METAL IONS IN BIOLOGY AND MEDICINE

LES IONS MÉTALLIQUES EN BIOLOGIE ET EN MÉDECINE

Sponsored by/*Sous le haut patronage de :*

American College of Nutrition
Anticancer Research
Food and Drug Administration
National Center for Toxicological Research
Institut National de la Santé et de la Recherche Médicale
Société Internationale pour le Développement des Recherches sur le Mg

The Organizing Committee wishes to thank LABCATAL and MERAM for their support in the realization of this book

METAL IONS IN BIOLOGY AND MEDICINE

LES IONS MÉTALLIQUES EN BIOLOGIE ET EN MÉDECINE

Proceedings of the First International Symposium on Metal Ions in Biology and Medicine held in Reims (France) on May 16-19, 1990

Premier Symposium International sur les Ions Métalliques en Biologie et en Médecine, Reims (France) 16-19 mai 1990

Edited by
Philippe Collery
Lionel A. Poirier
Michel Manfait
Jean-Claude Etienne

John Libbey
EUROTEXT
LONDON · PARIS

British Library Cataloguing in Publication Data
Metal Ions in Biology and Medicine
Congrès International
1. Ph. Collery et al.

ISBN 0-86196-243-5

Editions John Libbey Eurotext
6, rue Blanche, 92120 Montrouge, France.
Tél. : (1) 47.35.85.52 – Fax : (1) 46.57.10.09

John Libbey and Company Ltd
13, Smith Yard Summerley Street, London SW18 4HR, England
Tél. : (01) 947.27.77

John Libbey CIC
Via L. Spallanzani, 11
00161 Rome, Italy
Tel. : (06) 862.289

Foreword

The role of metal ions in molecular cell biology, cell growth, differentiation, carcinogenesis, tumor growth and immune system function is still not fully appreciated. The same can be said of the clinical consequences of common ions like calcium, cobalt, copper, iron, manganese, magnesium, potassium, sodium and zinc. 180 researchers, scientists and physicians, from 25 different countries, came to Reims , from the 16th to the 19th of May 1990, to expose in a rigorous way their implications in aging, inflammation, diseases of the immune system and cancer, which were the major themes of the symposium. Their papers also describe the mechanism of action of toxic ions like aluminium, arsenic, cadmium, chromium, nickel, lead, mercury or vanadium. They show the far reaching therapeutic possibilities of certain metal ions like gallium, germanium, gold, lanthanum, palladium, platinum, rhodium, ruthenium, selenium, titanium and tin. Much is already known but many questions remain to be answered. Indeed, the study of metal ions is very complex : a metal ion may be deficient in one sector and in excess in an other. The same is true of ion and magnesium which can be found in low concentrations in the blood but in high concentrations in the tumor of the same patient, much more so than in the surrounding tissue of the same organ. Concentrations can be low in the plasma and high in the red blood cells or in the leukocytes. Wide variations in intratissular distribution can occur depending on the way in which a metal ion is administered : orally or parenterally, once only or repeatedly. It is also important to take into consideration the interactions, which can be either agonistic or antagonistic, between metal ions, in order to understand the outcome of the studies performed as well as to exploit therapeutically the influence that one metal ion has on another. It is therefore necessary to acquire an overall picture of all metal ions so as to understand how they function in enzyme systems, the regulation of genes, membrane permeability, cytoskeleton formation, the synthesis of nucleic acid, protein and high energy compounds, ... and their clinical effects. It is hoped that this symposium will serve as a catalyst to promote further studies comparing the effects of each salt formed from metal ions with the same molar concentration and exposed for the same length of time. As result of the international cooperation fostered by the symposium, studies of the interactions of several metal ions in biology and medicine may also be promoted.

Philippe COLLERY
Jean-Claude ETIENNE

Avant-Propos

Le rôle des ions métalliques en biologie moléculaire, dans la croissance cellulaire, la différenciation, la carcinogenèse, la croissance tumorale et les fonctions du système immunitaire est encore insuffisamment apprécié. Il en est de même des conséquences cliniques de la dysrégulation des ions métalliques habituellement présents dans l'organisme comme le calcium, le cobalt, le cuivre, le fer, le manganèse, le magnésium, le potassium, le sodium et le zinc. 180 chercheurs, scientifiques et cliniciens de 25 pays différents, venus à Reims, du 16 au 19 mai 1990, entendent faire le point avec le maximum de rigueur sur les certitudes et les probabilités de leurs implications possibles dans les processus de vieillissement, les maladies inflammatoires ou dysimmunitaires et le cancer, thèmes majeurs du symposium. Il décrivent également le mécanisme d'action des ions toxiques comme l'aluminium, l'arsenic, le cadmium, le chrome, le nickel, le plomb, le mercure ou le vanadium. Ils montrent les larges possibilités thérapeutiques de certains ions métalliques comme le gallium, le germanium, l'or, les lanthanides le palladium, le platine, le rhodium, le ruthénium, le selenium, le titane et l'étain. Ces certitudes sont déjà bien établies mais de nombreuses questions restent posées. Les études sont en effet fort complexes : un ion métallique peut être en déficit dans un secteur et en excès dans un autre. Ainsi en est-il du fer et du magnésium dont les concentrations sanguines peuvent être diminuées alors qu'ils sont retrouvés en grande quantité dans la tumeur du même organe. Les concentrations peuvent être abaissées dans le plasma et élevées dans les érythrocytes ou les leucocytes. Les variations dans la distribution intratissulaire peuvent être grandes, fonction du mode d'administration, orale ou parentéral, unique ou répété. Les interactions entre les ions métalliques qu'elles soient agonistes, ou antagonistes conditionnent la compréhension de nombreux phénomènes biologiques et peuvent laisser espérer certaines implications thérapeutiques. Il est donc nécessaire d'avoir une connaissance globale de l'ensemble des ions métalliques pour mieux comprendre leur mécanisme d'action sur les systèmes enzymatiques, de régulation des gènes, de perméabilité membranaire, de formation du cytosquelette, de synthèse des acides nucléiques, des protéines, des composés riches en énergie... et leurs effets cliniques. Ce symposium doit servir de catalyseur pour promouvoir des études comparant les effets des sels formés à partir des ions métalliques à des concentrations et des temps d'exposition équivalents. Il doit aussi permettre des études coopératives internationales sur les interactions des ions métalliques en biologie et en médecine.

Philippe COLLERY
Jean-Claude ETIENNE

List and addresses of authors

Philippe Collery, CHRU, médecine interne/cancérologie, Département des maladies respiratoires, hôpital Maison Blanche, 45 rue Cognacq Jay, 51092 Reims Cedex, France.

Lionel A. Poirier, Food and Drug Administration, National Center for Toxicological Research, Jefferson AR 72079, USA.

Michel Manfait, Université de Reims, Laboratoire de spectroscopie biomoléculaire, Faculté de Pharmacie, 51 rue Cognacq Jay, 51100 Reims, France.

Jean-Claude Etienne, CHRU, médecine interne, hôpital Robert Debré, rue Alexis Carrel, 51092 Reims Cedex, France

CONTENTS/SOMMAIRE

A COMMON METAL IONS : *calcium, cobalt, copper, iron, magnesium, manganese, potassium, sodium, zinc*

I. IMMUNOLOGY AND INFLAMMATION

II. CARCINOGENESIS

III. AGING

IV. VARIOUS EFFECTS AND MISCELLANEOUS DISEASES

B

TOXIC METAL IONS : aluminium, arsenic, baryum, beryllium, cadmium, ceryum, chromium, lead, mercury, nickel, titanium, vanadium, yttrium

I. TISSUE DISTRIBUTION

II. EPIDEMIOLOGY, ANALYSIS

III. IMMUNOLOGY, INFLAMMATION

II. GOLD

III. GALLIUM

IV. NEW METAL COMPLEXES

V. RUTHENIUM

VI. SELENIUM

D INTERACTIONS METAL IONS/BIOLOGICAL SUBSTRATES

A

COMMON METAL IONS : CALCIUM, COBALT, COPPER, IRON, MAGNESIUM, MANGANESE, POTASSIUM, SODIUM, ZINC

I. Immunology and Inflammation

Metal Ions in Biology and Medicine. Eds. Ph. Collery, L.A. Poirier, M. Manfait, J.C. Etienne. John Libbey Eurotext, Paris © 1990, pp. 5-7

Speciation studies on the copper-salicylate system in relation to its antiinflammatory activity

V. Brumas, G. Berthon

INSERM U.305, University Paul Sabatier, 38, rue des Trente-Six-Ponts, 31400 Toulouse, France

During the past decade, reports have accumulated on the beneficial effects of copper compounds in controlling inflammation in various animal and human models (Sorenson, 1987).
Inflammation is a very complex process whose origin remains unclear, but which would result from a breakdown of the patient's autoimmune system (Jackson **et al.**, 1977). Among other effects, total serum copper levels are raised during acute as well as chronic inflammation states (Milanino **et al.**, 1985). Copper concentrations increase in association with the onset and persistence of the active disease but return to normal with remission (Sorenson, 1982). According to some authors, ceruloplasmin copper levels are raised (Milanino **et al.**, 1985). Others contend that it is the ultrafiltrable fraction of the metal which is enhanced (Linder, 1983). First considered a cause of the disease, these copper metabolic changes are now thought to proceed from the natural antiinflammatory response of the host (Sorenson, 1982). This new interpretation is consistent with the fact that copper complexes of antiinflammatory substances have been observed to be more effective than their parent drugs; these may indeed help restore normal serum levels of copper by promoting its tissue distribution (Sorenson, 1982).

Apart from D-penicillamine, however, the ability of all antiinflammatory drugs to interfere with copper metabolism in inflammatory conditions is still largely questionable (Milanino **et al.**, 1985).
In this connection, the prime objective is to identify the low-molecular-weight (l.m.w.) complexes present in the main biofluids (blood plasma at least) of both normal and inflamed organisms.
Unfortunately, to investigate the distribution of a given substance into its l.m.w. fractions in vivo is not possible by analytical techniques. Such speciation studies require the use of simulation models based on (i) analytical data relative to the biofluid under consideration, and (ii) formation constants of all the complexes potentially formed under the specific conditions involved.

Among nonsteroidal antiinflammatory drugs, salicylic acid has undoubtedly been the most often used in conjunction with copper. Also, implications of copper - salicylic acid coordination equilibria in the antiinflammatory activity of copper and salicylate have already been discussed (Arena **et al.**, 1977). The main conclusion of this work was that "antiinflammatory effects of metal salicylate complexes cannot act through the metal complexes themselves since their bonds dissociate in plasma".
Surprisingly however, these authors at the same time advocate that salicylate cannot successfully compete for the metal ion in plasma and claim that its antiinflammatory effect would stem from its capacity to directly attack copper ions bound to albumin (Fiabane and Williams, 1977).

Before taking up new studies on the reactivity of copper salicylate complexes, it was thus worth reexamining the above discussions in the light of new data obtained by reinvestigating copper salicylic acid interactions over larger concentration ranges. Histidine being the main l.m.w. ligand of copper in plasma (Berthon **et al.**, 1986), formation of copper salicylate ternary complexes with this aminoacid was also taken into account on this occasion.
Insufficient space is available here to report experimental details on these determinations. It is only worth noting that in comparison with the above study, the investigation of extended reactant concentration ranges effectively led to the characterisation of new copper salicylate complexes: in addition to the "classical" ML and ML_2 species, MLH and ML_2H were also definitely proved to exist. Ternary complexes have also been found with histidine, among which MLX and MLXH (where L stands for the dianionic form of salicylate and X for the histidinate anion) predominate.

A series of simulation studies was undertaken following the incorporation of the stability constants of these new species in our current databank. As far as concentration data are concerned, the results of these calculations remain basically the same as those previously reported (Arena **et al.**, 1977).

Simulations of the distribution of copper as well as salicylate in the gastrointestinal fluid confirm the synergistic effect existing in favour of the absorption of these two reactants: the fraction of salicylic acid not absorbed as such in the stomach is made bioavailable as copper monosalicylate in the small intestine; reciprocally, copper absorption is also favoured above pH 5.
Plasma simulations also confirm that, in contrast, a priori no favourable effect is to be expected from either of these reactants with respect to the tissue penetration of the other. For a concentration of $5x10^{-4}$ mol.dm^{-3} of salicylate, which roughly corresponds to the peak plasma level following the oral administration of 1gr of aspirin, concentrations of ML and MLXH neutral copper salicylate complexes respectively amount to $1.93x10^{-15}$ and $2.66x10^{-15}$ mol.dm^{-3}, which is quite negligible with respect to the pool of neutral copper complexes with histidine and other monoanionic aminoacids whose concentration reaches about 10^{-9} mol.dm^{-3}. On the other hand, the neutral fraction of salicylate to which these concentrations are to be compared is found around 10^{-7} mol.dm^{-3} under the same conditions.

The above results do indeed tend to exclude any possibility of mutual interaction between copper and salicylate with respect to their antiinflammatory activities. However, care should be taken to the fact that even species present in seemingly negligible concentrations may play determining parts in living systems (Berthon **et al.**, 1987).
First, their possible lipophilic properties may overcome their apparent concentration insufficiencies to make them produce significant physiological effects.
Secondly in the present case, the efficiency of nonsteroidal antiinflammatory agents used in conjunction with copper may not be limited to their potential capacity to improve the tissue penetration of Cu^{2+} ions. In addition to their potential role of circulating forms of the metal required to activate copper-dependent enzymes, l.m.w. copper complexes may also display chemical reactivities likely to facilitate the healing process by themselves.

For example, they have been claimed to be capable of dismutating superoxide radicals in a SOD-like manner (de Alvare **et al.**, 1976). Generated as a part of the inflammatory process, O_2^- radicals are believed to contribute to tissue damage and to perpetuate inflammation (McCord, 1982).
Besides, it has lately been suggested that hydroxyl free radicals, which are the most likely candidates for the toxic reactant in oxidative stress, would induce molecular damage at the very site of the metal binding where the Fenton reaction occurs, and that chelators capable of removing metals from biologically sensitive sites might prevent oxidative damage. In particular, free radical trapping by salicylic acid would be directly related to the capacity of this agent to chelate and shift from its original binding site the metal ion involved in the Fenton reaction (Saltman, 1989).

Thus, if we accept the hypothesis of a synergistic effect between salicylic acid and copper as antiinflammatory agents, the question still remains as to know whether copper salicylate complexes simply act as carriers of Cu^{2+} ions to help these diffuse towards the therapeutic target, or are directly involved in the antiinflammatory process either by dismutating superoxide radicals or by "luring" the Fenton reaction.
Work is currently in progress in our group to check each of these hypotheses.

REFERENCES

Arena, G., Kavu, G. and Williams, D.R. (1977): Metal-ligand complexes involved in rheumatoid arthritis - V Formation constants for calcium(II)-, magnesium(II)- and copper(II)-salicylate and acetylsalicylate interactions. *J. Inorg. Nucl. Chem. 40*, 1221-1226.

Berthon, G., Hacht, B., Blais, M.J. and May P.M. (1986): Copper-histidine ternary complex equilibria with glutamine, asparagine and serine. The implications for computer-simulated distributions of copper(II) in blood plasma. *Inorg. Chim. Acta Bioinorg. Chem. 125*, 219-227.

Berthon, G., Varsamidis, A., Blaquiere, C. and Rigal, D. (1987): Histamine as a ligand in blood plasma. Part 7. Malate, malonate, maleate and tartrate as adjuvants of zinc to favour histamine tissue diffusion through mixed-ligand coordination. *In vitro* tests on lymphocyte proliferation. *Agent. Action. 22*, 231-247.

de Alvare, L.R., Goda, K. and Kimura, T. (1976): Mechanism of superoxide anion scavenging reaction by bis(salicylato)copper(II) complex. *Biochem. Biophys. Res. Comm. 69*, 687.

Fiabane, A.M. and Williams, D.R. (1977): Metal-ligand complexes involved in rheumatoid arthritis - II Acidic drug-serum albumin-copper(II) interactions investigated using visible spectrophotometry and molecular filtration. *J. Inorg. Nucl. Chem. 40*, 1195-1200.

Jackson, G.E., May, P.M. and Williams, D.R. (1977): Metal-ligand complexes involved in rheumatoid arthritis - I Justifications for copper administration. *J. Inorg. Nucl. Chem. 40*, 1189-1194.

Linder, M.C., (1983): Changes in the distribution and metabolism of copper in cancer: a review. *J. Nutr. Grow. Cancer 27*, 1.

McCord, J.M. (1982): Roles of superoxide in inflammation and ischemic shock. In *Inflammatory Diseases and Copper*, ed J.R.J. Sorenson, pp. 255-266. Clifton, New Jersey: Humana Press.

Milanino, R., Conforti, A., Franco, L., Marrella, M. and Velo, G. (1985): Review: Copper and inflammation - a possible rationale for the pharmacological manipulation of inflammatory disorders. *Agent. Action. 16*, 504-513.

Saltman, P. (1989): Oxidative stress: a radical view. *Semin. Hematol. 26*, 249-256.

Sorenson, J.R.J. (1982): The anti-inflammatory activities of copper complexes. In *Metal Ions in Biological Systems, Vol. 14*, ed H. Sigel, pp. 77-124. New York: Marcel Dekker.

Sorenson, J.R.J. (1982): Copper complexes as the active metabolites of antiinflammatory agents. In *Inflammatory Diseases and Copper*, ed J.R.J. Sorenson, pp. 289-301. Clifton, New Jersey: Humana Press.

Sorenson, J.R.J. (1987): A physiological basis for pharmacological activities of copper complexes: an hypothesis. In *Biology of Copper Complexes*, ed J.R.J. Sorenson, pp. 3-16. Clifton, New Jersey: Humana Press.

Metal Ions in Biology and Medicine. Eds. Ph. Collery, L.A. Poirier, M. Manfait, J.C. Etienne. John Libbey Eurotext, Paris © 1990, pp. 8-10

The effect of D-penicillamine and phenylbutazone on the «turn-over» of endogenous copper during inflammation

D.E. Auer, *J.C. Ng, A.A. Seawright

*Department of Veterinary Pathology and Public Health, University of Queensland, St. Lucia, Queensland, 4067 Australia; *Present address : Drug Detection Laboratory, Racing Services, Bogan Street, Albion, Queensland, 4010 Australia*

INTRODUCTION

Numerous studies have demonstrated modification of copper metabolism in inflamed animals (Nev ***et al.***,1988) leading to speculation that endogenous copper modulates and contributes to the maintenance of homeostasis. Body stores of copper are sometimes supplemented as pharmacological control of inflammation despite availability of conventional therapies such as phenylbutazone or D-penicillamine. Sorenson (1976) hypothesised the mechanism of anti-arthritic and anti-inflammatory drugs was linked to endogenous copper through formation of copper complexes ***in vivo.*** The effect of D-penicillamine and phenylbutazone upon the "turnover" of endogenous copper in the inflamed tissues of rats was investigated in this study.

MATERIALS AND METHODS

Two subcutaneous polyurethane sponges (15x15x6mm) were surgically implanted on either side of the dorsal midline of 63 male Wistar rats (200±20grams). In 21 rats D-penicillamine (50mg/kg in 1 ml of normal saline) was administered daily for 5 days beginning on the day of sponge insertion, the last dose 30 minutes before the intracardiac injection of ^{64}Cu labelled copper acetate (2.75µg in 0.05ml normal saline) under ether anaesthesia. In another 21 rats, phenylbutazone (120mg/kg in 0.5ml DMSO) was administered intraperitoneally 30 minutes before injection of the labelled copper and a control group of 21 rats were not treated but were administered the labelled copper. The rats were exsanguinated from the jugular veins and carotid arteries in groups of three, at 10, 30 and 60 minutes and 6, 12, 24 and 48 hours after the ^{64}Cu injection. The plasma was separated from the blood, the liver, sponges and sponge capsules were weighed and their copper concentration and radioactivity measured. For each tissue, the uptake ratios and the relative specific activity (RSA), a measure of labelled to unlabelled copper were calculated (McIntosh & Lutwak-Mann, 1972). The RSA was analysed using 2-way analysis of variance and the uptake ratio and copper concentrations of the sponge and capsule at 1 and 24 hours were compared with the t-test for unpaired observations.

RESULTS

Figure 1 shows the RSA of plasma, liver, sponge and capsule. Irrespective of the time, the sponge RSA from D-penicillamine treated rats was always greater compared to PBZ treated ($p<0.01$) and untreated rats ($p<0.01$). Irrespective of the treatment, the liver RSA was greater and the sponge RSA less, between earlier and later times for all treatments ($p<0.01$).

Significant interactions between time and treatment occurred in the plasma and sponge capsule. The plasma RSA was higher for D-penicillamine treated rats compared to untreated rats at 1 ($p<0.05$) and 6 hours ($p<0.05$) and the plasma RSA of PBZ treated rats was less compared to untreated rats at 6 hours ($p<0.05$). Capsular RSA for D-penicillamine and PBZ treated rats were higher than controls at 30 and 60 minutes ($p<0.01$) and 60 minutes respectively ($p<0.05$) and for both groups at 24 hours ($p<0.01$). The sponge uptake ratio was greater after D-penicillamine treatment at 24 hours ($p<0.05$), in the capsule at 1 and 24 hours ($p<0.005$) but significantly less in the capsule after PBZ treatment at 24 hours ($p<0.005$). The sponge copper concentrations after D-penicillamine and PBZ treatment was less than controls at 1 ($p<0.05$) and 24 hours ($p<0.05$) respectively and in the capsule after PBZ treatment at 24 hours ($p<0.005$).

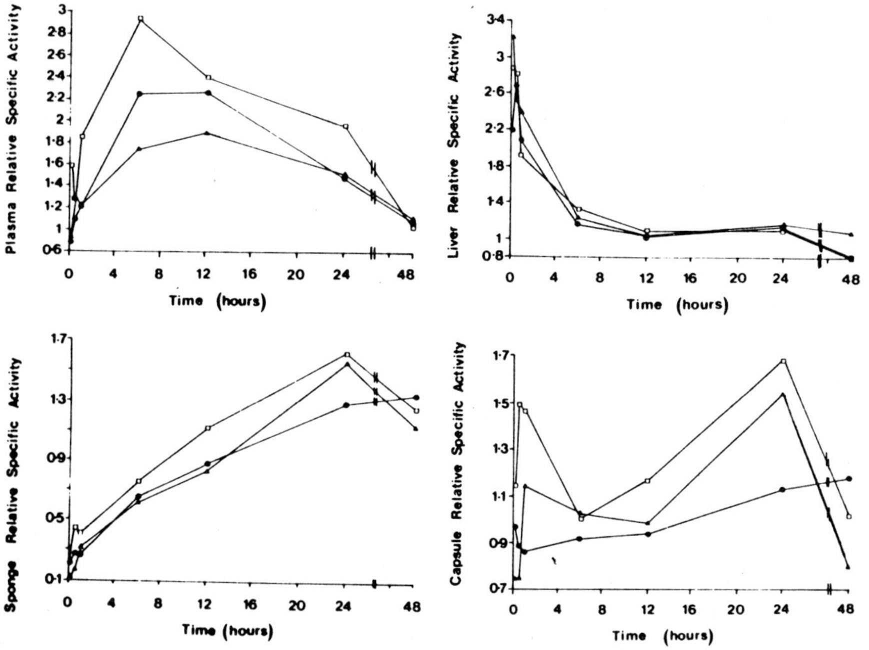

Fig 1. The relative specific activity (RSA) in the plasma, liver, sponge and sponge capsule at various times after administration of ^{64}Cu in rats that received no treatment (●——●), D-penicillamine (▫——▫) or phenylbutazone (▲——▲).

DISCUSSION

A comparison of the plasma and liver RSA suggests the ^{64}Cu initially bound to albumin or chelated to low molecular weight peptides and amino acids (Jackson ***et al.,*** 1978) was rapidly removed by the liver. Between 6 and 12 hours later, the mean plasma RSA was highest for all treatments due to secretion of newly labelled caeruloplasmin (Owen, 1965). The similar RSA-time profiles of the liver for each treatment suggests differences in the RSA of inflamed tissues were independent of the liver.

Since D-penicillamine would be present in blood and bound to body tissues, including the capsule at the time of ^{64}Cu administration, the excess plasma RSA of D-penicillamine treated rats at 10 minutes was probably circulating copper D-penicillamine. D-penicillamine chelates copper, promotes cupresis and binds to connective tissues (Patzschke & Wegner, 1977) with a half-life of 90 hours (Planas-Bohne, 1972). Since copper penicillamine complexes form ***in vivo*** (Cashin ***et al.***, 1985), the increased capsular uptake ratio of D-penicillamine treated rats at 1 hour suggests more labelled copper was retained in the capsule, either delivered as copper penicillamine or formed there from tissue-bound penicillamine. The decline in capsular RSA at 6 hours in D-penicillamine treated rats suggests mobilization of the labelled copper occurred. By comparison, the sponge RSA progressively increased because of isolation of the sponge interstices from the vascular supply. At later times, ^{64}Cu at the inflamed site was probably derived from labelled caeruloplasmin made available for chelation by penicillamine following degradation of the caeruloplasmin molecule (Gutteridge & Stocks, 1981).

A similar biphasic capsular RSA pattern was apparent after phenylbutazone treatment. Although the proportion of labelled to total copper was increased at times subsequent to increased copper availability in the blood, the decreased capsular uptake ratio and copper concentration suggests diminished vascular leakage reduced diffusion of labelled and unlabelled copper to the inflamed site. The increased capsular RSA at 24 hours apparently resulted from a reduction in the capsule copper concentration. These results suggest D-penicillamine and phenylbutazone influence the "turn-over" of copper in inflamed tissues. Similar capsular RSA patterns were observed after treatment with both drugs. However, phenylbutazone appeared to restrict diffusion of both labelled and unlabelled copper to the inflamed site whereas D-penicillamine possibly enhanced both its delivery, retention and removal, the effects evident in the proliferating connective tissues rather than the exudate. These effects may be germane to the mechanism of successful therapy of rheumatoid arthritis with D-penicillamine since copper-D-penicillamine complexes with superoxide dismutase activity are stable in the presence of endogenous copper binding ligands (Roberts & Robertson, 1985). Conversely, copper supplementation and simultaneous therapy with phenylbutazone would not appear to enhance delivery of copper to the inflamed site. Although the ratio of labelled to total copper increased soon after supplementation and later after caeruloplasmin synthesis, the inflamed tissues contained relatively less labelled and total copper at these times.

REFERENCES

Cashin, C.H., Lewis, E.J. & Burden, T. (1985): The development of screening tests for D-penicillamine like anti-rheumatic activity based on the ***in vivo*** interactions with copper. *Br. J. Rheumatol.*, 24,137-146.

Gutteridge J.M.C. & Stocks J. (1981): Caeruloplasmin: Physiological and pathological perspectives. *CRC Crit. Rev. Clin. Lab. Sci*, 14,257-329.

Jackson, G.E., May, P.M. & Williams, D.R. (1978): Metal-ligand complexes involved in rheumatoid arthritis-VI. *J. Inorg. Nucl. Chem.*, 43,825-829.

McIntosh, J.E.A. & Lutwak-Mann, C. (1972): Zinc transport in rabbit tissues. *Biochem.J.* 126,869-876.

Neve, J., Fontaine, J., Peretz, A & Famaey, J.P. (1988): Changes in zinc, copper and selenium status during adjuvant-induced arthritis in rats. *Agents and Actions.*, 25,147-155.

Owen, C.A. (1965): Metabolism of radiocopper (^{64}Cu) in the rat. *Am. J. Physiol.*, 209(5),900-904.

Patzschke Von K. & Wegner, L.A. (1977): Pharmakokinetische untersuchungen nach applikation von 14C-D-penicillamine an ratten. *Arzneim. Forsch. (Drug Res.).*, 27(1)NR.6,1152-1158.

Planas-Bohne Von (1972): Pharmakokinetische untersuchungen an 14C-markiertem penicillamin. *Arzneim. Forsch. (Drug Res.).*, 22(9),1426-1433.

Roberts, N.A. & Robinson, P.A. (1985): Copper chelates of anti-rheumatic and anti-inflammatory agents: Their superoxide dismutase-like activity and stability. *Br. J. Rheumatol.*, 24,128-136.

Sorenson, J.R.J. (1976): Some copper chelates and their anti-inflammatory and anti-ulcer activities. *Inflammation*, 1(3),317-331.

Metal Ions in Biology and Medicine. Eds. Ph. Collery, L.A. Poirier, M. Manfait, J.C. Etienne. John Libbey Eurotext, Paris © 1990, pp. 11-13

Modulating leukotriene syntheses by copper and zinc

A.M. Badawi[1], G.N. Al-Sousi[2], K.E. El-Tahir[3]

Department of application[1], Petroleum Research Institute, Nasr City, Cairo, Egypt; Department of Chemistry[2], College of Teachers, Taif, PO, Box 1070 and Department of Pharmacology[3], College of Pharmacy, King Saud University, PO, Box 2457, Riyadh, Saudi Arabia

INTRODUCTION

A variety of pathological lung disorders are accompanied by obstructive pulmonary disease (OPD) which is associated by altered copper and zinc metabolism. Serum copper and zinc values are significantly elevated and lowered respectively in chronic OPC (Sullivan et aL., 1979) suggesting that this is a physiological response to overcome this disease. Zinc blocked histamine release from sensitized mast cells (Hogberg & Uvnos, 1960) and inhibited 12-lipoxygenase (12-LO) in vitro (Wallach & Brown, 1981). Copper complexes of 3,5-diisopropylsalicylic acid and of chromone-2-carboxylic acid were suggested to inhibit 5-Lipoxygenase (Badawi et al., 1987).

M = Cu or Zn

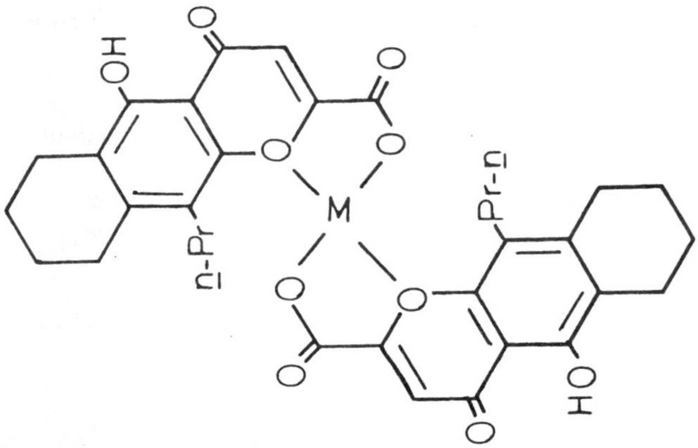

Fig. 1 Metal Complex of THP-CC.

To investigate the possibility that prophylactic agents against bronchial asthma can be potentiated by chelating them with either copper or zinc, it occurred to us that the antiallergic drug tetrahydro-5-hydroxy-8-propyl-chromone-2-carboxylic acid (THP-CC) might be complexed with either copper or zinc to give Cu(II) $(THR-CC)_2$ and Zn (II)$(THP-CC)_2$ respectively (Fig.1). Both metal complexes were found to inhibit 12-LO more than the parent drug.

MATERIALS AND METHODS

THP-CC was a gift from Fison Ltd, Pharmaceutical Divisions, England. Details of synthesising the copper (II) and zinc (II) chelates will be described elsewhere. These test compounds were solubilized in DMSO at 10 mM concentration and diluted to assay concentration. Activities of the lipoxygenase derived product 12-hydroxy-5,8,10,14 eicosatetraenoic acid (12-HETE) released by the rat testis were determined by TLC/UV spectrophotometric method (El-Tahir, 1986).

Data presented in Table 1 show inhibition of 12-HETE syntheses at 10^{-4} M concentrations which suggests that the copper (II) and zinc (II) complexes have 12-LO inhibitory activity .

Table 1: Effect of THP-CC and its metal complexes on 12-HETE syntheses by the rat testis

Treatment & Dose(M)		12-HETE syntheses ng/mg wet tissue					%
	Expt.No.	1	2	3	4	Mean	Change
Control		20.16	20	19.2	19.76	19.75	
THP-CC	$3.1x10^{-4}$	12.4	17.1	16.3	16.2	15.5	↓21.63
	$6.2x10^{-4}$	8.64	13.6	14.0	12.8	12.26	↓38.01
Control		22.4	21.6	21.1	22.12	22.00	
Cu(II)$(THP-CC)_2$							
	$1.5x10^{-4}$	16.8	15.55	15.1	17.1	16.13	↓26.68
	$3.0x10^{-4}$	14.1	13.9	10.24	12.8	12.76	↓42.0
Control		18.8	17.2	17.8	18.9	18.17	
Zn(II)$(THP-CC)_2$							
	$1.5x10^{-4}$	14.3	13.7	14.8	14.55	14.33	↓21.13
	$3.0x10^{-4}$	8.8	8.4	13.2	10.12	10.13	↓44.24

The ID-50 value of the parent drug (defined as that concentration of inhibitor producing 50% of the enzyme as calculated by linear regression) was found 9.88 x 10^{-4} M. The ID-50 values of Cu(II) $(THP-CC)_2$ and Zn(II) $(THP-CC)_2$ were $3.91x10^{-4}$ and $3.26x10^{-4}$ M respectively.

DISCUSSION

It seemed that both the Cu (II) $(THP-CC)_2$ and Zn (II) $(THP-CC)_2$ showed approximately from two and half to three times more inhibitory 12-LO activity than the parent drug on the bases of ID_{50} values.

The activity of inorganic forms of zinc and copper was studied before with the human platelet 12-LO system (Wallach & Brown, 1981). At a concentration of $3.7x10^{-3}M$ zinc chloride inhibited 100% 12-HETE syntheses. At a 10-fold lower concentration zinc chloride inhibited approximately 30% 12-HETE syntheses and its ID_{50} value was 1.34×10^{-3} M. The inorganic copper salt caused a dramatic increase in oxygen consumption which has been speculated to be due to auto-oxidation of the substrate (Wallach & Brown, 1981).

On the bases of ID_{50} values it seems that chelating THP-CC by copper or zinc increased its 12-LO modulating effect, which indicates that both metal ions have a role in normal function of the respiratory system, presenting a new immunopharmacological approach to the treatment of allergy.

REFLERENCES

Badawi, A.M.., Summan H.D, Sarau, H.M., Foly, J.J. and Ali, F. (1987): Antiasthmatic activity of some copper complexes. In Biology of Copper Complexes, ed. J.R.J. Jorensun, pp 573-579. Clifton, New Jersey: The Humana Press.

El-Tahir, K.E. (1986): Spectrophotometric quantitation of the lipoxygenase derived conjugated 1,3-diene (12-HETE) in biological systems. Oriental J. Chem,. 2: 141-144.

Hogberg, B., and Uvnas (1960): Further observations on the disruption of rat mesentery mast cells caused by compound 48/80, antigen-antibody reaction, Lecithinase A, and decylamine. Acta Physiol. Scan. 48: 133-145.

Sullivan, J.F., Blotcky, A.J., Jetton,; M.M., Hahn, H.K., and Burch, R.E. (1979): Serum levels of selenium, calcium, copper, magnesium, manganese and zinc in various human diseases. J. Nutr. 109: 1432-1437.

Wallach, D.P., and Brown, V.R.(1981): A novel preparation of human platelet lipoxygenase, characteristics and inhibition by a variety of phenylhydrazones and comparisons with other lipoxygenases. Biochem. Biophys. acta 663: 361-372.

Metal Ions in Biology and Medicine. Eds. Ph. Collery, L.A. Poirier, M. Manfait, J.C. Etienne. John Libbey Eurotext, Paris © 1990, pp. 14-17

Anti inflammatory action of zinc related to cutaneous pathology

B. Dreno*, H.L. Boiteau**, P. Litoux*

** Department of Dermatology, Hôtel Dieu, 44035 Nantes Cedex 01, France; ** Laboratory of toxicology, Hôpital St-Jacques, Nantes, France*

Cutaneous pathology associated with zinc deficiency is characterised by inflammatory lesions with pustules.This treatment of acne with zinc is only efficienty on inflammatory lesions and has no effect on retentional non inflammatory lesions. However, the mechanism by which zinc produces anti-inflammatory action remains uncertain.

In this context, we studied the changes induced by zinc therapy on granulocyte and erythrocyte zinc levels in 15 patients with inflammatory acne, 21 with retentional acne and 15 controls. Zinc analysis was performed before and after 2 months of zinc gluconate treatment (200 mg/per day; Labcatal) using atomic absorption spectrophotometry. Erythrocytes and granulocytes were isolated by Ficoll gradient. Before treatment, the zinc granulocyte level was high in the inflammatory group compared to retentional and control groups ($\alpha < 0.05$),whereas the zinc erythrocyte level was low in the inflammatory group compared to the other 2 groups ($\alpha < 0.01$). After 2 months of zinc therapy, the zinc granulocyte level increased in retentional and control groups, but decreased in the inflammatory group; the zinc erythrocyte level increased in the inflammatory group. At least 2 mechanisms could govern zinc action on cutaneous inflammatory lesions : firstly, by inhibiting granulocyte functions (bactericide, chemotaxis) and secondly by enhancing the activity of superoxide dismutase.

Evidence indicates that trace elements, particularly zinc, are among factors regulating the activity of various inflammatory cells. Thus, the cutaneous pathology associated with a zinc deficiency (acrodermatitis enteropathica, prolonged use of total parenteral nutrition without zinc) is characterized by inflammatory cutaneous lesions with pustules.In 1976 Fitzherbert (2) first postulated zinc deficiency to be causally related to acne vulgaris. Subsequently, Michaelsson & Juhlin (4) and Hillstrom & Pettersson (3) reported from double-blind studies a beneficial effect of oral zinc administration on acne lesions. Our recent study (1) confirmed the selective effect on inflammatory lesions.

However the mechanism by which zinc produces anti-inflammatory lesions is still uncertain. The relation between serum zinc deficit and inflammatory cutaneous lesions remains unclear since results are conflicting (6). Nor has a cutaneous zinc deficit ever been demonstrated.

In this context, we first investigated erythrocyte and granulocyte zinc levels in retentional (zinc therapy inefficient) and inflammatory acne (zinc therapy efficient), and then studied the variations induced by zinc therapy on granulocyte and erythrocyte zinc levels in these 2 groups of acne patients (1).

MATERIALS AND METHOD

Twenty-one retentional acne (more than 20 comedos on the face), 15 inflammatory acne (more than 10 pustular lesions on the face) and 15 controls were included in this study.

For each patient, analysis was performed between 9. and 10 a.m.before treatment and then after 2 months of zinc gluconate therapy (200mg per day; Labcatal), using atomic absorption spectrophotometry. Blood was collected by venous puncture, and erythrocytes and granulocytes where then isolated by Ficoll gradient. Starting with the collection of specimens, general precautions for microanalysis of zinc (contamination) were observed throughout the entire procedure of zinc estimation.

RESULTS

PRIOR TO THIS TREATMENT, ZINC GRANULOCYTE LEVELS were elevated in the inflammatory group ($\alpha < 0.05$) compared to retentional and control groups (Table 1), and the **ZINC ERYTHROCYTE LEVEL** was decreased in the inflammatory group ($\alpha < 0.001$) compared to retentional and control groups (Table 1)

	controls	retentional acne	inflammatory acne
zinc granulocyte level pg/10^6 cells	3.78 ±0.2	7.72 ±7.08	13.43 α <0,05 ± 9.2
zinc erythrocyte level µg/l	11901 ± 1212	11498 ± 1409	8672.19 α <0,001 ± 1497

<u>Table 1</u> Zinc granulocyte and erythrocyte levels (pg/ 10^6 cells) in control, retentional acne and inflammatory acne groups.

After 2 months of zinc gluconate (200mg/per day), the <u>zinc granulocyte level</u> increased in the control group (Table 2) remained unchanged in the retentional group BUT DECREASED (α<0.05) in the inflammatory group (Table 2)

	controls µg/10^6 cells	retentional acne	inflammatory acne
before therapy	3.78 ±0.12	7.72 ±7.08	13.43 ± 9.2
after 2 months of zinc gluconate	6.96 ± 1.46	10.9 ±5.2	6.41 ±7.8
	$\alpha<0.01$	ns	$\alpha<0.05$

Table 2 Variations in zinc granulocyte (pg/10^6 cells) after 2 months of zinc gluconate (200mg/per day) in the 3 groups.

On the contrary, after 2 months of therapy, zinc erythrocyte levels increased in the inflammatory acne group ($\alpha<0.01$) and in the other 2 groups (not significantly);

	controls prg/l	retentional acne	inflammatory acne
before therapy	11901 ±1212	11498 ±1409	8672.19 ± 1497
after 2 months of zinc gluconate	12720 ±1410	11722 ±1218	10678.33 ±1225
	ns	ns	$\alpha<0.01$

Table 3 : Variations in zinc erythrocyte levels (µg/l) after 2 months of zinc gluconate therapy in the 3 groups.

DISCUSSION

This study demonstrates that erythrocyte and granulocyte zinc levels have different values between inflammatory and retentional acne. Indeed, in inflammatory acne, zinc granulocyte levels were increased, whereas zinc erythrocyte levels were decreased. Moreover, it is of particular interest to note that this biological difference was associated with a different clinical effect . Indeed, zinc salts were only efficient for inflammatory acne.

In view of these results, it is possible that zinc acts on inflammatory lesions by at least 2 mechanisms. By decreasing the zinc granulocyte level, zinc salts could inhibit granulocyte functions (bactericide, chemotaxis). Indeed, Chapvil & Stankova (2) have shown in vitro that the addition of zinc in a medium has an inhibitory effect on granulocyte functions by washing zinc from cells.

In the same manner, the increase of the zinc erythrocyte level after zinc therapy could induce the activity of superoxide dismutase enzyme which plays a crucial role in the elimination of superoxide ions.

It would thus appear that the anti-inflammatory action of zinc results from several mechanisms, especially on granulocyte and erythrocyte functions.

REFERENCES

1 - Dreno, B.& Amblard, P. (1989): Low doses zinc gluconate in inflammatory acne. Acta Derm. Venerol... 69, 541-543

2 - Chapvil, M.& Stankova, L. (1977): Inhibition of some functions of polymorphonuclear leukocytes by in vitro zinc. J Lab Clin. 89, 135-146

3 - Fitzherbert, JC. (1976): Acne vulgaris: zinc deficiency ? Med J Aust. 1, 848-850

4 - Hillstrom, L. & Pettersson, L. (1977): Comparison of oral treatment with zinc sulphate and placebo in acne vulgaris. Br J Derm. 97, 681-683

5 - Michaelsson, G. & Juhlin, L. (1977): A double blind study of the effect of zinc and oxytetracycline in acne vulgaris. Br J Derm. 97, 561-564

6 - Pohit, J. & Saha, KC. (1985): Zinc status of acne vulgaris patients. Journal of applied nutrition. 37, 18-25

Metal Ions in Biology and Medicine. Eds. Ph. Collery, L.A. Poirier, M. Manfait, J.C. Etienne. John Libbey Eurotext, Paris © 1990, pp. 18-20

The effects of lanthanum and cobalt ions on enzyme release from neutrophils

J.G.R. Elferink

Department of Medical Biochemistry, University of Leiden, POB 9503, 2300 RA Leiden, The Netherlands

INTRODUCTION

The neutrophil contains granules in which a large number of cytotoxic products, mainly enzymes, are stored. Release of these enzymes can be either beneficial for the host, when they are used to kill microorganisms, or can be detrimental when they are released extracellularly and cause inflammation and tissue destruction. The release of enzymes from the neutrophil may occur in two ways.

A cytotoxic enzyme release occurs when the plasma membrane is damaged, which process is followed by cell desintegration and granule enzyme release. The release of the cytoplasmic enzyme lactate dehydrogenase (LDH) is a measure for plasma membrane damage. A number of physiologic important substances, such as hydroxyapatite microcrystals, asbestos, and cationic proteins cause cytotoxic enzyme release via a pattern of positive charges on their surface.

The second way of enzyme release is by exocytosis, which occurs after activation of the neutrophil with phagocytosable particles or with soluble agents. A commonly used activator is the chemotactic peptide formyl-methionyl-leucyl-phenylalanine (FMLP), which activates exocytosis in cytochalasin B-treated neutrophils. Exocytosis can be measured as the release of the granule enzyme lysozyme, in the absence of significant LDH release.

We studied the effect of La^{3+} and Co^{2+} on cytotoxic enzyme release, induced by poly-L-arginine or chrysotile asbestos, and on FMLP-induced exocytotic lysozyme release from rabbit neutrophils.

MATERIALS AND METHODS

Rabbit peritoneal neutrophils were isolated as described previously (Elferink *et al.*, 1985). During the experiments 3×10^6 neutrophils in a final volume of 1 ml were preincubated with or without metal ions for 10 min at 37°C and subsequently incubated with reagents for 20 min at 37°C. Thereafter the suspension was centrifuged and enzyme release was determined in the supernatant.

LDH, lysozyme and superoxide release were determined as described by Elferink and Deierkauf (1989). Enzyme release was expressed as a percentage of a maximum value, obtained by treating the cells with 0.05 % Triton X-100.

A homogeneous stock suspension of small asbestos particles was obtained by sonicating chrysotile asbestos in medium for 10 min. Poly-L-arginine (M=40000) and FMLP were from Sigma Chemical Co. 5×10^{-6} M cytochalasin B was present in the medium when exocytosis or superoxide production was induced by 10^{-8} M FMLP.

RESULTS AND DISCUSSION

Chrysotile asbestos and poly-L-arginine induce extensive LDH release (Fig.1),

indicating that they damage the plasma membrane and cause cytotoxic enzyme release. Both La^{3+} and Co^{2+} inhibit LDH release induced by these agents. La^{3+} however, is much more effective than Co^{2+} in antagonizing the membrane-damaging effect of poly-L-arginine or asbestos (Fig.1). In the absence of Ca^{2+} (Fig.1) there is little lysozyme release. In the presence of extracellular Ca^{2+} the situation becomes more complex because then an activating effect is superimposed on the cytotoxic effect, due to an influx of Ca^{2+} into the cell. The resulting strong lysozyme release is inhibited by La^{3+} and by Co^{2+}, but this may be a consequence of the inhibition of the membrane damaging effect.
Exposure of neutrophils to FMLP results in a strong lysozyme release while there is no LDH release. This implies that with FMLP the release of lysozyme is a direct measure for exocytosis. FMLP-induced exocytotic enzyme release is inhibited by Co^{2+}. The inhibitory effect of Co^{2+} is antagonized by extracellular Ca^{2+}, hence Co^{2+} acts as a calcium antagonist. The effect of La^{3+} on exocytotic enzyme release is different from that of Co^{2+}: in the absence of extracellular

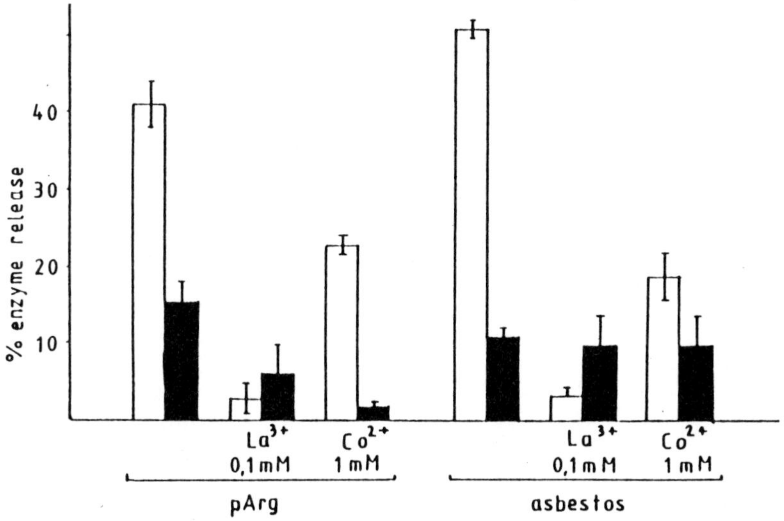

Fig.1. The effect of 0.1 mM La^{3+} or 1 mM Co^{2+} on cytotoxic enzyme release, induced by 5 µg poly-L-arginine (pArg) or by 100 µg asbestos, in the absence of extracellular Ca^{2+}. [open bar] : LDH release; [filled bar] : lysozyme release.

Table 1

Effect of La^{3+} and Co^{2+} on chemotactic peptide-induced enzyme release, in the absence or in the presence of 1 mM Ca^{2+}

	% enzyme release			
	–		+ Ca^{2+}	
	LDH	lys	LDH	lys
–	2 ± 1	4 ± 2		
0.1 mM La^{3+}	1 ± 0	6 ± 3		
1 mM Co^{2+}	2 ± 1	0 ± 1		
FMLP	4 ± 2	47 ± 2	3 ± 2	80 ± 1
FMLP, 0.1 La^{3+}	4 ± 2	65 ± 3	2 ± 1	74 ± 1
FMLP, 1 mM Co^{2+}	1 ± 1	8 ± 2	2 ± 1	37 ± 3

Ca^{2+} it does not inhibit exocytosis, but in concentrations up to 10^{-4} M it causes an enhancement of FMLP-induced lysozyme release. High concentrations (1mM La^{3+}) are inhibitory. Activation of neutrophils with FMLP not only causes exocytosis, but also the production of superoxide by these cells. The effect of Co^{2+} and La^{3+} on FMLP-induced superoxide production is the same as that on exocytosis: Co^{2+} inhibits superoxide production while La^{3+}, in concentrations up to 10^{-4} M, causes an enhancement of superoxide production. In the absence of FMLP there is no activating effect of La^{3+}. The effects are not specific for FMLP-induced activation, but also occur when the phorbol ester PMA is used to activate superoxide production.

Table 2
Effect of La^{3+} and Co^{2+} on superoxide generation induced by FMLP or by PMA

	Superoxide production (nmol/3x10^6 cells) in the presence of			
	FMLP		PMA	
	-	+ Ca^{2+}	-	+ Ca^{2+}
-	45.5 ± 3.7	71.8 ± 2.1	37.7 ± 1.2	42.4 ± 2.2
0.1 mM La^{3+}	66.1 ± 0.8	65.6 ± 1.8	56.4 ± 0.7	55.0 ± 0.8
1 mM La^{3+}	56.9 ± 0.9	57.6 ± 0.3	57.6 ± 0.8	58.8 ± 0.5
0.1 mM Co^{2+}	3.8 ± 1.6	55.7 ± 0.4	25.2 ± 4.9	41.0 ± 3.8
1 mM Co^{2+}	0.7 ± 0.4	15.2 ± 0.8	23.0 ± 2.9	27.1 ± 4.1

While the effect of La^{3+} and Co^{2+} on cytotoxic enzyme release is only quantitatively different, the effects on activation of enzyme release by exocytosis differ for the two ions, when they are present at low concentrations. La^{3+} has an a potentiating effect that is absent for Co^{2+}.

FMLP-induced activation of neutrophils is accompanied by an increase of calcium-association with the cell. Both Co^{2+} and La^{3+} inhibit FMLP-induced ^{45}Ca association with neutrophils (Boucek and Snyderman, 1976; Elferink and Deierkauf, 1989). It is unlikely however, that the inhibitory effect of either La^{3+} or Co^{2+} is associated with their ability to inhibit the influx of Ca^{2+} through the plasma membrane (Hagiwara and Byerly, 1981) because their effects are also present in the absence of extracellular Ca^{2+}. Co^{2+} is able to penetrate into the neutrophil (Elferink and Deierkauf, 1989). It is therefore conceivable, that some of the effects of La^{3+} and Co^{2+} are due to an interference with intracellular Ca^{2+} movements.

REFERENCES

Boucek, M.M. and Snyderman, R. (1976). Calcium influx requirement for human neutrophil chemotaxis: inhibition by lanthanum chloride. *Science* 193, 905-907.

Elferink, J.G.R., and M.Deierkauf. (1985). The effect of quin2 on chemotaxis by polymorphonuclear leukocytes. *Biochim. Biophys. Acta* 846, 364-369.

Elferink, J.G.R. and Deierkauf, M. (1989). The suppressive action of cobalt on exocytosis and respiratory burst in neutrophils. *Amer. J. Physiol.* 257, C859-C864.

Hagiwara,S., and Byerly, L.(1981). Calcium channel. *Ann. Rev. Neurosci.* 4,69-125.

Metal Ions in Biology and Medicine. Eds. Ph. Collery, L.A. Poirier, M. Manfait, J.C. Etienne. John Libbey Eurotext, Paris © 1990, pp. 21-23

Manganese complexes as superoxide radical anion scavengers

F.J. Arnaiz Garcia[1], C. Capul[2], P. Castan[3], D. Deguenon[3], Ph. Derache[4], F. Nepveu

[1] Universidad de Valladolid, Colegio Universitario de Burgos, S-09002 Burgos, Spain; [2] Laboratoire de Chimie Bioinorganique, INSERM U-305, 38, rue des 36-Ponts, F-31400 Toulouse, France; [3] Université Paul Sabatier, Laboratoire de Chimie Inorganique, 118, route de Narbonne, F-31062 Toulouse Cedex, France; [4] Université Paul Sabatier, Faculté des Sciences Pharmaceutiques, F-31062 Toulouse Cedex, France; [5] Laboratoire de Chimie Analytique, 31, allées J. Guesde, F-31400 Toulouse, France

INTRODUCTION

Reactive oxygen species in biological systems allow redox reactions by oxygen radicals. Under normal circumstances, intracellular enzymes (superoxide dismutases (SOD), catalases) control the levels of superoxide radical and hydrogen peroxide and prevent oxidative tissue damage. Under perturbed conditions, such as infectious diseases, oxygen metabolism is increased. The subsequent formation of reactive oxygen free radicals which are not detoxified is believed to account for much of the tissue damage observed in inflammatory disorders, aging and cancer (Halliwell, 1985). Investigations have demonstrated that superoxide radicals are scavenged not just by specific metalloproteins but also by free hydrated metal ions and by low molecular mass metal complexes (Aust ***et al.***, 1985). A number of results have been reported for the copper(II) ion and its complexes (Sorenson, 1982). Manganese compounds have been less thoroughly investigated (Archibald & Fridovich, 1982). Interest in developing small molecular mass Mn(II) complexes as models for Mn oxidase enzymes and preparing more active antiinflammatory agents led to the present study.

In a previous work (Nepveu ***et al.***, 1988) conventionnal methods, such as cytochrome-C or chemiluminescence, have been used to compare the superoxide scavenging properties of some manganese compounds; these methods were unefficient to observe significant difference of reactivity between the tested compounds, due to the presence of the additional redox couple, the manganese complex. In this paper, the scavenging of superoxide radical by manganese(II) complexes has been investigated by electron spin resonance (ESR) using 5,5-dimethyl-1-pyrroline-N-oxide (DMPO) and α-(4-pyridyl-1-oxyde)-N-t-butyl-nitrone (POBN) as spin traps.

MATERIELS AND METHODS

Chemical compounds were obtained from commercial sources and used without further purification. DMPO, POBN and enzymes were obtained from Sigma Chemical C°. The DMPO solution was purified with activated charcoal and checked for absence of any ESR contaminating signals. The following experimental conditions have been taken to minimize artifacts and secondary reactions. The acetaldehyde/xanthine oxidase (Ach/XO) reaction has been used as superoxide generating system; diethylenetriamine pentacetic acid (DETAPAC) was included in the phosphate buffer to remove metal ion impurities like iron and prevent hydroxyl radical formation and DMPO oxidation (Archibald & Fridovich, 1982; Finkelstein, 1980).

The ESR spectra were recorded at room temperature using a Brucker ER200D X-band spectrometer. The ESR spectra were obtained within 5 minutes (DMPO) or 15 minutes (POBN) after adding Ach. The instrument settings were: power 20 mW, microwave frequency 9.72 GHz, central field set 3500 G, field scan 200 G, modulation of amplitude 1 Gpp, modulation frequency 100 kHz, time constant 0.5 s.

The manganese compounds have been prepared from manganese(II) chloride or manganese(II) acetate and various substituted carboxylic acids. Analytical data obtained for the studied complexes are consistent with the following formulas: lactic acid ($(C_3H_5O_3)_2Mn.2H_2O$), salicylic acid ($(C_7H_5O_3)_2Mn.3H_2O$), malic acid ($(C_4H_4O_5)Mn.2H_2O$), chelidonic acid ($(C_7H_2O_6)Mn.4H_2O$), gluconic acid ($(C_6H_{11}O_7)_2Mn.2H_2O$), squaric acid ($(C_4O_4)Mn.2H_2O$), croconic acid ($(C_5O_5)Mn.3H_2O$).

RESULTS AND DISCUSSION

Fig. 1 illustrates ESR spectra of POBN spin adducts. The ESR spectrum (Fig. 1a) of the superoxide adduct of POBN, POBN-OOH, was recorded 15 min after adding Ach to the reaction mixture. To confirm the formation of POBN-OOH signal from the superoxide radical, 0.75 U/ml SOD was added to the solution; the formation of POBN spin adducts was prevented (spectrum not shown). In following experiments, manganese lactate complex ($(C_3H_5O_3)_2Mn.2H_2O$) was added at various concentrations (Fig. 1, spectra b), c), d)). The POBN-OOH signal gradually disappeared as the concentration of the complex was increased, indicating a superoxide dismutase-like activity of this compound. Since the typical six lines signal of POBN-OOH is not easily distinguishable from the six lines ESR signal of POBN-OH, it was not possible to observe the eventual decomposition of POBN-OOH into POBN-OH.

The superoxide scavenging properties of the manganese complexes cited in this work have been compared using DMPO as spin trap in the Ach / XO reaction system. Fig. 2 illustrates ESR spectra of DMPO spin adducts. The ESR spectrum of a test solution consisted mainly of DMPO-OOH signal (Fig. 2a). After 5 min, both spin adducts DMPO-OOH and DMPO-OH were observed. In order to verify that generation of DMPO-OH was due to a decomposition of DMPO-OOH, 0.75 U/ml SOD was added to the test solution and formation of both spin adducts was decreased (Fig. 2b); addition of excess SOD prevented DMPO-OOH and DMPO-OH signal. This result showed that DMPO-OH was due to the spontaneous decomposition of DMPO-OOH. When the manganese croconate complex ($(C_5O_5)Mn.3H_2O$) (40 μM) was added to the reaction system, DMPO-OOH and DMPO-OH spin adducts disappeared and an unidentified radical was observed (Fig. 2c). However the signal intensity of this species decreased when SOD was added to the reaction mixture. Similar results were observed with all the tested manganese compounds. DMPO-OH intensity was never increased in presence of these manganese complexes.

These results demonstrate that all the tested manganese compounds do not generate the hydroxyl radical (OH°) by catalyzing the decomposition of H_2O_2 but have superoxide scavenging properties. Quantitative comparison could not be made due to the signal intensity of the unidentified radical; this radical species could be due to a decomposition of DMPO spin adduct by superoxide radical in our reaction mixture (Samuni *et al*, 1989).

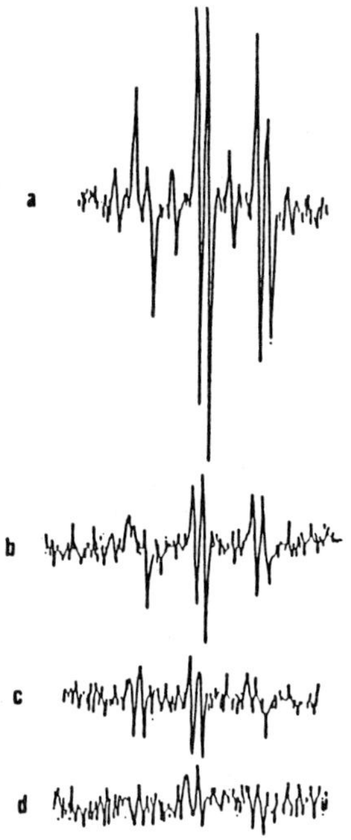

Fig. 1. ESR spectra of POBN spin adducts: Experimental conditions: 100 mM Ach, 60 mU/ml XO, 15 µM DETAPAC, 50 mM POBN, final volume 1 ml, phosphate buffer 0.1M, pH 7.4. a) without manganese compound; b), c), d) with 10, 20, 40 µM manganese lactate complex respectively.

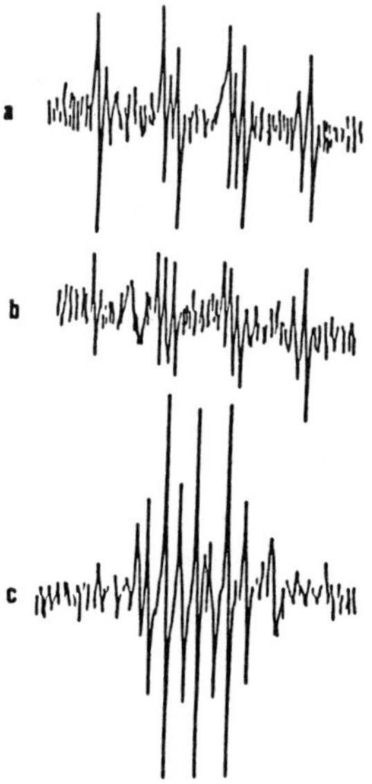

Fig. 2: ESR spectra of DMPO spin adduct: Experimental conditions: 20 mM Ach, 32 mU/ml XO, 15 µM DETAPAC, 50 mM DMPO, final volume 1 ml, phosphate buffer 0.1M pH 7.4. a) without manganese compound; b) with 0.75 U/ml SOD and without manganese compound; c) with 40 µM manganese croconate complex.

REFERENCES

Archibald, F.S., and Fridovich, I. (1982): The scavenging of superoxide radical by manganous complexes: in vitro. Arch. Biochem. Biophys. 214: 452-463.

Aust, S.D., Morehouse, L.A., and Thomas, C.E. (1985): Role of metals in oxygen radical reactions. J. Free Rad. Biol. Med. 1: 3-25.

Finkelstein, E., Rosen, G.M., and Rauckman, E.J. (1980): Spin trapping of superoxide and hydroxyl radical: practical aspects. Arch. Biochem. Biophys. 200: 1-16.

Halliwell, B., and Gutteridge, J.M.C. Eds (1985). In Free Radicals in Biology and Medicine. Oxford: Clarendon Press.

Nepveu, F., Ducrocq, A., and Derache, P. (1988): Determination of the superoxide scavenging activity of various manganese compounds by use of optical and ESR techniques. In Trace Element Analytical Chemistry in Medecine and Biology, eds P. Brätter and P. Schramel. Berlin: Walter de Gruyter.

Samuni, A., Krishna, M., Riesz, P., Finkelstein, E., and Russo, A. (1989): Superoxide reaction with nitroxide spin adducts. J. Free Rad. Biol. Med. 6: 141-148.

Sorenson, J.R.J. (1982): Copper complexes as the active metabolites of the antiinflammatory agents. In Inflammatory Diseases and Copper, ed J.R.J. Sorenson; pp. 289-301. Clifton, New-Jersey, Humana Press.

Metal Ions in Biology and Medicine. Eds. Ph. Collery, L.A. Poirier, M. Manfait, J.C. Etienne. John Libbey Eurotext, Paris © 1990, pp. 24-26

Metallothionein as a scavenger of free radicals

S. Cheng, C.S. Tang*

*Department of Biophysics and *Department of Pathophysiology Beijing Medical University, Beijing, 100083 Republic of China*

Metallothioneins (MTs) are nonenzymic low molecular weight proteins and rich in sulfur and metals. They are widely found in living system. According to the remarkable struture similarity of amino acid sequence in mammalian MTs and exact consistency of cystein position in all the proteins, we believe that MTs should be exerting some basic physiological functions except for metabolism and detoxification of metal ions. At frist we begin to investigate their ability to scavenge free radical and protect cells.

1. Influence of MT (Sigma Co) on postischemic reperfusion injury of isolated rat hearts. Isolated rat hearts were kept at 37°C for 30 min and then reperfused with oxygenated Krebs-Hnseleit (H-K) buffer. After 15 sec of reperfusion, hearts were immediately cut into strips and put in plastic tubes. These tubes were placed in liquid nitrogen and then transferred to the resonnance cavity. ESR spectra were recorded (Fig 1 and 2). There are three signals of ESR spectra for reperfused hearts, g=2.024, 2.004 and 1.935. Signals with g=2.024 and 2.004 are related with axial asymmetry $g_{\parallel}$ and $g_{\perp}$ of oxygen free radicals and g=2.004 and g=1.935 with simiquinone and transitional metal ions free radicals respectively. Signal of g=2.024 of global ischemic heart was remarkably reduced and of reperfused hearts protected with MT diappeared although signals of simiquinone and transitional metal ions did not change a lot. At the time of making specimen for ESR, a strip of myocardium was taken to make specimen for TEM. Mitochondria are the most sensitive organelle to ischemia. EM morphometric studies showed that specific surface (the ratio of the surface to the volume) of mitochondria in reperfused hearts remarkably decreased but of hearts protected eith MT approched that of normal control (Tab 1). It means that MT can alleviate mitochondria swelling and protect cardiac muscle cells from oxygen

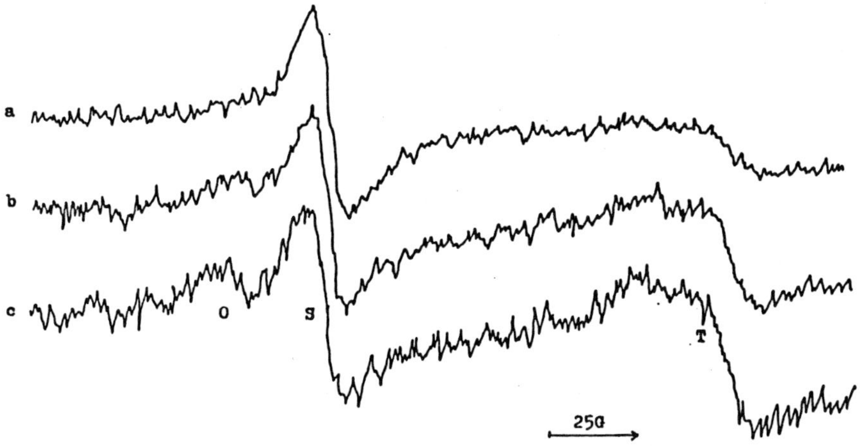

Fig 1 ESR spectra of heart after 30 min ischemia followed by 15 sec of reperfusion showing the power saturation behavior of O,S and T signals. Trace a.with microwave power 1mw. Trace b.with microwave power 10mw. Trace C.with microwave power 20mw.

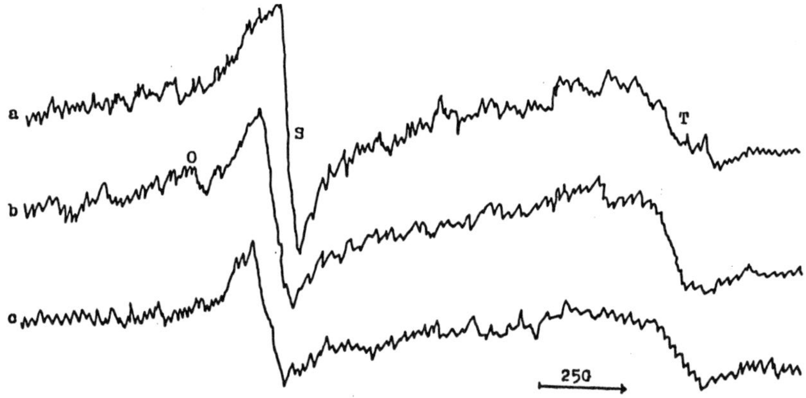

Fig 2 ESR spectra of heart at microwave power 20mw showing the changes in intensity of signal at G = 2.024. Trace a.sample from rat myocardium after 30 min ischemia. Trace b.sample from myocardium after 30 min ischemia followed by 15 sec of reperfusion. Trace c. sample from rat myocardium protected by metallothionein after 30 min ischemia followed by 15 sec of reperfusion.

Tab 1 specific surface (δ) of mitochondria in rat myocardium

	δ (M ± SE)
1 control	7.17±0.14*
2 ischemia	5.83 ±0.13**
3 reperfusion	5.33 ±0.12
4 reperfusion with MT	6.46 ±0.14*

n=50, the number of pictures; * $P<0.001$, ** $P<0.01$

paradox. It is interesting that changes in mitochondria specific surface and appearance and disappearance of oxygen free radical signals were sychronized, indicating oxygen free radicals are involved in postischemic reperfusion injury, and MT (10^{-5}mol/L) is possessed of a capacity to scavenge oxygen free radicals and protect myocardium.

2. Influence of MT on deoxidized/reoxygenated injury in rat cardiac muscle cells. By perfusing isolated rat hearts with K-H buffer containing collagenace, cardiac muscle cells were separated and collected. Intact cells were incubated either in this buffer bubbled at first with 95% N_2-5% CO_2 and then 95% O_2-5% CO_2 or with $FeCl_2$-ascorbic acid (free radical generating system, FRGS) at room temperature. Results are as Tab 2 and 3. In cells hurt by deoxidization/reoxigenation or FRGS MDA increased at double and rate of cells in rod-shape decreased. Oppositely, protected cells were not damaged so seriously, MAD production was restricted and this restraint was dependent on the dosage of MT.

In conclusion, the present results offer evidence that MT is able not only to clean away free radicals in vitro, but also to quench them produced in some pathologic processes, in which free radical overproduction occurs.

Tab 2 protection of MT on cardiac muscle cells from deoxidized/reoxygenated injury

	control	injury	MT protection (mol/L)		
			10^{-5}	10^{-6}	10^{-7}
rate of living cell (%)	73.2+0.7**	56.4±2.1	64.9±1.4**	60.3±1.5*	54.4±1.9
MDA content (nmol/mg protein)	0.040±0.04**	0.79±0.06	0.55±0.04**	0.72±0.05*	0.77±0.06

n=6, comparing with injury group * $P<0.05$, **$P<0.01$

Tab 3 protection of MT on cardiac muscle cells from FRGS injury

	control	injury	MT protection (mol/L)		
			2×10^{-5}	2×10^{-6}	2×10^{-7}
rate of living cell(%)	72.8±0.7**	56.4±3.6	66.4±3.6*	62.1±2.8*	57.8±2.1
MDA content (nmol/mg protein)	0.45±0.04**	0.94±0.07	0.76±0.08**	0.80±0.10*	0.84±0.14

n=4, comparing with injury group * $P<0.05$, ** $P< 0.01$

Metal Ions in Biology and Medicine. Eds. Ph. Collery, L.A. Poirier, M. Manfait, J.C. Etienne. John Libbey Eurotext, Paris © 1990, pp. 27-29

Efficacy of the Treatment with TIOMAG in Allergic Rhinites

P.J. Porr[1], J. Szántay[2], E. Tomescu[3], O. Marineanu[2], M. Rusu[2]

[1] *Inst. Hyg. publ. Hlth, Cluj, Romania;* [2] *3rd med. Clin., Cluj, Romania;* [3] *ORL Clin., Cluj, Romania*

Starting from the casual observation of a patient with Mg deficit (MD) treated by TIOMAG (original product of Mg gluconate and methionine) in whom besides the disappearance of spasmophilia phenomena the symptoms of a coexisting allergic rhinitis (AR) also disappeared, we proposed to study the therapeutical influence of AR by a Mg substitution treatment.

Material and Method

31 patients (18 women and 13 men) aged between 13 - 70 years (average 37 years) with MD and AR were taken into study. Their symptoms were characteristic: rhinorrhea, sneezing by fits, stifled nose. The diagnosis was confirmed by the ENT Department. The performed allergology tests were positive for home dust, pollen, mould, feathers or cat hair. Two of the patients also had allergic bronchial asthma. In all the cases serum Mg, erythrocytic Mg as well as serum Ca have been determined by complexometric methods. 7 patients also showed a simultaneous Ca deficit.

All the patients were treated exclusively with TIOMAG, vitamins B1, B6 and when necessary lactic Ca. The given dozes were 300-600 mg TIOMAG 3 times per day, 3 times of 2 vitamin B1 tablets per day and 3 times of vitamin B6 tablet per day. Lactic Ca was given 1 g per day. The mean period of treatment was 6.2 weeks.

Results

It was noticed that at the same time with MD correction (fig.1) in all cases the clinical manifestations of AR disappeared. The patients were followed up between 2-32 months (in average 8 months), the evolution being generally good. In 8 of them the symptoms reappeared during spring if they were not under TIOMAG treatment.

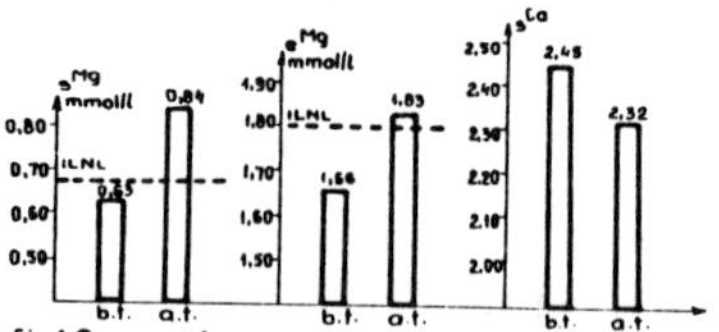

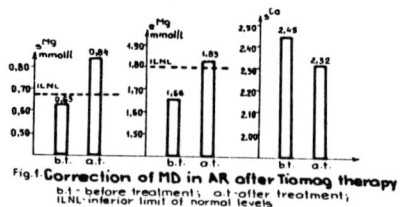

Fig.1. Correction of MD in AR after Tiomag therapy
b.t.- before treatment; a.t.-after treatment;
ILNL-inferior limit of normal levels

Discussion

As to the correlation between MD and various allergical forms, three pathogenetic rings are accepted:

1. By decreasing the cAMP level, MD may favour mastocytes degranulation and respectively the basophylic one, with the increase of histamine and serotonin release.

2. By this effect as well as by its action on essential fatty acids metabolism, MD favours the inflammatory effect of leukotrienes (1).

3. By its effect on the protein synthesis and on T lymphocytes the IgE production is increased (6).

A recent study analyses the Mg effect on bronchoconstriction in allergic bronchial asthma (7). In this case it seems that the main effect of Mg is to antagonize the Ca at the Ca channels level, blocking the Ca^{2+} passage and decreasing the bronchial muscle hyperactivity.

Taking in study a group of 405 patients with noninfectious rhinites, French authors (5) found a rate of 52% with coexistent MD, against 48% without MD. Among those with MD, one third presented AR, two thirds pseudoallergic vasomotor rhinites. However there are sometimes mixed forms of AR as well as pseudoallergical ones (9).

A large proportion of men is to be noted, much larger than usually encountered with MD. (fig.2)

A similar case was also noticed in diabetes mellitus (1).

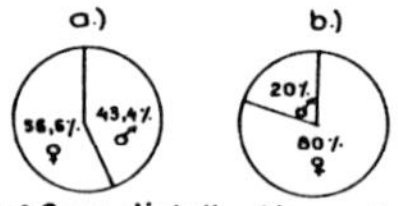

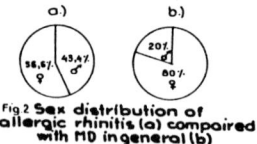

Fig.2 Sex distribution of allergic rhinitis (a) compaired with MD in general (b)

Conclusions

1. RA, if within a MD, can be treated efficiently exclusively by Mg substitution therapy.

2. TIOMAG is suitable in the treatment of these AR forms as well as in the prophylaxis of recurrences.

References

1. Durlach, J. (1988): Magnesium in clinical practice, p.101. London: J. Libbey Co.
2. Elin, R.J. (1987): Overview of Problems in the Assessment of Magnesium Status. In "Magnesium in Cellular Processes and Medicine", ed. B.T. Altura, J.Durlach, M.Seelig, p.67. Basel: Karger.
3. Golf, S.W. et al. (1988): Zur Bedeutung des Magnesiums in der Labormedizin. Med.Welt, 39, 241.
4. McCoy, J.H., Kennedy, M.A. (1987): Magnesium and Immune Function: A Review. In "Magnesium in Cellular Processes and Medicine", ed. B.T.Altura, J. Durlach, M.Seelig, p.196. Basel: Karger.
5. Moneret-Vautrin, D.A. et al. (1980): Profils cliniques différentiels des rhinites allergiques, rhinites vasomotrices non allergiques et polyposes nasales. Med.et Hyg. 38, 119.
6. Prouvost-Danon, A. et al. (1975): Données sur le taux sérique d'anticorps réaginiques (IgE) chez le souris en carence magnésique. Rev.Franç. Allergol., 15, 147.
7. Rolla, G. et al. (1988): The effect of Magnesium on bronchial hyperreactivity of asthmatics, V.Internat.Magnesium Sympos., Kyoto, F-7-12.
8. Shils, M.E. (1988): Magnesium in health and disease. Ann.Rev.Nutr., 8, 429.
9. Wayoff, M. et al. (1978): Rhinites vasomotrices et tests vasomoteurs. Ann. Oto-Laryngol., 95, 211.

II. Carcinogenesis

Metal Ions in Biology and Medicine. Eds. Ph. Collery, L.A. Poirier, M. Manfait, J.C. Etienne. John Libbey Eurotext, Paris © 1990, pp. 33-34

Prevention of cytotoxicity by divalent cations : from medicine to biophysics

C.A. Pasternak

Department of Cellular and Molecular Sciences, St George's Hospital Medical School, Cranmer Terrace, London SW17 ORE, England

Divalent cations (Zn^{2+} >Ca^{2+} >Mg^{2+}) prevent cytotoxicity caused by pore-forming agents of exogenous (certain viruses, bacterial and animal toxins), endogenous (cytolytic immune molecules) or synthetic (some detergents) origin. The mechanism by which these common divalent metal ions protect against membrane damage induced by such diverse agents has been studied in purely lipidic systems (electrical conductivity across planar bilayers and leakage of fluorescent dyes from small unilamellar vesicles) as well as in cells (collapse of membrane potential, leakage of ions, phosphorylated metabolites and larger molecules): it is in every case compatible with a "gating" of agent-induced pores by divalent cations.

Zinc, which has been used therapeutically for more than a thousand years, will be taken as an example of a metal ion with several defined biological roles: (prosthetic group in enzymes and in other proteins (zinc fingers); membrane stabilizing action, etc. By comparing the concentration-dependency of zinc in vitro and in vivo, it is concluded that a major action of zinc in medicine is to protect cellular membranes against certain types of damage.

REFERENCES

Bashford, C.L., Alder, G.M., Menestrina, G., Micklem, K.J., Murphy, J.J. and Pasternak, C.A. (1986): Membrane damage by hemolytic viruses, toxins, complement and other cytotoxic agents: a common mechanism blocked by divalent cations. J. Biol. Chem 261, 9300-9308.

Bashford, C.L., Menestrina, G., Henkart, P.A. and Pasternak C.A. (1988) Cell damage by cytolysin. Spontaneous recovery and reversible inhibition by divalent cations. Journ. Immunol. 141, 3965-3974.

Bashford, C.L., Alder, G.M., Graham, J.M., Menestrina, G. and Pasternak, C.A. (1988) Ion modulatioin of membrane permeability: effect of cations on intact cells and on cells and phospholipid bilayers treated with pore-forming agents. J. Membr. Biol 103, 79-94.

Bashford, C.L., Rodriques, L., and Pasternak, C.A. (1989) Protection of cells against membrane damage by haemolytic agents: divalent cations and protons act at the extracellular side of the plasma membrane. Biochim. Biophys. Acta. 983, 56-64.

Mahadevan, D., Ndirika, A., Vincent, J., Bashford, C.L., Chambers, T.J., and Pasternak, C.A. (1990) (to follow) Protection against membrane-mediated cytotoxicity by calcium and zinc. (in press). J. Path. (in press)

Menestrina, G., Bashford, C.L., and Pasternak, C.A. (1990) Pore-forming toxins: experiments with S. aureus toxin and E. coli haemolysin in lipid bilayers, liposomes and intact cells. Toxicon (in press).

Pasternak, C.A. (1987) A novel form of host defence: Membrane protection by Ca^{2+} and Zn^{2+}. Biosc. Rep. 7, 81-91

Metal Ions in Biology and Medicine. Eds. Ph. Collery, L.A. Poirier, M. Manfait, J.C. Etienne. John Libbey Eurotext, Paris © 1990, pp. 35-39

Dietary minerals and colorectal cancer; a review

R.L. Nelson

Department of Surgery, University of Illinois College of Medicine and Epidemiology/Biometry Program, University of Illinois School of Public Health, Box 6998, M/C 957 Chicago, Illinois 60680 USA

Despite the prevalent opinion that dietary fat and fiber are the most important nutrient factors that determine colorectal cancer risk, there is a growing belief that nutritional studies done to date do not provide enough evidence for this (1,2). This has led Shimizu to state:

" Our findings stress even more strongly that the effects of migration (e.g. Japanese from Japan to Hawaii) on cancer occurring are complex and that the westernization of diet is a gross oversimplification; and by itself it provides an inadequate explanation for the gradients of cancer incidence in migrant populations" (3).

Recently, in light of this disillusionment with established theories of colorectal cancer causation, attention has been more focused on mineral nutrition and cancer risk. The mineral that has received the most attention has been selenium because of the hope that it might prove to be an anticancer substance. Though there is good evidence that many of these minerals have a role to play in colorectal cancer risk, there are not nearly as much data available as there are for fat and fiber. The conclusions drawn from this section are therefore considered best as preliminary findings.

SELENIUM

The epidemiologic evidence that selenium might decrease cancer risk is broadly based, coming from cross cultural, case/control and prospective cohort studies (4). There have however been recent negative studies related specifically to colorectal cancer (5,6). The mechanisms by which selenium might protect against colon tumor induction or incidence include augmentation of glutathione peroxidase activity, alterations in hepatic heme metabolism, alterations in carcinogen metabolism, stimulation of the immune surveillance mechanism, or a cytotoxic effect on the colonic mucosa resulting in decreased mitosis and cell turnover. In vitro studies have shown that selenium decreases the effect of mutagen in the Ames Salmonella microsomal assay and the incidence of sister chromatid exchange induced in mammalian cells, also by mutagens (4).

Numerous animal studies have demonstrated the protective effect of selenium both on the incidence of spontaneously arising breast cancers in mice, on the growth of transplanted tumors on the incidence of tumors induced by chemical carcinogens (4). Many investigators have demonstrated a protective effect of inorganic and organic selenium compounds in colorectal carcinogenesis in for peroxidation. Roughly half of the population of the United States ingest vitamin supplements which almost uniformly contain iron. Yet adult humans absorb less than 10% of dietary iron, the remainder of which remains within the lumen of the colon with lipids from many sources (iron is very constipating) before defecation (13).

Augmentation of DMH colorectal tumor induction occurs in rats, as measured by tumor yield with parenteral iron supplementation and tumor incidence with oral iron supplementation. The alteration in tumor induction associated with oral iron supplementation of a high fat, low fiber diet was reversed by the addition of phytic acid to that diet (13). In an additional study, augmentation of DMH colorectal tumor induction was seen in mice with an iron supplemented diet. Mechanisms investigated in this study included the effect of dietary iron on enzyme activities of those enzymes that metabolize DMH to an active carcinogen, though no significant variation was seen in this regard (14).

The mechanism by which iron may increase colorectal cancer risk relates to iron's effect on oxygen radical metabolism. In the presence of divalent cations, particularly iron, a series of reactions occur, referred to as the Haber-Weiss cycle (or superoxide driven Fenton chemistry), which result in the synthesis from superoxide of the hydroxyl radical and other oxidants. This cycle has been demonstrated in-vitro and in-vivo and also has been found to be sensitive to iron availability in-vivo, for iron delivered either parenterally or enterally. Products of these reactions have been shown to be genotoxic as manifested both through DNA strand breaks and mutagenicity, associating iron with the initiating and promoting events of carcinogenesis (13).

SODIUM/POTASSIUM

Observational epidemiology and case control studies from Japan, France, Norway and of the Japanese in Hawaii have correlated increased sodium intake with increased mortality from cancer of the stomach and of the colon. Similar epidemiologic data also exist for a decreased overall cancer mortality with increased potassium intake in Evans County, Georgia, and Iran. Case control studies in France demonstrate decreased colon cancer mortality with increased potassium or decreased sodium intake. An increased K/Na (dietary or intracellular) ratio is felt to decrease risk, though the mechanism by which this association occurs is unknown (15). One study has focused on the effect of dietary sodium deficiency and vascular volume depletion on tumor proliferation (16). Factors that increase K/Na ratio include increasing altitude of habitat, vitamin A intake, vitamin C intake, and crude fiber intake. Factors that decrease the K/Na ratio include increasing age, stress, polyunsaturated fatty acids in the diet and DMH injection. Increasing the K/Na ratio in the diet in experimental animals has been found to decrease DMH colon cancer induction in those animals. The current K/Na ratio in the American diet is approximately 0.8. This represents a 20-fold drop in the ratio from paleolithic times and it has been hypothesized that this changing ratio may be a causative factor for increasing cancer mortality.

ZINC

Zinc is a trace element with many functions in mammals, being required for the activity of over 100 enzymes, including enzymes involved in DNA synthesis, those essential for the normal immune function of the host and superoxide dismutase, an enzyme that clears cells of toxic and mutagenic oxygen radicals. In contrast to esophageal tumor-associated zinc deficiency, zinc excess has also been linked in epidemiologic studies with tumor incidence of many organs. Though zinc can be primarily carcinogenic in very toxic levels, the mechanism by which elevated zinc is associated with these cancers is clearly not due to this effect. Zinc excess rodents (7). A variety of different rodent species and strains have been used and several carcinogens, though most of them related to 1,2 dimethylhydrazine (DMH) and its metabolites (azoxymethane and azoxymethanol).

A factor common to many studies of selenium in colon carcinogenesis is the supplementation of selenium in the rodent diet in gross excess of what would ever be encountered in human diets. Thus the physiologic relevance of these studies must be questioned, as they have not reproduced the situation in which selenium has been found to be protective in human populations. The protective effect may have been mediated through a cytotoxic effect of selenium excess.

CALCIUM

Epidemiologic evidence that calcium or vitamin D intake diminishes the risk of intestinal cancer is broadly based and more definitive than for any other element except selenium, including prospective cohort as well as case/control studies (4,8,9). A protective effect of dietary calcium has been seen in experimental studies, though not consistently (4,10,11).

A problem with many experimental animal studies relates to calcium supplementation of a rat diet that already contains calcium far in excess of what is seen in human diets. All commercially available rodent diets, from crude rat chow to purified solid and liquid diets, contain 50 times more calcium than is found in human diets, when calculated on a milligram per kilogram diet basis (1% by weight), and 8 times human consumption when calculated on the basis of nutrient density (1.5 mg Ca/kcal) (12). In addition the calcium-phosphate ratio is inverted in rodent diets when compared to human consumption (from 0.26 to 1.4 on a molar basis), a significant factor in determining the bioavailability of calcium.

It has been proposed that calcium has a protective effect in colon cancer by the following mechanism: since fat thought to be a significant risk factor in colon cancer and since fat consumption increases bile salt and cholesterol synthesis in the liver, calcium will bind bile salts and free fatty acids within the lumen of the colon and render them insoluble. The tumor promoting effect of these compounds would then be prevented by oral calcium supplementation (9).

There are certainly other factors in milk, the principal source of dietary calcium, that might alter colon cancer risk besides calcium and vitamin D, including vitamin A, which is often supplemented in milk, and lactose, which may result in a more acid fecal pH. Such acidification has been proven to be protective both epidemiologically and experimentally in colon cancer. An additional mechanism by which milk may diminish gastrointestinal cancer risk is that milk has been found to decrease nitrite concentration and nitrosamine synthesis in-vitro. Once again, it is not at all certain whether the protective effects of milk are mediated through its calcium content (4).

IRON

It has been suggested that the protective effect of dietary fiber observed in human colorectal cancer studies may not be due to alterations in fecal bulk, water content, transit time or pH, but to the chelation by the phytic acid within dietary fiber of dietary iron. When fiber consumption is corrected for phytate content, certain epidemiologic inconsistencies observed related to fiber consumption and colorectal cancer risk are eliminated. It has also been demonstrated in the National Health and Nutrition Examination Survey (NHANES- I) that body iron stores correlate with subsequent cancer risk, particularly cancer of the colon. Dietary iron levels also were positively associated with colon cancer risk (though no other cancer) in that report. Diets that are high in red meat and fat content are iron rich and also provide copious lipid substrate can abolish the protective effect of selenium on experimental tumor induction in rodents, and this may be due to zinc binding of selenicals, thus reducing their bioavailability (4).

Experimental studies of zinc in cancer of various organs can best be characterized by their contradictory results. This inconsistency may be due to a protective effect of zinc in the earliest phases of carcinogenesis, both through its effect on the immune system and on oxygen radical clearance. Yet zinc may augment tumor proliferation though its effect on tissue growth later in the proliferative phase of carcinogenesis. The epidemiologic work on zinc is too scant, and the experimental work too contradictory to draw any definite conclusions about the role of zinc in cancer except for esophageal cancer in certain endemic areas of severe zinc deficiency.

FLUORIDE

The association of fluoride with cancer risk uniquely first appeared in a political rather than scientific forum. The consequences of this technique of scientific revelation were predictable and unfortunate. Crude mortality rates from cancer in ten cities in which water supplies were supplemented with fluoride were compared with ten non-fluoridated cities in the United States and it was found that cancer mortality was high in fluoridated cities. This study failed to control adequately for age, sex, socioeconomic and ethnic background in the populations concerned. It was also not noted in the material that was read into the Congressional Record that the mortality rates in the ten fluoridated cities were in fact higher prior to the institution of fluoride supplementation than subsequent to fluoridation. Extensive reexamination of cancer mortalities in the United States, in areas of high and low natural fluoride content, as well as differential mortalities in fluoridated and non-fluoridated areas have demonstrated no increased cancer risk associated with fluoride consumption or artificial supplementation of water supplies with fluoride (4).

However one interesting finding was revealed in these studies concerning cancer of the colon and rectum. Rectal cancer mortality was found to be decreased in areas with high natural fluoride levels or was found to decrease subsequent to the institution of fluoride supplementation, when compared either with areas with low fluoride levels or cities not yet instituting fluoride supplementation. In addition it appears that proximal colon cancer mortality in women increased subsequent to fluoride supplementation (4).

DIET IN COLORECTAL CANCER; A SUMMARY

In the two-step theory of carcinogenesis, it is felt that initiation is an irreversible mutational event caused by a chemical carcinogen. This is followed by promotion, which is a reversible alteration in gene expression caused by chemicals that, by themselves, are not carcinogenic. Promotion is mediated through the production of oxygen radicals, the rare unstable by-products of oxidative metabolism. None of the dietary items associated with colorectal cancer risk epidemiologically have proven to be on their own carcinogenic, but rather alter colorectal cancer risk by effecting promotion (17). Dietary fat increases the induction of oxygen radicals, as may dietary iron. Those items found both epidemiologically and experimentally to decrease cancer risk do so principally through their effect on lowering the concentration of oxygen radicals, that is vitamins A, E, C, (18) and selenium.

REFERENCES;*

1. Stemmerman G. Geographic Epidemiology of Colorectal Cancer: Role of Dietary Fat. <u>In Colorectal Cancer: From Pathology to Prevention?</u> (Ed.s Seitz HK, Simanowski UA and Wright NA) . Springer Verlag. Berlin 1989. pp 3-23.

2. Byers T. Diet and Cancer, Any Progress in the Interim?

Cancer. 1988. 62. 1713-24.
3. Shimizu H, TM Mack, RK Ross, and BE Henderson. 1987. Cancer of the Gastrointestinal Tract among Japanese and White Immigrants in Los Angeles County. JNCI. 78:223-33.
4. Nelson RL. 1987. Dietary Minerals and Colon Carcinogenesis. Anticancer Res. 7. 259-70.
5. Dworkin BM, WS Rosenthal, A Mittelman, L Weiss, L Applebee-Brady and Z Arlin. 1988. Selenium Status and the Polyp-Cancer Sequence: A Colonoscopically Controlled Study. Am. J. Gastroent. 83:748-51.
6. Coates RJ, N Weiss, JR Daling, JS Morris and RF Labbe. 1988. Serum Levels of Selenium and Retinol and the Subsequent Risk of Cancer. Am. J. Epid. 128:515-23.
7. Reddy BS, S Sugie, H Maruyama, K El-Bayoumy and P Marra. 1987. Chemoprevention of Colon Carcinogenesis by Dietary Organoselenium, Benzylselenocyanate, in F344 Rats. Cancer Res. 47:5901-04.
8. Slattery ML, Sorenson AW and Ford MH. Dietary Calcium Intake as a Mitigating Factor in Colon Cancer. Am. J. Epid. 128. 504-14. 1988.
9. Alder RJ and McKeown-Eyssen G. Calcium Intake and Risk of Colorectal Cancer. Front. Gastroint. Res. 14. 177-87. 1988.
10. Appleton GVN, PW Davies, JB Bristol and RCN Williamson. 1987. Inhibition of Intestinal Carcinogenesis by Dietary Supplementation with Calcium. Br. J. Surg. 74:523-25.
11. Pence BC and Buddingh F. 1988. Inhibition of dietary Fat-promoted Colon Carcinogenesis in Rats by Supplemental Calcium or Vitamin d. Carcin. 9. 187-90.
12. Newmark HL. 1987. Nutrient Density: an Important and Useful Tool for Laboratory Animal Studies. Carcinogen. 8:871-73.
13. Nelson RL, SJ Yoo, JC Tanure and G Andrianopoulos and Misumi A.. 1990. The Effect of Iron on Experimental Colorectal Carcinogenesis. Anticancer Res. in press
14. Siegers CP, D Bumann, G Baretton and M Younes. 1988. Dietary Iron enhances the Tumor Rate in Dimethylhydrazine Induced colon Carcinogenesis in Mice. Can. Let. 41:251-55.
15. Tuyns A. 1988. Salt and Gastrointestinal Cancer. Nutr. & Cancer. 11:229-37.
16. Fine BP, Ponzio NM, Denny TN, Maher E and Walters TR. 1988. Restriction of Tumor Growth by Sodium-deficient Diet. Can. Res. 48. 3445-48.
17. Ames BN. 1983. Dietary Carcinogens and Anticarcinogens. Science 221:1256-64.
18. Colacchio TA, VA Memoli and L Hildebrandt. 1989. Antioxidants vs. Carotinoids: Inhibitors or Promoters of Experimental Colorectal Cancer. Arch. Surg. 124:222-26.
* Limitations in space have necessitated abbreviating the list of references. Other sources can be found in the above citations.

Metal Ions in Biology and Medicine. Eds. Ph. Collery, L.A. Poirier, M. Manfait, J.C. Etienne. John Libbey Eurotext, Paris © 1990, pp. 40-44

Cancer and magnesium status

J. Durlach*, A. Guiet-Bara**, M. Bara**, Y. Rayssiguier***, Ph. Collery****

* *Président de la SDRM, 64, rue de Longchamp, 92200 Neuilly, France; ** University P. et M. Curie, 7 Quai Saint-Bernard, 75252 Paris Cedex 05, France; ***CNRZ, INRA Theix, 63122 Ceyrat, France; **** Hôpital Maison Blanche, F-51100 Reims, France*

Relationship between magnesium and cancer are complex: both Mg load and Mg deficit may produce either carcinogenic or anticarcinogenic effects. Carcinogenesis modifies the Mg status inducing Mg disturbances which may frequently associate a tumor Mg load with Mg depletion in non neoplastic tissues.

INTERACTION BETWEEN CARCINOGENESIS AND Mg STATUS

Magnesium disturbances by carcinogenesis

During chemical carcinogenesis, it is possible to observe structural and functional alterations of the plasma membrane (increase of fluidity and permeability) with an increase of the extracellular cations Ca and Na and a decrease of the intracellular cations Mg and K (Anghileri et al., 1981; Durlach, 1985). These drastic changes are grossly similar to the cellular alterations observed in Mg deficiency. In fact, the disorders in Mg distribution in carcinogenesis are far more complex than those observed in simple Mg deficiency. During the neoplastic state, Mg bound in intracellular structures decreases while its concentration in the cytosol increases. This type of disturbance indicates an alteration of the intracellular Mg regulation (Durlach et al., 1987). In man, the disturbance of Mg distribution associates an increased Mg level in the tumor and in blood cells (erythrocytes and lymphocytes) with some stigmata of extra-tumoral Mg depletion which differ according to the clinical type and treatment (Collery et al., 1981; Durlach et al., 1987; 1990).

Effect on carcinogenesis induced by disturbances of Mg status

Mg deficiency may produce either tumorogenic or anticarcinogenic effects (Fig.1). In the rat only, a prolonged Mg deficiency may induce tumor-like connective proliferation in the intestine or thymic/lymphoma and myeloid leukemia. Genetic factors are prominent: Mg deficiency does not induce oncogenicity in any species other than the rat, or in any rat except in a minority of rats of the same particular strains. The anticarcinogenic properties of Mg observed on this particular model provide only poor support to the general notion of the anticarcinogenic properties of Mg. It is very important to emphasize that the preventive effects of Mg exists only at the early stage of tumorogenesis. On the other hand, it has been well established for almost 25 years that Mg deficiency antagonizes tumor implantation and inhibits growth of induced or spontaneous tumors in the rat. In the man, Mg-deficiency-associated with K-deficiency, induced by diet and hemodialysis has been used as an anticancer treatment to slow the progression of inoperable cancer (Durlach et al., 1987).

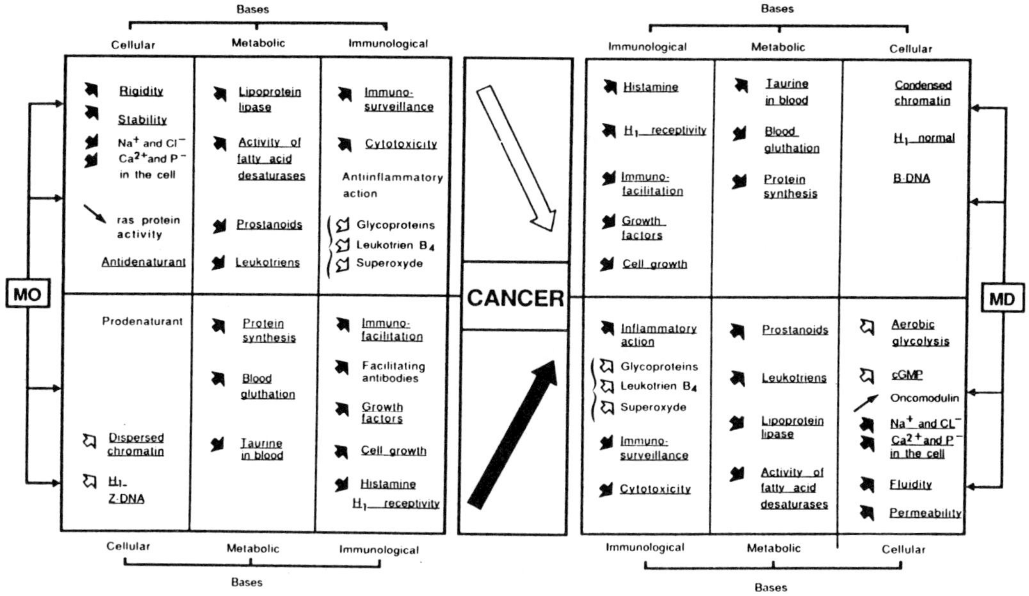

Fig.1. Magnesium metabolism and cancer

CELLULAR INTERACTION BETWEEN CANCER AND Mg STATUS

The concentration of intracellular free Mg is mainly regulated by adaptations of Mg efflux. But tumor cells, rapidly reaccumulate Mg.

Plasma membrane, Mg and cancer (Fig.1)
It is very important to point out the similarities between the plasma membrane impairment due to cancer and Mg deficiency. The stabilizing properties of Mg are due first to its structural role (Heaton et al., 1989). Conformational changes induced by Mg often depend on Mg-dependent transitions. For example, the Mg ion exerts a central role in the regulation of membrane receptors, promoting interconversion from monomer to dimer forms. However, this molecular basis for membrane Mg actions does not exclude enzymatic pathways. Neoplastic transformation of Ehrlich ascites tumor cells reduces the activity of numerous membrane Mg dependent enzymes (Ca-Mg ATPases, Flatman, 1989).
Our in vitro studies on the human amniotic membrane model (Bara et al., 1988; Durlach et al., 1986) showed that Mg induces a biphasic effect on the maternal and fetal sides: ionic fluxes first decrease, then increase, but without alterations of the flux ratio. In the other hand, the carcinogenic metals reduce the fluxes, with alterations of the flux ratio. The reverse actions of the effect of Mg might act as a general antidote of these pollutants. Our earlier and recent data have well established the erroneous character of these hypothesis. Mg may sometimes be a competitive (vs Pb and Cd) or a noncompetitive antagonist (vs Ni), but it may be also inactive on some pollutants (As, Be, Cr, Co, Fe) and sometimes even act as a specific or a non specific agonist (vs Al) (Fig.2). For instance, various cations such as Mg, Ca, Zn, Pb may either compete with or replace each other which may have anticancer or carcinogenic implications.

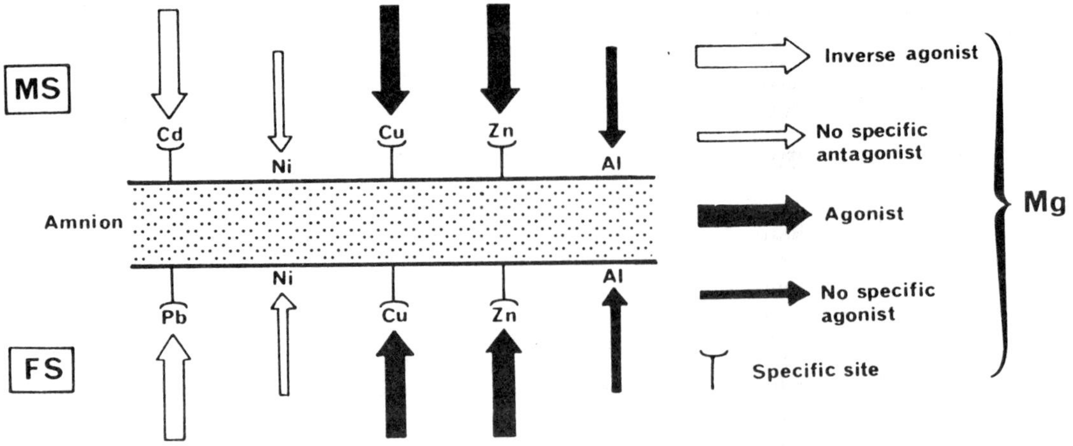

Fig.2. Agonist or antagonist effects of Mg towards carcinogenic metals on the human amnion

Postmembrane and prenucleus Mg and cancer interactions (Fig.1)
In cancer cells, the Ca/Mg ratio-dependent aerobic glycolysis enhancement is mediated by phosphofructokinase activation. A basic effect caused by Mg deficiency is an increased cGMP level and cGMP seems capable of stimulating normal and malignant cell growth (Durlach et al., 1990).
Mg^{++} may be an antagonist of oncomoduline and stabilize the least active form of the transforming protein induced by ras genes. In these reactions, Mg acts as an anticarcinogenic agent.

Nucleus, Mg and cancer (Fig.1)
The most important recent advances concerning the relationship between Mg and cancer rest on the findings concerning the relations with chromatin, histones and nucleic acids, which are most often biphasic. With a laser microfluorimeter coupled to a phase constrast analyzer, it is possible to identify a shrinking-swelling process induced by Mg. It seems that the shrinking-swelling phenomenon has a molecular correspondence at the genome level. The process of acridine orange intercalation can be considered as a stimulation of the accessibility of the genome to biological molecules. During the shrinking phenomenon a high level of chromatin condensation is preserved and during the swelling phenomenon chromatin is dispersed, leading to an increased accessibility to the cell genome and to the higher activation of cell functional processes. Among the main cellular cations, Mg is the one with the most influence on the chromatin structure (Durlach et al., 1990).
Histones, particularly histone H1, stabilize the structure of chromatin. When the concentration of Mg increases, the histones exhibit an increased content of elongated left-handed helices of the poly-L-proline II type. Mg deficiency in the rat at an early stage alters the histone H1, but this disorder is compensated by a subtle subcellular regulation due to a release of Mg from other sites. It seems to be equally important to maintain normal levels of Mg and histone H1 in chromatin in order to benefit from their critical role in higher order structures. Mg can generate the transition from right-handed conformation of DNA (B-DNA) to its left-handed conformation (Z-DNA). For example, a synthetic polynucleotide shows a left-handed form when the ratio Mg/nucleotides is $>$ 190/1000 and a right-handed conformation when the same ratio is $<$ 45/1000. The hexahydrated Mg ion

binds directly to the guanine N7 and O6 atoms and to the negative oxygen of the phosphate. The Mg atom near the N7 site prevent this site from denaturants: a normal Mg status is preventive against cancer. At the contrary, an increased concentration of Mg^{++} induces and stabilizes the transition from B-DNA to Z-DNA, it may stimulate carcinogenesis.

SYSTEMIC MECHANISMS OF THE RELATION BETWEEN Mg AND CANCER

Metabolic basis (Fig.1)

Among the drastic changes in cancer and Mg-deficient cells, an increase of Na (and Cl) and of Ca (and P) involve grossly similar features. Mg deficiency produces several variations in lipid metabolism which may participate in carcinogenic processes, e.g., reduced activity of lipoproteinlipase system and disturbances of fatty acids metabolism, involving inhibition of several steps of desaturation, increased synthesis of prostanoids and leukotriene B4 (Hanada et al., 1989). All these metabolic actions of Mg are relevant to its carcinogenic effects. The metabolic anticancer properties of Mg may come from its antagonism against carcinogenic or cocarcinogenic agents.

Among the numerous effects of Mg on protein metabolism, which plays a prominent part in mechanisms of cell growth, its role in sulfur amino acid metabolism is important. The inhibition of tumor growth due to Mg deficiency in the rat may depend on mobilization of taurine and inhibition of the biosynthesis of blood glutathione. These metabolic actions of Mg agree with some carcinogenic powers of this ion.

Immunological basis (Fig.1)

The anticarcinogenic action of Mg rests on Mg stimulation of cancer immunosurveillance and inhibition of immunofacilitation. Reverse mechanisms concern the carcinogenic effects of Mg deficit.

Immunostimulation and immunosuppression may be either useful or noxious in cancer treatment. The beneficial or detrimental effects of Mg load and Mg deficit agree with this notion. But, in fact, hemolymphoreticular malignancies are usually under immunosurveillance: in such cases the Mg-stimulating effects are useful. However, most often solid tumors are immunostimulated. It is therefore advisable to be careful with all types of immunostimulation and particularly with Mg treatment. This rule admits some exceptions, e.g., the control of a badly tolerated Mg deficit, especially when it is induced by a side effect of an efficient cytolytic treatment like Cis-Pt.

CONCLUSIONS

The data described may provide a basis for further developments in the physiological, experimental and therapeutic fields.

1. Through numerous cellular and systemic mechanisms, the relation between Mg status and cancer appears very complex: both Mg load and Mg deficit sometimes induce anticancer and sometimes carcinogenic effects. The following general scheme may be retained. Established carcinogenesis induces Mg disturbances which associate tumor Mg load due to Mg mobilization through blood cells with Mg depletion in non neoplastic tissue.
In experimental and clinical cancer and anticancer research, it appears very important to investigate magnesium status and the homeostatic or perturbating factors which regulate or disturb cellular, subcellular and systemic Mg metabolism.

2. Mg supply may be used as an adjuvant in hemolymphoreticular malignancies, particularly when symptomatic Mg deficit appears, aggravated by iatrogenic factors, i.e., use of therapies inducing Mg deficit: cyclosporine, amino-

glycosides.... Prophylactic anticancer action requires a balanced, normal Mg metabolism, with sufficient Mg intake in particular.

3. In the future, the ideal treatment for Mg disturbances in cancer should control the systemic and local factors which regulate cellular and subcellular Mg distribution in tumoral cells and those which prevent Mg depletion in nonmalignant cells.

REFERENCES

Anghileri, L.J., Collery, P., Coudoux, P. and Durlach, J. (1981): Experimental relationship between magnesium and cancer. Mag. Bull. 3, 1-5.

Bara, M., Guiet-Bara, A. and Durlach, J. (1988): Relationship between magnesium, carcinogenic metals and ionic permeability of the human amnion. Mag. Res. 1, 231-237.

Collery, P., Anghileri, L.J., Coudoux, P. and Durlach, J. (1981): Magnésium et cancer: données cliniques. Mag. Bull. 3, 11-20.

Durlach, J. (1985): Le magnésium en pratique clinique. 412pp. Paris: J.B. Baillère.

Durlach, J., Bara, M., Guiet-Bara, A. and Collery, P. (1986): Relationship between Mg, cancer and carcinogenic or cocarcinogenic metals. Anticancer Res. 6, 1353-1362.

Durlach, J., Bara, M., Guiet-Bara, A., Rinjard, P. and Collery, P. (1987): Données nouvelles sur les rapports entre Mg et cancer. In Magnesium Physiologische Aspekte für die Praxis, Ed. B. Lasserre, pp 26-45. Hedingen-Zurich: Panscienta Verlag.

Durlach, J., Bara, M. and Guiet-Bara, A. (1990): Magnesium and its relation to oncology. In Metals ions in biological systems. Ed. H. Sigel. New-York: M. Dekker Publ.

Flatman, P.W. (1989): Magnesium and ion transport in red cells. In Magnesium in health and disease, Ed. Y. Itokawa & J. Durlach, pp 43-51. London: John Libbey & Company.

Hanada, K., Suzuki, S., Hasimoto, I., Sone, K. and Ishida, K. (1989): Aetiological role of leukotriene B4 and superoxide anion on magnesium deficiency dermatis. Mag. Res. 2, 19.

Heaton, F.W., Tongyai, S. and Rayssiguier, Y. (1989): Membrane function in magnesium deficiency. In Magnesium in health and disease, Ed. Y. Itokawa & J. Durlach, pp 27-33. London: John Libbey & Company.

Metal Ions in Biology and Medicine. Eds. Ph. Collery, L.A. Poirier, M. Manfait, J.C. Etienne. John Libbey Eurotext, Paris © 1990, pp. 45-48

Iron in cancer. A review of the present knowledge

E. Harju

Department of Surgery. Central Hospital. Central Finland. SF-40620 Jyväskylä. Finland

Iron deficincy and body iron lack

Lack of body iron is common in cancer patients and it associates with complications in surgery and in animal experiments. From 42 to 51 per cent of 171 patients with gastrointestinal cancer had a low serum ferritin concentration indicating lack of body iron stores (Harju & Lindberg, 1985). Iron lack has also shown to be persistent e.g. in patients with stomach cancer after total gastrectomy (Harju, 1985). Among 228 patients with iron lack 38 had postoperative complications,whereas the respective number among 220 patients with normal body iron was 7 patients (Harju, 1988). The patients with lack of body iron stores may have only a slightly lower blood hemoglobin concentration than patients with normal iron stores. The detection of lack of body iron is not possible only from blood hemoglobin concentration value (Harju, 1988).

Empty iron stores have shown as a risk in prospective series of abdomino perineal resection for rectum carcinoma (Harju et al., 1990). The patients with postoperative complications had the lowest serum ferritin, metabolic wound healing was decreased in patients with low serum ferritin, and amounts of collagen and OH-proline were lower than in patients with normal serum ferritin. Rats with body iron lack showed increased gastric mucosal bleeding and a higher frequency of respiration under stress, increased mortality and blood lactic acid concentration in peritonitis, and decreased strength of gastrotomy agains pressure when compared with controls (Harju & Lindberg, 1988;Harju et al., 1988).

Anemia is a risk for oral cancer (Altini et al., 1989;Maresky et al., 1989). Epidemiologic data indicate a lower risk of lung cancer in iron-depleted women, but not in men (Selby & Friedman, 1988). Occupational exposure in iron foundry (Perera et al., 1988) and to iron seems as a risk factor for lung cancer.

High serum ferritin concentration

High serum ferritin level seen frequently in cancer may predict a poor prognosis in head and neck malignancies (Maxim & Veltri,1986; Bhatavdekar et al., 1987),leukemia (Aulbert & Schmidt, 1985;Potaznik et al., 1987), Hodgkin's disease (Bezwoda et al., 1985) and in abdominal malignancies (Harju, 1985;Yokoyama et al., 1985).High serum ferritin level is often present withother risk factors like malnutrition and icterus. High serum ferritin level predicts complications in bone marrow transplantation (Or et al., 1987). There are contradictionary results about the value of high serum ferritin level in hepatocellular, testicular (Szymendera et al., 1985; Ockhuizen et al., 1985), gastrointestinal (Harju & Lindberg, 1985; Harju, 1985), and in bronchial malignomas (Rapellino et al., 1986; Cox et al., 1986).

Basic ferritin is increased in acute nonlymphosytic leukemia, Hodgkin's disease, breast and lung cancer (Cox et al., 1986). As the mechanisms are suggested adverse effects on host immune response, hydroxyl radical formation, lipid peroxidation and inhibition of hematopoesis (Cazzola et al., 1985;O'Connell & Peters, 1987).The combination of the information about serum ferritin with other markers seems reasonable in prognostic sense e.g. with CEA and alfa-antitrypsin in lung, gastrointestinal and breast cancer (O'Connell & Peters, 1987), with immunosuppressive acid protein in gastric cancer (Yokoyama et al., 1985;Saji et al. 1986) and with alfa-fetoprotein in hepatocellular carcinoma and with CA125, TPA, IAP, CEA in ovarian cancer (Harju,1990b).

Hemochromatosis

Untreated hemochromatosis may produce liver cancer; among 163 patients with hemochromatosis during follow-up of 10.5+/-5.6 years liver cancer was 219 times more frequent than death rates expected for an age-matched normal populations (Strohmeyer et al.,1988). Iron overload may affect the structure and function of cellular DNA, for a marked reduction of the proliferative capability after a mitogenic stimulus and a dramatic decrease of the capacity to repair DNA damages were found in lymphocytes from iron overloaded rats (Pietrangelo et al.1988).Tumor growth of colon adenocarcinoma, hepatoma and mammary adenocarcinoma in mice is slower on low iron than on normal diet (Hannet al.,1988). A combination of N-hydroxy-N-aminoguanidine and iron chelators synergistically can inhibit the growth of L1210 cells (Weckbecker et al.,1988).The dependency of tumor growth on iron is also shown when transferrin synthesis by human leukemic cells and small cell lung cancer has been found to relate to iron requirement for cellular proliferation, which is inhibited by gallium salts, iron uptake inhibitors (Bergamaschi et al.,1988;Vostrejs et al.,1988). Expression of the human transferrin reseptor and its mRNA is strongly induced by iron deprivation (MÜllner & Kuhn, 1988).

Iron and sytotoxicity.

In human chronic myeloid leukemia cells Fe++ partially reversed the cytotoxicity of caracemide in combination with 2,2-bipyridine, while it had no effect on the cytotoxicity of caracemide alone (Satyamoorthy et al.,1988). The cardiotoxicity of adriamycin may be mediated by its OH formation. This has been inhibited by iron chelator in perfused rat heart (Rajagopalan et al., 1988). The cytotoxcicity to breast tumor cells of dihydroxy derivate of VP-16 may involve iron-dependent free radicals (Sinha et al., 1988). Mycotoxin ochratoxin, a nephrotoxic carcinogen, enhances the rate of lipid peroxidation in rat liver microsomes; a process requireing the presence of iron (Rahimtula et al., 1988). It has been shown that gamma-linolate kills human breast cancer cells. Iron accelerates the rate of cell death (Begin et al., 1988).

Iron and mutations. Iron and immune function.

Oxygen free radicals damage cellular constituents and have been causally implicated in the pathogenesis of many human diseases. In isolated DNA oxygen free radicals generated by Fe++ in aqueous solution seems to be mutagenic (Loeb et al., 1988).Iron chelator can produce an irreversible inhibition of DNA biosynthesis under hyperthermia in intact P388 murine lymphocytic leukemia cells in vitro (Satyamoorthy & Chitnis,1988).

Iron overload impairs antigen-specific immune responses and reduces the number of functional helper precursor cells, and impairs the generation of cytotoxic T-cells, enhances suppressor T-cell activity and reduces the proliferative capacity of helper T-cells. In iron deficiency impaired T- and B-cell function have been demonstrated (Good et al., 1988).

References

Altini, M., Peters, E. & Hille, J.J. (1989): The causation of oral precancer and cancer.S.Afr.Med.J.Suppl 6-10.

Aulbert, E. & Schmidt, C.G. (1985): Ferritin - a tumor marker in myeloid leukemia. Detect.Prev. 8,297-302.

Begin, M.E., Ells, G. & Horrobin, D.F. (1988): Polyunsaturated fatty acid-induced cytotoxicity against tumor cells and its relationship to lipid peroxidation. J.Natl.Cancer Inst. 80,188-94.

Bergmaschi, G., Cazzola, M., Dezza, L., Savion, E., Consonni, L. & Lappi, D. (1988): Killing of K562 cells with conjugates between human transferrin and a ribosome-inactivating protein (SO-6). Br.J.Haematol.68,379-84.

Bhatavdekar, J.M., Vora, H.H., Goyal, A., Shah, N.G., Karelia, N.H. & Trivedi, S.N. (1987): Significance of ferritin as a marker in head and neck malignancies. Tumori 73,59-63.

Cazzola,M., Arosio, P., Bellotti, V., Bergamaschi, G., Dezza, L., Iacobello, C., Ruggeri, G., Zappone, E., Albertini, A. & Ascari, E. (1985): Immunological reactivity of serum ferritin in patients with malignancy. Tumori 71,547-54.

Cox, R., Gyde, O.H. & Leyland, M.J. (1986): Serum ferritin levels in small cell lung cancer. Eur.J.Cancer clin.Oncol.22,831-5.

Good, M.F., Powell, L.W. & Halliday, J.W. (1988): Iron status and callular immune competence. Blood Rev.2,43-9.

Hann, H.W., Stahlhut, M.W. & Blumberg, B.S. (1988): Iron nutrition and tumor growth: decreased tumor growth in iron-deficient mice. Cancer.Res.48,4168-70.

Harju, E. (1985): Lack of body iron after total gastrectomy for adenocarcinoma of the stomach: a serum ferritin follow-up. Int. Surg.70,319-22.

Harju, E. (1988): Empty iron stores as a significant risk factor in the abdominal surgery. JPEN 12,282-5.

Harju, E. (1990): Serum ferritin concentration in cancer. An overview. Alim.Nutr.Metab. in press.

Harju, E., Huttunen J. & Ylä-Herttuala S. (1990): Clinical and metabolic effects and wound healing metabolism in controlled total parenteral nutrition with high vs. low nitrogen content for seven days after abdominoperineal rectum resection for carcinoma. J.Int.Med.Res. inpress.

Harju, E. & Lindberg, H. (1985): Lack of iron stores is common in patients with gastrointestinal disease. Surg.Gynecol.Obstet. 161,362-6.

Harju, E. & Lindberg, H. (1988): Total depletion of iron stores with moderate anemia brings about a significant risk in abdominal operations. An experimental study on rats. Current Surg. 45,185-8.

Harju, E, Lindberg, H. & Niittymäki, S. (1988): High serum lactic acid concentration after laparotomy and peritonitis in iron depletion. An experimental study on rats. In Vivo 2,385-8.

Loeb, L.A., James, E.A., Waltersdorph, A.M. & Klebanoff, S.J. (1988): Mutagenesis by the autoxidation of iron with isolated DNA. Proc.Natl.Acad.Sci.USA.85,3918-22.

Maresky, L.S., deWaal, J., Pretorius, S., vanZyl, A.W. Wolfaardt, P. (1989): Epidemiology of oral precancer and cancer. S.Afr. Med.J.Suppl 18-20.

Maxim, P.E. & Veltri, R.W. (1986): Serum ferritin as a tumor marker in patients with squamous cell carcinoma of head and neck. Cancer 57,305-11.

MÜllner, E.W. & Kuhn, L.C. (1988): A stem-loop in the 3'untranslated region mediates iron-dependent regulation of transferrin receptor mRNA stability in the cytoplasm. Cell.53,815-25.

Ockhuizen,T., Kok, A.J., Sleijfer, D.T., deBruijn, H.W., Koops, H.S. & Marrink, J. (1985): Serum ferritin determinations are of no value in the management of patients with differented non seminomatous germ cell tumors. Eur.J.Cancer Clin.Oncol.21,931-4.

O'Conell, M.J. & Peters, T.J. (1987): Ferritin and hemosiderin in free radical operation, lipid peroxidation and protein damage. Chem.Phys Lipids 45,241-9.

Or, R., Matzner, Y. & Konijn, A.M. (1987): Serum ferritin in patients undergoing bone marrow transplantation. Cancer 60,1127-31.

Perera, F.P., Hemminki, K., Young, T.L., Brenner, D., Kelly, G. Santella, R.M. (1988): Detection of polycyclic aromatic hydrocarbon-DNA adducts in with blood cells of foundry workers. Cancer Res.48,2288-91.

Pietrangelo, A., Cossarizza, A., Monti, D., Ventura, E. & Françeschi, C. (1988): DNA repair in lymphocytes from humans and rats with chronic iron overload .Biochem.Biophys.Res. Commun. 154,698-704.

Potaznik, D., Groshen, S., Miller, D., Bagin, R., Schwartz, M. & deSousa, M. (1987): Association of serum iron, serum transferrin saturation, and serum ferritin with survival in acute lymphocytic leukemia. Am.J.Pediatr.Hematol.Oncol.9,350-5.

Rahimtula, A.D., Bereziat, J.C., Bussacchini-Griot, V. & Bartsch, H. (1988): Lipid peroxidation as a possible cause of ochratoxin A toxicity. Biochem Pharmacol.37,4469-77.

Rajagopalan, S., Politi, P.M., Sinha, B.K. & Myers, C.E. (1988): Adriamycin-induced free radical formation in the perfused rat heart: implications for cardiotoxicity. Cancer.Res.48,4766-9.

Rapellino,M., Baldi, S., Obert, R., Pecchio, F., Oliaro, A. & Mancuso, M. (1986): Usefulness of measuring serum ferritin in the diagnosis and staging of non-microcytoma bronchial carcinoma. Minerva Med.77,2221-2.

Saji, S., Yokoyama, Y., Niwa, H., Takao,H., Kida, H., Kawata, R., Tanemura, H. & Sakata, K. (1986): Clinical studies on serum immunosuppressive acidic protein (IAP) and ferritin in gastric cancer patients: with special reference to preoperative value and infulence of surgical stress. J.Surg.Oncol.33,215-22.

Satyamoorthy, K. & Chitnis, M.P. (1988): Hyperthermia blocks reversibility of hydroxurea and biopyridine induced synergistic inhibition of DNA biosynthesis in P388 murine leukemia cells. Neoplasma 35,153-9.

Satyamoorthy, K., Chitnis, M.P. & Advani, S.H. (1988): In vitro cytotoxicity of caracemide alone and in combination with hydroxyurea or iron-chelating agents in human chronic myeloid leukemia cells and nurine tumors. Neoplasma 35,27-35.

Selby, J.V. & Friedman, G.D. (1988): Epidemiologic evidence of an association between body iron stores and risk of cancer. Int.J. Cancer 41,677-82.

Sinha, B.K., Eliot, H.M. & Kalayanaraman, B. (1988): Iro-dependent hydroxyl radical formation and DNA damage from a novel metabolite of the clinically active antitumor drug VP-16. FEBS Lett. 227,240-4.

Strohmeyer,G., Niederau,C. & Stremmel, W. (1988): Survival and causes of death in hemochromatosis. Observations in 163 patients. Ann.N.Y.Acad.Sci.526,245-57.

Szymendera, J.J., Kozlowicz-Gudzinska, I., Madej, G., Sikorowa, L., Kaminska, J.A. & Kowalska, M. (1985): Clinical usefulness of serum ferritin measurements in patients with testicular germ cell tumors. Oncology 42,235-8.

Vostrejs, M., Moran, P.L. & Seligman, P.A. (1988): Transferrin synthesis by small cell lungcancer cells acts as an autocrine regulator of cellular proliferation. J.Clin.Invest.82,331-9.

Weckbecker, G., Lien, E.J. & Cory, J.G. (1988): Properties of N-hydroxy-N'-aminoguanidine derivates as inhibitors of mammalian ribonucleotide reductase. Biochem.Phamacol.37,529-34.

Yokoyama, Y., Saji, S., Misao, H., Kunii, Y., Kajima, T., Kunieda, K., Nishiwaki, T., Adachi, Y., Kawata, R. & Sakata, K. (1985): Clinical study on serum immunosuppressive factors in gastric cancer patients - with special reference to preoperative ferritin levels and influences of surgical stress. Gan.No Rinsho.31,1387-92.

Metal Ions in Biology and Medicine. Eds. Ph. Collery, L.A. Poirier, M. Manfait, J.C. Etienne. John Libbey Eurotext, Paris © 1990, pp. 49-53

On the experimental role of copper in cancer

A.M. Badawi

Department of Application, Petroleum Research Institute, Nasr City, Cairo, Egypt

INTRODUCTION

Copper is recognized as an essential metalloelement and is primary associated with copper-dependent cellular enzymes (Sorenson, 1987). Cytochrome C oxidase is an enzyme required for cellular utilization of oxygen. Cytosolic and extracellular superoxide dismutase (SOD) are required for disproportionation of superoxides. Tyrosine is required for synthesis of dopa from tyrosine. Dopamine-B-hydroxlase is required for synthesis of norepinephrine from dopamine. Lysyl oxidase is required for synthesis of collagen and elastin from procollagen and proelastin. Amine oxidases are required for oxidation of primary amines to aldehydes in chatecholamine and other primary amine metabolism. Ceruloplasmin which represents the bulk of serum copper is required for mobilization and utilization of stored iron (Frieden, 1986).

The essentiality of copper is now understood upon its recognised need for copper-dependent enzymes that may have a role in cancer development and inhibition.

COPPER AND CELL GROWTH

The plasma tripeptide glycyl-L-histidyl-L-lysine (GHL) stimulates the growth of a wide group of cultured systems (Pickard et al., 1980). During the isolation of GHL, it was found the compound to co-isolate with copper. It is suggested that GHL may act as a copper transport factor and when added to culture media has been found to: enhance the growth of hepatoma cells; stimulate the growth of lymphocytes; and raise the viability of kidney cells, macrophages and mast cells.

COPPER AND CELL DIFFERENTIATION

Solid Ehrlich tumors taken from mice treated with Cu (II) (2,5-diisopropyl-salicylate)$_2$, Cu (II) (3,5-DIPS)$_2$ contained differentiated epithelial cells in duct arrangement, suggesting that Cu (II) (3,5-DIPS)$_2$ treated did not kill tumor cells but caused them to differentiate to normal duct cells (Leuthauser, 1979). Cu (II) (3,5-DIPS)$_2$ added to neuroblastoma culture medium caused differentiation of these neoplastic cells to normal neuronal cells in a concentration related manner (Sahu, 1979).

There is now evidence which suggests that SOD is intimately involved in cell division (Oberley et al., 1981).

ALTERED COPPER METABOLISM IN CANCER

Copper metabolism has been studied in a variety of neoplastic diseases. Elevated plasma copper concentration has been reported in adults and children with active Hodgkin's disease (Asbjornsen, 1979) and non-Hodgkin's lymphomas, cervical carcinoma, mammary carcinoma, as well as several other carcinomas (Sorenson, 1982).

ALTERED COPPER METABOLSIM IN CARCINOGENESIS

Rat, with implanted tumor undergo change in serum copper and ceruloplasmin oxidase activity with various forms of cancer (Linder, 1983). There is a decrease in the percentage of copper in ceruloplasmin and an increase in the percentage of low molecular weight forms of copper in serum.

Copper content increased in mitochondrial and microsomal fractions, but decreased in nuclear, lysosomal, and supernate fractions in tissues of chemically induced tumor bearing strain A mice. The subcellular distribution of ceruloplasmin was lower in cell fractions (Chakravarty et al., 1984).

EFFECT OF COPPER ON CARCINOGENESIS

Although a large number of inhibitors of chemical carcinogenesis have been described, copper complexes represent a novel class of cancer chemoprotective agents (Wattenberg, 1985).

There is ample evidence to suggest reactive oxygen species participate in cell stages of carcinogenesis (Kensler, 1984). Reactive oxygen detoxifiers and scavengers such as copper complexes and SOD inhibit the biochemical and biological action of tumor promoters.

It was reported that a single application of tumor promoter to the dorsal skin of mice led to a rapid and mark diminution of epidermal SOD activity (Solanski, 1981). Repetitive treatment with the phorbol diester TPA, as occurs during tumor promotion, caused a substaind 40 to 50 per cent decrease in total epidermal SOD activity. This reduction in activity was specific to the copper zinc form of SOD.

Application of Cu (II) $(3,5\text{-DIPS})_2$ before each biweekly treatment of TPA to CD-1 mice initiated with 7,12-dimethylbenz [a] anthracene (DMBA) reduced tumor development 87 per cent in terms of tumor multiplicity, after 24 weeks (Egner et al., 1985). Analogs lacking SOD-activity, namely DIPS and ZnDIPS did not inhibit promotion (Kensler et al., 1983; Egner & Kensler, 1985).

EFFECT OF COPPER ON EXPERIMENTAL TUMOR GROWTH

A large variety of different classes of copper complexes were reported to have antitumor activity. Of the recent copper complexes reported to have antitumor activities in rodents are : Cu (II) glycyl-glycyl-histidinate (Kimoto et al., 1983), Cu (II) 2,9-dimethyl-1,10-phenanthroline (Mohindru, 1983), Cu (II) trans-bis-salicylaldoximate (Paavo & Elo, 1985), Cu (II), 3,4-dihydroxybenzohydroxamate (Basosi et al., `1987) and Cu (II) bis-acetato-bis-imidazole (Tamura et al., 1987). Mechanistic studies provided a great deal of information for these copper complexes concerning their possible inhibition of DNA synthesis.

A reduction in tumor growth was reported (Oberley et al., 1982) when, CF1 mice were injected with a single dose of Cu Zn-SOD one hour after they implanted Sarcoma 180 tumor cells.

It was first reported that copper complexes of antiinflammatory drugs, including Cu (II) salicylate had SOD-mimetic activity (de Alvare et al., 1976). It was then thought that small molecular weight copper complexes might have anticancer activity. Cu (II) $(3,5\text{-DIPS})_2$ was compared with several other copper complexes and found to be a very effective anticancer agent in CBA/J mice implanted with Ehrilich carcinoma cells (Leuthauser et al., 1981; Pottathil & Lang, 1983).

It is not known how Cu (II) $(3,4\text{-DIPS})_2$ inhibitis tumor growth. If tumor cells are sensitive to hydrogen peroxide then addition of Cu (II) $(3,5\text{-DIPS})_2$ could increase hydrogen peroxide generation from superoxide and kill the cell. The protective effect of glutathione and sensitizing effect of BCNU when given with Cu (II) $(3,5\text{-DIPS})_2$ supports this idea. Tumor cells generate superoxide which is detoxified via either removing SOD activity in the cell or by reaction with other intracellular components such as glutathione or sulfhydryl (Leuthauser et al., 1984). A second possibility is that cancer patients have elevated levels of serum aldehydes, possibly generated by lipid peroxidation. Aldehydes inhibited immune cell function by altering membrane fluidity or surface receptors (Loven et al., 1985). Nonimmune cells might also be affected by such products (Henle et al., 1986). Addition of lipophilic compound with SOD-like activity could prevent lipid peroxidation and protect cells from aldehydes, thus allowing the host immune system to respond to the tumor. The third mechanism might involve the facilitation of syntehsis of copper-dependent enzymes in biological systems (Willingham & Sorenson, 1986).

CONCLUSION

After consideration of the experimental role of copper in cancer development and inhibition, the future use of copper chelates with SOD-like activity to treat neoplastic diseases without killing transformed cells has some exciting possibilities .

REFERENCES :

Asbjornsen, G. (1979): Serum copper compared to erythrocyte sedimentation rate as indicator of disease activity in Hodgkin's disease. Scand J. Haematol._22: 193-196.

Basosi, R., Trabalzini, L., Pogni, R., and Antholine, W.E. (1987): Biomolecular dynamics and electron spin resonance spectra of copper complexes of antitumor agnets in solution. J. Chem. Soc., Faraday Trans. 83: 151-159.

Chakravarty, P.K., Ghosh, A., and Chowdhury, J. (1984): Subcellular distribution of copper and ceruloplasmin in chemically induced tumor tissue. J. Comp Path. 94:607-609.

de Alvare, L.R., Goa, K., and Kimura,T. (1976): Mechanism of superoxide anion scavenging reaction by bis (salicylato) copper (II) complex. Biochem. Biophys. Res. Comm. 69:687-694.

Egner, P.A., and Kensler, T.W. (1985): Effects of a biomimetic superoxide dismutase on complete and multistage carcinogenesis in mouse skin. Carcinogenesis 6: 1167-72.

Egner, P.A., Taffe, B.G., and Kensler, T.W. (1985): Effect of copper complexes on multistage carcinogenesis. In Biology of Copper Complexes, ed. J.R.J. Sorenson, pp. 413-422. Clifton, New Jersey : Humana Press.

Frieden, E. (1986): Perspective on copper biochemistry. Clin. Physiol. Biochem. 4:11-19.

Henle, K., Moss, A., and Nagle, W. (1986): Mechanism of spermidine cytotoxicity at 37°C and 43°C in Chinese hamster cells . Cancer Res. 46: 175-182.

Kensler, T.W., Bush, D.M., and Kuzumbo, W.J. (1983): Inhibition of tumor promotion by a biomimetic superoxide dismutase. Science 221: 75-77.

Kensler, T.W., and Trush, M.A. (1984): Role of oxygen radicals in tumor promotions. Envir. Mut. 6: 593-616.

Kimoto, E., Tananaka, H., Gyotoku, J., Morishige, F., and Pauling, C. (1983): Enhancement of antitumor activity of ascorbate against Ehrlich ascites tumor cells by the copper:glycylglycylhistidine complex. Cancer Res. 43 : 824-828.

Leuthauser, S.W.C. (1979): Antitumor activities of superoxide dismutase and copper coordination compounds. University Microfilms International. Thesis No. 8, 012,387.

Leuthauser, S., Oberley, L., Oberley, T., Sorenson, J.R.J., and Ramakrishna, K. (1981): Antitumor effect of a copper coordination compound superoxide dismutase-like activity J. Nat. Cancer Inst. 66:1077-1081.

Leuthauser, S., Oberley, L., Oberley,T., and Loven, D. (1984): Lowered superoxide dismutase activity in distant organs of tumor bearing mice. J. Natl. Cancer Inst. 72: 1065-1074.

Linder, M.C. (1983): Change in the distribution and metabolism of copper in cancer: A review. J. Nutr. Grow. Cancer 1 : 27-38.

Loven, D., Leeper, D., and Oberely, L. (1985): Superoxide dismutase levels in Chinese hamster ovary cells and ovarian carcinoma cells after hyperthermia or exposure to cycloheximide. Cancer Res. 45:3029-3033.

Mohindru, A., Fisher, J., and Rabinovitz, M. (1983): 2,9-dimethyl-1, 10-phenanthroline (neocuproine): A potent, copper-dependent cytotoxin with antitumor activity. Biochem. Pharmacol. 32: 3627-3632.

Oberley, L.W., Oberley, T.D., and Buettner, G.R. (1981): Cell division in normal and transformed cells: The possible role of superoxide and hydrogen peroxide. Med. Hypotheses 7: 21-42.

Oberley, L.W., Leuthauser, S.W.C., Buettner, G.R., Sorenson, J.R.J., Oberley, T.D., and Bize, I.B. (1982): The use of superoxide dismutase in the treatment of cancer. In Pathology of Oxygen, ed. A.D. Autor, pp. 207-218, New York: Academic Press.

Paavo, O.L., and Elo, H.O. (1985): Antitumor activity and metal complexes, a comparison. Inorg. Chim. Acta 107:L15-L16.

Pickart, L., Freedman, J.H., Loker, W.J., Peisach, J., Perkins, Ch. M., Stenkamp, R.E., and Weinstein, B. (1980): Growth-modulating plasma tripeptide may function by facilitating copper uptake into cells. Nature 288: 715-717.

Pottathil, R., and Lang, D. (1983): Interferon-induced biochemical changes in cell membranes : Possible role of cellular enzyme superoxide dismutase. Prog. Clin. Biol. Res. 135:275-297.

Sahu, S.K. (1979): Effects of dexamethasone of neuroblastoma cell differentiation, superoxide dismutase activity, and possible role of negative oxygen ion on cell differentiation. University Microfilm International, Thesis No. 8, 012,417.

Solanki, V., Rana, R.S., and Slaga, T.J. (1981): diminution of mouse epidermal superoxide dismutase and catalase activities by tumor promoters. Carcinogenesis 2: 1141-1146.

Sorenson, J.R.J. (1982): The anti-inflammatory activities of copper complexes In Metal Ions in Biological Systems XIX, ed. H. Siegel, p. 77. New York and Basel : Dekker.

Sorenson, J.R.J (1989):A physiological basis for pharmacological activities of copper complexes : An hypothesis. In Biology of Copepr Complexes, ed. J.R.J. Sorenson, pp. 3-10, Clifton, New Jersey: Humana Press.

Tamura, H., Imai, H., Kuwahara, J., and Sugiura, Y. (1987): A new antitumor complex: Bis (actetato) bis (imidazole) copper (II) J. Amer. Chem. Soc. 109: 6870-6871.

Wattenberg, L.W. (1985): Chemoprevention of cancer.Cancer Res. 45: 1-8.

Willingham, W.M., and Sorenson, J.R.J. (1986): Physiologic role of copper complexes in antineoplasia. Trace Elements in Med. 4: 139-152.

III. Aging

Metal Ions in Biology and Medicine. Eds. Ph. Collery, L.A. Poirier, M. Manfait, J.C. Etienne. John Libbey Eurotext, Paris © 1990, pp. 57-61

The influence of oral zinc supplementation on aging processes in humans. Preliminary research outline

D. Garfinkel, M. Jedwab

Department of Medicine «C», The E. Wolfson Medical Center, Holon 58100 and Sackler School of Medicine, Tel Aviv University, Tel Aviv, Israel

INTRODUCTION

The significant progress in medical treatment during the last decades has brought a substantial increase in the average lifespan which is now around 75 years in the developed western countries. However, as the maximal lifespan has not changed at all since antiquity, we are facing a rapid increase in the percentage of elderly people in our population. In other words, there is a significant increase in the ratio between the older population with its medical, nursing, economical and social problems and the younger population who carries this burden. The western society is trying to correct this imbalance between non-contributing/contributing populations by directing economical and technical resources for specific care for the elderly. But, it seems that the need for specific services for the old is much higher than the availability and the problem is only going to get worse - a vicious circle with the continuing trend of aging in the population.

However, there is another way of attacking this problem. The main morbidity and disability of the aged are the end stage of chronic diseases which are age dependent or at least time dependent. These age related diseases include non insulin-dependent diabetes mellitus (Type II), most types of cancer, osteoporosis, senile cataract, senile dementia and probably atherosclerosis.. Some authors have speculated that if we succeed in slowing down the aging processes which underlie the age-related phenomena, the latter will be delayed and their consequences (the disabling diseases) postpone to an older age (Hayflick,1975; Walford,1980; Garfinkel,1986). Nowdays, we realize that some phenomena regarded as "aging processes" may be slowed down. The best example is the postponement of osteoporosis by preventive medications, another being the delay of aging processes by nutiritional means (Holehan & Merry,1986; Newsome et al.,1988). The main challenge of gerontology today is the finding of new factors by which aging processes may be delayed and as a result, postpone the age related degenerative diseases. The more factors found, the more improved will be the individual's medical condition, thus will the beneficial, economical and social implications which were mentioned earlier.

This philosophy provides the basis for the proposed research which is to test the influence of zinc on aging processes in humans.

Background

In the next paragraphs we will summarize the reasons for our assumption that zinc deficiency plays a significant role in the development of aging processes and the rational of prolonged administration of zinc compounds to humans in an attempt to slow down aging processes (Garfinkel,1986).

In various species, zinc is a co-factor of more than 200 enzymes including carbonic anhydrase, alkaline phosphatase, lactic, malic and alcohol dehydrogenase, carboxypeptidase A, B and C, retinal reductase and super-oxide dismutase. Furthermore, most enzymes concerned with DNA replication, transcription and repair (DNA polymerases, DNA-dependent RNA polymerases, TdT and thymidine kinase) are zinc enzymes (Vallee & Galdes,1984; Wu,1987). Zinc may also have an important role in maintaining the normal structure and function of biomembranes (Bettger & O'Dell,1981). The importance of zinc for growth is unquestionable. In the rat, zinc deficiency causes growth retardation, testicular atrophy, skin changes and immune deficiency. Zinc deficienty during pregnancy in female rats was associated with a very high incidence of chromosomal aberations and congenital malformations (Hurley,1977; Hurley et al.,1982). Fetal growth retardation and an increased rate of congenital malformations have also been found in the offspring of women who had zind deficiency during their pregnancy (Bergmann et al.,1980; Meadows et al.,1981). It may be concluded that zinc is essential for normal growth, development and function of every living cell. In humans, clinical syndromes caused by zinc deficiency are rare. Symptoms of zinc deficiency include growth retardation, dermatitis, alopecia, lethargy, mental depression, neuropathy, ophtalmic damage and immune deficiency (Prasad' et al.,1961; Moynahan,1974). During the last 3 decades, an increasing body of evidence has indicated that zinc deficient diets are much more common than previously suspected particularly in low income populations and in conditions of increased body requirements for zinc such as pregnancy, lactation and old age (Food and Nutrition Board NRC,1980; Prasad,1979; Wagner et al.,1980; Butte et al,1981; Sanstead et al,1984; Duncan,1988). There is no simple and reliable index of zinc status (McKenzie,1979; Tulikoura & Vuori,1986; King,1986) and in conditions of marginal zinc deficiency, its serum concentration may be normal (Walravens et al.,1983; Hambidge,1986). Walravens et al. (1983) gave zinc supplementation to normal children who had a relatively poor growth but normal serum zinc concentration and showed a significant improvement in growth in comparison to the control group. Therefore, some experts suggest that the only way of detecting chronic zinc deficiency is a therapeutic trial of zinc administration (Editorial, 1981). Lately, zinc concentration in leucocytes was shown to be a good indicator of intracellular zinc status (Wells et al.,1987). There are several clinical conditions in which zinc administration was proven to be beneficial; zinc deficiency causes severe impairment in collagen production and in wound healing, probably because zinc has a major role in the cross linking of collagen (McClain et al.,1973). Zinc administration has been reported to improve the healing of surgical wounds, leg ulcers and acne (Husain,1963; Pories et al.,1967; Michaelsson,1980; Fell,1985). The importance of zinc in maintaining the acuity of the special senses has been reviewed by Russell et al (1983); zinc administration improves taste and smell acuity especially in patients with chronic dialysis and in the elderly, even when serum zinc concentration is normal. Zinc concentration in the retina is relatively high; zinc administration improves the impairment of dark adaptation which is commonly found in cirrhotic patients. This improvement probably results from restoration of retinal reductase activity (a zinc enzyme)(Vallee et al.,1959). Zinc administration also improves the activity of vitamin A and increases its serum concentration (Cassidy et al.,1978). Zinc metabolism has been shown to be changing in Aging Macular Degeneration (AMD) - a retinal age related disease which is the leading cause of blindness in the aged in the developed countries (Silverstone et al.,1985). Recently, Newsome et al. (1988), have found that zinc administration to patients with AMD caused a significant delay in this degenerative retinal process with the accompanying postponement of blindness. Zinc deficiency may cause impairment of immune responses, particularly those mediated by T-lymphocytes. Zinc deficiency was associated with decreased mass of the thymus and spleen, impairments in neutrophil and monocyte chemotaxis and in the production of immunoglobulins (Moynahan,1974; Pekarek et al.,1979; Beach et al.,1980; Fraker et al.,1987). In vitro supplementation of zinc restored antibody formation of aged spleen cells (Winchurch et al.,1984); an increase in the number of T-lymphocytes and improved IgG-antibody responses were observed in old people after oral zinc supplementation (Duchateau et al.,1981). Furthermore, zinc supplementation caused an increase in serum Thymulin concentration (Fabris et al.(1984); Travaglini et al.,1989).

Zinc is practically nontoxic and doses of 660 mg of zinc salts were given to humans for more than a year without serious complications. Only in a few patients taking megadoses of zinc for more than a year, microcytic anemia of copper deficiency was noted, a condition easily corrected by copper administered for several days (Czerwinski et al.,1974; Prasad et al.,1978).

THE RESEARCH PLAN

1) THE SUBJECTS: The volunteers, healthy, nonsmoking men and women aged 40-65, participate in a longitudinal study of at least 5 years. A special subgroup of volunteers come from the eye clinics in which people suffering from early stages of AMD or cataracts are encouraged to join the research. Another source is the Dermatological clinic. People suffering from chronic diseases such as diabetis mellitus, high blood pressure or ischemic heart disease are being excluded.

2) THE EXAMINERS: Physicians of different specialties working at the E. Wolfson Medical Center, participate in the examination, follow up and assessment of the subjects, each according to his own speciality i.e. internal medicine, ophthalmology, plastic surgery, dermatology, E.N.T. and audiology, lung functions and histopathology. In some cases zinc concentration is being assessed in leucocytes.

3) THE PILL: Each volunteer is asked to swallow one pill every morning on an empty stomach. The pill contains one of three preparations, a) zinc sulfate - 300mg, b) zinc sulfate - 300 mg, with the additionn of an amino acid - 50 mg, and c) a placebo.

4) EVALUATION AND FOLLOWUP OF VOLUNTEERS: Each volunteer is being examined by a physician before entering the study; routine hematological and biochemical blood tests are taken as well. Those volunteers who meet our criteria are being examined by an opthalmologist and undergo audiometry battery and lung function tests. A skin biopsy from the buttocks is being performed on each volunteer. A systematic followup is done for each volunteer every 3 months.

5) EVALUATION OF AGE RELATED PARAMETERS: The influence of zinc supplementation on the aging process is being evaluated by comparing objective parameters which are changing with age, between the study and the placebo group. The following age related parameters will be evaluated:

a. The development of senile cataracts
b. The development of AMD (including fluorescine angiography of the retina).

c. Histological skin changes
d. Lung function tests
e. Audiometry Battery
f. Kidney functions

6)STUDY DESIGN AND EVALUATION OF THE RESULTS: This research project is a randomized double blind study. The zinc tablets and the placebo are being supplied in containers which are labeled "A", "B", or "C". The volunteers as well as all the physicians participating in the research do not know which container is zinc and which is the placebo.

A preliminary statistical evaluation of the results will be performed at the end of two years of research.

REFERENCES

Beach RS, Gershwin ME, Makishima RK, Hurley LS. (1980): Impaired immunological ontogeny in postnatal zinc deprivation. J. Nutr. 110,805-15.
Bergmann KE, Makosch G, Tews KH. (1980): Abnormalities of hair zinc concentration in mothers of newborn infants with spina bifida. Am. J. Clin. Nutr. 33,2145-50.
Bettger W, O'Dell BL (1981): A critical physiological role of zinc in the structure and function of biomemebranes. Life Sci. 28, 1425-38.
Butte NF, Calloway DH, Van Duzen JL. (1981): Nutritional assessment of pregnant and lactating Navajo women. Am. J. Clin. Nutr. 34,2216-28.
Cassidy AC, Brown ED, Smith JC, Jr. (1978): Alteration in zinc and vitamin A metabolism in alcoholic liver disease: a review. In: Zinc and Copper in Clinical Medicine (ed. Hambidge KM & Nichols BLG) Spectrum Publication, Inc. pp. 59-79.
Czerwinski AW, Clark ML, Serafetinides EA et al.(1974): Safety and efficacy of zinc sulfate in geriatric patients. Clin. Pharmacol. Therap. 15,436-41.
Duchateau J, Delepesse G, Vrijens R et al.,(1981): Beneficial effects of oral zinc supplementation on the immune response of old people. Am. J. Med. 70,1001-4.
Duncan Jr. (1988): Zinc nutriture in pregnant and lactating women in different population groups. S. Afr. Med. J. 73,160-2.
Editorial (1981): Another look at zinc. Brit. Med. J. 282,1098-9.
Fabris N, Mocchegiani E, Amadio L et al.(1984): Thymic hormone deficiency in normal aging and Down's Syndrome: is there a primary failure of the thymus? Lancet I,983-86.
Fell GS. (1985): The link with zinc. Brit. Med. J. 290, 242 (letter).
Food and Nutrition Board National Research Council, Recommended Dietary Allowances. (1980): National Academy of Science, 9th ed. Washington, D.C..
Fraker PJ, Jardieu P, Cook J. (1987): Zinc deficiency and immune function. Arch. Dermatol. 123,1699-701.
Garfinkel D. Is Aging Inevitable? The intracellular zinc deficiency hypothesis of aging. (1986): Medical Hypotheses 19,117-137.
Hambidge KM.(1986): Zinc deficiency in the weanling - how important? Acta. Paediatr. Scand. (suppl) 323,52-8.
Hayflick L.(1975): Aging in American's future (a symposium in dedication of the Pharmaceutical Research Center & Medical Administration Building). Hoechst-Roussel Pharmaceuticals, Inc., Somerville, New Jersey.
Holehan AM, Merry BJ. (1986): The experimental manipulation of aging by diet. Biol. Rev. Cambridge Philosophic Soc. 61,329-68.
Hurley LS.(1977): Zinc deficiency in prenatal and neonatal develompment. In: Zinc Metabolism: Current Aspects in Health and Disease (GJ Brewer, AS Prasad, eds) Alan R. Liss, Inc. New York, New York, pp. 47-58.
Hurley LS, Keen CL, Lonnerdal B.(1982): Trace elements and teratology: an interactinal perspective. Basic Life Sci. 21,189-99.
Husain SL. (1969): Oral zinc sulphate in leg ulcers. Lancet 1,1069-71.

King JC.(1986): Assessment of techniques for determining human zinc requirements. J. Am. Diet Assoc. 86,1523-28.
McClain PE, Wiley ER, Beecher GR, Anthony WL. (1973): Influence of zinc deficiency on synthesis and cross-linking of rat-skin collagen. Biochim. Biophys. Acta 304,457-65.
McKenzie JM. (1979): Content of zinc in serum, urine, hair and toenails of New Zealand adults. Amer. J. Clin. Nutr. 32,570-79.
Meadows NJ, Ruse W, Smith MF et al. (1981) Zinc and small babies. Lancet II:1135-37.
Michaelsson G. (1980): Oral zinc in acne. Acta Dermatovener (suppl) 89,87-93.
Moynahan EJ. (1974): Acrodermatitis enteropathica-a lethal inherited human zinc deficiency disorder. Lancet II,399-400.
Newsome DA, Swartz M, Leone NC et al.(1988): Oral zinc in macular degeneration. Arch. Ophtalmol. 106,192-198.
Pekarek RS, Sanstead HH, Jacob RA et al. (1979): Abnormal cellular immune-responses during acquired zinc deficiency. Am. J. Clin. Nutr. 32,1466-71.
Pories WJ, Henzel JH, Rob CG et al.(1967): Acceleration of wound healing in man with zinc sulphate given by mouth. Lancet I,121-124.
Prasad AS, Halsted JA, Nadimi M. (1961): Syndrome of iron deficiency anemia, hepatosplenomegaly, hypogonadism, dwarfism and geophagia. Am. J. Med. 31,532-46.
Prasad AS, Brewer GJ, Scoonmaker EB, Rabbani P. (1978): Hypocupremia induced by Zinc Therapy in Adults. J. Am. Med. Assoc. 240,2166-68.
Prasad AS. (1979): Clinical, biochemical and pharmacological role of zinc (review). Ann. Rev. Pharmacol. Toxicol. 19,393-426.
Russell RM, Cox ME, Solomons NW. (1983): Zinc and the special senses. Ann. Intern. Med. 99,227-39.
Sanstead HH, Henriksen LK, Greger JL, Prasad AS.(1982): Zinc nutriture in the elderly in relation to taste acuity, immune response and wound healing. Am. J. Clin. Nutr. 36,1046-59.
Silverstone BZ, Landal L, Berson D et al.(1985): Zinc and copper metabolism in patients with senile macular degeneration. Ann. Ophtalmol. 17,419-22.
Travaglini P, Moriondo P, Togni E et al.(1989): Effect of oral zinc administration on prolactin and thymulin circulating levels in patients with chronic renal failure. J. Clin. Endocrinol. Metab. 68,186-90.
Tulikoura I, Vuori E.(1986): Effect of total parenteral nutrition on the zinc, copper, and manganese status of patients with catabolic disease. Scand. J. Gastroenterol. 21,421-7.
Vallee BL, Wacker WEC, Bartholomay AF et al.(1959): Zinc metabolism in hepatic dysfunction. Ann. Intern. Med. 50,1077-91.
Vallee BL, Galdes A. (1984): The metallobiochemistry of zinc enzymes. Adv. Enzymol. 56,283-430.
Wagner PA, Krista ML, Bailey LB et al.(1980): Zinc status of elderly Black Americans from urban low-income households. Am. J. Clin. Nutr. 33,1771-77.
Walford RL. (1980): Immunology and aging. Am. J. Clin. Path. 74,247-53.
Walravens PA, Krebs NF, Hambidge KM.(1983): Linear growth of low-income preschool children receiving a zinc supplement. Am. J. Clin. Nutr. 38,195-201.
Wells JL, James DK, Luxton R, Pennock CA. (1987): Maternal leucocyte zinc deficiency at start of third trimester as a predictor of fetal growth retardation. British Med. J. 294,1054-1056.
Winchurch RA, Thomas DJ, Adler WH et al.(1984): Supplemental zinc restores antibody formation in cultures of aged apleen cells. J. Immunol. 133,569-71.
Wu FYH, Wu CW.(1987): Zinc in DNA replication and transcription - Ann. Rev. Nutr. 7,251-72.

Metal Ions in Biology and Medicine. Eds. Ph. Collery, L.A. Poirier, M. Manfait, J.C. Etienne. John Libbey Eurotext, Paris © 1990, pp. 62-66

Aging and magnesium status

Y. Rayssiguier[1], J. Durlach[2], A. Guiet-Bara[3], M. Bara[3]

[1] *Laboratoire des Maladies Métaboliques, INRA, Theix 63122 Ceyrat, France;* [2] *SDRM, 64, rue de Longchamp, Neuilly, France;* [3] *Université Pierre et Marie Curie, Paris, France*

Etiology of Mg deficit

The requirements and the consequences of suboptimal intakes of Mg for the elderly have been recently reviewed (Mountokolakis, 1987 ; Durlach, 1988 ; Durlach et al., 1989). Mg deficit can be induced by insufficient intake (Mg deficiency) or by dysregulation of the metabolism (Mg depletion). Elderly subjects are particularly susceptible to Mg deficit because of decreased intake or absorption, internal redistribution or increased urinary losses. The metabolism is affected by a variety of disease states and medications.

Dietary surveys indicate that diet in industrialized countries provides marginal Mg levels for a significant percentage of adults. In fact, these surveys have not included biochemical or other health indicators of Mg status. Studies upon which current adult recommendations are based, have primarily used young or middle aged subjects with extrapolation of results to older persons. Dietary intake of Mg decreases with age and the elderly may not be as adaptable as the young in maintaining adequate status on low intake. There is an obvious need for more quantitative data on the needs of elderly subjects using adequate analytic and more precise and prolonged balance techniques. A daily intake of 6 mg/kg/day of Mg has been recommended to maintain balance of this mineral (Durlach, 1989). This level of Mg intake is not readily obtainable from the usual diet (Kant et al., 1989 ; Thomas et al., 1989). This may be particularly true for the older groups of the population.

Several experiments indicate the influence of normal aging on Mg status. Old animals are more suceptible to a negative Mg balance. Mg status is maintained primarily through the absorption of dietary Mg by the intestine and tubular reabsorptive processes in the nephron. An age-related decline in the capacity of the intestine to absorb dietary Mg has been suggested but further quantification of the endogenous losses via feces is needed to clarify the effect of aging on Mg absorption. Renal function may decline with aging and the age-related decrement in kidney function could either increase magnesuria through reduction of the ability to maintain Mg homeostasis (Durlach, et al., 1989), either decrease magnesuria with positive relation with the creatinine clearance (Lowik et al., 1989). Part of the increased urinary losses may also be attributable to the muscle atrophy that occurs in elderly subjects and the negative Mg balance may be the result of an accelerated rate of skeletal catabolism.

In the notion that aging will influence Mg status, the pathological aspect of aging must also be considered. Diseases are more frequent in the aged. Elderly patients are particularly susceptible to Mg deficit because they have a high incidence of illness predisposing to Mg depletion (diabetes mellitus, alcoholism). The elderly also often take drugs such as diuretics to cause hypomagnesemia and conditions associated with hypercatabolic state usually exhibit hypermagnesuria (Mountokalakis, 1987 ; Durlach et al., 1989).

Mg deficit in experimental animals

There is a large volume of literature showing that Mg deficit contributes to the aging process and to the vulnerability of age-related diseases (Mountokalakis, 1987 ; Durlach, 1988 ; Durlach et al., 1989). Mg generally declines with aging in the tissues of rats, with higher rates of decline occurring in the shorter-lived Fisher rats (Mc Broom and Weiss, 1983). Neuromuscular abnormalities and irritability are prominent findings in animals fed Mg-deficient diet. Magnesium plays a role within the pathogenesis of stress-related disease and Mg deficiency was found to be connected with modifications in the storage, release and action of chemical neurotransmitters. Mg-deficient animals were shown to be less tolerant to stress and have a decreased lifespan. Even a moderate degree of Mg deficiency markedly aggravates diverse stressor effects. Stress may further contribute to the increased need for Mg since stress has been recognized as a potential cause of Mg deficit.

The secondary effects of Mg deficit on cell constituents are well known. These include loss of potassium, accumulation of sodium and calcium. Several results support the hypothesis that defective membrane function could be the primary lesion underlying the cellular disturbances that occur in Mg deficiency (Rayssiguier, et al., 1989). Fluorescence polarization was used to compare the fluidity of membrane preparations from Mg-deficient and control rats. Plasma and subcellular membranes from Mg-deficient animals were more fluid than those of control rats. An increased permeability of the membrane to ions may be caused by looser packing among the molecules of the bilayer. The loss of Mg from the membrane may contribute to the increased fluidity owing to the direct binding of the phospholipid headgroups but metabolic alterations of the lipid composition are also involved in the modification of membrane fluidity that occurs during Mg deficiency. One of the consequences of damage of the membrane in Mg-deficient animals is the inflow of relatively large amounts of calcium. When calcium homeostasis falls, cell viability is threatened and uncontrolled calcium inflow may be a central and common event in the aging process. The increased permeability of the membrane to Ca forces the mitochondria to sequester the extracalcium resulting in inhibition of ATP production and mitochondrial failure due to Ca overload. Other mechanisms have been proposed based on the ability of Ca to activate lipases. Such activation if excessive could be detrimental because toxic products of phospholipid hydrolysis such as lysophospholipids, free fatty acids, or leucotrienes adversely affect membrane structure and function. Many diverse cytotoxic agents damage cells by rendering their plasma membrane leaky to the passage of ions and low molecular weight metabolites. This increase in non specific permeability can be prevented by raising extracellular Mg. There are other studies showing that Mg is implicated in host defense processes. The evidence is in favor of increased lipid peroxidation in Mg deficiency, an occurence that is also observed during aging (Mahfouz and Kummerow, 1989). Low Mg status is associated with a decrease in collagen resorption leading to an irreversible accumulation of connective tissue.

Mg deficiency has been discussed as a possible contributory factor in the development of atherosclerosis, myocardial damage and arterial hypertension (Rayssiguier and Gueux, 1986). When rats were fed a Mg-deficient diet, hypertriglyceridemia and dyslipoproteinemia characterized by an increase in VLDL and LDL and a decrease of HDL fraction were reported. Mg-deficient rats showed an increased plasma level of apo B and a decreased plasma level of apo A1. Mg deficiency perturbs essential fatty acid metabolism as shown by the increase in plasma linolic acid and decrease in arachidonic acid (Mahfouz and Kummerow, 1989). Other results demonstrate altered platelet function and predisposition to thrombosis in Mg-deficient rats. The arterial damage resulting from Mg deficiency has been extensively reviewed. This includes internal thickening, thinning and fragmentation of elastic membranes, calcification and increased lipid deposition in animals on atherogenic diets. Blood pressure is one of many physiological parameters affected by dietary Mg deficiency. Magnesium deficiency in vitro is thought to lead to an increase in vascular smooth muscle tone and reactivity by modulating uptake content and distribution of Ca in the smooth muscle cell. However the effects of experimental Mg deficiency on blood pressure have been much debated. Mg deficiency in rats results in no blood pressure modification or after an initial decrease to an increase in blood pressure (Luthringer et al., 1988 ; Nishiyama et al., 1990). We have recently observed increased blood pressure in rats fed a Mg-deficient diet containing 0.080 g Mg/kg, for 21 weeks with a decrease in responsiveness to contractile agonists. It has been suggested that senescence is due to functional hemodynamic changes brought about by alterations in the mechanical and functional properties of blood vessels. In general, vascular smooth muscle cells become hyporesponsive to contractile agonists in old age. A similar situation was noted for NA-induced contractile response in Mg-deficient rats i.e. the responses were lower in Mg-deficient rats than in control rats. Arterial damage and increased level of stress hormones may be contributory factors (Rayssiguier et al., unpublished results).

Mg deficiency in rats has been associated with the development of brittle bones and a possible role for marginal magnesium deficiency in bone disease was suggested by these findings. Mg deficiency results in lower rates of bone formation and bone resorption. The resistance to vitamin D is likely to be due to the impaired skeletal responsibility to 1-25$(OH)_2$D. Extensive reviews on the relationship of Mg to immune function are available showing that dietary Mg deficiency leads to abnormal immune response. Another area of research which would be relevant to Mg nutrition in the elderly is the influence of Mg status on the evolution or resistance to cancer (Durlach et al., 1990). Because of the essentiality of Mg in tumor cell, deprivation of Mg in normal diet has been demonstrated to markedly decrease solid tumor growth rate. However Mg is known to be a fundamental regulator of cell cycle. Increased incidence of lymphoma was observed in rats kept on a deficient diet and the inhibitory effect of Mg on tumor induction by heavy metal has been reported.

Clinical aspects of Mg deficit

It is difficult to document Mg deficit in the elderly since depletion of intracellular Mg can occur in the absence of detectable changes in plasma or RBC Mg content. Regardless of age, the variation in plasma and RBC values has been known to be small in healthy people. Some studies underline a large prevalence of Mg deficit in elderly subjects and age-related decreases of Mg contents in cerebrospinal fluid, lymphocyte and skeletal muscle may indicate a decrease in Mg status with age.

Mg deficit in humans leads to neuromuscular dysfunction manifested by hyperexcitability and is sometimes accompanied by behavioral disturbances. It has therefore been speculated that Mg deficit plays a role in age-related muscular and neurological disorders. Cardiovascular disease is the major cause of morbidity and mortality in old age. Mg deficit has been discussed as a possible contributory factor in the development of atherosclerosis, myocardial damage, cardiac arrhythmias and perhaps arterial hypertension. Several observations indicate that patients with ischemic heart disease are often Mg-deficient and that Mg deficit may, together with several other factors, be involved in the development of IHD. However documentation of Mg deficit and well controlled trials of Mg therapy are needed to clarify the role of Mg in human hyperlipemia and cardiovascular damage. In postmenopausal osteoporotic women, trabecular bone Mg content is lower than normal and magnesium deficit was identified in these women, but the importance of dietary Mg deficiency in osteoporosis remains incertain. The hypothesis that Mg is part of the etiology of degenerative neurological syndromes is controversial. A recent study has demonstrated that the brain of patients with Alzheimer's disease contained a normal content of Mg (Durlach, 1988 ; Durlach et al., 1989).

References

Durlach, J. (1988): In Magnesium in clinical practice. London: John Libbey & Company.

Durlach, J. (1989): Recommended dietary amounts of magnesium: Mg RDA. Magnesium Res. 2: 195-203.

Durlach, J., Rayssiguier, Y., Guiet-Bara, A. and Bara, M. (1989): Magnésium et Vieillissement. J. Med. Esth. et Chir. Derm. 16: 283-287.

Durlach, J., Bara, M. and Guiet-Bara, A. (1990): Magnesium and its relationship to oncology. In Metal ions in biological systems, Magnesium and its role in Biology, Nutrition and Physiology, vol. 26, ed. Siegel H., New York, Dekker Inc.

Kant, A.K., Moser-Veillon, P.B. and Reynolds, R.D. (1989): Dietary intakes and plasma concentrations of zinc, copper, iron, magnesium and selenium of young, middle aged and older men. Nutr. Res. 9: 717-724.

Löwik, M., Schrijer, J., Odink, J., Vander Berg, Wedel, M. and Hermus, R. (1990). Nutrition and aging: Nutritional status of "apparently healthy" elderly (Dutch Nutrition Surveillance System). JACN 9: 18-27.

Luthringer, C., Rayssiguier, Y., Gueux, E. and Berthelot, A. (1988): Effect of moderate magnesium deficieny on serum lipids, blood pressure and cardiovascular reactivity in normotensive rats. Br. J. Nutr. 59, 243-250.

Mc Broom, M.J. and Weiss, A.K. (1983): Soft tissue potassium, magnesium, sodium and ash contents throughout the life of highly inbred rats. Comp. Biochem. Physical 74A: 529-533.

Mahfouz, M.M. and Kummerow, F.A. (1989). Effect of magnesium deficiency on delta 6 desaturase activity and fatty acid composition of rat liver microsomes. Lipids 24, 727-732.

Mountokalakis, T.D. (1987): Effect of aging, chronic disease and multiple supplements on magnesium requirements. Magnesium 6: 5-11.

Nishiyama, S., Saito, N. and Konishi, Y. (1989): Effect of severe and moderate magnesium deficiency on blood pressure, cardiac function and regional blood flow in male rats. In Magnesium in Health and Disease. ed. Y. Itokawa, J. Durlach, pp. 253-260, J. Libbey.

Rayssiguier, Y. and Gueux, E. (1986): Magnesium and lipids in cardiovascular disease. J.A.C.N. 5: 507-519.

Rayssiguier, Y., Gueux, E. and Motta, C. (1989): Evidence for a membrane modification in magnesium nutritional deficiency in the rat: fluorescence polarization study. In Biomembranes and Nutrition. Vol. 195, ed. C.L. Léger, G. Béréziat, pp. 441-452, Paris: INSERM.

Thomas, A.J., Bunker, V.W, Sodha, N. and Clayton, B.E. (1989): Calcium, magnesium and phosphorus status of elderly in patients: dietary intake, metabolic balance studies and biochemical status. Brit. J. Nutr. Vol. 62: 211-219.

IV. Various effects and miscellaneous diseases

Metal Ions in Biology and Medicine. Eds. Ph. Collery, L.A. Poirier, M. Manfait, J.C. Etienne. John Libbey Eurotext, Paris © 1990, pp. 69-74

The concept (intra and extra cellular minerals)

M. Nishimuta

The National Institute of Health and Nutrition, 1-23-1 Toyama, Shinjuku-ku, Tokyo (162), Japan

INTRODUCTION

Nowadays, it has been demonstrated that some of minerals were deficit under an inadequate diet because of low mineral intakes. Our recent studies, however, revealed that in some commom living circumstances, such as overeating or stress and so on, excretions of some minerals were increased, with or without a decrease in their intestinal absorption.

Adding to these phenomena, one of minerals, calcium(Ca) is known to move from bone to some soft tissues with age (age related osteoporosis), and with diseases such as atherosclerosis (to the smooth muscle of blood vessels), stone diseases (kidney, gall-bladder) and abnormal calcification.

At the latter case, although bone calcium content was decreased, that of calcified tissues were undoubtedly increase. So, it is confusing if dietary calcium supplement is effective to patients suffered from osteoporosis together with pathological calcification. A possible illustration for this problem may need a new concept appeared at the below.

After human mineral balance studies done in this decade, the concept [intra- and extra- cellular minerals] was introduced to mineral nutrition. This concept may permit us to develope new theories for the aetiology of chronic degenerative diseases, describing as follows.

THE CONCEPT [INTRA- AND EXTRA- CELLULAR MINERALS]

'Intracellular minerals'(ICM) is proposed to be defined as minerals whose content in cells are higher than those in plasma. 'Extracellular minerals'(ECM) is also proposed to be defined as minerals whose content in plasma are higher than those in cells.

Potassium(K), magnesium(Mg), phosphorus(P), zinc(zn), iron(Fe) and so on, are ICM according to the above definition. In contrast, sodium (Na), calcium (Ca), chloride (Cl) and so on, are ECM.

Copper(Cu) does not belong to neither minerals. For its serum and whole blood levels are essentially the same (unpublished observation).

BONE MINERALS

Bone minerals (BM) consist of both ICM and ECM. It was reported that bone was pool of minerals, and that 99% of Ca, 80% of P, 65% of Mg, and 50% of Na was present in the bone in adult human.

Some minerals such as manganese(Mn), strontium(Sr), lead(Pb), and so on, are accumulated into bone when they are appeared in blood.

CONTROLLED VARIABLE OF MINERALS

Plasma levels of essential minerals (PM: Na, K, Cl, Ca, Mg, P, Zn, Fe and Cu) are considered to be regulated, physiologically. Actually, PM are maintained within narrow range both after their ingestion and after their uresis.

However, the mechanism to regulate PM has not yet been fully understood. If PM are controlled variables of ICM, as is proposed by the author, cellular contents of ICM must not be constant(Nishimuta et al., 1989a). Human muscle contents of K and Mg were illustrated to be decreased in some diseases by biopsy.

For the nature of this plasma homeostasis of these minerals, informations from serum chemistry for minerals, usually failed to reveal the intracellular deficit of ICM in clinical practice.

CHARACTERISTICS OF INTRACELLULAR MINERALS

Common characteristics of ICM are as follows. 1)When PM of ICM are under normal ranges, intracellular deficit of ICM can be presumable. 2)These are widely distributed in foodstuffs except in energy mass of lipid and carbohydrates nor in refined products. 3)Most of ICM has been treated as toxic substances or nutrients never to be deficit.

Today, nutritional problems of ICM have been focused at only those of low intake. However, hereafter, they should be also focused at those of absorption and excretion, which were revealed to be changed by a lot of factors usually meet in common living circumstances.

<u>Potassium</u>
Intake of K in common Japanese diet is ranged 1-4 g/d. Most of K is absorbed in intestine, and, apparant absorption of K was ranged 80-90 % in our laboratory(Kodama et al., 1989). Urine K is correlated with the levels of intake, usually. However, that after low K diet was not reflected the levels of intake,

and, was kept low levels even after K load. Almost all plasma K is ultrafiltrable. More than 90 % of K filtered in the glomeruli in the kidney is reabsorbed. Intake of high dose of sodium chloride increased in the urine K (Nishimuta, 1983). An ingestion of any kind of energy source temporarily decreased in urine K (Nishimuta et al., 1986ab, 1989b). Sweat content of K is constant (about 200 mg/l)(Nishimuta et al, 1985). Balance of K in our studies usually showed positive(unpublished observation), suggesting shortage of K in the subjects before the studies. Cerebrospinal fluid K is lower than that of plasma(Hattori et al., 1987).

Magnesium
Intake of Mg in common Japanese diet is ranged 200-300 mg/d. Some of Mg ingested absorbed in intestine, and apparant absorption of Mg was widely ranged 30-70 % of the intake(Suzuki & Nishimuta, 1984), depend on the conditions of individuals. A two third of plasma Mg is ultrafiltrable. Ultrafiltrable fraction of plasma Mg is positively correlated with age (Nishimuta et al., 1989d). This may suggest a possible factor of Mg deficit with age. More than 90% of Mg filtered in the glomeruli is reabsorbed. Urine Mg temporarily increased by the risk factors for chronic degenerative diseases with isomolar excretion of urine Ca, i.e., after excess ingestions of any kind of energy sources(Nishimuta et al., 1986ab, 1989b, Yamada & Nishimuta, 1987), mental and emotional stress(Nishimuta et al., 1988), and physical exercise(Nishimuta & Kobayashi, 1988). However, energy restriction did not decrease in urine Mg(Kodama et al., 1989). Dietary Mg restriction did not decrease in urine Mg, neither(Suzuki et al., 1989). Dietary restriction of P, one of BM, (800 mg/d) increased urine Mg and Ca(unpublished observation). Dietary restriction of Na, another BM, (6 g/d as NaCl) increased sweat content of Mg and Ca, about ten times as much as was not restricted(Nishimuta et al., 1985). Sweat contents of Mg and Ca were usually about 1 mg/l for Mg, and about 10 mg/l for Ca. Cerebrospinal fluid Mg was higher than that of, not only ultrafiltrable fraction, but also total of blood plasma(hattori et al., 1987).

Phosphorus
Optimal P intake in Japanese was reported to be 1300 mg/d. Apparant absorption of P were relatively narrowly ranged 60-80 % in our studies (unpublish obsevation. Some of stress decreased in urine P in the morning(Nishimuta et al., 1988). However, some of stress increased in urine P in the afternoon(Nishimuta et al., 1989b). Physical exercise increased in urine P with the increased in urine Mg and Ca(Nishimuta & Kobayashi, 1988). Dietary restriction of P did not affect urine P. Energy restriction increased urine P. Sweat P was not detectable in our laboratory, and was considered to be very low.

Zinc
Intake of Zn in common Japanese diet was estimated ranged 6-15 mg/d. Main excretory pathway of Zn is via intestine. By this reason, apparant absorption of Zn is affected by the intestinal excretion. Urine Zn is ten times as much as that of Fe, although intake levels of Zn and Fe is almost the same. Energy restriction and physical exercise increased urine Zn(Kitajima et al., 1989). Diabetic patients and aged people excreted more Zn into urine than those of young adults(Nishimuta et al., 1986b). Sweat Zn is considerably high (about 1 mg/l)(Nishimuta et al., 1985), and sweat loss of Zn is one of the aetiology of Zn deficiency in people suffer from heat exposure or in athletes(Nishimuta et al., 1989e). Serum Zn is negatively, but, cerebrospinal fluid Zn is positively cor-

related with age(Nishimuta et al., 1989c).

Sign and symptoms of Zn deficiency in adult population has not been well understood. Theoretically, Zn is essential in DNA and protein synthesis, so, incomplete synthsis of DNA, disfunction of immunity, atrophy of organs whose function are maintained with transmigration, and so on, should be symptoms of Zn deficiency.

Iron

Iron is known to be one of nutrients whose deficiency in human is recognized. Intake of Fe in common Japanese diet was estimated 10 mg/d. Dermal and urine output of Fe is so small as to neglect in our balance studies (less than 0.2 mg /d). However, another group estimated urine Fe as 0.5 mg/d. Animal experiment suggested that Zn deficiency caused uresis of Fe. Absorption of Fe has been considered 10% for normal and 40% for anaemic people. Intestinal output is main path of Fe excretion (0-5 mg/d, presumption by the author). In the case that dietary intake: 10 mg/d, intestinal excretion: 5 mg/d, and fecal output: 9 mg/d, absorption is (10+5-9)/(10+5)=6/15 (40 %). And in the other case that dietary intake: 10 mg/d, intestinal excretion: 0 mg/d, and fecal output: 6 mg/d, absorption is also 40 % [(10-6)/10].

Estimation of plasma levels of Fe has not yet been established. The value obtained after ashing is usually higher than that after acid preparation. This suggested the presence of iron, not bounded to transferrin, but tightely bounded to other protein.

TRANS TISSUE TRAVELING OF MINERALS

There are two faces of tissue or cell function including the bone, physiologically. One of them is a proper action of each tissue, contraction in muscle, for example. The other is acted as reserver of nutrients. And by this action, honeostasis of plasma components may be maintained when dietary supply of nutrient was restricted.

Tissues whose function are maintained without transmigration, such as brain, muscle and so on, may atrophy under chronic deficit of ICM, as donators. On the other hand, tissues, whose functions are maintained with transmigration, such as blood cells, liver, mucous membrane, skin, gland and so on, may cause incomplete multiplication under chronic deficit of intracellar minerals.

It has been beliebed that Mg (ICM & BM) was supplied from the bone when plasma levels of Mg was decreased. However, Mg uresis by risk factors for chronic degenerative diseases is accompanied with isomolar Ca uresis. And Ca density in the bone is far higher than that of Mg, ten to hundred times as much. Furthermore, mineral supplement from the bone is done by way of osteolysis made by osteophagocytosis, total bone absorption.

If plasma Mg is compensated by bone Mg, excess Ca must be released from the bone, inevitably associated with Mg. Therefore, most of Mg must be derived from soft tissues such as liver, muscle and so on, where the minority of body Mg is thought to be located heterogeneously.

The actual conditions may be supposed that excess Ca, released accompanied with Mg from the bone, invades into soft tissues whose content of Mg was decreased by the deprivation. This hypothesis may not only explain symptoms taking place after exposure to stress (shoulder stiffness by muscle contracture and hypertension by blood vessel shrinkage), but also provide an explanation why osteoporosis could often be accompanied with pathological calcification of soft tissues.

CONCLUSION

The concept [intra- and extracellular minerals] was proposed to understand the actual results of human mineral balance studies. It was shown that balances of intracellular minerals tended to be negative at some common living condition in human, although those conditions were different from each minerals.

The aetiology of chronic degenerative diseases such as cardiovascular, metabolic and neoplastic diseases was proposed as chronic deficiency of intracellular minerals.

To confirm these hypotheses, results of human mineral studies, where risk factors for chronic degenerative diseases were taken into acount, were shown to illustrate the presence of negative mineral balance of minerals.

REFERENCES

Hattori,M. et al.(1987):Serum and cerebrospinal levels of sodium,potassium, calcium and magnesium in patients with neurological disorders. J. Jpn. Soc. Magnesium Res. 6, 25-31.

Kitajima,H. et al.(1989):Imcreased urine zinc excretion at low energy intake in human. J.Trace Elements in Exp.Med. 2,.78.

Kodama,N. et al.(1989):Positive mineral balances(Na,K,Ca & Mg) in young Japanese females consuming high fat but low energy diet with additional physical exercise. Magnesium Res. 2, 135.

Nishimuta,M. et al.(1983):Post exercise depression of serum magnesium in human. J. Jpn. Soc. Magnesium Res. 2, 49-56.

Nishimuta,M.(1983):Effects of oral saline administration on urinary sodium and potassium excretion in Japanese young men. J. Jpn. Soc. Nutr. Food Sci. 36, 367-371

Nishimuta,M. et al.(1985):Mineral contents in arm sweat at a low mineral diet with special reference to the onset of physical exercise. J. Jpn. Soc. Magnesium Res. 4, 13-21

Nishimuta,M. et al.(1986a):Magnesiuresis after butter and egg rich diet in young females. J. Jpn. Soc. Magnesium Res. 5, 53-60.

Nishimuta,M. et al.(1986b):Renal handling of minerals after 100 g oral glucose tolerance test. Trace Metal Metabolism. 14, 95-101

Nishimuta,M. and Kobayashi,S.(1988):Mineral requirement for long distance runner Sports Science Report VIII. pp. 46-47. Sports Science Committee. JAAA.

Nishimuta,M. et al.(1988):Stress induced magnesiuresis in human. J. Jpn. Soc. Magnesium Res. 7, 123-132

Nishimuta,M. et al.(1989a):Magnesium uresis by risk factors for chronic degenerative diseases. In Magnesium in Health and Disease, ed. Y. Itokawa & J. Durlach, pp. 279-284. London: John Libbey & Co.

Nishimuta,M. et al.(1989b):Fatigue and mineral metabolism in human. Hiro to kyuyo no kagaku. 4, 55-60
Nishimuta,M. et al.(1989c):Trace metal shift with age in human.Trace Metal Metabolism, 17, 61-64.
Nishimuta,M. et al.(1989d):Age related increase in serum ultrafiltrable calcium and magnesium in human.Magnesium Res. 2, 129
Nishimuta,M. et al.(1989e):Zinc status under risk factors for chronic degenerative diseases in human - "zinc deficient theory" for the aetiology of diseases -. J.Trace Element in Exp.Med. 2, 75
Suzuki,K. and Nishimuta,M.(1984):Magnesium requirment in Japanese young woman. J.Jpn.Soc.Magnesium Res. 3, 7-12
Suzuki,K. et al.(1989):A decrease in faecal magnesium output, without a decrease in urine magnesium excretion, under a relatively low mineral intake in young females.Magnesium Res. 2, 134.
Yamada,H. & Nishimuta, M.(1987):Some energy sources shift mineral metabolism in young rat. J.Jpn.Soc.Magnesium Res. 6, 185-191.

Metal Ions in Biology and Medicine. Eds. Ph. Collery, L.A. Poirier, M. Manfait, J.C. Etienne. John Libbey Eurotext, Paris © 1990, pp. 75-79

Life in Oxygen with iron and copper catalysts. Problems of molecular damage

J.M.C. Gutteridge

NIBSC, Blanche Lane, South Mimms, Potters bar, Herts., EN6 3QG UK

INTRODUCTION TO OXYGEN TOXICITY

Free molecular oxygen probably appeared on the Earth's surface some 2 x 10^9 years ago as a result of photosynthetic micro- organisms acquiring the ability to split water. Oxygen is now the most abundant element in the Earth's crust and the second most abundant element in the biosphere. The concentration of molecular oxygen (hereafter referred to as oxygen) in dry air has risen to 21%. Iron is the second most abundant metal in the Earth's crust whereas copper is considerably less abundant.

The author defines a free radical as an atom or molecule with one or more unpaired electrons. This definition includes the oxygen molecule, a hydrogen atom and most of the transition metal ions (Halliwell and Gutteridge, 1989). To avoid a spin, and possibly, orbital restriction oxygen prefers to accept electrons one at a time. This considerably slows down the reaction of oxygen with the majority of covalent molecules, which are non-radicals, and can be considered an advantage for life in oxygen . A major disadvantage is, however, that electrons when added singly to oxygen lead to the formation of reactive intermediates, two of which are free radicals. The 4-electron reduction of oxygen to water gives rise sequentially to the superoxide radical (O_2^-), hydrogen peroxide (H_2O_2) and the hydroxyl radical ($^{\bullet}OH$).

The superoxide radical is produced in numerous biological processes, particularly the electron transport chains of mitochondria and the endoplasmic reticulum, where electrons can 'leak' onto oxygen. The production of superoxide by activated phagocytic cells is one of the most studied radical-producing systems (Babior, 1978). Radicals produced by phagocytic cells are usually beneficial to the host because they participate in the killing of certain ingested micro-organisms.

Many studies have shown that the superoxide radical is not a particularly reactive species in aqueous solution and cannot account for most of the damage observed in systems in which it is generated (Fridovich, 1975). However, there are a few specific sites within cells at which O_2^- can do some direct damage (Fridovich, 1986) During the last 20 years it has been well established that the superoxide radical can assist the formation of a more damaging radical species by 'Fenton Chemistry'.

The superoxide radical dismutes to form hydrogen peroxide and also reduces, to the ferrous state, iron complexes present. This reduction proceeds with the intermediate formation of perferryl, an iron-oxygen complex with a resonance structure intermediate between that of $Fe^{2+}-O_2$ and $Fe^{3+}-O_2^-$. Ferrous ions and hydrogen peroxide then give the 'Fenton Reaction' to form hydroxyl radicals (·OH). It is likely that simple Fenton chemistry is an over simplification of the process of ·OH formation since it is now known that iron (IV) (ferryl states of iron) are reactive intermediates in the pathway to ·OH formation. The hydroxyl radical is one of the major products of the radiolysis of water and is responsible for much of the damage to tissue done by ionizing radiation. It is highly reactive and able to damage biological molecules at an almost diffusion-controlled rate.

The formation of ·OH radicals from O_2^- and H_2O_2 is a transition metal-catalysed reaction (reviewed in Halliwell and Gutteridge, 1989) and the reaction of ferrous ions with oxygen is an example of an electron transfer reaction in which the initial free radical is the ferrous ion.

Fenton-type Reactions in which a cuprous salt, ferrous salt, or an iron or copper complex, react with H_2O_2 to yield a highly-aggressive oxidant has been the subject of much debate during the past 40 years. Opinion has often shifted away from the view that the Fenton reaction produces hydroxyl radical and that an alternative metal-oxygen species such as ferryl (FeO^{2+}) or cupryl ($Cu(OH)_2^+$) are formed instead. Ferryl species would be expected to be less reactive than ·OH to show different second order rate constants with ·OH radical scavengers and, would not be expected to react with spin traps nor to hydroxylate aromatic compounds in exactly the same way as ·OH radicals do. As already discussed, it seems likely that both ferryl and ·OH are produced in Fenton systems and each might cause damage under different circumstances (Gutteridge et al., 1990).

The oxygen radicals so far discussed (HO_2·, O_2^-, ·OH) are inorganic oxygen radicals. However, organic oxygen radicals are equally important in biological systems. Probably the most studied organic oxygen radical reactions are the autoxidations of polyunsaturated fatty acids. Lipid peroxidation, as this process is now more widely known, proceeds by a chain reaction. When a hydrogen atom is abstracted from an unsaturated lipid (LH) by a free radical (R·), the resulting carbon-centred radical (L·) reacts rapidly with oxygen to give a peroxy radical (LO_2·) and this is its most likely fate under aerobic conditions. LO_2· has sufficient energy to abstract a hydrogen atom from another lipid molecule to perpetuate the chain, converting itself into a lipid hydroperoxide (LOOH).

The reaction will end when the substrate is exhausted or when radicals annihilate each other. Lipid hydroperoxides (LOOH) can be decomposed by iron and copper complexes to give alkoxy (LO·) and peroxy radicals (Equation) and eventually they yield numerous aldehydic products (Schauenstein et al., 1977).

$$2LOOH \xrightarrow[\text{complexes}]{\text{iron/copper}} LO^{\cdot} + LO_2^{\cdot} + H_2O$$

This decomposition pathway (equation) probably accounts for much of the observed stimulation of lipid peroxidation seen when iron salts or iron complexes are added to bulk lipids, isolated subcellular fractions or tissue extracts since they are certain to contain detectable amounts of lipid peroxides (Gutteridge and Kerry, 1982). Whenever tissues are damaged metal complexes are released from vacuoles, organelles or other sequestered sites within the cell. Released metal ions can then interact with cell lipids and lipid peroxides produced in cyclooxygenase and lipoxygenase pathways. This can account for the frequent observation that lipid peroxidation accompanies many disease processes and is associated with cellular injury by toxins. The common interpretation that cellular damage and disease processes are often caused by lipid peroxidation has recently been questioned.

Aldehydes, produced from oxidised lipids, are reactive molecules and will thus not get far from their site of formation, and, although they are not free radicals, they must be considered as part of the toxicity of 'oxidant stress' resulting from free radical damage.

ARE IRON AND COPPER IONS AVAILABILITY IN VIVO FOR RADICAL REACTIONS ?

The normal adult human contains some 4g of iron and 80 mg of copper with about two-thirds of the iron present in haemoglobin. Iron is preciously retained by the body and constantly re-cycled for use as there are no specific physiological mechanisms for iron excretion. Subtle changes in the delicate balance between iron absorption and iron loss can readily lead to iron-overload or to iron-deficiency diseases. Iron and copper salts never exist free in biological systems; they will be bound to biological molecules such as proteins, lipids, nucleic acids, carbohydrates and low-molecular-mass chelators. The question then arises as to which of these metal complexes are able to catalyse radical damage in biological systems?

Loosely-bound iron and copper are the most likely promoters of ·OH radical damage in vivo and attempts have been made to detect and measure them using specific DNA recognition techniques (Gutteridge, et al., 1981; Gutteridge, 1984). Bleomycin is a metal-binding antibiotic which also binds to the DNA molecule. In the presence of iron,oxygen and a suitable reducing agent the DNA molecule is cleaved to release base propenals which break down to give malondialdehyde. The release of malondialdehyde can be measured and related to the amount of iron in biological samples present in the low micromolar range. A similar approach has been adopted using the chelator 1,10-phenanthroline which, only in the presence of complexable copper, releases TBA-reactive material from DNA. Ferritins and haemoglobin appear to promote ·OH formation when present in superoxide generating systems. The reason for this appears to be the ability of superoxide or hydrogen peroxide to release small amounts of iron from the protein-metal centres and it is this low-molecular-mass chelatable iron which is the true catalyst in the Fenton reaction (Gutteridge, 1986; Puppo and Halliwell, 1988). In this way, superoxide radicals can provide all the necessary ingredients in a biological system for ·OH radical formation.

The key role of iron and possibly copper in biological radical formation offers the potential for protective intervention by the use of chelation therapy. Unfortunately, very few of the chelators well known to chemists are suitable for clinical use because they usually facilitate stages of radical formation by altering the redox potential of iron and increasing its solubility. However, desferrioxamine, an iron-chelator widely used clinically in humans, is usually effective at inhibiting iron-dependent radical reactions. The in vivo use of desferrioxamine has given considerable evidence consistent with the biological importance of iron-dependent radical reactions (reviewed in Halliwell and Gutteridge, 1989).

WHAT ARE THE IMPORTANT BIOLOGICAL ANTIOXIDANTS

Specialised antioxidant mechanisms have evolved to deal with specific problems of oxidant stress in different environments. Inside the cell oxygen is reduced to water and, as we have seen, superoxide and hydrogen peroxide are produced. The superoxide dismutase enzymes function to catalytically remove superoxide radicals, bringing about a dismutation reaction at a considerably faster rate than occurs in the absence of enzyme. The product of this reaction is hydrogen peroxide and two different enzymes exist to remove it, namely catalase and glutathione peroxidase (selenium enzyme). Nature considers it worth removing superoxide at the expense of making hydrogen peroxide. Providing superoxide and hydrogen peroxide are effectively removed, no ·OH radicals will be formed and a low molecular mass iron pool can safely exist within the cell to provide iron for the synthesis of iron-containing proteins. The inside of the cell membrane is a lipophilic region in which lipid-soluble radicals are formed. Protection depends to a large extent on lipid-soluble scavenging antioxidants such as vitamin E or β-carotene. Despite their lack of involvement in oxygen metabolism, extracellular fluids are, nevertheless, constantly exposed to the production of O_2^- and H_2O_2 especially from activated phagocytic cells. Extracellular fluids contain little, if any, of the protective enzymes found inside the cell yet protection is required against oxygen radicals. Extracellular antioxidant protection depends largely on mechanisms that prevent metal complexes participating in radical reactions. Iron binding by transferrin and lactoferrin, the binding of haemoglobin and haem by haptoglobins and haemopexin, copper binding by albumin, and ferrous ion oxidation by the copper-containing protein caeruloplasmin (ferroxidase I) contribute a substantial antioxidant potential to extracellular fluids (reviewed in Gutteridge and Halliwell, 1988) Several of these proteins are part of the body's 'acute-phase' response, increasing in concentration whenever tissue damage occurs. Tissue damage will of course amplify oxidative reactions because it will lead to 'mal-placed' iron and copper becoming available for the type of radical reactions described in this review. Many low molecular mass antioxidants have been described in biological extra-cellular fluids such as, urate, bilirubin, glucose and ascorbate. However, their biological contribution is sometimes difficult to assess since the assays used to demonstrate scavenging are often not physiological pro-oxidant systems (reviewed in Halliwell and Gutteridge, 1990). Prevention of radical formation is likely to be a primary role of the proteins described above with secondary protection by scavenging through to the low molecular mass antioxidants. Survival, however, ultimately depends on constant and effective repair of continuous oxidative damage.

REFERENCES

Babior, B.M. (1978): Oxygen-dependent microbial killing by phagocytes. N. Engl J. Med. 298: 659-668.

Fridovich, I. (1975): Superoxide dismutases. Annu. Rev. Biochem. 44: 147-152.

Fridovich, I. (1986): Biological effects of the superoxide radical. Arch. Biochem. Biophys. 247: 1-11.

Gutteridge, J.M.C. (1984): Copper-phenanthroline induced site-specific oxygen radical damage to DNA. Detection of loosely bound trace copper in biological fluids. Biochem. J. 218: 983-985.

Gutteridge, J.M.C. (1986): Iron promotors of the Fenton reaction and lipid peroxidation can be released from haemoglobin by peroxides. FEBS Lett. 201:

291-295.
Gutteridge, J.M.C., and Halliwell, B. (1988): The antioxidant proteins of extracellular fluids. In Cellular Antioxidant defense mechanisms Vol. II. ed. C.K. Chow. pp 1-23 CRC Press : Boca Raton.
Gutteridge, J.M.C., and Kerry, P. (1982): Detection by fluorescence of peroxides and carbohydrates in samples of arachidonic acid. Br. J. Pharamcol. 76: 459-461.
Gutteridge, J.M.C., Maidt, L., and Poyer, L. (1990): Superoxide dismutase and Fenton chemistry : Reaction of ferric-EDTA and ferric-dipyridyl with hydrogen peroxide without the apparent direct formation of iron (II). Biochem. J. (In press)
Gutteridge, J.M.C., Rowley, D.A., and Halliwell, B. (1981): Detection of 'Free' iron in biological symptoms by using bleomycin-dependent degradation of DNA. Biochem. J. 199: 263-265.
Halliwell, B., and Gutteridge, J.M.C. (1989): Free Radicals in Biology and Medicine. 2nd Ed. Oxford University Press: Oxford.
Halliwell, B., and Gutteridge, J.M.C. (1990): The antioxidants of human extracellular fluids. Arch. Biochem. Biophys. (In press).
Puppo, A., and Halliwell, B. (1988): Formation of hydroxyl radicals from hydrogen peroxide in the presence of iron. Biochem. J. 249: 185-190.
Schauenstein, E., Esterbauer, H., and Zollner, H. (1977): Aldehydes in Biological Systems. Pion : London.

Metal Ions in Biology and Medicine. Eds. Ph. Collery, L.A. Poirier, M. Manfait, J.C. Etienne. John Libbey Eurotext, Paris © 1990, pp. 80-83

Metabolic Pathways of dietary carbohydrates play a major role in the expression of copper deficiency

M. Fields, Ch.G. Lewis

Vitamin and Mineral and Carbohydrate Nutrition Laboratories, Beltsville Human Nutrition Research Center, U.S. Department of Agriculture, Agricultural Research Service, Beltsville, Maryland 20705 and Division of Endocrinology, Georgetown University Medical Center, Washington, DC 20007 USA

Copper deficiency in experimental animals results in a reduced body weight, anemia, hypercholesterolemia, hypertriglyceridemia, pancreatic atrophy, impaired glucose tolerance, depressed immune response, heart hypertrophy and heart pathology that includes histopathological changes such as necrosis, degeneration and inflammation (Fields et al., 1983; Fields et al., 1984; Klevay & Viestenz, 1981; Reiser et al., 1983; Redman et al., 1988). However, the most dramatic effect of copper deficiency is on rat mortality. From the fifth week of being fed their respective diets from weaning the copper deficient rats began to die. By the end of the ninth-twelfth week the majority of the rats die of the deficiency (Fields et al., 1983). Critical examination of the diets that have been used to induce copper deficiency in a variety of experimental animals reveals that the diets always contain simple sugars such as fructose, sucrose or powder milk or milk based diets (Fields et al., 1983; Fields et al., 1984; Klevay & Viestenz, 1981; Reiser et al., 1983; Redman et al., 1988; Weissman et al., 1963). When the copper deficient diets contain starch, the animals are protected against the severity of the deficiency (Fields et al., 1984; Redman et al., 1988; Reiser et al., 1983). The animals are not anemic, they gain weight, they do not exhibit pancreatic atrophy nor abnormal glucose tolerance. In addition, their heart is normal in appearance, it is not atrophied and it is devoid of pathological changes. Furthermore, the animals survive. Thus, it has been established that it is the type of dietary carbohydrate that plays a major role in the expression and severity of copper deficiency.

Copper deficient rats when fed fructose containing diets or starch containing diets are similarly copper deficient. This has been established by measuring direct copper indices in blood and tissues. Thus, the dramatic differences seen in the expression of copper deficiency between animals fed starch and those fed fructose could not be due solely to copper status. Therefore it has been suggested that certain metabolic pathways of fructose may be responsible for the severity of copper deficiency. These pathways may not be activated when starch is the main dietary carbohydrate.

A series of studies were conducted at the Beltsville Human Nutrition Research Center, Beltsville, Maryland, U.S.A., in order to understand the mechanisms that are responsible for this copper-carbohydrate interaction. This review summarizes the results of some of these studies in which differences in metabolic pathways between fructose and starch could be responsible for the copper-carbohydrate interaction in experimental animals.

a) Polyol Pathway

The consumption of a diet containing fructose gives rise to an abundance of fructose particularly in the liver. An elevation of fructose in extrahepatic tissues has also been observed (Bellomo et al., 1987; Jeffrey & Jornvall, 1983). This results in the conversion of fructose into its corresponding sugar alcohol sorbitol (Jeffrey & Jornvall, 1983). Sorbitol is poorly diffusible and its accumulation increases intracellular osmolarity and cellular hydration (Jeffrey & Jornvall, 1983). Sorbitol also chelates copper (Briggs et al., 1981) and could make copper unavailable for utilization. In addition, during the process of the conversion of fructose into sorbitol, reduced nucleotides are generated. These reduced nucleotides can provide a reduced environment for numerous transition trace metals (Cantoni et al., 1989; Rowley & Halliwell, 1982; 1985). Changes in the redox state of transition metals can be toxic to tissues due to the generation of free radicals (Cantoni et al., 1989; Rowley & Halliwell, 1982; 1985).

b) Fatty Liver

Unlike starch that in the process of digestion is hydrolyzed to glucose which is available to all body tissues, fructose either as the free sugar or as the fructose moiety of sucrose, is primarily metabolized in the liver (Hers, 1955). Thus, extrahepatic fructose metabolism is very low (Hers, 1955). The consumption of diets containing fructose induce fatty liver and stimulate enzymes involved in lipogenesis (Fields et al., 1985). The activities of these enzymes are not affected by copper deficiency (Fields et al., 1985). In contrast, starch feeding does not stimulate lipogenesis and its consumption does not induce fatty liver (Fields et al., 1985). During the process of lipogenesis, reduced nucleotides are generated. As previously noted, they could alter the oxidation-reduction of transition metals. In addition, during the biosynthesis of hepatic lipogenesis, from fructose, glyceraldehyde is generated (Fields et al., 1989). Glyceraldehyde, a precursor in lipogenesis is considered as a toxic substance since it has been shown to generate free radicals (Thoranelly et al., 1984).

If indeed fatty liver may play a role in the exacerbation of copper deficiency, then by preventing lipogenesis, those rats that are fed fructose should be protected against the severity of the deficiency. However, massive doses of choline, a well established lipotropic agents that has been shown to prevent the accumulation of fatty liver, failed to prevent the formation of fatty liver in copper deficient rats fed fructose (Fields et al., 1988).

c) Copper Deficiency and Alcohol Consumption

If our hypothesis is correct, and the reasons for the exacerbation of copper deficiency by the consumption of diets containing fructose are due to certain metabolic pathways of fructose that in turn create a specific environment for the copper-carbohydrate interaction, then any nutrient other then fructose that is metabolized in a manner that is similar to fructose should be able to mimick the fructose effect and should exacerbate the signs associated with copper deficiency. Alcohol should be such a nutrient. Both alcohol and fructose induce the formation of a fatty liver. Both generate reduced nucleotides and aldehydes. Indeed, the consumption of a 20% ethanol drink for six weeks by copper-deficient rats fed starch resulted in the exacerbation of the deficiency similar to that exerted by fructose (Fields & Lewis, 1990). The signs associated with the deficiency in both alcohol and fructose consumption included anemia, heart hypertrophy with gross pathological changes and mortality. In contrast, none of the copper deficient rats that consumed starch and drank water exhibited any signs of copper deficiency and all survived (Fields & Lewis, 1990). The data from that study support the contention that the combination of certain metabolic pathways of carbohydrate metabolism with copper deficiency are responsible for the exacerbation of the deficiency.

d) Copper Deficiency is Gender Dependent

The sex of the rat plays a major role in the expression of the deficiency (Fields et al., 1987). Unlike their male counterparts, female rats that consume a copper deficient diet containing fructose are protected against the signs associated with copper deficiency and survive (Fields et al., 1987). Endogenous sex hormones do not seem to play a role in this process, since both intact and castrated males die of the deficiency. In contrast, female rats regardless of ovariectomy are protected against the severity of the deficiency and survive (Fields et al., 1987). However, unlike their male counterparts, female rats do not develop fatty liver, they do not generate glyceraldehyde and they do not stimulate the polyol pathway (Fields et al., 1989). Thus, it is suggested that female rats metabolize fructose differently than males. These differences in metabolic pathways that exist between males and females when fructose containing diets are consumed may be responsible for the protection of the copper deficient female rats against the pathology of copper deficiency.

REFERENCES

Bellomo, G., Comstock, J.P. and Wen, D. (1987): Prolonged fructose feeding and aldose reductase inhibition: effect on the polyol pathway in kidneys of normal rats. *Proc. Soc. Expt. Biol. Med. 186:* 348-354.

Briggs, J., Finch, P. and Matulewicz, C. (1981): Complexes of copper (II) calcium and other metal ions with carbohydrates: their layer ligand-exchange chromatography and determination of relative stabilities of complexes. *Carbohydrate Res. 97:* 181-188.

Cantoni, O., Fumo, M. and Cattabeni, F. (1989): Role of metal ion in oxidant cell injury. *Biol. Trace Elem. Res. 21:* 277-281.

Fields, M., Ferretti, R.J., Judge, J.M., Smith, J.C. and Reiser, S. (1985): Effects of dietary carbohydrates on hepatic enzymes of copper deficient rats. *Proc. Soc. Expt. Biol. Med. 178:* 362-366.

Fields, M., Ferretti, R.J., Reiser, S. and Smith, J.C. (1984): The severity of copper deficiency is determined by the type of dietary carbohydrate. *Proc. Soc. Exp. Biol. Med. 175:* 530-537.

Fields, M., Ferretti, R.J., Smith, J.C. and Reiser, S. (1983): Effect of copper deficiency on metabolism and mortality in rats fed sucrose or starch diets. *J. Nutr. 113:* 1335-1345.

Fields, M. and Lewis, C.G. (1990): Alcohol consumption aggravates copper deficiency. *Metabolism 39:* in press, 1990.

Fields, M., Lewis, C.G. and Beal, T. (1988): High dietary choline and copper deficiency. *Nutr. Rep. Intl. 37:* 1281-1287.

Fields, M., Lewis, C.G. and Beal, T. (1987): Sexual differences in the expression of copper deficiency in rats. *Proc. Soc. Expt. Biol. Med. 186:* 183-187.
Fields, M., Lewis, C.G. and Beal, T. (1989): Accumulation of sorbitol in copper deficiency: dependency on gender and dietary carbohydrate. *Metabolism 38:* 371-375.
Hers, H.G. (1955): The conversion of fructose-1-^{14}C and sorbitol-1-^{14}C to liver and muscle glycogen in the rat. *J. Biol. Chem. 214:* 373-381.
Jeffrey, J. and Jornvall, H. (1983): Enzyme relationship in sorbitol pathway that bypasses glycolysis and pentose phosphates in glucose metabolism. *Proc. Natl. Acad. Sci. USA 80:* 901-905.
Klevay, L.M. and Viestenz, K.E. (1981): Abnormal electrocardiogram in rats deficient in copper. *Am. J. Physiol. 240:* H185-H189.
Lee, V.M., Szepesi, B. and Hansen, R.J. (1986): Gender linked differences in dietary induction of hepatic glucose-6-phosphate dehydrogenase, 6 phosphogluconate dehydrogenase and malic enzyme in the rat. *J. Nutr. 116:* 1547-1554.
Redman, R.S., Fields, M., Reiser, S. and Smith, J.C. (1988): Dietary fructose exacerbates the cardiac abnormalities of copper deficiency in rats. *Atherosclerosis 74:* 203-214.
Reiser, S., Ferretti, R.J., Fields, M. and Smith, J.C. (1983): Role of dietary fructose in the enhancement of mortality and biochemical changes associated with copper deficiency in rats. *Am. J. Clin. Nutr. 38:* 214-222.
Rowley, D. and Halliwell, B. (1982): Superoxide-dependent formation of hydroxyl radicals from NADH and NADPH in the presence of iron salts. *FEBS Lett. 142:* 39-41.
Rowley, D.A. and Halliwell, B. (1985): Formation of hydroxyl radicals from MADH and NADPH in the presence of copper salts. *J. Inorg. Biochem. 23:* 103-108.
Thoranelly, P., Wolff, S. and Crabbe, J. (1984): The autooxidation of glyceraldehyde and other simple monosaccharides under physiological conditions catalyzed by buffer ions. *Biochem. Biophys. Acta 797:* 276-287.
Weissman, N., Shields, G.S. and Carnes, W.H. (1963): Cardiovascular studies on copper-deficient swine. IV. Content and solubility of the aortic elastin, collagen and hexosamine. *J. Biol. Chem. 238:* 3115-3118.

Metal Ions in Biology and Medicine. Eds. Ph. Collery, L.A. Poirier, M. Manfait, J.C. Etienne. John Libbey Eurotext, Paris © 1990, pp. 84-88

Copper and endocrine and neuroendocrine function

S.J. Bhathena, L. Recant

USDA, ARS, Beltsville Human Nutrition Research Center, Carbohydrate Nutrition Laboratory, Beltsville, MD 20705, USA and Diabetes Research Laboratory, V.A. Medical Center, Washington, DC 20422, USA

The essentiality of copper in human and animal nutrition is now firmly established. Though the actual quantitative requirements of copper in humans is not established, the average intake of copper in humans is 1.0 mg/day, far below the estimated safe and adequate daily intake of 1.5-3.0 mg/day. Both excess as well as very low intake of copper affect the normal functions of the central nervous system (CNS) in humans and animals. In experimental animals low copper intake has been shown to alter biochemical, metabolic, and endocrine functions of several organs besides brain, notably, heart, pancreas, pituitary, adrenal, thyroid, and gonads. Copper ions play a major role in hormone release, catecholamine synthesis, opiate processing and in the modulation of opiate receptors either directly or through copper-containing and copper-dependent enzymes. This review is limited to the role of copper in endocrine and neuroendocrine functions in several tissues in humans and experimental animals.

BRAIN

Among all tissues, brain is studied most extensively for the role of copper (Bhathena, 1989). Bennetts (1932) first described the role of copper in the CNS. He observed ataxia in lambs grazing on pastures grown on land with low copper content. Though perturbations in brain copper metabolism have been reported in several species, neurological consequences of copper deficiency are observed only in humans, lambs, and mice. These include neonatal enzootic ataxia, defective myelination (demyelination, hypomyelination and amyelination), motor incoordination, and brain and spinal cord degeneration. Although most symptoms of dietary copper deficiency are readily reversed by feeding high levels of copper, dietary copper treatments should be started very early in life to reverse neurological changes. A defect in catecholamines has been reported to be involved in these neurological disturbances (Morgan & O'Dell, 1977). It is also possible that alterations in neuropeptides, opiates and/or their receptors may also be involved in many of these disorders. In this regard, copper plays a role in modulating peptides, opiates and their receptors in brain (Marzullo & Hine, 1980; Chapman & Way 1980; Sadee *et al.*, 1982). In humans, an interaction between copper metabolism and B-endorphin has been suggested in schizophrenia (Kolb & Sadee, 1983). Similarly, levels of peptidyl-glycine-α-amidating monooxygenase PAM are increased in anterior pituitary homogenates in copper deficient rats (Loh *et al.*, 1984), and decreased in brain and CSF in patients with dementia and Alzheimer's disease. PAM is involved in amidation of several neuropeptides and peptide hormones which require post-translational processing to be biologically active. Copper is also implicated in the synthesis of many biologically active neuropeptides such as peptide histidine isoleucine (PHI), peptide YY, neuropeptide Y, galanin, neuropeptide K, and pancreastatin (Eipper & Mains 1988).

Opiate receptors: That copper is involved in modulating opiate receptors is evidenced from both *in vitro* as well as *in vivo* studies. The binding of opiate agonists as well as antagonists to brain membranes is inhibited by copper (Sadee *et al.*, 1982). In a recent study, no changes in the binding of an opiate agonist, etorphine to brain membrane were observed in copper-deficient rats (Bhathena *et al.*, 1988a). However, changes in muscarinic, dopaminergic and benzodiazepine receptors have been observed in different areas of the brain in copper-deficient rats (Farrar & Hoss, 1982). *In vivo*, ICV injection of copper in mice produced analgesia (Marzullo & Hine, 1980), and IP administration of copper in mice enhanced morphine-induced analgesia (Stern, 1968). Similarly, B-endorphin-induced analgesia was reversed by the copper chelator penicillamine (Chapman & Way, 1980). Thus, higher brain and CSF levels of copper and elevated B-endorphin in CSF in schizophrenic subjects may explain their poor responses to pain stimuli (Kolb & Sadee, 1983).

HEART

Copper is intimately involved in normal cardiac function. Absolute or relative copper deficiency has been recognized as one of the important factors in ischemic heart disease (Klevay, 1984). Cardiac abnormalities including rupture of the heart at the apex (resulting in death) in severe experimental copper deficiency, have been reported by Klevay (1984) and Fields (1985). Heart and atria have now been recognized as endocrine tissues. Thus, B-endorphin is present in rat heart and is modulated by gonads (Forman *et al.*, 1989), and atrial natriuretic peptides (ANP) are present in atria. Further, enkephalins are cardioactive and act either directly or through CNS in regulating myocardial function (Holaday, 1983). Recently, significant decrease in atrial ANP has been shown in both male and female rats fed copper-deficient diets (Bhathena *et al.*, 1988b). Plasma ANP were elevated in only male rats fed copper-deficient diets with fructose. No change was observed in plasma ANP in copper-deficient rats fed starch, suggesting a protective effect of starch (Bhathena *et al.*, 1988b). Significant increases in plasma ANP have been reported in patients with congestive heart failure and hypertension (Tang *et al.*, 1986). Since opiates, catecholamines and atrial peptides are all involved in cardiac regulation, it is important to study their interactions and relative importance in cardiac function in severe copper deficiency (Bhathena & Recant, 1987; Bhathena *et al.*, 1986b). It is important to note that only male rats fed copper-deficient diets with fructose die of cardiac rupture (Fields *et al.*, 1986). Studies in gonadectomized male and female rats indicate that gonadal steroids are not directly involved in this sexual dimorphism (Bhathena *et al.*, 1988b). However, it is possible that brain and pituitary sex hormones may be involved. These relationships need to be further studied.

PANCREAS

Studies in copper deficient rats have elucidated that cellular functions dependent on copper are involved in pancreatic hormone secretion. Copper deficiency produced by either dietary means (Recant *et al.*, 1986) or by use of penicillamine (Weaver *et al.*, 1986) causes significant atrophy of exocrine pancreas. Significant morphologic changes also occur in endocrine pancreas when copper deficiency is produced by feeding low copper diets (Bhathena & Recant, 1987) but not when the deficiency is induced by penicillamine (Weaver *et al.*, 1986). In rats fed a copper-deficient diet, there was a significant decrease in plasma insulin, but not plasma glucagon (Bhathena *et al.*, 1986a), and a significant increase in pancreatic insulin content (Recant *et al.*, 1986). Significant increases in glucagon as well as somatostatin were seen in copper-deficient rats fed fructose but not starch when expressed per unit of protein content or per weight of the pancreas (Bhathena *et al.*, 1986a). Perfusion of the pancreas with glucose showed a significantly reduced response in copper-deficient rats, indicating a possible defect in insulin secretion. However, in rats made copper deficient with penicillamine, insulin secretion in response to glucose is not altered (Weaver *et al.*, 1986).

Pancreas also contains enkephalins in B cells and β-endorphin in A cells. It is thus possible that copper status of the animals also affect the content and secretion of these peptides. A significant decrease in total free leu- as well as met-enkephalin was observed when copper-deficient rats were fed fructose (Recant *et al.*, 1986). It is important to note that copper deficiency causes significant atrophy of the pancreas. When the enkephalins were expressed per unit of protein, there was an increase in free enkephalins in copper-deficient rats fed fructose compared to control animals with a concomittant increase in precursor met- and leu-enkephalin. The increase was more marked when the dietary carbohydrate was fructose rather than starch. This indicates a possible defect in proenkephalin processing in copper deficiency. However, no specific role for copper has been suggested in the conversion of precursors to free peptides. It is interesting to note that a significant negative correlation was observed between free met-enkephalin and insulin and a signifcant positive correlation between insulin and total leu- and met-enkephalin. Though β-endorphin levels were lower in copper-deficient rats compared to controls, the changes were not significant. Thus, changes in enkephalins appear to be related to changes in insulin while those of β-endorphin are related to changes in pancreatic glucagon. Since changes in pancreatic enkephalin precede changes in insulin in copper-deficent rats, it is possible that enkephalins my affect insulin release secondarily via a paracrine effect within the islet (Recant *et al.*, 1986). However, independent alterations in insulin and enkephalin secretion cannot be ruled out.

ADRENALS, PITUITARY AND HYPOTHALAMUS

Besides catecholamines, and steroid hormones, adrenals also contain opiates. Copper status of the animals in general does not appear to affect endocrine function of the adrenals to the same extent as brain, heart or pancreas. No significant effect of copper deficiency was observed in free or precusor enkephalin (Recant *et al.*, 1986), or catecholamines (Kennedy, *et al.*, 1986). Pituitary and hypothalamus are also less responsive to changes in copper status. Copper deficiency in rats did not produce any significant changes in either β-endorphin or free enkephalins in the anterior or posterior pituitary. Leu-enkephalin precursor was lower in the posterior pituitary of copper-deficient rats fed fructose (Recant *et al.*, 1986). Somatostatin was higher in the posterior pituitary but not in the anterior pituitary. Somatostatin in the hypothalamus was not affected by dietary copper (Bhathena & Recant, 1987).

Copper, however, plays a significant role in neuroendocrine secretion from pituitary and hypothalamus. Incubation of bovine pituitary with copper promotes the release of growth hormone, prolactin, luteinizing hormone (LH), ACTH and thyrotropin stimulating hormone (TSH) (LaBella *et al.*, 1973). However, TSH secretion in respones to thyrotropin releasing hormone is normal in copper-deficient rats (Allen, *et al.*, 1982). Intravenous administration of copper has been shown to decrease pituitary LH concentration in rats (Jelinek *et al.*, 1970). *In vivo* release of luteinizing hormone releasing hormone (LHRH), LH, prolactin and FSH by copper has been reported in rabbits (Pau &

prostaglandin E_2 and forskolin-stimulated LHRH release from hypothalamic neurons (Barnea, 1987).

COPPER AND ENDOCRINE REGULATION IN HUMANS

There are only a few studies which describe the role of copper in endocrine regulation and the role of hormones on copper homeostasis in humans. Kolb and Sadee (1983) have described an interaction between copper and β-endorphin in schizophrenic subjects. Significant decreases in plasma enkephalins have been reported in humans fed low copper diets (Bhathena *et al.*, 1986b). However, this observation needs to be confirmed. Exogenous estrogen and testosterone but not progesterone reversibly increase serum copper in males and premenopausal women (Johnson *et al.*, 1959) and in subjects with Wilson's desease (German & Bearn, 1961). Changes in serum coper have been reported during pregnancy and in women taking oral contraceptives (Henkin, 1976, Ettinger, 1984). Injections of epinephrine as well as thyroid hormone also elevate serum copper (Ettinger, 1984). Elevated serum copper levels have been observed in acromegalics and a reduction of growth hormone in these subjects decreased serum copper (Henkin, 1976). Though no endocrine changes have been reported in subjects with Menke's or Wilson's disease, it is possible that altered copper metabolism in these subjects may have affected endocrine and neuroendocrine function in addition to altered neurotransmitters. This needs to be explored.

REFERENCES

Allen, D.K., Hassel, C.A., and Lei, K.Y. (1982): Function of pituitary-thyroid axis in copper-deficient rats. *J. Nutr.* 112, 2043-2046.

Barnea, A. (1987): Copper - a modulator of peptide release in the brain. In *Biology of Copper Complexes*, ed J.R.J. Sorenson, pp. 81-93. New Jersey: The Humana Press.

Bennetts, M.W. (1932): Enzootic ataxia of lambs in Western Australia. *Aust. Vet. J.* 8, 137-142; 8, 183-184.

Bhathena, S.J., Voyles, N.R., Timmers, K.I., Fields, M., Kennedy, B.W., and Recant, L. (1986a): Effect of copper deficiency on the content and secretion of pancreatic islet hormones. *Fed. Proc.* 45, 235.

Bhathena, S.J., Recant, L. Voyles, N.R., Timmers, K.I., Reiser, S., Smith, J. C., Jr., and Powell, A.S. (1986b): Decreased plasma enkephalins in copper deficiency in man. *Am. J. Clin. Nutr.* 43, 42-46.

Bhathena, S.J., and Recant, L. (1987): Peptides and opiates in copper deficiency. In *Biology of Copper Complexes*, ed J.R.J. Sorensen, pp. 315-328. New Jersey: The Human Press.

Bhathena, S.J., Katki, A.G., Kennedy, B.W., Fields, M., and Rockwood, G. (1988a): Effect of copper deficiency and dietary carbohydrates on brain opiate receptors. *Med. Sci. Res.* 16, 945-946.

Bhathena, S.J., Kennedy, B.W., Smith, P.M., Fields, M., and Zamir, N. (1988b): Role of atrial natriuretic peptides in cardiac hypertrophy of copper deficient male and female rats. *J. Trace Elem. Exp. Med.* 1, 199-208.

Bhathena, S.J. (1989): Recent advances on the role of copper in neuroendocrine and central nervous system. *Med. Sci. Res.* 17, 537-542.

Chapman, D.B., and Way, E.L. (1980): Metal ion interactions with opiates. *Ann. Rev. Pharmacol. Toxicol.* 20: 553-579.

Eipper, B.A., and Mains, R.E. (1988): Peptide α-amidation. *Ann. Rev. Physiol.* 50, 333-344.

Ettinger, M.J. (1984): Copper metaboism and diseases of copper metabolism. In *Copper Proteins and Copper Enzyme*. ed R. Lontie, pp. 175-229. Boca Raton:CRC.

Farrar, J.R. and Hoss, W. (1984): Effect of copper on the binding of agonists and antagonists to muscarinic receptors in rat brain. *Biochem. Pharmacol.* 33, 2849-2856.

Fields, M. (1985): Newer understanding of copper metabolism. *Int. Med.* 6, 91-98.

Fields, M., Lewis, C., Scholfield, D.J., Powell, A.S., Rose, A.J., Reiser, S., and Smith, J.C. (1986): Female rats are protected against the fructose induced mortality of copper deficiency. *Proc. Soc. Exp. Biol. Med.* 183, 145-149.

Forman, L.J., Estilow, S., and Hock, C.E. (1989): Localization of β-endorphin in the rat heart and modulation by testosterone. *Proc. Soc. Exp. Biol. Med.* 190, 240-245.

German, J.L., III, and Bearn, A.G. (1961): Effect of estrogens on copper metabolism in Wilson's disease. *J. Clin. Invest.* 40, 445-453.

Henkin, R.I. (1976): Trace metals in endocrinology. *Med. Clin. North Am.* 60, 779-797.

Holaday, J.W. (1983): Cardiovascular effects of endogenous opiate systems. *Ann. Rev. Pharmacol. Toxicol.* 23, 541-594.

Jelinek, J.M., Markan, O., and Seda, M. (1970): The effect of copper sulphate on pituitary LH in rats. *Endocrinol. Exptl.* 4; 37-43.

Johnson, N.C., Kheim, T., and Kountz, W.B. (1959): Influence of sex hormones on total serum copper. *Proc. Soc. Exp. Biol. Med.* 102, 98-99.

Kennedy, B.W., Bhathena, S.J., Fields, M., Voyles, N.R., Timmers, K.I., and recant, L. (1986): Effect of copper deficiency on plasma and adrenal catecholamines. *Fed. Proc.* 45, 356.

Klevay, L.M. (1984): The role of copper, zinc, and other chemical elements in ischemic heart disease. In *Metabolism of Trace Metals in Man*, eds O.M. Rennert and W.Y. Chan, pp. 128-157. Boca Raton: CRC.

Kolb V.M., and Sadee, W. (1983): Copper-endorphin schizophrenia hypothesis. *Res. Commun. Psychol. Psychiatry Behav.* 8, 343-347.

LaBella, F., Dular, R., Vivian, S., and Queen, G. (1973): Pituitary hormone releasing or inhibiting activity of metal ions present in hypothalamic extracts. *Biochem. Biophys. Res. Commun.* 52, 786-791.

Loh, Y.P., Brownstein, M.J. and Gainer, H. (1984): Proteolysis in neuropeptide processing and other neural functions. *Ann. Rev. Neurosci.* 7, 189-222.

Marzullo, G. and Hine, B. (1980): Opiate receptor function may be modulated through an oxidation-reduction mechanism. *Science* 208, 1171-1173.

Morgan R.F., and O'Dell, B.L. (1977): Effect of copper deficiency on the concentration of catecholamides and related enzyme activity in the rat brain. *J. Neurochem.* 28, 207-213

Pau, K.Y.F., and Spies, H.G. (1986): Effect of cupric acetate on hyphothalamic gonadotrophin-releasing hormone release in intact and ovariectomized rabbits. *Neuroendocrinology* 43, 197-204.

Recant, L. Voyles, N.R., Timmers, K.I., Zalenski, C., Fields, M., and Bhathena, S.J. (1986): Copper deficiency in rats increases pancreatic enkephalin-containing peptides and insulin. *Peptides* 7, 1061-1069.

Sadee W., Pfeiffer, A., and Herz, A. (1982): Opiate receptor: multiple effects of metal ions. *J. Neurochem.* 39, 659-667.

Stern, P. (1968): Mode of action of histamine in central nervous system. *Klin. Wockenschr.* 80, 181-185.

Tang, J., Song, D.L., Suen, M.Z., Xie, C.W., Chang, D., and Chang, J.K. (1986): Alpha-human atrial natriuretic polypeptide (alpha-hANP) in normal volunteers and patients with heart failure or hypertension. *Peptides* 7, 33-37.

Weaver, Fr.C., Sorenson, R.L., Kaung, H.L.C. (1986): An immunohistochemical, ultrastructral, and physiologic study of pancreatic islets from copper-deficient, penicillamine treated rats. *Diabetes* 35, 13-19.

Metal Ions in Biology and Medicine. Eds. Ph. Collery, L.A. Poirier, M. Manfait, J.C. Etienne. John Libbey Eurotext, Paris © 1990, pp. 89-91

Evaluation of increased copper levels in the cerebrospinal-fluid of Parkinson's patients

J.F. Belliveau[1,2], J.H. Friedman[2], G.P. O'Leary Jr[1], D. Guarrera[3]

[1] *Departments of Chemistry and Biology, Providence College, Providence, RI 02918;* [2] *Department of Medicine, Roger Williams General Hospital/Brown University, 825 Chakstone Avenue, Providence, RI 02908 and* [3] *Chemistry Laboratory, Texas Instruments, MS10-16, Attleboro, MA 02703, USA*

The development of analytical methodologies for determining trace copper levels in cerebrospinal-fluid (CSF) has spanned the last two decades. Normal/control mean CSF copper levels reported in the literature have decreased by an order of magnitude during this period. Goody *et al.* (1974) reported an average level of 240 ng/mL, Bodgen *et al.* (1977) reported 95 ng/mL, Tyrer *et al.* (1985) reported 130 ng/mL and Pall *et al.* (1987) reported 20 ng/mL. These differences appear to be a function of inadequate methodology (i.e., sample handling, contamination, matrix enhancement effects, etc.) and recent reports have established that the average copper levels in CSF are reliably on the order of 20-40 ng/mL (Pall *et al*, 1987; Kapaki *et al, 1989*).

Pall *et al.* (1987) reported that the copper levels in the CSF of patients with Parkinson's disease (PD) were significantly higher than those of healthy controls. CSF copper levels were measured by electrothermal atomization/atomic absorption spectrophotometry (ETA-AAS) and both of the data distributions for the control and Parkinson's disease groups were non-parametric. This present study sought to duplicate these previous results and to further parameterize this effect. Copper levels were measured by d-c argon plasma emission spectroscopy (DCP-AES), a methodology reportedly having less matrix interferences than (ETA-AAS). In addition, sample storage conditions and the relative magnitudes of matrix effects between ETA-AAS and DCP-AES copper CSF measurements were evaluated.

MATERIALS AND METHODS

measurement of CSF samples

CSF samples were received from the Division of Neurology at Roger Williams General Hospital and frozen at -20°C until analyzed. The CSF samples came from Parkinson's patients who had stages I and II, mild PD, and who had been untreated for at least one month prior to the spinal tap. Control samples were from non-PD patients or from persons undergoing myleograms. The spinal fluid was taken as the approximate 20th cc. All samples were introduced into a conventional 3-electrode, d-c argon plasma source contained in a Spectraspan III spectrometer (Applied Research Laboratories, Sunland, CA). Copper was measured at 327.3 nm, zinc at 206.2 nm, iron at 259.9 nm and silicon at 251.6 nm. Zn, Fe and Si data were used to assess storage conditions (glass versus plastic storage containers) and sample integrity (presence or absence of yellow protein color in samples). All chemicals were analytical grade or better. Because of matrix-interferences with the analyte signal, all calibration standards were made up in an artificial CSF matrix (138 meq/L of Na, 2.8 meq/L of K, 2.1 meq/L of Ca and 2.3 meq/L of Mg - using the chloride salts). These calibration standards had the analyte concentration ranges of 0-100 ng/mL for Cu, 0-500 ng/mL for Zn and Fe, and 0-1000 ng/mL for Si. Distilled/deionized water (DDW) was run through an anion exchange column as well as the usual cation exchange column to decrease the background silicon level (as silicates) to a negligible level. Calibration curves for the DCP-AES methodology, because of their linearity over the analyte concentration range of interest, were evaluated using the Linear Least Squares technique.

<u>analytical comparison of DCP-AES and ETA-AAS methodologies</u>

Calibration curves for copper (0-200 ng/mL) in various matrices (deionized/distilled water, artificial CSF matrix and 0.2% nitric acid) were established for both the DCP-AES and ETA-AAS methodologies to evaluate the magnitude of the respective matrix effects. Protein (human albumin) was also added (0-1%) to a 50 ng/mL standard to evaluate its matrix effect with these two methodologies. The atomic absorption spectrometer was a Varian SpectrAA with Zeeman background correction and copper was measured at 324.8 nm. Samples were desolvated at 120°C, ashed at 900°C and atomized/measured at 2300°C.

RESULTS AND DISCUSSION

CSF samples stored in glass had abnormally high zinc and silicon levels (474±250 & 2108±578 ng/mL, respectively) compared to those stored in plastic (33±16 & 47±39 ng/mL, respectively). Those samples showing a yellowish protein color (xanthochromic) had high zinc and iron levels (87±12 & 282±70 ng/mL, respectively) compared to those without any yellowish tint (33±16 & 26±17 ng/ml, respectively). Copper levels in the samples that were stored in glass (21±6 ng/mL) or which were xanthochromic (23±4 ng/mL) were not elevated compared to those stored in plastic and non-xanthochromic (19±6 ng/mL) as measured by the DCP-AES methodology. However, because of the abnormalities in the zinc, iron and silicon levels, the copper data from these samples stored in glass or which were xanthochromic were not included in the comparison between Parkinson's patients and controls.

There was no statistically significant difference between the CSF copper levels in PD (18.7±6.3 ng/mL, range 7-30 ng/mL, N = 16) and controls (19.2 ±5.8 ng/mL, range 10-33 ng/mL, N = 21). Our data was more normally distributed and did not show the extended concentration ranges (PD 6-107 ng/mL, controls 3-67 ng/mL) of the previous study (Pall *et al.*, 1987). More importantly, there was a statistically significant difference (p = 0.009, Mann-Whitney Rank-sum test) between our controls (19.2±5.8 ng/mL, median = 19.0) and those of the previous study (19.4±19.2 ng/mL, median = 11.4). It is unclear whether the differences between these studies are due to different sample populations or to artifacts of analytical methodologies (DCP-AES with the present study versus ETA-AAS with the previous study).

In an initial attempt to distinguish between these two possible explanations (different sample populations versus artifacts of analytical methodologies), we measured the magnitude of the matrix enhancement effects between the two methodologies. The effect of differing ionic strengths was determined by evaluating calibration curves for copper (0-200 ng/mL) in various matrices (deionized/ distilled water, an artificial CSF matrix and 0.2% nitric acid). The DCP-AES methodology is less prone to matrix interferences in that the variation among copper calibration curves was approximately three times smaller for the DCP-AES methodology compared to the ETA-AAS methodology. For example, given the range of our experimental copper calibration curves, a sample measured by DCP-AES with a relative emission intensity of 14 would have a [Cu]-range of 20-37 ng/mL whereas a sample measured by ETA-AAS with an absorbance of 1.0 would have a [Cu]-range of 20-78 ng/mL. The effect of abnormal amounts of protein was also evaluated by addition of protein (0-1%) to a 50 ng/mL copper standard. The results also showed that the DCP-AES methodology is less prone to matrix enhancement in that the relative emission intensity for copper increased by 85% going from 0 to 1 % protein compared to a 227% increase in absorbance for the ETA-AAS methodology. Thus the differences between the two studies (i.e., the extended copper concentration ranges of the ETA-AAS measurements) could be due to larger matrix enhancement effects with some samples measured by the ETA-AAS methodology. The definitive experiment, which is presently being scheduled, consists of analyzing an extended set of Parkinson/control samples from both the Rhode Island, USA and Birmingham, England areas by both methodologies using identical instrumental conditions employed in the measurement of previous data.

ACKNOWLEDGEMENTS

This research was supported in part by NIH Grant NS24778 and an American Parkinson Disease Association Grant.

REFERENCES

Goody, W., Williams, T.R. and Nichols, D. (1974): Spark-source mass spectrometry in the investigation of neurological disease. *Brain* 97, 327-336.

Bodgen, J.D., Troiano, R.A. and Joselow, M.M. (1977): Copper, zinc, magnesium and calcium in plasma and CSF of patients with neurological diseases. *Clin. Chem.* 23, 485-489.

Tyrer, S.P., Delves, H.T. and Weller, M.P.I. (1985): CSF copper concentrations. *Am. J. Psychiatr.* 142, 143.

Pall, H.S., Williams, A.C., Blake, D.R., Lunec, J., Gutteridge, J.M., Hall, M. and Taylor, A. (1987): Raised cerebrospinal-fluid copper concentration in Parkinson's disease. *Lancet* ii, 238-241.

Kapaki, E., Segditsa, C. and Papageorgiou, C. (1989): Zinc, copper and magnesium concentration in serum and CSF of patients with neurological disorders. *Acta Neurol. Scand.* 79, 373-378.

Metal Ions in Biology and Medicine. Eds. Ph. Collery, L.A. Poirier, M. Manfait, J.C. Etienne. John Libbey Eurotext, Paris © 1990, pp. 92-94

Zinc whole blood determination in a normal finnish population and dose response to long-term zinc administration

K. Jaakkola, J. Laakso, I. Ruokonen, K. Mahlberg

Mineral Laboratory Mila Ltd, Sirrikuja 4 B, SF-00940 Finland

INTRODUCTION

The biochemical basis of the body's need for Zn is well established. Among other cellular processes, Zn is required for the maximal activity of over 70 enzymes and is also essential for the integrity of cell membranes. Interest in the determination of nutritional Zn status has increased ever since Zn deficiency was established as a primary factor in growth retardation, retarded sexual maturation, skin lesions and anemia in Egyptian children by Prasad in 1963 (Prasad et al. 1963). Later Zn has also been found essential for growth and cell development, and for the integrity of host defence mechanisms (Pekarek et al. 1979). Zn deficiency has been reported in severe metabolic dysfunctions, such as cancer (Feustel et al. 1989).

Numerous laboratory tests have been proposed for determining Zn status. Plasma and, in spite of the critics, even hair Zn determinations are still used (Dorea et al. 1988, Diez et al. 1989). Zn determination from lymphocytes and neutrophils has also been suggested, although the isolation of these cells is tedious and feasible for routine purposes (Goode et al. 1989). Since the major part of the whole blood Zn is located in the erythrocytes (over 80%), whole blood Zn (B-Zn) levels are sensitive to changes in hematocrit (hcr) values, causing errors in the interpretation of the results. Therefore when B-Zn determinations are used to assess the individual and interindividual changes in Zn status, the results should be corrected with hcr. In the present work, plasma Zn values were compared to whole blood values with or without hcr standardization. The effect of Zn supplementation on B-Zn levels was also studied.

MATERIALS AND METHODS

Patients: To determine reference intervals for B-Zn in the normal population, blood samples were collected from 101 healthy volunteers (36 men aging 39.6 +/- 14.2 and 65 women aging 37.0 +/- 11.1 years). For the Zn supplementation study, 54 patients expressing B-Zn levels below the reference mean were selected (21 men and 33 women, mean age 42.3 +/- 12.7 years, range 4 - 64 years).

Oral Zn administration: Individual Zn supplementation was started in the selected group and followed for 10 to 14 months. The doses were determined according to patient's original B-Zn levels. The mean daily Zn dose was 42.0 +/- 11.6 mg (as Zn(II), in gluconate form). The patients were examined and the laboratory tests taken at the beginning and at the end of the study.
Samples: Blood was collected by venopuncture into 10 ml stoppered acid-washed polypropylene tubes containing heparin as anticoagulant, after discarding the first 5 ml of the blood. The samples were centrifuged at 2150 g for 10 min, and the plasma was transferred to acid-washed polypropylene tubes. Samples were kept at -20 C until assayed.

Analytical methods: An IL-Video 22 flame atomic absorption spectrophotometer equipped with autosampler, FASTAC-II nebulizer and one slot burner (Instrumentation Laboratory,GB) was used for the Zn determinations. An aliquot of 200 ul of whole blood, serum or erythrocytes was diluted with 1.0 ml water. 75 ul of concentrated nitric acid (Merck, Darmstadt, GDR, suprapur quality) was added and mixed in order to precipitate proteins. After a short centrifugation, 200 ul of the clear supernatant was transferred into sample cups and aspirated into the flame AAS. The absorption signal was recorded for 6.0 seconds and the peak area integrated.The intra-assay imprecision was 2.5, and 2.3% for serum and whole blood. The respective inter-assay imprecisions were 2.7 and 2.3%. The B-Zn values were corrected for hcr by dividing the B-Zn level with the respective hcr.

Hematological parameters were determined employing a Coulter Counter T-540 (Coulter Electronics Inc, USA), following the manufacturer's instructions.

Statisticalmethods: The SPSSPC + package (Statistical Package for the Social Sciences, SPSS Inc, Illinois, USA) was employed for the statistical analysis.

RESULTS

In the normal population ($n = 101$) the mean Hb level was 138.43 +/- 13.07 g/l and hcr 0.43 +/- 0.04. The mean B-Zn level was 99.07 +/- 10.9 μmol/l. The mean P-Zn level was 14.52 +/- 2.0 μmol/l. Men and women revealed no significant differences in these parameters.

B-Zn levels were found to correlate to erythrocyte count ($r=0.67$ $p<0.001$), hcr ($r=0.45$, $p<0.05$) and Hb (0.46, $p<0.05$). P-Zn levels did not correlate to B-Zn, the erythrocyte count, Hb or hcr.

A significant increase from the base B-Zn levels, 85.4 +/-11.9 μmol/l, to 90.6 +/- 13.4 μmol/l ($p<0.01$) was detected after the follow-up time in the supplemented group. The significance of the difference was markedly improved by hcr standardization ($p<0.001$), although no significant changes in either Hb or hcr values were found. In 30.2 % of the group no change or a slight decrease in B-Zn levels in response to oral Zn therapy was detected. No side- effects during the Zn therapy was encountered.

DISCUSSION

With respect to the biochemical importance of Zn and the deleterious effects of Zn deficiency, a simple and reliable method for assessing nutritional Zn status is needed. Serum or plasma Zn determinations are widely used, although it is doubtful if these tests really reflect the Zn status in humans (Hess et al. 1977, Prasad 1978). B-Zn determinations have been suggested, but these studies are hampered by the ignorance of the effect of hcr on B-Zn levels. The normal levels of uncorrected B-Zn in the present study were in agreement with previous studies (Manser et al 1989).

In the present work B-Zn showed a clear response to oral Zn supplementation. The relatively high amount of non-responders can be explained by the low Zn dose we used, individual homeostatic control of the Zn absorption , or poor compliance.

In conclusion , the determination of hcr corrected B-Zn levels provides a reliable and simple method to assess the individual long-term nutritional status of Zn.

REFERENCES:

Diez, M., Cerdan, F.J., Arroyo, M., Balibrea, J.L. (1989): Use of the copper/zinc ratio in the diagnosis of lung cancer. *Cancer* 63: 726-730.

Dorea, J.G., Paine, P.A. (1988): Making sense of the hair zinc literature: where do we go from here?. *Arch. Latinoam. Nutr.* 33: 93-112.

Feustel, A., Wennrich, R., Schmidt, B. (1989): Serum Zn levels in prostatic cancer. *Urol.* Res. 17: 41-42.

Goode, H.F., Kelleher, J., Walker, B.E. (1989): Zinc concentrations in pure populations of peripheral neutrophils, lymphocytes, and monocytes. *Ann. Clin. Biochem.* 26: 89-95.

Hess F.M., King, J.C., Margen, S. (1977): Zinc excretion in young women on low zinc intake and oral contraceptive agents. *J. Nutr.* 107: 1610-1620.

Manser, W.W., Kahn, M.A. (1989): Trace element studies on Karachi population, Part 1: Normal ranges for blood copper, zinc and magnesium for adults. *JPMA* 39: 43-49.

Pekarek, R.S., Sandsteadt, H.H., Jacob, R.A., Barcome, D.F. (1979): Abnormal cellular immune response during acquired zinc deficiency. *Am. J. Clin. Nutr.* 32: 1466-1471.

Prasad, A.S., Miale, A.jr., Farid, Z., Sandsteadt, H.H., Schulert, A.R. (1963): Zinc metabolism in patients with syndrome of iron deficiency anemia, hepatosplenomegaly, dwarfism, and hypogonadism. *J. Lab. Clin. Med.* 61: 537-549.

Prasad, A.S., Rabbani, P., Abbasii, A., Bowersox, E., Fox, M.S. (1978): Experimental zinc deficiency in humans. *Ann. Intern. Med.* 89: 483-490.

Metal Ions in Biology and Medicine. Eds. Ph. Collery, L.A. Poirier, M. Manfait, J.C. Etienne. John Libbey Eurotext, Paris © 1990, pp. 95-99

Nutritional role of chromium in glucose and lipid metabolism of humans

R.A. Anderson

Vitamin and Mineral Nutrition Laboratory, Beltsville Human Nutrition Research Center, U.S. Department of Agriculture, ARS, Beltsville, Maryland 20705, USA

SUMMARY

Dietary intake of chromium appears to be suboptimal and numerous studies have reported beneficial effects of supplemental Cr on hyper- and hypoglycemic subjects. Chromium is only of benefit to subjects whose signs or symptoms are due to suboptimal dietary Cr and has no detectable effects on subjects with good glucose tolerance and normal serum lipids. Chromium binds to nucleic acids and effects may be either beneficial or toxic depending upon form and amount of Cr. Various forms of stress including high sugar diets, exercise, infection and physical trauma increase Cr losses. Recent advances in chromium nutrition strengthen the association of insufficient dietary Cr and risk factors associated with maturity-onset diabetes and cardiovascular diseases and further document the role of chromium in the maintenance of optimal health.

INTRODUCTION

Chromium is an essential element required for normal glucose and lipid metabolism. Insufficient dietary Cr leads to signs and symptoms often associated with maturity-onset diabetes and(or) cardiovascular diseases.

Signs and symptoms of marginal chromium deficiency in normal free-living people are shown in Table 1. These signs of Cr deficiency are widespread and several studies have reported that 75-90% of the subjects with one or more of the signs or symptoms relating to glucose tolerance listed in Table 1 may respond to supplemental Cr. Chromium supplementation of control subjects (those not displaying any signs or symptoms listed in Table 1) is without effect. Chromium is a nutrient, not a therapeutic agent, and therefore will be of benefit only to those whose signs and symptoms are due to marginal Cr deficiency.

Overt signs of chromium deficiency, in addition to those listed in Table 1, including unexpected weight loss, impaired nerve conduction and abnormal respiratory quotient that were refractory to insulin, were reversed by supplemental chromium in a female patient receiving total parenteral nutrition (Jeejeebhoy et al., 1977). These results have been confirmed (Freund et al., 1979; Brown et al., 1986).

While Cr toxicity is usually limited to industrial settings and is due to hexavalent Cr, Cr deficiency is widespread and is due to Cr in the trivalent state. Animals and humans can reduce limited amounts of toxic hexavalent Cr to the nutritionally beneficial trivalent form. Reduction occurs primarily in the stomach and if the stomach is bypassed by injecting Cr directly into the small intestine, Cr is absorbed primarily in the hexavalent state.

Since the discovery of a biological role of trivalent Cr in experimental animals, several studies have reported beneficial effects of Cr primarily on glucose tolerance and related variables of hyperglycemics, hypoglycemics and diabetics (see review, Anderson, 1987). Chromium has also been shown to be involved in nucleic acid and lipid metabolism. This review will be an update of the nutritional role of chromium in primarily glucose and lipid metabolism of human subjects from studies published in the past decade.

Table 1: Common Signs and Symptoms of Cr Deficiency Observed in the General Population

Fasting hyperglycemia	Elevated serum cholesterol
Impaired glucose tolerance	Elevated serum triglycerides
Elevated circulating insulin	Decreased insulin binding
Glycosuria	Decreased insulin receptor number

DIETARY CHROMIUM INTAKE

The recommended safe and adequate adult daily intake for chromium is 50-200 ug. However, most reliable reports indicate that females and males are only consuming roughly 40 and 60 percent of the minimum suggested intake, respectively. Anderson and Kozlovsky (1985) reported a mean daily intake (seven day average) for 22 female subjects of 25 $\pm$ 1 ug (Mean $\pm$ SEM) and 33 $\pm$ 3 ug for 10 male subjects. Similar values have been reported from England, Finland and New Zealand (see review, Anderson, 1987). Dietary daily chromium intake of 18 two-months postpartum women was 41 $\pm$ 4 ug and that for control women was 27 $\pm$ 2 ug (Anderson et al., 1988). Greater intake of mothers appeared to be due primarily to increased caloric intake rather than differential selection of foods. In both the control and postpartum subjects, chromium absorption was inversely related to dietary intake. However, the absorption of the two months postpartum women was considerably greater than that of the control females at dietary chromium intakes of less than 30 ug per day.

There are no comprehensive data bases to calculate dietary chromium intake, therefore chromium intake needs to be measured and cannot be calculated. Anderson and Bryden (1983) determined the chromium content of individual brands of beer, as well as individual lots of the same beer, and reported that there is not only a large variation among different brands of beer but also different lots of the same beer. Similar results were observed for different brands and lots of breakfast cereals (Anderson et al., 1988). It is unlikely that tables containing values to calculate Cr content of foods will be useful due to the large variations observed not only among different varieties or brands of the same food but also variations within the same brand or specific lot of foods. Seasonal and geographical variation may further increase the variability.

CHROMIUM SUPPLEMENTATION OF HUMAN SUBJECTS

The overwhelming majority of the studies published since 1980 demonstrate significant improvements in glucose metabolism following Cr supplementation.

Subjects with slightly elevated blood glucose (90 min glucose greater than 100 mg/dl) show consistent improvements following Cr supplementation. The effects of Cr on diabetic subjects appear to depend upon the amount of supplemental Cr. One hundred fifty micrograms of Cr seems to be insufficient (Rabinowitz et al., 1983). The study of Mossop (1983) which reported large effects of supplemental Cr on diabetics, gave 600 ug. Improvements in fasting glucose, amount of exogenous insulin or hypoglycemic drugs and HDL-cholesterol all improved following supplementation with 600 ug of Cr. Urberg and Zemmel (1987) reported that nicotinic acid and Cr both had to be consumed to observe beneficial effects of 200 ug of supplemental Cr. In the study of Polansky et al. (1990), a vitamin supplement was given to all subjects during the entire study to ensure proper vitamin nutrition while consuming a low Cr controlled diet. In that study, supplemental Cr led to significant improvements in glucose tolerance, circulating insulin, glucagon and serum cortisol.

Anderson et al. (1983), reported that not only did supplemental Cr lead to a decrease in the blood glucose of subjects with slightly elevated blood glucose but caused an increase in blood sugar of subjects with tendencies towards hypoglycemia. In a follow-up study, blood sugar, hypoglycemic symptoms, insulin binding and insulin receptor number of hypoglycemic subjects improved following Cr supplementation (Anderson et al., 1987). Clausen (1988), also reported beneficial effects of supplemental Cr on hypoglycemic symptoms and glucose values of hypoglycemics. The beneficial effects of chromium on both low and high blood sugar are due likely to a normalization of insulin. Chromium potentiates the action of insulin in hyperglycemic patients and with a more effective insulin, blood sugar levels decrease. With increased insulin efficiency, less insulin is required and insulin levels also decrease leading to decreased likelihood of hypoglycemia.

The effects of Cr on serum lipids are not as persuasive as those for glucose metabolism. Effects of Cr on serum lipids seem to be greater for subjects with higher serum lipids at the onset of the study. Two studies (Riales and Albrink, 1981; Mossop, 1983) reported significant improvements in HDL cholesterol following Cr supplementation but other studies have not been able to reproduce these results. However, other studies have reported decreased blood cholesterol and(or) triglycerides following Cr supplementation (Offenbacher and Pi-Sunyer, 1980; Saner et al., 1983; Riales and Albrink, 1983; Wang et al., 1989).

CHROMIUM AND NUCLEIC ACIDS

Besides its role in glucose and lipid metabolism, Cr is also postulated to play a role in nucleic acid metabolism. Chromium is postulated to be involved in the structural integrity of nucleic acids. Wacker and Vallee (1959) reported that beef liver fractions high in nucleic acids are also high in Cr. The interaction among Cr and nucleic acids was quite strong since precipitation of beef liver RNA six times from EDTA solutions did not reduce the amount of Cr associated with RNA while the concentration of all other metals tested decreased dramatically.

Chromium appears to be involved in gene expression by binding to chromatin causing an increase in the number of initiation sites and thereby enhancing RNA synthesis. This enhancement is caused by induction of a nuclear bound protein of 70,000 daltons and by activation of nuclear chromatin (Okada et al., 1989).

Chromium has also been shown to be mutagenic and carcinogenic. However, these effects are due primarily to chromates or other forms of hexavalent Cr. As stated, the form of Cr involved in normal glucose, lipid and nucleic acid metabolism is trivalent.

The toxic and beneficial effects of Cr on nucleic acids seem to be closely related. For example, Cr binds to DNA causing an increase in the number of initiation sites and thus enhances RNA synthesis (Okada et al., 1989) but Cr may also activate DNA polymerase which may lead to nuclear infidelity and mutagenesis. Similarly, a Cr glutathione complex is postulated as the biologically active form of chromium (Mertz, 1974) but Cr is also postulated to form a glutathione-chromium-DNA adduct that may be important in chromium genotoxicity (Borges and Wetterhahn, 1989). Chromium may also play both a toxic and beneficial role in zinc-finger-loops. It is possible that the exchange inert properties of Cr could lead to very stable finger-loop domains leading to the required reactions and products. However, Cr could also form distorted stable complexes leading to metal carcinogenesis.

CHROMIUM AND STRESS

Stress often leads to significant increases in glucose metabolism. Accompanying these changes in glucose metabolism there are usually significant changes in Cr metabolism. Increased utilization of glucose leads to increased mobilization of chromium; once Cr is mobilized it is not reabsorbed but excreted in the urine. Therefore, urinary Cr losses are a rough estimate of Cr mobilization.

Infection, glucose challenge, high sugar diet, exercise and physical trauma have all been shown to lead to increased mobilization of Cr and subsequent increased urinary losses (Anderson, 1987). The intensity and duration of the stress can be correlated with the increases in urinary Cr losses. For example, glucose loading, a mild stress leads to lower Cr losses than more severe stresses such as strenuous exercise or physical trauma. Duration and intensity of exercise can also be correlated with Cr losses (Anderson et al., 1982).

Increases in urinary Cr losses are usually associated with not only increases in glucose utilization but also increases in circulating insulin. However, during exercise insulin remains constant or decreases yet Cr losses may increase several-fold. Basal urinary Cr losses do not appear to correlate with circulating insulin levels. However, urinary Cr losses of control subjects do increase with the insulinogenic properties of individual carbohydrates or in combination with fructose (Anderson et al., 1990). Subjects with slightly elevated insulin levels appear to lose this ability to mobilize Cr to combat increases in carbohydrate intake and subsequent increases in insulin. This lack of an increase in Cr mobilization may be due to marginal Cr deficiency in which Cr stores may be depleted.

REFERENCES

Anderson, R.A. (1987): Chromium In *Trace Elements in Human and Animal Nutrition,* Vol. 1, ed. W. Mertz, Academic Press, Inc., NY 225-244.

Anderson, R.A. and Bryden, N.A. (1983): Concentration, insulin potentiation and absorption of chromium in beer. *J. Agric. Food Chem. 31:* 308-311.

Anderson, R.A. and Kozlovsky, A.S. (1985): Chromium intake, absorption and excretion of subjects consuming self-selected diets. *Am. J. Clin. Nutr. 41:* 1177-1183.

Anderson, R.A., Polansky, M.M., Bryden, N.A., Roginski, E.E., Patterson, K.Y. and Reamer, D.C. (1982): Effect of exercise (running) on serum glucose, insulin, glucagon and chromium excretion. *Diabetes 31:* 212-216.

Anderson, R.A., Polansky, M.M., Bryden, N.A., Roginski, E.E., Mertz, W. and Glinsmann, W.H. (1983): Chromium supplementation of human subjects: Effects on glucose, insulin and lipid parameters. *Metabolism 32:* 894-899.

Anderson, R.A., Polansky, M.M., Bryden, N.A., Bhathena, S.J. and Canary, J. (1987): Effects of supplemental chromium on patients with symptoms of reactive hypoglycemia. *Metabolism 36:* 351-355.
Anderson, R.A., Bryden, N.A., Polansky, M.M., Reynolds, R., Andon, M. and Moser-Veillon, P. (1988): Elevated chromium intake, absorption and urinary excretion of lactating women. *Fed. Proc. 2:* A1102.
Anderson, R.A., Bryden, N.A., Polansky, M.M. and Reiser, S. (1990): Urinary chromium excretion and insulinogenic properties of carbohydrates. *Am. J. Clin. Nutr. 51:* 864-8.
Borges, K.M. and Wetterhahn, K.E. (1989): Chromium cross-links glutathione and cysteine to DNA. *Carcinogenesis 10:* 2165-2168.
Brown, R.O., Forloines-Lynn, S., Cross, R.E. and Heizer, W.D. (1986): Chromium deficiency after long-term total parenteral nutrition. *Dig. Dis. Sci. 31:* 661-664.
Clausen, J. (1988): Chromium induced clinical improvement in symtomatic hypoglycemia. *Biol. Trace Elem. Res. 17:* 229-236.
Freund, H., Atamian, S. and Fischer, J.E. (1979): Chromium deficiency during total parenteral nutrition. *J. Am. Med. Assoc. 241:* 496-498.
Jeejeebhoy, K.N., Chu, R.C., Marliss, E.B., Greenberg, G.R., Bruce-Robertson, A. (1977): Chromium deficiency, glucose intolerance, and neuropathy reversed by chromium supplementation in a patient receiving long-term total parenteral nutrition. *Am. J. Clin. Nutr. 30:* 531-538.
Mertz, W. (1974): Chromium as a dietary essential for man. In: *Trace Element Metabolism in Animals-2.* Baltimore, University Park Press, 185-198.
Mossop, R.T. (1983): Effects of chromium (III) on fasting glucose, cholesterol and cholesterol HDL levels in diabetics. *Cent. Afr. J. Med. 29:* 80-82.
Offenbacher, E.G. and Pi-Sunyer, F.X. (1980): Beneficial effect of chromium-rich yeast on glucose tolerance and blood lipids in elderly subjects. *Diabetes 29:* 919-925.
Okada, S., Tsukada, H., Tezuka, M. (1989): Effect of chromium(III) on nuclear RNA synthesis. *Biol. Trace Elem. Res. 21:* 35-39.
Polansky, M.M., Bryden, N.A., Canary, J.J. and Anderson, R.A. (1990): Beneficial effects of supplemental chromium (Cr) on glucose, insulin and glucagon of subjects consuming controlled low chromium diets. *FASEB J.* A2964.
Rabinowitz, M.B., Gonick, H.C., Levine, S.R., Davidson, M.B. (1983): Clinical trial of chromium and yeast supplements on carbohydrate and lipid metabolism in diabetic men. *Biol. Trace Elem. Res. 5:* 2670-2678.
Riales, R., Albrink, M.J. (1981): Effect of chromium chloride supplementation on glucose tolerance and serum lipids including high density lipoprotein of adult men. *Am. J. Clin. Nutr. 34:* 2670-2678.
Saner, G., Yuzbasiyan, V., Neyzi, O., Gunoz, H., Saka, N. and Cigdem, S. (1983): Alterations of chromium metabolism and effect of chromium supplementation in Turner's syndrome patients. *Am. J. Clin. Nutr. 38:* 574-578.
Urberg, M. and Zemmel, M.B. (1987): Evidence for synergism between chromium and nicotinic acid in the control of glucose tolerance in elderly humans. *Metabolism 36:* 896-899.
Wacker, W.E.C. and Vallee, B.L. (1959): Nucleic acids and metals. *J. Biol. Chem. 234:* 3257-3262.
Wang, M.M., Fox, E.A., Stoecker, B.J., Menendez, C.E., Chan, S.B. (1989): Serum cholesterol of adults supplemented with brewer's yeast or chromium chloride. *Nutr. Res. 9:* 989-998.

Metal Ions in Biology and Medicine. Eds. Ph. Collery, L.A. Poirier, M. Manfait, J.C. Etienne. John Libbey Eurotext, Paris © 1990, pp. 100-103

A study of cognitive effect of mild iron deficiency anemia in school age children

X.Y. Liu, J. Cao

Department of Nutrition, Research Institute of Child and Adolescent, Beijing Medical University, People's Republic of China

Iron deficiency anemia (IDA) is the commonest nutritional deficiency in Children in China. Most cases are mild. It is well recognized that children with IDA are lower in physical work Capacity, but the effect of mild iron deficiency for mental function is not clear. The present study with double blind method compared the IDA children with noniron deficiency anemia children (NIDA) before iron treatment (T_1) and after iron treatment (T_2) in congnitive function.

Subject and Method

Subjects: Primary school children in Beijing City, chosen from 1023 8-10 year-old children. Medical history, anthropometry, hematological examination, socioeconomic status survey were made for them.

Blood samples were obtained on finger skin puncture. Hemoglobin (Hb) was measured by Cyanmethemoglobin method, Free Erythrocyte Protoporphrin (FEP) by fluorescent test, Serum Ferritin (SF) by Immuno-Radiometric Assay. The operation of measurement was carefully done with quality control, fixed operator and instrument. The criteria of IDA are Hb $<$ 120g/L, FEP $>$ 35.0 µg/dl or FEP $>$ 2.8 µg/Hb, SF $<$ 20 ng/ml. The selected IDA children were free of acute and chronic illness, birth complications and mental defect. 61 IDA children were chosen and 61 NIDA children were matched with age, sex, height, weight, parents socialeconomic status. 30 of IDA and 30 of NIDA children selected were treated with $FeSO_4$ Tablets at dosage of 0.3g/day for 2 months, designated as group IDA-I and group NIDA-I respectively. The remaining 31 of IDA and 31 of NIDA children selected were treated with placebo Tablets, designated as group IDA-P and group NIDA-P respectively. The characteristic features of the group receiving iron treatment were similar with those of the group to be treated with placebo. The children do not know what tablets they take. Cognitive function of all selected children was tested, fefore treatment (T_1).

Hematological examination was made after taking tablets for 2 and 4 months. Cognitive function was reevaluated 4 months after being treated with iron or placebo (T_2). The cognitive test used was designed by Department of Psychology, Beijing Normal University, which is authoritative and adaptable to Chinese school children. Includes: 1. Number span test (forward and backward memory). 2. efficiency and accuracy test. 3. Visual recognition test. 4. General reasoning test. 5. Discrimination and response test (with letter matching test and number scanning memory test computerized).

Results and Discussion

The results of biochemical determination of IDA and NIDA group children are Hb (g/L) 110.7 ± 6.8, 130.7 ± 5.7; FEP (μg/dl) 39.7 ± 10.6, 28.8 ± 7.73; FEP (μg/Hb g) 3.6 ± 1.0, 2.2 ± 0.6; SF (ng/ml) 16.7, 26.0. There are significant difference between two groups in biochemical determination ($P < 0.001$). The characteristic feature of IDA and NIDA group children are: age (year) 9.49 ± 1.02, 9.46 ± 0.9; sex (male/female) 27/34, 29/32; height (cm) 133.0 ± 8.6, 134.9 ± 6.3; weight (Kg) 28.3 ± 5.5, 30.3 ± 5.9; Total children in family 1.6 ± 0.6, 1.6 ± 0.6. There are no significant difference between two groups in characteristic feature.

Hemological change observed after treatment two and four months. The results of Hemoglobin are shown in Fig. 1. Hemoglobin was obviously higher than before in IDA-I group and reached normal (Hb > 120 g/L) Hb in IDA-P group increased too, it might be influenced by dietary factor, but it was still below normal. The results of FEP, SF in IDA-I group after two months iron treatment improved.

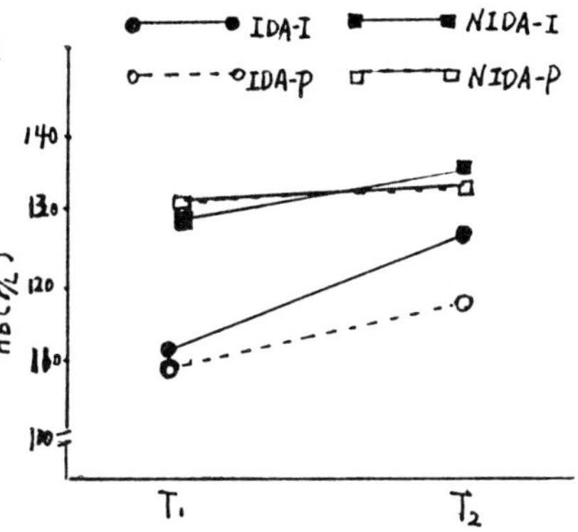

Fig.1. Hb change after iron treatment

The results of cognitive test befor iron treatment showed that the scores of visual recognition; response time of letter matching; number-scanning memory and general reasoning test at T_1 of IDA children were significantly lower than those of NIDA children ($P < 0.05$)

The results of cognitive test after iron treatment and placebo treatment (T_2) compared with the results of before treatment (T_1) are shown in Fig.2 .
Results of visual recognition test are shown in Fig. 2-1. The scores of IDA-I after iron treatment (T_2) are higher than those of (T_1) having significant difference ($P < 0.001$). The remaining group showed no change between T_1, T_2. The score of IDA-P is the lowest of all groups.

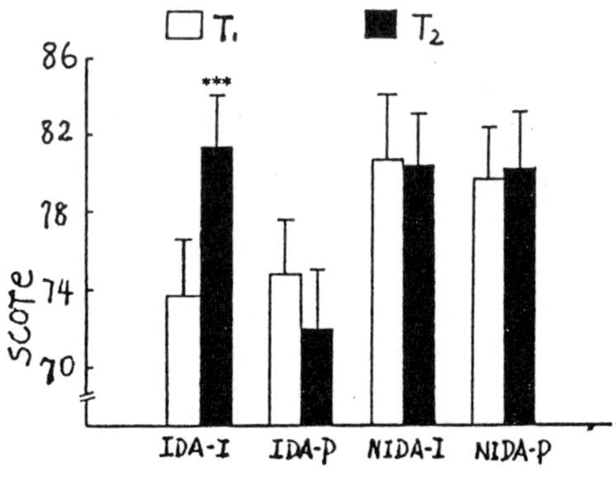

Fig.2-1. Results of visual recognition test

Results of letter matching test are shown in Fig. 2-2. The scores of IDA-I, NIDA-I after iron treatment (T_2) increased ($P < 0.01$, $P < 0.05$), the remaining two groups also showed increase but not significants.

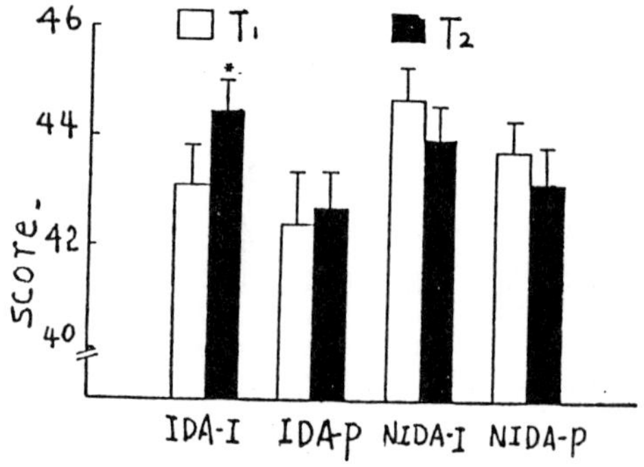

Fig. 2-2. Results of letter matching test

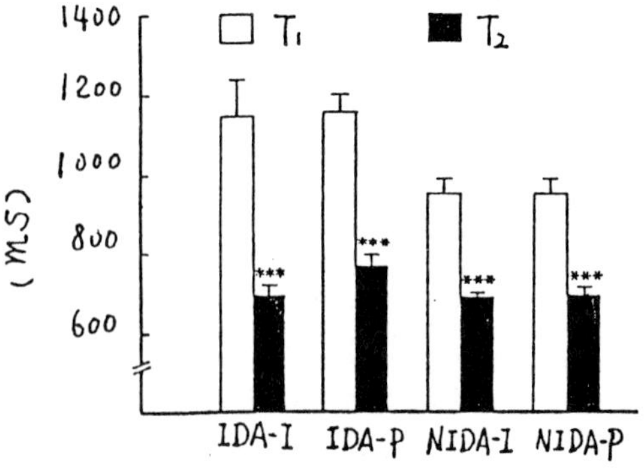

Fig. 2-3. Results of response time of letter matching test

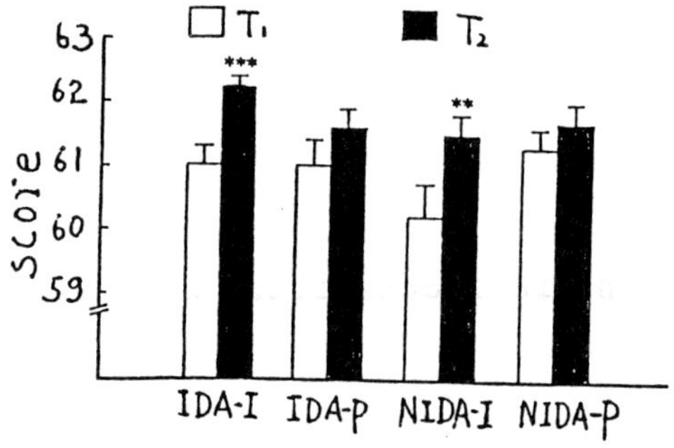

Fig. 2-4. Results of number scanning memory test

Fig.2. Results of cognitive test before iron treat (T_1) compared with after iron treatment (T_2) *** $P < 0.001$, ** $P < 0.05$, * $P < 0.1$.

Results of response time of letter matching test are shown in Fig.2-3. The response time of all groups are improved in (T_2). It was probably caused by experience. IDA-I (T_2) had the most improvement in all groups.

Results of number scanning memory test are shown in Fig. 2-4 only IDA-I (T_2) had slight improvement. IDA-P was the lowest of all groups.

The results of cognitive test in any else showed no significant change between (T_1 and T_2).

This study shows that in congnitive function test of visual recognition the score of NIDA group were higher than those in IDA group after iron treatment, showes obvious improvement. This test was assess the shorttime memory and attention. The letter matching test and number scanning memory test were used to assess not only the memory but also the response as shown in Fig. 2-2 and Fig. 2-4 the group of iron treatment better than the control groups. However iron treatment showed no significant improvement on general reasoning, for which is higher cognitive process affected by various factors.

These findings suggest that, IDA has delecterious effect on special function in school children. The results obtained in the present study are identical to these gained by pollitt's[1] research which means that iron deficiency has adverse effect on cognitive function but it is reversible following iron repletion.

REFERENCES

Pollitt,Ernesto. et al. (1986) : Iron deficiency and behavioral development in infants and preschool children. Am. J. Clin. Nutr. 1986, Vol. 43, p555.

Soemantri, A.G., et al (1985):Iron deficiehcy anemia and educational achievement. Am. J.Clin. Nutri. Vol. 42, p1221.

Oski,F.A. et al (1983) : Effect of iron therapy on behavior performence in nonanemic iron deficient infants. Pediatrics, Vol. 71. p877-880.

Walter, T.F. et al (1983) : Effect of mild iron deficiency on infant ment scores. J. Pediatr. Vol. Lo2 p519-522.

Webb, T.E.,et al. (1973) Iron deficiency anemia and scholastic achievement in young adolescents. J. Pediatr. vol. 82. p827-830.

Thomas,F.Massaro (1982) :Effect of iron deficiency on learning. in ; Pollitt,E, et al. Iron deficiency, Brain Biochemistry and Behavior. New York. Raven Press p125

Youdim,M.B.H., et al: (1980) The effect of iron deficiency on brain biogenic monoamine biochemistry and function in rats. Neuropharm. Vol. p 295.

Metal Ions in Biology and Medicine. Eds. Ph. Collery, L.A. Poirier, M. Manfait, J.C. Etienne. John Libbey Eurotext, Paris © 1990, pp. 104-107

The detection of iron in tissue : analytical and histochemical investigation of the perls iron-hexacyanoferrate reaction

J.P. Cassella*, J. Hay**

Department of cell pathology, St-Thomas's Hospital, London, SE1 7EH, England School of Pharmacy, Leicester Polytechnic, Leicester, Lei 9BH, England***

Iron occurs in the body in two oxidation states, (+2 and +3). It is associated with haemosiderin and ferritin which act as iron storage complexes. Iron is bound to B-globulin and transported as transferrin in blood plasma; it is associated with the respiratory pigment haemoglobin. A number of diseases are associated with either a deficiency, or an excessive build-up of iron, in the body tissues.The parameter of iron status chosen for any given clinical situation, will depend on the precise objective and whether iron overload or deficiency is suspected. Iron status is best determined by measuring the amount of iron present in the three main iron containing compartments of the body: storage iron, transport iron, and erythrocyte iron. Following routine haematological and biochemical investigations, tissue biopsies are a common method of determining for the presence of iron. Liver and bone marrow biopsies not only permit the detection of iron, but also a view of the pathological status of the tissue.

Histological assessment of haemosiderin in bone marrow particles showed good agreement with chemical estimation over a wide range of values (Davidson and Jennison,1952) . This confirmed original findings of Rath and Finch (1948) that the type of anaemia and amount of stainable iron in bone marrow smears and histological sections were closely related. Reactions involving hexacyanoferrate ions are often used in diagnostic histopathology for the detection of iron (Kiernan,1981) . The so-called 'Prussian blue' reaction is considered to demonstrate iron in its ferric (III) oxidation state by reaction of these ions with acidified potassium ferrocyanide. The so-called 'Turnbull's blue' reaction is suggested to involve reduction of ferric (III) ions to ferrous (II) ions by treatment with aqueous ammonium sulphide followed by reaction of the generated ferric (II) ions with acidified potassium ferricyanide. Sulphydryl groups can be demonstrated by virtue of their reducing properties; a solution containing ferric (III) iron and ferricyanide ions produces a blue insoluble, product often cited as "Prussian blue' , when acted upon by reducing groups.

The chemistry of iron-hexacyanoferrate complexes (IHC) is well documented (Cotton and Wilkinson 1988). There is interaction between the individual complexes as a result of oxidation-reduction reactions. The latter is of considerable interest to the histochemist; in a tissue section chemical components capable of acting as oxidants or reductants, may be present in the immediate vicinity of a forming IHC; this may make difficult the determination of the precise nature of the reaction and its product. A further complication for the histochemist is the wide range of differing formulae and chemical structures available for each IHC. Furthermore, various synonms are assigned to each IHC (Hawker and Twigg, 1987).

These discrepancies in nomenclature and structure of individual IHCs and of equations presented for the chemical reactions between iron and hexacyanoferrates, is therefore of considerable relevance; the primary aim of histochemical methods is to utilise chemical reactions in order to determine unequivocally the presence or absence of a particular component, such as iron, in a tissue section.

Thus an investigation was undertaken in order to establish the effect of modifying IHC reaction conditions in an attempt to mimic those which might be present in a tissue section. The product of each reaction was then subjected to chemical analysis. The reaction between ferric (III) iron and potassium ferrocyanide (Perls, 1867) was of special interest since this is the most widely used IHC complex in diagnostic histopathology.

For this purpose a simple qualitative technique was developed (Cassella et al, 1990). This involved addition of the various ingredients required for IHC formation, as well as possible interfering substances, at different concentrations and under differing physical conditions, to the wells of microtitre plates. The same reactions were performed concurrently on sections of pathological spleen known to contain haemosiderin. In some instances IR or UV spectroscopy, and X-ray powder diffraction studies, were undertaken on appropriately prepared reaction products.

The microtitre plate studies clearly indicated that the reaction between ferric (III) iron and potassium ferrocyanide was hydrogen ion dependent. If ferrocyanide was used in conjunction with no acid, weak organic acid or very dilute mineral acid, ferric ferrocyanide was not formed in sections known to contain haemosiderin.sections of This finding was confirmed in sections of spleen known to contain ferric (III) iron. Excess, or prolonged incubation of ferric (III) iron with potassium ferrocyanide, in the absence of acid, yielded a gelatinous yellow-brown product; in the tissue sections this was observed as a yellow, diffuse, non-granular precipitate (FIG 1). The method used by Singh (1968) for staining bone marrow fragments in previously diagnosed megaloblastic anaemia resulted in diffuse non-granular staining. This was due to an excessive incubation period in a hyperconcentrated potassium ferrocyanide solution. In the present study acid present with excess ferrocyanide resulted in the formation of a blue-coloured soluble product; this was readily removed from a section simply by washing with water. The soluble blue product was so-called 'soluble Prussian blue'; this species was rapidly precipitated with excess potassium ions. If ferric (III) iron was present in excess, a precipitate of blue granular material was noted; this was so-called 'insoluble Prussian blue'. Use of over-dilute mineral acid or weak organic acid in the reaction between ferric (III) iron and potassium ferocyanide resulted in the formation of a green non-granular reaction product; in the section leaching of haemosiderin was observed. The blue coloured product formed on the reaction between ferric (III) iron , potassium ferrocyanide and a suitable defined concentration of HCl ,was rendered green or colourless in the presence of relatively low concentrations of either oxidising or reducing agents respectively. In sections, addition of such reagents to the incubation mixture, resulted in no product being observed microscopically.

Analytical studies showed that so-called 'Prussian blue' and 'Turnbull's blue' were chemically identical, thus confirming the findings of Nicholls, (1973). Ferric (III) ions are surrounded octahedrally by 6 nitrogen atoms and iron (II) having 6 carbon atoms as neighbours. The structure can be represented as $Fe_4^{III}[Fe^{II}(CN)_6]_3 . xH_2O$ where x =14-16 (Cotton and Wilkinson, 1988). It is likely that reports suggesting that these are separate species result from differences in preaparation, and presence of varying amounts of metal ions, water and other components in the crystal lattice.

On the basis of the microtitre plate test with concomitant staining of sections known to contain ferric (III) iron an improved Perls' reaction protocol was developed (Cassella et al,1989). The method was used with success on tissues which had previously been fixed in alcohol based fixatives, or those fixed in neutral formalin. It has been suggested that such fixatives are inferior to acid formalin which contains calcium for demonstration of ferric (III) iron in tissue sections (Hadler et al, 1969). The modified method was not associated with any obvious release of haemosiderin granules from pathological spleen; the associated ferric (III) iron was clearly and discretely demonstrated in the tissue (FIG 2). There was no requirement for pre-incubation of section with potassium ferrocyanide prior to addition of HCl. This was suggested by Gomori (1952) to compensate for the differing diffusion rates of the two compoounds. The modified method was successfully used for the demonstration of ferric (III) iron in tissue sections derived from cold blooded vertebrates, or from invertebrates; it has been suggested that the simple Perls method is inadequate for this as a result of insensitivity (Gabe, 1976). The modified methodology also overcame the problems stated to be associated with 'Prussian blue' staining of whole human brain sections (Barnett et al , 1980).

It is our contention that the imprecise nomenclature, and lack of understanding of chemical reactions involving iron-hexacyanoferrates, at least in a histochemical setting, may lead to misinterpretation of histological sections subjected to existing methodologies. The use of the simple microtitre plate test to determine the optimum reaction conditions in the Perls' technique allowed better and more reliable results to be obtained in sections containing ferric (III) iron; it also overcame problems associated with non-optimal fixation. This method can easily be used for other IHC reactions in histochemistry. It may also be useful when used in conjunction with analytical techniques for determination of the role of iron in the formation of coloured chelation products in histochemical reactions.

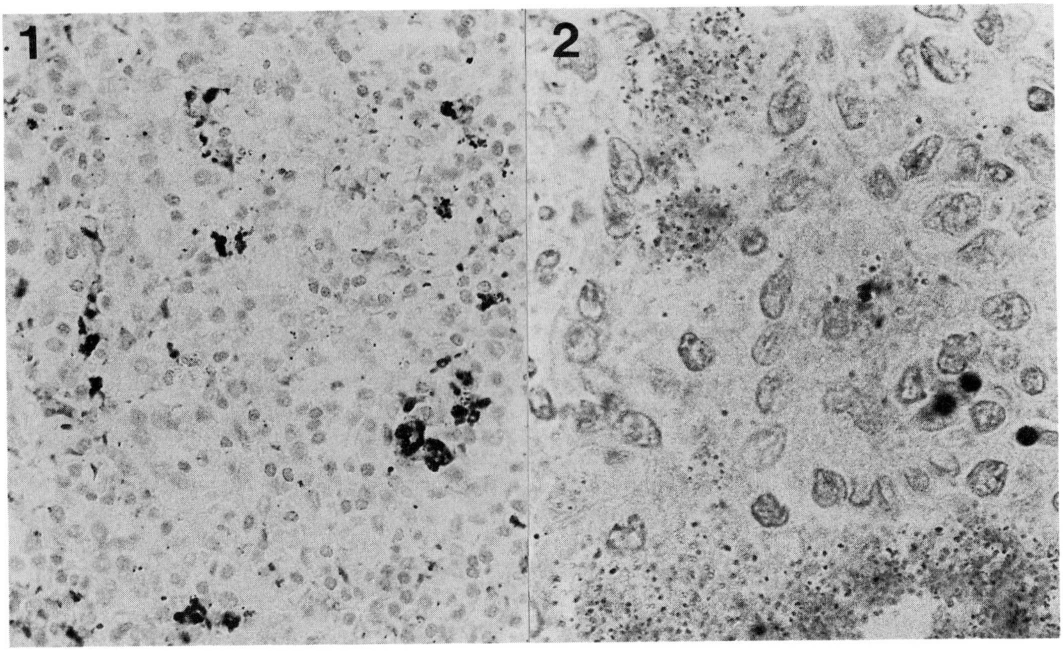

Human patholological spleen showing;

FIG 1 Diffuse extracellular staining after prolonged incubation with aqueous potassium ferrocyanide (1000:1);

FIG 2 Discrete intacellular staining of haemosiderin (1000:1).

REFERENCES:

Barnett R.I, Lyons G.W, Driscoll J.D and Forrest W.J, 1980, Improved sectioning and Berlin blue staining of whole human brain,*Stain Technol*, **55** 235.

Cotton F.A and Wilkinson G, 1988, *Advanced Inorganic Chemistry*, New York : Wiley.

Cassella,J.P, Hay,J, Ball,M, 1990, A modified method for improved demonstration of ferric iron in histological sections, Royal Microscopical Societies *Proceedings*, **25**: (1) 41.

Cassella, J.P, Hay,J, Cairns, D. 1989, Analytical and histochemical investigationof the Perls' iron-hexacyanoferrate reaction, *Inst. Phys .Conf .Ser. No 98 Chapter 16* p 747, EMAG-MICRO.

Davidson, W.M; Jennison, R.F, 1952, Relationship between iron storage and anaemia, *J. Clin Path* **5** 287.

Ed Jacobs A; Worwood M, 1976, Iron in biochemistry and medicine, Academic Press, New York.

Gabe M, *Histological Techniques*,Paris: Masson.

Gomori G ,1952, *Microscopic Histochemistry,* Chicago : Chicago University Press.

Hadler W A , Lucca O, de Ziti L M and Patelli A S ,1969, An analysis of the effect of some fixatives on the histochemical detection of nonhaem ferric iron in spleen sections,*Revista Brasileira de Pesquisas Medicas Medicas e Biologicas,* **2** 378 .

Hawker P N and Twigg M V ,1987, *Comprehensive Coordinate Chemistry,* London : Pregammon Press.

Kiernan J. A , 1981 ,*Histological And Histochemical Methods,* Toronto : Pergammon Press .

Nicholls D, 1973, *Comprehensive Inorganic Chemistry,* Oxford : Pergammon Press.

Perls M,1867, Nachweis von eisenoxyd in gewissen pigmente,*Virchows Arch Pathol Anat,* **79** 42.

Rath, C.E; Finch, C.A, 1948, Sternal marrow haemosiderin; Method for the determination of available iron stores in man, *J. Lab & Clin Med* , **33** 81.

Singh, A.K, 1968, Studies on iron metabolism with particular reference to iron store, PhD, University of London.

ANALYTICAL AND HISTOCHEMICAL INVESTIGATION OF THE PERLS IRON-HEXACYANOFERRATE REACTION

J P CASSELLA AND J HAY

Department of Cell Pathology, Institute of Dermatology, St Thomas's Hospital, Lambeth Palace Road London, SW1, England

Reactions involving hexacyanoferrate ions are often used in diagnostic histopathology. The 'Prussian blue' reaction is used for the demonstration of iron in its ferric (III) oxidation state. The 'Turnbull's blue' reaction involves reduction of iron (III) ions to ferrous (II) ions by treatment with ammonium sulphide followed by reaction of the generated iron (II) ions with potassium ferrocyanide. Sulphydryl groups can be demonstrated by virtue of their reducing properties; a solution containing iron (III) and ferricyanide ions produces a blue insoluble product when acted upon by reducing groups.

The chemistry of iron-hexacyanoferrate complexes (IHC) is well documented. There is interaction between the individual complexes as a result of oxidation-reduction reactions. The latter is of considerable interest to the histochemist; in the tissue section chemical components capable of acting as oxidants or reductants may be present in the immediate vicinity of the forming IHC; this may make difficult the determination of the precise nature of the reaction and its product. A further complication for the histochemist is the wide range of different formulae and chemical structures available for the IHC. Furthermore various synonyms are assigned to each IHC.

An investigation was undertaken in order to establish the effect of modifying the IHC reaction conditions in an attempt to mimic those which might be present in a tissue section. The products of the reactions were then subjected to chemical analysis.

A simple qualitative testing procedure was developed. This involved addition, to the wells of the microtitre plate, of the various ingredients required for IHC formation as well as possible interfering substances. This allowed visual assessment of the reaction products. The same reactions were performed concurrently on sections of pathological spleen known to contain haemosiderin. In some instances IR or UV spectroscopy and X-ray powder diffraction studies were undertaken on appropriately prepared reaction products.

Findings clearly indicated that the reaction between ferric (III) iron and potassium ferrocyanide was hydrogen ion dependent.

On the basis of the microtitre plate test with concomitant staining of sections known to contain ferric (III) iron an improved Perl's reaction protocol was developed.

It is our contention that the imprecise nomenclature of and lack of understanding of chemical reactions involving iron and hexacyanoferrates at least in a histochemical setting may lead to misinterpretation of histological sections subjected to existing methodologies. The use of a microtitre plate test could easily be used for other IHC reactions in histochemistry. It may also be useful when used in conjunction with analytical techniques for determination of the role of iron in the formation of coloured chelation products in histochemical reactions.

Metal Ions in Biology and Medicine. Eds. Ph. Collery, L.A. Poirier, M. Manfait, J.C. Etienne. John Libbey Eurotext, Paris © 1990, pp. 108-110

Vine sensitivity to frost and manganese

C. Choisy*, G. Lemoine*, J.P. Clément*, R. Fay**, B. Itier***, O. Brun****, P. Choisy*

* *Laboratoire de Microbiologie, UFR Pharmacie, Reims, France.* ** *Laboratoire de Pharmacologie, UFR, Médecine, Reims, France.* *** *Station de Bioclimatologie, INRA, Thivernal Grignon, France.* **** *GCEV, Division Recherche MUMM, Epernay, France*

Late frosts in Champagne often cause significant economic losses by destroying buds (Vazart 1988). Since Lindow's studies (1983), it has been acknowledged that Pseudomonas syringae can be inductors of ice-nucleation in plant tissues. The active substance has been identified (Green 1985): it is a protein bound to the bacterial cell for activity. In the aim of studying the different factors of frost damage, a multidisciplinary group was set up to survey a champagne vineyard (MUMM) at AVIZE. During 2 consecutive springs, the following parameters were measured : i) bud sensitivity to frost, ii) P. syringae number, iii) stage of bud development. We had hoped to cut off subjective description ambiguity by measuring tissue hydration and the amounts of Al, Zn, Cu, Mn, Mg. The curves for sensitivity to frost number of ice-nucleating Pseudomonas and concentrations of Mg and Mn in plant tissues seen to be roughly parallel. It induced us to seek wether ice-nucleating Pseudomonas syringae are dependent of MnII and MgII concentrations.

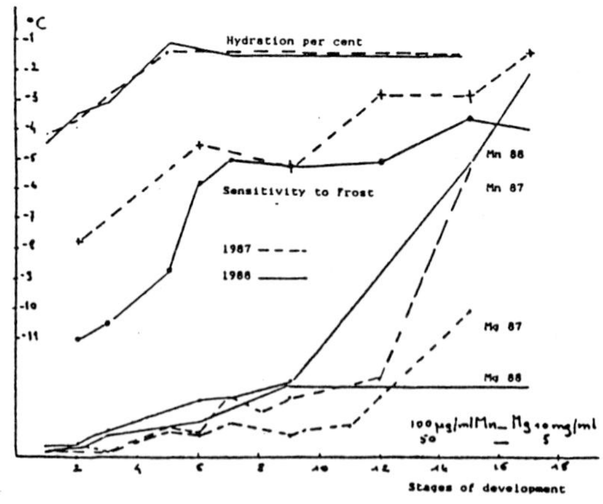

Fig 1 : Frost sensitivity of grapevine, amounts of MnII, MgII and water in buds in 87-88

MATERIAL AN METHODS

Sensitivity to frost was measured by setting up convective chambers around vine plant in situ and by recording the difference in temperature between buds and a piece of dead vine branch. Bud freezing was clearly shown by an exotherm. Amounts of mineral ions were assayed by Atomic Absorption spectrophotometry (A.A.S). Bacterial strains came from grape-vine samples and were identified according to Lelliott (1966). Before assay, they were transplanted 3 times each 24h. Strains dependence was studied in the following sterile medium : Difco Casaminoacids 14g ,Glucose 6 g, K2HPO4 0,5 g, water 1 l, extemporaneously added with polyvitamins 0,02 ml per l. Ice-nucleating power was evaluated from young suspensions in distilled water (#10^8 cells/ml) in a Huber Cryostat (Bioblock). The effect of dehydration of medium was assessed by seeding P. Syringae with a Spiral (Interscience) automatic seeding machine plates weighed before and after dehydration.

RESULTS

1 - Manganese, Magnesium and Pseudomonas growth

8 strains of *P. Syringae* and 4 strains of *P. fluorescens* were submitted to a range of 12 MnII concentrations from 2 to 50 µg/ml and 12 MgII concentrations from 20 to 1 000 µg/ml. The maximum of growth was determined by nephelemetry.

Table 1 : Fluorescent *Pseudomonas* sentivitiy to MnII and MgII

conc. g/ml souches	MnII	MnII+small set conc	Mg II
P.Syringeas	35 ± 8	30 ± 11	69 ± 26
P. fluorescens	42.5	45	85

2 - Manganese, Magnesium and ice-nucleation power.

Table 2 : Strain ice-nucleating temperature variations as a function of MgII and MnII concentrations.

conc.g/l / strains	control	ManganeseII					Magnesium II				
	0	0.05	0.1	0.25	0.5	1	0.01	0.1	0.25	0.5	1
SAB 10	- 3.5°C	-3.3	-4	-4	-3.3	-3.3	-4	-4	-4.6	-5	-8
SAB 2	- 6.6°C	-6.3	-6.3	-6.6	-33	-10	-6.6	-6.6	-9	-10	-10

3 - Hydratation and P.syringae growth

Table 3: loss of Colony Forming Units in relation to dehydration

Duration	5days	4days	3days	2days	1day
lost water %	11,22	6,88	5,88	1,88	0
lost Cfu %	15	14	3,33	2	0

DISCUSSION AND CONCLUSION

Amoung the 12 assessed strains only one of P. syringae and one of P. fluorescens seemed MnII independent, all are MgII dependent. It can be inferred that MnII has a favorable action on the development of P. syringae and P. fluorescens. Since ice-nucleation is due to a protein, the factors favouring protein synthesis most probably favour ice-nucleating power. Mg II effect on protein synthesis is well known. MnII is a constituant of some essential enzymes and especially Superoxide dismutase (Keele 1970) particularly important to a strict aerobic genus as Pseudomonas. Therefore, MnII is also necessary to non ice-nucleating species. Protein synthesis inhibition by MnII at high concentration was demonstrated by Korc (1984) in animal cells. Table 2. Shows that medium amounts of Mn and Mg favoured ice-nucleation and highest amounts inhibit it.

At this present state of our knowledge, it can be noted that numerous factors contribute to favour freezing at slightly negative temperature : i) the increase in amounts of MnII and MgII enhance P.syringae growth and ice-protein synthesis; ii) hydration of plant tissues increases Pseudomonas population (the aw value of which being among the highest (Beuchat 1981) and the development of plant organs is linked to this hydration ; according to Fig.1, the phenologic stage plays a major role in frost sensitivity; iii) besides, our group experimented that outer water (dew) probably has prime importance in sensitivity to frost (Itier 1989).

REFERENCES

Beuchat, L.B. and Golden, D.A. (1981) : Microbial activity as affected by water activity. Cereal Foods World, 26, 7.

Green, R.L. and Warren, G.J. (1985) : Physical and functional repetition in a bacterial ice-nucleation gene. Nature, 317, 645-648.

Itier, B., Flura, D., Brun, O. Luisetti, J., Gaignard, J.L., Choisy, C. and Lemoine, G. (1989). Journées de Bioclimatologie INRA, PARIS, 21 Novembre.

Keele, B.B., McCord, J.M. and Fridovich, I. (1970) : Superoxide dismutase from Escherichia coli , a new manganese containing enzyme. J. Biol. Chem., 245, 6176-6181.

Korc, M. (1984) : Manganese action on proteins synthesis in diabetic rat pancreas evidence for a possible physiological role.Amer. J. Clin. Nutr., 27, 2119-2126.

Lederer, J. (1989) : L'iode et le manganèse. Paris. Maloine.

Lelliott, R.A., Billing, E. and Hayward, A.C. (1966) : A determinative scheme for fluorescent plant pathogenic Pseudomonas. J. Appl. Bacteriol. 29, 3, 470-489.

Lindow, S.E. (1983) : The role of bacterial ice-nucleation in frost injury to plants. Ann. Rev. Phyto Pathol., 21, 363-384.

Vazart, V. (1988) : Les gelées de printemps de 1875 à 1986 dans le vignoble champenois : historique et conséquences, relation avec les productions de raisin. Thèse Doct. Etat Pharmacie. Reims.

Metal Ions in Biology and Medicine. Eds. Ph. Collery, L.A. Poirier, M. Manfait, J.C. Etienne. John Libbey Eurotext, Paris © 1990, pp. 111-113

Effect on exercice on copper, zinc and ceruloplasmin levels in blood of athletes

M. Marrella, * F. Guerrini, * P.L. Tregnaghi, * S. Nocini, G.P. Velo, R. Milanino

*Istituto di Farmacologia, Università di Verona, 37134, Italia and * Istituto di medicina dello Sport, 37122, Verona, Italy*

Summary

Eleven healthy male triathletes (26.7 ± 3.77 years) underwent submaximal cycle ergometric exercise test (1 hour, 70% VO_2 max) to determine the effect of exercise on blood copper and zinc (plasma and total blood cells) and on plasma ceruloplasmin. Blood was collected four times: before the exercise (PRE), at the end (POST), after one and two hours of resting (REST1 and REST2 respectively).

Mean basal values of plasma copper and zinc concentration, in the athletes, were lower than those measured in a control group of healthy blood donors (Cu: -15.7%; $p<0.01$ and Zn: -23.8%; $p<0.01$) but no differencies were found in the basal values of ceruloplasmin and in total blood cell (TBC) copper and zinc concentration.

After the exercise a slight but significant decrease in plasma copper concentration was observed (-6.1%, $p<0.01$) that returned to basal values at REST1. The ceruloplasmin and plasma zinc concentrations showed no significant decrease during the trial. There was no change in TBC copper and zinc concentration.

Introduction

Aerobic exercise leads to many metabolic, cardiovascular and muscolar changes and also the trace elements metabolism is modified by either acute exercise or long training (Campbell W.W. & Anderson R.A. 1987). The changes of plasma copper and zinc concentrations have been studied in different sports and training conditions with different results (Dowdy R.P. & Burt J. 1980, Lukaski H.C. et al. 1983, Olha A.E. et al. 1982).

However, it is generally accepted that there is a mobilization of these minerals and a lost of them with physical work (Campbell W.W. & Anderson R.A. 1987). The negative balance that occur in the athletes if their diet is inadequate in minerals, and the subsequent marginal deficiency which may ensue, can induce a direct effect on athletic performance (McDonald R. & Keen C.L. 1988).

Although the plasma copper concentration can be take as an index of the copper status (only in health and resting conditions) (Underwood E.J. 1977) the plasma zinc concentration is not a reliable index of the status of zinc (Prasad A.S. 1983).

The purpose of this study was to evaluate the effects of exercise (1 hour, 70% VO_2 max) on plasma copper, zinc and plasma ceruloplasmin levels and total blood cells copper and zinc concentrations. All these data were also compared with those obtained from a control group of healthy blood donors.

Material and methods

The investigation was conducted, one month after the end of the competitive season, on eleven healthy male triathletes who gave their informed consent to participate in this study.

VO_2 and experimental work loads were determined for each subjects during a preliminary test session by an automatic analyzer (Oxitest, Biotec) on a cycle ergometer (Cardioline STS3). Subsequently, all subjects performed an exercise of 60 min. duration at a constant work load corresponding to 70% VO_2 max. Each test was conducted 3-4 hours after a standardized light meal (boiled rice, and vegetables). At the end of the test, the subjects observed a two hours rest, during which eating or drinking was not allowed.

Blood was drawn from an antecubital vein and collected into heparinized tubes four times: before the exercise (PRE), at the end (POST), after one and two hours of resting (REST1 and REST2 respectively).

Red blood cells count (RBC), haemoglobin (HGB) and haematocrit (HCT), were assayed automatically by a Sysmec CC 780 instrument. For copper and zinc determinations, all the chemical used were A.R. grade free from copper and zinc contamination, and only copper and zinc free glassware was used. Non-haemolytic plasma was deproteinized by addition of an equal amount of 15% trichloroacetic acid solution. Total blood cells were acid digested following a procedure which has been previously described (Marrella M. & Milanino R., 1986). Copper and zinc determinations were carried out by flame atomic absorption spectrophotometry (Perkin-Elmèr 3030 SAA)(Marrella M. & Milanino R., 1986). Ceruloplasmin was assayed by immunological method (Beckman ICS analyzer II).

The values obtained after exercise for plasma copper, zinc and ceruloplasmin, were corrected because of the changes in blood volume according to Dill D.B. & Costill D.L. (1974).

Statistics was made by means of paired and umpaired Student's t-test when appropiated.

Results

The results are reported in Table 1, in the last column are reported the data obtained from a control group of 85 healthy blood donors.

Table 1

	PRE	POST	REST 1	REST 2	DONORS
RBC ($\cdot 10^6$/ml)	4.89 ± 0.4	5.23 ± 0.4 **	4.95 ± 0.4 ◇◇	4.97 ± 0.4 ◇◇	4.86 ± 0.3
HGB (g/dl)	14.90 ± 0.6	16.20 ± 0.8 **	15.31 ± 0.7 ** ◇◇	15.41 ± 0.8 ** ◇◇	14.81 ± 1.0
HCT (%)	45.05 ± 2.7	48.00 ± 2.0 **	45.74 ± 1.8 ◇◇	45.94 ± 2.0 * ◇◇	45.63 ± 2.7
Cp (mg/dl)	33.22 ± 6.6	32.69 ± 10.8	31.53 ± 4.9	29.35 ± 4.7	29.50 ± 6.5
Cu Pl (μg/dl)	88.18 ± 16.4 ✦✦	82.77 ± 15.9 **	87.15 ± 17.7 ◇◇	85.57 ± 14.9 **	104.60 ±14.0
Zn Pl (μg/dl)	77.60 ± 13.2 ✦✦	74.62 ± 14.2	71.03 ± 8.9	71.10 ± 8.3	101.90 ± 9.3
Cu Cell (μg/dl)	91.55 ± 11.9 ✦✦	91.10 ± 14.8	88.44 ± 10.9	94.52 ± 12.7	83.40 ± 9.1
Zn Cell (μg/dl)	1304 ± 120	1323 ± 140	1300 ± 111	1361 ± 140	1337 ± 165

Paired Student t-test: * $P < 0.05$, ** $P < 0.01$, versus PRE; ◇◇ $P < 0.01$, versus POST.
Unpaired Student t-test ✦✦ $P < 0.01$, versus DONORS

Mean basal values of plasma copper and zinc concentration in the athletes were lower than those measured in the control group but no differencies were found in the basal values of ceruloplasmin and in the total blood cell copper and zinc concentration. Basal value of TBC copper concentration is higher in the athletes than controls.
As axpected RBC, HGB, HCT increases after exercise. At REST2, HGB and HCT were still higher than at PRE. The exercise induced a slight but significant decrease in plasma copper concentration (-6.1%, $p<0.01$) at POST that returned to basal values at REST1. A significant decrease of plasma copper concentration is also observed at REST2.
The ceruloplasmin and zinc plasma concentration showed no significant decrease during the trial. There was no change in TBC copper and zinc concentration.

Discussion

Results from the present study show that triathletes are a population in which basal plasma values of copper and zinc are lower than normal.
However TBC copper and zinc concentrations, which represent a more reliable index of the status of these elements in blood, are not reduced in the athletes, compared with control group, apparently indicating that a condition of overt copper and zinc deficiency does not exist.
Nevertheless, the observed decrease in plasma copper and zinc concentrations may indicate a loss or a change in the distribution of these elements during exercise. If the former alternative is correct it would be desirable to supplement with copper and zinc the athletes during competition period in order to equilibrate their metal balance and prevent the appearance of a deficiency status. Further work is in progress to verify this hypothesis.

References

Campbell, W.W. & Anderson, R.A. (1987): Effects of aerobic exercise and training on the trace minerals chromium, zinc and copper. Sport Medicine 4, 9-18.

Dill D.B. & Costill D.L. (1974): Calculation of percentage changes in volume blood, plasma and red cells dehydration. J.Appl.Physiol.2, 247-248.

Lukaski H.C., et al. (1983): Maximal oxigen comsumption as related to magnesium, copper and zinc nutriture. Am.J.Clin.Nutr. 37, 407-415.

Marrella M. & Milanino R. (1986): Simple and riproducible method for acid extraction of copper and zinc from rat tissue for determination by flame atomic spectroscopy. Atomic Spectroscopy 7,1, 40-42.

McDonald R. & Keen C.L. (1988): Iron, Zinc and Magnesium nutrition and athletic performance. Sport Medicine. 5, 171-184.

Olha A.E., et al. (1982): Effect of exercise on concentration of elements in the serum. Journal of Sport Medicine and Physical Fitness. 22, 414-425.

Prasad A.S. (1983) The role of zinc in gastrointestinal and liver disease. Clinics in Gastroenterology 12,3, 713-741. .

Underwood E.J. (1977), Copper. In trace elements in human and animal nutrition. 4th ed. pp. 56-108. Academic Press, New York.

Metal Ions in Biology and Medicine. Eds. Ph. Collery, L.A. Poirier, M. Manfait, J.C. Etienne. John Libbey Eurotext, Paris © 1990, pp. 114-118

Electron microscopic studies on the intracellular translocation of calcium ions during the contraction-relaxation cycle in muscle

H. Sugi, S. Suzuki

Department of Physiology, School of Medicine, Teikyo University, Itabashi-ku, Tokyo 173, Japan

REGULATION BY CALCIUM IONS OF THE CONTRACTION-RELAXATION CYCLE IN MUSCLE

Muscle contraction results from the alternate formation and breaking of cross-links between the thick and thin filaments coupled with ATP hydrolysis. In skeletal muscle at rest, the cyclic interaction between the filaments is inhibited by the regulatory proteins, tropomyosin and troponin, on the thin filaments. When calcium ions (Ca^{2+}) bind to troponin, the inhibitory action of the regulatory proteins is removed to result in muscle contraction (Ebashi & Endo, 1968). Conversely, detachment of Ca^{2+} from troponin causes relaxation. In smooth muscles, on the other hand, the cyclic interaction between the filaments takes place when Ca^{2+} binds to calmodulin in the myoplasm to form Ca-calmodulin, which in turn activates myosin light chain kinase to cause phosphorylation of myosin light chain in the thick filament; relaxation is produced by dephosphorylation of myosin light chain after detachment of Ca^{2+} from calmodulin (Sovieszek, 1977). In spite of the above difference in the mode of activation of contraction between skeletal and smooth muscles, the contraction-relaxation cycle is regulated by the change in free Ca^{2+} concentration in the myoplasm in all kinds of muscle.

In resting muscle fibers, the myoplasmic free Ca^{2+} concentration is very low, being of the order of 10^{-7}M. This condition is maintained by the intracellular membranous structures accumulating Ca^{2+} and/or the action of the surface membrane, i.e. the uphill Ca^{2+} transport to the extracellular space and the Na-Ca exchange mechanism. Thus, the Ca^{2+} activating contraction (activator Ca) may originate from the intracellular Ca^{2+}-accumulating structures and/or from the extracellular fluid. In physiological conditions, the translocation of activator Ca initiated by neural stimulation is generally mediated by surface membrane depolarization, and the resulting increase in the myoplasmic free Ca^{2+} concentration (Ca-transients) can be detected with various intracellular Ca indicators (for a review, see Blinks *et al.*, 1982). Since the Ca-transients do not tell us about the sources of activator Ca, the mechanism of translocation of activator Ca regulating the contraction-relaxation cycle in muscle can only be studied at the level of muscle ultrastructure. This article deals with electron microscopic studies for detecting the intracellular Ca translocation in skeletal and smooth muscles.

INTRACELLULAR TRANSLOCATION OF CALCIUM IONS IN SKELETAL MUSCLE FIBERS

Skeletal muscle fibers contain two different types of membranous structures, the transverse tubules (T-tubules) and the sarcoplasmic reticulum (SR). The T-tubules are tubular invaginations of the surface membrane, forming a continuous tubular network surrounding each myofibril. The SR consists of three continuous elements, i.e. the terminal cisternae, the longitudinal tubules and the fenestrated collar, and envelopes each myofibril along its long axis as shown diagrammatically in Fig. 1. In frog muscle fibers, the T-tubules are located at the level of Z-line, while the SR runs longitudinally from Z-line to Z-line, so that the terminal cisternae of the SR are closely apposed to the T-tubules (Fig. 1A). The SR accumulates Ca^{2+} in its lumen by the active transport operated by Ca-activated ATPase at the SR membrane. The influence of surface membrane depolarization is transmitted inwards along the T-tubule membrane, and further to the SR *via* bridge-like structures between the T-tubules and the SR.

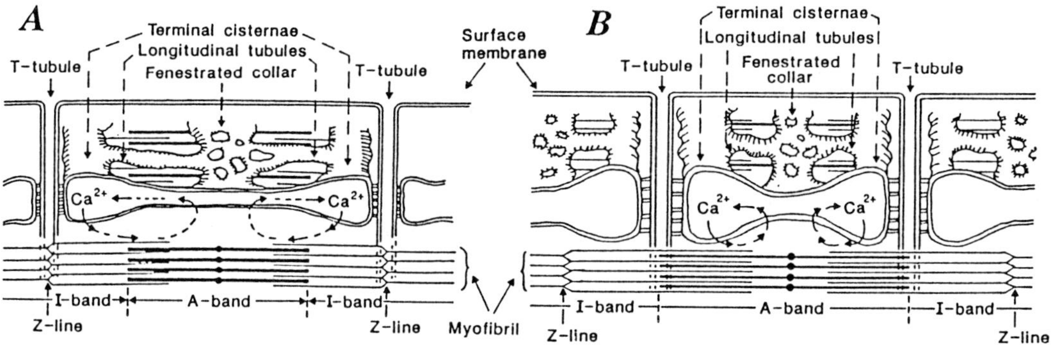

Fig. 1. Diagrams showing the organization of the T-tubules and the SR and the direction of Ca translocation during the contraction-relaxation cycle in frog skeletal muscle fibers (A) (Winegrad, 1968) and in fish swimbladder muscle fibers (B) (Suzuki *et al.*, 1990).

Using ^{45}Ca autoradiography combined with quick freezing technique, Winegrad (1965, 1968) first studied the intracellular translocation of Ca^{2+} by observing the change in ^{45}Ca distribution in frog muscle fibers under a light microscope not only at rest, but also during and after a tetanus. In resting fibers, ^{45}Ca was mostly localized around the Z-line, while in the fibers fixed during a tetanus, ^{45}Ca originally present around the Z-line decreased in amount to distribute toward the A-band region, indicating the release of Ca^{2+} from the terminal cisternae into the myoplasm. When the fibers were fixed at various times after a tetanus, the change in ^{45}Ca distribution suggested that the released Ca^{2+} is first taken up at the longitudinal tubules, and then slowly return to the terminal cisternae. The above direction of Ca^{2+} translocation during the contraction-relaxation cycle is also indicated in Fig. 1A.

Recent development of electron probe X-ray microanalysis technique has made it possible to measure the intercellular localization of various elements including Ca under an electron microscope on cryosections of quickly-frozen tissues. Thus, Somlyo *et al.* (1981, 1985) performed the X-ray microanalysis of cryosections of frog muscle fibers prepared at rest and during and after a tetanus. Contrary to the report of Winegrad, they found no evidence that Ca^{2+} was taken up by the longitudinal tubules, insisting that the Ca^{2+} was released from, and again taken up by, the terminal cisternae during the contraction-relaxation cycle. Though the spatial resolution of the X-ray microanalysis is much better than that of the ^{45}Ca autoradiography, the detection of Ca in the SR elements other than the terminal cisternae seems to be extremely difficult because of poor preservation of the membranous structures (especially the longitudinal tubules) in cryosections, and the above debate remains to be settled.

We have studied the intracellular Ca translocation during the contraction-relaxation cycle in the swimbladder muscle fibers of a teleost fish *Sebasticus marmoratus* (Suzuki *et al.*, 1990). The muscle contracts very rapidly to produce sound and contains extremely well developed SR, providing a material suitable for detecting Ca localization within the elements of the SR. As shown in Fig. 1B, the T-tubules of the swimbladder muscle are mostly located around each boundary between the A- and I-bands unlike frog muscle fibers. In resting fibers, Ca was mostly localized around the A-I boundary, indicating the Ca localization in the terminal cisternae. During a tetanus, Ca was distributed in other regions in each sarcomere with a decrease in the amount of Ca around the A-I boundary, indicating the release of Ca^{2+} into the myoplasm. At 100ms after the onset of relaxation, on the other hand, a large amount of Ca was seen around the level of the fenestrated collar and the longitudinal tubules, strongly suggesting that, during relaxation, the Ca^{2+} released from the terminal cisternae can be taken up at the fenestrated collar and the longitudinal tubules.

INTRACELLULAR TRANSLOCATION OF CALCIUM IONS IN SMOOTH MUSCLE FIBERS

Smooth muscles exhibit wide variations in both ultrastructure and physiological properties, so that the experimental results obtained from one kind of muscle can not be directly applied to the other. Figure 2 is a diagram showing the membranous structures observed in most kinds of smooth muscle. The SR is only poorly developed compared to that in skeletal muscle, and can be seen in the form of vesicles in fiber cross-sections. Both the SR and the mitochondria are sometimes closely apposed to the surface membrane. The surface membrane exhibits bottle shaped invaginations called the caveolae. Evidence has been accumulating that many kinds of smooth muscle contain intracellularly stored activator Ca (for reviews, see Sugi & Suzuki, 1982; Suzuki & Sugi, 1982). A number of histochemical studies on the intracellular Ca localization in smooth muscles have been made using various techniques, such as Ba and Sr to replace Ca, oxalate and pyroantimonate to precipitate Ca, and Ca loading. It has been shown that Ca is localized in various membranous structures, i.e. the SR, the mitochondria, the nucleus and the surface membrane, though these results do not give information about the source of activator Ca in physiological contraction.

Since 1976, we have succeeded in visualizing the intracellular Ca translocation in various types of vertebrate and invertebrate smooth muscles under an electron microscope, by fixing muscles with OsO_4 solution containing K-pyroantimonate. The results on a mammalian visceral smooth muscle and a mammalian vascular smooth muscle are shown in Figs. 3 and 4. As shown in Fig. 3, electron-opaque pyroantimonate precipitate is localized along the inner surface of the surface membrane as well as at the mitochondria and the SR in resting muscle fibers of guinea-pig taenia coli (A and B), while the precipitate diffusely distributes in the myoplasm in contracted fibers (C), indicating the release of Ca^{2+} from these structures into the myoplasm (Sugi & Daimon, 1977). The X-ray microanalysis of the pyroantimonate precipitate always indicates the presence of Ca in the precipitate. Similar results are obtained in some invertebrate somatic smooth muscles (Atsumi & Sugi, 1976; Suzuki & Sugi, 1978). In the case of dog coronary artery smooth muscle, on the other hand, the pyroantimonate precipitate containing Ca is localized only in the lumen of the caveolae, i.e., a part of the extracellular space, in resting fibers (Fig. 4A, B), while the precipitate diffusely distributes in the myoplasm in contracted fibers (Fig. 4C), indicating the extracellular origin of activator

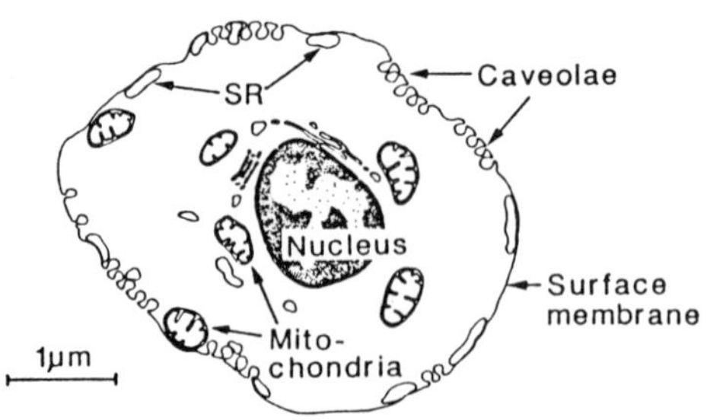

Fig. 2. Diagram showing the membranous structures in smooth muscles.

Ca in this muscle (Suzuki & Sugi, 1989). The above difference in the resting Ca distribution between the two kinds of smooth muscle is well correlated to the fact that K- or drug-induced contraction persists after the removal of external Ca^{2+} in taenia coli but not in coronary artery smooth muscle.

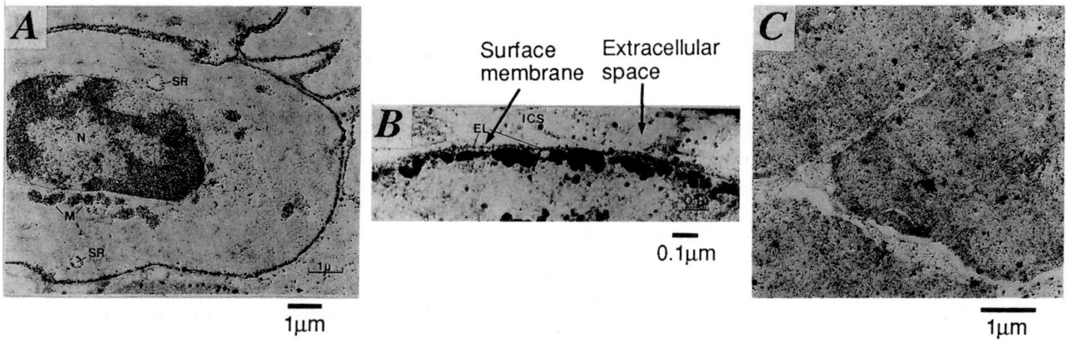

Fig. 3. Change in distribution of the pyroantimonate precipitate in resting (A and B) and contracted (C) smooth muscle fibers of guinea-pig taenia coli (Sugi & Daimon, 1977).

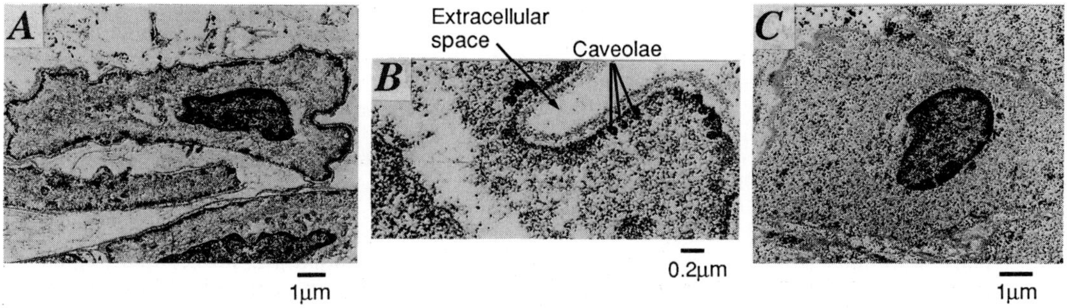

Fig. 4. Change in distribution of the pyroantimonate precipitate in resting (A and B) and contracted (C) smooth muscle fibers of dog coronary artery (Suzuki & Sugi, 1989).

In our work with pyroantimonate, the amount of the precipitate in the nucleus shows no appreciable change between resting and contracted states. The rate of Ca^{2+} uptake by the mitochondria is known to be too slow to account for the rate of relaxation. These results preclude the possibility that these structures serve as sources of activator Ca in smooth muscles during physiological contraction. Thus, the SR and the inner surface of the surface membrane are regarded to be the main sources of activator Ca. Recently, we succeeded in obtaining a high molecular weight (450kD) Ca-binding protein from the surface membrane-enriched fraction of a molluscan smooth muscle, the anterior byssal retractor muscle of *Mytilus edulis* (Yamanobe *et al.*, 1988), supporting the view that the inner surface of the surface membrane is the source of activator Ca; its Ca-binding capacity might be readily influenced by membrane potential changes to release and reuptake Ca^{2+}. In the case of the SR, it is necessary to consider some mediators including inositol triphosphate to transmit signals to the SR located distant from the surface membrane.

At present, the X-ray microanalysis of cryosections is not so effective in studying the intracellular Ca translocation in smooth muscles, because of poor preservation of the intracellular membranous structures as well as the surface membrane. Though the situation is the same in cryosections of skeletal muscle, the poor preservation of the T-tubules and the SR is compensated by their regular and periodic structures in each sarcomere. The problem concerning the relative contribution of the Ca-accumulating structures to the source of activator Ca in smooth muscles would be clarified when techniques in making cryosections are very much improved.

REFERENCES

Atsumi, S. & Sugi, H. (1976): Localization of calcium-accumulating structures in the anterior byssal retractor muscle of *Mytilus edulis* and their role in regulation of active and catch contractions. *J. Physiol. (Lond.)* 257: 549-560.

Blinks, J.R., Wier, W.G., Hess, P. & Prendergast, F.G. (1982): Measurement of Ca^{2+} concentrations in living cells. *Prog. Biophys. Mol. Biol.* 40: 1-114.

Ebashi, S. & Endo, M. (1968): Calcium ion and muscle contraction. *Prog. Biophys. Mol. Biol.* 18: 123-183.

Somlyo, A.V., Gonzalez-Serratos, H., Shuman, G., McClellan, G. & Somlyo, A.P. (1981): Calcium release and ionic changes in the sarcoplasmic reticulum of tetanized muscle: an electron-probe study. *J. Cell Biol.* 90: 577-594.

Somlyo, A.V., McClellan, G., Gonzalez-Serratos, H. & Somlyo, A.P. (1985): Electron probe X-ray microanalysis of post-tetanic Ca^{2+} and Mg^{2+} movements across the sarcoplasmic reticulum *in situ*. *J. Biol. Chem.* 260: 6801-6807.

Sovieszek, A. (1977): Ca-bind phosphorylation of a light chain of vertebrate smooth muscle myosin. *Eur. J. Biochem.* 73: 477-483.

Sugi, H. & Daimon, T. (1977): Translocation of intracellularly stored calcium during the contraction-relaxation cycle in guinea pig taenia coli. *Nature* 269: 436-438.

Sugi, H. & Suzuki, S. (1982): Physiological and ultrastructural studies on the intracellular calcium translocation during contraction in invertebrate smooth muscles. In *Basic Biology of Muscles*, ed. B.M. Twarog, R.J.C. Levine & M.M. Dewey, pp. 359-370. New York: Raven.

Suzuki, S., Hino, N. & Sugi, H. (1990): Intracellular calcium translocation during the contraction-relaxation cycle of fish swimbladder muscle. *J. Muscle Res. Cell Motil.* (In press).

Suzuki, S. & Sugi, H. (1978): Ultrastructural and physiological studies on the longitudinal body wall muscle of *Dolabella auricularia*, II. Localization of intracellular calcium and its translocation during mechanical activity. *J. Cell Biol.* 79: 467-478.

Suzuki, S. & Sugi, H. (1982): Mechanisms of intracellular calcium translocation in muscle. In *The Role of Calcium in Biological Systems. Vol. 1*, ed. L.J. Anghileri & A.M. Tuffet-Anghileri, pp. 201-217. Boca Raton: CRC.

Suzuki, S. & Sugi, H. (1989): Evidence for extracellular localization of activator calcium in dog coronary artery smooth muscle as studied by the pyroantimonate method. *Cell Tissue Res.* 257: 237-246.

Winegrad, S. (1965): Autoradiographic studies of intracellular calcium in frog skeletal muscle. *J. Gen. Physiol.* 48: 455-479.

Winegrad, S. (1968): Intracellular calcium movements of frog skeletal muscle during recovery from tetanus. *J. Gen. Physiol.* 51: 65-83.

Yamanobe, T., Mimura, T. & Sugi, H. (1988): Detection of a Ca-binding proteins in the plasma-membrane-enriched fraction of a molluscan smooth muscle. *J. Muscle Res. Cell Motil.* 9: 286.

Metal Ions in Biology and Medicine. Eds. Ph. Collery, L.A. Poirier, M. Manfait, J.C. Etienne. John Libbey Eurotext, Paris © 1990, pp. 119-122

The application of Proton-induced X-ray-emission to the analysis of Ca^{2+} distribution in vascular tissue

C. Spieker, W. Zidek, K. Kisters, D. Heck, D.B. von Bassewitz, K.H. Rahn

Medizinische Universitäts-Poliklinik, Albert-Schweitzer-Straße 33, D-4400 Münster, FRG

Introduction

Studies by Zidek et al. (1982), Erne et al. (1984), and other authors (Erne et al., 1984 and Furspan and Bohr, 1986) demonstrated an elevation of intracellular free calcium in essential hypertension. The available data have been obtained in blood cells, whereas only few and indirect determinations of cellular calcium have been performed in vascular smooth muscle (Zidek et al. 1983). Therefore calcium concentrations in arterial smooth muscle cells are of interest. To analyse the calcium content in spontaneously hypertensive rats (SHR) and in normotensive animals of arterial vessels, the method of proton-induced X-ray emission (PIXE) was used in this study to differentiate changes in calcium metabolism in hypertensive arteries from secondary degenerative phenomena due to the incipient arteriosclerosis, spontaneously hypertensive rats have been examined in the earliest stage of hypertension. PIXE qields a correlation of the calcium distribution within the structure of the arterial muscle. In this method an ion beam is used containing H^+, ^{2+}H or He which is characteristic for each element (Heck 1979). The data show that aortic tissue calcium was significantly elevated in SHR aged 4 weeks and in adult SHR whereas there was no difference in the aortic tissue calcium in the SHR aged one week as compared to normotensive rats at this stage of age.

Methods

Experiments were performed on male spontaneously hypertensive rats (SHR) aged one (n=9), four (n=9) and twelve weeks (n=15) of the Münster strain. 8 normotensive Wistar Kyoto rats aged one week, 12 four weeks old WKY rats and 12 WKY rats aged 3 months served as controls. Systolic blood pressure as measured with a tail cuff was 163,2 $\pm$ 10,8 mmHg in four weeks SHR and 192,5 $\pm$ 10,2 mmHg in 12 weeks SHR. The analysis were performed in the aorta. Frozen tissue blocks were cut at -25°C using a cryomicrotome (adjusted at 6 μm). The cryosections were placed on Formvar backing foils (thickness rang 20-40 ug/cm^2) and dried in vacuo. Before irradiation the serial section were transferred to microscope slides for the routine histological investigation. Adequate fields for micro-PIXE analysis were chosen, and photographs were taken both before and after irradiation in order to correlate the histological structures with the micro-PIXE measurements. Principally, the PIXE method involves the following procedures: A target usually consisting of a sample placed on suitable backing is iiradiated in vacuo with protons (energy 2-3 MeV), thereby inducing the emission of X-rays from the target atoms. The X-rays are detected by a Si (Li) X-rax detector, and a spectrum is obtained and analysed by a computer. The PIXE measurements were performed with the 3,5 MeV van de Graff accelerator of the Karlsruhe Research center (Heck 1979). The proton beam was directed perpendicularly throught the arterial wall layers. Among the scanning line all elements were recorded simultaneously to convert the X-Rays intensities which reflect only the number of atoms hit, to element concentrations. To determine the mass of the irradiated specimen, the C, N, and O content of the probe was determined, since beside hydrogen these three elements form the main constituents of organic tissue, and contribute to 90 % of its dry weight. The data on Ca^{2+} concentrations given in the results were obtained by integrating the Ca^{2+} distribution curve within the media. The extension of the media was determined by photographs. The Mann-Whitney test was used for statistical analysis.

Results

Ca^{2+} concentrations above the limit of detection was only found in the tunica media of the aortas. The Ca^{2+} content was not elevated in the aortic smooth muscle of SHR aged one week as compared to the normotensive WKY rats in this age group (186,8 ± 89,9 µg/g Ca^{2+}/g tissue versus 254,0 + 173,7 µg Ca^{2+}/g tissue, mean ± SD; n.s.). Ca^{2+} content was significantly raised in the aortic smooth muscle of SHR aged 4 weeks compared to 4 weeks old WKY rats (726,0 ± 130,4 µg Ca^{2+}/g tissue versus 440,3 ± 214,4 µg Ca^{2+}/g tissue, mean ± SD, $p<0,01$) and in SHR aged 3 months as compared to normotensive WKY-rats respectively (3317,0 ± 734,0 ug Ca^{2+}/g tissue versus 1632,0 ± 569,6 µg Ca^{2+}/g tissue; $p<0,05$).

Discussion

The method of proton induced X-ray emission allows to determine the Ca^{2+} content selectively in the tunica media, and hence in arterial smooth muscle, excluding other Ca^{2+}-depots such as adventitia and intima or atheromatous plaques. Aortic Ca^{2+} analysis with this method is not influenced by the procedures for cell isolation, which has often been a point of criticism of studies on isolated cells. A disadvantage of the PIXE method in the form used in this study is that due to the present limits of dissolution only the total amount of the vessel calcium can be evaluated, whereas intracellular studies on isolated smooth muscle cells are not performed up to now. Initial studies have been performed with ion-selective electrodes by Zidek et al. in red blood cells (Zidek et al. 1982), subsequently the results were confirmed by measurements of Erne et al (1984) and Bruschi et al. (1985) in lymphocyts and platelets using the fluorescent dye, quin 2 to determine cytoplasmatic free Ca^{2+} concentrations (Erne et al. 1984 and Bruschi et al. 1985). Both methods have considerable elements of uncertainity, for quin 2 in concentrations necessary to obtain satisfactory signals probably acts as a calcium puffer. Quin 2 is known to induce a depletion of the cells from ATPase thus possible affecting cellular Ca^{2+} handling. In blood cells ion selective electrodes have been used after disrupting the cells. Thereby cellular Ca^{2+} distribution is inevitably altered. Arterial Ca^{2+} content in the whole vessel in SHR had been measured previously. In a serious of experiments on SHR and normotensive controls Massingham and Shevde measurd a marked difference in the total calcium ion content of aortai from the two study groups. In this study the hypertensive tissue contained approximately 65 % more calcium than the normotensive tissue. Furthermore they did not measured any difference in the aortic water content in hypertensive rats as compared to normotensive controls (Massingham and Shevde 1973). The Ca^{2+} measurements were carried out in dried material from the whole vessel with atomabsorption spectometry, the rats had been nephrectonized 4 hours prior to the experiments. In other forms of experimentally induced secondary hypertension similar findings were obtained. Postnov and Orlov (1985) described an elevated Ca^{2+} content in the aortic wall of hypertensive rats. Fleckenstein et al. (1986) carried out experimental studies with various models of arterial calcinosis or artherosclerosis in rats. The most interesting results were obtained in aging rats. These normotensive rats showed a statistically significant increase in arterial calcium content with the age from the second to the 10 month of life (Fleckenstein et al. 1986). Our results confirm the age related increase in the arterial Ca^{2+} content in normotensive rats and demonstrate additionally that this age related rise in arterial Ca^{2+} content is accelerated in spontaneously hypertensive rats. In SHR aged 4 weeks both blood pressure and arterial Ca^{2+} content were found elevated. Therefore it may be speculated, that the arterial Ca^{2+} content rises in parallel with blood prssure. But it cannot be decided whether the rise in arterial Ca^{2+} preceeds the increase in blood pressure, since measurements of blood pressure at the age of one week were subjected to methological difficulties.

Our results and the cited studies give hints that changes of cellular clacium in hypertensice rats developed simultaneously with the rise in arterial blood pressure. A more direct confirmation of the crucial role of Ca^{2+} metabolism in the pathogenesis of primary hypertension would be possible, if Ca^{2+} determinations with PIXE in arteriolar smooth muscle cells could be performed.

References

1. Bruschi G., Bruschi M.E., Caroppo M., Orlandini G., Spaggiori M., Vovatora A. (1985): Cytoplasmatic free (Ca^{2+}) is increased in platelets of spontaneously hypertensive rats and essential hypertensive patients. Clin. Sci 68, 179-184
2. Erne P., Bolli P., Burgisser E., Bühler F.R. (1984): Correlation of platelet calcium with blood pressure. N. Engl. J. Med. 310, 1084-1088

3. Fleckenstein A., Frey M., Fleckenstein-Grün G. (1986): Antihypertensive and arterial anticalcinosis effects of calcium antagists. Am. J. Cardiol. 57, 1d-10d
4. Furspan P.B., Bohr D.F. (1986): Calcium-related abnomalities in lymphocytes from genetically hypertensive rats. Hypertension 8 (suppl. II), I.123-II.126
5. Heck D. (1979): The Karlsruhe Proton Microbeam System. Beiträge Elektronenmikroskop. Direkt abb Oberfl. 12, 259-262
6. Massingham R., Shevde S. (1973): The ionic composition of aortic smooth muscle from A.S. Hypertensive rats. Br. J. Pharmacol. 48, 422-424
7. Postnov Y.V. and Orlov S.N. (1985): Physiological Reviews Vol 65, 904-944
8. Zidek W., Losse H., Dorst K.G., Zumkley H., Vetter H. (1982): Intracellular sodium and calcium in essential hypertension. Klin. Wschr. 60, 859-862
9. Zidek W., Vetter H., Dorst K.G., Zumkley H. and Losse H. (1982): Intracellular Na^+ and Ca^{2+} activities in essential hypertension. Clin. Sci. 63, 41s-43s
10. Zidek W., Kerenyi T., Losse H., Vetter H. (1983): Intracellular Na^+ and Ca^{2+} in aortic smooth muscle cells after encymatic isolation in spontaneously hypertensive rats. Res. Exp. Med. 183, 129-132

Metal Ions in Biology and Medicine. Eds. Ph. Collery, L.A. Poirier, M. Manfait, J.C. Etienne. John Libbey Eurotext, Paris © 1990, pp. 123-125

Electron probe X-Ray microanalysis studies on the intracellular calcium translocation during the contraction-relaxation cycle of a fish sound-producing muscle

S. Suzuki, N. Hino*, H. Sugi

*Department of Physiology, School of Medicine, Teikyo University, Itabashi-ku, Tokyo 173, Japan and * Department of Physiology, School of Medicine, Juntendo University, Bunkyo-ku, Tokyo 113, Japan*

The contraction-relaxation cycle in the skeletal muscle is controlled by the release of Ca^{2+} from, and its uptake by, the sarcoplasmic reticulum (SR) (Ebashi & Endo, 1968). By use of ^{45}Ca autoradiography, Winegrad (1965, 1968) found that, in frog skeletal muscle, Ca^{2+} released from the terminal cisternae of the SR into the myoplasm to cause contraction is first taken up at the longitudinal tubules of the SR during relaxation, and then slowly returns to the terminal cisternae. On the other hand, Somlyo et al. (1985) claimed that the released Ca^{2+} is again taken up directly by the terminal cisternae during the relaxation, based on the electron probe X-ray microanalysis of cryosections of frog skeletal muscle fibers, though the preservation of the SR structures in cryosections is very poor. We studied the intracellular Ca translocation during the contraction-relaxation cycle by the electron probe X-ray microanalysis of cryosections of a fish swimbladder muscle, which contracts rapidly to produce sounds for communication and contains extremely well developed SR, providing a material suitable for detecting Ca localization within the SR components. It will be shown that Ca^{2+} released from the terminal cisternae can be taken up at the fenestrated collars and the longitudinal tubules of the SR, thus supporting the view of Winegrad.

MATERIALS AND METHODS

Bundles of 3-5 fibers with attached tendons were dissected from the swimbladder muscle of a teleost fish, *Sebasticus marmoratus* (16-20 cm body length), at 4°C in a fish Ringer's solution of the following composition (mM): NaCl, 167.5; KCl, 4.4; $CaCl_2$, 2.2; $MgCl_2$, 1.3 (pH adjusted to 7.2 by 10 mM Tris-malate buffer). The fiber bundle was stimulated to contract isometrically at slightly stretched length (sarcomere length, 2.5-2.7 μm) with a 1.0-sec train of supramaximal sinusoidal voltages (100 Hz), and rapidly frozen at rest, during sustained contraction and at 0.1 and 1.0 sec after the onset of relaxation with a dual-jet liquid propane freezing device (Suzuki and Pollack, 1986). Cryosections (200 nm thick) were cut from the frozen fiber bundle at -110°C on a LKB NOVA cryoultramicrotome, and placed on thin carbon supporting films on Ni-grids. Then the cryosections were freeze-dried at 10^{-6} torr and below -80°C. Electron probe X-ray microanalysis was performed on a liquid N_2-cooled Be-stage at -130°C (JEOL EM-CTH10 CRYO transfer holder) in a JEOL 2000FX electron microscope operated at 80 kV and equipped with a Tracor Northern TN5500 energy dispersive X-ray microanalyzer. The elemental concentrations were calculated from the X-ray spectra in mmol/kg dry wt based on the Hall's quantitative equation (Hall, 1971; Shuman et al., 1976). Conventional electron microscopy of the fibers were also made by prefixing resting fibers with 2.5% glutaraldehyde and post-fixing them with 2% OsO_4.

RESULTS AND DISCUSSION

The swimbladder muscle fibers contain extremely well developed SR, which in most cases showed triadic contacts with the transverse tubules (T-tubules) at the boundary of the A- and I-bands (Fig.1A), though triadic contacts were also occasionally seen at the Z-band (Fig.1B). Examination of the SR along the entire length of a single muscle fiber revealed that the triadic contacts at the Z-band were only localized near both ends of the fibers. Since the cryosections were only obtained from the middle portion of the frozen fibers, they only contained the triadic contacts at the boundary of the A- and I-bands. In the cryosections of about 200 nm thick, the SR and myofilaments generally overlaped each other (Fig.2A). As illustrated in Fig.2, spot analysis of the cryosections (beam diameter, 0.16 μm) were performed on five differenr regions (named as Z, I, A1, A2 and H) in each half sarcomere examined. The concentrations of K^+, Na^+ and Cl^- were almost uniform in these regions, and were comparable to the values reported for frog skeletal muscle fibers (Somlyo et al., 1981, 1985), while Ca concentrations in these regions were not uniform and changed during the contraction-relaxation cycle.

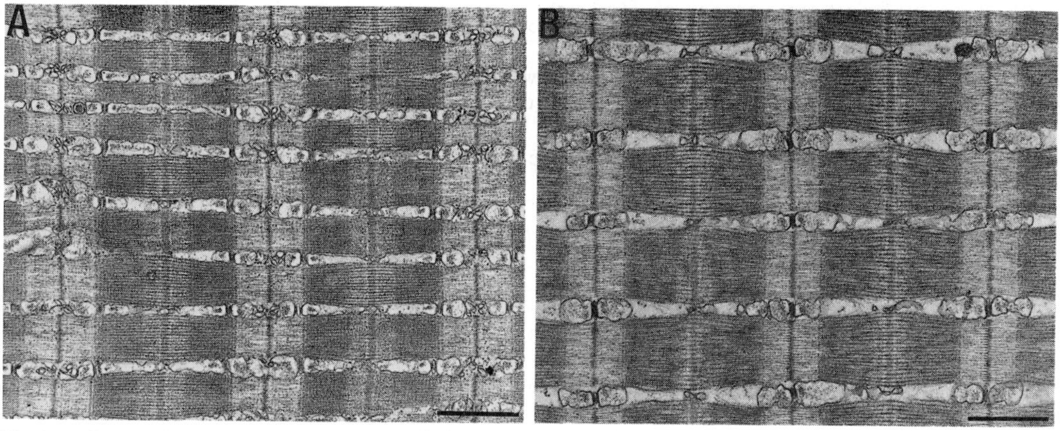

Fig.1. Conventional electron micrographs showing the triadic contacts between the SR and T-tubules at the boundary of the A- and I-bands (A) and at the Z band (B) in the swimbladder muscle fibers. Scale bars, 1 μm.

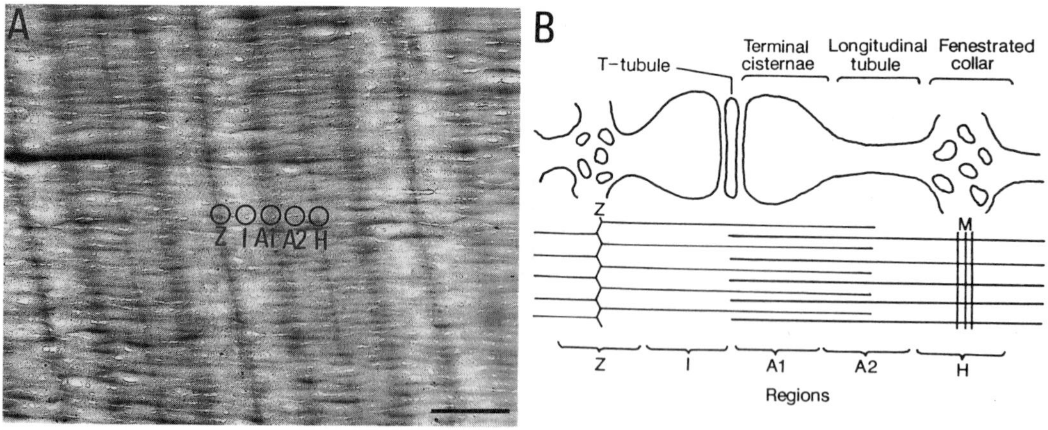

Fig.2. Method of the spot analysis of the cryosections. (A) Electron micrograph of part of a cryosection showing the five regions for the spot analysis in a half sarcomere examined. Scale bar, 1μm. (B) Diagram showing the relation between the five regions and the SR structures.

In the resting fibers, the Ca concentration was highest at the I and A1 regions (42.7 and 46.6 mmol/kg dry wt, respectively), indicating the Ca localization in the terminal cisternae of the SR located at these regions. The Ca concentrations in the Z, A2 and H regions were also fairly high (23.8, 13.3 and 9.9 mmol/kg dry wt, respexctively), probably because these regions sometimes include a part of terminal cisternae. The fact that the Ca concentration at the H region was the lowest may support this view.

In the fibers frozen during the sustained contraction, on the other hand, the Ca concentration decreased significantly at the I and A1 regions (10-26 mmol/kg dry wt), and increased significantly at the A2 and H regions (3-6 mmol/kg dry wt). This change in the Ca distribution in each sarcomere indicates that Ca stored in the terminal cisternae in the resting fibers is released into the myoplasm to cause contraction. At 0.1 sec after the onset of relaxation, the Ca concentration at the A2, H and Z regions further increased to about twice of those in the contracting fibers, though the isometric tension already decreased to a half of the maximum tension. This result indicates that Ca^{2+} released from the terminal cisternae can be taken up at the fenesterated collars and the longitudinal tubules, which are located at the Z, H and A2 regions. At 1.0 sec after the sustained contraction, the fiber relaxed completely. In this relaxed fiber, the Ca concentration at the I and A1 regions increased to a level nearly as high as that in the resting fibers, with corresponding decrease in the Ca concentrations at the A2, H and Z regions. The decrease in Ca concentration was larger at the H region than at the A region, indicating the translocation of Ca from the fenestrated collars to the terminal cisternae passing through the longitudinal tubules. The present results strongly suggest that, during relaxation, Ca^{2+} can be taken up mainly at the fenestrated collars and longitudinal tubules, and then slowly return to the terminal cisternae, being consistent with the view of Winegrad (1968).

REFERENCES

Ebashi, S. & Endo, M. (1968): Calcium ion and muscle contraction. *Prog. Biophys. Mol. Biol.* 18: 123-183.

Hall, T. A. (1971): The microprobe assay of chemical elements. In *Physical techniques in biological research*, ed G. Oster, pp.157-275. New York: Academic Press.

Shuman, H., Somlyo, A. V. & Somlyo, A. P. (1976): Quantitative electron probe micro-analysis of biological thin sections: Methods and validity. *Ultramicroscopy* 1, 317-339.

Somlyo, A. V., Gonzalez-Serratos, H., Shuman, G., McClellan, G. & Somlyo, A. P. (1981): Calcium release and ionic changes in the sarcoplasmic reticulum of tetanized muscle: an electron-probe study. *J. Cell Biol.* 90: 577-594.

Somlyo, A. V., McClellan, G., Gonzalez-Serratos, H. & Somlyo, A. P. (1985): Electron probe X-ray microanalysis of post-tetanic Ca^{2+} and Mg^{2+} movements across the sarcoplasmic reticulum *in situ*. *J. Biol. Chem.* 260: 6801-6807.

Suzuki, S. & Pollack, G. H. (1986): Bridgelike interconnections between thick filaments in stretched skeletal muscle fibers observed by the freeze-fracture method. *J. Cell Biol.* 102, 1093-1098.

Winegrad, S. (1965): Autoradiographic studies of intracellular calcium in frog skeletal muscle. *J. Gen. Physiol.* 48: 455-479.

Winegrad, S. (1968): Intracellular calcium movements of frog skeletal muscle during recovery from tetanus. *J. Gen. Physiol.* 51: 65-83.

Metal Ions in Biology and Medicine. Eds. Ph. Collery, L.A. Poirier, M. Manfait, J.C. Etienne. John Libbey Eurotext, Paris © 1990, pp. 126-128

The Ca^{2+} metal binding site to proteolytic enzymes «A novel method to ellucidate the amino acid residues involved in binding to metal ions in proteins»

B. Farzami*, F. Jordan

** Department of Biochemistry, School of Medicine, Tehran University of Medical Sciences, PO Box, 14155-5399 Tehran, Iran and Department of Chemistry, Rutgers, the State university of New Jersey, Newark, NJ, 07102, USA*

The role of metal ions bound to proteins provide diverse functions in biological systems which varies from structural role to catalysis and capability as messengers (Farsen 1989). The use of different techniques have shed lights on the nature of these interactions,and only the crystallographic methods have had a definite significance in determining the exact location of metal ions in proteins and the residues involved. The Ca^{2+} binding sites in proteins were mainly detected using crystallographic methods(Kretsinger R.H.,1980),Einspahr H.(1980,81,84). Other techniques such as spectroscopy has been used and thermodynamic data on binding of Ca^{2+} to enzymes are investigated (Sipos,T.,Merkel J.) There are several new deductions regarding Ca^{2+} binding sites obtained from crystal structure. It is known that the majority of Ca^{2+} binding sites contribute seven to six oxygen ligands to metal ion. In these inctances the oxygen is located at the seven vertices of a pentagomal bipyramidal structure with the Ca^{2+} in the center.Many of these binding ligands have one or more water molecules in their structure. It is suggested also that glutamate and aspartate residues are mostly involved in binding to Ca^{2+} as well as residues such as Imidazol and serine (Strynadka and James N.G.1989) In present study, we have employed a simple spectrophotometric technique which could identify the PKa of the functional proups involved in binding of proteolytic enzymes to Ca^{2+} this method could be extended to identify binding sites of diverse metal ions to proteins. As an example we have introduced along with the result from proteolytic enzymes, the results obtained from binding of Mg^{2+} ion bound to pyruvate decarboxylase,in wich the binding locality had been unknown.

The enzymes used in these studies were trypsin, chymotrypsin, subtilizin , Elastase and pyruvate decarboxylase. A dipeptide tryptophanyl-glutamic acid was also used as a model to confirm the liability of the method for estimation of the PKa of the functional groups involved in proteins binding to metal ions. The PKa determinations on enzyme were based on the observation that binding of metals to proteins caused changes at 280 mm, which was pH dependent. Thise changes were found to correlate the shifts in wavelengths from which the energy changes could be estimated. The changes in energy in the form of wave number ($\bar{\nu} = 1/\lambda$),correlated closely the degree of ionization of the specific group involved in binding as could be estimated from the Henderson-Hasselbach equation. In our studies the proteolytic enzymes were depleted of metal ions, through the use of LBTI and sephadex G-50 calumns. The bakers yeast apo-pyruvate decarboxylase was prepared using Morey and Junie

(1968) techniques. The difference spectra of the enzyme solutions were prepared at specific PH's using buffers in their range of action with strict ionic strength. The blank contained all the components except the metal ion which was replaced by an inert salt. The shifts in the wave number obtained from the printed data, were plotted against the pH. The plot showed a sharp minimum corresponding to the PKa of the group involved in binding. To test further, the reliability of the assumption, the logarithmic dependency of the changes of abosrption were plotted against the pH. A straight line was obtained that intercepted the pH axis at the PKa of the group bound to metal ion. The use of the dipeptide, tryptophanyl-glutamic acid as a probe to test the reliability of the method gave a PKa=4.30. Corresponding to the PKa of glutamic acid (Fig.1).

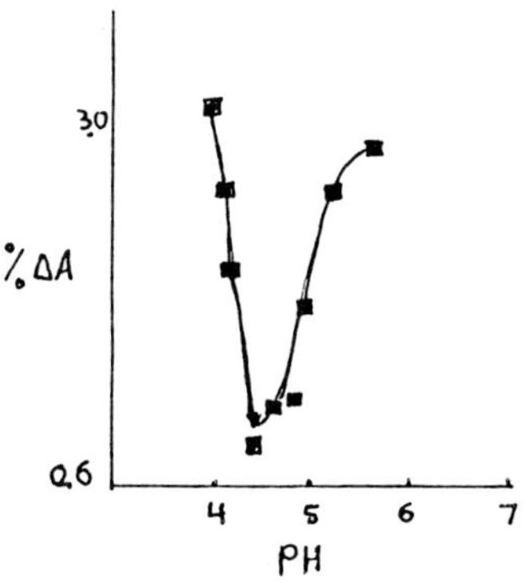

Fig.1: The change in absorbance at 280 vs pH for tryp-glutaminc acid bound to calcium;minimum at 4.30

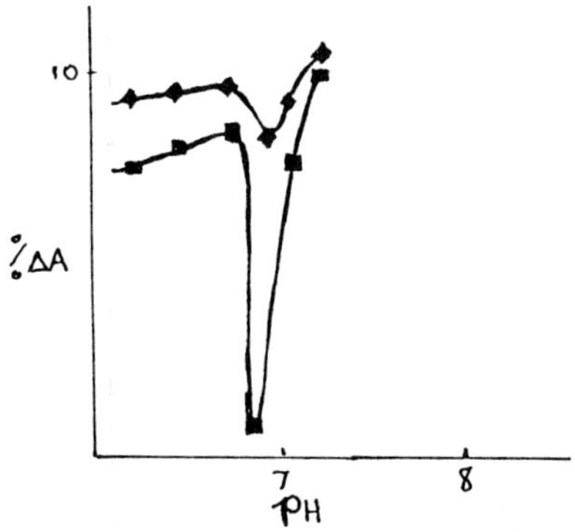

Fig.2: The change in absorbance vs pH at two wavelengths,250 ◆ and 280,■ ; for pyruvate decarboxylase bound to Mg^{2+}; minimum at 6.93

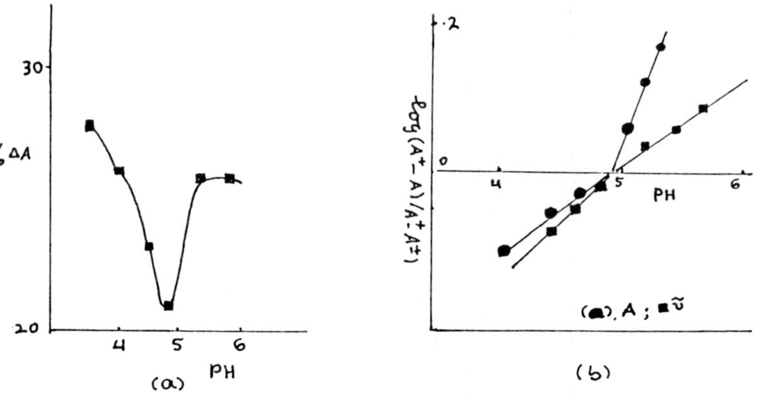

Fig.3: a:the change in absorbance at 280 vs. pH for trypsin bound to calcium, minimum at 4.97 b: the log of absorbance change and the shifts in terms of wave number vs pH. for trypsin bound to Ca^{2+};the intercept is at 4.97.

The PKa's obtained for the functional groups bound to Ca^{2+} in enzymes under the study are depicted in (table1). The low PKa may be assigned to glutamate. PKa's appear from 6.5-7.5 may be assigned to Imidazole group of histidine. In pyruvate decarboxylase a sharp minimum at pH=6.39 was obtained when Mg^{2+} ion and thiamine pyrophosphate (TDP), as coenzyme were used with the apoenzyme. (Fig.2) Our results indicate that the method could also be used for estimating other properties of the binding site such as the environment of the metal ion as well as its position. i.e. wether the binding is direct or through other factors such as coenzymes.

Table 1. PKa's of amino acid residues involved in binding to metal ions

Proteins and Peptides	PKa_1	PKa_2	Metal ion
Trypsin	4.97	6.49	Ca^{2+}
Chymotrypsin	4.87	6.40	"
Elastase	4.84	-	"
Subtilizin	4.65	-	"
Pyruvate decarboxylase	6.93	-	Mg^{2+}
Tryp-Glu	4.30	-	Ca^{2+}

References:

1. Einspahr H. Bugg, C.E., (1986, Acta Crystallogr. Sec. B36; 246-71.
2. Einspahr H. Bugg, C.E., (1981) Acta Crystallogr. Sect. B37, 1049-52.
3. Einspahr H. Bugg. C.E., (1984) In Metal ions Biological Systems ed H. Sigel 17:51-97, New York/Basd : Dekker.
4. Forsen, S. (1989). In Inorganic Biochemistry ed. 1, Bertini, Mill Valley, Calif. Univ. Sci. Books.
5. Kretsinger, R.H. 1980. CRC Crit Rev. Biochem. 8:119-74.
6. Morey and Juni (1968) J. Biol. Chem., 243, 2009-3019.
7. Sipos, T., Merkel, J.R. (1970) Biochemistry, 9, 2766.
8. Strynadka, N.C.J., James, M.N.G. (1989). Ann. Rev. Biochem. 1989, 58:951-98.

Metal Ions in Biology and Medicine. Eds. Ph. Collery, L.A. Poirier, M. Manfait, J.C. Etienne. John Libbey Eurotext, Paris © 1990, pp. 129-131

The effect of Ca^{2+} ion in proteolytic enzymes against the action of inhibitors

F. Jordan, B. Farzami*

* *Department of Biochemistry, School of Medicine, Tehran University of Medical Sciences, PO Box, 14155-5399 Tehran, Iran and Department of Chemistry, Rutgers, the State university of New Jersey, Newark, NJ, 07102, USA*

It has been known for several years that the activity of most serine proteases, as well as their other properties such as activation of zymogen to the active enzyme (McDonald and Kunitz 1941) the protection of enzyme from autolysis(Gorini 1951, Bier and Nord,1951), and activation of several component of the blood coagulation (Shore, J.D. et al,1987) are enhanced by the presence of Ca^{2+}. The X-ray crystallography and the nuclear magnetic resonance techniques(Hinrichs & Saenger 1988, Adebodun and Jordan 1989), have shown that the site of binding on some serine proteases is located at some distance away from the active center. On the basis of crystallographic evidences, it is suggested that the Ca^{2+}ions triggers a conformational change leading to the formation of the substrate recognition site, via secondary structure hydrogen bonds between Ca^{2+} binding site and the substrate recognition site (Bajorath,Hinrichs and Saenger 1988).

We have made the first observation regarding the protective effect of Ca^{2+} in proteolytic enzymes under the study against the action of inhibitors whose implications for Inhibition/drug design are important. The several fold enhancement may evolve in inactivation of these enzymes, employing a covalently linking Ca^{2+} chelator to an active site directed inactivator. In our studies the enzyme used were trypsin, chymotrypsin, subtilizin, elastase. The use of inhibitors was based on specific affinity for the active site or the affinity for the specific groups on the active site, thus PMSF, TPCK, TLCK,DIFP, were used as inhibitors. The specific substrates for the enzymes were chosen and the products of acylation reaction were measured at their specific wavelenghts, the substrates used were PNPA, BAEE, CBZ-Ala-PNA and N-Succ (Ala) 3PNA. The studies were carried out using LBTI, or Sephadex G50 column to strip of Ca^{2+} from the enzymes. In parrallel experiments the enzymes used were incubated for at least 15 minutes with about hundred fold concentration of Ca^{2+}. The inhibition studies were carried out using several concentration of inhibitor in each run incubated with the enzyme at an appropirate pH. Aliquots were then taken at different time intervals and tested for the activity. Kitz and Wilson plots were then used to estimate the inhibition rate constant (ki) and the inhibition constant (Ki), from which the ratio of (ki/Ki) was estimated. The results showed that the enzymes such as trypsin, chymotrypsin and subtilizin are strongly protected against the action of inhibitor when Ca^{2+} ion was present an increase in ki and Ki was observed when Ca^{2+} was present and the decrease in both constants occured when the enzymes were stripped off Ca^{2+}. These results suggest that when Ca^{2+} was not present the initial binding between the inhibitor and the enzyme (depicted by Ki) was more favorable. This stage was accompanied by a low rate of inactivation (low ki). The reverse effect was found in experiments when Ca^{2+} was reacted with the enzyme.

Table 1. The Ca^{2+} ion effect on inactivation parameters of proteolytic enzymes against the action of inhibitors

enzyme	Inh.	Sub.	pH	T	ki min^{-1}	$Ki.10^{-5}$ mol	$ki/Ki.10^{-4}$ min^{-1} mol^{-1}	fold inactiv. increase
1 trypsin+ Ca^{2+}	PMSF	PNPA	8.1	30	1.89	86	0.217	17.6
2 trypsin Ca^{2+} + depl.	"	"	"	"	1.54	4	3.82	
3 trypsin Ca^{2+}	"	BAEE	7.6	"	1.40	40	0.35	4.52
4 trypsin Ca^{2+} + depl.	"	"	7.6	"	1.34	8.5	1.58	
5 trypsin Ca^{2+}	TLCK	PNPN	8.1	15	0.78	9.2	0.85	3.17
6 trypsin Ca^{2+} dept	TLCK	PNPA	8.1	15	0.53	2.0	2.7	
7 Chymo.+ Ca^{2+}	TPCK	PNPA	8.1	15	1.14	5.40	2.11	2.94
8 Chymo.+ Ca^{2+} + dept	TPCK	PNPA	8.1	15	0.84	1.35	6.22	
9 Elastase Ca^{2+}	PMSF	CBZ-(Ala) PNA	7.6	30	0.77	0.138	55.6	0.62
10 Elastase NO Ca^{2+}	"	"	7.6	30	0.54	0.158	34.2	
11 Subtitizin .025Ca^{2+}	DIFP	BAEE	7.8	23	-	-	-	5.0
12 Sublitizin .05Ca^{2+}	"	"	"	"	"			

Abreviations: PMSF,phenylmethyl sulfonyl floride; PNPA, para nitrophehyl acetate BAEE,benzoyl arginine ethyl ester;TLCK,Tosyl lysine chloromethyl keton;

The comparisons of total reactivity of inhibitors in cases were Ca^{2+} was present and the cases were enzyme did not contain any metal ions were made using the ratios of (ki/Ki)1:(ki/Ki)2 for the two sets. The results showed that this ratio changed few to several folds in some cases under study signifying the role of Ca^{2+} in protecting the proteases against the action of inhibitors. The above trend was not detected for elastase for which the dependency of the rate of enzymatic hydrolysis on Ca^{2+} was also negligible indicating the active participation of Ca^{2+} in both inhibition and activation of Ca^{2+} dependent proteases and their impartial role in non Ca^{2+} dependent ones.

Considering the results obtained from the crystallographic studies the Ca^{2+} bound to these enzymes may induce a conformational change which effect the active site in such a way to provide a more relaxed conformation when the Ca^{2+} is not present leading to a better affinity for the inhibitior, with a lesser rate of inactivation,possibly because of the fact that in such a state the appropirate functional groups are not fully in place. The overall increases in inactivation due to the cooperative effect of both inhibition factors were in favor of the state when Ca^{2+} was not bound to the enzyme.

References:

1. a: Adebodun,F.,Jordan,F.(1989) Biochemistry,28,7524.
   b: Adebodun, F.,Jordan,F.,(1989) J.Cell. Biochem. 40:249-260.
2. Bajorath,J., Hinrichs,B.W., Saenger,W.,(1988) Biochem,76,441-447.
3. Bier,M.,Hord,E.F.(1951) Arch. Biochem. Biophys. 33,320.
4. Gorini,L. (1951) Biochim. Biophys. Acta. 7,318.
5. McDonald,M.R., Kunitz,M.(1941) J.Gen. Physiol. 25,53.
6. Shore,J.D. et al (1987) Biochemistry,26,2251.

Abreviations: TPCK, Tosyl phenyl chloromethyl keton; CBZ-(Ala)-PNA,Carbobenzoxy alanine para nitro anilid; DIFP, diisopropyl fluorophosphate.

Metal Ions in Biology and Medicine. Eds. Ph. Collery, L.A. Poirier, M. Manfait, J.C. Etienne. John Libbey Eurotext, Paris © 1990, pp. 132-134

Transient Ca^{2+} increase during fertilization and its role in cortical granule breakdown in the sea urchin egg

Y. Hamaguchi, M.S. Hamaguchi, T. Mohri

Biological Laboratory, Faculty of Science, Tokyo Institute of Technology. O-Okayama, Meguro-ku, Tokyo 152, Japan

Upon fertilization, a number of events occur within seconds of sperm-egg attachment, including a transient increase in intracellular Ca^{2+} concentration (Ca_i), cortical granule breakdown and fertilization envelope (FE) elevation (for a review; Epel, 1978). It is thought that the stimulation of sperm attachment is transmitted by the second messengers of Ca^{2+} and inositol 1,4,5-trisphosphate (IP_3) to the cortical granule (Turner et al., 1986; Whitaker et al., 1989). After the cortical granule breakdown, FE is elevated by colloid osmotic pressure generated by a component of cortical granule inclusion which protrudes into the space between the plasma membrane and the vitelline coat (Moser, 1939; Hiramoto, 1955). In this study, the timing of sperm-egg attachment, the increase in Ca_i, and FE elevation were determined using a microscope which allows us to observe both epifluorescence and differential interference contrast (DIC) images simultaneously without any changes in the optical setup throughout the observation period. Furthermore, the reaction times when the cortical granules broke down were determined in order to investigate the sequence of steps in which Ca-EGTA buffers, IP_3, and GTPγS introduced by microinjection act in echinoderm eggs.
The eggs of the sand dollar, C. japonicus were injected with a Ca^{2+} indicator, Fluo-3 (Minto et al., 1987) as previously reported (Hamaguchi & Hiramoto, 1981; Hamaguchi, 1982) and observed with a Nikon microscope equipped with a simultaneously observable system of both DIC and epifluorescence using a B-2 filter cassette. In order to distinguish the fluorescent image from the DIC image, the latter was formed by red transmitted light with wavelength greater than 580 nm which was obtained by adding a long-pass barrier filter to the illuminating system, whereas the former was green (the peak wavelength of fluorescence emission is 526 nm (Minto et al., 1987)). The microscopic images were recorded through a color video camera with a video cassette recorder and analyzed by reviewing the records. During fertilization, the attachment of the sperm and FE elevation were observable as red in DIC images, and simultaneously, Ca_i increase was detected as an increase in the green fluorescence of Fluo-3, which had been injected into the egg. Therefore, the Ca_i-increased area in the egg became yellow as the result of mixing red and green light, as shown in Fig. 1. Ca_i increase was detected accurately beneath the sperm-attached site on the egg surface (Fig. 1a); the yellow area of the egg cytoplasm spread gradually as a concentric sphere (Fig. 1b), and then its spreading front became planar rather than spherical when passing near the egg center. Finally, the Ca_i-increased area spread throughout the egg, but Ca_i near the attached site of the sperm began to decrease before the arrival of the increase at the opposite site (Fig. 1c). These results confirm that Ca^{2+} increasing around the attached site in the egg cytoplasm does not simply diffuse

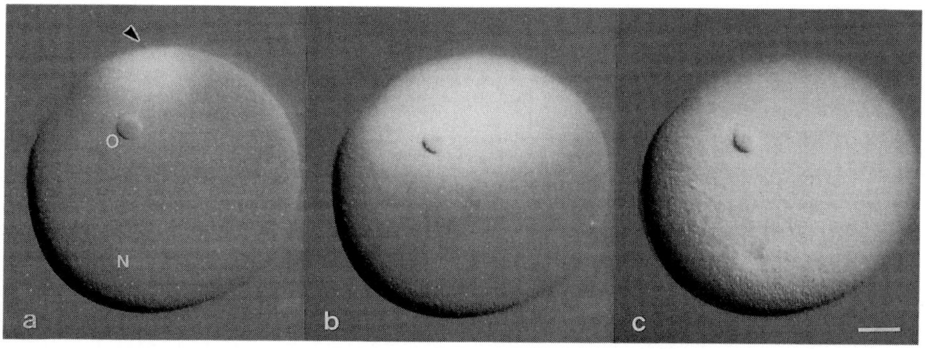

Fig. 1 Time course of Ca_i increase after sperm-egg attachment. The increase was detected 8 sec after the attachment of the sperm (yellow cytoplasm in **a**) and spread throughout the entire egg (**c**). FE began to elevate in **b**. The arrowhead indicates the fertilizing sperm and **O** is an oil drop introduced at the time of injection. **N**; nucleus. Scale bar is 20 μm.

from the site into the surrounding cytoplasm, but that Ca^{2+} is released and then sequestered gradually even through all parts of the egg cytoplasm as well as the initial site. FE began to elevate when the increase in Ca_i reached the egg center (Fig. 1b). When the results of 15 samples which were suitable for analysis are summarized, Ca_i increased 8.6±2.6 sec (no. of sample=5) after the sperm attached to the egg surface. Ca_i increase spread throughout the egg within 24.2±3.5 sec (n=10). FE became elevated 10.3±3.3 sec (n=12) after the initiation of Ca_i increase and 21.2±4.4 sec (n=5) after sperm attachment.

The breakdown of cortical granules was observed with a DIC microscope equipped with a 100X/1.25 objective and recorded by a videosystem, after microinjecting Ca-EGTA buffers, IP_3 and GTPγS into the eggs of _S. mirabilis_, _T. hardwicki_ and _C. japonicus_ using an injection apparatus (Mohri & Hamaguchi, 1989). The reaction time when the first breakdown of a cortical granule was detected after injection was measured by reviewing the recorded image at slow speed. The cortical granules in _C. japonicus_ eggs, which were 1.0-1.2 μm in diameter, were larger in size and smaller in number than those in _S. mirabilis_ and _T. hardwicki_. The mean reaction times of the breakdown of the cortical granule by injection of Ca-EGTA buffers, IP_3 and GTPγS were determined to increase in this order and to be 0.4 sec, 1.0 sec and 2.7 sec, respectively, in _C. japonicus_ as shown in Fig. 2; minor differences in these values existed in other species. In all three species of echinoderm used in this study, the reaction times were shortened in the order of GTPγS, IP_3 and Ca-EGTA buffer. These results indicate that GTP-binding protein, IP_3 and Ca^{2+} may react in this order in the egg during fertilization, if the corresponding reactions are activated successively in the egg cytoplasm by them. Turner et al. (1986) concluded that they successively activated corresponding reaction steps in the egg cytoplasm in a different sea urchin species.

FE elevation is caused by the osmotic pressure in the perivitelline space generated after cortical granule breakdown (Moser, 1939; Hiramoto, 1955). The close relationship between Ca_i increase and cortical granule breakdown was found, and the first breakdown of cortical granules was observed about 0.5 sec after the increase by means of a microinjection of calcium buffer solutions at high Ca^{2+} (see also Mohri & Hamaguchi, 1989). However, FE elevation was not observed until 10.3 sec after Ca_i increase in the present study. Because it was found that 60% of the total cortical granules broke down within 10 sec after the first breakdown was recognized (manuscript in preparation), the osmotic pressure might become sufficient for FE elevation when more than 60% of the total cortical granules in the corresponding egg cortical area break down.

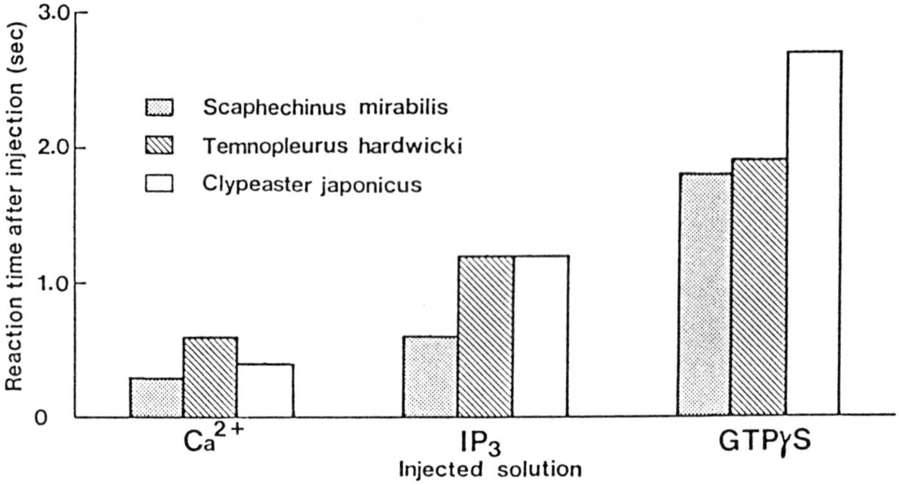

Fig. 2 Summary of injection of Ca^{2+}, IP_3, and GTPγS into sea urchin eggs of three species. This figure summarizes the mean reaction times of the breakdown of the cortical granule after injections of Ca-EGTA buffers (2.8-5.7 μM Ca^{2+}), IP_3 (0.1-20 μM), and GTPγS (10-50 mM).

Acknowledgment

We thank Nikon Co. Ltd. for the generous loan of the simultaneous observation microscope used in this study. Supported partly by a grant from Japan Ministry of Education, Science and Culture to Y. H.

REFERENCES

Epel, D. (1978): Mechanisms of activation of sperm and egg during fertilization of sea urchin gametes. Curr. Top. Dev. Biol. **12**, 185-246.

Hamaguchi, M. S. (1982): The role of intracellular pH in fertilization of sand dollar eggs analyzed by microinjection method. Develop. Growth Differ. **24**, 443-451.

Hamaguchi, Y., & Hiramoto, Y. (1981): Activation of sea urchin eggs by microinjection of calcium buffers. Exp. Cell Res. **134**, 171-179.

Hiramoto Y. (1955): Nature of the perivitelline space in the sea urchin eggs. III. On the mechanism of membrane elevation. Annot. Zool. Japon. **28**, 183-193.

Minto, A., Harootunian, A.T., Kao, J.P.Y., & Tsien, R.Y. (1987): New fluorescent indicators for intracellular sodium and calcium. J. Cell Biol. **105**, 89a.

Mohri, T., & Hamaguchi, Y. (1989): Analysis of the breakdown of cortical granules in echinoderm eggs by microinjection of second messengers. Cell Struct. Funct. **14**, 429-438.

Moser, F. (1939): Studies on a cortical layer response to stimulating agents in the Arbacia egg. I. Response to insemination. J. Exp. Zool. **80**, 423-445.

Turner, P.R., Jaffe L.A., & Fein, A. (1986): Regulation of cortical vesicle exocytosis in sea urchin eggs by inositol 1,4,5-trisphosphate and GTP-binding protein. J. Cell Biol. **102**, 70-76.

Whitaker, M.J., Swann, K., & Crossley, I. (1989): What happens during the latent period at fertilization. In Mechanisms of Egg Activation eds. R. Nuccitelli, G.N. Cherr, & W.H. Clark, Jr. pp. 157-171, New York: Plenum Press.

Metal Ions in Biology and Medicine. Eds. Ph. Collery, L.A. Poirier, M. Manfait, J.C. Etienne. John Libbey Eurotext, Paris © 1990, pp. 135-137

Effects of buprenorphine on mitochondrial protein bound Ca^{2+} and ultrastructural distribution of Ca^{2+} in the some brain regions*

X.N. Zhao, H. Liao, S.L. Shi, J.J. Wang, J. Xing, J. Chen Z.X. Zhang, R.S. Chen**

*Medical School of Nanjing University, Nanjing 210008, P.R. China and ** Chemistry Department, Nanjing University, Nanjing 210008, People's Republic of China*

INTRODUCTION

Opiate effects have often been associated with Ca^{2+}, which appears to have multiple functions in the central nervous system. Recently the significant role of Ca^{2+} in the mechanism of action of narcotic drug–induced analgesia and the development of tolerance to morphine has aroused great interest (Carafoli, 1987). It has been reported(Zhang, 1987) that both morphine analgesia and acupuncture analgesia could be antagonized by Ca^{2+}, and La^{3+}, EDTA,EGTA agents presumed to prevent Ca^{2+} influx into cells, can themselves produce analgesia (Harris, 1975). In the present study, we employed fluorescent probe Tb^{3+} and transmission electron microscope to observe the changes of the mitochondrial protein bound Ca^{2+} in vitro and distribution of intracellular Ca^{2+} in the some brain regions after treatment with nonopiate buprenorphine, and to know the role of Ca^{2+} in drugs–induced analgesia.

EXPERIMENTAL DETAILS

The animals rapidly were perfused by intracardiac catheterization with saline containing 90 mM potassium oxalate, then 3% glutaraldehyde containing 90 mM potassium oxalate for 20 minutes. In accordance with stereotaxic atlas of the albino mouse forebrain (Slotnick,1975), the periaqueductal gray (PAG) and hypothalamus were dissected, and cut into 1 mm^3 pieces. The samples for electro–microscope observation were preparaed according to the procedures (Shi,1988). For identifing the chemical constituents of the electron–dense precipitates, unstained sections mounted on Formvar–coated copper grids. The analysis of the deposits were performed in Link–860 X–ray energy dispersive spectrography operated at 120 kV, count 200 seconds.

The mitochondria was prepared by the method of Moldave (Moldave,1967),and protein was measured by the method of Lowry et al.(Lowry,1951). Mitochondria (40μg protein / ml) was incubated at 37℃ for 30 minutes in the incubation medium containing 0.4 ml of hexamine and 0.2 ml of 0.02 M Tb^{3+}, pH 6.9. The fluorescent spectrum of mitochondrial protein–Tb^{3+} was determined.

RESULTS AND DISCUSSION

Data in Table 1 show that the lower dose (0.4 mg) of buprenorphine had no effects on Ca^{2+}–uptake by mitochondria in the three brain regions under addition of 1 μmol / L of Ca^{2+}. When the initial Ca^{2+} content in the incubation medium was 2 or 3 μmol / L, there was an increase of mitochondrial protein bound Ca^{2+}, i.e. a decrease of Tb^{3+} relative fluorescent intensity in PAG region under observation. Addition of verapamil 5 minutes before buprenorphine induced a reversed changes of Ca^{2+}–uptake by mitochondria of the PAG region.

* The Project Supported by National Natural Science Foundation of China

Table 1. Effects of the lower dose of buprenorphine on Ca^{2+}–uptake by mitochondria in discrete brain regions in vitro

added Ca^{2+} (μmol / L)	Group	Tb^{3+} relative fluorescent intensity		
		PAG	hypo	hip
1.0	Control	43.6 ± 2.5	45.8 ± 3.2	53.1 ± 4.8
1.0	Bup	38.9 ± 2.7	44.8 ± 2.3	49.8 ± 3.3
1.0	Ver+Bup	42.3 ± 2.1	48.2 ± 5.7	52.3 ± 6.6
2.0	Control	39.8 ± 4.0	41.0 ± 4.6	46.3 ± 4.8
2.0	Bup	31.8 ± 6.0^{a}	34.7 ± 4.5^{a}	47.5 ± 8.2
2.0	Ver+Bup	48.4 ± 1.6^{ac}	40.6 ± 5.6	50.9 ± 4.8
3.0	Control	38.1 ± 3.9	32.6 ± 2.6	49.8 ± 4.8
3.0	Bup	25.0 ± 4.5^{a}	27.4 ± 5.1^{a}	45.2 ± 8.5
3.0	Ver+Bup	56.8 ± 5.2^{bc}	32.4 ± 3.7	50.6 ± 8.6

To each assay tube was added 40μg of protein / ml of mitochondria in discrete brain regions and incubated at 37℃ for 30 minutes, pH 6.9, in the basic medium containing different calcium loadings and 0.4mg of buprenorphine. Verapamil (4μg) was added 5 minutes before buprenorphine.

Mitochondrial protein bound Tb^{3+} fluorescent intensity (arbitrary unit) was determined by a Hitachi RF–540 fluorescence spectrophotometer.

Bup = buprenorphine; Ver = verapamil; PAG = periaqueductal gray Hypo = hypothalamus; Hip = hippocampus

a) Significantly different from the control, $p < 0.05$

b) Significantly different from the control, $p < 0.01$

c) Significantly different from Bup group, $p < 0.05$ or $p < 0.01$

Moderate dose (0.8 mg) of buprenorphine had clearly effects on Ca^{2+}–uptake by mitochondria in both PAG and hypothalamus regions under 1, 2 or 3 μmol / L of Ca^{2+} loading, after addition of verapamil, the changes of buprenorphine–induced Ca^{2+}–uptake by mitochondria in the two regions were clearly antagonized.

Table 2. Effects of the moderate dose of buprenorphine on Ca^{2+}–uptake by mitochondria in discrete brain regions in vitro

added Ca^{2+} (μmol / L)	Group	Tb^{3+} relative fluorescent intensity		
		PAG	Hypo	Hip
1.0	Control	42.6 ± 1.3	46.1 ± 1.0	52.8 ± 2.5
1.0	Bup	37.0 ± 3.0^{a}	39.5 ± 1.5^{a}	48.2 ± 2.5
1.0	Ver+Bup	46.1 ± 3.0^{c}	47.6 ± 3.3^{c}	50.5 ± 7.0
2.0	Control	37.0 ± 2.4	40.2 ± 2.3	45.5 ± 2.5
2.0	Bup	27.5 ± 3.3^{b}	32.2 ± 1.4^{a}	42.2 ± 2.0
2.0	Ver+Bup	47.7 ± 3.5^{bc}	40.0 ± 2.1^{c}	45.6 ± 3.3
3.0	Control	28.1 ± 2.3	33.2 ± 2.5	42.8 ± 2.3
3.0	Bup	20.1 ± 2.2^{b}	25.6 ± 4.2^{b}	36.5 ± 5.4
3.0	Ver+Bup	28.6 ± 2.5^{c}	35.9 ± 5.0^{c}	41.2 ± 8.0

To each assay tube was added 0.8 mg of buprenorphine, and other conditions and notes are the same as in table 1.

The analogous results were obtained with the same treatment in a higher dose (1.6 mg) of buprenorphine or verapamil plus buprenorphine.

The above treatments do not affect the Ca^{2+}–uptake by mitochondria in the hippocampus region, indicating Ca^{2+} in the PAG and hypothalamus regions for buprenorphine have a relative speciality.

The changes of ultrastructural distribution of calcium ions in the some brain regions after administration of buprenorphine were also studied (Fig. 1). The analogous results were obtained in hypothalamus.

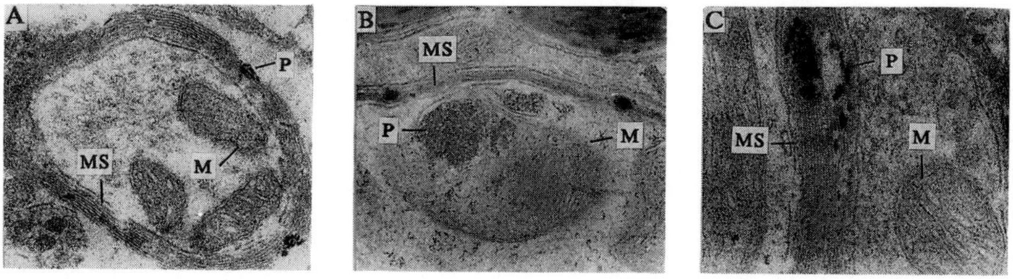

Fig.1. Ultrastructural distribution of calcium in PAG induced by i.p. of buprenorphine(Bup) in mice. A) was the case of 30 min following i.p. saline, which showed that PAG was normal. B) showed a great number of the precipitate pellets at myelin sheeth and in mitochondria after i.p. Bup. C) showed significant decrease of the precipitate pellets at these sites induced by i.p. of ruthenium red before Bup. × 50,000(A.C) and 40,000(B)
MS = myelin sheath; M = mitochondeia; P = precipitate pellets

The above results suggest that the various drugs or other ways, such as acupuncture which may produce analgesia, could lead to corresponding changes of distribution of Ca^{2+} in the some brain regions, i.e. they are likely to share a mutual ion basis and mechanism of action. These findings also suggest that Ca^{2+} transport across the neuroplasmic membranes could play a mediate role in drugs–induced analgesia, and from the distribution of Ca^{2+} ions in myelin sheath under observation, extracellular Ca^{2+} is likely to flow into the intracellular through the myelin sheath.

REFERENCES

Carafoli, E.(1987): Intracellular calcium homeostasis. Ann. Rev. Biochem. 56,395–433.

Harris, R. A. et al.,(1975): Analgesic effects of lanthanum:cross–tolerance with morphine. Brain Res. 100,221–225.

Lowry, O.H. et al.,(1951): Protein measurement with the folin phenol reagent. J. Biol. Chem. 193,265–275.

Moldave, K.(1967): Preparation of RNA from mammalian ribosomes. In Methods in Enzymology, ed.Grossman, L. and Moldave, K., Vol.12, Part A, pp. 607–632. NY : Academic Press.

Shi Shan–ling et al., (1988): The effects of electroacupuncture, morphine and Tb^{3+} on theultrastructural distribution of calcium ions in the discrete brain regions of mice. J.Nanjing University. 24, 495–540.

Slotnick, B.M. et al.,(1975): A stereotaxic atlas of the albino mouse forebrain. pp. 54–146. Maryland: Rockville Press.

Zhang Zu–zuan. et al.,(1987): Role of calcium in electroacupuncture analgesia and the development of analgesic tolerance to electroacupuncture and morphine. Sci. Sin. 30,974–985.

Metal Ions in Biology and Medicine. Eds. Ph. Collery, L.A. Poirier, M. Manfait, J.C. Etienne. John Libbey Eurotext, Paris © 1990, pp.138-140

Intralymphocytic Ca^{2+} content in essential and renal hypertension

K. Kisters, W. Zidek, C. Spieker, T. Fetsch*, F. Wessels**, K.H. Rahn

Medizinische Poliklinik & Medizinische Klinik C der Universität 4400 Münster, Albert-Schweitzer-Str. 33, West Germany, St. Franziskus Hospital Essen**, West Germany*

Introduction

Intracellular free Ca^{2+} content has been found to be elevated in essential hypertensives and in spontaneously hypertensive rats by several authors. The measurements have been performed with ion-selective electrodes and with the fluorescent dye, quin 2. Both methods yielded free Ca^{2+} concentrations. Whereas the electrode measurements have been performed in red blood cells (Zidek et al., 1982; Zidek et al., 1982) and in arterial smooth muscle cells (Zidek et al., 1983), the quin 2 method was applied to lymphocytes and platelets (Erne et al., 1984; Bruschi et al., 1985; Furspan & Bohr, 1986; Bing et al., 1986). As to the mechanisms by which intracellular Ca^{2+} is elevated, disturbances in both intracellular binding processes and transmembraneous Ca^{2+} transport systems come into account. To clarify this question, measurements of total intracellular Ca^{2+} concentration are required. In this study an approach to determine intracellular total Ca^{2+} content in lymphocytes is demonstrated, using lymphocytic protein as a reference for the lymphocytic Ca^{2+} content.

Furthermore it is still open whether the changes in intracellular Ca^{2+} content reported in essential hypertension are specific for this disorder. In the aforementioned studies only essential hypertensives have been compared with normotensives, so that the changes in intracellular free Ca^{2+} could also be a consequence of arterial hypertension per se, e. g. by altered membrane properties due to enhanced mechanical stress. In the present study a group of renal hypertensives was included to test the specificity of changes in intracellular Ca^{2+} content for essential hypertension.

Methods

17 essential hypertensives (10 males, 7 females, aged 51,2$\pm$9,7 years, blood pressure 176,2$\pm$14,1/105,0$\pm$13,4 mm Hg, means $\pm$ SD), 8 renal hypertensives with chronic glomerulonephritis (4 males, 4 females, aged 48,6$\pm$8,2 years, blood pressure 181,4$\pm$11,3/110,0$\pm$7,7 mm Hg) and 11 normotensives (6 male, 5 female, aged 47,2$\pm$13,7 years, blood pressure 115,0$\pm$10,7/85,2$\pm$7,2 mm Hg) were studied.
20 ml heparinized blood was drawn of each patient. Lymphocytes were isolated by

the density gradient method using Lymphoprep[R]. The suspension of lymphocytes was then washed twice in bidistilled water, which has been passed through a Chelex column to remove Ca^{2+} ions, containing 150 mmol/l NaCl Suprapur[R] (Merck, Darmstadt, West Germany). Ca^{2+} concentration in this solution was below 1 nmol/l.
After washing the lymphocytes twice, 1 ml of bidistilled water treated with the Chelex resin was added to the lymphocytes. For lysis of the cells, the suspension was then frozen to -18°C and then rethawed. Thereafter, in the sample Ca^{2+} concentration was determined by atomic absorption photometry.
Protein concentration was measured using the Coomassie Blue method. Lymphocytic Ca^{2+} content was then expressed as µmol/g protein.

Results

Lymphocytic Ca^{2+} content in normotensives, essential hypertensives and renal hypertensives was 31 ± 12, 53 ± 23 and 43 ± 34 µmol/g protein respectively (Fig. 1).

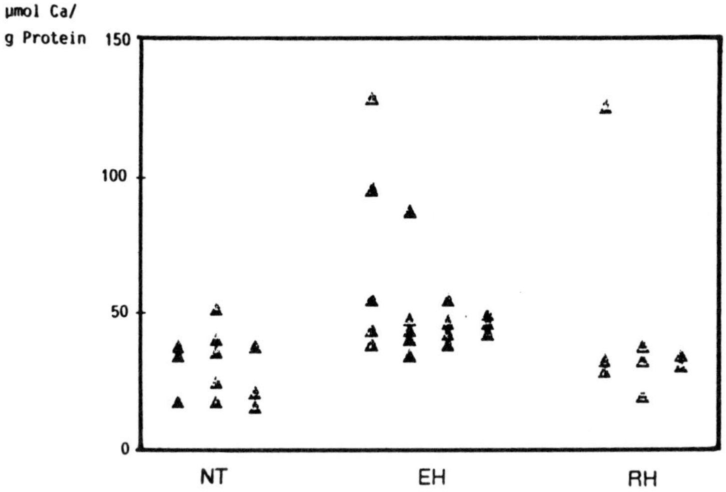

Fig. 1. Lymphocytic Ca^{2+} content in normotensives (NT), essential hypertensives (EH) and renal hypertensives (RH) in µmol/g protein.

In renal hypertensives there was a relatively wide distribution of calcium values, some values lying in the range of the normotensive subjects and some in the hypertensive range.

Discussion

The results show that in lymphocytes from essential hypertensives not only free calcium content is elevated, but also total calcium content. This implies that the underlying defect of calcium transport should be an increased net calcium influx.

On the other hand, an increased release of calcium from intracellular stores, e. g. induced by IP_3 release, seems unlikely to be the primary defect in cellular calcium metabolism in essential hypertension.

The results obtained in renal hypertensives show that the elevation of lymphocytic calcium content is not an unspecific effect associated with arterial hypertension per se, but may be the consequence of a specific defect in transmembrane calcium transport.

References

Bing, R.F., Heagerty, A.M., Jackson, J.A., Thurston, H., Swales, J.D. (1986): Leukocyte ionized calcium and sodium content and blood pressure in humans. In Hypertension 8: 483-488.

Bruschi, G., Bruschi, M.E., Caroppo, M., Orlandini, G., Spaggiari, M., Cavatorta, A. (1985): Cytoplasmic free (Ca^{2+}) is increased in the platelets of spontaneously hypertensive rats and essential hypertensive patients. In Clin. Sci. 68: 179-184.

Erne, P., Bolli, P., Bürgisser, E., Bühler, F.R. (1984): Correlation of platelet calcium with blood pressure. Effect of antihypertensive therapy. In N. Engl. J. Med. 310: 1084-1088.

Furspan, P.B., Bohr, D.F. (1986): Calcium-related abnormalities in lymphocytes from genetically hypertensive rats. In Hypertension 8 (Suppl. II): II 123- II 126.

Zidek, W., Vetter, H., Dorst, K.G., Zumkley, H., Losse, H. (1982): Intracellular Na^+ and Ca^{2+} activities in essential hypertension. In Clin. Sci. 63: 41s-43s.

Zidek, W., Losse, H., Dorst, K.G., Zumkley, H., Vetter, H. (1982): Intracellular sodium and calcium in essential hypertension. In Klin. Wschr. 60: 859-862.

Zidek, W., Kerenyi, T., Losse, H., Vetter, H. (1983): Intracellular Na^+ and Ca^{2+} in aortic smooth muscle cells after enzymatic isolation in spontaneously hypertensive rats. In Res. Exp. Med. 183: 129-132.

Metal Ions in Biology and Medicine. Eds. Ph. Collery, L.A. Poirier, M. Manfait, J.C. Etienne. John Libbey Eurotext, Paris © 1990, pp. 141-144

Determination of intracellular Ca^{2+} in primary and various types of hypertension

C. Spieker, W. Zidek, K. Kisters, K.H. Rahn

Medizinische Universitäts-Poliklinik Albert-Schweitzer-Straße 33, D-4400 Münster, FRG

Introduction

Numerous investigations in the last two decades have shown that in essential hypertension, as well as in the spontaneously hypertensive rat, characteristic deviations in cellular electrolyte metabolism occur. Various studies revealed raised intracellular free Ca^{2+} concentrations in blood cells from essential hypertensives (Bruschi et al., 1985; Erne et al., 1984; Zidek et al. 1982) with the quin 2 method or with ion-selective electrodes. Many attempts have been made to clarify the nature of the underlying cellular defect. Principally either a disturbed transmembranous Ca transport through one of the several Ca^{2+} channels, or shifts of intracellular Ca^{2+} between the cellular compartments come into account. In the latter situation, total intracellular Ca^{2+} would remain unchanged. Therefore, in addition to studies on free intracellular Ca, measurements on total Ca content are required to elucidate the principal source of increased intracellular Ca^{2+}. Furthermore, the methods to study intracellular free Ca may not be suitable for studying large groups of patients. In this study an approach to determine intracellular total Ca^{2+} by the Ca_i^{2+}/K_i^+ ratio, i.e., the ratio between intracellular Ca^{2+} and K^+ content, is used. Since intracellular K^+ concentration is not much altered in essential hypertension, the Ca_i^{2+}/K_i^+ ratio should yield information about changes of total intracellular Ca^{2+} concentration. The measurements of intracellular free Ca^{2+} concentrations in the above mentioned studies were mainly obtained with methods of a comparatively high methodological demand. In our study we used a relatively simple method for assessing total intracellular Ca^{2+} using intracellular K^+ as an "internal standard". This procedure can be used because intracellular K+ concentrations are not altered in essential hypertension and since establishing another variable as a reference would have the disadvantage of introducing another type of measurement with additional sources of error. The method was used to dertermine total intracellular Ca^{2+} in red cells from essential hypertensives, patients with phaeochromocytoma and with renal hypertension.

Methods and materials

The investigations were performed in essential hypertensives (178.4 $\pm$ 19.8/110.3 $\pm$ 9.4) (n=13), renal hypertenives (170.4 $\pm$ 16.5/115.8 $\pm$ 8.1) (n=15), patients with renal artery stenosis (159.1 $\pm$ 2.9/110.5 $\pm$ 5.3) (n=7) (angiography of the renal artery) and patients with phaeochromocytoma (169.1 $\pm$ 5.2/119.1 $\pm$ 2.3) (n=16) (raised urine catecholamines; histological diagnosis of the tumour postoperatively). These patients were either untreated or inufficiently treated.
Normotensive patients (n=21) served as controls. The renal hypertensives had a mean serum creatinine of 4.76 $\pm$ 3.21 mg/dl. The underlying disorders were chronic glomerulonephritis (n=8), polycystic renal disease (n=4) and diabetic glomerulosclerosis (n=4). From each subject, 10 ml heparinized blood was drawn. The measurements of intracellular Ca^{2+} were performed with ion-selective electrodes. (Zidek et al. 1982) The results are given as ion activities, referring to free cell water.
Total intracellular Ca^{2+} concentration, as determined by flame photometry, was measured as described in earlier studies using the same preparation of cells [4]. The haemolysate obtained by freezing to -18°C and rethawing was ashed at 1,100°C. To the ashed material 5 ml 5 mmol/l LiCl solution, which was used to preare samples for flame photometry, was added. After thorough mixing, the insoluble content was separated with a Milley HA filter SCHA 025 BS. The filtered fluid was analysed for Ca^{2+} and K^+ with an automatized Eppendorf flame photometer type FCM 6341. The volume of distilled water added to the ash was chosen so as to avoid Ca^{2+} concentrations below the lower limit of detection of the flame photometer.
In another sample of haemolysate, K^+ concentration was determined by flame photometry without previous ashing.
Statistical analysis was performed using Students's test for unpaired samples.

Results

In normotensives (n=21) the mean $Ca_i^{2+}/K_i^+ \ 10^{-3}$ ratio was 2.01 $\pm$ 0.91, in essential hypertensives 4.58 $\pm$ 1.68 ($p<0.01$), in renal hypertensives 3.73 $\pm$ 1.90 ($p<0.05$), in patients with phaeochromocytoma 2.15 $\pm$ 1.62 and in patients with renal artery stenosis 2.25 $\pm$ 1.25. It appeared that the range of Ca_i^{2+}/K_i^+ ratios was much wider in essential and renal hypertensives than in normotensives. Using the intracellular K^+ concentrations, intracellular total Ca^{2+} concentration of 437.1 $\pm$ 223.7 µmol/l ($p<0.01$) in essential hypertensives, of 352.0 $\pm$ 197.0 µmol/l ($p<0.05$) in renal hypertensives, of 198.2 $\pm$ 149.1 µmol/l in patients with phaeochromocytoma, of 201.1 $\pm$ 139.1 µmol/l in patients with renal artery stenosis and of 183.9 $\pm$ 88.7 µmol/l in normotensive subjects was calculated.

Discussion

The data obtained in this study show that the Ca^{2+}/K^+ ratio in red blood cells is elevated in essential hypertension as

compared to normotensives. As intracellular K^+ concentrations were unchanged in these patients, the measurements indicate that total intracellular Ca^{2+} concentrations are elevated. Using ion-selective electrodes, Zidek et al. (1983) found elevated intracellular free Ca^{2+} concentrations in both spontaneously hypertensive rats and essential hypertensives. Erne et al. (1984) and Bruschi et al. (1985) confirmed these findings by the quin 2 method. Both methods involved studies in blood cells or isolated arterial smooth muscle cells.
However, there is still considerable debate as to the mechanism by which intracellular free Ca^{2+} is elevated. Blaustein (1977) suggested an increased Na^+-Ca^+ exchange across the cell membrane to cause an elevation of intracellular free Ca^{2+} in primary hypertension, but further work on the cellular electrolyte metabolism could not verify an altered Na^+-Ca^{2+} exchange in essential hypertension. Postnov et al. (1979) and Postnov and Orlov (1980) described a reduced Ca^{2+} binding to the cell membrane of different types of cells and furthermore a decreased maximum activity to the Ca^{2+} ATPase of the cell membrane. Devynck et al. (1981) confirmed a decreased Ca binding in plasma membranes from spontaneously hypertensive rats. It was found that intracellular total Ca^{2+} was elevated to a similar or a slightly lesser extent as compared to the increase in intracellular free Ca^{2+} transport rather than a change in the bound/free ratio of intracellular Ca^{2+} exchange or potential dependent Ca^{2+} extrusion might be altered in red cells from essential hypertensives. Another possibility to explain the change of the Ca^{2+} distribution in hypertensives is an increased "leak" which causes the passive inward influx of Ca^{2+}. With respect to the above mentioned theories, the elevation of intracellular total Ca^{2+} in the red blood cells of renal hypertensives might allow the conclusion that beside the specific mechanisms, a nonspecific change to the cell membrane could also alter the electrolyte composition, perhaps as a result of uremia or an altered rheology. However, the findings in renal hypertensives could as well be explained assuming that some of the patients suffering from renal insufficiency have either essential hypertension or a genetic disposition to hypertension thich may be contributory to the elevated Ca^{2+} concentration. In the patients with renal artery stenosis there was only a tendency to elevation of Ca^{2+}, but the study group was small, so that final conclusions cannot be drawn. There was no significant elevation of the e concentration in patients with phaeochromocytoma as compared to normotensive subjects. This could be due to the fact that most of the patients of our study with renal-renal artery stenosis and essential hypertension were mainly treated with diuretics and/or ß-blockers. The patients with phaeochromocytoma were included in the study before operation and received a-blocking agents.

References

1. Blaustein M.P. (1977): Sodium ions, calcium ions, blood pressure regulation, and hypertension: a reassessment and a hypothesis. Am. J. Physiol. 232, C165-C173

2. Bruschi G., Bruschi M., Caroppo M., Orlandini G., Spagigiari, M.; Cavatorta, A. (1985): Cytoplasmic free (Ca^{2+}) is increased in the platelets of spntaneously hypertensiv rats and essential hypertensive patients. Clin. Sci. 68, 179-184
3. Devynck M.A., Pernollet M.-G., Nunez A.-M., Meyer P. (1981): Analysis of calcium handling in erythrocyte membranes of genetically hypertensive rats. Hypertension 3, 397-403
4. Erne P., Bolli P., Bürgisser E., Bühler F. (1984): Correlation of platelet calcium with blood pressure. Effect of antihypertensive therapy. New Engl. J. Med. 310, 1084-1088
5. Postnov Y.V., Orlov S.N., Pokudin N.J. (1979): Decrease of calcium binding by th red blood cell membrane in spontaneously hypertensive rats and in essential hypertension. Pflügers Arch. 379, 191-195
6. Postnov Y.V., Orlov S.N. (1980): Alteration of membrane control over intracellular calcium in essential hypertension and in spontaneously hypertenive rats; in Zumkley, Losse, intracellular electrolytes and arterial hypertension, pp. 144-171 (Thieme Stuttgart)
7. Zidek W., Losse H. Dorst K.G., Zumkley H., Losse H. (1982): Intracellular sodium and calcium in essential hypertension. Klin. Wschr. 60, 859-862
8. Zidek W., Vetter H., Dorst K.G., Zumkley H., Losse H. (1982): Intracellular Na^+ and Ca^{2+} activities in essential hypertension. Clin. Sci. 63, 41S-43S
9. Zidek W., Kerenyi T., Losse H., Vetter H. (1983): Intracellular Na^+ and Ca^{2+} in aortic smooth muscle cells after enzymatic isolation in spontaneously hypertensive rats. Res. exp. Med., 129-132

Metal Ions in Biology and Medicine. Eds. Ph. Collery, L.A. Poirier, M. Manfait, J.C. Etienne. John Libbey Eurotext, Paris © 1990, pp. 145-147

Calcium ion involvement in analgesic tolerance to acupuncture and morphine*

Z.X. Zhang, X.N. Zhao, J. Xing, X. Li, J. Chen, R.S. Chen

*Medical School of Nanjing University, Nanjing 210008, P.R. China and ** Chemistry Department, Nanjing University, Nanjing 21008, People's Republic of China*

INTRODUCTION

It has been reported in recent years that electroacupuncture, morphine and lanthanides–induced analgesia not only were very similar, but had a good parallel to the changes in the brain Ca^{2+} concentrations as well, and electroacupuncture,morphine, and lanthanides could produce analgesia by analogous mechanism (Zhang, 1983,1984). Therefore, it is reasonable that the mechanism underlying the production of analgesic tolerance to morphine and acupuncture is probably the same. A preliminary account of this work has been presented previously(Zhang, 1987).

EXPERIMENTAL DETAILS

Mice weighting 25–30g were used.They were divided at random into a variety of experimental groups.Analgesia was determined by the heat radiation tail–flick procedure, the tests of the development of analgesic tolerance to electroacupuncture and morphine in mice were made by the method which we used previously.

Analgesic tolerance was estimated by comparing the daily analgesic effect induced by electroacupuncture or morphine for 5 consecutive days. The free calcium concentration and mitochondrial protein bound calcium in discrete brain regions were measured with a specific ion electrode and fluorescent probe Tb(III) respectively (Xie,1988). Anisomycin was a gift from Charles Pfizer Co., Groton, Connecticut. Solutions were prepared at appropriate concentration in 0.9% NaCl. In order to dissolve anisomycin, an approximately equal molar amount of HCl was added,and the pH was finally adjusted to 6–7.

RESULTS AND DISCUSSION

The pain threshold decreased on the fourth day after the daily administration of electroacupuncture or morphine, and the significant analgesic tolerance to electroacupuncture or morphine was elicited on the fifth day. Thus,in the following experimental groups the electroacupuncture and morphine–induced analgesic tolerant models were used.

As can be seen in Fig. 1 , a significant increase of the free Ca^{2+} concentration and a decrease of mitochondrial protein bound Ca^{2+} were observed in the two brain regions during analgesic tolerance to morphine. The analogous results were obtained with electroacupuncture. The results suggest that Ca^{2+} transport across the neuroplasmic membrane of PAG and hypothalamus seems to play an important role in mediating analgesic tolerance to morphine and electroacupuncture.

* The Project Supported by National Natural Science Foundation of China

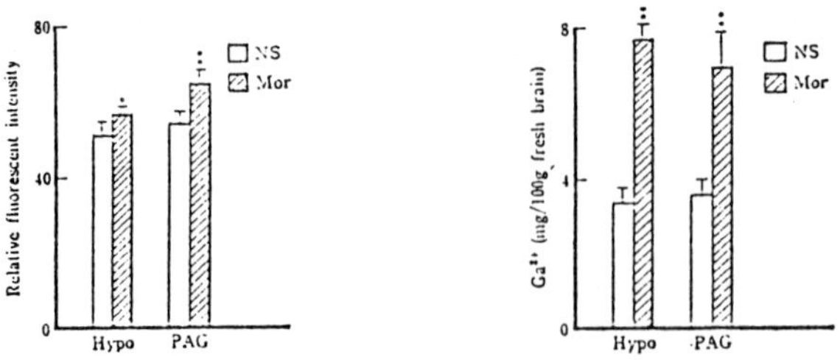

Fig.1 Changes of Tb^{3+}–mitochondrial fluorescence (left) and free Ca^{2+} concentration (right) in discrete brain regions of mice during analgesic tolerance to morphine
The free Ca^{2+} was detected by Ca^{2+}–selective electrode. Vertical lines indicate means ± S.D. of 7–9 experiments.
*** $P<0.05$, as compared with saline ** $P<0.01$, as compared with saline Hypo = hypothalamus PAG = periaqueductal gray NS = saline Mor = morphine**

The test animals were divided into 5 groups, each having 9 or 10 animals.Anisomycin was injected i.p. 30 min before morphine on the first day. Then each group was given daily repeated treatments.
Data in Table 1 show that anisomycin itself has no analgesic effect. The lower dose of anisomycin (0.4 μg / mouse) had no effect on the development of analgesic tolerance to morphine. However, Table 1 also shows a definite trend toward decreasing tolerance development with increasing dose of anisomycin.
In the next test, larger doses of anisomycin were tried using the same procedure. The results in Table 2 again indicate antagonism of tolerance development related to the dose of anisomycin; and with increasing dose such as 40μg / mouse,pain threshold had no significant difference between the anisomycin–treated and the control animals ($P>0.05$), i.e.the development of analgesic tolerance to morphine was fully antagonized.

Table 1. Effects of anisomycin on morphine tolerance development in mice

	DA (%)				
Group	1	2	3	4	5(days)
Anisomycin 4μg	3.6 ± 3.8	2.4 ± 4.4	4.1 ± 3.4	3.8 ± 4.9	3.4 ± 5.8
Morphine 10 mg / kg	91.5 ± 12.9	85.4 ± 12.1	56.4 ± 17.5	36.3 ± 5.7	17.2 ± 6.4
+ anisomycin 0.4μg	87.4 ± 15.8	89.3 ± 18.4	61.3 ± 19.4	44.6 ± 10.5	27.4 ± 7.2
+anisomycin 0.8μg	83.7 ± 17.6	81.5 ± 14.8	72.6 ± 12.0*	65.3 ± 11.2**	37.3 ± 6.7*
+ anisomycin 1.6μg	84.6 ± 18.5	82.9 ± 15.4	74.8 ± 8.9*	70.6 ± 12.4**	39.8 ± 8.9*

Anisomycin was injected intraperitoneally (i.p.) 30 minutes before the daily morphine (i.p.). Values in the table are means ± S.D. for 9–10 experiments.
* $p<0.05$, as compared with morphine control
** $p<0.01$, as compared with morphine control

Table 2. Effects of anisomycin on morphine tolerance development in mice

	DA(%)				
Group	1	2	3	4	5(days)
Morphine 10 mg / kg	92.7 ± 14.8	83.8 ± 10.4	58.6 ± 9.8	24.7 ± 5.1	12.1 ± 5.6
+ anisomycin 5μg	89.4 ± 15.6	81.2 ± 10.9	72.4 ± 8.4*	70.3 ± 11.0**	44.5 ± 11.8**
+ anisomycin 20μg	88.5 ± 16.1	85.4 ± 16.4	76.3 ± 10.1*	76.5 ± 10.4**	62.1 ± 15.6**
+ anisomycin 40μg	84.3 ± 11.6	89.6 ± 11.8	83.4 ± 8.7**	75.8 ± 7.9**	76.4 ± 12.8**

Values in the table are means ± S.D. for 10 experiments, other notes are the same as in table 1.

The above two tests show that the diminution in morphine tolerance obviously related to the dose of anisomycin; and Fig.2 shows the dose–response relationship for the pooled results of the last 2 experiments. Ordinate indicates the degree of analgesia (%) on the fifth day, and abscissa, dose in μg per mouse on log scale.

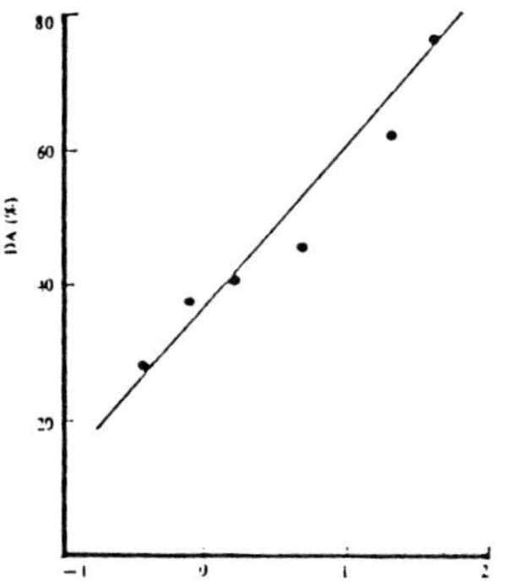

Fig.2 Decrease in morphine tolerance by increasing doses of anisomycin Abscissa: dose in μg per mouse on log scale; ordinate: degree of analgesia (%).

Our previous study(Zhang, 1987) has demonstrated that protein synthesis inhibitor anisomycin could block the development of analgesic tolerance to electroacupuncture, and electroacupuncture tolerance may be related to the production of protein or peptide in CNS. In the present study, it is observed that the inhibitor also blocks the development of analgesic tolerance to morphine and displays the obvious dose–dependent relationship. Moreover, from the changes of mitochondrial protein bound Ca^{2+} and free Ca^{2+} concentration in the discrete brain regions of mice, electroacupuncture and morphine appeared to share the same sites for antinociceptive activity. The hypothalamus and the PAG of midbrain, especially the latter, could be the most sensitive site for morphine and acupuncture. This site is reportedly also the brain region of greatest opiate receptor concentration (Kuhar, 1973). These results suggest that the tolerant effects of both morphine and acupuncture may be related to reduced influx and / or binding of Ca^{2+} in brain tissue, especially in the PAG region.

REFERENCES

Kuhar, M.J. et al., (1973): Regional distribution of opiate receptor binding in monkey and human brain. Nature. 245, 447–450.

Xie Xin–ying, et al.,(1988): The change of mitochondrial protein bound Ca^{2+} in some brain regiona under the conditions of electroacupuncture or morphine analgesia and analgesic tolerance. Acta. Physiol. Sin. 40, 553–560.

Zhang Zu–xuan, et al., (1983): The similarity of influences of calcium on electroacupuncture analgesia and morphine analgesia. Acta Physiol. Sin. 35,172–177.

Zhang Zu–xuan, et al., (1984): Enhancement of electroacupuncture analgesia by intraventricular praseodymium chloride. Kexue Tongbao. 29,537–542.

Zhang Zu–xuan, et al.,(1987): Role of calcium in electroacupuncture analgesia and the developments of analgesic tolerance to electroacupuncture and morphine. Sci. Sin. (B Series) 30,974–985.

Metal Ions in Biology and Medicine. Eds. Ph. Collery, L.A. Poirier, M. Manfait, J.C. Etienne. John Libbey Eurotext, Paris © 1990, pp. 148-150

Ca^{2+} and the starting of silicosis*

Y. Mao, R.S. Chen, A.B. Dai, Z.X. Zhang**

*Department of Chemistry, Nanjing University, Nanjing 210008, P.R. China and ** Medical School of Nanjing University, Nanjing 210008, People's Republic of China*

SUMMARY

In this paper, the beginning of the formation of silicosis has been studied. In the silicotic rats, with ion–selective electrode, it was found that concentration of free Ca^{2+} in their lungs increased significantly. Along with this, by means of Tb^{3+}–fluorescent probe, $^{45}Ca^{2+}$–tracer, Raman and NMR spectra, the process of increase and effect of Ca^{2+} in lungs of silicotic rats were studied and the difference between α–quartz and amorphous silica in potency to cause silicosis was demonstrated.

INTRODUCTION

As to the cause of silicosis, studies have usually emphasized the destructive reaction of SiO_2 crystalline dusts with lung tissues resulting in the development of silicotic lesion. As an extension of our previous studies on chemical behavior of SiO_2 in solution and biological effect of metallic ion on acupuncture analgesia (Dai,1981; Zhang,1987), the variation of Ca^{2+} concentration in silicotic organ has been studied. Some interesting results were obtained and preliminary explanation was attempted.

EXPERIMENTS AND RESULTS

1.The concentration of Ca^{2+} in lung of rats which had been affected with experimentally silica–induced silicosis by intratracheal injection of crystalline SiO_2 dusts were measured by ion–selective electrode. It is found for the first time that concentration of Ca^{2+} in lung of rats increased significantly from the third day after injection of dusts of α–quartz. The increase continued the second month after which the increase

Table 1. Change of fluorescence intensity of Tb^{3+}–mitochondria in lung of rats after injection of α–quartz

days after injection of α–quartz	fluorescence intensity control(I_o)	treatment(I_t)	$(I_t-I_o)/I_o$
3	22.8±0.93	33.7±1.78**	0.478
7	22.9±0.80	35.0±1.00**	0.528
14	22.5±0.69	36.7±1.35**	0.631
60	23.2±0.60	34.8±0.48**	0.500
90	23.6±0.26	35.4±0.67**	0.500
150	22.5±0.68	42.0±0.74**	0.868

note: each value represents the mean ± S.D. n = 8–10 rats / group
** significantly different from control, $p < 0.01$

* The Project Supported by National Natural Science Foundation of China

slowed down and almost stopped at the fifth. This shows that a relationship exists between the increase of Ca^{2+} in the lung and development of silicosis.

2.The change of Ca^{2+} concentration in the mitochondria of lung of rats was studied by the Tb^{3+}–fluorescent probe. The results are shown in Table 1. The fluorescence intensity of the treatment group was higher than that of the control. It indicates that the content of Ca^{2+} of mitochondria in lung of silicotic rats was lower than that of the control group.

3. The macrophage obtained from the alveolus of rabbits was incubated at 37℃ with α–quartz, then the content of Ca^{2+} on macrophage membrane and in mitochondria of macrophage was studied by Tb^{3+}–fluorescent probe and $^{45}Ca^{2+}$–tracer. The results are shown in Table 2 and Table 3. It was seen that the content of Ca^{2+} of the treatment group was lower than that of the control.

Table 2. Effect of α–quartz on the fluorescence intensity of Tb^{3+}–macrophage in alveolus

sample	incubation for 40 minutes		incubation for 80 minutes	
	fluorescence intensity	$(I_t-I_o)/I_o$	fluorescence intensity	$(I_t-I_o)/I_o$
control	19.6±1.03		19.5±1.53	
+o.5mg α–quartz / ml	34.1±1.74**	0.740	32.8±0.65**	0.682
+1.0mg α–quartz / ml	33.0±0.62**	0.684	33.4±0.72**	0.713

note: each value represents the mean ± S.D. n = 8 rabbits / group

* * significantly different from the control, $p<0.01$

Table 3.1 Effect of α–quartz on the fluorescence intensity of Tb^{3+}–mitochondria in macrophage of alveolus

sample	incubation for 40 minutes		incubation for 80 minutes	
	fluorescence intensity	$(I_t-I_o)/I_o$	fluorescence intensity	$(I_t-I_o)/I_o$
control	19.7±1.27		19.7±1.81	
+0.5mg α–quartz / ml	26.8±1.50**	0.360	28.9±0.88**	0.467
+1.0mg α–quartz / ml	29.7±0.79**	0.508	31.7±0.94**	0.609

note: as in table 2.

Table 3.2 Effect of α–quartz on Ca^{2+} of mitochondria in macrophage of alveolus

sample	$^{45}Ca^{2+}$(cpm / 40μg protein)
mitochondria in macrophage (I_o)	379±39 (8)
incubated at 37 ℃ with α–quartz for 40 minutes (I_t)	796±74** (8)
$(I_t-I_o)/(I_o)$	417 / 379 = 1.10

note: as in Table 2.

All the above finding suggest that the increased amount of free Ca^{2+} in the lung came mainly from the mitochondria of macrophage.

DISCUSSION

1. Replacement of membrane Ca^{2+} by the H^+ of silanol group of α–quartz

H^+ from the silanol group of α–quartz may replace $Ca^{2+}$from the membrane of macrophage, as indicated by the equation:

$$R{-}Ca + (HO)_2Si \rightleftharpoons R{-}H_2 + CaO_2Si$$

here R denotes the membrane of macrophage, and $(HO)_2Si$, the 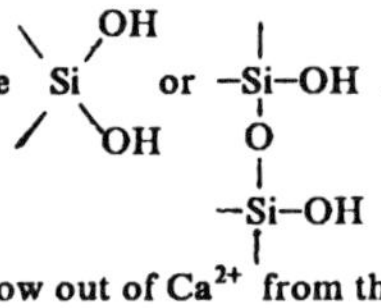 . As a result,

Ca^{2+} displaced from the membrane of macrophage gave rise to the flow out of Ca^{2+} from the cell and thus Ca^{2+} in the mitochondria was released from the macrophage.

2. Bond of α–quartz with membrane of macrophage

The results of NMR and Raman spectra showed that the α–quartz particle had affility to $N(CH_3)_4Cl$, which suggested that silanol and siloxan groups on the surface enabled bonding with the protein or phospholipid molecules of the cell membrane, and definite amount of α–quartz was adsorbed to the membrane of macrophage.

The α–quartz causes silicotic lesion but amorphous silica does not. Such difference is due to (1) the higher degree of dissociation of H^+ from the silanol group of α–quartz and (2) the more regular arrangement of silanol and siloxan groups on the surface of α–quartz which enables more complete bonding with the protein or phospholipid of the cell membrane.

CONCLUSION

It is found that concentration of Ca^{2+} in lung of silicotic rats increased significantly from the third day after injection of crystalline dusts of α–quartz. The increased amount of free Ca^{2+} in the lung came mainly from the mitochondria of macrophage. H^+, dissociated from the silanol group of α–quartz, displaces Ca^{2+} from the membrane of macrophage, which results in the flow out of Ca^{2+} from the cell and the release of Ca^{2+} in the mitochondria. The stronger adsorbability of α–quartz to the membrane of macrophage and the higher degree of dissociation of H^+ from the silanol group of α–quartz are probably the primary cause of the detrimental effect.

REFERENCES

Dai An–bang & Chen Rong–san,(1981): A theory of polymerization of silicic acid in aqueous solution. In 30 year's review of China's science and technology. Chapter 10, ed. Ziran Zazhi. World's Science Press. Sigapore.

Zhang Zu–xuan et al.,(1987): Role of calcium in electroacupuncture analgesia and the developments of analgesic tolerance to electroacupuncture and morphine. Sci. Sin. (B Series) 30,974–985.

Metal Ions in Biology and Medicine. Eds. Ph. Collery, L.A. Poirier, M. Manfait, J.C. Etienne. John Libbey Eurotext, Paris © 1990, pp. 151-155

Electronprobe X-ray microanalysis of Na^+ content in vascular smooth muscle cells from spontaneously hypertensive and normotensive rats

K. Kisters, W. Zidek, E.R. Krefting*, C. Spieker, K.H. Rahn

*Medizinische Poliklinik and * Institut für Medizinische Physik der Universität 4400 Münster, Albert-Schweitzer-Str. 33, West Germany*

Summary

In aortic smooth muscle cells from spontaneously hypertensive rats (Münster strain n=12, SHR) and of normotensive Wistar- Kyoto rats (n=11, NTR) intracellular Na^+ content was measured by electronprobe microanalysis. Measurements were performed in aortic cryosections of a thickness of 3 µm. Na^+ content was 12,5±2,4 g/kg dry weight in SHR versus 6,96±1,1 g/kg dry weight in NTR (p<0,01). Thus aortic smooth muscle cells from SHR are characterized by markedly elevated intracellular Na^+ content compared to normotensive cells. This may either be due to genetically determined disturbances in transmembrane Na^+ transport or to a circulating factor affecting Na^+ transport. In aortic smooth muscle of spontaneously hypertensive rats cellular Na^+ handling may be disturbed similarly as in hypertensive blood cells.

Key words: Na^+, vascular smooth muscle, spontaneously hypertensive rats, hypertension

Introduction

In primary hypertension disturbances of intracellular electrolyte concentrations and metabolism have often been described. In earlier studies mainly disturbances in cellular Na^+ transport have been investigated (D' Amico, 1958; Losse et al., 1960), whereas the role of intracellular Ca^{++} handling received growing interest in the last years (Kaplan, 1988; Zidek & Vetter, 1987). In most studies blood cells of essential hypertensive patients were used. As to possible analogies between electrolyte metabolism in any type of blood cells and arterial smooth muscle cells, comparatively few data are available. To test cellular Na^+ handling in primary hypertension, electronprobe microanalysis was performed in aortic smooth muscle cells from spontaneously hypertensive and normotensive rats.

Methods

Aortae from 12 spontaneously hypertensive rats and from 11 normotensive rats (systolic pressure 118,3±6,8 mm Hg, mean value ± SD) aged 3 months were used. The spontaneously hypertensive rats from the Münster strain had reached a systolic blood pressure of 192,3±10,7 mm Hg at this age. The aortae were freed of surrounding connective tissue and immediately thereafter frozen in liquid propane cooled with liquid nitrogen at a temperature of about - 190 °C. Then cryosections with a thickness of 3 μm were made and lyophilized (Fig. 1.).

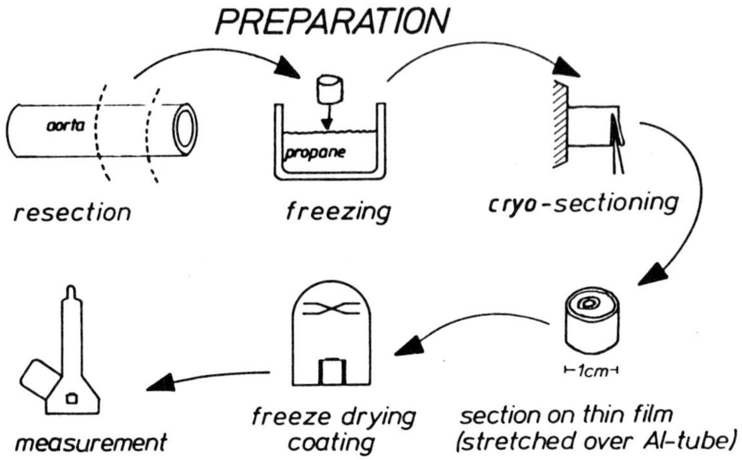

Fig. 1. Preparation of aortic tissue for electronprobe microanalysis.

For electronprobe microanalysis an electron microscope with an X-ray detector system is used. When the electrons of the incoming beam strike an atom in the specimen, they can knock an electron out of the kernel. If this hole is in an inner shell, it is filled with an electron of a higher shell and an X-ray photon with a discrete energy corresponding to the difference between the two atomic shells is emitted simultaneously. The energy of these X-rays is characteristic for each element (Fig. 2.).

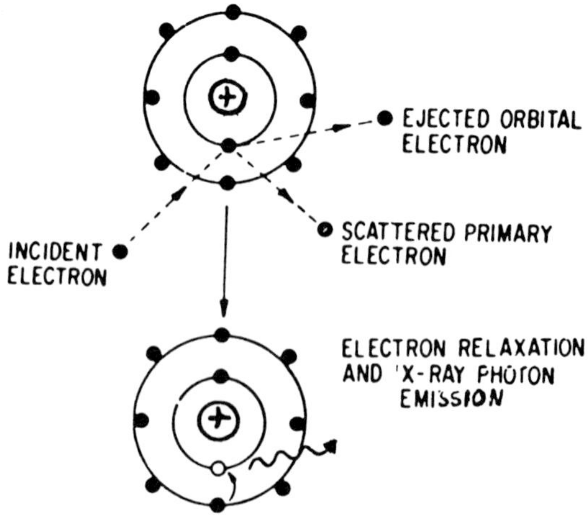

Fig. 2. Principle of electronprobe microanalysis. Schematic illustration of the process of inner-shell ionization and subsequent deexcitation by X-ray photon emission.

For quantification the continuum method developed by Hall (Goldstein et al.,1981) was used. Intracellular sites of measurements were identified by the morphology obtained by electron microscopy and by simultaneous measurement of sulfur and phosphorus, the concentrations of which were markedly elevated in the intracellular compared to the extracellular space. In each aorta, mean values of at least 5 intracellular measurements at different sites were calculated. Na^+ content was expressed in mg/kg dry weight of the tissue. For statistical analysis unpaired Student's t test was used.

Results

In spontaneously hypertensive rats intracellular Na^+ content was $12,5 \pm 2,4$ g/kg dry weight as compared to $6,96 \pm 1,1$ g/kg dry weight in normotensive rats (mean values $\pm$ SD, $p<0,01$, Fig. 3.).

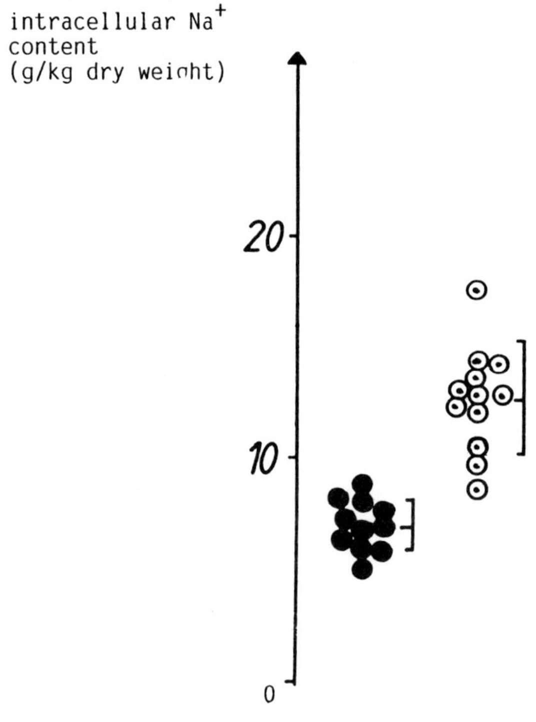

Fig. 3. Intracellular Na^+ content in aortic smooth muscle cells from normotensive (•) and spontaneously hypertensive rats (⊙). Mean values and standard deviations are noted.

Variations in intracellular Na^+ content in aortic smooth muscle cells of one animal were 20,5 % of the mean value in spontaneously hypertensive rats and 12,4 % in normotensive rats.

Discussion

There are only sparse data on intracellular Na^+ content in arterial smooth muscle cells from hypertensive animals. Tobian & Binion (1952) found an elevation of intracellular Na^+ content in arterial smooth muscle cells in one kidney-one clip hypertensive rats. On the other hand, Massingham and Shevde (1973) described principally unchanged total Na^+ content in aortae from genetically hypertensive rats, but did not distinguish between intracellular and extracellular Na^+ concentration.

More indirect evidence has been presented demonstrating a circulating Na, K ATPase inhibitor in plasms of essential hypertensives (De Wardener & Clarkson, 1985). The inhibition of Na, K ATPase has been postulated to cause hypertension as a consequence of elevated intracellular Na^+ concentrations. Increased intracellular Na^+ concentrations have been claimed to cause vasoconstriction by increasing Na^+ - Ca^{++} exchange (Blaustein, 1977), by increasing sensitivity of arterial smooth muscle to noradrenaline (Aronson, 1984), or by decreasing lumen/wall ratio of resistance vessels (Folkow, 1978).
Elevated intracellular Na^+ concentration of vascular smooth muscle cells in primary hypertension may be due to disturbances in Na^+ - K^+ cotransport (Garay, 1980), Na^+ - H^+ exchange (Livne, 1987) and Na^+ - K^+ countertransport (Poston, 1987). Changes in Na^+ - K^+ ATPase were related to the action of a still hypothetical natriuretic hormone (De Wardener & Clarkson, 1985), the secretion of which may be enhanced in primary hypertension. Although from the present results it cannot be decided between these possible mechanisms,the findings demonstrate that in aortic smooth muscle of spontaneously hypertensive rats similar changes in Na^+ handling can be detected as in blood cells from hypertensive animals or subjects.

References

Aronson, J.K. (1984): The role of the Na^+, K^+ - ATPase in the regulation of vascular smooth-muscle contractility and its relationship to essential hypertension. In Biochem. Soc. Trans. 12: 943-945.

Blaustein, M.P. (1977): Sodium ions, calcium ions, blood pressure regulation and hypertension: a reassessment and a hypothesis. In Am. J. Physiol. 232: C165-C173.

D' Amico, G. (1958): Red cell sodium and potassium in congestive heart failure, essential hypertension and myocardial infarction. In Am. J. Med. Sci. 236:156-161.

De Wardener, H.E., Clarkson, E.M. (1985): Concept of natriuretic hormone. In Physiol. Rev. 65: 658-759.

Folkow, B. (1978): Cardiovascular structural adaption: its role in the initiation and maintenance of primary hypertension. The Fourth Volhard Lecture. In Clin. Sci. Mol. Med. 55: 3-22.

Garay, R.P., Dagher, G., Pernollet, M.G., Devynck, M.A., Meyer, P. (1980): Inherited defect in a Na^+, K^+, co-transport system in erythrocytes from essential hypertensive patients. In Nature 284: 281-283.

Goldstein, J.T., Newbury, D.E., Echlin, P., Joy, D.C., Fiori, C., Lifshin, E. (1981): Scanning electron microscopy and X-ray microanalysis. In New York: Plenum Press.

Kaplan, N.M. (1988): Calcium and blood pressure. In Cardiovasc. Drugs Ther. 2: 269-274.

Livne, A., Balfe, J.W., Veitch, R., Marquez-Julio, A., Grinstein, S., Rothstein, A. (1987): Increased platelet Na^+ - H^+ exchange rates in essential hypertension: application of a novel test. In Lancet I: 533-536.

Losse, H., Wehmeyer, H., Wessels, F. (1960): Der Wasser- und Elektrolytgehalt von Erythrozyten bei arterieller Hypertonie. In Klin. Wschr. 38: 393-395.

Massingham, R., Shevde, S. (1973): The ionic composition of aortic smooth muscle from A.S. - hypertensive rats. In Br. J. Pharmacol. 47: 422-424.

Poston, L. (1987): Endogenous sodium pump inhibitors: a role in essential hypertension? In Clin. Sci. 72: 647-655.

Tobian, L. Jr., Binion, J.T. (1952): Tissue cations and water in arterial hypertension. In Circulation 5: 754-758.

Zidek, W., Vetter, H. (1987): Calcium and primary hypertension. In Nephron 47 (suppl. 1): 13-20.

Metal Ions in Biology and Medicine. Eds. Ph. Collery, L.A. Poirier, M. Manfait, J.C. Etienne. John Libbey Eurotext, Paris © 1990, pp. 156-158

Morphine enhancement of glutamate and kainic acid neurotoxicity*

X.N. Zhao, X. Li, J.J. Wang, J. Chen, Z.X. Zhang, X.J. Lou**

*Medical School of Nanjing University, Nanjing 210008, P.R. China and ** Shanghai Institute of Materia Medica, Shanghai 200031, People's Republic of China*

Monosodium glutamate (MSG), the main ingredient of glutamate, has been extensively used in food industry and clinically in the treatment of hepatic damage or ammonia poisoning induced by the some behavioral disorders. Olney et. al.(Olney, 1969) successively reported that MSG not only damaged the retinas, but also the some brain regions, especially the arcuate hypothalamic nucleus, and resulted in endocrine disorders in animals. The neurotoxicity of MSG could be enhanced by morphine (Wang, 1986). At the present work we have continued to observe the correlation of brain Ca^{2+} with neurotoxicity of MSG emphasis being placed on a comparison of the morphine enhancement of MSG and KA.

EXPERIMENTAL DETAILS

Mice of both sexes, weighing around 25g, were divided into a variaty of experimental groups at random. Pain was tested by the heat radiation tail–flick procedure. The average of three–time tests before administration of drug was taken as normal pain threshold. After administration of drug, pain was also measured three times, and the analgesic index was calculated (Xie, 1988). Each mouse was decapitated after the final pain test, and the front segment brain (including hypothalamus) was immediately isolated and homogenated in a Teflon bar glass homogenizer containing 1ml of 0.15M NaCl. After centrifugation at 1000 × g for 10 min, the precipitate was resuspended in the same buffer solution. Each assay tube had its duplicate, and protein was measured. 100μg of brain protein was put into a vial with 0.3ml of H_2O_2 and 0.4ml of formic acid, incubated at 80℃ for 1 h and cooled, 2ml of scintillation fluid (PPO,POPOP) and 1.5ml of ethyleneglycol ether were added in each vial. The radioactivity in the transparent solution was determined with a FJ–2101 double channel liquid scintillation spectrometer.In order to prepare brain mitochondria, the front segment brain was immediately homogenated in a Taflon bar glass homogenizer containing 2ml of sucrose solution, and the mitochondria was prepared by the method of Moldave (Moldave, 1967). The radioactivity in 50μg mitochondria protein from each tube was measured according to the above–mentioned procedure.

RESULTS AND DISCUSSION

As can be seen in Table 1, MSG own had mild analgesic effects and reduced the brain ^{45}Ca concentration. After pretreatment of morphine, pain threshold was markedly raised and the brain ^{45}Ca levels, compared with MSG control, obviously lowered. These effects could be partially antagonized by naloxone.
The results of replacing MSG with KA showed that KA also raised pain threshold and reduced the brain ^{45}Ca levels. The enhancing effects of morphine on KA were also reversed by naloxone.(Table 2).

* The Project Supported by National Natural Science Foundation of China

Table 1. Effects of naloxone onmorphine enhancement of MSG–induced analgesia and brain ^{45}Ca levels

Group	DA(%)	^{45}Ca (cpm x 10^3 / mg protein)
control	5.9 ± 3.3	147.30 ± 10.19
MSG	21.8 ± 11.0^{a}	114.43 ± 8.34^{a}
Mor+MSG	91.3 ± 10.8^{bc}	89.14 ± 4.91^{bc}
Nal+Mor+MSG	10.6 ± 8.4^{d}	136.84 ± 9.42^{d}

MSG (20μg / mouse) i.cv., saline (0.2ml / mouse), morphine(10mg / kg) or naloxone (5mg / kg) containing 9.25 x 10^3Bq / kg of ^{45}Ca i.p.20 minutes before administration of MSG. $x \pm$ S.D., n = 10. MSG, monosodium glutamate; Mor, morphine; Nal, naloxone

a) $p<0.05$; b) $p<0.01$ vs control; c) $p<0.01$ vs MSG group; d) $p<0.01$ or $p<0.001$ vs Mor+MSG group, $p<0.05$ vs MSG

Table 2. Effects of naloxone on morphine enhancement of MSG–induced analgesia and brain ^{45}Ca levels

Group	DA(%)	^{45}Ca(cpm x 10^3 / mg protein)
Control	4.8 ± 3.3	136.30 ± 8.44
KA	49.5 ± 10.2^{a}	108.47 ± 6.81^{a}
Mor+KA	94.6 ± 11.5^{bc}	79.64 ± 5.48^{bc}
Nal+Mor+KA	28.4 ± 3.7^{ad}	129.86 ± 10.24^{d}

KA (0.5μg / mouse) i.cv., saline (0.2ml / mouse), morphine (10mg / kg) or naloxone (5mg / kg) containing 9.25 x 10^3Bq / kg of ^{45}Ca i.p. 20 minutes before administration of KA. $\bar{x} \pm$ S.D., n = 8–10. KA, kainic acid, other abbreviations are the same as in table 1.

a) $p<0.05$; b) $p<0.01$ vs control; c) $p<0.001$ vs KA group; d) $p<0.001$ vs Mor+KA group, $p<0.05$ vs KA group

Data in Table 3 and Table 4 showed that administration of terbium prior MSG or KA remarkably enhanced pain threshlod and reduced the brain ^{45}Ca levels in correspondence with morphine, and these effects could also be antagonized by naloxone.

Table 3. Effects of naloxone on terbium (III) enhancement of MSG–induced analgesia and brain ^{45}Ca levels

Group	DA(%)	^{45}Ca(cpm x 10^3 / mg protein)
Control	5.3 ± 13.4	131.41 ± 15.15
MSG	28.7 ± 12.3^{a}	114.30 ± 9.34^{a}
Tb^{3+}+MSG	73.4 ± 15.7^{bc}	86.17 ± 8.40^{bc}
Nal+Tb^{3+}+MSG	9.1 ± 11.8^{d}	139.44 ± 11.87^{d}

MSG (20μg / mouse) i.cv., saline (0.2ml / mouse), Tb^{3+}(8μmol / kg) or naloxone (5mg / kg) containing 9.25 x 10^3Bq / kg of ^{45}Ca i.p. 20 minutes before administration of MSG. $x \pm$ S.D., n = 8–10. Tb^{3+}, terbium, other abbreviations are the same as in table 1.

a) $p<0.05$; b) $p<0.01$ vs control; c) $p<0.01$ vs MSG group; d) $p<0.001$ vs Tb^{3+}+MSG group, $p<0.05$ vs MSG group

Table 4. Effects of naloxone on terbium (III) enhancement of KA–induced analgesia and brain ^{45}Ca levels

Group	DA(%)	^{45}Ca(cpm x 10^3 / mg protein)
Control	5.4 ± 11.2	148.40 ± 14.71
KA	47.8 ± 18.2^{b}	124.84 ± 9.40^{a}
Tb^{3+}+KA	71.4 ± 14.4^{bc}	69.75 ± 5.44^{bc}
Nal+Tb^{3+}+KA	29.8 ± 13.6^{ad}	151.63 ± 7.98^{d}

KA (0.5μg / mouse) i.cv., saline (0.2ml / mouse), Tb^{3+}(8μmol / kg) or naloxone (5mg / kg) containing 9.25 x 10^3Bq / kg of ^{45}Ca i.p. 20 minutes before administration of KA. x ± S.D., n=8–10. The abbreviations are the same as in table 2 and table 3.

a) $p<0.01$; b) $p<0.001$ vs control; c) $p<0.05$ vs KA group; d) $p<0.01$ or $p<0.001$ vs Tb^{3+}+KA group, $p<0.05$ vs KA

The changes in the pain thershold and mitochondrial protein bound Ca^{2+} in the front segment brain after administration of MSG or KA were observed. The animals were divided into two test groups, the tests consisted of 6 phases. Each phase involved 8–9 mice. The latency was measured once every 10 min on the three occasions, and the average was taken as the baseline threshold. After measurement of the final pain threshold, each of the animals was decapitated immediately, then mitochondria of the front segment brain was prepared.

Fig. 1 shows that the degree of analgesia by MSG was enhanced gradually with time, the peak time was at 50th min. The effect of KA is more powerful and prolonged than that of MSG. In correspondence with the results in Fig.1, the enhancement of pain threshold was accompanied by an increase of mitochondrial protein bound Ca^{2+} of the brain regions (Fig. 2).

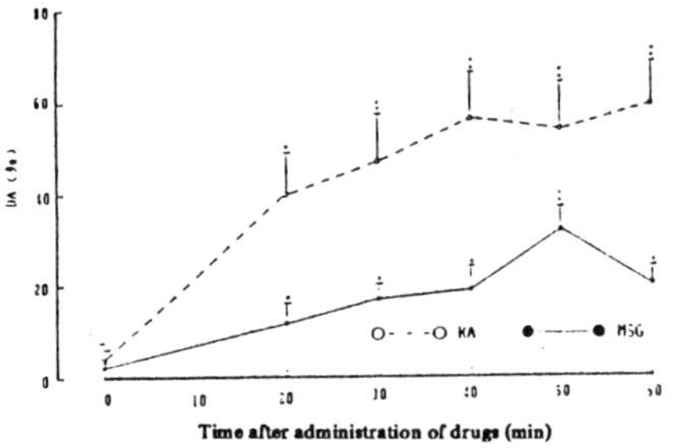

Fig. 1. Comparison of MSG and KA–induced analgesia MSG (20μg / mouse) or KA (0.5 μg / mouse) was injected i.cv., saline (0.2ml / mouse) containing 9.25 × 10^3Bq / kg of ^{45}Ca was injected i.p., x ± S.D. n = 8–9 mice / group * $p<0.05$ vs control, * * $p<0.01$ vs control

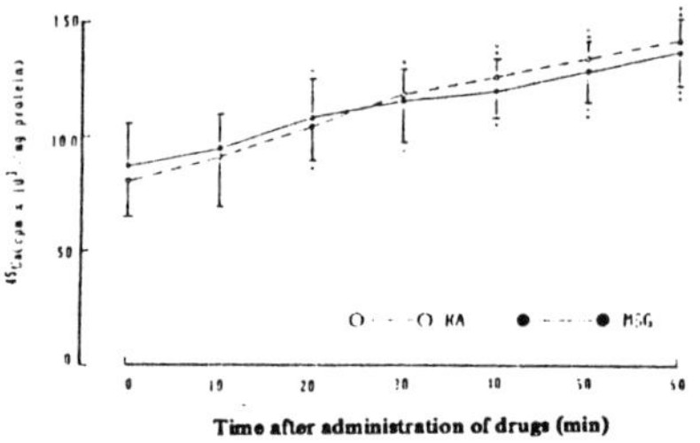

Fig. 2. Effect of MSG and KA on ^{45}Ca uptake by mitochondria. The treatment of drugs and notes are same as in Fig.1

The increase in Ca^{2+} binding to mitochondrial membrane, caused by MSG or KA, can be due to increase of cytoplasmic free Ca^{2+} concentration and a large uptake of Ca^{2+} by mitochondria. Cytoplasmic free Ca^{2+} overloading will lead to cellular injury. When morphine, which stimulates influx of extracellular Ca^{2+}, is added, the cytoplasmic free Ca^{2+} overloading will be aggrevated. The Ca^{2+} accumulation has become irreversible and causes structural damage to the cell. This type of precipitation appears to be an important part of the sequence of cell injury even death induced by some excitatory neurotoxins, such as MSG and KA.

REFERENCES

Moldave, K.(1967): Preparation of RNA from mammalian ribosomes. In Methods in Enzymology. eds. Grossman, L. and Moldave, K., Vol. 12, Part A, pp. 607–632. NY: Academic Press.

Olney, J.W.(1969): Brain lesions, obesity, and other disturbances in mice treated with monosodium glutamate. Science. 164,719–721.

Wang Guang–jian et al.,(1986): Enhancing effects of morphine on glutamate neurotoxicity in mice. Chinese J. Pharmacol. Toxicol. 1,57–62.

Xie Xin–ying et al.,(1988): The change of mitrochondrial protein bound Ca * 2+ in some brain regions under the conditions of electroacupuncture or morphine analgesia and the analgesic tolerance. Acta. Physiol. Sin. 40,553–560.

Metal Ions in Biology and Medicine. Eds. Ph. Collery, L.A. Poirier, M. Manfait, J.C. Etienne. John Libbey Eurotext, Paris © 1990, pp. 159-161

Abnormalities of membranes of myocardial mitochondria and erythrocytes from patients with keshan disease*

F.Y. Yang, Z.H. Lin, W.H. Wo, Q.R. Xing, W.W. Chen, S.Y. Wang, J.F. Wang, S.G. Li, B.Q. Guo

Institute of Biophysics, Academia Sinica Beijing, 100080 China

INTRODUCTION

The main pathological feature of Keshan disease(KD), an endemic cardiomyopathy of unknown cause, is the multifocal necrosis of heart muscle. In the present paper we describe changes of structure and function of myocardial mitochondria obtained postmortem from patients with subacute KD. Our results suggest that KD might be classified as a " mitochondrial cardiomyopathy " endemic in China. Besides, abnormalities of erythrocyte membrane from patients with latent KD have also been presented.

MATERIALS AND METHODS

Materials

Myocardial tissue (left ventrile) was obtained from 19 patients with Keshan disease (11 subacute, 8 chonic) in Chu-xiong County, Yun-nan Province, Southwestern China and was stored in liquid N_2 before use. Control samples were collected from healthy victims of accidents and treated with the same procedure as above. In the case of latent KD, samples were collected both from patients and control children from the same endemic region. Another 20 control children(Group3) from families of staff in Forestry Bureau, located in that region but their food was supplied by other Se non-deficient region.

Methods

Isolation of cardiac mitochondria. Mitochondria were isolated from myocardial tissue by previously described procedures(Yang,F.Y.et al.1980). Enzyme measurements. Activities of succinic oxidase, succinate dehydrogenase, cytochrome c oxidase, H'-ATPase as well as its sensitivity to oligomycin were determined as described. Se determination. Se concentration was determined by the diaminonaphthalene method. Fluidity measurements. Lipid fluidity measurements were carried out by the fluorescence polarization method using diphenylhexatriene as a probe. The degree of association of spectrin with erythrocyte membrane and its association state were analyzed as reported in (Liu, S.C. et al. 1984). ^{31}P-NMR measurement was carried out by the procedure described.

RESULTS AND DISCUSSION

Keshan disease —an endemic mitochondrial cardiomyopathy in China

1.Changes in the activities of respiratory enzyme

*This work was partly supported by the National Natural Science Foundation of China (NSFC)

Table 1 shows that striking changes in the activities of respiratory enzymes were found in myocardial mitochondria of Keshan disease patients. However, the spctrum of reduced-minus-oxidized cytochromes showed no obvious difference in cytochrome c oxidase (aa_3) content between patients and controls (not shown). This may indicate that the decrease in cytochrome c oxidase activity in Keshan disease patients is not due to a biosynthetic defect but to some subtle change of structure correlated to enzymic activity.

Table 1. Changes in succinic oxidase, succinate dehydrogenase and cytochrome c oxidase activities of myocardial mitochondria from patients with KD

Samples	Cases	Succinic oxidase		Succinate dehydrogenase		Cytochrome c oxidase	
		Activity nmol O_2 per min and per mg protein	Comparison vs control	Activity μ mol succinate oxidized per min and per mg protein	Comparison vs control	Activity K per min and per mg protein	Comparison vs control
Control	9	61.4±6.3*		0.64±0.01		8.73±0.78	
Subacute Keshan disease	7	20.6±3.2	p <0.001	0.28±0.03	p <0.001	3.37±0.49	p <0.001
Chronic Keshan disease	4	36.8±5.0	p <0.001	0.31±0.55	p <0.001	4.87±0.35	p <0.001

* $\bar{x}$ ±S.D.

2. Changes in H^+-ATPase activity and sensitivity to oligomycin in patient mitochondria

It is well known that H^+-ATPase is a crucial enzyme complex for the coupling of oxidative phosphorylation. Table 2 shows that mitochondrial H^+-ATPase activity in subacute patients was significantly lower than in controls ($p<0.01$) or in chronic Keshan disease patients ($p <0.05$). On the other hand, oligomycin sensitivity of H^+-ATPase was decreased in both subacute and chronic Keshan disease patients.

Table 2. Changes in H^+-ATPase activity as well as its sensitivity to oligomycin of patient myocardial mitochondria

Samples	Cases	Activity		Oligomycin sensitivity	
		μ mol ATP hydrolyzed per min and per mg protein	comparison vs control	Inhibition (%)	comparison vs control
Control	9	0.440±0.079		77.64±5.11	
Subacute Keshan disease	7	0.157±0.019	p <0.01	44.80±4.33	p <0.001
Chronic Keshan disease	4	0.335	NS	52.71	p <0.05

3. Changes in lipid fluidity of myocardial mitochondrial membrane from patients with Keshan disease

It is recognized that the activities of membrane enzymes are influenced by the physical state of the surrounding phospholipid. The lipid fluidity of myocardial mitochondria membrane in subacute KD patients appears to be decreased. However, no obvious difference could be detected between chronic patients and controls.

Table 3. Lipid fluidity of patient myocardial mitochondria membrane

Samples	Cases	DPH fluorescent polarization p	Comparison vs control
Control	8	0.142±0.007	
Subacute Keshan disease	7	0.176±0.011	p <0.02
Chronic Keshan disease	4	0.150±0.014	NS

4. Se content of patient myocardial mitochondria

The results shown in Table 4. indicate that the Se content of myocardial mitochondria from subacute Keshan disease patients was 1/8 of the control ($p <0.001$). In chronic patients the Se content was about 1/4 of the control. Thus, there seems to be a close correlation between Se deficiency and dysfunction of myocardial mitochondria in patients with Keshan disease.

Table 4. Se content in myocardial mitochondria from patients with Keshan disease

Samples	Cases	Se content μg per mg protein	Comparison vs control
Control	9	1.59 ± 0.053	
Subacute Keshan disease	6	0.175 ± 0.085	p <0.001
Chronic Keshan desease	4	0.439 ± 0.096	p <0.001

Keshan disease may be classified as a "mitochondrial cardiomyopathy" not linked to genetic deficiencies, but resulting from malnutrition through some unknown pathogenic factors.

Changes in the erythrocyte membrane from patients with latent KD

1. Change in the tightness of association of spectrin with erythrocyte

It can be seen in Table 5 that the degree of tightness of spectrin linked with erythrocyte from patients with latent KD is less than that of children from families of staffs in Forestry Bureau in endemic region.

Table 5. The Degree of Tightness of Spectrin linked with Erythrocyte Membrane

Sample	number of Samples Determined	Spectrin/Band 3	t-test (Comparison of latent KD with control)
(1) Control in endemic region	14	0.738 ± 0.103	(2)—(1) NS
(2) Patients with latent KD	6	0.787 ± 0.081	(2)—(3) p <0.05
(3) Children from families of staffs in Forestry Bureau in endemic region	20	0.831 ± 0.099	(1)—(3) p <0.05

2. Change in the association state of spectrin

It was shown that SP-O/SP-D of the patients with latent KD was less than that of the control (Tab. 6) and much less than that of the children from families of staffs in Forestry Bureau in endemic region.

Table 6. The Association State of Spectrin in Erythrocyte membrane

Sample	Number of Sample Determined	SP-O/SP-D	t-test (Comparison of latent KD with control)
(1) Control in endemic region	14	6.65 ± 0.90	(2)—(1) p <0.01
(2) Patients with KD of latent type	6	4.67 ± 1.60	(2)—(3) p <0.001
(3) Children from families of staffs in Forestry Bureau in endemic region	20	7.04 ± 1.01	(1)—(3) p <0.05

3. Study on the lipid-protein interaction of erythrocyte membrane from patients with latent KD by using ^{31}P-NMR method

The result shows that the value of chemical shift anisotropy of ^{31}P-NMR $\Delta\sigma$ (40.5 ppm) of erythrocyte membranes from patients with latent KD is higher than that of the control (38.8 ppm). It is likely that Se deficiency may be related to the abnormalities of patients' erythrocyte membranes as well. Presumably, other factors in addition to Se deficiency are required in the pathogenesis of Keshan disease.

We are greatly indebted to Mr. Niu Cunlin and other medical doctors from Chu-xiong Institute of Prevention and Cure of Endemic Diseases for their cooperation in the present research.

REFERENCES

Tan, J.A., et al., (1987): Se ecological chemico-geography and endemic Keshan disease and Kaschin-Beck disease in China. In Selenium in Biology and Medicine, eds G.F. Combs, Jr.; J.E. Spallholz; O.A. Levander; J.E. Oldfield, Van Nostrand Reinhold, pp 859-876. New York.

Yang, F.Y. and Wo. W.H. (1987): Role of Se in stabilization of human erythrocyte membrane skeleton. Biochemistry Intern. 15: 475-478.

Metal Ions in Biology and Medicine. Eds. Ph. Collery, L.A. Poirier, M. Manfait, J.C. Etienne. John Libbey Eurotext, Paris © 1990, pp. 162-164

The role of magnesium on insulin secretion and insulin sensitivity

V. Durlach, H. Grulet, A. Gross, J.M. Taupin, M. Leutenegger

Medical Clinic, Robert Debré Hospital, Reims Medical University, rue Alexis Carrel, 51092, Reims, France

Insulin secretion after glucose binding on its beta islet cell receptor, involves coupling phenomenons such as ATP production secondary to intracellular glycolysis, closing of ATP dependant potassium channels, opening of voltage dependant calcium channels with a raise of intracellular calcium (6). Cellular action of the hormone after binding to its specific receptor is followed by complex phenomenons where calmodulin phosphorylation and raise in cytosolic calcium can be partially implicated (5).

Intracellular magnesium is essentially linked to cell membranes, it controles the Na+/K+ dependant ATP ase activity and behaves mainly as a calcium channel blocker. (4,14). Thus Magnesium deficit (MD) can be responsible for a stimulation of insulin secretion as shown in animal models (1,3,12), even in experimental stress conditions where insulin remains unuasually high(8) ; conversely high magnesium levels suppress insulin production. Nevertheless the type (oral, parenteral), the dose (pharmacological, physiological) and the duration (acute, chronical) of MD have to to be taken into consideration to interprete insulin modification ; a prolonged chronical MD induced hyperinsulinism can thus exhausts itself and lead to insulinopenia in the animal (3) . Insulin sensitivity can also be altered by magnesium modifications, however the peripheral site of action of the ion (receptor, post-receptor) is still on discussion (15) ; both insulin-like (through a raise of intracellular calcium) and/or diabetogenic effects (through a reduction of trans-phosphorylation reactions) have been described.

In healthy man the diabetogenic effect seems to be predominant : nutritionnal changes such as saccharose and lipid loads both reduce the number of insulin erythrocyte receptors (10) and aggravates MD. An impaired glucose tolerance is noted in severe

hypomagnesemic patients which is reversible with the correction of hypomagnesemia (9) with no modification of the number and affinity of insuline erythrocyte receptors.This suggests a post receptor effect, although the erythrocyte could not be considered as a main target of insulin.
Besides clinical studies (2,7,11,13) converge to make insulin dependent and non insulinodependant diabetes responsible for secondary MD; in insulin dependant diabetic patients a magnesium oral load reduces exogenous insulin needs (11).
Then as far as insulin secretion is conserved a vicious circle can be created : MD induces an impaired glucose tolerance (excessive insulin release and insulin resistance) which can create a secondary deficit . Nevertheless further study are necessary to confirm this data.

REFERENCES

1 - Andersen, T., Berggren, P.O. et al(1982) : Amount and distribution of intracellular Magnesium and Calcium in pancreatic beta. cells. Acta Physiol. Scand. 114 : 235-241.

2 - De Leeuw, I. (1989) : Magnesium status in diabetes mellitus : old and new facts. Magnesium in health and disease. Y. Itokawa and J. Durlach. J. Libbey Ed : 343-348.

3 - Durlach J. , Rayssiguier, Y. (1983) : Données nouvelles sur les relations entre magnésium et hydrates de carbone. Magnésium 2 : 174-191

4 - Durlach J. (1980) : Les contrôles neuro-hormonaux du métabolisme du magnésium et leurs conséquences cliniques. Revue Fr. Endocr. Clin. 21 : 507-524.

5 - Kahn, G.R., White, M.F. (1988) : The insulin receptor and the molecular mecanism of in of insulin action. J. Clin. Invest. 82 : 1151-1156.

6 - Meglasson, M.D., Matschinsky, F.M. (1986) : Pancreatic islet glucose metabolism and regulation of insulin secretion. Diab. Metab. Rev. 2 : 163-214.

7 - Paschen. K., Bachem, M.G. et al (1981) : Magnesium stoffwechsel beim diabetes mellitus. Magnesium Bull. 3 : 307-313.

8 - Porta. S, Ersewhuber, W. et al (1990) : Reversed stress response of the endocrin rat pancreas in Mg++ depletion. Magnesium Res. 3,1 : 51.

9 - Rapado. A, Rovina, A. et al (1990) : Insulin resistance with patients with hypomagnesemia. Magnesium Res. 3,1 : 56.

10 - Rizkalla, S.W. ,Le Bouc, Y. et al (1982) : Influence des facteurs nutritionnels sur la liaison de l'insuline aux

récepteurs érythrocytaires. Journée de Diabètologie de l'Hôtel Dieu : 53-59 (Flammarion. Paris).

11 - Sjoegren, A. (1989) : Magnesium deficiency in diabetes mellitus : metabolic consequences, and effects of treatment with magnesium hydroxyde. Magnesium in health and disease. Y. Itokawa and J. Durlach. J. Libbey Ed : 355-362.

12 - Taljedal, I.G. (1974) : Interaction of Na++ and Mg++ with Ca++ in pancreatic islets as visualized by chlortetracycline fluorescence. Biochim. Biophys. Acta. 372 : 154-161.

13 - Vanroelen, W.F., Vangaal , L.F. et al (1985) : Serum and erythrocyte Mg level in type I and type I and type II diabetics. Acta. Diabetol. Lat. 22.

14 - Wollheim, CB. ; Sharp. (1981) : Regulation of insulin release by Ca. Physiol. Rev. 61 : 914-973

15 - Yajnik, C.S., Smith R.F. et al (1984) Fasting plasma Mg concentrationand glucose disposal in diabetes. Br.Med.J. , 228, 1032-1034.

Metal Ions in Biology and Medicine. Eds. Ph. Collery, L.A. Poirier, M. Manfait, J.C. Etienne. John Libbey Eurotext, Paris © 1990, pp. 165-167

Influence of magnesium supplementation on atherogenic risk factors

K. Kisters, W. Zidek, C. Karoff, K.H. Rahn

Medizinische Poliklinik der Universität 4400 Münster, Albert-Schweitzer-Str. 33, West Germany

Summary

In the present study the effect of oral magnesium supplementation on serum lipids in patients with hyperlipidemia of Frederickson type IV and IIb and on blood pressure was examined. In 65 patients with normal renal function, on cholesterol-poor and calorie restricted diet, blood pressure, serum cholesterol, triglyceride, HDL- and LDL- cholesterol as well as plasma and intracellular magnesium concentration were measured before and 4 weeks after therapy. 35 patients received 500 mg Mg^{++} daily additionally.
The results of our study show that magnesium supplementation in addition to the usual dietary measures is beneficial with regard to serum triglycerides (values, means$\pm$SD, decreased from 197,57$\pm$47,12 to 163,80$\pm$40,73, $p<0,05$), but exerts no positive effect on blood pressure and serum cholesterol.
Furthermore intracellular magnesium concentration increased significantly under magnesium supplementation (values, means$\pm$SD, increased from 1,73$\pm$0,23 to 1,90$\pm$0,19, $p<0,05$).

Introduction

Animal experiments have shown that magnesium deficiency causes atherogenic alterations in blood lipid composition (Rayssiguier, 1981; 1984).
Furthermore, some of the most common causes of magnesium deficiency are diabetes mellitus, long-term treatment with loop- and/or thiazide diuretics, and chronic alcoholism (Flink, 1981; Kisters et al., 1989; Ryan et al., 1981); it is well established that these conditions are associated with atherogenic alterations in blood lipid composition (Grimm et al., 1981).
For this reasons, magnesium deficiency may influence blood lipid composition and furthermore repletion of this deficiency could change the blood lipid composition such as to protect against atherosclerosis. To test this hypothesis our study was performed.

Methods

In 65 patients with normal renal function, on cholesterol-poor and calorie restricted diet blood pressure, serum cholesterol, triglyceride, HDL- and LDL- cholesterol as well as plasma and intracellular magnesium concentration were measured before and 4 weeks after therapy. 35 patients received 500 mg Mg^{++} daily additionally. Plasma magnesium concentrations and intracellular (red blood cells) magnesium were determined by atomic absorption spectroscopy, using a Video 12 apparatus (Thermo Electron).

Results

The following table shows the data of 65 patients without and with magnesium substitution before and after treatment.

		Mg-		Mg+	
blood pressure (mm Hg)	b	130,0±11,6/82,0±3,5		133,2±16,8/81,8±3,7	
	a	121,0± 6,5/80,0±2,5		123,2±12,3/78,6±5,1	
cholesterol (mg/dl)	b	286,44±27,55	*	282,03±35,24	*
	a	238,11±41,71		233,71±22,17	
triglyceride (mg/dl)	b	186,89±66,15		197,57±47,12	*
	a	187,44±64,25		163,80±40,73	
HDL-cholesterol (mg/dl)	b	57,33±15,61		49,43±12,60	
	a	50,88±10,73		45,66±10,03	
LDL-cholesterol (mg/dl)	b	168,33±52,36	*	206,11±35,26	*
	a	134,56±49,14		167,74±55,89	
plasma Mg^{++} (mmol/l)	b	0,88± 0,14		0,85± 0,12	
	a	0,86± 0,13		0,99± 0,15	
intracellular Mg^{++} (mmol/l)	b	1,80± 0,17		1,73± 0,23	*
	a	1,84± 0,14		1,90± 0,19	

Tab. 1. Means±SD of blood pressure, cholesterol, triglyceride, HDL- and LDL-cholesterol, plasma and intracellular magnesium concentrations of 30 patients without (Mg-) and of 35 patients with (Mg+) magnesium supplementation before (b) and after (a) therapy (cholesterol-poor and calorie restricted diet). *=$p<0,05$.

The results show that a magnesium supplementation in addition to the usual dietary measures is beneficial with regard to serum triglycerides, but exerts no positive effect on blood pressure and serum cholesterol. Furthermore intracellular magnesium concentration increases significantly under magnesium substitution.

Discussion

This sdudy indicates that long-term peroral magnesium supplementation changes blood lipid composition. In addition to dietary measures magnesium supplementation causes a significant decrease in triglycerides, whereas no positive effect on

serum cholesterol and blood pressure was noted in patients with normal renal function and hyperlipidemia of Frederickson type IV and IIb.
Only a few studies have evaluated the effect of peroral magnesium therapy on blood lipid composition in humans. Haywood and Sylvester, 1962, found that treatment over a period of 19 months with a magnesium combined therapy lowered the ß and α lipoproteins. Davis et al., 1984, found that treatment of 16 patients with hyperlipidemia for a period of 118 days significantly reduced total cholesterol.
The physiologic and biochemical background of the effect of an oral magnesium supplementation has not beeb studied in detail yet. Two enzymes, both of which are essential in lipid metabolism may be involved, namely the lecithin cholesterol acyl transferase (LCAT) and the lipoprotein lipase. From animal studies, it has been suggested that magnesium is an important cofactor for both of these enzymes (Gueux et al., 1984).
This study has shown that peroral magnesium therapy can alter blood lipid composition and therefore be of use in reducing atherogenic risk factors, but further investigations, especially with regard to a long-term magnesium therapy, still remain to be settled.

References

Davis, W.H., Leary, W.P., Reyes, A.J. (1984): Monotherapy with magnesium increases abnormally low high density lipoprotein cholesterol: a clinical assay. In Curr. Ther. Res. 36: 341-346.

Flink, E.B. (1981): Magnesium deficiency: etiology and clinical spectrum. In Acta Med. Scand. 169(Suppl. 647): 125-137.

Grimm, R.H., Leon, A.S., Hunningshake, D.B. (1981): Effects of thiazide diuretics on plasma lipids and lipoproteins in mildly hypertensive patients. In Ann. Intern. Med. 94: 7-11.

Gueux, E., Alcindor, L., Rayssiguier, Y. (1984): The reduction of plasma lecithine cholesterol acyl transferase activity by magnesium deficiency in the rat. In J. Nutr. 114: 1479-1483.

Haywood, J., Sylvester, R. (1962): Effect of oral magnesium and potassium on serum lipids. In Clin. Med. 10: 67.

Kisters, K., Zidek, W., Rahn, K.H. (1989): Wirkung von Trichlormethiazid und Amilorid auf die zelluläre Na^+-, K^+- und Mg^{2+}- Konzentration. In Schweiz. med. Wschr. 119: 1837-1839.

Rayssiguier, Y., Gueux, E., Weiser, D. (1981): Effect of magnesium deficiency on lipid metabolism in rats fed a high carbohydrate diet. In J. Nutr. 111: 1876-1883.

Rayssiguier, Y. (1984): Role of magnesium and potassium in the pathogenesis of arteriosclerosis. In Magnesium 3: 226-238.

Ryan, M.P., Ryan, M.F., Counihan, T.B. (1981): The effect of diuretics on lymphocyte magnesium and potassium. In Acta Med. Scand. 169(Suppl. 647): 53-61.

Metal Ions in Biology and Medicine. Eds. Ph. Collery, L.A. Poirier, M. Manfait, J.C. Etienne. John Libbey Eurotext, Paris © 1990, pp. 168-171

Relationship between plasma and erythrocyte magnesium concentrations in dairy cattle

C.M. Mulei

University of Nairobi, Department of clinical studies, PO Box 29053, Nairobi, Kenya

Introduction

There is conflicting data on the relationship between plasma magnesium (PMg) and erythrocyte magnesium (EMg) concentrations. The concentrations of PMg and EMg in circulation have been said to be independent of each other, that is, there is no exchange of Mg ions across the red cell membrane (Salt 1950, McAleese et al 1961. Wallach et al, 1962, Hellerstein et al. 1970, Ross et al. 1980). A two to fourfold increase in PMg levels produced by intravenous magnesium infusion were not found to be accompanied by any changes in EMg levels (McAleese et al. 1961, Wallach et al. 1962). On the other hand, marked uptake of Mg by theerythrocytes following intravenous Mg injection has been reported (MacDonald et al 1959). However, most date indicate that htere is very little uptake of Mg by the erythrocytes (Zumoff et al. 1958, Aikawa et al. 1960, Ginsburg et al. 1965, Hilmy and Somjen 1968).
It has been suggested that EMg concentration reflect the magnesium status of an animal at the time of erythropoiesis (Tufts and Greenberg 1937, Salt 1950, Hellerstein et al. 1970, Elin et al. 1971), that is, erythrocytes produced during a period of low Mg status should have low Mg values and vice versa. In addition EMg values might also reflect the age of the erythrocytes in circulation at the time of blood collection (Bernstein 1959, Geinsberg et al. 1962). Young erythrocytes are richer in Mg that mature erythrocytes (Bernstein 1959, Ginsberg et al. 1962, Bunn et al. 1971), therefore their number in circulation could influence the EMg concentration. However, unless there is increased erythropoietic activity which provokes an increased liberation by the bone marrow of immature erythrocytes (Bernstein 1959, Ginsburt et al. 1962) the EMg values should reflect the Mg status of the animal at the time of erythropoiesis (Hellerstein et al. 1970, Elin et al. 1971).

Materials and Methods

The experimental animals and designs, and procedure for determination of erythrocyte and plasma magnesium concentrations of the four experiments discussed in this paper have been described (Mulei and Daniel 1988 a, b, c,). The four experiments were:- (1) a fourty hour starvation period of ten yearling dairy heifers, (2) feeding of low magnesium diet to seven young dairy calves, (3) intramuscular injection of 100 iu of ACTH to 10 dairy heifers and (4) intramuscular injection of a 20% solution of magnesium sulphate to seven young dairy calves.

Results

The results are shown in figure 1. The mean PMg concentrations decreased significantly ($P<0.05$) during starvation the lowest level recorded being at 16 hours after reintroduction of feed (Fig. 1). The EMg concentration did not show any significant change during starvation (Fig. 1). Plasma Mg concentration decreased significantly in the calves fed on low Magnesium diet and showed no significant changes in the control calves (Fig. 1). Similarly the EMg concentrations decreased significantly in the calves fed on low Mg diet ($P<0.01$) and the control ($P<0.05$) calves (Fig. 1). The decrease was similar in both groups up to the fourth week and then slowed down in the control group.

Following ACTH administration concentration of PMg decreased significantly ($P<0.05$) while that of EMg did not show any significant change (Fig. 1). Within 4 hours after Mg load there was a significant increase in PMg concentration in both normomagnesaemic ($P<0.05$) and the hypomagnesaemic ($P<0.01$) calves (Fig. 1). However, no significant changes in EMg concentration was observed in both groups following the Mg load (Fig. 1).

Discussion

Changes in PMg concentrations in cattle have been reported during starvation (Robertson _et al_. 1960, Herd 1966), following ACTH administration (Wegner and Stott 1972) and as a result of low Mg dietary intake (Economides _et al_. 1973, Herdt 1981).

The decrease in PMg concentrations during starvation could have been partly due to absence of Mg absorption from the gut into circulation (Herd 1966) and partly due to chelation of Mg by free fatty acids (FFA) (Flink _et al_. 1973). Increased production of FFA during starvation (Tilakaratne _et al_. 1980) could lead to a decrease in PMg concentration following the tissue uptake of FFA - Mg complex (Rayssigner 1977). Administration of ACTH to an animal induces release of aldosterone as well as glucocorticoids from the adrenal cortex (Crabble _et al_. 1959), Mulrow 1967). The released aldosterone stimulates renal secretation of Mg ions (Mulrow 1967, Massry 1981) which then could result in decreased PMg concentration. Regulation of plasma Mg concentration is not under any rigid homeostic machanism (Herdt and Stevens 1981). Therefore, animals on low Mg dietary intake will have decreased PMg concentrations. Likewise administration of Mg would lead to an increase in PMg concentrations.

The lack of any significant changes in EMg concentrations in response to the short term changes in PMg concentration would indicate that erythrocyte membrane permitted only small diffusion of magnesium. This would be in agreement with other observation (e.g. Wallach _et al_. 1962, Ginsburg _et al_. 1962, Aikawa 1965). In acute Mg deficiency EMg levels do not fall as quickly as does PMg levels (Tufts and Greenberg 1932, Salt 1950). Similarly acute two to fourfold increases in PMg levels induced by intravenous Mg infusions were not accompanied by changes in EMg levels (Wallach _et al_. 1962). Hemodialysis of hypermagnesaemic patients with Mg free or low dialysates also had very little or no effect on EMg levels (Seelig 1980).

The decrease in both PMg and EMg concentrations as a result of feeding low Mg diet indicate that PMg and EMg values would be expected to decrease in cases of prolonged Mg deficiency. The erythrocytes formed during the period of low Mg status will have low Mg values (Tufts and Greenberg 1937, Salt 1950, Hellerstein _et al_. 1970, Elin _et al_. 1971).

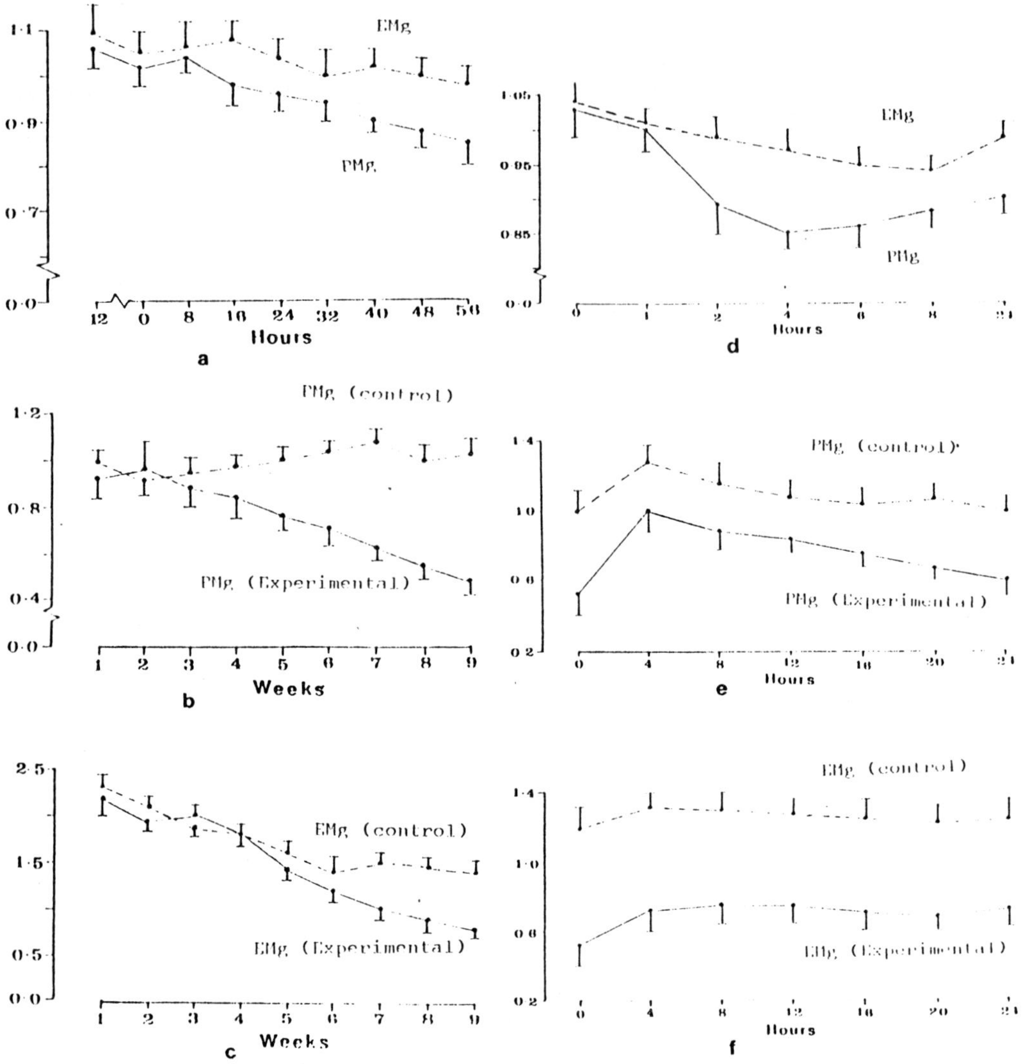

Fig. 1. Changes in EMg and PMg (mmol/l) during (a) starvation (b) and (c) feeding of low Mg diet (d) after ACTH administration and (e) and (f) after $MgSO_4$ administration.

References

Aikwa, J. K. 1960: Proc. Soc. Exp. Biol. Med. 104:461-463
Aikwa, J. K. 1965: In "Electrocytes and Cardiovascular Diseases". 1. pp 9-27.
Bernstein R. E. 1959: J. CLin. Invest. 38: 1572-1585.
Bunn, H. F., Ransil, B. J. and Chao A. 1971: J. Biol. Chem. 246: 5273-5279.
Crabble, J., Reddy, W. J. Ron, E. J. and Thorn G. W. 1957: J. CLin Endocr. Metab. 19:1185-1191.
Economides, S. J., Miller, T. B., Topps, J. H., Gelman, A. L. and Keith G. 1973. Br. Vet. J. 129:63-71.
Elin, R. J. Armstrong, W. D. and Singer, L. 1971: Am. J. Physiol. 220: 534-548.
Flink, E. B., Flink, P. F., Shane, S. R., Jones, J. E. and Steefs, P. E. 1973: Clin. Res. 21:880-891.
Ginsburg, S., Smith, J. G., Ginsburg, F. M. Readon, J. Z. and Aikwa, J. K. 1962: Blood. 20: 723-729.
Hellerstein, S., Spees, W. and surapathana, L. D. 1970: J. Lab. Clin. Med. 76: 10-24.
Henrotte, J. G. 1982: magnesium bull. 1:69-80.
Herd, R. P. 1966: Aust. Vet. J. 42. 269-272.
Herdt, T. H. and Stevens, J. B. 1981: Continuing education. Vol. 3. No. 1. pp 532-543.
Hilmy, M. G. and Somjen, G. G. 1968: Am. J. Physiol: 214: 406-413.
MacDonald, D. C., Care, A. D. and Nolan, E. 1959: Nature. 183: 736-737.
Massry, S. G. 1981: Magnesium Bull. 3 (1a) : 277-281.
McAleese, D. M., Bell, M. C. and Forbes, M. 1961; J. Nutr. 74: 505-514.
Mulei, C. M. and Daniel, R. C. W. 1988a: Vet. Res. Comm. 12: 289-293.
Mulei, C. M. and Daniel, R. C. W. 1988b: J. Vet. Med. A. 35: 516-521.
Mulei, C. M. and Daniel, R. C. W. 1988c: Indian J. dairy Sci. 41:53-56
Mulrow, P. J. 1967: In "The adrenal cortex" ed. Elsenstein, A. B. Little brown and Co. Boston pp. 293-313.
Rayssinger, J. 1977: Horm. and melab. Res. 9: 309-314,
Robertson, A., Paren, H. Barden, P. and Marr, T. G. 1960: Res. Vet. Sci. 1:117-124.
Ross, R. S. Seeling, M. S. and Berger, A. R. 1980: In "Magnesium in Health and Disease". pp. 365-369.
Salt, F. J. 1950: Lab. J. 8.357-367.
Seeling, M. S. 1980: Magnesium deficiency in the pathogenesis of disease. Plenum publiching corporation. p. 364.
Tilakaratne, H., Osmond, T. J., Carr, W. R., Alliston, J. C. Anim. Prod. 30: 327-340.
Tufts, E. V. and Greenberg, D. M. 1937: J. Biol. Chem. 122:693-727.
Wallach, S., Cahill, L. N., Rogan, F. H. and Jones, H. L. 1962: J. Lab. Clin. Med. 59:196-210.
Wegner, T. N. and Stott, G. H. 1972: J. Dairy Sci. 55. 1464-1469.
Zumoff, B. Bernstein E. H., Imarisio, J. J. and Hellman, L. 1958: Clin. Res. 6:260.

Metal Ions in Biology and Medicine. Eds. Ph. Collery, L.A. Poirier, M. Manfait, J.C. Etienne. John Libbey Eurotext, Paris © 1990, pp. 172-176

Mg^{2+} metabolism under diuretic treatment with the loop diuretic piretanide

K. Kisters, W. Zidek, K. Fehske, A. Kwapisz, K.H. Rahn

Medizinische Poliklinik der Universität 4400 Münster, Albert-Schweitzer-Str. 33, West Germany

Summary

Whereas changes of plasma Mg^{2+} concentration under diuretic treatment have often been described, comparatively few data on intracellular Mg^{2+} concentration are known. Therefore it was of interest to study the effect of the loop diuretic piretanide (6 mg/d) on intracellular free and total Mg^{2+} concentration in red blood cells in 18 patients with essential hypertension. Measurements were performed in red blood cells by atomic absorption spectroscopy. Free and total Mg^{2+} concentrationstions increased significantly ($p<0,05$) during diuretic treatment. Furthermore intracellular free Ca^{2+} decreased significantly. Intracellular Na^+ and K^+ concentrations did not change significantly. The results show that under diuretic treatment with the loop diuretic piretanide no losses of cellular Mg^{2+} occur.

Key words: Mg^{2+}, diuretic, hypertension, intracellular electrolytes

Introduction

The fundamental importance of magnesium to biological systems is well documented. Since magnesium acts as a co-factor in many enzyme-catalyzed reactions, magnesium deficiency in humans may result in nervous, gastrointestinal and muscular symptoms. Furthermore Mg^{2+} depletion can be associated with an increased frequency of cardiac arrhythmias, which is of particular clinical importance (Levine et al., 1982; Loeb et al., 1988; Topol & Cerma, 1983). It is well known that diuretic treatment can result in Mg^{2+} depletion, since many diuretics inhibit the reabsorption of filtered magnesium in the proximal and distal tubule (Dyckner & Wester, 1984). Changes in plasma magnesium concentrations have often been described (Brendan & Milligran, 1987; Fehske et al., 1989; Kisters et al., 1989), but comparatively few data on cellular Mg^{2+} concentration under diuretic treatment are known.
Therefore we studied the effect of the loop diuretic piretanide on intracellular electrolytes, especially on free and total Mg^{2+} concentration.

Patients and methods

Patients: In 18 patients (10 male, 8 female) with essential hypertension, who gave informed consent to participate in the study, intracellular Na^+, K^+ and Mg^{2+} concentrations and intracellular free Ca^{2+} and Mg^{2+} concentrations were measured in red blood cells before and 6 and 12 weeks after starting treatment with 6 mg piretanide daily.

Methods: 10 ml heparinized blood were drawn and centrifuged with 2500 g. Then plasma was removed. The red blood cells were washed twice with an isotonic lithium acetate solution. Measurements were performed in the hemolysate after freezing and rethawing the red blood cell pellet. Intracellular total Mg^{2+} concentrations in red blood cells were measured with atomic absorption spectroscopy, using the Thermo Jarrell Ash Video 12 apparatus. Calibration curves were established using solutions with known Mg^{2+} concentrations, in the lower, upper and intermediate range (Seronorm charge No 176, Merck, Pathonorm H charge No 21, Nyegaard, Pathonorm L charge No 20, Nycomed). For each sample a mean value was calculated from 3 measurements. All measurements were done in one series at room temperature. Intra-assay variability was 5,7% in 10 subsequent measurements. Intracellular free Ca^{+} and Mg^{2+} concentrations were measured with a Ca^{+} and Mg^{2+} selective electrode. Intracellular Na^{+} and K^{+} concentrations were measured by flame-photometry (Zidek et al., 1982).

Results

Table 1 shows mean values and standard deviations of blood pressure (syst./diast., in mm Hg), free and total intracellular Mg_i^{2+} concentrations (in mmol/l, * =p<0,05) before and after 6 and 12 weeks under treatment with the loop diuretic piretanide.

Tab. 1:

	before piretanide	after 6 weeks piretanide	after 12 weeks piretanide
blood pressure	147,5 ± 16,5 101,0 ± 8,8	139,5 ± 16,9/ 91,5 ± 9,4	132,0 ± 9,5/ 88,0 ± 7,9
free Mg_i^{2+}	0,36 ± 0,16	0,50 ± 0,19*	0,45 ± 0,21*
total Mg_i^{2+}	1,78 ± 0,16	1,90 ± 0,19*	1,98 ± 0,22*

Durind treatment with piretanide blood pressure decreased from 147,5 ± 16,5/101,0 ± 8,8 to 139,5 ± 16,9/91,5 ± 9,4 and 132,0 ± 9,5/88,0 ± 7,9 mm Hg after 6 and 12 weeks, respectively. Intracellular free and total Mg_i^{2+} concentration increased significantly(Table 1, * =p<0,05).
Intracellular Na^{+} and K^{+} concentrations did not change significantly (Figure 1, 2). In contrast intracellular free Ca^{2+} concentration decreased significantly (* = p< 0,05, Figure 3).

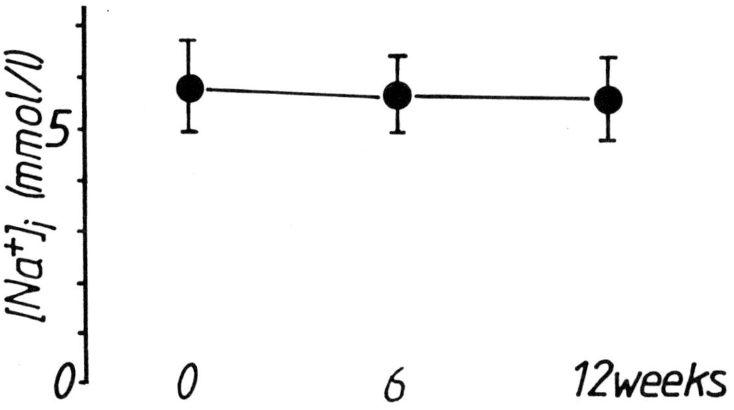

Figure 1. Mean values and standard deviations of intracellular sodium concentrations (in mmol/l) before and after 6 and 12 weeks during treatment with the loop diuretic piretanide.

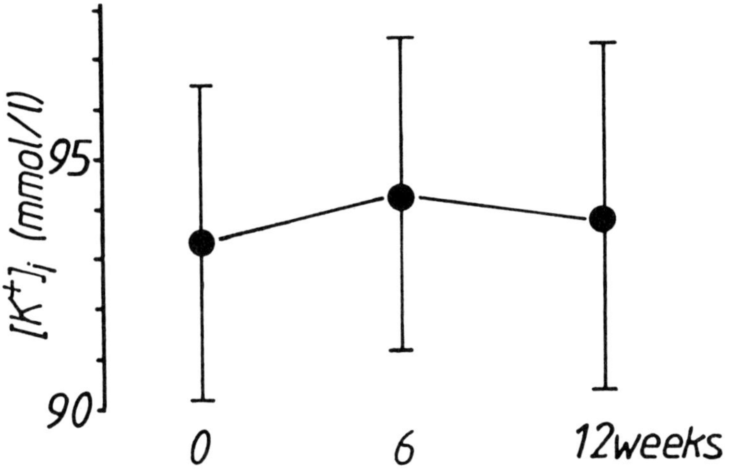

Figure 2. Mean values and standard deviations of intracellular potassium concentrations (in mmol/l) before and after 6 and 12 weeks during treatment with the loop diuretic piretanide.

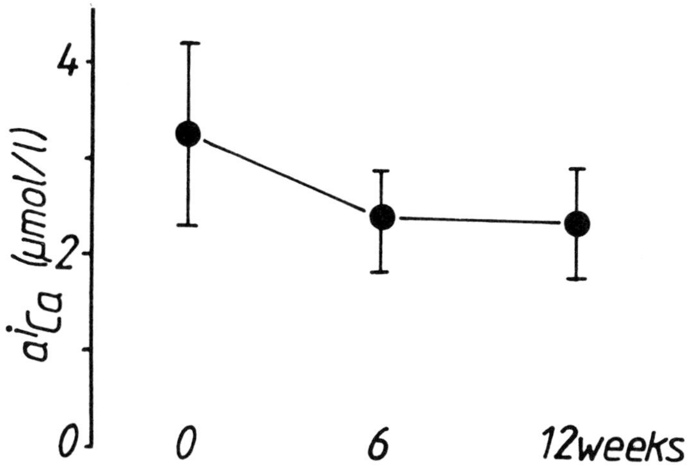

Figure 3. Mean values and standard deviations of intracellular Ca^{2+} activity (in µmol/l) before and after 6 and 12 weeks during treatment with the loop diuretic piretanide.

Discussion

In contrast to plasma Mg^{2+} concentration under diuretic treatment, intracellular Mg^{2+} concentration has not been studied in detail yet. Extracellular Mg^{2+} constitutes only about 3% of total body Mg^{2+}. Therefore plasma Mg^{2+} concentration may not be adequate for assessing the Mg^{2+} status of the body. To this purpose, Mg^{2+} determinations in blood cells may be more useful, although a satisfactory correlation between Mg^{2+} concentration in red blood cells and other tissues has not been established yet.
Whereas several authors found a decrease in intracellular Mg^{2+} concentration during treatment with thiazides (Hollifield, 1986), combined treatment with potassium-sparing diuretics and thiazides did not induce a decrease in intracellular Mg^{2+} concentration (Ryan, 1987). With respect to loop diuretics, divergent results were obtained. During treatment with furosemide a decrease in plasma and intracellular Mg^{2+} concentration was observed. In contrast a significant decrease in plasma Mg^{2+} concentration under diuretic treatment with the loop diuretic piretanide was not found (Verho et al., 1987).
The findings in the present study correspondend with the data by Verho et al., as a small, but significant Mg^{2+} sparing effect was found also with respect to intracellular magnesium.

References

Brendan, J.M., Milligran, K. (1987): Diuretic-associated Hypomagnesemia in the Elderly. In Arch. Intern. Med. 147: 1768-1771.

Dyckner, T., Wester, P.O. (1984): Intracellular Magnesium Loss After Diuretic Administration. In Drugs 28 (Suppl.1): 161-166.

Fehske, K.J.,Kisters, K., Kwapisz, A., Zidek, W. (1989): Cellular magnesium handling during treatment with piretanide. In III. International Conference on Diuretics, Mexico City 1989 (in press).

Hollifield, J.W. (1986): Thiazide Treatment of Hypertension. In Am. J. Med. 80 (Suppl. 4A): 8-12.

Kisters, K., Zidek, W., Rahn, K.H. (1989): Wirkung von Trichlormethiazid und Amilorid auf die zelluläre Na^{+}-, K^{+}- und Mg^{2+}- Konzentration. In Schweiz. med. Wschr. 119: 1837-1839.

Levine, S.R., Crowley, T.J., Hai, H.A. (1982): Hypomagnesemia and ventricular tachycardia. In Chest 81: 244-246.

Loeb, H.S., Pietras, R.J., Gunnar, R.M., Tobian, J.R. (1988): Paroxysmal ventricular fibrillation in two patients with hypomagnesemia; treatment by transvenous pacing. In Circulation 37: 210-216.

Ryan, M.P. (1987): Diuretics and Potassium/Magnesium Depletion. In Am. J. Med. 82 (Suppl. 3A): 38-47.

Topol, F.J., Cerma, B.B. (1983): Hypomagnesemia, Torsades de pointes. In Amer. J. Cardiol. 52: 1367-1368.

Verho, M., Irmisch, R., de Looze, S., Rangoonwala, B. (1987): Lack of Effect of Piretanide (A Potassium-Stable Diuretic) on Serum Magnesium. In Int. J. Clin. Pharm. Res. 7: 433-442.

Zidek, W., Losse, H., Dorst, K.G., Zumkley, H., Vetter, H. (1982): Intracellular sodium and calcium in essential hypertension. In Klin. Wschr. 60: 859-862.

Metal Ions in Biology and Medicine. Eds. Ph. Collery, L.A. Poirier, M. Manfait, J.C. Etienne. John Libbey Eurotext, Paris © 1990, pp. 177-179

Magnesium and calcium content of drinking water, fruit juices, salt and saffron of Greece

P. Tarantilis, S. Haroutounian, M. Polissiou

Agricultural University of Athens, lab. General Chemistry, Iera Odos, 75, 118 55 Athens, Greece

Introduction

The inverse correlation between water hardness and the development of several diseases has been studied thoroughly for a long period of time (Masiorini R.,1978). Recent studies has associated this beneficial effect with the presence and content of magnesium in hard water (Marier J.R.,1985;1986). More specifically numerous epidemiological studies have confirmed that high total daily magnesium intake plays crucial role for the prevention of development of various diseases as cardiovascular fatalities (Marier J.R.,1978), cancer (Durlach J. et al,1987), urolithiasis (Johansson G. et al,1982) etc. On the other hand it has been prouved that balance of Mg/Ca ratio is correlated with circulatory diseases (Hiraike et al,1989).

It has also been prouved that unlike calcium and other electrolytes (whose dietary daily requirements are obtained almost wholly from food), magnesium is not found in modern foods in sufficient amounts (Masiorini R.,1978). Although the RDA (Recommended Dietary Allowance) for magnesium is 300 - 500 mg/day or 6 mg/Kg day (Durlach J.,1989) at industrialized countries the daily magnesium intake from all sources (except water) is about 238 milligrams (Marier J.R.,1978). Thus the attention has been focused to the waterborn magnesium, which can make a sizeable contribution to the total daily intake.

Materials and Methods

The content of magnesium and calcium were determined by Atomic Absorption Spectrometry using a Perkin - Elmer model 380 spectrrophotometer. The samples were prepared by diluting drinking water and salt in deionised water. In the case of fruit juices and Saffron, the samples were prepared as follows: 20 g and 0.725 g respectively of these substances were calcinated with 5 ml HCl, 15 ml HNO_3 and 10 ml $HClO_4$ and subsequently were diluted in deionised water.

Results and Discussion

As a part of our efforts to optimize the amount of the daily magnesium intake in Greece, we have determined the magnesium and calcium content of drinking water of various cities of Greece, along with their concentration in most common bottled waters of Greece. The results, the molar ratio of Mg/Ca, as well as the estimated magnesium intake from drinking water per day, are presented in Table I.

Table I. Compositional diversity of water from different Greek cities and bottled waters.

Location (City)	Mg mg/l	Ca mg/l	Mg/Ca molar ratio	Mg intake from 2.4 l water/day[1] (mg)
Athens	10	54	0.30	24
Thessaloniki	12	60	0.33	28.8
Kerkyra (Corfu)	40	270	0.24	96
Kalamata	27	88	0.51	64.8
Chania (Crete)	9	40	0.37	21.6
Larissa	18	69	0.43	43.2
Loutraki (bottled)	86	5	29.5	206.4
Sariza (bottled)	17	116	0.57	40.8
Kimi (bottled)	24	50	0.34	57.6

[1] A daily water intake of 2.4 l (in all forms is in aggreement with the findings of ICRP (1975) and several researchers (Schroeder et al,1969; Spencer et al,1970).

It is note worthy that the Mg/Ca molar ratio of the drinking water for all parts of Greece, is relatively stable, varying from 0.24 - 0.57 and is independent from the partial concentrations of magnesium and calcium in water. On the other hand, except the bottled water of Loutraki and partially the water of Kerkyra (Corfu), the other drinking waters are not capable to fulfil the magnesium deficiency between the RDA value and the average daily intake.

In an attempt to determine the content of the said elements in other food sources, such as fruit juices and Saffron of Greece, we have observed that the balance of the molar ratio is entirely different. Thus we have found that the content of magnesium is 17.8 + 5.8 ppm in fruit juices and 565 mg/100 g in Saffron, while the content of calcium is 24.8 + 9 ppm and 150 mg/100 g, leading to molar ratio values of Mg/Ca to 1.18 and 6.22 respectively.

In order to suppress this deficiency and balance the magnesium daily intake with the RDA value, we have explored the possibility to use the table salt as a potential source of magnesium in the daily diet. Thus we have determined the loss of magnesium during the table salt refining/ processing. The results are depicted in Table II indicating that the magnesium was removed, almost quantitatevely, during this process (95% loss of magnesium).

Table II. Loss of magnesium during salt refining / processing.

	Magnesium mg/Kg weight	Magnesium lost, %
Crude salt (from salt sea)	6,000	-
after refining (before drying)	850	86
table salt	300	95

Research concerning the increase of the daily magnesium intake, in relation to the addition of mineral magnesium ($MgCl_2.6H_2O$ or $MgCO_3$) at the softening process of drinking water, as well as the refining/processing of the table salt is currently underway.

REFERENCES

Altura, B. M.; Altura, B. T. (1985): New perspectives on the role of magnesium in the pathophysiology of the cardiovascular system. Magnesium. 4: 226-244.

Durlach, J.; Rinjard, P.; Bara, M.; Guiet - Bara, A.; Collery, P. (1987): Donnees nouvelles sur les rapports entre magnesium et cancer. Magnesium, Physiologische Aspekte fur die Praxis. Eds. B. Lasserre and Petit - Lancy Panscientia verlag, pp. 26-45.

Durlach, J. (1989): Recommended dietary amounts of magnesium: Mg RDA[1]. Magnesium Res. 2: 195-203.

Hiraike, H.; Kimura, M.; Yokoi, K.; Sekine, K.; Ahmed, S. M.; Sata, F.; Itokawa, Y. (1989): Magnesium, Calcium and other minerals in human placenta. Magnesium Res. 2: 90-91.

Johansson, G.; Backman, U.; Danielson, B. G.; Fellstrom, B.; Ljunghall, S.; Wikstrom, B. (1982): Effects of magnesium hydroxide in renal stone disease. J. Am. Col. Nutr. 1: 179-185.

Marier, J. R. (1978): Cardio - protective contribution on hard water to magnesium intake. Rev. Can. Biol. 37 (2): 115-125.

Marier, J. R.; Neri, L.C. (1985): Quantihying the role of Magnesium in the interrelationship between Human mortality/morbidity and water hardness. Magnesium. 4: 53-59.

Marier, J. R. (1986): Magnesium Content of the food supply in the monder day word. Magnesium 5: 1-8.

Masiorini, R. (1978): How trace elements in water contribute to health. WHO Chronicle 32: 382-385.

Schroeder, H. A.; Nason, A. P.; Tipton, I. H. (1969); Essential metals in man - magnesium. J. Chrom. Dis. 21: 815.

Spencer, H.; Lewin, I.; Wiatrowski, E.; Samachsom, J. (1970): Fluoride metabolism in man. Am. J. Med. 49: 807.

Metal Ions in Biology and Medicine. Eds. Ph. Collery, L.A. Poirier, M. Manfait, J.C. Etienne. John Libbey Eurotext, Paris © 1990, pp. 180-182

Effects of chronic alcoholism on trace elements status. Influence of withdrawal program

P. Pirollet, F. Paille, M.F. Hutin**, A.M. Corroy, F. Nabet-Belleville*, D. Burnel**

*Centre d'alcoologie. Hôpital Fournier. CO n° 34, 54035 Nancy Cedex; * Laboratoire Central de Chimie A. Hôpital de Brabois, 54511 Vandœuvre; ** Service de Chimie Générale. Faculté de Médecine. B.P. 184, 54500 Vandœuvre, France*

INTRODUCTION

Trace elements are involved in alcohol metabolism, in the defense mechanisms against its toxic effects, and in the genesis of histologic lesions. A low serum concentration of Zinc (Zn) has long been observed in cirrhotic alcoholic patients. Many studies have been published concerning Zn metabolism in those patients, and later in all patients with chronic intake of alcohol without cirrhosis lesions. Copper (Cu) is suspected to take part in pathogenesis of hepatic histologic lesions as it described in Wilson's disease. Alcoholic beverages contain lead (Pb) and iron (Fe), and iron tissue overload is frequent. Glutathion peroxydase and selenium (Se) control free radicals intra-cellular pool. Excessive production of oxygen free radicals may lead to tissue damage. An excessive alcoholic consumption changes serum and tissue concentrations on many trace elements.
The aim of the present study is to measure the level of 10 trace elements in patients with chronic alcoholic consumption and to evaluate influence of withdrawal programm.

SUBJECTS AND METHODS

141 chronic alcoholics (117 men and 24 women) were studied before and after a 28 days weaning period in our detoxification unit of Nancy hospital. All patients were dependent on alcohol according to DSM III R dependance criteria. The mean age was 39 ± 9 years (from 20 to 70 years). None had severe somatic complications (chronic pancreatitis, encephalopathy, severe hepatic failure). The mean daily alcohol intake was 231 ± 161 g (from 38 to 1280 per day). Withdrawal program began immediately on admission in hospital unit and was respected all stay long. Tranquillizers and oral vitamins (vit. B1 750 mg/day ; vit. B6 750 mg/day) were systematically given. Patients had usual hospital meals. On admission and after 28 days of abstinence, the blood concentration of these trace elements were examined : Zinc (Zn), Copper (Cu), lead (Pb), Silicium (Si), Nickel (Ni), Manganese (Mn), Molybdene (Mo), Chrome (Cr). They were dosed by emission spectrophotometry on argon DCP plasma on serum or total blood for Pb. A colorimetric method was used for iron dosage, and spectrophotometry atomic absorption for selenium dosage. The results were compared with control subjects for Zn, Cu, Pb, Si and Fe. Usual hematologic, electrolytic, renal and hepatic tests were performed. Nutritional status was defined by retinol binding protein (RbP) and prealbumin dosage.

RESULTS

The main results are gathered in table I
On admission, Mo and Cr were undetected in all patients. Serum concentrations of Zn, Cu and Se were significantly lower in alcoholics when compared to controls (Zn : p = 0,0002 ; Cu : p = 0,001 ; Se : p = 0,03), whereas levels of Pb and Si were higher (Pb : p = 0,03 ; Si :

p = 0,003). Serum concentration of Fe was within the normal range. According to sex, Cu level was found higher and Pb level lower among women. Last, Zn level was significantly lower and Si level higher in patients with cytolytic hepatic lesions (ASAT ≥ 50 IU/l). A positive correlation was found between Zn, Mn and Se. Zn level was highly correlated with protein anabolism markers. On the opposite, there was a negative correlation between Cu and blood albumin levels. No correlation was established either between Pb blood concentration and ∂-amino-levulinique acid deshydase activity or between these trace elements and daily alcohol intake. However, Ni serum level was significantly lower in heavier drinkers (daily alcohol intake ≥ 200 g).
After one month of abstinence, Si and Fe levels had decreased (respectly p = 0,03, p < 0,0001), but those of Ni had increased (p = 0,002). Cu level had increased only in men (p = 0,02), whereas Zn/Cu ratio reduced. Zn and Se levels were lowered though this evolution was not significant. Mn and Pb levels remained unchanged.

	CONTROL		ADMISSION		AFTER ONE MONTH	
	n	level	n	level	n	level
Zn	40	1063±191	141	928,25±203,45	92	888,53±171,29
Cu	40	1164±436	141	1003,52±213,97	92	1037,58±183,24
Pb	50	221±51	138	256,43±115,26	92	268,81±118,60
Si	50	173±13	141	224,23±120,83	91	193,60±90,45
Se	125	71±11	54	45,22±14,08	39	43,11±11,67
Ni	/	/	54	3,02±1,79	30	4,70±1,78
Mn	/	/	48	25,79±2,97	30	25,73±2,88
Fe	/	1,30±0,35	124	1,37±0,52	112	0,86±0,25

Table I : Blood trace element levels in controls and alcoholics.
(serum values are mcg/l and mg/l for iron.)

DISCUSSION

The originality of this study is first based on the simultaneous study of 10 trace elements with similar subjects and then on their evolution after withdrawal program.
We confirm an important Zn level diminution during chronic alcoholism, even without cirrhosis (1, 4, 5, 6, 9). In fact, no hepatic biopsy for histologic analysis has been performed in those patients in so far as our recruitment and many previous studies made it useless. A few patients are cirrhotics, indeed, but without complications. In case of severe hepatic failure, however, Zn level gets lower. Even if nutrition is well balanced, Zn level keeps falling, even though, oral supplementation is quickly effective (10). The observation of markedly low levels of anabolism markers which is positively correlated to zinc in all alcoholics suggests that low serum zinc levels could be attributed to a decrease in circulating albumin. It has been suggested that the decreased circulating albumin levels enhance the binding of zinc to amino acids and other small molecular weight ligands which would facilitate its excretion in the urine.
A higher Cu level is observed in cirrhotic patients but Conri C et al. (2) could not find any significant variation of Cu blood level in chronic alcoholism. In our study, there is a significantly lower rate of Cu compared with control subjects.
In most authors, we notice a rise in lead level but no correlation is established either with delta-amino-levulinic-acid deshydrase (ALAD), or with daily alcohol intake or systemic arterial pressure. ALAD activity becomes apparent as an alcohol marker and not as a lead intoxication marker (3, 8). Its activity indeed, gets quickly higher after the weaning period without any significant alteration in the serum Pb level. The content in Pb of beverages, should influence serum lead level, but all our patients are heavy drinkers, and those lead level variations seem, thus, difficult to be analysed. Systolic and diastolic arterial pressure has significantly dropped during the weaning period and there is no correlation with serum Pb level (3).

We can state chronic alcoholism causes Se to decrease and it is more pronounced after withdrawal program (7).
We notice a significant diminution of Ni level in cirrhotic patients, and a normal Ni level in patients without cirrhosis. In this study no control subjects are available and our levels are as low as the sensitivity limits of the method. However it's worth noting that Ni serum level is significantly higher after 28 days of abstinence.
Silicium, initially higher, and much higher in case of hepatic cytolysis, reach the range of normal values after weaning.

CONCLUSION

This study points out that chronic alcoholism modifies most of the trace elements, especially Zn, Cu, Se (decreased), Si and Pb (increased). Owing to the high daily alcohol intake, we could not find any correlation between the latter and the levels of trace elements. Nonetheless, Zn is well correlated with protein energy markers and cytolysis. Thus, Zn requirements seem to be increased in all alcoholics, especially in those with severe liver disease. The evolution of serum trace elements levels during the first month of abstinence is variable : some of them tend to reach the range of normal values whereas others are more disturbed in spite of adequate dietary intake. These data suggest a metabolic origin for these alterations rather than a dietary deficiency.

REFERENCES

1.Atukorala, T.M.S. et al, (1986) : Zinc and vitamin A status of alcoholics in a medical unit in Sri Lanka. Alcohol Alcoholism 21, 269 - 275.
2.Conri, C. et al (1988) : serum and erythrocyte selenium, serum zinc and copper in alcoholics with or without abnormal liver fonction tests. Alcohol toxicity and free radical mechanisms. In advances in the biosciences, 71 ed. Pergamon Press, Oxford, 107 - 110.
3.Dally, S. et al (1986) : Elévation de la plombémie au cours de l'alcoolisme. Relation avec la pression artérielle. Presse Med, 15, 1227 - 1229.
4.McClain, C.J. et al (1983) : Zinc metabolism in alcoholic liver disease.Alcoholism Clin Exp Res, 10, 582 - 589.
5.Mills, P.R. et al. (1983) : A study of zinc metabolism in alcoholic cirrhosis. Clin Sci, 64, 527 - 535.
6.Riitta Hartoma, T. et al (1977) : Serum zinc and serum copper and indices of drug metabolism in alcoholics. Eur J Clin Pharmacol, 12, 147 - 151.
7.Thuong, T. et al (1986) : Sélénium sanguin et alcoolisme chronique. Rev Alcol , 31, 87 - 94.
8.Vives, J.F. et al (1980) : Alcoolisme chronique et intoxication saturnine. Gastroenterol Clin Biol, 4, 119 - 122.
9.Zarski, J.P. et al (1985) : Oligo-éléments (zinc, cuivre, manganèse) dans la cirrhose alcoolique : influence de l'alcoolisme chronique. Gastroenterol Clin Biol, 9, 664 - 669.
10.Zarski, J.P. et al (1987) : Evolution des concentrations sériques et tissulaires de zinc après supplémentation orale chez des alcooliques chroniques avec et sans cirrhose. Gastroenterol Clin Biol, 11, 856 - 860.

Metal Ions in Biology and Medicine. Eds. Ph. Collery, L.A. Poirier, M. Manfait, J.C. Etienne. John Libbey Eurotext, Paris © 1990, p. 183

Un laboratoire pharmaceutique pour les oligo-éléments

Laboratoires Labcatal, 7, rue Roger Salengro, BP 305, 92541 Montrouge Cedex, France

Introducteur, il y a 40 ans, de l'oligothérapie catalytique à doses modérées - les OLIGOSOLS® - LABCATAL développe également aujourd'hui l'oligothérapie à doses fortes comme le lithium en psychiatrie - NEUROLITHIUM® - et le zinc en dermatologie - RUBOZINC® -.
Le domaine de la nutrition le concerne également, et il a initié de vastes programmes d'études et d'enquêtes épidémiologiques.

Développer des spécialités pharmaceutiques irréprochables est une des vocations du Laboratoire LABCATAL. En effet, les oligo-éléments sont des médicaments à part entière.

Quatre préoccupations sont essentielles à LABCATAL : la recherche fondamentale, la collaboration hospitalo-universitaire, le goût de communiquer avec ses prescripteurs, le dialogue au niveau international.

Avec un chiffre d'affaires qui a progressé de plus de 300 % en dix ans, des implantations, non seulement dans les pays de la CEE, mais également au Canada, LABCATAL a un essor international qui le mène au premier rang des producteurs d'oligo-éléments. Son usine d'Annemasse a été restructurée entièrement selon les normes des Bonnes Pratiques de Fabrication et LABCATAL affirme par cette usine de haute technologie son dynamisme et son désir d'appliquer, dans le domaine des oligo-éléments, les techniques les plus performantes.

Entre 1953, date de mise sur le marché des premiers OLIGOSOLS®, et 1990 c'est une expérience consacrée aux métaux.

B

TOXIC METAL IONS : ALUMINIUM, ARSENIC, BARYUM, BERYLLIUM, CADMIUM, CERYUM, CHROMIUM, LEAD, MERCURY, NICKEL, TITANIUM, VANADIUM, YTTRIUM

I. Tissue distribution

Metal Ions in Biology and Medicine. Eds. Ph. Collery, L.A. Poirier, M. Manfait, J.C. Etienne. John Libbey Eurotext, Paris © 1990, pp. 189-191

Ion microprobe mass resolved imaging of metal ions in biological tissue

P. Hallegot[1], C. Girod[1], J.M. Chabala[1], R. Levi-Setti[1], J.P. Berry[2] and P. Galle[2]

[1] *The Enrico Fermi Institute, The University of Chicago, 5640 S. Ellis Avenue, Chicago, Illinois 60637, U.S.A.;* [2] *Laboratoire de Biophysique, SC 27 INSERM, Faculté de Médecine de Créteil, 94010 Créteil, France*

Secondary Ion Mass Spectrometry (SIMS) (Castaing and Slodzian, 1962) is a microanalytical technique used to image elemental distributions. In SIMS analysis, the sample is bombarded by an energetic ion beam which sputter erodes the specimen surface; constituent atoms of the sputtered volume are consequently emitted as neutral, negative or positive particles. Electrically charged particles are collected and subsequently mass separated by a mass spectrometer in order to identify the different elements present in the tissue and to determine their lateral distributions. Shortly after its development, SIMS has been applied to biological studies and since extensively used in this field (for reviews see (Burns, 1982; Galle, 1982)). The appeal of the technique lies in its ability to discriminate between isotopes of any element and in its high sensitivity.

SCANNING ION MICROPROBE AND SAMPLE PREPARATION

The SIMS instrument we are presently using at the University of Chicago is a newly developed scanning ion microprobe (University of Chicago Scanning Ion Microprobe: UC-SIM) (Levi-Setti et al., 1985) with a lateral resolution 10 to 50 times better than that of conventional ion microscopes (a 20 nm resolution has been recently achieved) (Levi-Setti et al., 1988). This high lateral resolution is obtained by eroding the sample with a finely focused Ga^+ beam extracted from a high brightness Liquid Metal Ion Source. For our biological applications we routinely used a 30 to 90 nm diameter probe carrying a 10 to 70 pA current. Ions emitted under Ga^+ bombardment are mass filtered through an RF quadrupole and the resulting signal used to construct isotopic maps in a KONTRON image processing system.

Best biological structure recognition is performed from analysis of perfectly flat tissue sections (to prevent edge effects) that are covered prior to analysis by a thin (5 nm) gold layer (to prevent charging effects). We used chemical fixation to investigate Be and Ag insoluble precipitates and cryofixation when analyzing diffusible ions in bone.

EXAMPLES AND COMMENTS

The first example deals with silver precipitates in basal membrane during experimental renal argyria (Berry and Galle, 1982). Figure 1A is a 160µm x 160µm analytical map at mass 26^- (CN^- ions mainly originating from proteins) which reveals with great detail the histology of a rat glomerulus transverse section. In Fig.1B (mass 107^+) silver underlines the basal membrane where the metal, ingested by the animal as silver nitrate, has precipitated. This example demonstrates our instrument's capability to quickly acquire readily interpretable metal distribution maps: total time to record the two pictures (Fig.1A (the reference picture) and Fig.1B (the investigated element map)) is as short as 17 minutes.

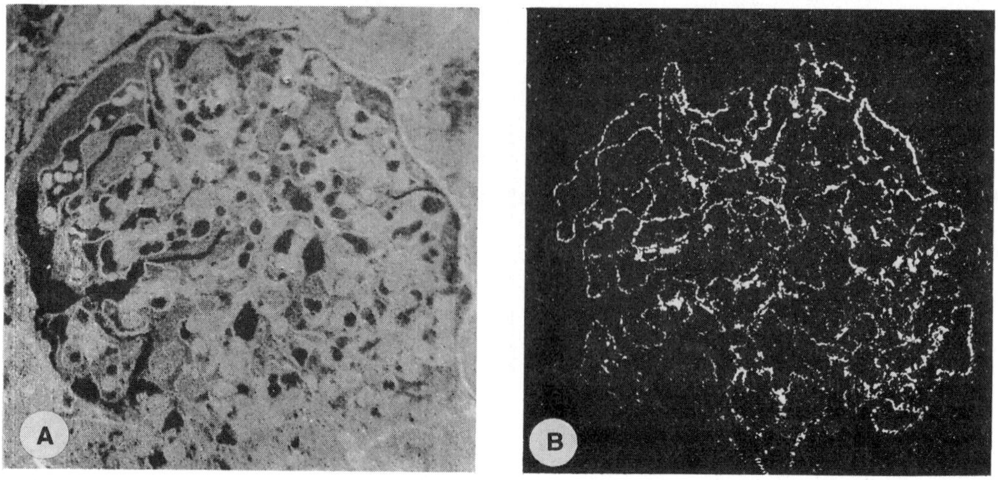

Fig.1. Rat kidney Epon section, 160µmx160µm, 1024x1024 pixels, probe diameter 70 nm: experimental intoxication with soluble silver nitrate. (**A**), $^{26}CN^-$ map showing tissue histology (13.10^6 counts collected in 512s). (**B**), 107^+ elemental map, silver has concentrated in the basal membrane (11700 counts in 512s).

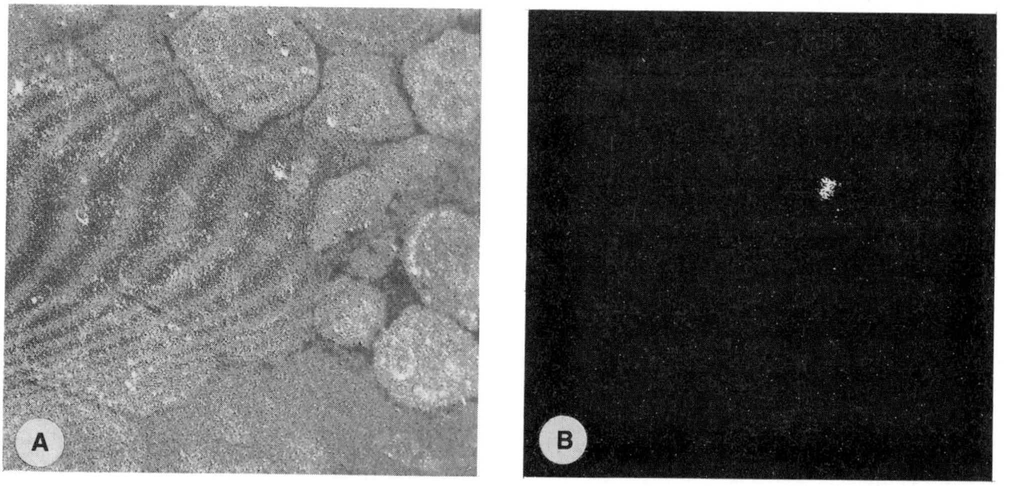

Fig.2. Rat bone marrow Epon section, 30µmx30µm, 512x512 pixels, probe diameter 30 nm: experimental intoxication by i.p injection of $BeSO_4(H_2O)_4$. (**A**), CN^- map describing the cell population (4,6.10^6 counts in 512s). (**B**), mass 9^+ distribution; a beryllium accumulation is present in the cytoplasm of a macrophage seen in (A) (941 counts in 512s).

The UC-SIMS ability to localize light metals at the ultrastructural level is an outstanding feature that we illustrate with an example of beryllium experimental intoxication. Figure 2A (30µm x 30µm) is a CN^- histological map from a rat bone marrow section where a macrophage is seen among the cell population; a beryllium accumulation (Fig.2B, same area) has formed in the macrophage cytoplasm following injection of $BeSO_4$ to the animal (Hallégot et al., 1989).

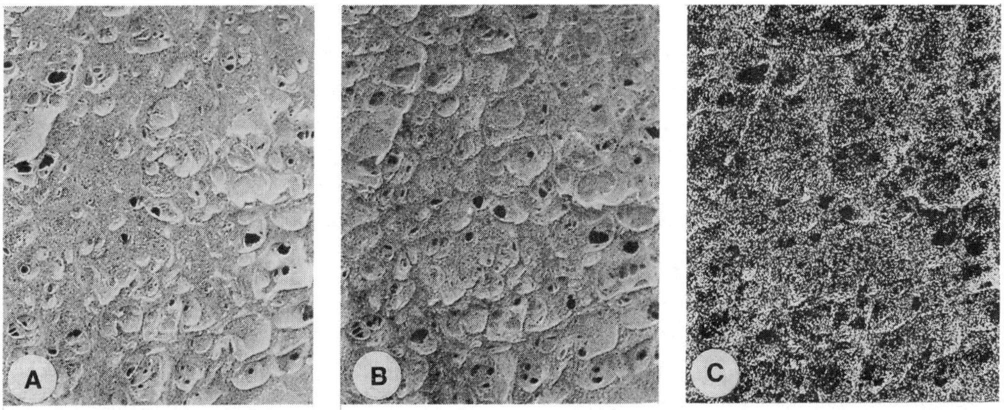

Fig.3. Frozen dried skull bone from neonatal mouse, 40μmx30μm, 512x384 pixels, probe diameter 70 nm. (**A**) and (**B**), Na^+ and K^+ distribution respectively, the high SIMS signal permits to reach a lateral resolution of 70 nm. (**C**), Ca^+ map exhibiting a lower resolution than (**A**) and (**B**) due to a lower counting statistics.

Given a probe diameter, elemental map resolution depends on the counting statistics (Chabala et al., 1988). Figure 3A and Fig.3B are respectively Na^+ and K^+ distributions (40μm x 30μm) from neonatal mouse frozen dried skull bone (Bushinsky et al., 1986); because of the intense SIMS signal the lateral resolution reaches the probe diameter size. Figure 3C (Ca^+) from the same area shows a lower lateral resolution due to a lower SIMS signal.
The above examples represent only a small part of the growing field of applications of the UC-SIM which is presently intensively used for microquantitation and detection of tracer isotopes.

ACKNOWLEDGEMENTS

This research is based on work supported by the National Science Foundation under Grant DIR-8610518.

REFERENCES

Burns M.S. (1982): Applications of secondary ion mass spectrometry (SIMS) in biological research: a review. *Journal of Microscopy*, 127, 237-258.
Berry J.P., Galle P. (1982): Selenium and kidney deposits in experimental argyria. Electron microscopy and microanalysis. *Path. Biol.* 3, 136-140.
Castaing R., Slodzian G. (1962): Microanalyse par émission ionique secondaire. *J. Microscopie*, 1, 395-410.
Bushinsky D., Levi-Setti R., Coe F. (1986): Ion microprobe determination of bone surface elements: effects of reduced medium pH. *Am. J. Physiol.* 250, F1090-F1097.
Chabala J.M., Levi-Setti R., Wang Y.L. (1988): Practical resolution limits of imaging microanalysis with a scanning ion microprobe. *Appl. Surf. Sci.* 32, 10-32.
Galle P. (1982): Tissue localization of stable and radioactive nuclides by secondary ion microscopy. *J. Nucl. Med.* 23, 52-57.
Hallégot P., Berry J.P., Levi-Setti R., Chabala J.M., Galle P. (1989): Metabolism of beryllium: a scanning ion microprobe analysis. *Trace Elements in Medicine*, 6, 3, 96-103.
Levi-Setti R., Crow G., Wang Y.L. (1985). Progress in high resolution scanning ion microscopy and secondary ion mass spectrometry imaging microanalysis. *Scanning Elec. Microsc.* 2, 535-551.
Levi-Setti R., Chabala J.M., Wang Y.L. (1988): Aspects of high resolution imaging with a scanning ion microprobe. *Ultramicroscopy*, 24, 97-114.

Metal Ions in Biology and Medicine. Eds. Ph. Collery, L.A. Poirier, M. Manfait, J.C. Etienne. John Libbey Eurotext, Paris © 1990, pp. 192-196

Localization of toxic ions in tissues by a laser micropobe mass analyser LAMMA 500

P.F. Schmidt, R. Barckhaus, B. Winterberg*

*Institut für Medizinische Physik, Hüfferstraße 68, and * Medizinische Universitäts-Poliklinik, Albert Schweitzerstraße 33, Westfälische Wilhelms-Universität, D-4400 Münster, Federal Republic of Germany*

In medical research and medical practice the importance of exactly determining trace elements has increased, for deviations in trace element concentrations may be the cause or the consequence of diseases. Usually estimations on concentrations of trace elements are based on integrating analyzing techniques like atomic absorption spectrometry. Since the increase or decrease of trace element content within cells is probably of pronounced importance for the effect of trace elements in toxic concentrations and thereby for pathologic symptoms, the analysis of intracellular trace elements contents is of particular significance. Information on the distribution of toxic ions at cellular or subcellular level is given by the laser microprobe mass analyzer (LAMMA 500).

METHOD

The laser microprobe mass analysis is based on a laser beam induced ion production by evaporating microvolumes of histological sections and the mass spectrometric analysis of these ions. The basic components of the laser microprobe mass analyzer LAMMA 500 are a combination of an optical microscope with a laser source and a mass spectrometer. By means of the optical microscope the specimen can be observed and a laser pulse can be focused to a small spot and evaporate the area of interest. The minimum diameter of the spot and therewith the lateral resolution is on the order of 0.5-1µm. The ionized part of the evaporated material is analyzed in a time-of-flight mass spectrometer in which ions of different m/e ratios are discriminated according to their different flight times. Thus elements can be detected in one spectrum. One of the most important features of the laser microprobe is the very low detection limit, 10^{-18} g, which enables the detection of toxic ions in trace concentrations (Kaufmann et al., 1979; Schmidt, 1984). This should be demonstrated by two examples:

LOCALIZATION OF ALUMINUM ACCUMULATIONS IN BONE TISSUE FROM PATIENTS WITH DIALYSIS OSTEOMALACIA

An increased concentration of aluminum in tissue and blood has been found in patients with dialysis osteomalacia. One of the consequences of this Al intoxication is an accumulation of Al in bone, combined with a defect bone mineralization that results in an increased osteoid volume and a reduced mineralization front (Al-induced osteomalacia (Drüeke, 1980)). To obtain more information on the pathogenic mechanisms between Al intoxication and development of dialysis osteomalacia, the Al content of various structural elements of bone tissue of patients with dialysis osteomalacia has been determined by laser microprobe mass analysis.

For these investigations bone samples were taken by needle puncture from the pelvic bone. In order to prevent any redistribution of Al, the tissue had been shock frozen by immersion into melting N , freeze dried in a freeze drying device at below - 80° C, plastic embedded in an epoxy resin and sectioned dry on a ultramicrotome. The thickness of the sections were 0,5 µm. The sections were mounted on 3mm grids used in electron microscopy.

By LAMMA it was possible to detect aluminum accumulations in following structural elements of bone:

a) In bone tissue from patients with dialysis osteomalacia, LAMMA revealed an increased Al content in the mineralization front of the calcified tissue and the osteoid (Schmidt et al., 1984).

b) In addition, LAMMA detected an increased Al content in osteocytes of patients with dialysis osteomalacia, especially in osteocytes near the mineralization front (Schmidt et al., 1984).

c) LAMMA investigations also demonstrated that osteoblasts from patients with dialysis osteomalacia contain an increased concentration of Al. Histological sections from bone tissue of a patient with dialysis osteomalacia show two areas of different mineralization phases: normal mineralized bone with a thin osteoid seam and elongated osteoblasts (Fig. 1a), and an area with a defect mineralization front, a wide osteoid seam, and cubic osteoblasts (Fig. 1b). The integral Al concentration was 47 µg/g.

In the area of intact mineralization, no Al was detected in the junction of mineralized bone and osteoid and in the osteoblasts (Fig. 1a).

In the area of defect mineralization: LAMMA analyses indicate a considerable accumulation of Al in the junction of mineralized bone and osteoid seam, and in the cytoplasma of osteoblasts (Fig. 1b). In particular, a comparison of osteoblasts with different Al concentrations indicates, that the content of Mg and Ca may decrease with increasing Al content (Schmidt et al., 1989).

These results have shown that by ultrastructural localization of Al it was possible to get more information on the correlation between Al intoxication and dialysis osteomalacia, in such a manner that the hypothesis of a direct toxic effect of Al on bone cells with consequent development of osteomalacia can be supported.

Many investigations made by different methods (such as histochemical staining, x-ray microanalysis, and secondary ion mass spectrometry) have shown that an Al accumulation could be found at the junction of mineralized bone and osteoid in patients with dialysis osteomalacia

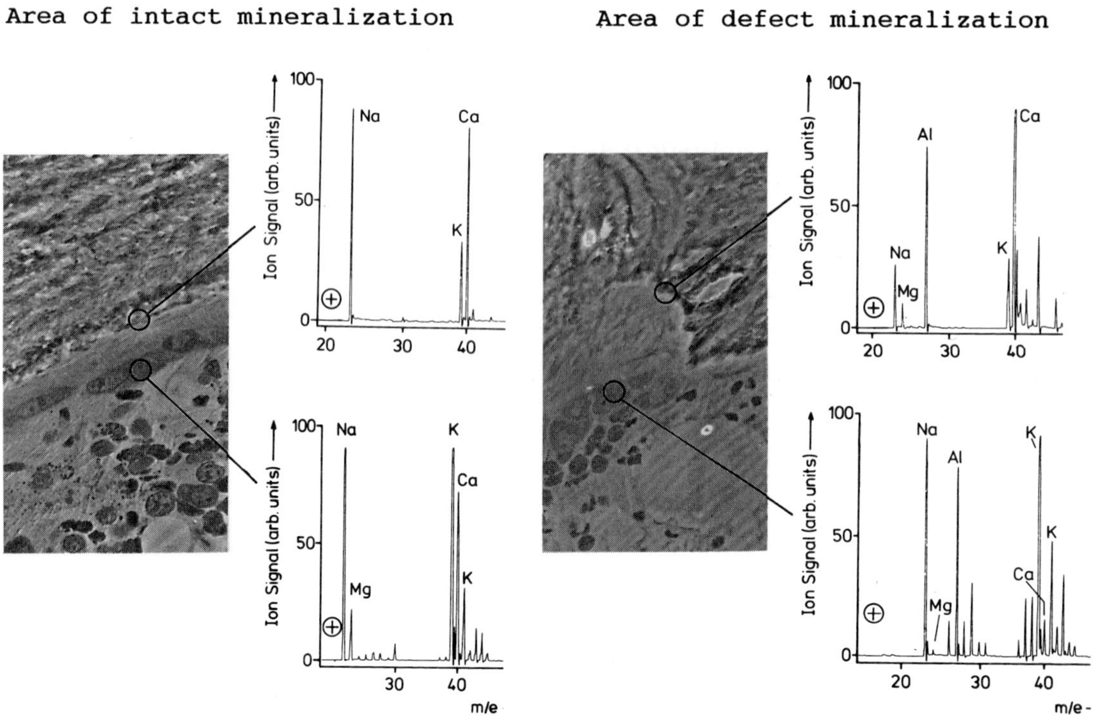

Fig. 1: LAMMA analyses of the mineralization front and of osteoblasts within the area of intact (a) and within the area of defect bone mineralization (b). Bone tissue of an Al intoxicated patient with dialysis osteomalacia.

(Buchanan et al., 1981; Cournot-Witmer et al., 1981; Plachot et al., 1984). These results have been confirmed by laser microprobe mass analysis. Moreover, LAMMA can also detect Al in cases, in which no staining could be observed although AAS measurements indicate an elevated Al level in bone (Schmidt et al., 1989). The findings on Al accumulations in osteoblasts of patients with dialysis osteomalacia are in accord with results obtained by x-ray microanalysis which demonstrate the presence of Al in the mitochondria of osteoblasts (Plachot et al., 1984; Cournot-Witmer et al., 1986).

Considering that Al ions show phosphate binding properties and that Ca ions can be replaced by Al ions in biochemical reactions (van de Vyver and de Broe, 1985), the presence of Al at the junction between osteoid and the mineralized matrix suggests that the mineralization of osteoid is inhibited by Al accumulations, so that available Ca cannot be taken up by bone. The finding that osteoblasts in front of areas of defect bone mineralization contain Al supports the supposition that one of the possible causes of Al-induced osteomalacia is given by Al accumulations which may inhibit the mineralization of the osteoid. The calcification process of bone occurs under control of osteoblasts. The presence of Ca aggregates in osteoblast mitochondria makes it evident that osteoblasts regulate the transport of Ca ions from the blood vessels towards the calcifying matrix.

This mainly occurs through the accumulation of Ca and phosphate ions in mitochondria followed by their release (Bonucci, 1984). The accumulation of Al in osteoblasts therefore suggests that Al inhibits the normal mineralization process by interfering with the Ca fluxes.

LOCALIZATION OF CADMIUM IN RENAL CORTEX OF RATS AFTER LONG TERM EXPOSURE TO CADMIUM

By laser microprobe mass analysis cadmium accumulations have been localized in different structural elements of the kidney cortex of rats after cadmium rich diet. Normotensive Wistar-rats and rats with spontaneous hypertension (SH-rats, Münster-strain) were used for these experiments. The rats were fed a cadmium poor diet or a cadmium rich diet, each group over a period of 40 weeks. The cadmium rich diet amounted in 10 mg cadmiumsulfat in 1 l drinking water.

Fig 2. shows a light micrograph of a histological section of renal tissue and one typical LAMMA spectrum of a proximal tubule cell (cytoplasm).

In order to quantify the results we have used a method for quantitation by referring the height of mass lines of cadmium to the height of mass lines of the organic background. By this variations in evaporated volume can be corrected by the use of suitable internal standards, in which the ion signal of the element to be analyzed will be referred to the ion signal of the reference element as internal standard (Schmidt et al., 1986).

In general the LAMMA measurements indicate an inhomogeneous distribution of cadmium in different elements of renal cortex. The proximal renal tubule has the highest accumulation of cadmium. The glomeruli, the distal tubules and the collecting tubes are showing lower cadmium content compared with the cadmium content in proximal tubule cells. The LAMMA results are in accordance with the conception on the mechanism of the cadmium absorption in the kidney. The cadmium metallothionein complex is transported to the renal tubules by means of glomerular filtration and a subsequent tubular reabsorption. Ultrastructural investigations by transmission electron microscopy have shown degeneration changes in proximal tubule cells of rats with a

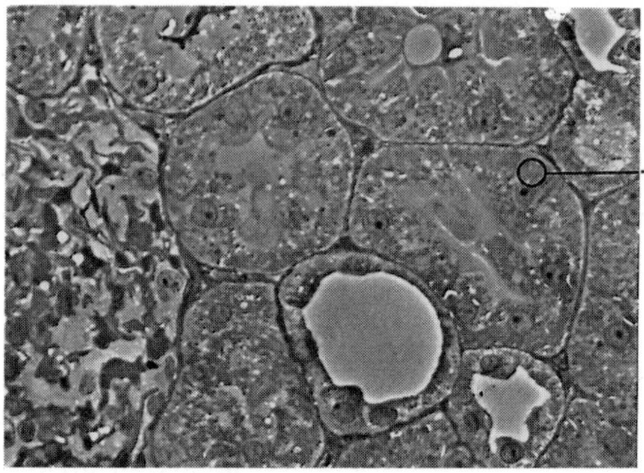

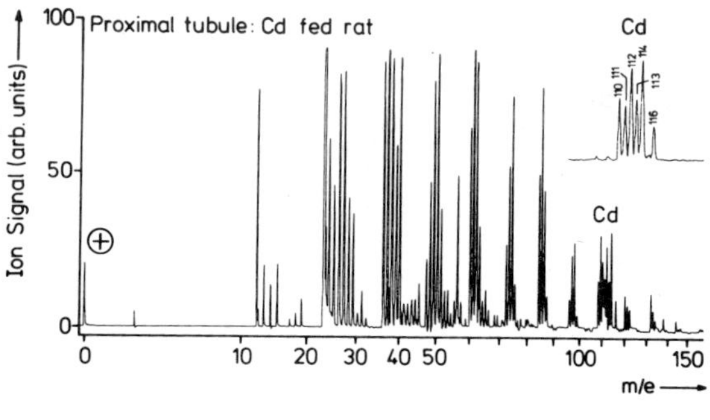

Fig. 2 Typical LAMMA spectrum of a proximal tubule cell (cytoplasm) of a cadmium fed rat demonstrating the occurance of cadmium

cadmium rich diet. The high cadmium concentration in the presumable early segment of proximal tubules estimated by LAMMA is in accord with observations that renal effects of cadmium were most severe in cells of the early segments of proximal portions of tubules.

CONCLUSIONS

The advantage of laser microprobe mass analysis is the feasibility of a morphologically controlled analysis of small specimen volumes with an outstanding low limit of detection and a relatively high sensitivity. Therefore LAMMA gives information on the local distribution and local concentrations of trace elements in cellular or subcellular range. Thus estimation of local toxicity of trace elements can be correlated to pathological alterations of tissues.

REFERENCES

Bonucci, E. (1984): The structural basis of calcification. In Ultrastructure of the connective tissue matrix, eds A. Ruggeri and P.M. Motta, Martinus Nijhoff Publishers, Boston

Buchanan, M. R. C., Ihle, B. U.,and Dunn, C. M. (1981): Haemodialysis related osteomalacia: A staining method to demonstrate aluminum. J. Clin. Pathol. 34: 1352-1354

Cournot-Witmer, G., Zingraff, J., Plachot, J.-J., Escaig, F., Lefevre, R., Boumati, P., Bourdeau, A., Garabedian, M., Galle, P., Bourdon, R., Drüeke, T., and Balsan, S. (1981): Aluminum localization in bone from hemodialyzed patients: Relationship to matrix mineralization. Kidney International 20: 375-385

Cournot-Witmer, G., Plachot, J.-J., Bourdeau, A., Lieberherr, M., Jorgetti, V., Mendes, V., Halpern, S., Hemmerle, J., Drüeke, T., and Balsan, S. (1986): Effect of Al on bone and cell localization. Kidney International 29, Suppl. 18: S37-S40

Drüeke, T. (1980): Dialysis osteomalacia and aluminum intoxication. Nephron 26: 207 - 210

Kaufmann, R., Hillenkamp, F., and Wechsung, R. (1979): The laser microprobe mass analyzer (LAMMA), a new instrument for biomedical microprobe analysis. Med. Prog. Technol. 6: 109-121

Plachot, J.-J., Cournot-Witmer, G., Halpern, S., Mendez, V., Bourdeau, A., Drüeke, T., Galle, P., and Balsan, S. (1984): Bone ultrastructure and X-ray microanalysis of aluminum-intoxicated hemodialyzed patients. Kidney International 25: 796-803

Schmidt, P.F. (1984): Localization of trace elements with the laser microprobe mass analyzer. Trace elements in medicine 1: 13-20.

Schmidt, P.F., Zumkley, H., Barckhaus, R., and Winterberg, B. (1984): Localization of aluminum in bone cells from patients with dialysis osteomalacia. Trace elements in medicine 1: 167-171

Schmidt, P.F., Barckhaus, R., and Kleimeier W. (1986): Laser microprobe mass analyzer investigations on the localization of cadmium in renal cortex of rats after long-term exposure to cadmium. Trace elements in medicine 3: 19-24

Schmidt, P.F., Zumkley, H., Barckhaus, R., and Winterberg, B. (1989): Distribution patterns of aluminum accumulations in bone tissue from patients with dialysis osteomalacia determined by LAMMA. In Microbeam Analysis-1989, ed. P.E. Russell, pp. 50-54. San Francisco: San Francisco Press

Van de Vyver, F. L. , and de Broe, M. E. (1985): Aluminum in tissues. Clin. Nephrol. 24 (Suppl. 1): 37-57

Metal Ions in Biology and Medicine. Eds. Ph. Collery, L.A. Poirier, M. Manfait, J.C. Etienne. John Libbey Eurotext, Paris © 1990, pp. 197-199

Intracellular beryllium distribution in rat tissue study by ion micropobe

J.P. Berry*, Ph. Hallegot**, P. Mentre***, R. Levi-Setti**, P. Galle*

** Centre de Microanalyse Appliquée à la Biologie du CNRS-SC 27 INSERM Faculté de médecine 94010 Créteil, France; ** Enrico Fermi Institute - University of Chicago Il. 60637 U.S.A.; *** Laboratoire de Biologie Cellulaire, 94205 Ivry-sur-Seine, France*

Beryllium and its compounds were used in industry to increase the hardness and conductivity of alloys. In the meantime this element has carcinogenic properties. Animal experiment has shown that it can induce malignant bone and lung tumors. After administration in the rat of beryllium sulfate (a soluble form) at long term, the intracellular distribution of beryllium was studied in following tissus of rat : kidney, spleen, liver, lung, bone marrow, nodes. These tissus were observed by following methods:

- electron microscopy : embedding were performed in epon or directly in GMA (Glycol methacrylate without dehydratation). Moreover, cytochemical study has been studied.
- electron microprobe : the apparatus used for analysis of ultrathin section was a Camebax (Cameca) electron probe microanalyzer.
- high resolution imaging ion microprobe proposed by Levi-Setti et al. (1975) (University of Chicago - Hughs Research Laboratories). Secondary ion were emitted following bombardment with a 40 Kev 40 pA scanning Ga^+ probe lateral resolution is about 70 nm.

Results shown that this element was concentrates in abnormal intranuclear structures distinct from the nucleoli and perichromatine grains. After a 4 month intoxication, these structures were generally grouped in one simple inclusion localized in the central part of the nucleus constituted of dense masses about 2 or 3 µm in diameter and composed of clusters of small spherules about 20 or 30 nm in diameter (Fig. 3)(Berry, Escaig et al. 1987). The distribution of these granules is not regular and there is not crystalline organization. These abnormal structures were observed in the proximal convoluted tubule of kidney cell, in alveolar wall cell, and spleen cell. (Fig. 2)
Ionic microscopy at high resolution has shown that these structures were rich in beryllium and in protein (Fig. 1). Electron probe microanalysis shows that the nuclear inclusions rich in beryllium yield marked peaks for the K lines of phosphorus and calcium. Cytochemical reactions (Berry, Mentré et al. 1989) have precised that beryllium was bind neither DNA, nor RNA, but with acidic protein which could have a fundamental role in the regulation of certain nuclear function. Hallegot, Levi-Setti et al. (1989) have also shown that beryllium can be localized in lysosome of others organs : liver and kupffer cells, mesangial cells, bone marrow macrophages. These data were in relation with transport of beryllium by blood either in soluble form or in insoluble form (Vacher et Stoner, 1988). The soluble form could be the most toxic and will be involved in nuclear structure rich in beryllium.

Concentrations of granules in nucleus could be due to the accumulation of a complex formed between beryllium and acidic protein. The high amount of calcium in the granules could be explaned by the affinity for the complex very likely for the phosphate group. The protein could have a fondamental role in the regulation of certain nuclear functions (Parker et Stevens 1978) (Marcotte 1972), since the outcome of beryllium intoxication is cellular transformation.

These results elucidate the mechanism of beryllium absorption, according to the chemical form injected by different types of cells and organelles. These findings contribution to a better understanding of toxic and carcinogenic effect of this element.

REFERENCES

Berry, J.P., Mentré, P. Hallegot, P. et al. (1989) : Cytochemical study of abnormal intranuclear structures rich in beryllium. Biology of the Cell 67: 147-157.

Berry, J.P., Escaig, F., Galle P. et al. (1987) : Etude de la localisation intracellulaire du beryllium par microscopie ionique analytique. C.R. Acad. Sc. III.10: 239-243.

Hallegot, P., Berry, J.P. Levi-Setti, R. et al.(1989) : Metabolism of beryllium: a scanning ion microprobe analysis. Trace Elements in Medicine 3: 96-103.

Levi-Setti, R., Crow, G., and Wang, Y.L. (1985) : Progress in high resolution scanning ion microscopy and secondary ion mass spectrometry imaging microanalysis. Scanning Electron Microsc. II: 535-551.

Marcotte, S., and Witachi, H.E. (1972): Synthesis of RNA and nuclear proteins in early regenerating rat livers exposed to beryllium. Res. Commun. Clin. Pathol. Pharmacol. 1: 97-104.

Parker, V.H. and Stevens C. (1979) : Binding of beryllium to nuclear acidic proteins. Chem. Biol. Interactions 26: 167-177.

Vacher, J., and Stoner, H.B. (1968) : The transport of beryllium in rat blood. Biochem. Pharmacol 17: 93-107.

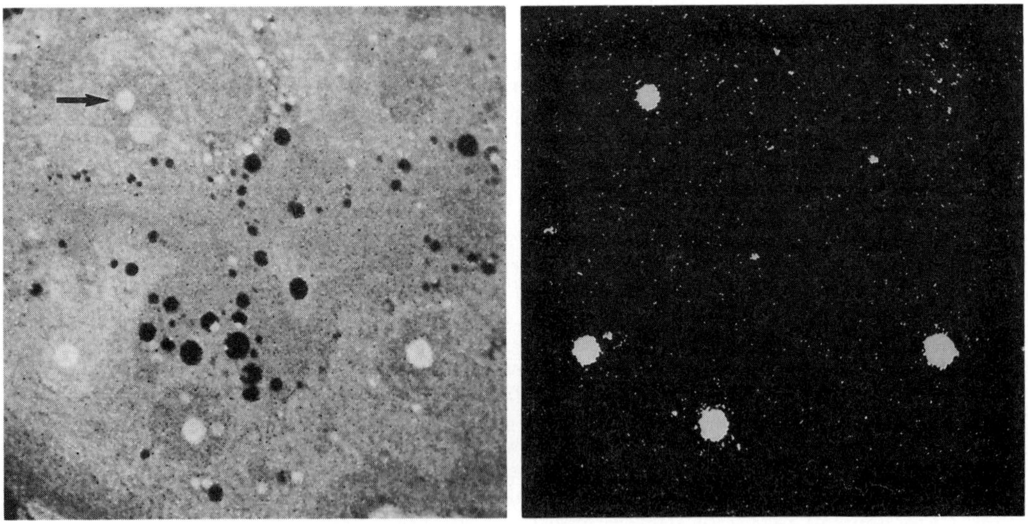

Fig. 1 Scanning ion microprobe.
Rat kidney proximal tubule. Distribution maps CN^- (A) and Be+ (B) in the same area. Strong accumulation of beryllium localized in the some of the nuclei. A nucleus contains, near its nucleus, an inclusion (arrow) which is emissive in Be+ and CN^- (representative of protein distribution)

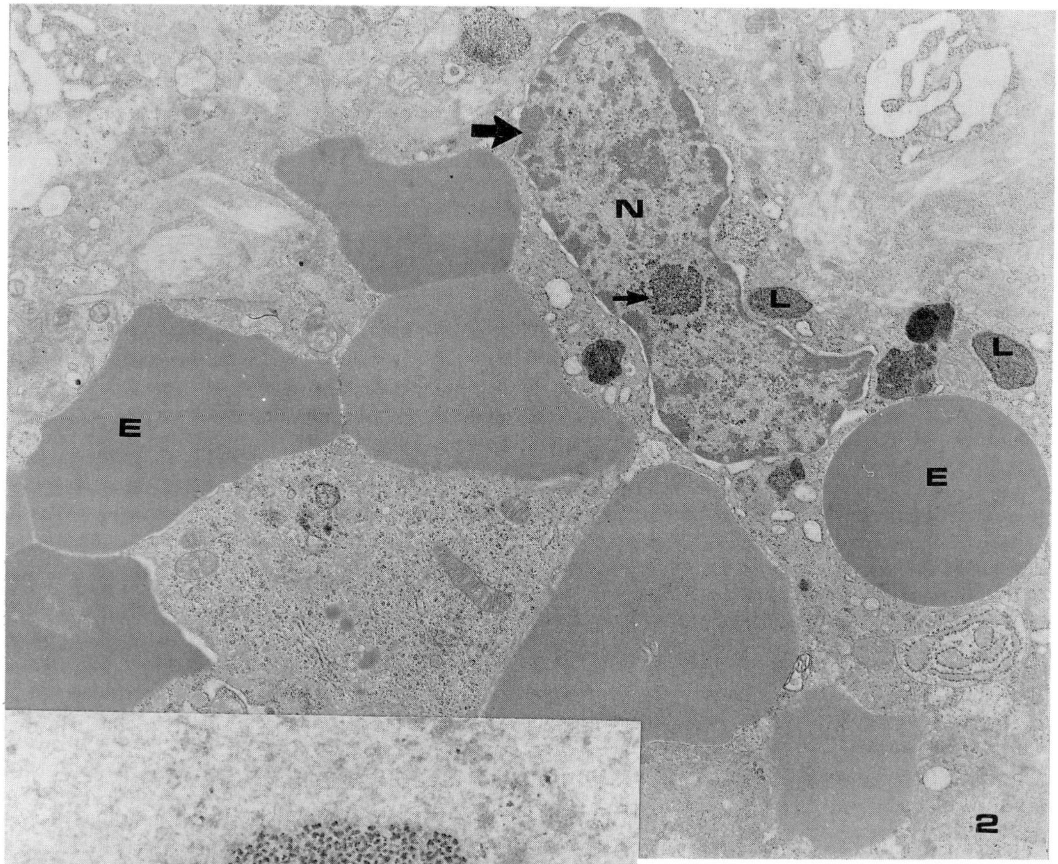

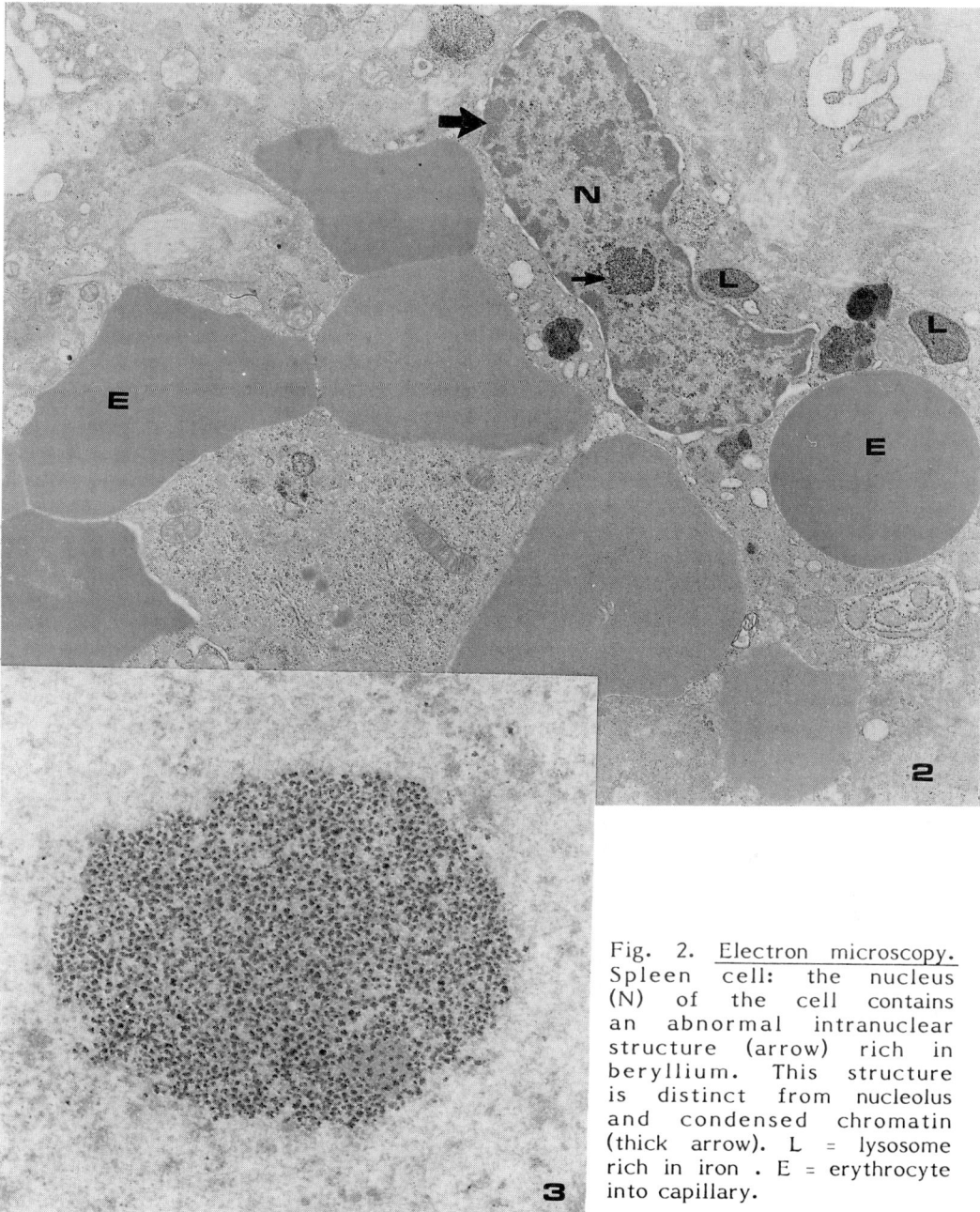

Fig. 2. <u>Electron microscopy.</u> Spleen cell: the nucleus (N) of the cell contains an abnormal intranuclear structure (arrow) rich in beryllium. This structure is distinct from nucleolus and condensed chromatin (thick arrow). L = lysosome rich in iron . E = erythrocyte into capillary.

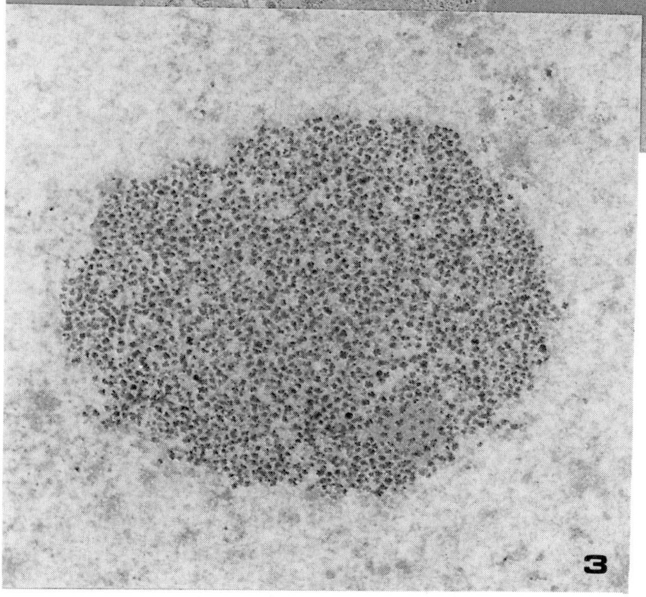

Fig. 3. High magnification showing an abnormal intranuclear structure with mean size of 3 μm containing clusters of electron-dense granules about 20 nm in diameter. The distribution of these granules is not regular and there is not crystalline organization.

Metal Ions in Biology and Medicine. Eds. Ph. Collery, L.A. Poirier, M. Manfait, J.C. Etienne. John Libbey Eurotext, Paris © 1990, pp. 200-202

Ultrastructural observations in mouse lungs after short term inhalation of cadmium compounds

I. Paulini, K.U. Thiedemann, C. Dasenbrock

Fraunhofer Institute for Toxicology and Aerosol Research, Dept. of Ultrastructure Research, Nikolai-Fuchs-Str. 1, D 3000 Hannover 61, FRG

Chronic environmental or workplace inhalation exposure to Cadmium (Cd) compounds are suspected to cause lung cancer in man. Chronic inhalation exposure to Cd was found to be carcinogenic in rats (Takenaka 1983). The morphological changes induced by short term Cd inhalation exposure have not been investigated extensively. In this study, the acute morphological changes in mouse lungs due to short term Cd exposure and the persistence of the alterations after a clean air recovery period were investigated by correlative light (LM), scanning (SEM) and transmission electron microscopy (TEM) in an attempt to gain insight into the sequence of morphological alterations and the events leading to the development of preneoplastic and neoplastic lesions.

MATERIALS AND METHODS

Six-week-old female SPF mice (Han: NMRI) were exposed to cadmium oxide dust (CdO), cadmium chloride ($CdCl_2$) and cadmium sulfide (CdS) (270 μg Cd/m^3; 18 hrs/d; 5 d/wk) for 3 d, 10 d, 30 d and 30 d plus 16 wks clean air (recovery groups). Control mice inhaled filtered clean air (Cd content 20 ng/m^3). A separate group of mice was sacrificed after 20 wks (recovery control).

Animals were sacrificed 2 d after the end of the exposure period. Lungs were immediately fixed by intratracheal instillation of 2 % glutardialdehyde in 0.1 M Na-cacodylate buffer (pH 7.3, 190 mOsmol) under a pressure of 20 cm water column. For LM/TEM, vibratome sections were prepared from the right middle lung lobe, postfixed in 1 % osmium tetroxide, and routinely embedded in Polarbed 812. Semithin sections were stained with Stevenel's blue, ultrathin sections with aqueous uranyl acetate and basic lead citrate, and viewed in an EM 10C (Zeiss, Oberkochem, FRG). Areas adjacent to the samples for LM/TEM were critical point dried from liquid CO_2 and sputtered with approx. 30 nm gold for SEM. They were viewed in a Stereoscan S 360 scanning electron microscope (Cambridge Instruments, Heidelberg, FRG).

RESULTS

Cadmium oxide: After 3 d of inhalation, CdO caused extensive type-I cell damage and necrosis. Most type-II pneumocytes as well as some bronchiolar epithelial cells (Clara and ciliated cells) showed morphological evidence of damage. Proliferated undifferentiated and hyperplastic type-II cells replaced the necrotic type-I epithelium. After longer exposure periods (9 d), a distinct type-II cell hyperplasia had developed. This hyperplastic cell population contained lamellar bodies unusual in size and density. This, as well as intraepithelial lamellar body accumulations suggested a disturbed surfactant production. Alveoli lined by hyperplastic type-II cells often contained serous/fibrinous fluid, neutrophilic granulocytes and macrophages. The interstitium in centroacinar regions was infiltrated with plasma cells, neutrophilic granulocytes, macrophages and fibroblasts. Around blood vessels, very small alveoli with thick and solid septae contained interstitial inflammatory cells and fibroblasts and were lined by type-I cells. Alveolar bronchiolization (extension of bronchiolar epithelium into alveolar ducts) and some rarification of septae (increased number and size of alveolar pores) were observed. Bronchiolar epithelial cells were damaged and ciliogenesis was observed more frequently. After 4 wks of CdO exposure, the type-II cell damage had decreased. In focally occurring very small and perfectly round alveoli, some elongated darkened type-II cells were found. The shape of these cells reminded of metaplastic cells, and they might represent intermediate stages in the premetaplastic cellular differentiation. The number of free alveolar cells (macrophages, leucocytes) and the amount of alveolar debris had decreased. The alveolar bronchiolization was more pronounced and the interstitial inflammatory reaction had regressed when compared to lungs exposed for shorter periods. Extremely large multinucleated macrophages with numerous phagocytosed lamellar bodies, cholesterol cristals, and fat vacuoles were often completely filling alveolar spaces. During the clean air recovery period, the interstitial inflammatory infiltration had regressed. Type-II cell hyperplasia, alveolar bronchiolization, and giant multinucleated macrophages, however, had persisted. Septal rarification was focally very pronounced and often alveolar septae were retracted so much, that only delicate net-like structures remained. In other areas of the same lungs, where the parenchyma was extremely condensed (thick and solid septae; type-II cell hyperplasia) the alveoli were clogged with giant foamy macrophages. The elongated dark, possibly premetaplastic, type-II cells still occurred, but they were less distinct.

Cadmium Chloride: Compared to the lesions caused by CdO after 3 d of inhalation, $CdCl_2$ caused less epithelial damage. However, more macrophages and leucocytes were found in alveolar spaces and alveolar bronchiolization had developed. After 9 d of $CdCl_2$ inhalation, extensive type-II cell hyperplasia, especially in alveolar ducts, and a pronounced alveolar bronchiolization had developed. The occurrence of elongated type-II cells and an increase in number and size of alveolar macrophages were observed. After 4 wks of $CdCl_2$ inhalation, the centroacinar interstitial infiltration with inflammatory cells had resolved, but persisted perivascularly. In centroacinar alveoli, cellular debris, lamellar bodies, serous exudate, neutrophilic granulocytes, and giant, frequently disrupted, macrophages were found. Large parts of the epithelial lining consisted of damaged, proliferated and hyperplastic type-II cells. Subpleurally, many disrupted giant macrophages had accumulated. Like CdO, $CdCl_2$ also induced septal rarification. The septae exhibited large and numerous pores. The unusually small and very round perivascular alveoli had persisted. The results for the $CdCl_2$ recovery group were very similar to the ones found in the CdO recovery group.

Cadmium Sulfide: After 3 d, CdS inhalation caused less severe type-I cell necrosis and accumulation of cellular debris than CdO and $CdCl_2$, but more macrophages had accumulated in alveolar spaces. Intraepithelial lamellar body accumulations were frequently found. After 9 d of CdS inhalation, the lining of centroacinar regions consisted of hyperplastic type-II cells with large lamellar bodies, and of an extensive bronchiolo-alveolar hyperplasia. Alveolar macrophages and neutrophilic granulocytes were abundant. They had neither cholesterol nor fat inclusions and were of medium electron density. The alveolar clusters of leucocytes were always surrounded by hyperplastic type-II cells. After 4 wks of CdS inhalation, contrary to the alveolar bronchiolization, the type-II cell hyperplasia had persisted. Elongated dark type-II cells in small round alveoli and giant macrophages with cholesterol cristals and lamellar bodies were observed. Large amounts of surfactant located in alveolar lumina might have originated from excessive surfactant production by damaged type-II pneumocytes. The lungs of animals from the CdS recovery group showed a less pronounced septal rarification and parenchymal clogging with giant macrophages than tissue from the CdO and $CdCl_2$ recovery groups. Large areas of epithelium consisted of damaged or rugged hyperplastic type-II cells with irregular lamellar bodies. In condensed parenchyma the alveolar lumina either were filled with lamellar bodies, cellular debris, disintegrating macrophages, and fibrin, or with multinucleated giant macrophages containing cholesterol cristals, fat vacuoles and many lamellar bodies.

DISCUSSION

Acute Cd inhalation causes severe lung damage in mice with initial necrosis of type-I cells and subsequent repair by proliferation of hyperplastic type-II cells. The persistence of type-II cell hyperplasias in this study suggests that the regeneration of type-I cells (Strauss 1976) is impaired under the continuous influence of the toxic agent. The persistent type-II cell hyperplasias possibly give rise to the alveolar bronchiolization and after prolonged periods of Cd exposure may undergo transformation.

A recovery period considerably longer than the Cd exposure period did not lead to a regression of the morphological alterations caused by Cd compounds. In this study, the possibly premetaplastic type-II cells had persisted throughout the recovery period and the septal rarification became even more pronounced. The latter observation might indicate an early stage of an emphysemateous lesion. Septal rarification was also found in hamsters chronically exposed to Cd compounds (Thiedemann 1990). We assume that even short term Cd exposure causes irreversible lung damage which might even be augmented over time.

Different Cd compounds evoke different types of morphological alterations, CdO causing more, CdS less extensive tissue damage. This varying biological action was previously reported in Cd inhalation studies (Grose 1987) and may be due to the differences in bioavailability, deposition, and water solubility of the compounds under study (Oldiges 1986). These data are in agreement with our observations.

REFERENCES

Grose, E.C., Richards, J.H., Jaskot, R.H., Menache, M.G., Graham, J.A., Dauterman, W.C. (1987) A comparative study of the effects of inhaled cadmium chloride and cadmium oxide: pulmonary response. J. Toxicol.Environ. Health, 21:219-232.

Oldiges, H., Glaser, U. (1986) The inhalative toxicity of different cadmium compounds in rats. Trace elements in medicine 3(2):72-75.

Strauss, R.H., Palmer, K.C., Hayes, J.A. (1976) Acute lung injury induced by cadmium aerosol. Am. J. Pathol. 84:561-578.

Takenaka, S., Oldiges, H., König, H., Hochrainer, D., Oberdörster, G. (1983) Carcinogenicity of cadmium chloride aerosols in W. rats. J. Nat. Cancer Inst. 79:367-373.

Thiedemann, K.-U., Kreft, A., Abel, U., Dasenbrock, C., Fuhst, R., Peters, L., Heinrich, U., Mohr, U. (1990) Qualitative and quantitative ultrastructural investigations in hamster lungs after chronic inhalation of Cadmium compounds, this volume.

ACKNOWLEDGMENT: The authors greatly appreciate the skilful technical assistance of R. Griebel, A.v. Malotki, and F. Müller.
This work was supported by the BMFT/DFVLR-HdA Nr. 01 VD 1838.

Metal Ions in Biology and Medicine. Eds. Ph. Collery, L.A. Poirier, M. Manfait, J.C. Etienne. John Libbey Eurotext, Paris © 1990, pp. 203-205

Histological observations and morphometric analysis in lungs of hamsters after long term inhalation of cadmium compounds

S. Rittinghausen, M. Aufderheide*, H. Ernst, C. Dasenbrock, R. Fusht, U. Heinrich, L. Peters, U. Mohr

*Fraunhofer Institute of Toxicology and Aerosol Research, Nikolai-Fuchs-Str. 1, D-3000 Hannover 61, FRG and * Institute of Experimental Pathology, Hannover Medical School, Konstanty-Gutschow-Str. 8, D-3000 Hannover 61 FRG*

INTRODUCTION

Cadmium compounds are suspected of being a hazard of human health, since epidemiological studies demonstrated that their occupational inhalation is associated with increased risk of lung cancer (Lemen *et al.*, 1976; Thun *et al.*, 1985, Soharan, 1987). These findings are confirmed by studies on rats that were chronically exposed to different cadmium aerosols and had high frequencies of lung tumours (Takenaka *et al.*, 1983; Oldiges *et al.*, 1985). Neoplasms observed in rats after cadmium chloride inhalation included squamous cell carcinomas and adenocarcinomas. Primary lung carcinomas were induced in 71.4 % of the animals exposed to 50 μg Cd/m^3, in 52.6 % of the animals exposed to 25 μg Cd/m^3, and in 15.4 % of the animals exposed to 12.5 μg Cd/m^3 (Takenaka *et al.*, 1983). Since no information is available on pathological lesions resulting from chronic exposure to cadmium aerosols in other laboratory animals, long-term inhalation studies with Syrian hamsters were performed to evaluate the effects of different cadmium compounds on the lung of this species.

MATERIALS AND METHODS

Male and female Syrian hamsters [Hoe: SYHK (SPF Ars)] were exposed to different concentrations of aerosols of CdS, $CdSO_4$, $CdCl_2$ and CdO (as dust and fume) in a long-term inhalation experiment. The animals were exposed 19 hours daily, 5 days a week up to 15 months. Additional groups of hamsters were exposed for only 8 hours a day and 5 days a week over a period of 15 months to 90 μg Cd/m^3 as CdO fume and to 270 μg Cd/m^3 as CdS. After termination of the exposure they were kept in clean air for another 6 to 12 months till death or at least till the final sacrifice (Table 1). During the inhalation periods all animals, including the controls, were kept individually in stainless steel wire mesh cages (type III) in horizontal flow inhalation chambers. The Cd-aerosols were generated by nebulizing solutions of $CdCl_2$ and $CdSO_4$ or a suspension of CdS. CdO-dust aerosols were produced by atomizing Cd-acetate solutions with subsequent pyrolization of the aerosol at 750 °C; CdO-fume was produced by burning metallic Cd from Cd-electrodes in an electric arc. The mass median aerodynamic diameter of the different Cd-particles was in the range of 0.2 - 0.6 μm, determined by cascade impactor measurements. At the end of the exposure time the animals were transferred to macrolon cages. The inhalation laboratories and the animal rooms were maintained under standard laboratory conditions (23 ± 2 °C, relative humidity 55 ± 10 %, 12 : 12 hours light-dark sequence starting at 7 a.m.). The animals were given a commercial cereal-based diet (RMH-TM, Hope Farms, Woerden, The Netherlands) and water ad libitum.

Dead or moribund animals were necropsied as soon as possible after they were detected. Surviving hamsters were killed at the end of the study by exsanguination after deep anaesthesia using Nembutal. A complete necropsy was performed on all hamsters. The lungs and trachea were removed from the thorax in toto and weighed. The trachea was cannulated and the lungs were fixed by intratracheal instillation of 10 % buffered formalin. Samples of the lung were embedded in paraffin wax. Sections (5 μm) were stained routinely with hematoxylin and eosin. Additional sections were provided with Masson-Goldner or Turnbull blue staining for proof of collagen fibres or haemosiderin.

For morphometrical analysis the lungs of five animals of distinct groups (marked with "*" in Table 1) were instilled with Karnovsky's fixative (0.7 % glutaraldehyde and formaldehyde in 0.08 M sodiumcacodylate buffer, pH 7.3, 190 m Osmoles) instead of formalin. The volumes of the fixed lungs were estimated by fixative displacement. The left lung was trimmed into maximally five 3 mm thick slices in the dorsal plane, which were then processed as described above. Ten randomly selected fields of view of five serial step-cut sections per tissue block were evaluated. To determine the volume density of proliferative areas an automatic image analyser (IBAS, Zeiss) was used. The percentage area of the lung occupied by proliferative tissue in the peribronchial region was determined and calculated. Morpho-

Table 1. Experimental groups and aerosol concentrations

Cd-compound-concentration (μg Cd/m^3)		Cd-Dose (mg Cd/m^3• h)	Exposure Time (h/d)	(w)	Experimental Time (w) m / f	Number of animals m / f
Clean air	-	-	-	-	105 / 86	24* / 24*
$CdCl_2$	30	180	19	65	113 / 76	23* / 22*
$CdCl_2$	90	497	19	60	102 / 76	23 / 24
$CdSO_4$	30	180	19	65	113 / 77	24* / 23*
$CdSO_4$	90	497	19	61	103 / 76	24 / 24
CdS	90	552	19	64	111 / 82	20* / 21
CdS	270	278	8	26	101 / 87	22* / 23
CdS	270	1447	19	60	70 / 64	24 / 23
CdS	1000	3563	19	44	60 / 61	21 / 23
CdO-d	10	63	19	64	113 / 86	23* / 24
CdO-d	30	135	19	52	108 / 76	24 / 24
CdO-d	30	155	19	60	113 / 84	24 / 24
CdO-d	90	216	8	64	113 / 86	24* / 23
CdO-d	90	375	19	49	89 / 65	24 / 24
CdO-d	90	408	19	52	73 / 67	23 / 23
CdO-d	270	282	19	13	105 / 83	24 / 24
CdO-d	270	517	8	57	108 / 84	23* / 23
CdO-f	10	46	19	55	95 / 78	23 / 24
CdO-f	30	127	19	50	104 / 81	24 / 24
CdO-f	90	208	8	64	107 / 83	20 / 24

-d = dust, -f = fume, * = morphometrical analysis

metrical. and histopathological data were evaluated using a two-tailed *t*-test and *U*-test of Wilcoxon, Mann and Whitney taking $p \leq 0.05$ as the level of significance.

RESULTS AND DISCUSSION

No tumours were observed in the lungs of the hamsters. A lesion which was present in all high dose groups of the different cadmium aerosols, is the bronchiolization. This hyperplastic lesion is localized at the bronchiolar-alveolar junction. The walls of the respiratory bronchioles, alveolar ducts and adjacent alveoli become replaced by cuboidal ciliated cells. Significant frequencies of marked bronchiolization were found in lungs of several animals of the different high dose treatment groups, i.e. the male hamsters receiving $CdCl_2$ in a dose of 497 mg, the male and female hamsters after exposure to CdS in a dose of 3563 mg, or CdO-dust in a dose of 408 mg, respectively. A much lower grade of this lesion was also observed in the controls, the low doses of $CdCl_2$, $CdSO_4$, the low and the medium doses of CdS and CdO-dust, and in all CdO-fume groups; histopathological findings coincided with the results of the morphometrical analysis. It is suggested that the lesion represents an adaptive response to dust exposure by which the bronchiolar epithelium extends into the alveoli in an attempt to facilitate increased particulate removal.

Interstitial fibrosis was observed in a significantly high frequency in the males and females exposed to CdS aerosol in doses of 1447 and 3563 mg, in the males receiving CdO-dust aerosol in doses of 375, 408, and 517 mg, and in the females receiving CdO-dust in a dose of 282 mg. Fibrosis may be seen as a repair stage after injury of alveolar walls by cadmium compounds. Cholesterol clefts were present in the same dose groups. The cholesterol clefts caused in most cases a granulomatous inflammation which is one further aspect of pathogenesis of focal fibrosis.

Dose-dependent significant higher frequencies of inflammatory changes and activation of the bronchus-associated lymphoid tissue, as well as of haemorrhages and haemosiderin pigmentation were also observed in the medium and high dose groups after inhalation of CdS- and CdO-aerosols. The haemorrhages indicate damage of vascular epithelium by the toxic action of CdS and CdO-dust. Only males of a high CdS dose group developed alveolar oedema in a significant frequency. Activity of alveolar macrophages was exhibited in all dose groups. Significantly high frequency of increased macrophage accumulation was observed in all middle and high dose groups. No significant differences between the different treatment groups were observed for leucocytostasis, calcification of bronchial or alveolar epithelium, and alveolar emphysema.

REFERENCES

Lemen, R.A., L.S. Lee, J.K. Wagoner, and H.P. Blejer (1976): Cancer mortality among cadmium production workers. *Ann. NY Acad. Sci.* **271**, 273-279.

Oldiges, H., Hochrainer, D., Takenaka, S., Oberdörster, G., and König, H. (1985): Lung carcinomas in rats after low level cadmium inhalation. *Curr. Top. Environ. Toxicol. Chem.* **8**, 409-419.

Soharan, T. (1987): Mortality from lung cancer among a cohort of nickel cadmium battery workers: 1946-84. *Brit. J. Ind. Med.* **44**, 803-809.

Takenaka, S., Oldiges, H., König, H., Hochrainer, D., and Oberdörster, G. (1983): Carcinogenicity of cadmium chloride aerosols in W rats. *JNCI* **70**, 367-373.

Thun, M.T., Schnorr, M.T., Smith, A.B., Halperian, W.E., and Lemen, R.A. (1985): Mortality among a cohort of U.S. cadmium production workers - an update. *JNCI* **74**, 325-333.

Metal Ions in Biology and Medicine. Eds. Ph. Collery, L.A. Poirier, M. Manfait, J.C. Etienne. John Libbey Eurotext, Paris © 1990, pp. 206-208

Qualitative and quantitative ultrastructural investigations in hamster lungs after chronic inhalation of cadmium compounds

K.U. Thiedemann, A. Kreft, U. Abel, C. Dasenbrock, R. Fuhst, L. Peters, U. Heinrich, U. Mohr

Fraunhofer Institute of Toxicology and Aerosol Research, Department of Ultrastructure Research, Nikolai-Fuchs-Str. 1, D-3000 Hannover 61, FRG

Epidemiologic studies have suggested, that inhalation of Cadmium compounds may cause lung cancer in humans. This has been confirmed by experimental studies in rats, where chronic inhalation exposure to $CdCl_2$ at concentrations as low as 12.5 μg Cd/m^3 caused lung tumors (Takenaka et al., 1983). No data, however, are available on longterm inhalation toxicity of less watersoluble Cd-compounds such as CdO and CdS (which may be more relevant with regard to workplace exposure and environmental hygiene) and on pathological lesions occuring in animal species other than rat. The observations reported in the present communication are part of a joint study of the Fh-ITA and the Fh-IUCT where rats, mice and hamsters were exposed to various concentrations of $CdCl_2$, $CdSO_4$, CdO and CdS in a chronic inhalation carcinogenicity setup (Heinrich et al., 1987; Oldiges et al., 1987). Since the histopathological examination of the lungs revealed an increased tumor incidence in the rat but no carcinogenic effect in the hamster, one goal of this ultrastructural study was to obtain more detailed information on exposure related epithelial alterations occurring in the latter species.

MATERIALS AND METHODS

Animals used for ultrastructural investigation were derived from the longterm inhalation study in male and female Syrian Golden Hamsters [Hoe:SYHK (SPF Ars)] performed with $CdCl_2$ (30 μg Cd/m^3), $CdSO_4$ (30 μg Cd/m^3), CdO (10, 90, 270 μg Cd/m^3), and CdS (90, 270 μg Cd/m^3). The animals were exposed for 19 h/day, 5 days/week, for up to 15 months or 8 h/day, 5 day/week, for 6 to 18 months, respectively. After the end of the inhalation period the surviving animals were kept in clean air until sacrifice. Details of the exposure experiments have been described previously (Heinrich et al., 1987; Oldiges et al. 1987).

Immediately after sacrifice, the lungs of the animals were fixed by intratracheal instillation of a modified Karnovsky fixative (0.53 % Paraformaldehyde and 0.66 % Glutaraldehyde in 0.08M Na-Cacodylate buffer, pH 7.4) at a pressure of 20 cm water column. Tissue samples were dissected from the right middle lobe of the lung, routinely postfixed with OsO_4 and embedded in Epon. Appropriate areas for ultrastructural evaluation were selected by light microscopy from semithin sections (1 μm) of the embedded tissue samples.

For morphometric analysis appropriate samples were selected from semithin sections.

Morphometry was performed on semithin sections (light microscopy) and ultrathin sections (electron microscopy) according to methods suggested by (Weibel 1979).

RESULTS

The ultrastructural alterations oberserved during the present study were qualitatively similar in all exposure groups. None of the changes was specific for a certain Cadmium compound or a certain exposure regimen.

A common feature observed in all exposure groups was a bronchiolo-alveolar hyperplasia (Rittinghausen et al. 1990). This alteration was preferentially located in the centro-acinar region adjacent to terminal airways. Alveoli lined by hyperplastic epithelium were commonly found to be in luminal continuity with terminal bronchioli. Morphometric investigations revealed that the extent of the hyperplasias varied with the compound used, the applied concentration and the length of exposure (Aufderheide et al. 1987).

Ultrastructurally brochiolo-alveolar hyperplasias were mainly composed of Clara cells and ciliated cells. Ultrastructural morphometric analysis revealed, that in animals exposed to CdO (270 $\mu g/m^3$) the hyperplastic foci consisted of 53.3 Vol% of ciliated cells and 44.7 Vol% of Clara cells. These values are similar to those observed in terminal bronchioli of clean air control animals (52.9 Vol% of ciliated cells, 45.9 Vol% Clara cells).

Clara cells in bronchiolo-alveolar hyperplasias often contained lamellar bodies which normally are typical constituents of type-II alveolar epithelial cells. In animals exposed to CdO (270 $\mu g/m^3$) this population amounted to 12.1 Vol% of the Clara cells in hyperplastic foci. Occasionally cells with typical structural features of both, Clara cells (apical dome-shaped protrusion, rich content in smooth endoplasmic reticulum) and type-II pneumocytes (electron dense cytoplasm, rough endoplasmic reticulum, lamellar bodies) were observed. Clara cells containing lamellar bodies occurred only very rarely in terminal bronchioli of clean air control animals (2.4 Vol% of all Clara cells). When present they were located in the most distal part of the bronchiolar epithelium next to the border of the alveolar epithelium.

Epithelial cells in bronchiolo-alveolar hyperplasias often showed degenerative features such as lipid droplets in ciliated cells, giant secretory granules in Clara cells, or nuclear inclusions. In some animals a focal proliferation of neuro-endocrine cells was observed.

By light microscopy the peripheral lung tissue appeared fragmented. Upon electronmicroscopic examination interalveolar septae consisted of numerous short strands that were entirely lined by type-I epithelium. Scanning electron microscopy revealed that in the high dose groups the alveolar pores were enlarged rendering a basket-like appearance to the alveolar walls.

In lung tissue distant from the centro-acinar region a generalized distinct thickening of the diffusion barrier was observed. This thickening was due to both, an increased thickness of type-I pneumocytes and of the interstitium of alveolar septae. The alveolar surface of type-I epithelium was undulated and their cytoplasm contained abundant cell organelles. The thickening of the alveolar septal interstitium results from an increased content in connective tissue elements, thickening of basement membranes, and an increased cellularity of the interstitial space. Morphometric analysis of the peripheral lung tissue confirmed, that in animals exposed to 270 $\mu g/m^3$ the thickness of alveolar septae was increased significantly and the overall alveolar surface was drastically reduced.

CONCLUSIONS

The type-I alveolar epithelial cells are very sensitive to damage by toxic agents. Regeneration of alveolar epithelium takes place through proliferation of type-II cells and subsequent differentiation into postmitotic type-I cells (for review see Plopper and Dungworth 1987).

Our observations indicate, that this regenerative process may not only result in a repair of the initial epithelial damage, but may also give rise to bronchiolo-alveolar hyperplasias by way of a transdifferentiation of type-II pneumocytes into Clara cells and possibly also into ciliated cells. Evidence for this is the occurrence of cells which are morphologically intermediate between type-II alveolar epithelial cells and Clara cells and of Clara cells containing lamellar bodies. In this regard the situation in hamsters is different from that observed in rats where Type-II cell hyperplasias persist and presumably may give rise to epithelial metaplasias and neoplasias (Thiedemann et al. 1989). In the hamster transdifferentiation of type-II pneumocytes into Clara cells may be a normal process which is occasionally observed in ageing clean air control animals. Here, too, it may be related to the formation of bronchiolo-alveolar hyperplasias which are commonly found in these animals (Aufderheide 1987).

Alterations induced by chronic inhalation of Cadmium compounds are not limited to the centro-acinar region. The changes observed in the peripheral lung tissue may be interpreted as a result of a generalized epithelial damage, a chronic inflammatory reaction, and an early stage of pulmonary emphysema.

REFERENCES

Aufderheide, M., Thiedemann, K.-U., Riebe, M., Kohler, M. (1987): Quantification of proliferative lesions in hamster lungs after chronic exposure to cadmium aerosols. In: Proceedings "Heavy metals in environment", CEP Consultants Ltd., Edinburgh, UK

Heinrich, U., Peters, L., Rittinghausen, S., Dasenbrock, C., König, H. (1987): Investigations of the carcinogenic effect of various cadmium compounds after inhalation exposure in rodents. In: Proceedings "Heavy metals in environment", CEP Consultants Ltd., Edinburgh, UK

Oldiges, H., Glaser, U., Hochrainer, D. (1987): Longterm inhalation study with four different cadmium compounds in Wistar rats. In: Proceedings "Heavy metals in environment", CEP Consultants Ltd: Edinburgh, UK

Plopper, C.G., Dungworth, D.L. (1987): Structure, function, cell injury and cell renewal of bronchiolar-alveolar epithelium. In: Lung Carcinomas (E.M. McDowell, ed.), Churchill Livingstone, Edinburgh

Rittinghausen, S., Aufderheide, M., Ernst, H., Dasenbrock, C., Fuhst, R., Heinrich, U., Peters, L., Mohr, U. (1990): Histological observations and morphometric analysis in lungs of hamsters after long-term inhalation of Cadmium compounds, this volume

Takenaka, S., Oldiges, H., König, H., Hochrainer, D., Oberdörster, G. (1983): Carcinogenicity of cadmium chloride aerosols in W rats. JNCI 70: 367 - 373.

Thiedemann, K.-U., Lüthe, N., Paulini, I., Kreft, A., Heinrich, U., Glaser, U. (1989): Ultrastructural observations in hamster and rat lungs after chronic inhalation of cadmium compounds. Exp. Pathol. 37: 264 - 268

Weibel, E.R. (1969): Stereological principles for morphometry in electron microscopic cytology. Int. Rev. Cytol. 26: 235 - 302

ACKNOWLEDGMENTS

The authors gratefully acknowledge the expert technical assistance of R. Griebel, A. v. Malotki, and F. Müller. Supported by BMFT/DFVLR-HdA Nr. 01 VD 1838.

Metal Ions in Biology and Medicine. Eds. Ph. Collery, L.A. Poirier, M. Manfait, J.C. Etienne. John Libbey Eurotext, Paris © 1990, pp. 209-211

Ultrastructural localization of aluminium in kidneys and gills of trouts taken in vosges from acidified streams

C. Galle*, C. Chassard-Bouchaud**, J.C. Massabuau***, D. Pepin*

** Laboratoire d'Hydrologie - Institut Louise Blanquet - Faculté de Médecine et de Pharmacie de Clermont-Ferrand; ** Laboratoire de Biophysique - Faculté de Médecine de Créteil; *** laboratoire de Neurologie et Physiologie Comparée - Arcachon, France*

The intracellular concentration of aluminium has been studied in trouts "Trutta fario" taken from acidified streams of Cornimont (Vosges, France) in june 1988.

Trout's kidneys and gills were studied by electron microscopy and two microanalytical methods : 1) the study of thick sections by Secondary Ion Mass Microanalysis (Ion Microscopy) (fig.1) and 2) the study of ultrathin sections, using an Electron Probe X-ray microanalyzer equipped with a transmission electron microprobe (CAMEBAX) (figs.3,4,5).

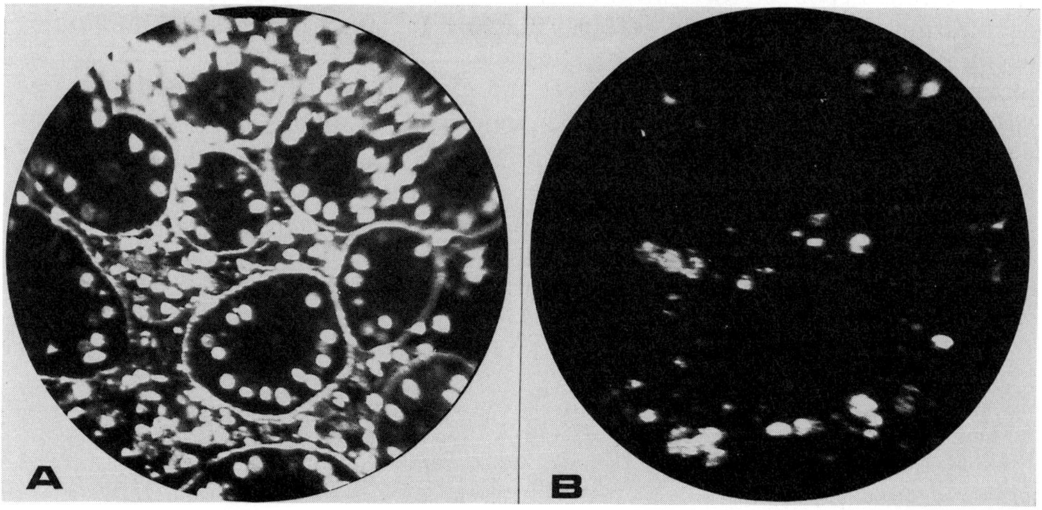

Fig. 1. The ionic image of $^{23}Na^+$ (A) reflects the structure of renal tissue and the ionic image of $^{27}Al^+$ (B) reveals the presence of many important spots of aluminium localized in the interstitial tissue.

Ion microscopy revealed an important emission of aluminium from the interstitial tissue of both kidneys and gills. In these trouts, no aluminium was found in the cytoplasm of tubular cells (fig. 1). The structure of trout's kidney observed by light microscopy is presented on figure 2.

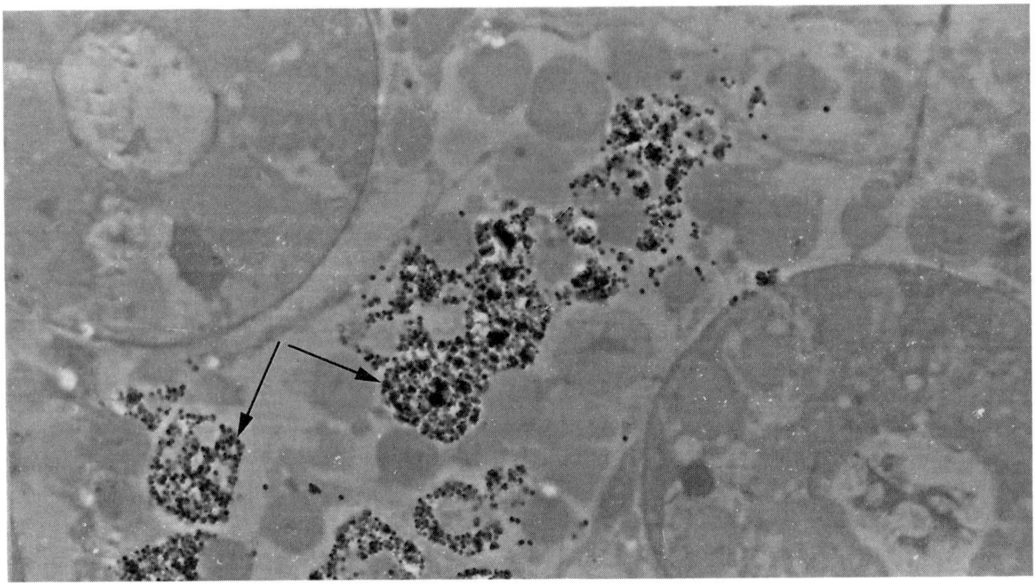

Fig. 2.: this picture of trout's kidney shows melanocytes which are numerous in these trouts kidneys. Their melanosomes (arrow) contain Al, P, Ca, and S.

And Electron Probe X-Ray allowed to determine the precise intracellular concentration of Aluminium (figs. 3,4,5). In the kidneys, Aluminium concentration is observed in melanosomes of melanocytes where it is associated with Phosphorus, Calcium and Sulfur. In gills, Aluminium is localized in melanosomes too (fig. 4) and also, in higher concentration in phagolysosomes of macrophages (fig. 5). Higher signals of Aluminium are observed for older trouts.

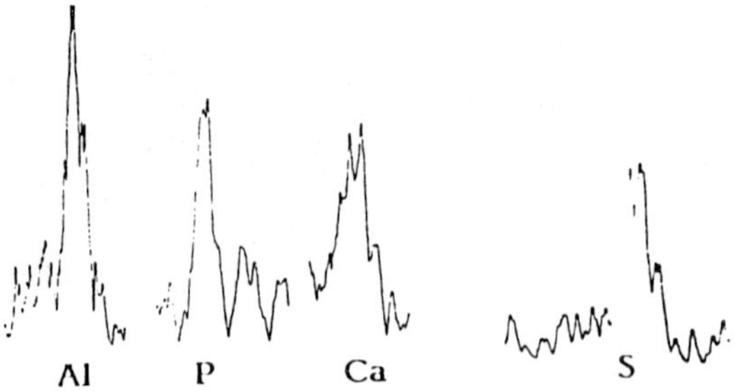

Fig. 3 : Kα lines of Al, P, Ca and S obtained during microanalysis of melanosome from trout's kidney. Aluminium peaks are bigger for gill's lysosomes.

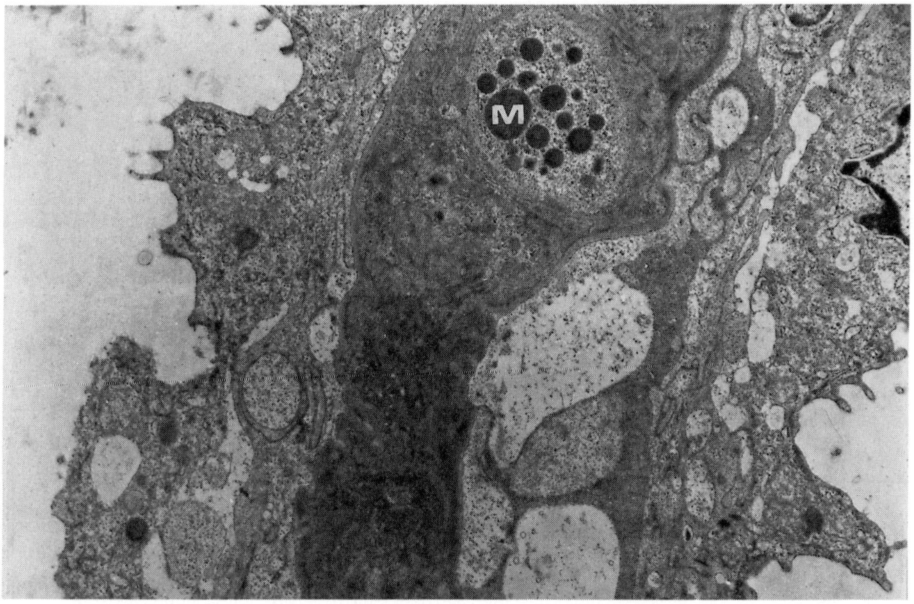

Fig. 4 : Gill's lamellae with melanosomes (M) (x 10 000)

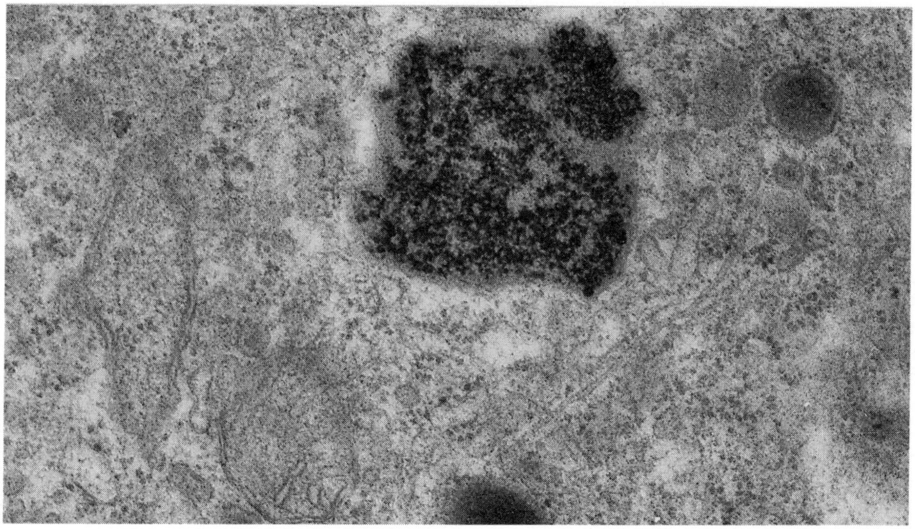

Fig. 5 : High magnification of a phagolysosome in the cytoplasm of a gill's macrophage.(x 50 000).

Metal Ions in Biology and Medicine. Eds. Ph. Collery, L.A. Poirier, M. Manfait, J.C. Etienne. John Libbey Eurotext, Paris © 1990, pp. 212-214

Comparative study of a Bioavailability of As^{5+} Administered to guinea pigs by an aerosol of thermal water from la bourboule and an other of sodium arsenate solution

D. Pepin*, F. Verdier*, N. Boscher**, M.P. Sauvant*

* *Laboratoire d'Hydrologie et d'Hygiène, Faculté de Pharmacie, B.P. 38, 63001 Clermont-Ferrand Cedex, France;* ** *Laboratoire d'Analyse par Activation, C.E.N. Saclay, 91191 Gif-sur-Yvette, France*

INTRODUCTION

At LA BOURBOULE spa (FRANCE) administration of aerosols of mineral water is one of the main treatments of some respiratory diseases. This mineral water (4,61 g/l) is characterized by its high arsenic content (7 mg/l), a mineral ion to which the therapeutic effect is attribued.

In previous study (IRIGARAY et al., 1988), we have use this element as a tracer to demonstrate the difference of behaviour between the mineral water and an ordinary sodium arsenate solution. The aims of this work were to prove in guinea pigs the pulmonary absorption of As present in aerosols, after repeated aerosols and the retention of As by different organs.

MATERIAL AND METHODS

Experimental protocol

This study was carried out on male HARTLEY guinea pigs (CHARLES RIVER, SAINT-AUBIN LES ELBOEUFS, FRANCE) (weight : 200 g), kept in metabolic cages and fed with a commercial standard diet. They were randomized in three groups : one exposed to the aerosols of mineral water from LA BOURBOULE (n = 10), another to the aerosols of sodium (7 mg/l) (n = 10) and the last to the aerosols of sodium chloride solution producing the same ionic strength as the thermal water (n = 4 - control group).

The guinea pigs were exposed to a specific aerosol in a collective enclosure for 30 minutes a day over a period of 21 days. For each animal, 24 hour urine samples were collected. At the end of the experiment, the guinea pigs were weighed and killed. Lungs, livers and kidneys were removed, weighed and freeze-dried.

As determinations

After extraction by the PUTTEMANS and MASSART' method (1982), urinary As were determined by G.F.A.A.S. on a PERKIN-ELMER spectrophotometer Model 5000, equiped with an HGA 500 furnace with a background correction Zeeman-effect.

In the lungs, livers and kidneys, As were determined by Neutron Activation Analysis using the DESCHAMPS et al.' method (1982).

Statistical analysis

Statistical analysis were carried out by KRUSKALL and WALLIS' test and by U-MANN-WHITNEY' test computed with Statview + on Macintosh II X. All data are representated as mean+/- S.E. .

RESULTS

Treated guinea pigs gained weight during the experiment and the rate of gain was the same as that observed in control animals.
Furthermore the response to the aerosols of sodium arsenate solution and to the aerosols of mineral water from LA BOURBOULE spa differed from each other and, moreover, from the control (aerosol of sodium chloride solution).
The results (ng As/g dry weight organ or ng As/ml urines) are in table 1.

ORGANS	GROUP 1 (n = 10) (La Bourboule mineral water)	GROUP 2 (n= 10) (Sodium arsenate solution)	GROUP 3 (n = 4) (Sodium chloride solution)
LUNGS	18.9 +/- 0.9*	15.7 +/- 0.6*	13.0 +/- 0.6
LIVERS	33.8 +/- 1.8*	26.2 +/- 2.2	29.4 +/- 2.5
KIDNEYS	26.7 +/- 2.2	36.9 +/- 1.8*	29.3 +/- 3.4
URINE (total quantity excreted over 21 days)	300.4 +/- 32.9 *	380.9 +/- 53.1*	216.1 +/- 21.5

TABLE 1 : ARSENIC CONCENTRATIONS IN LUNGS, LIVERS, KIDNEYS AND URINES AFTER 21 AEROSOLS
(Results expressed as mean +/- S.E.)
* $p < 0,05$ - Comparison with the control group

The groups 1 and 2 showed a amount of As in the **lungs** statistically greater from the control group. Moreover, As was increased statistically in the **livers** in the group 1 and in the **kidneys** in the group 2.
The total excreted As in urine over 21 days was increased statistically in the both groups.

DISCUSSION

This study proved the absorption and accumulation in guinea pigs of As present in aerosols. The pattern and magnitude of response to sodium arsenate solution and to LA BOURBOULE mineral water differed. This suggests a particular bioavailability relative to chemical species of As present in solution. It is to be noted however that after the aerosol of mineral water As was detected especially in the lungs and livers. This phenomenom can be due to a different absorption of As by the lungs according to the nature of the aerosol in the first time, then a passage in the circulation and an accumulation in the livers, before the urinary excretion ; the presence of As being noted first in the urine. The same particular distributions of chemical species As were noted by MARAFANTE and VAHTER (1987) in hamsters after intratracheal instillations.
The long retention time of As in the lungs following exposure to the aerosols of mineral water is interesting in view of the reported remanent therapeutic effects of As, and of thermal aerosols generally.

REFERENCES

DESCHAMPS N. et al. (1982) : Dosage des oligoéléments de l'émail dentaire par irradiation neutronique - J. Radioanal. Chemistry, 70, 109-116.
IRIGARAY J.L. et al. (1988) : Nuclear method applications in the study of arsenic and molybdenum reabsorption in the body - J. Trace Microprobe Technics, 6, 125-132.
MARAFANTE E. et VAHTER M. (1987) : Solubility, retention and metabolism of intratracheally and orally administered inorganic arsenic compounds in the hamster. - Environ. Res., 42, 72-82.
PUTTEMANS F. et MASSART D.L. (1982) : Solvent extraction procedures for the differential determination of arsenic (V) and arsenic (III) species by electrothermal atomic absorption spectrometry - Anal. Chim. Acta, 70, 109-116.

II. Epidemiology, analysis

Metal Ions in Biology and Medicine. Eds. Ph. Collery, L.A. Poirier, M. Manfait, J.C. Etienne. John Libbey Eurotext, Paris © 1990, pp. 217-219

Chemical speciation studies in relation to aluminium toxicity

S. Daydé, G. Berthon

INSERM U305, Université Paul Sabatier, 38, rue des Trente-Six Ponts, 31400 Toulouse, France

Long considered biologically inert, aluminium has recently been recognized as the causative agent of a series of diseases occurring in renal patients on long-term hemodialysis therapy. The so-called dialysis encephalopathy was the first of these to be characterized (Alfrey **et al.**, 1976), osteodystrophy and microcytic anemia being later shown to be associated with it.
The source of aluminium was initially attributed to the tap water of the dialysates (Ward **et al.**, 1978). However, the use of purified water did not lead to the complete disappearance of these disorders. Indeed, aluminium intoxication was also shown to affect uremic patients administered oral aluminium-containing phosphate-bindings gels only (Andreoli **et al.**, 1984). Such phosphate binders are now the main source of aluminium overload for dialysis patients treated with low aluminium dialysates (Molitoris **et al.**, 1989).

Iatrogenic aluminium toxicity does not affect uremic patients only. Aluminium poisoning has also been reported in those suffering from peptic ulcer disease treated with aluminium-containing antacids in the long run (Alfrey, 1984). Aluminium is also present in food, especially in nutrients prepared in aluminium cookware ustensils, and drugs. In particular, the frequent occurrence of this metal in nutritive solutions represents a serious hazard for infants on long-term total parenteral nutrition (Sedman **et al.**, 1985). In the same respect, aluminium has been shown to contaminate milk formulae (Watt and Ross, 1989) as well as blood products (Fell **et al.**, 1986).
The threat of aluminium scattering in everyday life is all the more preoccupating as this metal does seem to play an active role in the pathogenesis of Alzheimer's disease (McLachlan **et al.**, 1989).

Healthy individuals are generally believed to develop formidable barriers towards aluminium absorption. Nonetheless, this absorption may increase when large doses of aluminium-containing drugs are administered (Alfrey, 1984). In fact, the apparent protective effect of the gastrointestinal tract mainly stems from the low doses of metal usually ingested. Dose-dependent effects are not the only factor to condition aluminium toxicity, and the apparent innocuity of this metal is also largely due to the nonabsorbability of the forms under which it is administered.
Aluminium bioavailability does indeed largely depend on the nature of the salts with which it coexists in vivo. For example, citrate oral administration simultaneously to that of aluminium hydroxide tends to increase the concentration of aluminium in blood and to facilitate its tissue retention (Slanina **et al.**, 1986).
Once absorbed, aluminium is normally eliminated by urine, but dose-dependent effects may interfere with its excretion since it is unknown whether the normal kidney can cope with large amounts of this metal.

Clearly, aluminium prevention as well as detoxification treatments are an urgent need. The only drug currently available to remove aluminium from the tissues of patients with aluminium-associated diseases is desferrioxamine (Kruck and Kalow, 1985). Unfortunately, this compound suffers serious drawbacks such as lack of aluminium specificity, undesirable side-effects, not to mention its high cost.
Before strategies can be elaborated to design new aluminium sequestering agents, factors affecting its bioavailability must be elucidated. First, speciation studies are necessary to discriminate the chemical forms under which aluminium occurs in the main biofluids. In this respect, whereas protidic fractions of aluminium are experimentally analysable, its low-molecular-weight (l.m.w.) pools, though the most active in terms of bioavailability, are not attainable by analytical techniques. Computer simulation is thus the only possible approach.

The reliability of any simulation model crucially depends on that of the parameters on which it is based. Like direct analytical data relative to the biofluid investigated, formation constants of the complexes potentially formed *in vivo* must also be known with precision. Since blood plasma plays a central role in terms of bioavailability, our first objective was to determine the constants for its main aluminium l.m.w. complexes under appropriate *in vitro* conditions (Venturini and Berthon, 1989; Daydé **et al.**, 1990).
Simulations based on these first results have shown that aluminium monophosphate and aluminium trihydroxide are by far the most predominant species in the ultrafiltrable fraction of this metal (Daydé **et al.**, 1990).

Table 1: pH ranges within which aluminium-trihydroxide or/and monophosphate salts can dissolve in presence of average amounts of dietary acids.

Acid concentration ($mol.dm^{-3}$)	Aluminium concentration ($mol.dm^{-3}$)	pH range	
		$Al(OH)_3$	$AlPO_4$
Citric acid	0.005	1-8	1-8
(0.0312)	0.05	<3.7	<2.3/7.8<
	0.5	<3.2	<1.8
Malic acid	0.005	<7.0	1-8
(0.0466)	0.05	<3.9	<2.4/7.8<
	0.5	<3.2	<1.8
Oxalic acid	0.005	<6.4	1-8
(0.0278)	0.05	<3.8	<2.4/7.9<
	0.5	<3.2	<1.8
Succinic acid	0.005	<4.3	1-8
(0.0100)	0.05	<3.7	<2.3/7.9<
	0.5	<3.2	<1.8
Tartric acid	0.005	<5.8	1-8
(0.020)	0.05	<3.7	<2.3/7.8<
	0.5	<3.2	<1.8

As mentioned above, citric acid has recently been recognized as a major factor in the toxicity of orally administered aluminium compounds (Molitoris **et al.**, 1989).
Simulations of the distribution of aluminium in the presence of citrate in the gastrointestinal fluid corroborate this clinical observation: the neutral aluminium monocitrate complex forms near pH 2.5, which favours stomach absorption (Venturini and Berthon, 1989). In contrast, parallel simulations relative to plasma have predicted that high citric acid levels should induce aluminium urinary excretion since electrically charged aluminium citrate species are predominantly formed in this medium. This prevision has been validated on mice (Domingo **et al.**, 1988).

Like citric acid, other carboxylic acids are commonly present in food, but no data has been produced so far concerning their capacity to influence aluminium absorption. Computer simulations relative to gastrointestinal conditions have thus been run to investigate the possible effect of their conjugated anions on the bioavailability of aluminium from aluminium-containing antacids.

Two different phenomena may a priori favour aluminium absorption:

(i) aluminium complexes with dietary acids may be stable enough to induce such a decrease in free aluminium concentration that the therapeutic salts administered can dissolve;

(ii) the above complexes may be electrically neutral, hence membrane diffusible. It may also be that these acids do not form any neutral species with aluminium, but that once the aluminium salt dissolved, its released free ions can be coordinated into neutral forms by others nutrients.

Table 1 shows the pH range within which the dissolution of aluminium trihydroxide or monophosphate salts appears thermodynamically possible between pH 1 and pH 8.

REFERENCES

Alfrey, A.C., Legendre, G.R., and Kaehny, W.D. (1976): The dialysis encephalopathy symdrome. Possible aluminium intoxication. *N. Engl. J. Med. 294*: 184-188.

Alfrey, A.C. (1984): Aluminium intoxication. *N. Engl. J. Med. 310*: 1113-1115.

Andreoli, S.P., Bergstein, J.M. and Sherrard, D.J. (1984): Aluminium intoxication from aluminium-containing phosphate binders in children with azotemia not undergoing dialysis. *N. Engl. J. Med. 310*: 1079-1084.

Daydé, S., Filella, M. and Berthon, G. (1990): Aluminium speciation studies in biological fluids. Part 3. Quantitative investigation of aluminium-phosphate complexes and assessment of their potential significance in vivo. *J. Inorg. Biochem.* in press.

Domingo, J.L., Gomez, M., Llobet, J.M. and Corbella, J. (1988): Parenteral citric acid for aluminium intoxication. *Lancet ii*: 1362-1363.

Fell, G.S., Shenkin, A. and Halls, D.J. (1986): Aluminium contamination of intravenous pharmaceuticals, nutrient, and blood products. *Lancet i*: 380.

Kruck, T.P.A., and Kalow, W. (1985): Determination of desferrioxamine and a major metabolite by high-performance liquid chromatography. Application to the treatment of aluminium related disorders. *J. Chromatogr. 341*: 123-130.

MacLachlan, D.R.C., Lukiw W.J. and Kruck, T.P.A. (1989): New evidence for an active role of aluminium in Alzheimer's disease. *Can. J. Neurol. Sci. 16*: 490-497.

Molitoris, B.A., Froment, D.H., Mackenzie, T.A., Huffer W.H. and Alfrey, A.C. (1989): Citrate: a major factor in the toxicity of orally administered aluminium compounds. *Kidney Int. 36*: 949-953.

Sedman,A.B., Klein, G.L., Merritt, R.J., Miller N.L., Weber, K.O., Gill, W.L., Anand, H. and Alfrey, A.C. (1985): Evidence of aluminium loading in infants receiving intravenous therapy. *N. Engl. J. Med. 312*: 1337-1342.

Slanina, P., Frech, W., Ekström, L.-G., Lööf, L., Slorach, S. and Cedergren, A. (1986): Dietery citric acid enhances absorption of aluminium in antacids. *Clin. Chem. 32*: 949-953.

Venturini, M. and Berthon, G. (1989): Aluminium speciation studies in biological fluids. Part 2. Quantitative investigation of aluminium-citrate complexes and appraisal of their potential significance in vivo. *J. Inorg. Biochem. 37*: 69-90.

Ward, M.K., Feest, T.G., and Ellis, H.A. (1978): Osteomalacic dialysis osteodystrophy: Evidence for a water-bone aetiological agent, probably aluminium. *Lancet 1*: 841-845.

Watt, S.J. and Ross, J.A.S. (1989): Aluminium and infant formulae. *Lancet i*: 614-615.

Metal Ions in Biology and Medicine. Eds. Ph. Collery, L.A. Poirier, M. Manfait, J.C. Etienne. John Libbey Eurotext, Paris © 1990, pp. 220-222

Accumulation of metals in teeth in the Population of Barcelona

J. Corbella, M. Luna, M. Torra, J. To-Figueras

Departament de Toxicologia. Facultat de Medicina, Hospital Clínic. Universitat de Barcelona. Villarroel 170. 08036 Barcelona, España

INTRODUCTION

Some heavy metals are ubiquitous environmental pollutants and tend to accumulate in tissues of animals and man. Several toxic effects caused by these metals are well known including carcinogenesis (1). The accumulation is produced in some cases preferably in hard tissues, and the analysis of organs like bone and teeth can be used as useful monitoring tools .

One of the most old references about the accumulation of metals in teeth can be found in the manuscript of Devergie (2).No other data can be found until the thirties , when the elemental composition of teeth was studied. Some of the most important references are those of Lowater et al (3), and Drea et al (4). The last reported the presence in teeth of Pb, Al, Ba, V, Cr, Mn and others . Some active research groups in this field are, actually, those of Needelman & Shapiro and others, mainly in Philadelphia and Boston in the USA (5). They have reported the relationship between the accumulation of lead in teeth and the appearence of behavior disorders during childhood. Losee et al (6) reported an analysis of the levels of several heavy metals in teeth. Other research groups can be found in the USA, UK, FRG, South Africa, Scandinavia, New Zeeland, Etc .

There are important variations when the different set of results are compared. However there seems to exist an agreement about the significant accumulation of lead in relation to age. In children , the reported concentrations of lead in teeth range from relatively low levels : p.e. 2.5 ppm in Norway (7) to the higher levels (about 51 ppm) found in the urban area of Philadelphia (8) or the levels up to 100 ppm in children exposed (9) . There have been also found differences between enamel and dentine (9) .

In this work the accumulation of Pb, Cr, Ni, Co and Mn in teeth of the human population of Barcelona has been studied and has been compared with the levels of these metals in other tissues and organs.

MATERIAL AND METHODS

Specimens of human tissues (liver,kidney, lung, heart, pancreas, rib and teeth) were collected during necropsies or odontologic activities . They were washed with distilled water and stored in polyethylene tubes. The analyses were performed treating the specimens with concentrated nitric acid (Suprapur, Merck 441) at 70ºC during 12 hours. After complete digestion the samples were analyzed for Pb,Cr,Ni, Co, and Mn concentration in a Perkin Elmer 3030 spectrophotometer with Zeeman background correction, and L´vov plataform according to the S.T.P.F (stabilized temperature platform furnace) concept (10).

RESULTS and DISCUSSION

The concentrations of metals in teeth can be found in table I (total) and table II (groups of age). The concentration of metals in other tissues can be found in table III.

TABLE I : Metals in teeth (μg / g).

	Pb	Cr	Ni	Co	Mn
N	196	154	193	152	154
X:	26.14	0.597	1.008	0.355	0.207
s.d.	40.94	0.655	0.704	0.178	0.103
s.d / X	1.56	1.09	0.73	0.50	0.49
Correlation with age					
r :	0.426	0.363	0.257	0.267	0.144
p:	<0.01	<0.01	<0.01	<0.01	--------

TABLE II : Metals in teeth according to groups of age (μg/g).

Age (years)	Pb	Cr	Ni	Co	Mn
1-20	9.29	0.321	0.724	0.314	0.184
21-40	15.49	0.453	1.090	0.327	0.200
41-60	26.26	0.612	1.156	0.340	0.219
61-80	56.84	0.985	1.033	0.406	0.227
t 1-20 / 61-80.	4.6358	4.130	2.662	2.621	1.850
p:	< 0.001	< 0.001	< 0.01	< 0.01	------

TABLE III : Metals in tissues (μg / g wet tissue)

	Pb	Cr	Ni	Co	Mn
Bone (rib)	6.92	0.43	2.03	0.50	0.23
Liver	0.82	0.19	0.29	0.23	2.11
Kidney	0.47	0.18	0.23	0.16	0.87
Lung	0.49	0.15	0.22	0.14	0.24
Heart	0.23	0.19	0.17	0.28	0.26
Pancreas	0.34	0.23	0.24	0.23	0.99

Teeth seem to be one of the best markers of human body burden of Pb and Cr with levels higher than those found in the bones. In the case of Co and Ni the teeth is the second most concentrated tissue after the bone. Mn accumulates mostly in soft tissues as liver.

The ratio standar deviation : mean is superior to 1 for lead and chromium. 0.75 for Ni, and around 0.50 for Co and Mn. This suggest that in the case of lead and chromium there are more external factors as a source of variation.

There is a positive correlation between the levels of Pb and Cr in teeth and age. There is also a positive correlation with age in the case Ni and cobalt but less strong, and there is no correlation in the case of Mn.

There is a strong correlation between Pb levels and Cr levels in teeth ($p: < 0.01$) in all the sub-samples. The correlation between Pb levels and Co levels is less strong. ($p : < 0.05$).

REFERENCES

1.-Sunderman, F.W. (Jr). " Metal carcinogenesis". in Goyer,R.A; Mehlman, M.A : " Toxicology of trace elements" . N.York (S.W.Ley). 1977. 257-295.

2.-Devergie, A : Médecine Légale théorique et pratique". Paris (G.Baillière edit). 3ème ed. 1852. (t.III).

3.-Lowater, F ; Murray, M.M : " Chemical composition of teeth .V. Spectrographic analysis ". Biochem. 1937. 31, 837-841.

4.-Drea, W.F. " Spectrum analysis of dental tissues for trace elements " .J. Dent.Res. 1936. 15, 403-406.

5.-Needelman, H.L.; Shapiro, I.M.. " Dentine lead levels in asymptomatic Philadelphia school-children : subclinical exposure in high and low risk groups" Environ.Health Perspect. 1974. 7, 24-31.

6.-Losee, F.L.; Cutress, T.W : Brown, R : " Natural elements of the periodic table in human dental enamel". Caries. Res. 1974. 8, 123-134.

7.-Fosse, G ; Justesen, N.P.B. : " Lead in deciduous teeth of norvegian children". Arch. Environ.Health. 1978. 33, 166-177.

8.-Needleman, H.L. ; Tuncay, O.C ; " Lead levels in deciduous teeth of urban and suburban american children" . Nature. 1972. 235, 111-112.

9.-Burde, B. de la ; Shapiro, I.M. : "Dental lead, blood lead and pica in urban children ". Arch. Environ. Health. 1975. 30, 281-284.

10.-Slavin, W ; Carnick G.R. : " The possibility of standerless furnace atomic absrption spectroscopy ". Spectrochimica Acta. 1984. 39B, 271-282.

Metal Ions in Biology and Medicine. Eds. Ph. Collery, L.A. Poirier, M. Manfait, J.C. Etienne. John Libbey Eurotext, Paris © 1990, pp. 223-225

Cadmium, magnesium, zinc and copper blood concentrations in non-smokers, healthy smokers and lung cancer smokers

D. Jolly[1], Ph. Collery[2,4], H. Millart[3], Ph. Betbeze[2], G. Bechambes[4], C. Cossart[2], D. Perdu[2], H. Vallerand[2], G. Barthes[4], H. Choisy[3], F. Blanchard[1], J.C. Etienne[4]

[1] *Service de Médecine interne à orientation Gérontologique, Hôpital Sébastopôl, 48 rue de Sebastopol, 51092 Reims Cedex (France);* [2] *Clinique des Maladies respiratoires, Hôpital Maison Blanche, 45, rue Cognac Jay, 51092 Reims Cedex (France);* [3] *Laboratoire de Pharmacologie, Hôpital Maison Blanche, 45 rue Cognac Jay, 51092 Reims Cedex, (France);* [4] *Clinique Médicale, Hôpital Robert Debré, rue Alexis Carrel, 51092 Reims Cedex, (France)*

SUMMARY :

We determined plasma Cadmium and Copper concentrations , plasma and red blood cell Zinc and Magnesium concentrations in 29 healthy non smoking men, in 16 healthy smokers and in 14 smoking men with non small cell lung cancer. We noted several significant level variations by comparison between the three groups. Some hypotheses could be proposed for a Magnesium, Gallium or Zinc substitution for the treatement or for the prevention of lung cancer , particulary in smokers.

INTRODUCTION :

We determined plasma Cadmium concentration (P Cd), plasma Copper concentration (P Cu), plasma Zinc concentration (P Zn), red blood cell Zinc concentration (R.B.C. Zn), plasma Magnesium concentration (P Mg), and red blood cell Magnesium concentration (R.B.C. Mg) in 29 healthy non smoking men, in 16 healthy smokers, and in 14 smoking men with non small cell lung cancer.

PATIENTS AND METHODS :

Lung cancer patients came from the hospital, where they were treated. The other participants, smokers and non smokers were healthy volunteers.
Plasma concentrations of Cadmium were routinely monitored by electrothermal atomic absorption spectrometry (ET-AAS). Cadmium levels were determined with a Perkin Elmer atomic absorption spectrophotometer (Model 5000) equipped with a graphite furnace atomizer with the Zeeman correction and a HGA 500 programmer. The instrument was operated in the peak height mode at 228.8 nm with a 0.7 nm slit band width. Red blood cells Magnesium and Zinc concentrations as well as plasma Magnesium, Zinc and Copper levels were measured by Flame atomic absorption spectrophotometry (Perkin Elmer, Model 5000). Wavelenght selection was 324.8 nm for Copper, 285.2 for Magnesium and 213.9 for Zinc.
The results obtained on the three groups were compared using the Mann and Whitney rank-sum test, because of little sample sizes.

RESULTS :

The study of age repartition between the three groups did not show any statistical différence. The age mean, in the 29 healthy non smoking men was 55 years (range : 21-84 years). In the 16 healthy smokers, the age mean was 49 years (range : 31-78 years), and in the 14 men with non small cell lung cancer the age mean was 61 years (range : 48-74 years). So the observed variations, in ion concentrations, discribed below, couldn't result from an age effect.

In table 1 are sum up, means and standard deviations of ion concentrations for each group, and all the comparisons two by two between groups.
The mean of plasma Cadmium concentration was 1.35 µg/l, in healthy men. We noted a significant increase of P Cd in smokers (2.15 µg/l) and in patients with lung cancer (2.34 µg/l). There was no difference between smokers and cancer patients. Plasma Copper concentration was 1000 µg/l in healthy men. We noted a significant increase in P Cu in smokers (1067 µg/l) and in patients with lung cancer (1165 µg/l). The increase was significantly higher in lung cancer patients than in healthy smokers. Plasma Zinc concentration was 0.77 mg/l in healthy men. In smokers P Zn was significantly higher (0.95 mg/l). There was a non significant fall in patients with lung cancer (0.7 mg/l). R.B.C. Zn was 12.18 mg/l in healthy men. In smokers, the concentration was significantly diminished (10.89 mg/l) as well as in cancer patients (9.86 mg/l). The fall was more important in cancer patients than in smokers. The magnesemia was 0.83 mmol/l, in healthy men. In smokers P Mg was significantly higher (0.9 mmol/l). There was a non significant fall in patients with lung cancer (0.78 mmol/l).R.B.C. Mg was 2.21 mmol/l in healthy men. In smokers, this concentration was non significantly diminished (2.05 mmol/l). The R.B.C. Mg increase observed in cancer patients (2.50 mmol/l) was significant in comparison with healthy subjects, and also in comparison with healthy smokers.

V- TABLE 1 -

METAL IONS CONCENTRATIONS

	Results			Comparison			
	Healty non smoking men	Healty smokers	Lung cancer patients	Healty non smoking men	Healty smokers	Lung cancer patients	Mann Whitney tests
P Cd	1.35	2.15	2.34	■	■		p < 0.005
µg/l	0.68	0.71	2.44		■	■	N.S.
				■		■	p < 0.05
P Cu	1000	1067	1165	■	■		p <0.05
µg/l	282	164	320		■	■	p < 0.03
				■		■	p < 0.004
P Zn	0.77	0.95	0.70	■	■		p < 0.005
mg/l	0.17	0.22	0.16		■	■	p < 0.005
				■		■	N.S.
R.B.C. Zn	12.18	10.89	9.86	■	■		p < 0.05
mg/l	2.21	1.64	2.03		■	■	p < 0.04
				■		■	p < 0.002
P Mg	0.83	0.90	0.78	■	■		p < 0.02
mmol/l	0.06	0.11	0.14		■	■	p < 0.03
				■		■	N.S.
R.B.C. Mg	2.21	2.05	2.50	■	■		N.S.
mmol/l	0.34	0.23	0.23		■	■	p < 0.001
				■		■	p < 0.003

These results, from a transversal study are only descriptive and we must be aware of some biases such as hemolysis for smokers. But some hypotheses could be proposed for further studies, and if these hypotheses are supported by further facts, some therapeutic approaches could be proposed for clinical trials.

DISCUSSION :

Cadmium is a highly toxic and carcinogenic divalent metal, as reported by Sunderman (1971), Lucis et al (1972) and Nath et al (1984). Carcinogenic action has been demonstrated in testicular tumors in Wistar rats and its action is antagonised by Zinc. Respiratory and prostatic cancer have been found in excess among Cadmium workers. High level of Cadmium and low level of Zinc, in patients with brochogenic carcinoma, have been so reported by Morgan (1970) and Spicker (1988).

Poirier et al (1983) has demonstrated that testicular carcinoma induced by injections of Cadmium chloride, in Wistar rats can be prevented by simultanéous injection of Magnesium acetate. Magnesium is the obligatary ion for proper transcription of D.N.A. Cadmium, by biochimical competition can replace Calcium and Zinc in certain enzymatic and carrier proteins and decrease the fidelity of Nucleotide incorporation into R.N.A.; This infidelity can be overcome by increased concentration of Magnesium. So a Magnesium substitutive treatement could antagonize Cadmium effects, and perhaps could have a lung cancer protective effect in smokers.
But on the other hand, as demonstrated by Collery et al. (1981), at cancer stage, the red blood cell concentration of magnesium is high (as it is in the tumor in comparison with the peritumoral healthy tissue) and it could facilitate malignant cell development, so its prescription is contra-indicated.
A Gallium treatement could be proposed, for its antagonism with Magnesium, and its inhibition of tumor cell reproduction, as reported by Collery et al. (1986 & 1989) and Vistelle et al (1989).
A Zinc substitution could result in a stimulation of cellular immunity and in a correction of abnormal T cell function discribed in lung cancer, by Allen et all (1985) and Flynn (1983). In healthy smokers or in patients with lung cancer and low red blood cell Zinc concentration, a Zinc substitution could be favorable.

BIBLIOGRAPHY :

ALLEN,J.I., BELL, E., BOOSALIS, M.G., OKEN, M.M., McCLAIN, C.J., LEVINE, A.S., MERLEY, J.E. (1985) : Association between urinary zinc excretion and lymphocyte dysfunction in patients with lung cancer. *Am. J. Med.* 79, 209-215.
COLLERY, PH., ANGHILERI, L.J., COUDOUX, P., DURLACH, J. (1981) : Magnesium et cancer : données cliniques. *Magnesium Bulletin* 3, 11-20.
COLLERY, PH., MILLART H., PLUOT, M., ANGHILERI, L.J. (1986) : Effect of Gallium chloride oral administration on transplanted C3HBA mammary adenocarcinoma : Ga, Mg, Ca and Fe concentration and anatomopathological Characteistics. *Anticancer Research* 6,1085-1088.
COLLERY, PH;, MILLART, H., LAMIABLE, D., VISTELLE, R., BERTHIER, A., BETBEZE,P., COSSART, C., MULETTE, T., MASURE, F., GOURDIER, B., PECHERY, C., CHOISY, H., ETIENNE, J-C., DUBOIS DE MONTREYNAUD, J-M. (1989) : Magnesium alterations and pharmacocinetic data in Gallium-treated lung cancer patients. *Magnesium* 8, 56-64.
FLYNN, A. (1983) : Effects of Antigen Stimulation and Inter leukin-1 on In Vivo Splenic Zinc Changes in the A/J Mouse. *J. Am. Coll. Nut.* 2, 205-213.
LUCIS, O.J., LUCIS, R., ATERMAN, K. (1972) : Tumorigenesis by Cadmium. *Oncology* 26, 53-67.
MORGAN, J.M. (1970) : Cadmium and Zinc abnormalities in bronchogenic carcinoma. *Cancer* 25, 1394-1398
NATH, R., PRASAD, R., PALINAL, V.K., CHOPRA, R.K. (1984) : Molecular basis of Cadmium toxicity. *Progress in Food and Nutrition Science* 8, 109-163.
NORDBERG, G. F. & ANDERSEN, O. (1981) : Metal interactions in carcinogenesis : Enhancement, Inhibition. *Environmental. Health Perspectives* 40, 65-81.
POIRIER, L.A., KASPRAZAK, K.S., HOOVER, K.L., WENK, M.L. (1983) : Effects of Calcium and Magnesium acetates on the carcinogenicity of Cadmium chloride in Wistar rats. *Cancer Reasearch* 43, 4575-4581.
SPIEKER, C., BERTRAM, H. P., ACHATZKY, R., STRATMANN, T., KISTER, K., ZIDEK, W., ZUMKLEY, H. (1988) : Cadmium levels in blood samples and heart tissue of smokers and non-smokers. *Trace element in Medicine* 5, 35-37.
SUNDERMAN F.W. (1971) : Metal carcinogenesis in experimental animals. *Fd. Cosmet. Toxicol.* 9, 105-120.
VISTELLE, R., COLLERY, PH., MILLART, H. (1989) : In vivo distribution of Gallium in healthy rats after oral administration and its intreaction with Fe, Mg and Ca. *Trace Elements in Medecine* 6, 27-32.

Metal Ions in Biology and Medicine. Eds. Ph. Collery, L.A. Poirier, M. Manfait, J.C. Etienne. John Libbey Eurotext, Paris © 1990, pp. 226-228

Blood lead level reference values in Zaragoza (Spain). Study of the seasonal variations

M. Gonzalez, A. Garcia De Jalon, T. Abadia, M. Perez

Unidad de Nutrición y Metales, Servicio de Bioquímica Clínica, Hospital Miguel Servet, Insalud, Zaragoza (España)

A first stage in the knowledge of epidemiology of lead toxicity can be the study of the blood lead levels in an adult and "healthy" population.

We have made a survey in our local area in order to know which are blood lead levels in an adult and healthy population. The values have ranged from 7.61 µg/dl (0.37 µM/l) to 31.62 µg/dl (1.53 µM/l), mean 13.27 ± 3.47 µg/dl (0.64 ± 0.17 µM/l), with a Kurtosis index of 3.42 and a bevel of 1.47. Before applying the estadistical Kolmogprov-Smirnov test we have checked that our blood lead level curve is gaussiana.

These values agree with those recorded in other countries that belong to European Economic Comunity (13 and 16 µg/dl).

Blood lead level and sexes:

Blood lead level found in males has ranged from 7.61 µg/dl (0.37 µM/l) to 31.62 µg/dl (1.53 µM/l), mean 14.89 ± 3.98 µg/dl (0.72 ± 0.19 µM/l). Blood lead level found in females has ranged from 7.61 µg/dl (0.37 µM/l) to 21.43 µg/dl (1.03 µM/l). (Fig. 1)

Mean blood lead level found in males is significantly higher than blood lead found in females. This can be due to the different haematocrit value between both sexes. We must remind that lead is measured in whole blood and that erytrocites are the main support of this metal in blood.

Besides, it may also happen that most jobs conneted with a high risk of lead toxicity are mainly done by men.

FIG. 1

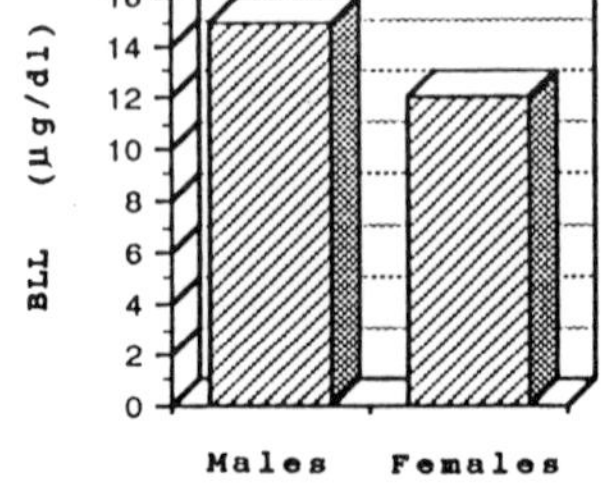

Blood lead level and age:

In figure 2 we can see how blood lead levels change according to age in both sexes, male and female. In both, there is an increase in blood lead until fifty, a light fall from fifty to seventy (over all in females); over eighty, values tend to reduce the difference between both sexes.

FIG. 2

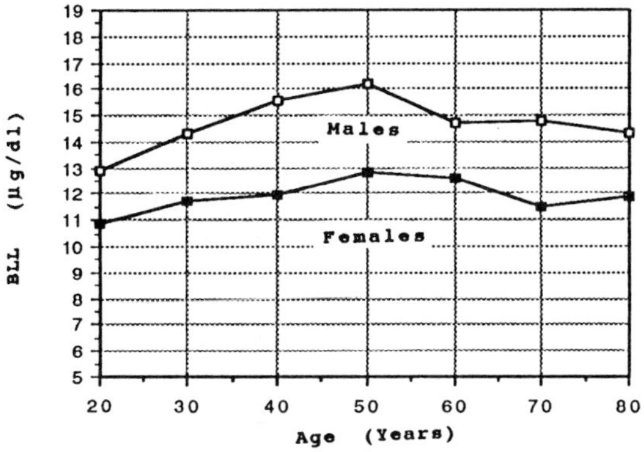

We have made another survey, from 1975 to 1987, trying to investigate if during all this time there has been a special behaviour in blood lead levels according to annual, monthly and seasonal variations.

Annual Variations:

From 1975 to 1977 blood lead levels show an increase. In 1977 and 1978 they go down; the maximun peak happened in 1979. Since then lead levels were getting lower and lower until 1987 when we finished our study.

Monthly Variations:

When we analysed monthly behaviour of blood lead levels (Fig. 3), we observe that values are smaller in December, January and February, and they are bigger in June, July, August and September.

FIG. 3

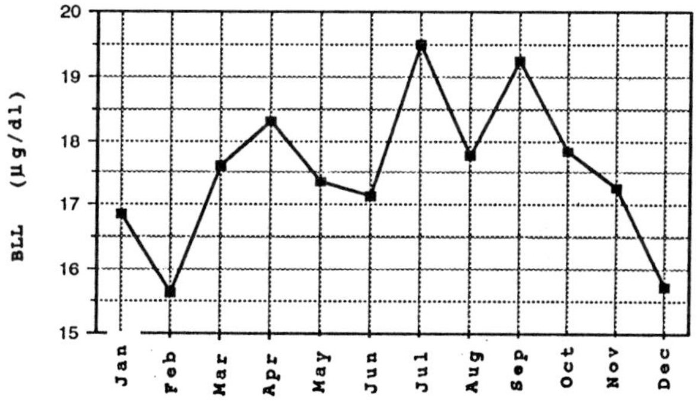

The upper level we found in July, 19.51 ± 7.15 µg/dl (0.94 µg/dl), followed by September, 19.25 ± 7.73 µg/dl (0.92 µg/dl). We found the lowest peak in February, 15.63 ± 6.07 µg/dl (0.75 µg/dl), followed by December, 15.72 ± 6.40 µg/dl (0.76 µg/dl).

Seasonal Variations:

In order to identify if there was a seasonal behaviour we gathered all values together in a cold season (December, January and February) and in a hot season (July, August and September) and compared, by applying the "t" Student test, the mean blood lead level between these two seasons.

During hot seasons, blood lead levels ranged from 6μg/dl (0.29 μM/l) to 35 μg/dl (1.69 μM/l), mean 18.93 ± 7.05 μg/dl (0.91± 0.3 μM/l). During cold seasons ranged from 2 μg/dl (0.10 μM/l) to 35 μg/dl (1.69 μM/l), mean 16.14 ± 6.28 μg/dl (0.78 ± 0.30 μM/l). (Fig. 4)

FIG. 4

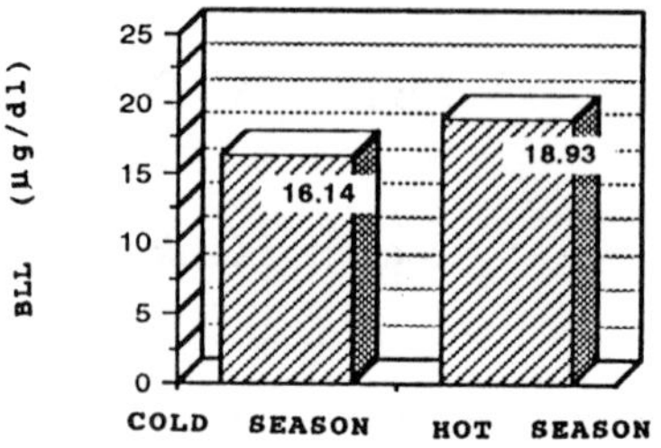

Blood lead level in hot seasons has been significantly higher that blood lead in cold seasons.

We can conclude that in 10 years there is a seasonal variation in the blood lead level in adults, since we have found that values in hot season are significantly higher than those we found in cold season.

The exact reasons for a seasonal pattern are not well known yet. There are several hypothesis:

1.- In hot season there is a sharp increase in comsumption of petrol, and a specific increase in use of motorcicles.

2.- In hot season, people remain longer outdoors, in the open air, and so, they are much more exposed to air pollution.

3.- A hot atmosphere can change chemical composition of lead compounds in cities air, the size of polluting particles, the time these particles stay in the air ... All these events can increase the contaminant effect of lead in atmosphere.

BIBLIOGRAPHY:

1.- ABADIA T.: Toxicidad por plomo. Determinación de la plumbemia. Valores de refencia en nuestro medio y estudio de su variacion temporal. Tesis Doctoral. Facultad de Medicina. Universidad de Zaragoza. Febrero 1989.

2.- CHISOLM J.Jr: Lead in red blood cells and plasma. J Ped 1974; 84 (1): 163-164.

3.- Comision European Communities. Directive on lead: The Safety. Practicioner 1983; 9: 48.

4.- DE ROSSA.; BRIGHENTI F.; ROSSI A.; CAROLDI S.; GORI GP.; CHIESURA P.: The ceramics industry and lead poisoning. Scand J Work Environ Health 1980; 6: 306.

5.- HARRIS R.W.; ELSEA W.R.: Ceramic glaze as a source of lead poisoning. J Am Med Ass 1967; 202: 544-546.

Metal Ions in Biology and Medicine. Eds. Ph. Collery, L.A. Poirier, M. Manfait, J.C. Etienne. John Libbey Eurotext, Paris © 1990, pp. 229-231

Modification of Hessel's method to measure blood lead. Methodology and quality control

A. Garcia De Jalon, M. Gonzalez, T. Abadia, M. Perez

Unidad de Nutrición y Metales, Servicio de Bioquínica Clínica, Hospital Miguel Servet, Insalud, Zaragoza (España)

SURVEY SAMPLE DESIGN

In order to stablish reference values in our local area we tested five hundred adults randomly selected, men and women, chosen one of five who came to samples extraction area of the hospital. We can accept that, if we previously exclude risk fellows, patients who came to surgeries of the hospital are similar to general population.

MATERIAL AND METHODS

To asses lead levels in blood we used a semi-micro method based on Hessel´s technique (modification of Hessel´s method). It consists in complexing directly blood lead with APDC, and then, to extract the complex APDC-Pb with MIC. Blood must be previously haemolized with Triton X-100.

We reduced to 2 ml serum samples without less of accuracy, sensibility and reproductibility of the method.

Lead concentrations in venous whole blood and quality control specimens were determined by atomic absorption spectrophotometry (E.A.A.) choosing a wavelength of 283 nm.

We have used a spectrophotometer Perkin-Elmer model 460 equipped with a hollow cathode lamp:

Wavelength : 283 nm
Crack : 0.7 nm
Flame : Acetylene/air
Current strenght of the lamp: 6 mA.
Reagents: APDC solution 8%.
Triton X-100.
Metilisobutilcetona (MIC).
Main solution 1 gr/l of lead.
Solution 5 mg/l of lead.
Solution 100 mg/l of lead.
Standard solutions.

FIDELITY OF THE ANALYTICAL METHOD

Specificity

We have chosen the 283 nm analytical emission line because although is less strong than 213 nm line, don´t show interferences with MIC spectrum. At this wavelength, the method is specific enough.

Accuracy

We determined the recovery ratio of lead added by a standard addition test. A same blood sample (pool from several individuals) was divided in 20 aliquots of 2 ml. We added 0, 25, 50 and 75 μg/dl of lead. Five repetitions per sample were made; the recovery ratio of lead added we´ve got has ranged from 97% to 99%.

Precision

A pool of heparinized blood from several individuals was prepared; we took 25 aliquots of 2 ml and measured them separately. The mean we got was 11.16 ± 0.78 μg/dl. The coefficient of variation was 6.98%.

Sensibility

According to methodology proposed (modified Hessel´s method), we can say that:

- Only 2 ml of heparinized whole venous blood are necessary. We can consider this method as a semi-micro one, able to be used in pediatrics.
- Reproductibility and accuracy of the method have been checked. Recovery ratio of lead added is 98%.
- Special accesories for flameless A.A.E. are not required.

QUALITY CONTROL

Blood lead level is a recent exposition ratio to lead, and it has a high prediction power when it is submitted into a Quality Control Program.

Measurements of this survey have been included into a QC-BLL Program interlaboratories. Every month three control samples are sent to each participant laboratory. They analyse blood lead by methodology they usually run, and send to the Program values they have got, and these are submitted into a statistical analysis in order to establish: valid results, variation ratio, mean variation ratio and centre values.

In figures below we compare the control values measured in our laboratory by AAE after organic extraction respected to:

- values given by laboratories that use the same method.
- the centre values.
- the rest of the values given by all laboratories that participate in this Program.

All these comparisons have been made with specimens collected from January 1987 through April 1988.

Figure 1. Comparison between control-1 values got in our laboratory by A.A.E. and the same given by other laboratories that also use this methodology (EAA-1).

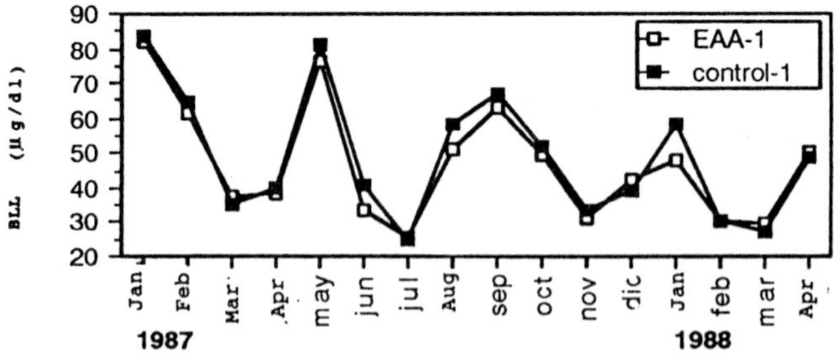

Figure 2: Comparison between control-1 values got in our laboratory and centre values given by all laboratories that participate in the Program

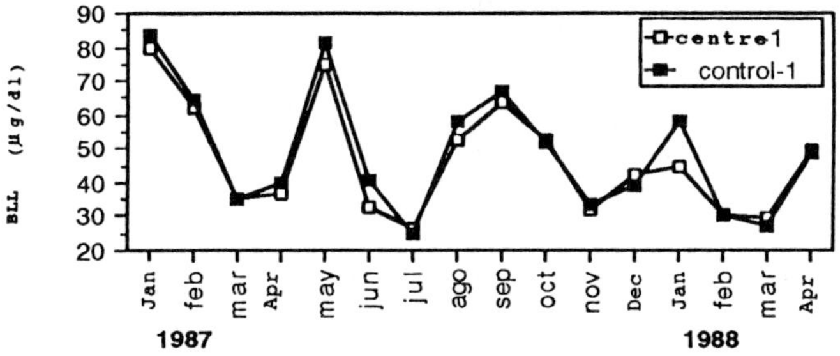

Figure 3: Comparison between control-1 values got in our laboratory and values given by all participants laboratories

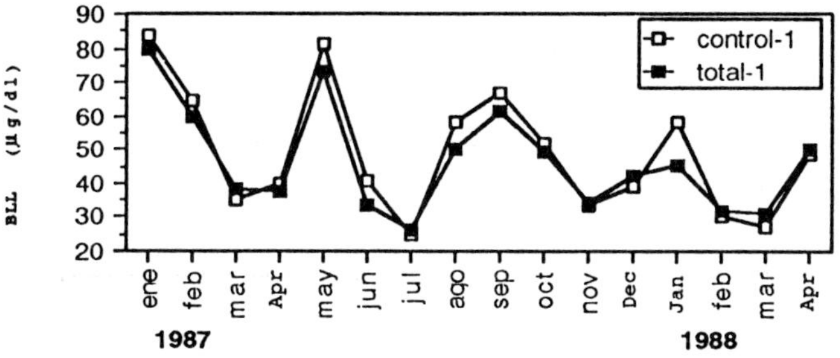

When we compare our outcome with these values we conclude that the analytical measurements done in this research by A.A.E. perfectly agree with: 1) all values given by other Laboratories that use non modified Hessel´s method, 2) the Centre Values and 3) Values given by other laboratories that participate in this Program and use all posible determination methods.

BIBLIOGRAPHY:

1.- ABADIA T.: Toxicidad por plomo. Determinación de la plumbemia. Valores de refencia en nuestro medio y estudio de su variacion temporal. Tesis Doctoral. Facultad de Medicina. Universidad de Zaragoza. Febrero 1989.

2.- ARROYO M.; XIMENEZ L.: Determinación analítica de plomo en sangre por Espectrofotometría de Absorción Atómica. Rev Clin Esp 1972; 2: 119.

3.- ARROYO M: La plumbemia como parámetro indicador en la exposición al plomo. Estudio comparativo de diferentes poblaciones. Med Seg Trab, 1977; 25 (97): 27-36.

4.- GARCIA DE JALON A.; SANCHEZ AGREDA M.; MARTINEZ M.P.; BASSECOURT M.: Aportación de la determinación en sangre y orina por E.A.A. (Niveles de plumbemia en una población normal). SANGRE 1976; 21 (2): 375-386.

5.- RICHTERICH R.;COLOMBO J.P.: Química Clínica. Teoría , práctica e interpretación. Ed. Salvat, 1983 Barcelona, España.

Metal Ions in Biology and Medicine. Eds. Ph. Collery, L.A. Poirier, M. Manfait, J.C. Etienne. John Libbey Eurotext, Paris © 1990, pp.232-234

Blood lead level : place of residence. Wine consumption. Laboral risk

T. Abadia, A. Garcia De Jalon, M. Gonzalez, M. Perez

Unidad de Nutrición y Metales, Servicio de Bioquímica Clínica, Hospital Miguel Servet, Insalud, Zaragoza (España)

We have made a survey in our local area with the purpose of assessing if there is any significantly association between values of blood level among adults in general population and some sociolaborals aspects: area of residence, wine consumption and laboral risk (individuals occupationally exposed).

AREA OF RESIDENCE

We have divided our area in:

1.- Urban area (city of Zaragoza); we have subdivided this study population into four subgroups:

a) Inner city
b) City center
c) Outskirts
d) Left bank of the Ebro river

2.- Rural area, including several smaller towns.

Schematic urban map of the city of Zaragoza

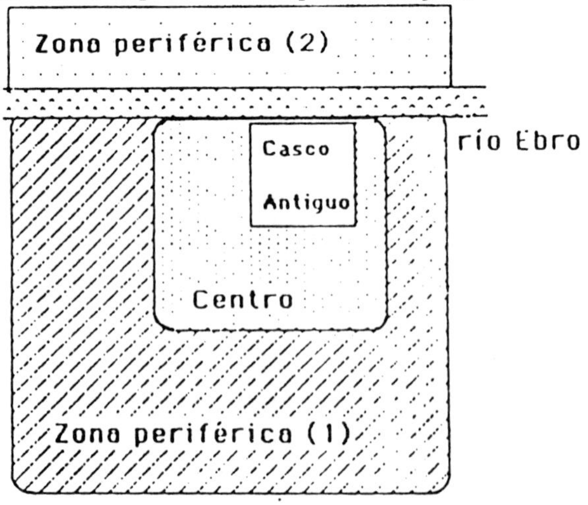

Althought there weren´t statistically significant differences, blood lead levels measured in people who live in rural area were higher than those of people who live in urban area.

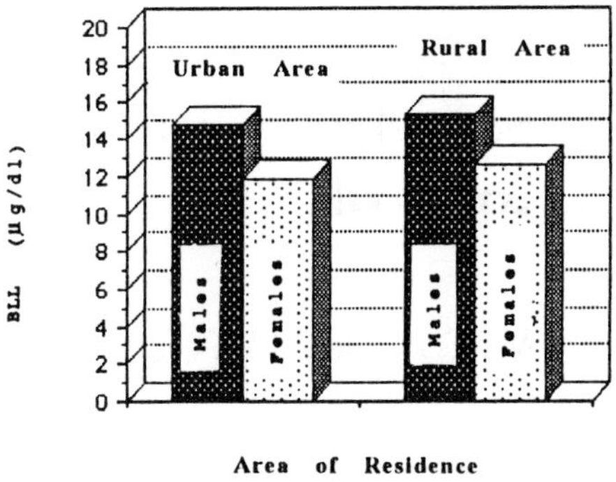

This fact can be explained because:

a) In our rural area is very common having old domiciliary lead piping systems.

b) In our rural area is extensivily diseminated the use of glazed pottery receptacles with lead salts used for storing home-made foods that take acid flavouring (vinegars, wine...) that melt lead salts of the varnishes of these receptacles contaminating foods.

MALES

Mean blood lead concentration in males who live in rural area was higher (15.26 µg/dl) than in males who live in the city (14.77 µg/dl). We haven´t found statistically significant association.

For each geographic subgroup tested, mean blood lead concentration in males was:

- BLL in males who live in inner city....15.63 µg/dl (0.75 µM/l).
- BLL in males who live in left bank ....15.28 µg/dl (0.74 µM/l).
- BLL in males who live in city center...14.37 µg/dl (0.57 µM/l).
- BLL in males who live in outskirts .. 13.92 µg/dl (0.67 µM/l).

We haven´t found statistically significant association.

FEMALES

Mean blood lead concentration in females who live in rural area is lightly higher (12.6 µg/dl) than in females who live in the city (11.8 µg/dl). We haven´t found statistically significant association.

For each geographic subgroup tested, mean blood lead concentration in females was:

- BLL in females who live in inner city..11.64 µg/dl (0.56 µM/l).
- BLL in females who live in left bank ..11.60 µg/dl (0.56 µM/l).
- BLL in females who live in city center.11.92 µg/dl (0.57 µM/l).
- BLL in females who live in outskirts ..12.04 µg/dl (0.58 µM/l).

We haven´t found statistically significant association.

BLOOD LEAD LEVEL AND WINE CONSUMPTION

BL levels in males that usually consume wine,(wine drinkers), and those ones abstemious have been:

-Wine drinkers: 16.96 ± 4.85 µg/dl (0.82 ± 0.23µM/l).
-Abstemious: 13.92 ± 3.45 µg/dl (0.67 ± 0.17 µM/l).

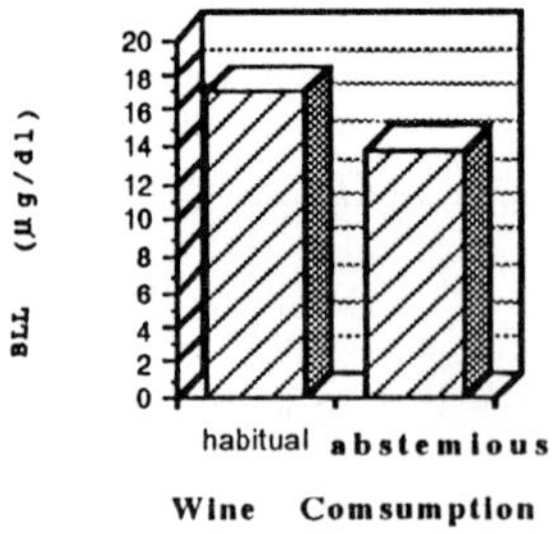

Before applying the "t" Student test of bilateral contrast we conclude that difference found between BLL in males that usually consume wine and that found in abstemious is statistically significant.

BLOOD LEAD LEVEL AND LABORAL RISK

BLL in males without laboral risk: 14.17 ± 3.46 µg/dl (0.68 ± 0.17 µM/l).

BLL in males with laboral risk: 18.7 ± 4.23 µg/dl (0.9 ± 0.2µM/l).

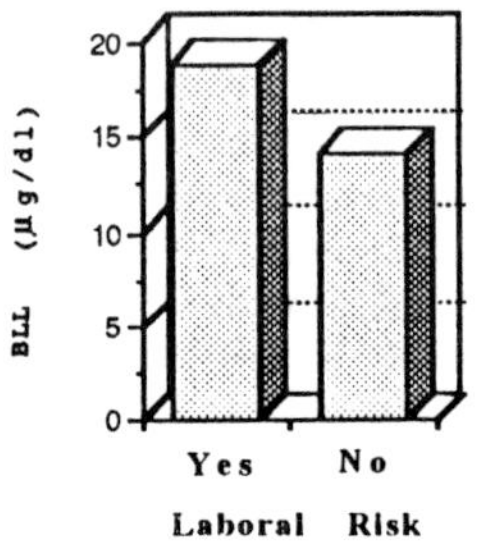

Mean BLL found in males occupationaly exposed to lead has been higher compared with BLL in males occupationally non exposed.

Applying the "t" Student test of unilateral contrast we conclude that BLL in males occupationally exposed to lead is statistically higher than the rest of the population ($p < 0.001$).

BIBLIOGRAPHY:

1.- ABADIA T.: Toxicidad por plomo. Determinación de la plumbemia. Valores de refencia en nuestro medio y estudio de su variacion temporal. Tesis Doctoral. Facultad de Medicina. Universidad de Zaragoza. Febrero 1989.

2.- GONZALEZ FERNANDEZ E.: Consideraciones para la evaluación fiable de la intoxicación laboral por plomo inorgánico. Med Seg Trab 1982; 30 (120): 232-247.

3.- NEEDLEMAN H.L.: Exposure to lead: sources and effects. N Engl J Med 1977; 297: 943-946.

4.- RABINOWITZ M.; NEEDLEMAN H.L.: Temporal trends in the lead concentrations of umbilical cord blood. Science 1982; 216: 429-431.

III. Immunology, Inflammation

Metal Ions in Biology and Medicine. Eds. Ph. Collery, L.A. Poirier, M. Manfait, J.C. Etienne. John Libbey Eurotext, Paris © 1990, pp.237-242

Metal influences on the incidence of autoimmunity and infectious disease

D.A. Lawrence, M.J. Mc. Cabe, M. Kowolenko

Department of Microbiology and Immunology, Albany Medical College, Albany, New York 12208, USA

Heavy metals are known to directly induce pathophysiologic changes which affect particular organ systems (reviewed by Goyer, 1986). Assuming the immune system is one of these systems, these target compounds may modulate the host immune system resulting in immune dysfunctions with subsequent immunopathophysiologic changes. These immunotoxicological effects of heavy metals can result from a direct toxic effect on a cellular component of the immune system or from a modulatory effect on a component of the immune system. Alternatively, the modulatory metal could indirectly modify the immune system via disruption of the bidirectional communication between the neuroendocrine and immune systems (Weigent & Blalock, 1987). Heavy metals do modulate both host resistance and select components of the immune system. Possible biochemical mechanisms of metal–induced alterations of immunoreactivity are discussed below.

The ability of the immune system to respond to foreign antigens while not reacting against "self" is the result of dynamic regulatory interactions between the cells of the immune system. Immunoregulation involves interactions between cells of a network of leukocytes with substantially different development and functional characteristics. The regulatory subpopulations (lymphocytes: helper and suppressor T cells and B cells; monocytes/macrophages; granulocytes) control overall immunity via cell–to–cell and factor–to–cell interactions. Divergence from a balanced control (as may occur after heavy metal exposure) generally manifests itself as a pathological disorder. The importance of proper immunoregulation is demonstrated by diseases in which there is a deficiency in one or several components of the immune system. There are several pathologic conditions characterized by defects in either T or B cells, currently, the most recognized being Acquired Immune Deficiency Syndrome (AIDS). Upon infection with HIV, previously normal individuals acquire a T helper cell defect which leads to increased morbidity and mortality due to an increase in opportunistic infections and/or neoplasms. HIV constituents concomitantly modulate B cell activities, bypassing T cell regulations (Lane & Fauci, 1985). Interestingly, the incidence of neoplasia and autoimmunity increases with age, in contrast to immunoregulated host resistance to infectious diseases which decline with age (Makinodan & Kay, 1980). It is important to note that accumulated body burdens of environmental toxicants such as heavy metals, especially lead (Pb), are also inherent with aging.

Most autoimmune diseases are of unknown etiology. Autoimmunity, hyperimmune responsiveness to self–constituents, is the result of altered T cell activitiesand/or hyperactive B cells. Autoimmune diseases encompass a variety of immunopathologic states and affect most organ systems. Genetic predisposition as well as environmental factors have been implicated in aberrant autoimmune responses which can progress to overt autoimmune disease (Schwartz & Rose, 1986). Numerous reports suggest that the incidence of autoimmune disorders has increased in recent years. In part, this increase could be attributed to better medical diagnosis or extended population longevity since autoimmune diseases are known to increase with aging; however, as stated above heavy metal exposures and body burdens also accumulate. Since there is an autoimmune disorder for each of the major organ systems affected by heavy metals, it is reasonable to postulate that heavy metal toxicity may, in part, be due to autoimmunity. This is only a hypothesis, but results with some heavy metals, especially mercury, support this hypothesis. In addition, metals are known inducers of allergic hypersensitivity. All four types of hypersensitivity are caused by sensitized lymphocytes which are activated by an endogenous or exogenous (environmental) factor and produce antibody responses (Type I, II, and III hypersensitivity) or cell mediated responses (Type IV hypersensitivity). The pathologic consequences of these processes could be autoimmune disease(s). Any metal known to produce an allergic response could be a promoter of autoimmunity.

Exposure to heavy metals has the potential of disrupting the body's normal immune homeostasis by either acting directly on the cells of the immune system or by interacting with other cells or organ systems so as to render them immunologically reactive. The direct effect of toxicant exposure on the immune system is quite obvious. There could be changes in lymphocyte subset ratios, altered morphology of immune tissue, decreases in total immune cell numbers or altered immune functions with or without the above changes (reviewed by Luster *et al.*, 1984). The more difficult modulations to assess are the subtle effects of toxicant exposure that cause an imbalance of immune homeostasis leading to immune activation directed at what the immune system views as "altered self" (Smith & Steinberg, 1983). An additional mechanism is a disruption of immunoregulatory circuits. Such alterations may manifest themselves as depressed suppressor cell activity or modification of idiotypic networks. Such defective circuits could allow the overproduction of autoantibodies or amplification of reactivities directed against "altered self".

Many metals have been associated with immunomodulation; however, in discussing the immunomodulatory effects of heavy metals, Au, Hg, and Pb serve as the prototype compounds. Many reports have implicated Au and Hg as inducers of autoimmune diseases. Mercuric chloride exposure of Brown–Norway rats induces lymphoproliferation, hypergammaglobulinemia, and the production of autoantibodies directed against the glomerular basement membrane as well as immune complex induced glomerulonephritis (Sapin *et al.*, 1981). In addition autoreactive T cell clones have been isolated from exposed rats which will induce a local graft vs. host reaction in syngeneic animals (Pelletier *et al.*, 1988; Pelletier *et al.*, 1985). Adoptive transfer of $W3/25^+$ T cells obtained from Hg exposed rats induced autoimmune disease in naive animals (Pelletier *et al.*, 1988). Transfer of B cells had no effect. Of particular importance was the observation that the recipients of autoreactive T cells responded with increased proliferation of $OX8^+$ T cells (cytotoxic/suppressor) cells. When $OX8^+$ cells were depleted by *in vivo* treatment with OX8 antibody, the severity of the autoimmunity increased. Animals exposed to Hg did not display increased numbers of $OX8^+$ cells. The data indicates that autoreactive T helper cells are induced while T cytotoxic/suppressor cells are inhibited from responding to the increase in autoreactivity. In the mouse, Hg caninduce a similar autoimmune process and is known to be restricted to the S haplotype (Robinson *et al.*, 1986). Hg is able to augment anaphylaxis, in that, Hg has been shown to enhance IgE production (Prouvost–Danon *et al.*, 1981). Au also modulates immune reactions (Harth, 1981). Au treatment for rheumatoid arthritis has been associated with allergic reactions and nephritic problems (Tornoth & Skrifuars, 1974; Geddes & Brostoff, 1976). Au also mediates immune complex–induced renal disease in rats (Nagi *et al.*, 1971) and guinea pigs (Ueda *et al.*, 1980).

MHC gene products are known to regulate Hg–induced renal immunopathologies in mice (Robinson *et al.*, 1986), rats (Sapin *et al.*, 1981), and man (Charpentier *et al.*, 1981). Au–induced nephritic syndrome has been associated with expression of the DR3w phenotype (Woody *et al.*, 1980). Au–induced thrombocytopenia also was reported to be immunologically mediated and regulated by MHC genes (Speerstra *et al.*, 1985). In tests of Pb–poisoned workers, it was reported that 21 of 57 individuals with high Pb levels had kidney disease; 15 of these 21 were described as having Pb nephropathy (Wedeen *et al.*, 1979). Eight of these 15 workers with nephropathy were examined for immune involvement and seven were positive. The study suggests that only 23% of the Pb burdened workers had Pb–induced glomerulonephritis. Polak *et al.* (1968) were the first to indicate that sensitization to inorganic metal compounds is under genetic control. Likewise, it has been indicated that many autoimmune diseases are influenced by the genetics of the MHC. Altered renal pathology, possibly the result of immune complexes, has been associated with Hg, Au, Cd, and Pb.

The hypersensitivities induced by heavy metals are proof of their ability to modulate immune reactivity; however, the mechanism by which they induce autoimmunity is unclear. Four possible mechanisms are outlined by which autoimmunity may be induced. First, the metal could cause tissue damage which results in the release of sequestered antigen or higher amounts of tissue antigens, and the immune system is activated by this new antigen or antigenic dose. Secondly, the metal may interact with a self–constituent altering its antigenic determinants which leads to an immune response to "altered self" and eventually to unaltered self as well. Metal–modified self antigens could induce a loss of tolerance to the modified antigens or bystander antigens. Antigenic challenges of this nature are due to activation of T cells which produce non–specific immunoregulatory products that exacerbate responses to modified antigen(s) and initiate responses to bystander antigens (Eastman & Lawrence, 1982). Alternatively, the metal may enhance morbidity, and the higher incidence of infections may aid in the establishment of autoimmunity. Finally, the metal may directly modulate lymphocyte activation; enhancement of B cell activity or T cell activity could lead to an autoimmunity. Direct interaction(s) with lymphocytes may alter their physiology resulting in polyclonal activation and/or in a hyperreactive state (Druet *et al.*, 1982; Hirsch *et al.*, 1982).

The direct activation of lymphocytes by heavy metals has been proposed. When human peripheral blood mononuclear cells were exposed to Pb, Ni, Zn, or Hg, there was increased ^{3}H–thymidine incorporation compared to control (no metal) cultures (Warner & Lawrence, 1986a). When B and T cells were isolated and exposed to either Pb or Zn, increased proliferation was only noted in T cell preparations. Like Hg effects on human lymphocytes, Hg can induce proliferation of rat lymphocytes; however, *in vitro* Hg is only inhibitory for mouse lymphocytes. Pb, Ni, and Zn have all been shown to induce progression of mouse lymphocytes from G_0 to G_1 of the cell cycle (Warner & Lawrence, 1986b). Pb also has been reported to induce proliferation of liver cells (Columbano *et al.*, 1983) and renal tubular cells (Choie & Richter, 1974). The mechanism by which heavy metals influence lymphocyte activation, proliferation, and effector phases of the lymphoid subsets are still unknown. There is evidence that *in vivo* Pb exposure can alter the C3b receptor on some lymphocytes (Koller & Brauner, 1977) and Pb and Hg are known to modulate numerous cellular enzymes (Vallee & Ulmer, 1972). Pb was shown to modulate cAMP and cGMP effects on B cells, both of which are implicated in lymphocyte activation. Hg's species–specific influences on lymphoproliferation appear to correlate with its elimination of cellular thiols. In a dose responsefashion, mouse cellular thiols were more sensitive to $HgCl_2$ treatment than the thiols of rat or human lymphocytes.

The association of metals with protein constituents resulting in altered self is an attractive hypothesis, however, data concerning metal–lymphocyte protein interactions has not materialized. Several metals are associated with proteins in the body. The metallothioneins, a group of cysteine rich, low molecular weight proteins have been demonstrated to bind several metals both *in vivo* and *in vitro*.

Both Zn and Cu–metallothionein (MT) complexes are found in man and may act as a storage depot for these essential trace elements. Interestingly, Interleukin 1 (IL–1) and gamma interferon (IFN) both upregulate the transcription rate of MT, indicating that this compound may be influenced by or affect immunoreactivity. IL–1 and IFN are associated with macrophage activation. The activation of macrophages can lead to the generation of reactive oxygen moieties which induce cellular damage. MT may act as a scavenger of free radicals that may work synergistically with glutathione to protect cells against oxidative damage. Heavy metals with their high affinity for thiols may interact within this system resulting in an enhanced inflammatory response. MT recently has been reported to have modulatory effects on lymphocytes and MT is endogenous to lymphocytes (Lynes *et al.*, 1990).

The ability of metals to react with thiols has been well documented. Pb, Hg, and Au have a high affinity for sulfhydryl groups (Vallee & Ulmer, 1972) and all are associated with dysregulation of the immune system. The thiol status of lymphocytes appears to be critical for cell activation. Experimental data has shown that modulation of thiols results in a decrease in cell activation by mitogen. In addition, the T suppressor cell subset has higher level thiol sensitivity than does the helper subset. The relationship between the lymphocyte subsets and macrophages is dynamic. Removal of one cell type alters homeostasis and may result in autoimmunity or impaired host resistance. The development of the Brown–Norway rat model of Hg induced autoimmunity appears to involve an imbalance between T helper and suppressor cells (Pelletier, *et al.*, 1985). T cytotoxic/suppressor cells have more cell surface thiols, are more sensitive to permeant thiol blockers, and are more radiosensitive than helper cells (Lawrence 1981b). One could then infer that T suppressor cells are more prone to the toxic effects of heavy metals. A disruption of helper:suppressor cell ratios could result in polyclonal cell activation, increased immunoglobulin formation to metal–protein antigens, immune–complex formation, and basement membrane destruction. Thus, heavy metals may accelerate the generation of autoimmune reactions due to the combined effects of lowered thiol status and increased body burden of metal. Clearly these are speculative mechanisms of how metals may induce immunopathologic changes.

Although the mechanisms by which heavy metals induce pathologic changes has not been delineated, immunologic parameters could be involved in pathologic effects of heavy metals, such as renal, hematopoietic, and central nervous system changes associated with Pb. A posited scenario for the ability of Pb to induce autoimmunity and yet concomitantly inhibit host resistance is presented herein. As for Hg and Cd, Pb lowers host resistance to a variety of pathogens (Reviewed by Lawrence, 1985). Mice orally exposed to Pb (0.4mM) in their drinking water for only two weeks had significantly enhanced mortality when injected with *Listeria monocytogenes* even though their blood Pb levels were only 20μg/dl. *Ex vivo* analyses of T cells, B cells, and macrophages from these mice (including phagocytosis, IL–1 production, lymphoproliferation, and antibody production) indicate that no obvious immune functions were inhibited with the exception of antigen presentation to some T cell clones/hybridomas. In addition, these mice had diminished numbers of bone marrow or splenic macrophage progenitors capable of responding to colony stimulating factor (M–CSF/CSF–1). Upon closer examination of the T cell clones affected by Pb in antigen–specific assays, we have determined that TH1 cells seem to be preferentially inhibited; whereas, the activation of TH2 cells is actually enhanced by Pb. These results provide an explanation for the apparent dichotomy in Pb's *in vivo* effects, in that, host resistance is suppressed and humoral immunity to foreign antigens is unaffected and yet we postulate enhancement of autoimmunity. The TH1 (inflammatory/DTH) cells produce less stimulatory factors (IL–2 and gamma–IFN) whenexposed to Pb; thus macrophages which are directly inhibited in development by Pb also are less activated. Although Pb can directly enhance B cell activation there are fewer TH2 (helper/inducer) cells activated to help these B cells via their production of IL–4 and IL–5 since macrophages normally present antigen to TH2 cells (Janeway *et al.*, 1988); however, syngeneic B cell:TH2 interactions may be enhanced leading to autoimmunity. It is important to note that Pb does not enhance Ia expression on macrophages but it does on B cells. Since Ia molecules are intimately involved in T cell—macrophage and T cell—B cell interactions these modifications also need to be considered in the immunomodulation by Pb. Hg also can enhance Ia expression on B cells, thus heavy metals may upset immunohomeostasis by alteration of Ia expression and other surface molecules which influence immunoregulation via cell—cell communication. Interestingly, the polypeptides of Ia molecules (α,β and invariant chains) have thiols which are acylated for translocation to the plasma membrane. In addition, many growth factor receptors contain thiol rich domains which aid in receptor/ligand binding. Thiols also are involved in cell—cell communication and adherence via the family of proteins known as integrins. Although cellular thiols are prime candidates as targets for heavy metals, the biochemical manifestations of heavy metals in modulation of immune reactivities are presently all speculative. However, the environmental influence on the incidence of autoimmune diseases may relate to accumulated body burdens of heavy metals.

Charpentier, B., *et al.* (1981): T lymphocyte functions in mercuric cholride—induced membranous glomerulonephritis in man. Evidence for a defect of presentation of the histocompatibility class II molecules at the cell surface. *Nephrologie* **2**, 153—157.

Choie, D.D. & Richter, G.W. (1974): Cell proliferation in mouse kidney induced by lead: I. Synthesis of DNA. *Lab Invest.* **30**, 647—651.

Columbano, A., *et al.* (1983): Liver cell proliferation induced by a single dose of lead nitrate. *Amer. J. Pathol.* **110**, 83—88.

Druet, P., *et al.* (1982): Immune dysregulation and autoimmunity induced by toxic agents. *Transplant. Proc.* **14**, 482—484.

Eastman, A.Y. & Lawrence, D.A. (1982): TNP—modified syngeneic cells enhance immunoregulatory T cell activities similar to allogeneic effects. *J. Immunol.* **128**, 926—931.

Geddes, P.M. & Brostoff, J. (1976): Pulmonary fibrosis associated with hypersensitivity to gold salts. *Brit. Med. J.* **1**, 1444.

Goyer, R.A. (1986): Toxic effect of metals in Klassen, C.D., Amdur, M.O., and Doull, J. (eds) Toxicology p.582—635 Macmillian, New York.

Harth, M. (1981): Modulation of immune response by gold salts. *Agents Actions* **8**, 465—474.

Hirsch, F., *et al.* (1982): Polyclonal effect of $HgCl_2$ in the rat, its possible role in an experimental autoimmune disease. *Eur. J. Immunol.* **12**, 620—625.

Janeway, C.A. *et al.*. (1988): $CD4^+$ T cells: Specificity and function. *Immunol. Rev.* **101**, 39—80

Koller, L.D. & Brauner, J.A. (1977): Decreased B lymphocyte response after exposure to lead and cadmium. *Toxicol. Appl. Pharmacol.* **42**, 621—624.

Lane, C.H. & Fauci, A.S. (1985): Immunologic abnormalities in the aquired immunodeficiency syndrome. *Ann. Rev. Immunol.* **3**, 477—500.

Lawrence, D.A. (1981a): Heavy metal modulation of lymphocyte activities. II. Lead, a mediator of B—cell activation. *Int. J. Immunopharmacol.* **3**, 153—161.

Lawrence, D.A. (1981b): Antigen activation of T—cells. In: *Handbook of Cancer Immunology*, edited by H. Waters, pp. 257—320. Garland Press, NY.

Luster, M.I., *et al.* (1984): Evaluation of immune functions in toxicology. In: Hayes, A.W. (ed) Principles and Methods in Toxicology, p. 561, Raven Press, New York.

Lynes, M.A. *et al.* (1990): Extracellular metallothionein effects on lymphocyte activites. *Mol. Immunol.* **27**: (in press)

Makinodan, T. & Kay, M.M.B. (1980): Age influence on the immune system. *Adv. Immunol.* **29**, 287—330.

Nagi, A.H., *et al.* (1971): Gold nephropathy in rats: light and electron microscopic studies. *Exp. Mol. Pathol.* **15**, 354—362.

Pelletier, L., *et al.* (1988): Autoreactive T cells in mercury—induced autoimmunity: ability to induce autoimmune diseases. *J. Immunol.* **140**, 750—754.

Pelletier, L., *et al.* (1985): *In vivo* self reactivity of mononuclear cells to T cells and macrophages exposed to HgCl2. *Eur. J. Immunol.* **15**, 460–465.
Polak, L., *et al.* (1968): The genetic control of contact sensitivity to inorganic metal compounds in guinea pigs. *Immunology* **14**, 707–711.
Prouvost–Danon, A. *et al.* (1981): Induction of IgE synthesis and potentiation of anti–OVA IgE antibody response by $HgCl_2$ in the rat. *J. Immunol.* **126**, 699–702.
Robinson, C.J.G. *et al.* (1986): Mercuric chloride–, gold sodium thiomalate–, and d–penacillamine–induced antinuclear antibodies in mice. *Toxicol. Appl. Pharmacol.* **86**, 159–169.
Schwartz, R.S. & Rose, N.R. (1986): Autoimmunity: Expermental and Clinical Aspects. *Ann. NY Acad. Sci.* **475**, 1–423.
Smith, H.R. & Steinberg, A.D. (1983): Autoimmunity–A perspective. *Ann. Rev. Immunol.* **1**, 175–210.
Speerstra, F. *et al.* (1985): HLA associations in aurothioglucose and D–penicillamine induced haematotoxic reactions in rheumatoid arthritis. *Tissue Antigens* **26**, 35–40.
Tornroth, T. & Skrifuars, B. (1974): Gold nephropathy prototype of membranes glomerclonephritis. *Am. J. Pathol.* **75**, 573–584.
Warner, G.L. & Lawrence, D.A. (1986a): Effect of heavy metals on human PBL proliferation and cellular thiols. *Fed. Proc.* **46**, 1322.
Warner, G.L. & Lawrence, D.A. (1986b): Cell surface and cell cycle analysis of metal–induced murine T cell proliferation. *Eur. J. Immunol.* **16**, 1337–1342.
Weeden, R.P. *et al.* (1979): Detection and treatment of occupational lead nephropathy. *Arch. Intern. Med.* **139**, 53–57.
Weigent, D.A. & Blalock, J.E. (1987): Interactions between the neuroendocrine and immune systems: common hormones and receptors. *Immunol. Rev.* **100**, 79–108.
Wooley, P.H. *et al.* (1980): HLA–DR antigens and the toxic reactions to sodium aurothiomalate and d–penicillamine in patients with rheumatoid arthritis. *N. Engl. J. Med.* **303**, 300–303.

Metal Ions in Biology and Medicine. Eds. Ph. Collery, L.A. Poirier, M. Manfait, J.C. Etienne. John Libbey Eurotext, Paris © 1990, pp. 243-245

Effect of a thermal water on human lymphocyte mitogenesis *in vitro* and its comparison with arsenite and arsenate

P. Mercier, B. Rouveix

Département de Pharmacologie Clinique, INSERM U13, Hôpital Claude Bernard, 10 Avenue de la Porte d'Aubervilliers, 75944 Paris Cedex 19, France

INTRODUCTION

Spa treatment using thermal water remains in Europe an important medical activity. In France particularly, more than 600 000 patients go every year to the country's 101 spa stations. Spa treatment and indications are based mostly on tradition, empirismus and isolated observations. Some experimental studies in animals and human begin however to support the therapeutic activity of certain thermal treatment (Chevance & Prevost, 1984; Fourot-Bauzon et al., 1988; Laroche, 1987; Magnin et al., 1989). The compositions of the thermal waters are quite different and complex. Physical parameters of the waters (temperature, pH) and natural gas of the spring could sometimes account for the therapeutic efficiency. Nevertheless, numerous trace, elements which undoubtly act on the human metabolism particularly on the immunological response (Gershwin et al., 1985; Srinivas et al., 1988), often exist in the thermal waters. This trace elements could be partly responsible for some waters activities. In this regard, arsenic (6,8 ppm, in the pentavalent (+5) oxidation state) is the most striking and abundant trace element found in a chlorobicarbonated thermal water (Table 1), indicated in the treatment of allergic respiratory diseases and skin aliments. The effect of this mineralized water was thus investigated on human lymphocyte mitogenesis, in comparison with arsenite [As (+3)] and arsenate [As (+5)].

IMMUNOLOGICAL ACTIVITY OF AN ARSENICAL THERMAL WATER

Experimental design

Thermal water was obtained from the spring (La Bourboule, France) and conserved in sealed glass ampoules. Its osmolarity was adjusted to 280 mOsm with NaCl and it was diluted from 1:2 to 1:16, in lymphocyte culture medium, corresponding to arsenate concentration of 3.4 to 0.425 ppm respectively. NaCl (0.9%; 1:4 diluted) was used as control. $NaAsO_2$ [As (+3)] and Na_2HAsO_4 [As (+5)] were employed at elemental concentrations : 10, 50, 100 ppb for arsenite (0.13 , 0.66 , 1.33 x 10^{-6} M) and 0.02 , 0.1 , 0.5 ppm (0.27 , 1.34, 6.67 x 10^{-6} M) for arsenate. Higher concentrations (500 ppb for arsenite and 1.5 ppm for arsenate) were found to be the threshold values before inhibition of mitogenesis. Lymphocytes from 17 healthy donors were stimulated with phytohemagglutinin (PHA), a mitogen, in the presence or not of arsenicals. The response was measured at 72 hr by 3H-thymidine incorporation.

Table 1. Chemical content of the thermal water. BRGM (Bureau de Recherche Géologique et Minière), 1978.

	mg/l		mg/l		μg/l		μg/l
Ca^{+}	35.00	HCO_3^-	1665.00	Al	<2	Mn	35
Mg^{2+}	8.80	Cl^-	1722.00	Cd	<1	Mo	190
Na^{+}	1752.00	SO_4^{2-}	128.00	Co	<10	Ni	10
K^{+}	96.80	PO_4^{3-}	0.23	Cr	14	Pb	14
Li^{+}	5.84	F^-	5.20	Cu	14	Sn	231
Fe^{2+}	0.21	CO_2	456.00	Ga	63	Ti	14
Sr^{2+}	3.31	SiO_2	115.00			Zn	114
NH_4^{+}	0.70	B	10.00				
		As	6.80				

Traces of : Ag, Ba, Be, Ge, Rb, Sb, V, W.

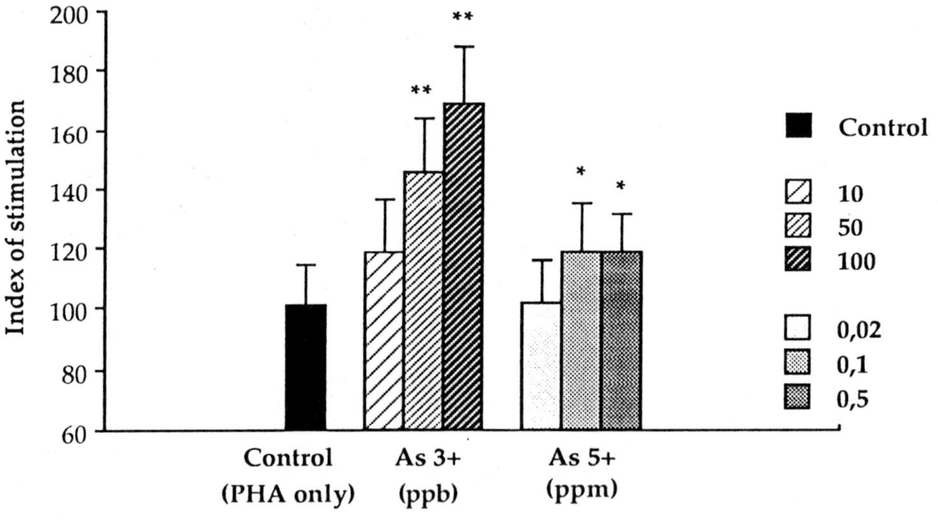

Fig.1. Effect of arsenic on PHA-induced proliferative response of human lymphocytes. Index of stimulation = count per minute in stimulated cells over count per minute in unstimulated cells. Statistics : two-way analysis of variance and Student's t-test. Bars represent mean values ± s.e.m. (n = 17; each point in triplicate).
* : $p < 0.05$; ** : $p < 0{,}001$.

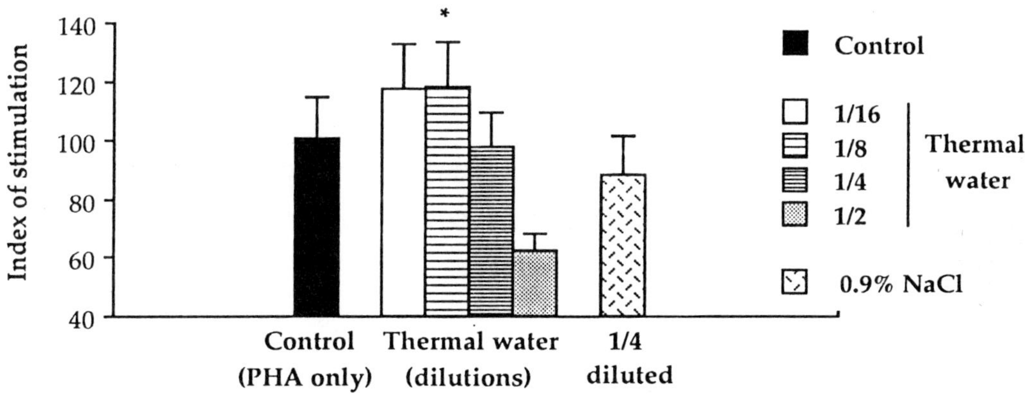

Fig.2. Effect of the thermal water on human lymphocyte mitogenesis. * : $p < 0.05$.

Results.

According to previous studies (Mc Cabe et al., 1983; Stewart et al., 1970), the present results show that arsenic enhances PHA-stimulated lymphocyte mitogenesis (fig. 1). However, this effect was observed over a narrow dose range. Although arsenate (As [+5]) is reduced in arsenite (As [+3]) *in vivo* (Vather & Envald, 1983) and *in vitro* (Bertolero et al., 1987) before being methylated, the latter is 10-fold more active as regard to dosage. In addition the mean enhancement of the 3H-thymidine incorporation is more important with arsenite (68 % augmentation) than with arsenate (18 %). Increase rates of lymphocyte mitogenesis (17 % and 18 % augmentation in mean; fig. 2) observed with 1:16 and 1:8 thermal water dilutions are mostly donor dependant.

It is concluded that this thermal water could have a biological activity in immune system, partly due to its arsenic content.

REFERENCES

Bertolero, F., Pozzi, G., Sabbioni, E. & Saffioti, U. (1987): Cellular uptake and metabolic reduction of pentavalent to trivalent arsenic as determinants of cytotoxicity and morphological transformation. *Carcinogenesis*, 8, 803-808.

Chevance, L.G. & Prevost, M.C. (1984): Etude de la regranulation in vivo des mastocytes de la muqueuse des voies aériennes supérieures chez le cobaye. Son inhibition par une eau thermale. *Rev. Fr. Allergol.*, 24, 7-18.

Fourot-Bauzon, M., Perrin, P. & Bedu, M. (1988): Effects of "ultrasoune aerosol" of natural thermal water (spa water) in treatment of upper airway disease in children. *J. Aerosol Medecine*, 1, 265.

Gershwin, M.C., Beach, R.S. & Hurley, L.S. (1985): Trace elements. In *Nutrition and Immununity*, pp.190-226. Academic Press.

Laroche, C. (1987): Le suivi d'une cohorte de 3000 curistes thermaux pendant trois ans par le service national du contrôle médical du régime général. *Bull. Acad. Natle. Méd.*, 171, 869-886.

Magnin, A., Kantelip, J.P., Drutel, P., Tran Minh, T. & Magnin P. (1989): Effects antispasmodiques d'une eau thermale administrée en aérosols et en injections parentérales. *C. R. Soc. Biol.* , 183, 180-185.

Mc Cabe, M., Maguire, D. & Novak, M. (1983): The effects of arsenic compounds on Human and bovine lymphocyte mitogenesis in vitro. *Environ Res.*, 31, 323-331.

Srinivas, U., Braconier, J.H., Jeppsson, B., Abdulla, M., Akesson, B. & Ockerman, PA. (1988): Trace element alterations in infectious diseases. *Scand. J. Clin. Lab. Invest.*,48, 495-500.

Stewart, T.H.M., Harris, J.E., Afghan, K.C. & Wise, P. (1970): The effect of inorganic arsenicals on the non-specific stimulation of lymphocytes by phytohemagglutinin in man. In *Proc. 5th Leucocyte Culture Conf.*, 75-95.

Vather, M. & Enval, J. (1983): In vivo reduction of arsenic in mice and rabbits. *Environ Res.*, 32, 14-24.

Metal Ions in Biology and Medicine. Eds. Ph. Collery, L.A. Poirier, M. Manfait, J.C. Etienne. John Libbey Eurotext, Paris © 1990, pp. 246-248

Arsenic and immune response : protective effect on oxygen-induced depression of lymphocyte proliferation

P. Mercier, B. Rouveix

Département de Pharmacologie Clinique, INSERM U.13, Hôpital Claude Bernard, 10 Avenue de la Porte d'Aubervilliers, 75944 Paris Cedex 19, France

INTRODUCTION

Arsenic is essentialy known and studied as an environmental toxic and carcinogenic heavy metal (Ishinishi etb al., 1986). Toxicity of arsenite (As [+3]) has been proposed to be a consequence of its affinity for dithiols (Knowles, 1982), whereas substitution of phosphate was supposed to be responsible for the uncoupling of respiratory chain phosphorylation by arsenate (As [+5]) (Crane & Lipman, 1952). Nevertheless, arsenic has still been used in therapeutic (for review see Goodman & Gilman, 1958). Indeed, as most of drugs, its activity is complex, dose and oxidation state dependant. In cell culture, high arsenic concentrations induce chromosomal damages, arsenite being about 10-fold more toxic than arsenate (Lee et al., 1986). On the opposite, lower doses of arsenic enhance cellular proliferation (Lee et al., 1985) particularly in immune cells (Stewart et al., 1970; Mc Cabe et al., 1983).

The present study investigates the effect of arsenic low doses on immune response using rat lymphocyte mitogenesis under hyperoxic conditions. This model was selected as it has been shown that normobaric oxygen could depress the proliferative response of lymphocytes by increasing the endogenous production of reactive oxygen metabolites (Kraus et al., 1985). Such an injury of immune cells by oxygen species has been shown to be involved in the suppressive effects of macrophage on lymphocyte responses (Hoffeld & al., 1981), and various immune stress: decline of the immune response with age (Harman et al., 1977), damages due to the ionizing radiations (Doria et al., 1982) and drug action or toxicity (Trush et al., 1982).

METHODS

Purified splenic rat lymphocytes stimulated with Concanavalin A (Con A, 2 µg/ml) were exposed to hyperoxic (40 %) normobaric oxygen for 24 hours and then replaced in a normal atmosphere (21 % oxygen). ^{3}H-thymidine incorporation was determined in cultures after 48 or 72 hr. Arsenic tri- and pentavalent were present in cultures at different concentrations. Arsenite [As (+3)] ($NaAsO_2$) 10 or 50 ppb (0.13 or 0.66 x10^{-6} M in element) or arsenate [As (+5)] (Na_2HAsO_4) 0.5 or 1.5 ppm (6.67 or 20 x10^{-6} M in element) were used. Besides, an arsenical (6.8 ppm arsenic, As (+5)) chloro-bicarbonated thermal water indicated in the treatment of allergic respiratory diseases was employed (La Bourboule, France). Its osmolarity was adjusted to 280 mOsm with NaCl and it was diluted in lymphocyte culture medium (dilutions ranging from 1:2 to 1:16). Viability was evaluated in cultures by LDH release measurement .

RESULTS AND DISCUSSION

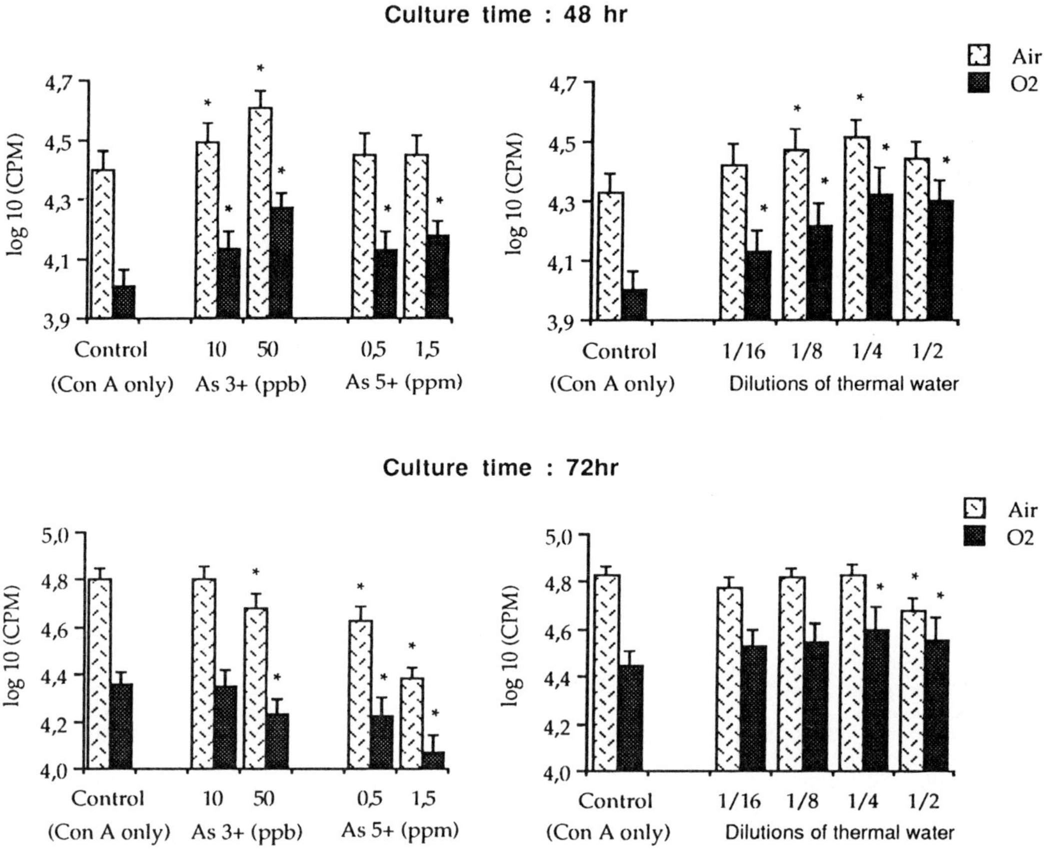

Fig. 1. Effects of arsenicals on Con A induced rat lymphocyte mitogenesis. CPM = count per min. in stimulated cells minus count per min. in unstimulated cells. Bars represent the mean values ± s.e.m., n = 14 for As and 9 for thermal water; each point was assayed in triplicate. Statistics: three-way analysis of variance and Student's t-test. * : significatly different from control under the same gas culture conditions, $p < 0.001$.

Table 1. Viability evaluated by LDH release measurement: (LDH activity in the supernatant / LDH activity in the cellular lysate) x 100. Results: Mean ± s.e.m., n = 12. Level of significance: $p < 0.001$: *

Time of culture		48 hr		72 hr	
		Air	O_2	Air	O_2
Control	(Con A only)	28.87 ± 2.16	26.30 ± 1.67	48.73 ± 5.85	43.35 ± 4.45
As $^{3+}$	50 ppb	34.25 ± 2.41 *	29.71 ± 2.38	62.20 ± 4.88 *	55.67 ± 5.51 *
As $^{5+}$	0.5 ppm	31.25 ± 2.18	29.01 ± 2.11	57.04 ± 5.38 *	53.26 ± 5.11 *
Thermal water	1/4 diluted	31.23 ± 2.31	29.53 ± 2.79	57.68 ± 5.23 *	52.80 ± 5.63 *
	1/8 diluted	31.89 ± 2.52	30.42 ± 2.37	52.93 ± 6.28	53.02 ± 5.90 *
0.9% NaCl	1/4 diluted	28.28 ± 2.11	27.78 ± 2.15	45.43 ± 5.32	46.44 ± 5.26

Results indicated that at 48 hr of culture, arsenical compounds enhance the thymidine incorporation in Con A stimulated lymphocytes, under normal conditions and partially restore the proliferative response after 40 % oxygen exposure (fig. 1). The arsenic and arsenate stimulating and protective effects disappear at 72 hr. Arsenicals impair viability, particularly at 72 hr of culture (Table 1).

Targets and effects of arsenic at cellular level are numerous. Thus, the finding that arsenic can protect rat lymphocyte mitogenesis against oxidant injury could be related to: interrelations between this element and glutathione metabolism (Lee et al., 1989), inhibition of free radical generating enzymes such as xanthine oxidase (Hille & Massey, 1981), synthesis induction of some stress proteins such as metallothionein (Maitani et al., 1987) or heat shock proteins (Levinson et al., 1980), or activity on DNA-repair systems (Nunoshiba et al., 1987).

REFERENCES

Crane, R.K. & Lipmann, F. (1952): The effect of arsenate on aerobic phosphorylation. J. Biol. Chem., 201, 235-243.

Doria, G., Agarossi, G. & Adorini, L. (1982): Selective effects of ionizing radiations on immunoregulatory cells. *Immunol. Rev.*, 65, 23-54.

Goodman, L.S. & Gilman, A. (1958): Arsenicals. In *The pharmacological basis of therapeutics.* Second ed. , pp. 948-969. Macmillan Company.

Harman, D., Heidrick, M.L. & Eddy, D.E. (1977): Free radical theory of aging: effect of free-radical-reaction inhibitors on the immune response. *J. Amer. Ger. Soc.*, 9, 400-407.

Hille, R. & Massey, V. (1981): Tight binding inhibitors of xanthine oxidase. *Pharmac. Ther.*, 14, 249-263.

Hoffeld, J.T., Metzger, Z. & Oppenheim, J.J. (1981): Oxygen-derived metabolites as suppressors of immune responses *in vitro.* In *Lymphokines,* ed. E. Pick, vol. 2, pp. 63-86. London: Academic Press.

Ishinishi, N., Tsuchiya, K., Vahter, M. & Fowler, B.A. (1986): Arsenic. In *Handbook on the toxicology of metals*, ed. L. Friberg, G.F. Nordberg & V. Vouk, 2nd ed., pp. 43-83. Elsevier.

Knowles, F.C. (1982): The enzyme inhibitory form of inorganic arsenic. *Biochem. Int.*, 4, 647-653.

Kraus, L., Lacombe, P., Fay, M. & Pocidalo J.J. (1985): High concentrations of oxygen modulate *in vitro* Con A responses of rat lymphoid cells. Effect of 2-mercaptoethanol. *Immunol. Lett.*, 11, 51-55.

Lee, T.C., Oshimura, M. & Barrett, J.C. (1985): Comparison of arsenic-induced cell transformation, cytotoxicity, mutation and cytogenetic effects in Syrian hamster embryo cells in culture. *Carcinogenesis*, 6, 1421-1426.

Lee, T.C., Wei, M.L., Chang, W.J., Ho, I.C., Lo, J.F., Jan, K.Y. & Huang, H. (1989): Elevation of glutathione levels and glutathione s-transferase activity in arsenic-resistant chinese hamster ovary cells. *In Vitro Cell. Develop. Biol.*, 25, 442-448.

Levinson, W., Oppermann, H. & Jackson, J. (1980): Transition series metals and sulfhydryl reagents induce the synthesis of four proteins in eukaryotic cells. *Biochim. Biophys. Acta*, 606, 170-180.

Maitani, T., Saito, N., Abe, M., Uchiyama, S. & Saito, Y. (1987): Chemical form-dependent induction of hepatic zinc-thionein by arsenic administration and effect of co-administered selenium in mice. *Toxicol. Lett.*, 39, 63-70.

Mc Cabe, M., Maguire, D. & Novak, M. (1983): The effects of arsenic compounds on Human and bovine lymphocyte mitogenesis *in vitro. Environ Res.*, 31, 323-331.

Nunoshiba, T. & Nishioka, H. (1987): Sodium arsenite inhibits spontaneous and induced mutations in *Escherichia coli. Mutat. Res.*, 184, 99-105.

Stewart, T.H.M., Harris, J.E., Afghan, K.C. & Wise, P. (1970) : The effect of inorganic arsenicals on the non-specific stimulation of lymphocytes by phytohemagglutinin in man. In *Proc. 5th Leucocyte Culture Conf.*, 75-95.

Trush, M.A., Mimmaugh, E.G., Gram T.E. (1982): Activation of pharmacologic agents to radical intermediates. Implications for the role of free radicals in drug action and toxicity. *Biochem. Pharmacol.*, 31, 3335-3346.

Metal Ions in Biology and Medicine. Eds. Ph. Collery, L.A. Poirier, M. Manfait, J.C. Etienne. John Libbey Eurotext, Paris © 1990, pp. 249-251

Differences in toxicology patterns of acute poisoning from arsenite and arsenate in the light of an extremely rare case of arsenic pentoxide ingestion

A. Gionovich

Johnson Matthey, Technology Centre, Blount's court, Sonning common, Reading, Berks, RG4 9NH, UK

A host of reports concerns the acute intoxication from arsenic trioxide (As_2O_3), but very few, if any at all, have been published on acute poisoning from arsenic pentoxide (As_2O_5). This remark originated from a literature examination following an extremely rare clinical case of intentional ingestion of As_2O_5 at a dose within the human lethal estimated range of As^{5+}. Therefore a critical review of literature was performed looking for affinity and differences between toxicological properties of As (III) and As (V). For example our attention was focused on: (a) the contribution of the amount of As (III) as resulting from "in vivo" reduction of As (V); (b) the influence of "in vivo" redox equilibrium after ingestion of high amount of As (III) or As (V) compounds. The main differences between toxicological properties of As_2O_3 and As_2O_5 concern: (a) distribution, (b) redox equilibrium, (c) excretion, (d) methylation, (e) lethal dose estimated for man, (f) clinical findings, prognosis and sequelae.

(a) Distribution. Once absorbed, both the compounds bind to the globin of heme in a range of 95-99%, but later the distribution is valence-dependent, i.e. arsenite rapidly distributes (Lerman, 1983) to both liver and kidney, whereas in the same time arsenate distributes to kidney only. The same author (Lerman, 1983+) explained the different uptake in hepatocytes with the different ionization at physiogical pH of the two forms: arsenite pKa=9,23 unionized, arsenate pKa=2,20 is charged. The first may gain entry to the hepatocytes by diffusion and the concentration gradient is maitained by rapid intracellular methylation.

(b) Redox equilibrium. "In vivo" redox equilibrium exists between the two forms, maintained by activation or inhibition of reduction-oxidation reactions (Benko, 1976) in the liver or kidney (Ginsburg 1976).

(c) Excretion. Kidney represents the main excretion route of arsenic, as shown by urinary elimination of As and its metabolites monomethylarsonic acid, MMA and dimethylarsinic acid, DMA. A facilitated renal elimination for As (V) was found in rabbit (Crawford, 1947) and recently in man too (Buchet, 1981).

(d) Methylation. It is extremely facilitated in the presence of reduced form and this biotransformation would be prevented by a large excess of thiolic groups which decrease the amount of free As^{3+} (Buchet, 1986). Metabolic equilibrium and kinetics of As (III) and As (V) are summarised in Fig. 1.

(e) Lethal dose estimated for man. In EPA (1985) the As_2O_5 lethal dose range in man was reported to be 0,3-3 g, i.e. 4-16 times higher than As_2O_3. The toxicological actions of As(V) and As(III) are summarised in Fig. 2.
(f) Clinical findings, prognosis and sequelae. The case of acute intoxication with As_2O_5 described in this report confirmed the higher severity and the worse prognosis of As_2O_3 intoxication, which is often followed by important neurological sequelae within weeks.

CASE REPORT. A 26-years old woman voluntary ingested one "Titrisol" (Merck) ampoule containing 1503 mg of arsenic pentoxide. This is a standard solution of As in H_2O (As 1,000 ± 0,001 g) currently used for atomic absorption spectrophotometry. After about 90 min from the ingestion, the patient came by herself to the local hospital and soon was admitted to I.C.U. , even if no tipical findings of arsenic acute intoxication were present. Nevertheless samples of urine, blood and gastric fluid were collected. In agreement with the Laboratory of Toxicology, gastric lavage was immediately undertaken followed by the administration of a suspension of activated charcoal. At the same time a mono-dithiolic treatment started: dimercaprol (BAL) 50 mg i.m. in six daily doses plus continuous i.v. infusion of N-acetylcysteine at the dose of 150 mg/kg in the first 2 h, subsequentely reduced to 50 mg/kg/24 h. Monitoring of vital signs and infusional-supportive therapy were instituted.
The results of toxicological investigations confirmed the presence of As in all the samples at the following concentrations: urine 12γ/ml, plasma < 2γ/ml, gastric fluid 15γ/ml. The treatment was integrated with haemodialysis on daily basis. Each dialysis was performed using a PAN membrane, dialysate flow was 500 ml/min in single pass, blood flow 280 ml/min. Dialysate fluid was added with $NaHCO_3$, each dialysis lasted 3 h. During the treatment neither pathological signs, nor changes of glycemia and other laboratory findings appeared. Periodical samples of urine and dialysate fluid were collected for toxicological monitoring; the faecal excretion was not evaluated as it usually corresponds to 6-9% of the ingested dose. Mono-dithiolic treatment plus haemodialysis were stopped after 3 days, in accordance with Dreisbach (1987) , when the total urinary excretion of As was below 50 /24 h. The patient was transferred from I.C.U. to medical division. The As removed by the treatment was distributed in the following percentages of the ingested dose: renal

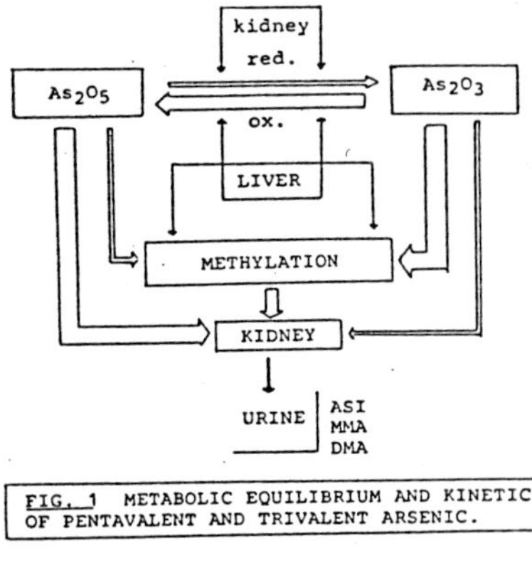

FIG. 1 METABOLIC EQUILIBRIUM AND KINETIC OF PENTAVALENT AND TRIVALENT ARSENIC.

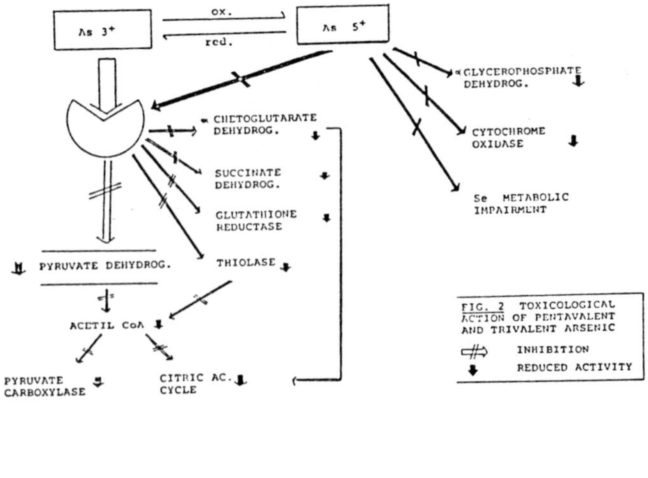

FIG. 2 TOXICOLOGICAL ACTION OF PENTAVALENT AND TRIVALENT ARSENIC

route 2,2%, haemodidlysis 24,9%, gastric lavage 15% (total = 42,1%).

CONCLUSIONS. This clinical case of voluntary ingestion of arsenic pentoxide confirms the low toxicity of arsenate and demonstrates, for the first time, the high efficacy of the mono-dithiolic treatment combined with haemodialysis. The same treatment was successfully applied in the same I.C.U. in an analogous case of mercuric chloride ingestion (Giunta, 1983). The mono-dithiolic compounds, promptly administered, have directly antagonised arsenic pentoxide and then probably neutralized the amount of As(III) resulted from "in vivo" reduction of As(V). This assumption was indirectly confirmed by the absence of any pathological findings. Likewise our clinical experience confirmed the only data in literature regarding the ability of BAL in preventing central and autonomic toxic effects of both arsenite and arsenate in the rabbit (Koppanyi, 1947).
Finally haemodialysis has likely provided a good protection for renal parenchyma, facilitated the removal of As and its chelates and prevented adverse reactions to BAL.

REFERENCES

Anon (1985): In *EPA chemical profiles*. Washington D.C. 2046 USA

Benko, V. et al. (1976): Biotransformation of As(III) to As(V) and arsenic tolerance. *Arch. Toxicol.* 36, 159-162.

Buchet, J.P. (1988): Role of thiols in vitro methylation of inorganic arsenic by rat liver cytosol. *Bioch. Pharm.* 37, 3149-3155.

Buchet, J.P. et al. (1981): Comparision of the urinary excretion of arsenic metabolites after a single oral dose of sodium arsenite, monomethylarsonate or dimethylarsinate in man. *Int. Arch. Occup. Envirom. Health.* 48, 71-79.

Crawford, T. and Lewy, G.A. (I947): Change undergone by phenyl-arsenious acid and phenyl-arsonic acid in animal body. *Biochem. J.* 41, 333-336.

Deisbach, R.H. and Robertson, W.O. (1987). In *Handbook of poisoning*. XII° Ed. pp. 223-224. Appleton and Lange, Norwalk.

Ginsburg, J.M. (1965): Renal mechanism for excretion and transformation of Arsenic in the dog. *Am. J. Physiol.* 208, 832-840.

Giunta, F. et al. (1983): Severe acute poisoning from the ingestion of a permanent wave solution of mercuric chloride. *Human Toxicol.* 2, 243-246.

Koppanyi, T. and Sperung, F. (1947): Central and autonomic effects following the combinated administration of sodium arsenite and 2-3 dimercaptopropanol (BAL). *J. Pharmacol. Exper. Therap.* 69, 350-355.

Lerman, S. and Clarhson, T. (1983): The metabolism of arsenite and arsenate by the rat. *Fundam. and Appl. Toxicol.* 3, 309-314.

Lerman, S. et al. (1983^{+}): Arsenic uptake and metabolism by liver cells is dependent on arsenic oxidation state. *Chem. Biol. Interaction.* 45, 401-406.

Metal Ions in Biology and Medicine. Eds. Ph. Collery, L.A. Poirier, M. Manfait, J.C. Etienne. John Libbey Eurotext, Paris © 1990, pp. 252-257

Toxic metals and irritant or allergic skin reactions

K. Nordlind

Department of Dermatology, Karolinska Hospital, S-104 01, Stockholm, Sweden

INTRODUCTION

Like other tissues the skin is vulnerable to the effects of toxic metals. Through interactions with different cell types in the epidermis such as keratinocytes and Langerhans cells, lymphocytes and cells of the connective tissue like endothelial cells, mast cells, granulocytes and histiocytes, both toxic and irritant skin reactions may appear. Irritant skin reactions are the most common of these. In the metal industry at least 50% of dermatoses are of an irritant type. As regards contact allergy to metals, nickel allergy is common in women, the number of nickel-sensitive women in the general population approximates 10% (Fisher, 1986a; Menné et al., 1982). Chromate sensitivity tends to be a problem in men and the chromate dermatitis tends to become chronic and lichenified particularly from industrial exposure (Fisher, 1986b).

The common way to diagnose contact allergy is by patch test. However, a problem which faces the clinician is to differentiate allergic and toxic reactions, and this is a common problem not only for metal compounds. There is also a need for in vitro tests which could replace the in vivo test.

Metal compounds have a tendency to bind to various chemical groups, like, e.g., mercury to sulfhydryl groups, being parts of immuno-regulatory molecules in addition to molecules important for morphological structure.

CONTACT ALLERGY

There are immediate and delayed type allergic skin reactions. Most interest has been focused on the delayed type reaction (cf. Andersen et al., 1987). After penetration of the corneal barrier, the antigen is believed to be presented to T lymphocytes by antigen-presenting cells of which Langerhans cells in the epidermis are the most important. The class II histocompatibility HLA-DR antigens on the cell surface of presenting cells and in association with the antigen, are necessary for lymphocyte-dendritic cell interactions in the induction of the immune response. After migration of antigen-presenting cells to

regional lymph nodes there is development of memory T lymphocytes. When there is a new challenge with the antigen, a proliferation of effector T lymphocytes occurs with release of different inflammatory mediators such as interleukins, interferons, neuropeptides and eicosanoids.
In addition, there is evidence that metals like nickel could bind directly and activate T lymphocytes without the participation of antigen-presenting cells and MHC-encoded molecules (Kapsenberg et al., 1988).

TOXIC REACTION

The mechanism for the toxic eczema is not fully understood, however, may be, e.g., damage to cellular membranes of keratinocytes, release of lysosomal enzymes, degranulation of mast cells, activation of polymorphonuclear leucocytes and release of eicosanoids and other inflammatory mediators (cf. Frosch, 1989). Individual susceptibility to irritants is extremely variable.

MORPHOLOGY

Allergic and irritant contact dermatitis are difficult to differentiate by clinical, light or electron microscopic examinations. Clinically, very strong patch test reactions may be differentiated, since the irritant reaction is sharply demarcated, promptly healing and "burn-like", compared to the spreading more slowly disappearing eczematous reaction to an allergen (Fisher, 1986c). However, there is no way to clinically distinguish a weak irritant from a weak allergic patch test reaction. Histologically, irritant reactions may show epidermal necrosis and less intercellular edema than allergic reactions (cf. Thestrup-Pedersen et al., 1989). Larger cell infiltrates are usually present in the allergic reaction. There is also focal distribution of the inflammatory cells in contrast to diffuse distribution in irritant reactions (Avnstorp et al., 1989). Of various ultrastructural changes, Langerhans cells, following application of allergens and irritants to the skin, show signs of both activation and damage. Irritants give degenerative changes as early as 2h after application (Picut et al., 1987) Immunohistologically, HLA-DR antigen expression on keratinocytes has been found in allergic but not in irritant reactions (Scheynius and Fischer, 1986). Investigation of qualitative differences by phenotype classification of the infiltrating cells has been unsuccessful (Scheynius et al., 1984; Avnstorp et al., 1987).

It should also be mentioned that lichenoid reactions may be found after application of, e.g., gold (Hjorthöj, 1977) and mercuric (Mobacken et al., 1984) compounds, giving a special appearance both clinically and histologically; a dense infiltrate of lymphocytes is found immediately beneath the epidermis.

IN VITRO TESTS

There has been much interest in trying to replace the patch test in vivo with an in vitro technique. The most used in vitro technique has been the lymphocyte transformation test (cf. von Blomberg-van der Flier et al., 1987). Other techniques also include test for leucocyte/macrophage migration inhibition factors. Antigen-stimulated T cells release factors which inhibit the migration of leucocytes/macrophages. A combination of lymphocyte transformation and macrophage/leucocyte migration inhibition tests has been suggested to be of value in diagnosis of contact allergy to nickel (von Blomberg-van der Flier, 1987).

The lymphocyte transformation test has been used for about 25 years, however, much remains to be modified before this test could be of clinical value. The suspected allergen is added at different concentrations at the start of the incubation of the lymphocytes, and after 5-6 days the cell proliferation is measured as uptake of tritiated thymidine, mitosis index or frequency of blasts. The main problem with toxic metals is the tendency for some of these to give an unspecific stimulated proliferation, since this proliferation occurs in every tested individual. Such unspecific effect has been reported for mercury (e.g., Nordlind, 1984), gold (e.g., Nordlind, 1985a), zinc (e.g., Nordlind, 1985b) and even nickel (von Blomberg-van der Flier et al., 1987). The proliferation occurs at concentrations which are close to cytotoxic ones. Addition of recombinant IL-2 (interleukin 2), the actual proliferative signal for T lymphocytes, may result in higher specific proliferation values (Prens et al., 1989).

INFLAMMATORY MEDIATORS

For the differentiation of allergic and irritant reactions in the skin or in vitro cultures, it is thus necessary to study other parameters and in that context we may consider the inflammatory mediators.

Among various cells, also mast cells and basophils are involved in contact dermatitis. These cells may release a variety of mediators such as histamin, serotonin and leukotrienes. LTB_4 (leukotriene B_4), which has potent neutrophil chemokinetic properties, has been found in low levels in some positive allergic patch tests but not in irritant reactions (Barr et al., 1984). The mast cells lie in close approximation with nerve fibers and various neuropeptides such as, e.g., substance P, may modulate experimental contact allergy (Wallengren and Möller, 1988). Of other mediators, gamma-IFN (gamma-interferon) induces the expression of class I and II antigens (Basham and Merigan, 1983), and ICAM-1 (intercellular adhesion molecule 1) (Nickoloff et al., 1987), being responsible for the adhesion between keratinocytes and T lymphocytes. The co-expression of adherence molecules and class II antigens creates the immunological background for an immune reaction (Boyd et al., 1988). IL-1 (interleukin 1) is released from a large variety of cells including keratinocytes. It has an ability to increase class II antigen expression on T lymphocytes (Oppenheim et al., 1986), a chemotactic activity (Miossec et al., 1984), induce T lymphocytes to produce IL-2 (Stanton et al., 1987) and is able to increase the expression of IL-2 receptors on T cells (Oppenheim et al., 1986).

According to a model (Kupper, 1989) a toxic stimulus on epidermal cells may lead to release of preformed IL-1 from epidermis, which leads to expression of adhesive molecules, giving trapping of leucocytes. The interaction with unimpaired keratinocytes also leads to release of chemotactic factors for leucocytes like, e.g., IL-8 (interleukin 8), and release of activating factors for inflammatory cells like, e.g., IL-6 (interleukin 6) and G-CSF (granulocyte colony stimulating factor) and GM-CSF (granulocyte/macrophage colony stimulating factor). There is also proliferation and chemokinetic activity which leads to reepitheliazation and, finally, IL-1 induces fibroblasts to proliferate and produce collagen and collagenase, which mediates the dermal components of wound healing.

With stimulation by antigens, gamma-IFN producing T cells are activated. Gamma-IFN induces adhesion molecule and also keratinocyte activation partly by inducing IL-1 production and release. There is release of chemotactic and activating factors and the following events such as epidermal hyperproliferation is initiated.

It has been reported that IL-1 is increased about 3-fold in epidermis overlying a positive allergic patch test (Larsen et al., 1988) but not an irritant skin reaction (Larsen et al., 1989).

TOXIC METALS AND INFLAMMATORY MEDIATORS

IL-1 has been shown to be produced by in vitro cultured peritoneal macrophages stimulated by cisplatin (Gupta and Sodhi, 1987) IL-2 is produced in the supernatants by human peripheral blood T lymphocytes after addition of cobalt chloride (Löfström and Wigzell, 1986) and nickel (Sinigaglia et al., 1985; Karttunen et al., 1988), and also by murine splenocytes after addition of nickel and zinc (Warner and Lawrence, 1988). Murine splenocytes produce interferon after addition of mercury and zinc (Reardon and Lucas, 1987). Interferon has also been shown for human peripheral blood lymphocytes after addition of nickel (Sinigaglia et al., 1985).

CONCLUSION

By using methods to detect expression of different inflammatory mediators such as interleukins and interferons in the skin and in supernatants of immune cell cultures, the problem of irritant or allergic effects of added toxic metal compounds, may be explored.

REFERENCES

Andersen, K.E., Benezra C., Burrows, D., Camarasa, J., Dooms-Goossens, A., Ducombs, G., Frosch, P., Lachapelle, J.-M., Lahti, A., Menné, T., Rycroft, R., Scheper, R., White I. & Wilkinson, J. (1987): Contact dermatitis. A review. Contact Dermatitis 16, 55-78.

Avnstorp, C., Ralfkiaer, E., Jörgensen, J. & Lange Wantzin, G. (1987): Sequential immunophenotypic study of lymphoid infiltrate in allergic and irritant reactions. Contact Dermatitis 16, 239-254.

Avnstorp C., Balslev, E. & Thomsen, H.K. (1989): The occurrence of different morphological parameters in allergic and irritant patch test reactions. In Current Topics in Contact Dermatitis, ed. P.J. Frosch, A. Dooms-Goossens, J.-M. Lachapelle, R.J.G. Rycroft & R.J. Scheper, pp. 38-41. Berlin Heidelberg: Lea & Febiger.

Barr, R.M., Brain, S., Camp, R.D.R., Cilliers, J., Greaves, M.W., Mallet, A.I. & Misch, K. (1984): Levels of arachidonic acid and its metabolites in the skin in human allergic and irritant contact dermatitis. Br. J. Dermatol. 111, 23-28.

Basham, T.Y., & Merigan, T.C. (1983): Recombinant interferon-gamma increases HLA-DR synthesis and expression. J. Immunol. 130, 1492-1494.

Boyd, A.W., Wanryk, S.O., Burns, G.F. & Feconds, J.V. (1988): Intercellular adhesion molecule 1 (ICAM-1) has a central role in cell-cell contact mediated immune mechanisms. Proc. Natl. Acad. Sci. (USA) 85, 3095-3099.

Fisher, A.A. (1986a): Nickel- the ubiquitous contact allergen. In Contact Dermatitis, pp. 745-761. Philadelphia: Lea & Febiger.

Fisher, A.A. (1986b): Chromate dermatitis and cement burns. In Contact Dermatitis, pp.762-772. Philadelphia: Lea & Febiger.

Fisher, A.A. (1986c): In Contact Dermatitis, p. 26. Philadelphia: Lea & Febiger.

Frosch, P.J. (1989): Irritant contact dermatitis. In Current Topics in Contact Dermatitis, ed. P.J. Frosch, A. Dooms-Goossens, J.-M. Lachapelle, R.J.G. Rycroft & R.J. Scheper, pp. 385-398. Berlin Heidelberg: Springer-Verlag.

Gupta, P. & Sodhi, A. (1987): Increased release of interleukin-1 from mouse peritoneal macrophages in vitro after cisplatin treatment. Int. J. Immunopharmacol 9, 385-388.

Hjortshöj, A. (1977): Lichen planus and acne provoked by gold. Acta Dermatovener. (Stockholm) 57, 165-167.

Kapsenberg, M.L., Van der Pouw-Kraan, T., Stiekema, F.E., Schootemeijer, A. & Bos, J.D. (1988): Direct and indirect nickel-specific stimulation of T lymphocytes from patients with allergic contact dermatitis to nickel. Eur. J. Immunol. 18, 977-982.

Karttunen, R., Silvennoinen-Kassinen, S., Juutinen, K., Andersson, G., Ekre, H-P.T. & Karvonen, J. (1988): Nickel antigen induces IL-2 secretion and IL-2 receptor expression mainly on $CD4^+$ T cells, but not measurable gamma interferon secretion in peripheral blood mononuclear cell cultures in delayed type hypersensitivity to nickel. Clin. Exp. Immunol. 74, 387-391.

Kupper, T.S. (1989): Mechanisms of cutaneous inflammation. Interactions between epidermal cytokines, adhesion molecules, and leukocytes. Arch. Dermatol. 125, 1406-1412.

Larsen, C.G., Ternowitz, T., Larsen, F.G. & Thestrup-Pedersen, K. (1988): Epidermis and lymphocyte interactions during an allergic patch test reaction. Increased activity of ETAF/IL-1, epidermal derived lymphocyte chemotactic factor and mixed skin lymphocyte reactivity in persons with type IV allergy. J. Invest. Dermatol. 90, 230-233.

Larsen, C.G., Ternowitz, T., Larsen, F.G., Zachariae, C. & Thestrup-Pedersen, K. (1989): ETAF/interleukin-1 and epidermal lymphocyte chemotactic factor in epidermis overlying an irritant patch test. Contact Dermatitis 20, 335-340.

Löfström, A. & Wigzell, H. (1986): Antigen specific human T cell lines specific for cobalt chloride. Acta Dermatovener. (Stockholm) 66, 200-206.

Menné, T., Borgan Ö. & Green A. (1982): Nickel allergy and hand dermatitis in a stratified sample of the danish female population: an epidemiological study including a statistic appendix. Acta Dermatovener. (Stockholm) 62, 35-41.

Miossec, P., Yu, C.L. & Ziff, M. (1984): Lymphocyte chemotactic activity of interleukin 1. J. Immunol. 133, 2007-2011.

Mobacken, H., Hersle, K., Sloberg, K. & Thilander, H. (1984): Oral lichen planus: hypersensitivity to dental restoration material. Contact Dermatitis 10, 11-15.

Nickoloff, B.J., Lewinsohn, D.M. & Butcher, E.C. (1987): Enhanced binding of peripheral blood mononuclear leukocytes to gamma interferon-treated cultured keratinocytes. Am. J. Dermatopathol. 9, 413-418.

Nordlind, K. (1984): Fractionation of human thymocytes and peripheral blood lymphocytes on Percoll density gradients and DNA synthesis stimulating effect of mercuric chloride. Int. Archs Allergy Appl. Immunol. 75,16-19.

Nordlind, K. (1985a): Stimulating effect of gold chloride on DNA synthesis of human lymphoid cells. Int. Archs Allergy Appl. Immunol. 77, 459-460.

Nordlind, K. (1985b): Stimulating effect of zinc chloride on DNA synthesis of human thymocytes. Int. Archs Allergy Appl. Immunol. 77, 461-462.

Oppenheim, J.J., Kovacs, E.J., Matsushima, K. & Durum, S.K. (1986): There is more than one interleukin 1. Immunology Today 7, 45-56.

Picut, C.A., Lee, C.S. & Lewis R.M. (1987): Ultrastructural and phenotypic changes in Langerhans cells induced in vitro by contact allergens. Br. J. Dermatol. 116, 773-784.

Prens, E.P., Benne, K., van Joost, T. & Benner, R. (1989): In vitro nickel-specific T lymphocyte proliferation: methodological aspects. In Current Topics in Contact Dermatitis, ed. P.J. Frosch, A. Dooms-Goossens, J.-M. Lachapelle, R.J.G. Rycroft & R.J. Scheper, pp. 578-583.

Reardon, C.L. & Lucas, D.O. (1987): Heavy-metal mitogenesis: Zn^{++} and Hg^{++} induce cellular cytotoxicity and interferon production in murine T lymphocytes. Immunbiol. 175, 455-469.

Scheynius, A., Fischer, T., Forsum, U. & Klareskog, L. (1984): Phenotypic characterization in situ of immunocompetent cells in allergic and irritant contact dermatitis in man. Clin. Exp. Immunol. 55, 81-90.

Scheynius, A. & Fischer, T. (1986): Phenotypic difference between allergic and irritant patch test reactions in man. Contact Dermatitis 14, 297-302.

Sinigaglia, F., Scheidegger, D., Garotta, G., Scheper, R., Pletscher, M. & Lanzavecchia, A. (1985): Isolation and characterization of Ni-specific T cell clones from patients with Ni-contact dermatitis. J. Immunol. 135, 3929-3932.

Stanton, G.J., Weigent, D.A., Fleischmann jr., W.R., Dianzani, F. & Baron, S. (1987): Interferon review. Invest. Radiol. 22, 259-273.

Thestrup-Pedersen, K., Larsen, C.G. & Rönnevig, J. (1989): The immunology of contact dermatitis. A review with special reference to the pathophysiology of eczema. Contact Dermatitis 20, 81-92.

Von Blomberg-van der Flier, M., van der Burg, C.K.H., Pos, O., van de Plassche-Boers, E.M., Bruynzeel, D.P., Garotta, G. & Scheper, R.J. (1987): In vitro studies in nickel allergy: Diagnostic value of a dual parameter analysis. J. Invest. Dermatol 88, 362-368.

Wallengren, J. & Möller, H. (1988): Some neuropeptides as modulators of experimental contact allergy. Contact Dermatitis 19, 351-354.

Warner, G.L. & Lawrence, D.A. (1988): The effect of metals on IL-2-related lymphocyte proliferation. Int. J. Immunopharmacol. 10, 629-637.

Metal Ions in Biology and Medicine. Eds. Ph. Collery, L.A. Poirier, M. Manfait, J.C. Etienne. John Libbey Eurotext, Paris © 1990, pp. 258-262

Nickel-induced immune reactions *in vitro* in cutaneous nickel allergy

S. Silvennoinen-Kassinen

Department of Medical Microbiology, University of Oulu, Kajaanintie 46 E, SF-90220 Oulu, Finland

SUMMARY

Nickel causes cutaneous hypersensitivity of the delayed type, often involving the hands, and eczema may become a chronic problem.

Nickel binds to serum and cellular proteins and forms an immunologically stimulatory antigen. The nickel allergic reaction can be studied in vitro using mononuclear cells from nickel sensitive subjects. It will then induce an allergic reaction in the cell cultures, so that immunological parameters reveal activated cells, soluble mediators and dividing cells.

Langerhans cells or monocytes present the nickel to T lymphocytes, which recognize it by means of a T cell receptor. The nickel must be in a complex with a HLA-DR molecule on the antigen presenting cell. The reacting cells are T helper cells with primed memory and activation markers. Many of them use αß genes for T cell receptors, and γδ genes are seldom used. IL-2, IL-2 receptors (IL-2R) and IL-6 are secreted during the nickel reaction.

Genetic susceptibility to nickel allergy has been mapped to chromosome 6 in the HLA-DQA region.

Nickel can be harmful because it causes allergic skin disease in genetically susceptible subjects.

BACKGROUND TO CUTANEOUS NICKEL ALLERGY

Metals cause contact allergies, and nickel is the most common cause of metal sensitivity (Cronin, 1983). About 10% of women and 1% of men suffer from nickel allergy in Finland (Peltonen, 1981), and the numbers are even higher in some other western countries. Nickel allergy is increasing in industrialized countries (Young & Howing, 1987) and is the most common cause of occupational metal sensitivity (Kanerva et al., 1988).

Nickel allergy is clinically annoying because it is often associated with chronic hand eczema and results in inflammatory and destructive skin eruptions. The mechanism is hypersensitivity reaction of the delayed type. The diagnosis is confirmed by skin testing.

AN *in vitro* MODEL FOR NICKEL ALLERGY

Nickel allergy is mediated by T lymphocytes (Silvennoinen-Kassinen *et al.*, 1986). The reacting cells are obtained from peripheral blood mononuclear cells and are cultured with culture medium, human serum proteins and nickel sulphate (6.25-12.5 µg/ml) (Silvennoinen-Kassinen, 1980). T cells specific for nickel initiate an allergic reaction in which they secrete soluble factors, undergo blast transformation and divide. Exponential growth is achieved within 7 days.

NICKEL AS AN ANTIGEN

Nickel binds to serum proteins and cellular components (Shelley & Juhlin, 1976; Abbracchio *et al.*, 1982). It forms an immunogenic complex with proteins and is regarded as a hapten. The exact form of this complex is not known.

Nickel is presented to T cells by another cell, the antigen-presenting cell. This is a monocyte (Silvennoinen-Kassinen, 1980) or a Langerhans cell (Res *et al.*, 1987). It can process the antigen, although the processing events involved are still unknown. They must present the nickel to the T cells together with a HLA-DR molecule, and carry HLA molecules on their cell membranes. The nickel-HLA complex is recognized by T cells which have a receptor molecule (T cell receptor specific for the antigen).

NICKEL-REACTING T CELLS AND SOLUBLE MEDIATORS (CYTOKINES)

The reacting T cells are predominantly helper T cells (CD4+, CD8-) (Silvennoinen-Kassinen *et al.*, 1986) with IL-2 receptors (Karttunen *et al.*, 1988) and HLA-DR molecules as activation markers. They are primed memory cells (CD45RO+). Two alternatives for T cell receptor genes exist, αß or γδ, with most nickel-reactive cells using the former. These genes have alternative variable regions for an antigen. None of the ß chain variables 5, 6 and 8 predominates in the nickel reaction. (Silvennoinen-Kassinen *et al.*, 1990a). The selection among the alternative variable α, ß, γ and δ genes in the nickel reaction is not yet understood.

Lymphocytes carry receptors which direct their homing to different organs. The expression of homing receptors in nickel allergy still requires investigation.

A subset of lymphocytes are killer cells and mediate cytotoxicity. They could contribute to the tissue damage in nickel allergy. This needs investigation.

Soluble mediators secreted during immune response promote or inhibit cellular reactions. Nickel induces IL-2 (Karttunen et al., 1988), IL-2R and IL-6 secretion, and a small quantity of tumor necrosis factor α is also secreted (Silvennoinen-Kassinen et al., 1990c), but usually no measurable amount of interferon γ (Karttunen et al., 1988). The induction of other soluble factors remains to be studied.

NICKEL-SPECIFIC T CELL CLONES

A clone of T cells is a population of similar cells which have all started from a single stem cell. Clones are established by limiting dilution and their growth requires IL-2 (T cell growth factor).

Nickel-specific T cell clones reflect the properties of nickel-specific blasts from nickel allergic subjects. They are helper T cells (CD3+, CD4+, CD8-) (Sinigaglia et al., 1985; Emtestam et al., 1989) and carry memory cell markers (CD45RA-, CD45RO+) and HLA-DR activation markers. They use αß T cell receptor genes, but do not select ß chain variables 5 or 8. (Silvennoinen-Kassinen et al., 1990b).

The nickel specific clones produce interferon γ (Sinigaglia et al., 1985), which is not found in cultures of peripheral blood lymphocytes with nickel (Karttunen et al., 1988). Our clones produced IL-4 and/or IL-6 and soluble IL-2R (Silvennoinen-Kassinen et al., 1990b). The clones help in antibody production, mainly for class IgG (Sinigaglia et al., 1985; Silvennoinen-Kassinen et al., 1990b).

Nickel-specific clones recognize nickel on the antigen-presenting cells, which may be monocytes from peripheral blood, although Langerhans cells from the skin sometimes present nickel more effectively (Kapsenberg et al., 1987). The clones recognize nickel only if HLA-DR molecules are also expressed on the antigen presenting cells. HLA-DQ and HLA-DP do not restrict the nickel reaction (Emtestam et al., 1989).

DISEASE SUSCEPTIBILITY GENES IN NICKEL ALLERGY

A Danish twin study suggests that heredity accounts for 60% in nickel allergy (Menné & Holm, 1983). Since the HLA genes (in chromosome 6) that take part in immune responses represent the most polymorphic gene cluster known so far in the human species, an association of different alleles with different immune aberrations can be reasonably expected.

No consistent association of nickel allergy with serologically defined HLA class I or class II antigens has been found (Lidén et al., 1978; Silvennoinen-Kassinen et al., 1979; Kapoor-Pillarisetti et al., 1981; Karvonen et al., 1984; White et al., 1986). HLA-DQ locus typing at the DNA level showed an association of nickel allergy with a polymorphic site on the HLA-DQA gene (Olerup & Emtestam, 1988), but this has not yet been confirmed by others.

T cell receptors recognize the antigen and may contribute to susceptibility. This should be investigated.

CONCLUSION

Nickel is harmful because it causes allergic skin eruptions. Its reaction shows genuine immunological activation. Susceptibility to nickel allergy is in part genetically determined.

ABBREVIATIONS: HLA = human histocompatibility complex; CD = cluster of differentiation; IL = interleukin

ACKNOWLEDGEMENTS: This work was supported by grants from the Finnish Allergy Foundation and the Paulo Foundation.

REFERENCES

Abbracchio, M. P., Evans, R. M., Heck, J. D., Cantoni, O., and Costa, M (1982): The Regulation of Ionic Nickel Uptake and Cytotoxicity by Specific Amino Acids and Serum Components. Biol. Trace Element Res. 4: 289-301.

Cronin, E. (1983): Contact dermatitis. pp. 338-367. Edinburgh: Churchill-Livingstone.

Emtestam, L., Carlsson, B., Marcusson, J. A., Wallin, J., and Möller, E. (1989): Specificity of HLA restricting elements for human nickel reactive T cell clones. Tissue Antigens 33: 531-541.

Kanerva, L., Estlander, T., and Jolanki, R. (1988): Occupational skin disease in Finland. An analysis of 10 years of statistics from an occupational dermatology clinic. Int. Arch. Occup. Environ. Health 60: 89-94.

Kapoor-Pillarisetti, A., Mowbray, J. F., and Cronin, E. A. (1981): HLA Dependence of Sensitivity to Nickel and Chromium. Tissue Antigens 17: 261-264.

Kapsenberg, M. L., Res, P., Bos, J. D., Schootemijer, A., Teunissen, M. B. M., and VanSchooten, W (1987): Nickel-specific T lymphocyte clones derived from allergic nickel-contact dermatitis lesions in man: heterogeneity based on requirement of dendritic antigen-presenting cell subsets. Eur. J. Immunol. 17: 861-865.

Karttunen, R., Silvennoinen-Kassinen, S., Juutinen, K., Andersson, G., Ekre, H.-P., and Karvonen, J. (1988): Nickel antigen induces IL-2 receptor expression mainly on CD4+ T-cells, but no measurable gamma interferon secretion in peripheral blood mononuclear cell cultures in delayed type hypersensitivity to nickel. Clin. Exp. Immunol. 74: 387-391.

Karvonen, J., Silvennoinen-Kassinen, S., Ilonen, J., Jakkula, H., and Tiilikainen, A. (1984): HLA antigens in nickel allergy. Ann. Clin. Res. 16: 211-212.

Lidén, S., Beckman, L., Cedergren, B., Göransson, K., and Nyquist, H. (1978): HLA antigens in allergic contact dermatitis. Acta Dermatovener. (Stockh) 79: 53-56.

Menné, T., and Holm, N. V. (1983): Nickel allergy in a female twin population. Int. J. Dermatol. 22: 22-28.

Olerup, O., and Emtestam, L. (1988): Allergic contact dermatitis to nickel is associated with a Taq I HLA-DQA allelic restriction fragment. Immunogenetics 28: 310-313.

Peltonen, L. (1981): Nickel sensitivity. An actual problem. Int. J. Dermatol. 20: 352-353.

Res, P. R., Kapsenberg, M. L., Bos, J. D., and Stiekema, F. (1987): The Crucial Role of Human Dendritic Antigen-Presenting Cell Subsets in Nickel-Specific T-Cell Proliferation. J. Invest. Dermatol. 88: 550-554.

Shelley, W. B., and Juhlin, L. (1976): Langerhans cells form a reticulo-endothelial trap for external contact allergens. Nature 261: 46-47.

Silvennoinen-Kassinen, S. (1980): Lymphocyte transformation in nickel allergy: Amplification of T-lymphocyte responses to nickel sulphate by macrophages in vitro. Scand. J. Immunol. 12: 61-65.

Silvennoinen-Kassinen, S., Ikäheimo, I., Kauppinen, M., and Karvonen, J. (1990a): Cell surface markers of nickel induced 7-day lymphocyte blasts in nickel sensitive subjects. Abstract to the 20th Meeting of the European Society for Dermatological Research.

Silvennoinen-Kassinen, S., Ikäheimo, I., Poikonen, K., and Kauppinen, M. (1990b): Nickel-specific T cell clones. Abstract to the XXIst Meeting of the Scandinavian Society for Immunology.

Silvennoinen-Kassinen, S., Ilonen, J., Tiilikainen, A., and Karvonen, J. (1979): No significant association between HLA and nickel contact sensitivity. Tissue Antigens 14: 459-461.

Silvennoinen-Kassinen, S., Jakkula, H., and Karvonen, J.(1986): Helper cells (Leu-3a+) carry the specificity of nickel sensitivity reaction in vitro in humans. J. Invest. Dermatol. 86: 18-20.

Silvennoinen-Kassinen, S., Karvonen, J., and Kauppinen, M.(1990c): Nickel induces IL-6, IL-2R and tumor necrosis factor α secretion from peripheral blood mononuclear cells of nickel-sensitive subjects. Abstract to the XXIst Meeting of the Scandinavian Society for Immunology.

Sinigaglia, F., Scheidegger, D., Garotta, G., Scherper, R., Pletscher, M., and Lanzavecchia, A. (1985): Isolation and charaterization of Ni-specific T cell clones from patients with Ni-contact dermatitis. J. Immunol. 135: 3929-3932.

White, S. I., Friedmann, P. S., and Stratton, A. (1986): HLA antigens and Langerhans cell density in contact dermatitis. Brit. J. Dermatol. 115: 447-452.

Young , E., and Howing, R. H. (1987): Patch test results with standard allergens over a decade. Contact. Derm. 17: 104-107.

Metal Ions in Biology and Medicine. Eds. Ph. Collery, L.A. Poirier, M. Manfait, J.C. Etienne. John Libbey Eurotext, Paris © 1990, pp. 263-265

Nickel salts and occupational respiratory allergy

F. Lavaud, C. Cossart, M.C. De Thésut, A. Prévost, Ph. Collery, J.M. Dubois De Montreynaud

Département des maladies respiratoires et Allergiques (Pr. S. Kockman), C.H.R. de Reims. 45, Rue Cognacq Jay – 51092 Reims Cedex – France

Occupational asthma caused by nickel is a rare phenomenon ; only few reports have been published (1, 3, 4, 5). We describe another case of asthma induced by nickel dust, where experimental challenge elicit an early asthmatic response.

Case report : A 27 year-old metal polisher with a history of contact dermatitis and recent episodes of shortness of breath was referred to our department. He was nonsmoker, and had no history of allergy and knew of none in his family. The patient had started to work in a metal plating factory 7 years ago, and had been a metal polisher since the age of 22 years. He had contact with nickel, copper, emery and chromium. His job consisted of grinding and polishing the surface of chromium or nickel taps. The atmosphere was dusty and insufficiently ventilated. He used gloves but rarely wore a mask. The patient develop cough with sputum, dyspnea and wheezing since few months. These symptoms recurred daily, and he experienced nocturnal attacks. Dyspnea seemed to be more noticeable ou mondays after a week-end away from work with an improvement on leaving work. On examination he had wheezing. The chest roentgenogram was normal, the Fev_1 was 2.7 1 (3.9 1 predicted) and FVC was 4.5.1 (4.8 1 predicted), with salbutamol inhalation, $Fe V_1$ reached 3.4 1. These findings strongly suggested that his asthma was related to his working environment. He had also a contact dermatitis to chromium with positive patchtests (negative for nickel) but inhalation challenge with chromium remained negative.

Monitoring of specific investigations :

- skin prick-tests showed a Weal of 4 x 5 mm at 10 mg./ml of nickel sulfate solution. In 3 control subjects this test remained negative.

- inhalation tests : Methacholine inhalation test was done after one month he had been absent from work. Initial FeV_1 was normal. The provocation concentration of methacholine producing a 20 % drop in FeV_1 (PC20) was 4,8 mg/ml (moderate bronchial reactivity).

Inhalation challenge test was performed 3 days after (fig 1), using sulfate solutions according to Malo (4). Nebulization of $NiSO_4$ at a concentration of 10 mg/ml for two consecutive periods of 30 seconds gave no significative fall in FeV_1. A further inhalation of $NiSO_4$, at a concentration of 100 mg/ml for 2 minutes gave an immediate reaction with a 40 % fall in FeV_1, and a syndromic response requiring administration of salbutamol and corticosteroids.

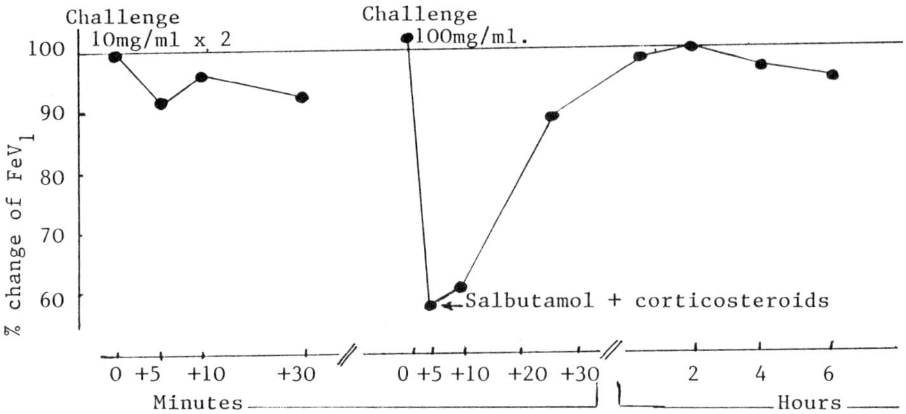

Fig. 1. Results of nickel sulfate challenges.

- immunologic studies : Human basophil degranulation tests were done with the patient's blood and 3 control subjects. This test, using $NiSO_4$-Human serum albumin (HSA) conjugates, according to Yeung-Laiwah (6), gave a positive specific degranulation only for the patient (degranulation index of 74 % at a $NiSO_4$ concentration of 0,1 mg/ml).

DISCUSSION :

Only few patients affected with occupational asthma attributed to nickel have previously been described (1, 3, 4, 5). More than 100 jobs are exposed to nickel and respiratory symptoms are very rare. However nickel is the first allergen for contact dermatitis. Atopy does not seen to be a predisposing factor. In 1971 Mc Connel (3) reported asthma due to nickel sulfate in a metal plating worker. Malo (4) reported another case in a metal plating factory and Block (1) a case of a metal polisher who was exposed to nickel dusts and electroplating of nickel. Nickel is involved by its salts and specially nickel sulfate, but nickel dusts may also give respiratory symptoms. This was the case in our observation. We have only performed skin tests and inhalation tests with nickel sulfate and not with nickel dusts, but Block (1) has observed the same response to inhalation with salts or dusts. This is in agree with a common reactivity to these 2 components, which is observed in contact dermatitis to nickel (2). In this case, nickel salts or nickel dusts may occure the same pathology. It is thougt that nickel act as a hapten and react with human proteins such as albumin. This reaction may be observed with the sweat or with respiratory tract proteins. Evidence of an immunologic response with nickel related specificity is well documented. Skin reactivity to $NiSO_4$ is presented is many studies (1, 4, 5) and in our observation. This reactivity is not comparable in control subjects. Evidence for specific antibodies was estabilited by Mc Connel (3) by hemagglutination of nickel-coated erythrocytes occuring on incubation in serum. The specificity of the IgG and IgM antibodies was skown by inhibition of the hemagglutination. However Block (1) failed to demonstrate the same results. Malo (4) and Novey (5) showed evidence for IgE, with a positive $NiSO_4$-HSA Radio Allergo sorbent Test (RAST) in their patient but failed to perform RAST inhibition. In our observation, a positive basophil degranulation test, ruling out a non specific response, is in agree for an IgE mediated reaction. Inhalation tests, with immediate reaction to nickel sulfate, act in the same way.

In conclusion, asthma induced by nickel remains a rare manifestation of the nickel induced pathology. It may be suggested by the clinical history and confirmed by results of nickel sulfate skin prick tests and bronchial challenge. The mechanism is very likely compatible with an immunological process.

Références

1 BLOCK G.T., YEUNG M. (1982)
Asthma induced by nickel.
J A M A. 247, 1 600-1 602.

2 LA CHAPELLE J.M., GROSHANS E. et al. (1988)
Eczemas de contact. In Allergologie, ed. J. Charpin,
pp 460-515- Paris : Flammarion.

3 Mc CONNELL L.H., FINK J.N. et al. (1973)
Asthma caused by metal sensitivity
Ann. Intern. Med. 78, 888-890.

4 MALO J.L., CARTIER A. et al. (1982)
Occupational asthma caused by nickel sulfate.
J. Allergy Clin. Immunol. 69, 55-59.

5 NOVEY H. (1983)
Asthma and IgE antibodies induced by chromium and nickel salts.
J. Allergy Clin. Immunol. 72, 407-412.

6 YEUNG-LAIWAH A.C., PATEL K.R. et al. (1984)
Evaluation of the human basophil degranulation test using the commercially available Basokit as a test of immediate-type hypersensitivity in hay-fever sufferers.
Clin. Allergy. 14, 571-579.

Metal Ions in Biology and Medicine. Eds. Ph. Collery, L.A. Poirier, M. Manfait, J.C. Etienne. John Libbey Eurotext, Paris © 1990, pp. 266-270

Intraocular metal foreign bodies : retinal intoxication, recovery after particle extraction and drug influences

J.G.H Schmidt

Eye Clinic, University of Cologne, 9 Joseph-Stelzmann-Str., D-5000 Köln 41, Federal Republic of Germany

INTOXICATION

Intraocular metal particles may cause an intoxication of the retina and other tissues of the eye which react with inflammatory responses in different kinds and intensities.
Special problems exist in case of patients injured by intraocular metal foreign bodies which can not be removed with a magnet. Aside from non-magnetic splinters the ophthalmologist has to deal with magnetic particles which are encapsulated and involve a higher risc of manual extraction (Schmidt, 1977).
Besides clinical examinations (recording of visual acuity, visual field, dark adaption) electrophysiological methods are appropriate to measure the retinal function. Mainly electroretinography (ERG) is applied representing the activity of the visual cell layer (a-wave) and the area of the bipolar layer (b-wave) of the retina.
To minimize the side effects of the anesthesia in animal experiments the mixture of halothane, nitrous oxide and oxygen was modified in additional investigations (Wasserschaff & Schmidt, 1986).

With the aid of animal experiments it was shown that the extent of the metallosis retinae is determined mainly by the particles chemical composition (1), the time lapse from invasion to removal (2), the surface area size (3) and the localization in the eye bulb (4).

ad (1) and (2). In three groups of rats (42 in total) we implanted pure iron or copper or lead wires with an active surface of 1.2 mm^2 into the center of the vitreous body. The intact fellow eyes were used as control. Lead caused a limited and iron a distinct decrease of the ERG-amplitudes, whereas copper destroyed these electrical potentials completely within 24 hours (Schmidt et al., 1972; Schmidt & Stute, 1973).
The toxic effect of copper alloys was strongly influenced by the other constituents (Schmidt, 1988). Intravitreal foreign bodies of 99.9 per cent copper and 0.1 per cent silver (A) destroyed the ERG faster than specially purified copper B (99.999%). In contrast,

copper-zinc alloys (C: Cu 85% + Zn 15% and D: Cu 64% + Zn 36%) affected the ERG more slowly than B. This effect is more obvious in alloys containing a large amount of zinc (D). The degree of complications in the anterior segment (iridocyclitis, corneal opacity, hypopyon) and in the vitreous body (intensity and rate of opacities) depends essentially on the type of alloy. The frequency and extent of inflammatory responses decreased in the following order: alloy A,B,C and D. Itoi (1937), Mielke (1940), Contzen and Broghammer (1964) have already stated that the rate of ionisation, and therefore of metal intoxication, depends on the metals position in the electrovoltaic series.
Interactions between copper and zinc are not only found in the physico-chemical, but also in the biochemical sphere (Schmidt et al., 1987).
The influence of intravitreal zinc particles (1.3 mm^2) in the eyes of rats caused the amplitudes of the electroretinogram to decrease to about 50 per cent of the initial values within 24 hours, whereas the prolongation of the peak times began only on the third day after particle implantation. The removal of the zinc particles after 30 days led to only partial recovery of the amplitudes, whereas the peak times reached normal values. Dissociated behaviour of the amplitudes and peak times as described here was not found in earlier experiments with intravitreal iron, copper or lead particles of the same size (Schmidt et al., 1987).

In additional in-vitro-experiments, the influence of H^+ on copper ionisation was explored. Wires of pure copper and copper-silver-alloys were placed into a polysaccharide paste which was prepared with a saline solution at pH 7.4. Only the immediate environs of these wires were bright green. Conducting the experiment with a saline solution at pH 6.4 led to a diffuse green coloring of the area surrounding the wires. This is explained by the change of the dispersion rate of ions in the presence of inflammation of the tissue and acidosis. Attention is drawn to corresponding observations made by Makiushi (1960), who found that in cases of cyclitis the decomposition of intraocular iron particles is accelerated.

ad (3). To obtain quantitative data about this factor we implanted different sized copper wires into the vitreous body of rabbits and rats (Schmidt et al., 1978). As basis for comparison we took the size of the foreign body in relation of the total weight of the eye bulb. In spite of the anatomical and physiological differences between the eyes of the two species a surprisingly good correspondance of the index figures was obtained regarding the toxicity of copper. The values of the 16 rabbits and ten rats coincided between 3 and 4 mm^2 per gram bulbus weight for the most part. Particles smaller than 1.5 mm^2 per gram bulbus weight affected the ERG potentials very slowly.
Performing corresponding procedures we gained quantitative data about retinal intoxication caused by intravitreal iron particles (Schmidt & Max, 1988).

ad (4). Copper-containing foreign bodies located in the anterior chamber of rat eyes did not cause a decrease of the ERG potentials for an observation time of up to 83 days (Schmidt & Weber, 1971).
The position of a copper particle in the vitreous body in rabbit eyes determined the time course of the inflammatory response (Rosenthal et al., 1974). Copper wires located in the center of

the vitreous body of rats (Schmidt et al., 1972) had a different influence on the ERG in comparison with those implanted close to the ora serrata (Schmidt & Wasserschaff, 1983).

RECOVERY

A recovery of the ophthalmoscopic findings after resorption of intravitreal metal particles was observed in a few clinical cases (Jess, 1929; Cordes & Harrington, 1935; Marner, 1945; Delaney, 1975). In view of the exceptional toxic effect of copper there is special interest in the question whether a functional recovery can be expected after removal of this metal.

In a series of experiments with rats we removed intravitreal copper wires (1.2 mm^2) two days after implantation and measured a 50 per cent loss of ERG amplitudes up to the end of the observation time of 260 days. In another group of animals we removed the copper particles after one day and found a loss of 25 per cent. Applying a copper wire with a surface of 0.3 mm^2 for one day we had a permanent loss of about 10 per cent (Schmidt et al., 1986; Schmidt et al., 1987). The importance of the time factor was also stated in experiments with intravitreal iron (Schmidt & Fischer, 1988), lead (Schmidt & Ehrgott, 1988) and zinc particles (Schmidt et al., 1987).

These experiments demonstrate that the four factors characterized above not only determine the intoxication rate of the tissue but also the extent of the functional recovery after particle removal. Just so, the ophthalmoscopic symptoms showed a dependence on surface area size and time lapse of intravitreal metal particles (Schmidt et al., 1987; Schmidt & Max, 1988). This was the case for the frequency and extent of vitreal opacities, retinal hemorrhages as well as retinal detachments.

DRUGINFLUENCES

The effects of drugs on metallosis and the following inflammatory response was investigated in several animal experiments and clinical observations.

Corticosteroids reduced the imflammatory responses and vitreal opacities of rabbits and rats, but increased the tendency of the foreign bodies to migrate (Rosenthal et al., 1976; Schmidt & Hummelsheim, 1986).

20 mg $CaNa_2$-EDTA injected into rat eyes subconjunctivally two times a day over a period of 17 days did not reduce the metalltoxic effect of lead wires under the applied experimental conditions. Furthermore this drug did not accelerate the recovery of the electric potentials after particle extraction (Schmidt & Ehrgott, 1988).

Intravitreal iron wires reduced the ERG amplitudes of male and female rats to the same extent. Subconjunctival injections of Desferal(R) over a period of 10 days retarded the iron intoxication of the ERG significantly in male rats. In contrast to this the drug damaged the amplitudes and affected the peak times of the retina of female rats besides the iron-established intoxication (Schmidt & Fischer, 1988).

REFERENCES

Cordes, F.C., and Harrington, D.O. (1935): Bilateral absorption of intraocular copper with chalcosis in one eye. *Am. J. Ophthalmol. 18*: 348-352.

Delaney, W.V. (1975): Presumed ocular chalcosis: A reversible maculopathy. *Ann. Ophthamol. 7*: 378-380.

Itoi, M. (1937): Über das Wesen der Verrostung des Eisens im Augeninneren. *Acta Soc. Ophthalmol. Jap. 41*: 669-670.

Jess, A. (1929): Das Verschwinden von Verkupferungserscheinungen des Auges. *Z. Augenheilk. 69*: 59-73.

Makiuchi, S. (1960): Clinical aspects and pathology of intraocular siderosis. *Acta Soc. Ophtahlmol. Jap. 64*: 2234.

Marner, E. (1945): A case of ocular chalcosis with spontaneous resorption of the copper, and disappearance of chalcosis. *Acta Ophthalmol. (Kbh) 23*: 171-174.

Mielke, S. (1940): Die Rolle der elektrochemischen Vorgänge bei der Entstehung der Linsen- und Hornhautverkupferung. *Albrecht von Graefes Arch. Ophthalmol. 141*: 644-654.

Rosenthal, A.R., Appleton, B., and Hopkins, J.L. (1974): Intraocular copper foreign bodies. *Am. J. Ophthalmol. 78*: 671-678.

Schmidt, J.G.H., and Weber, E. (1970): The Effect of intraocular copper alloys on the ERG of human and rat retinas. In *Symposium on electroretinography*, ed. A. Wirth, pp. 240-250. Pisa: Pacini.

Schmidt, J.G.H., Stute, A., and Weber, E. (1972): Elektroretinographische und ophthalmoskopische Befunde bei intraokularen Metallfremdkörpern der Ratte. *Ber. Dtsch. Ophthalmol. Ges. 71*: 391-396.

Schmidt, J.G.H., and Stute, A. (1973) Electroretinogram and ophthalmoscopic findings in intravitreous iron, copper and lead particles. *Docum. Ophthalmol. Proc. Ser. 15*: 85-90.

Schmidt, J.G.H. (1977): Metallosis retinae - Pathophysiologie und Klinik. *Ber. Dtsch. Ophthalmol. Ges. 75*: 670-674.

Schmidt, J.G.H., Micovic, V., and Stute, A. (1978): Surface area sizes of intravitreal copper particles: their effects on the ERG of rabbits and rats. *Docum. Ophthalmol. Proc. Ser. 15*: 63-68.

Schmidt, J.G.H., and Wasserschaff, M. (1983): On the recovery of the electroretinogram of rats after removal of intravitreal iron particles. *Docum. Ophthalmol.Proc. Ser. 37*: 293-299.

Schmidt, J.G.H., and Hummelsheim, St. (1986): Über den Einfluß von Decadron(R)-Phosphat auf die ophthalmoskopischen Veränderungen des Rattenauges bei intravitrealen Messingdrähten. *Klin. Mbl. Augenheilk. 188*: 234-238.

Schmidt, J.G.H., Mansfeld-Nies, R., and Nies, C. (1986): Über die Rückbildungsfähigkeit von Netzhaut- und Glaskörperveränderungen der Ratte nach Extraktion intravitrealer Kupferpartikel. *Klin. Mbl. Augenheilk. 189*: 39-43.

Schmidt, J.G.H., Mansfeld-Nies, R., and Nies, C. (1987): On the recovery of the electroretinogram after removal of intravitreal copper particles. *Docum. Ophthalmol. 65*: 135-142.

Schmidt, J.G.H., Nies, C., and Mansfeld-Nies, R. (1987): On the recovery of the electroretinogram after removal of intravitreal zinc particles. *Docum. Ophthalmol. 65*: 471-480.

Schmidt, J.G.H., and Max, M. (1988): Surface area sizes of intravitreal iron wires: their effects on the electroretinogram of rats. *Docum. Ophthalmol. 67*: 263-272.

Schmidt, J.G.H. (1988): Intravitreal cupriferous foreign bodies: electroretinograms and inflammatory responses. *Docum. Ophthalmol.* 67: 253-261.

Schmidt, J.G.H., and Ehrgott, H.J. (1988): Über die Erholbarkeit des Elektroretinogramms der Ratte nach Extraktion intravitreealer Bleipartikel und die Wirkung von $CaNa_2$-EDTA. *Spektr. Augenheilkd.* 2: 54-59.

Schmidt, J.G.H., and Fischer, U.M. (1988): On the recovery of the electroretinogram after removal of intravitreal iron particles and the effect of Desferal(R). *XXVI Symposium of the International Society for Clinical Electrophysiology of Vision (I.S.C.E.V.)*. Estoril/Portugal 1988.

Wasserschaff, M., and Schmidt, J.G.H. (1986): Electroretinographic responses to the addition of nitrous oxide to halothane in rats. *Docum. Ophthalmol.* 63: 347-354.

Metal Ions in Biology and Medicine. Eds. Ph. Collery, L.A. Poirier, M. Manfait, J.C. Etienne. John Libbey Eurotext, Paris © 1990, pp. 271-276

Aspects of Lead (Pb) potentiation of B lymphocytes responses and its relationship to immune disregulation

M.J. Mc Cabe Jr. and D.A. Lawrence

Department of Microbiology and Immunology, Albany Medical College, Albany, New York, 12208, USA

Lead (Pb) is a well known toxicant for the renal, hematopoietic and central nervous systems (Goyer, 1971). In addition to these organ systems, the immune system has been implicated as a target of Pb (Lawrence, 1981; 1985). The role of the immune system is to protect the host from disease causing pathogens or neoplasia. It achieves this task through a balanced network of cells which discriminate between self and non-self antigens. Lymphocytes and macrophages are the primary cell types involved in the regulation of responsiveness to self and non-self. Lymphocyte subsets have been characterized ontogenically, phenotypically, and functionally. T lymphocytes develop within the thymic microenvironment and, upon maturation, bear clonotypic receptors (TcR) which recognize specific antigenic fragments in association with appropriate major histocompatibility complex (MHC) molecules on the surfaces of antigen presenting cells. In addition to their antigen specific receptor, TcR, T cells express other surface markers which delineate functionally distinct subsets. For instance, $CD4^+$ T cells (ie. helper and inflammatory T cells) recognize antigen in association with MHC class-II (Ia) molecules, whereas, $CD8^+$ T cells (ie. cytotoxic and suppressor T cells) recognize antigen in association with MHC class-I molecules. B lymphocytes mature within the bone marrow in mammals. They express surface immunoglobulin (sIg) as their clonotypic antigen specific receptors. Unlike the TcR, sIg binds native antigen which upon appropriate stimulation of the B cell is secreted. An outstanding feature of both T cells and B cells is their extraordinary specificity. Specific responses are inducible and strongly amplified by antigen, cellular interactions, and lymphokines. Agents which interfere with immunohomeostasis by mimicking or modulating immunoregulation can promote disease either by reducing responses against non-self (ie. immunosuppression) or inducing hyperreactivity against self (ie. autoimmunity).

One strategy for assessing Pb-immunomodulation is to evaluate the influences of Pb on the various components of the immune system *in vitro*. This report describes studies which indicate that Pb can directly influence B cells by increasing cell surface Ia density, enhancing Ig production, and promoting $CD4^+$ T cell:B cell interactions. This Pb-facilitated T-B cell interaction, mediated by the increased density of B cell surface Ia, augments factor production by the T cells and B cell responsiveness to these factors. Modulation of the regulation of B cell responses by Pb is consistent with a posited role for the involvement of environmental toxicants in the induction of autoimmune diseases. In addition to the in vitro studies, some in vivo data indicating Pb alteration of lymphocyte trafficking which similarly could promote the induction of autoimmune disease is presented.

Previous studies have shown that Pb enhances B cell differentiation in vitro as measured by the hemolytic plaque forming cell (PFC) assay (Lawrence, 1981). In those studies the T cell dependent antigen sheep erythrocytes was used. To address whether Pb could directly influence B cell differentiation in the absence of T cells, the ability of Pb to modulate Ig secretion induced by a T cell independent polyclonal activator, LPS, was evaluated (Table 1). Pb significantly increased the amount of IgM secreted from LPS stimulated B cells comparable to the amount reported for the lymphokine, interleukin-5 (Tonkonogy et al., 1989). Pb did not augment LPS-induced B cell IgG production or proliferation.

Table 1. Effects of Pb on LPS-induced B cell responses.

IgM	IgG	proliferation
1.69	1.10	0.98

IgM and IgG were measured by ELISA, and the results are expressed as the ratio of the response from day 7 (peak) LPS-stimulated B cell cultures containing 10 μM $PbCl_2$ compared to cultures without Pb. Proliferation was measured by ^{3}H-thymidine incorporation of the day 3 (peak) response. The result is expressed as the ratio of the LPS plus Pb to LPS response.

In addition to its influence on B cell differentiation, Pb induces $CD4^+$ T cell proliferation in vitro (Warner & Lawrence, 1986); however, Pb-induced T cell proliferation requires the presence of Ia^+ cells in culture suggesting that Pb may enhance the autologous mixed lymphocyte response, a response against self Ia. The Ia molecules consist of two isotypic forms in the mouse designated I-A and I-E which are codominantly expressed and have a limited tissue distribution. Unlike the MHC class-I molecules (eg. H-2 K and D) which are found on all cells, MHC class-II molecules are found only on select cell types. B cells constituitively express Ia, but the density of B cell surface Ia molecules increases during B cell activation. Increased B cell surface Ia has been posited in immune response regulation by promoting B cell:$CD4^+$ T cell interaction. Since Pb increases responses where T-B cell collaboration is important, Pb's effect on B cell Ia expression was evaluated.

Pb's ability to increase B cell surface I-A expression was compared to other known B cell activators such as LPS, anti-Ig, and interleukin-4 (Noelle et al., 1986). These polyclonal activators have distinct effects on B cells yet they all have been shown to increase Ia. Flow cytofluorometry was used to quantitate B cell surface Ia density. 10 μM Pb substantially increased the median fluorescence intensities after staining with monoclonal antibodies against both I-A and I-E (Table 2). The increase caused by Pb was greater than 2-fold relative to the saline control and comparable to that induced by optimal doses of LPS, anti-Ig, or IL-4.

Table 2. Comparison between the effects of Pb and other B cell activators on B cell surface MHC molecule density.

Culture Additive	Ratio of Fluorescence Intensity I-A	I-E	K
10 μM Pb	2.50	2.61	0.98
20 μg/ml anti-Ig	2.86	3.85	0.99
10 μg/ml LPS	2.30	ND	ND
30 units/ml IL-4	3.10	ND	ND

Values represent the ratio of the median fluorescence intensity for each marker from B cells cultured with the indicated additives compared to the control (saline). ND = not done.

Pb elevation of I-E density was detected by 3 monoclonal antibodies which recognize distinct epitopes on I-E (data not shown). These results indicate that Pb increases the expression of both isotypic forms of MHC class-II molecules, I-A and I-E, and not a fortuitous cross reactive epitope recognized by one monoclonal antibody. Although MHC class-II molecule expression increases during B cell activation, MHC class-I expression does not change. Notably, Pb did not influence B cell MHC class-I expression.

Other relevant B cell activation markers were modulated by Pb. As indicated in Table 3, Pb consistently increased the expression of CD23 (the Fc epsilon receptor) and decreased surface IgD. These markers are known to be similarly modulated by the B cell activators IL-4 and LPS respectively.

Table 3. Pb modulation of B cell activation antigens.

Culture Additive	Ratio of Fluorescence Intensity CD23	sIgD	sIgM
10 μM Pb	1.32	0.54	0.80
30 units/ml IL-4	2.49	ND	ND
10 μg/ml LPS	1.61	0.27	0.84

Values represent the ratio of the median fluorescence intensity for each marker from B cells cultured with the indicated additive compared to the saline control.

10 μM Pb caused the maximum, 2-fold increase in I-E, although Pb concentrations ranging from 1 to 100 μM also significantly increased B cell surface I-E (data not shown). The effects of other metals on Ia expression were tested (data not shown). Similar to Pb, Hg increased Ia but neither Ni, Zn, nor Cd at doses known to be immunomodulatory affected B cell Ia expression indicating that metals influence immune responses by different mechanisms.

Since Pb's effects on B cells appear in some respects to be similar to IL-4's, namely increased expression of Ia and CD23, the possibility that residual T cells in our B cell preparations were stimulated by Pb to secrete IL-4 which could subsequently upregulate Ia expression was considered. Inclusion of an anti-IL4 antibody in sufficient quantity to completely abrogate I-A upregulation by exogenously added IL-4 had no inhibitory effect on Pb upregulation of B cell I-A (data not shown). These results strongly suggest that Pb does not increase B cell surface Ia via IL-4 secretion.

Our interpretation of these results is that phenotypically and functionally Pb behaves like a direct B cell activator.

Furthermore, by augmenting B cell surface Ia density Pb may facilitate T-B cell interaction resulting in enhanced responsiveness of both cell types.

The requirement for T-B cell contact for Pb enhancement of B cell responses to T cell dependent antigens was addressed. Membrane segregated cultures, designed so that the influences of direct T-B cell contact and of T cell factors on B cells could be independently evaluated, were used. Double chambered wells containing a 1.0 μm Nucleopore membrane were employed to separate B cells (upper chambers) from T + B cells (lower chambers). Both chambers were immunized with sheep erythrocytes $\pm$ 100 μM $PbCl_2$ or Pb-pretreated B cells in either chamber. As shown in Table 4, Pb consistently increased the number of antibody forming cells (AFCs) from the upper chambers. The magnitude of the enhancement was dependent upon the state of activation of the responding B cells (ie. resting B cells were influenced by Pb to a much greater extent than large non-resting B cells), and optimum enhancement (> 6-fold) was obtained when Pb was added directly to the cultures; however, the use of Pb-pretreated B cells in the lower or upper chambers also caused a 3-fold and 2-fold increase respectively. This suggests that Pb enhances factor mediated T cell help by both enhancing the production and responsiveness to putative T cell derived lymphokines. Studies are underway to identify these lymphokines.

Table 4. Effect of Pb on the AFC response of B cells segregated from direct T cell contact.

Culture Conditions: lower chamber	Pb	upper chamber	Pb	upper chamber AFC response (ratio)
T cells plus unseparated B cells	+	unseparated B cells	+	6.58
T cells plus unseparated B cells	-	unseparated Pb-B cells	-	1.94
T cells plus unseparated Pb-B cells	-	unseparated B cells	-	3.45
T cells plus resting B cells	+	resting B cells	+	38.43
T cells plus activated B cells	+	activated B cells	+	3.75

Values represent the ratio of the number of AFCs detected from upper chamber B cells cultured as indicated compared to the appropriate identical culture conditions without any Pb treatment.

Since Pb-pretreated B cells added to the lower chambers enhance the AFC response of upper chamber B cells, it is conceivable that Pb-B enhances T-B cell interaction (ie. cognate help). To substantiate this, splenocytes obtained from CBA/N mice which require cognate T cell help for B cell differentiation were studied. Comparable to results obtained with splenocytes from "normal" mouse strains, Pb significantly increased the AFC response of CBA/N splenocyte cultures (data not shown).

Since Pb increased B cell Ia and optimally enhanced Ig production under conditions where T-B cell interaction was favored, it was of interest to determine the nature of the T-B interaction in the presence of Pb (ie. Pb-altered-self or cognate). The nature and influence of Pb on cognate T-B cell interaction was evaluated using antigen specific T cell clones which would be unable to recognize Pb-altered self. Murine $CD4^+$ T cell clones are divided into two subsets based upon the repetoire of lymphokines that they produce. TH1 type clones produce IL-2 and gamma-IFN, whereas, TH2 type clones produce IL-4 and IL-5. TH1 like cells, termed inflammatory T cells, and TH2 like cells, termed helper T cells, have been described _in vivo_. Like their _in vitro_ counterparts they mediate different effector functions which in part may be attributed to the different lymphokines that they produce. As a result of these functional differences, $CD4^+$ T cell subsets differentially regulate immune responses and perturbation of either subset may preferentially be protective or immunopathologic (reviewed by Scott et al. 1989). We have noted a dichotomy with Pb's influence on TH1 vs. TH2 responses (Table 5). Although more extensive testing of TH1 and TH2 clones is necessary, we predict based on our available data that Pb enhances antigen specific T cell help for antibody responses but inhibits responses dominated by TH1 cells. In addition, since Pb enhances antigen specific TH2 cell help for antigen specific immunoglobulin production as well as the response to sheep erythrocytes using a SE primed T cell line, it is unlikely that Pb activates T cells by a Pb-B cell altered-self phenomenon but probably augments cognate T-B cell interaction.

Table 5. Effects of Pb on the activities of T cell clones.

clone	type	antigen specific Ig production	polyclonal Ig production	antigen presentation
D10.G4	TH2	increased	increased	increased
HDK-1	TH1	not applicable	not applicable	decreased
M264-37	TH1	not applicable	decreased	not done
DO11.10	TH1-like	not done	not applicable	decreased
SE line	?	increased	not applicable	not done

As an _in vivo_ corollary to the effects of Pb on immune parameters _in vitro_, the popliteal lymph node (PLN) assay was used. Injection of 100 µg of $PbCl_2$ into the foot-pad resulted in a marked 3-fold increase in the weight and cellularity of the draining PLN compared to the contralateral PLN or saline control (Table 6). Likewise, $HgCl_2$ injection into the foot-pad caused a marked increase in the size of the draining PLN, as well as the contralateral PLN and spleen suggesting a more systemic effect of Hg than Pb. Pelletier et al., 1985, have shown a similar increase in PLN size after Hg injection in the Brown-Norway rat model of metal-induced autoimmune disease. $NiCl_2$ injection resulted in a slight yet statistically significant increase in the size of the draining PLN. Like Pb, Ni had a more local effect and did not affect the size of the contralateral PLN or spleen. The maximum Pb dose causing PLN enlargement was 250 µg; however, 100 µg was used for subsequent assays due to solubility problems at higher $PbCl_2$ concentrations. The peak time of PLN enlargement after a single injection of the metals was day 7, but enlargement was detected as early as day 2. This perhaps was a hint that the increased PLN size induced by Pb represented altered lymphocyte trafficking (ie. nodal accumulation) rather than intra-nodal lymphocyte proliferation. In fact, by propidium iodide staining of cellular DNA content and flow cytometry, no cycling cells were detected in the lymph nodes from either the saline control or Pb injected mice (data not shown). Studies are currently underway to substantiate Pb alteration of lymphocyte trafficking or homing to the PLN. Although Pb increased the number of all cell types within these nodes, it preferentially increased the frequency of cells expressing B cell markers (eg. I-A, I-E, and sIgM) at the expense of $CD4^+/Ly1^+$ cells (data not shown). In addition, the density of I-A and I-E was slightly but consistently elevated on the lymphocytes obtained from the nodes of Pb injected mice.

Table 6. Effects of metal injection after 7 days on popliteal lymph node (PLN) weight and cellularity.

metal	draining PLN weight (mg)	draining PLN cell number ($x 10^6$)	contralateral PLN weight (mg)	spleen weight (mg)
saline	1.69 ± 0.30	1.59 ± 0.52	1.75 ± 0.21	96.67 ± 6.59
Pb	4.18 ± 1.70	4.93 ± 1.25	1.66 ± 0.31	92.70 ± 8.34
Hg	4.40 ± 0.60	8.25 ± 2.54	3.60 ± 1.06	123.47 ± 6.92
Ni	2.56 ± 0.88	2.90 ± 1.97	1.68 ± 0.41	89.97 ± 7.16

Features of the induction of many autoimmune diseases include aberrant expression of Ia and lymphokines as well as lymphoid infiltration of target tissues due to altered interactions between the lymphoid subsets. Although to date there is little data that directly implicates Pb in any autoimmune disease, our data suggests that Pb can provoke immune dysregulations of these types. In that little is known about the etiology of autoimmune diseases, a role for toxic agents such as Pb warrants further consideration.

Goyer, R.A. (1971): Lead Toxicity: A problem in environmental pathology. Am. J. Pathol. 64, 167-180.

Lawrence, D.A. (1981): _In vivo_ and _in vitro_ effects of lead on humoral and cell-mediated immunity. Inf. Immun. 31, 136-143.

Lawrence, D.A. (1985): Immunotoxicity of heavy metals. In: Immunotoxicology and Immunopharmacology, edited by J. Dean, M. Luster, A.E. Munson, and H. Amos, pp. 341-353. NY: Raven Press.

Noelle, R.J., et al. (1986): Regulation of the expression of multiple class II genes in murine B cells by B cell stimulatory factor-1 (BSF-1). J. Immunol. 137, 1718-1723.

Pelletier, L., et al. (1985): _In vivo_ self reactivity of mononuclear cells to T cells and macrophages exposed to $HgCl_2$. Eur. J. Immunol. 15, 460-465.

Scott, P., et al. (1989): Role of cytokines and $CD4^+$ T-cell subsets in the regulation of parasite immunity and disease. Immunol. Rev. 112, 161-182.

Tonkonogy, S.L., et al. (1989): Regulation of isotype production by IL-4 and IL-5. Effects of lymphokines on Ig production depend on the state of activation of the responding B cells. J. Immunol. 142, 4351-4360.

Warner, G.L. & Lawrence, D.A. (1986): Cell surface and cell cycle analysis of metal-induced murine T cell proliferation. Eur. J. Immunol. 16, 1337-1342.

IV. Carcinogenesis

Metal Ions in Biology and Medicine. Eds. Ph. Collery, L.A. Poirier, M. Manfait, J.C. Etienne. John Libbey Eurotext, Paris © 1990, pp. 279-283

Cadmium carcinogenesis in review

M.P. Waalkes

Inorganic carcinogenesis section, Laboratory of comparative carcinogenesis, National Cancer Institute, Frederick Cancer Research Facility, Building 538, Room 205E, Frederick, MD 21701-1013, USA

ABSTRACT

Cadmium is a metallic toxicant of great environmental and occupational concern. A potent carcinogen in animals, cadmium is also a suspected human carcinogen. Occupational exposure to cadmium has been associated with tumors of the lung and prostate in humans. The epidemiological data showing an association between cadmium exposure and pulmonary cancer has recently been reinforced while the association with prostatic carcinogenesis has perhaps weakened. Nevertheless recent studies in rats have shown that cadmium is an effective prostatic carcinogen after systemic exposure or intraprostatic application, confirming its potential within this tissue. Prostatic carcinogenesis in rodents after systemic cadmium exposure only occurs at cadmium doses below those that induce chronic degeneration and dysfunction of the testes, confirming the androgen dependency of this tumor type. Toxicokinetic data also indicate a clear androgen dependency of cadmium disposition and retention in the prostate, as more metal is deposited and retained within the rat prostate with androgen supplementation. Chronic inhalation of cadmium has been proven to be a very effective pulmonary carcinogen in rats, likewise confirming the potential for cadmium in this organ. Other targets of cadmium in rodents include the testes, injection sites, and recently the lymphatic system. In the testes, cadmium induces interstitial (Leydig) cell tumors which, though typically benign, can become malignant if the dose of cadmium is high enough or cadmium is given repeatedly. Cadmium treatment in rats has also been associated with very rare tumors of the testes including seminomas, a rete testis adenocarcinoma, and a mixed sertoli-Leydig cell tumor, although not in sufficiently high numbers to draw a firm conclusion with regard to cause and effect. Injection site sarcomas are common with cadmium and occur at subcutaneous and intramuscular sites. The malignancy of cadmium-induced injection site sarcomas is increased by repeated exposures. Incidence of leukemia can be markedly increased in rats or mice by cadmium. Various treatments can modify cadmium carcinogenesis. For instance, supplemental zinc prevents both cadmium induced injection site and testicular tumor formation while facilitating prostatic tumor formation. In contrast, dietary zinc deficiency increases testicular tumor formation but reduces prostatic carcinogenesis. In summary, cadmium is a potent carcinogen in many tissues but its carcinogenic effects are related in complex fashions to various factors. The mechanisms of cadmium carcinogenesis remain obscure but may well be tissue dependent.

INTRODUCTION

Cadmium is a very potent metallic toxicant of continuing environmental and occupational concern. Cadmium has an extremely long biological half life and thus must be considered a cumulative toxin. To date no proven effective treatment exists for cadmium intoxication. Several sources of human exposure to cadmium exist including primary metal occupations and tobacco smoking. Cadmium accumulates primarily in the liver and kidney and is generally found in association with the metal binding protein, metallothionein which is thought to be involved in detoxication of the metal through sequestration. Frequently the toxic effects of cadmium occur through a disruption of zinc mediated metabolic processes, and as zinc is essential in various critical cellular functions including DNA, RNA and protein synthesis such disruption is readily seen as highly detrimental. Treatment with excess zinc can often prevent or reduce the toxic effects of cadmium, including carcinogenesis. Among the non-carcinogenic toxic effects of cadmium nephrotoxicity is the most important, and has been frequently observed in environmentally or occupationally exposed individuals. Several comprehensive reviews exist concerning various aspects of cadmium toxicology (Friberg et al., 1986; Webb, 1979).

CADMIUM CARCINOGENESIS; EPIDEMIOLOGY

Occupational exposure to cadmium has been on occasion associated with development of tumors of the lung, prostate, kidney and stomach (IARC, 1976; for review see Oberdorster, 1989; and Waalkes and Oberdorster, 1990). However, the association of cadmium exposure with cancer of any site in humans has not been consistently observed in all studies. The most firmly established target site of cadmium in humans must be considered the lung and several studies has shown a positive association of occupational exposure and pulmonary carcinogenesis. Furthermore, recent epidemiological data have reinforced the concept of the lung as a target site of cadmium in humans (see Waalkes and Oberdorster, 1990). The production of lung tumors in rats chronically exposed to inhaled cadmium (Takenaka et al., 1983) strengthens the case for a role of cadmium in generation of lung tumors in humans. On the other hand the role of cadmium in human prostatic cancer is less well defined although cadmium has been associated with prostatic cancer in humans after occupational (Lemen et al., 1976) and presumed environmental (Bako et al., 1982) exposure. However, several studies not showing an association with prostatic cancer have appeared including some recent work (Kazantzis et al., 1988). It must be kept in mind that prostatic cancer is a very important and deadly form of human cancer with what is considered to be an extremely complex etiology. The complexity of causation in this case may make linkage to a single factor very difficult to discern. As is the case with lung cancer recent animal studies detecting prostatic cancer in rodents following systemic cadmium exposure add credence to the suppositions that cadmium may play a role in human prostatic carcinogenesis; they do not, however, allow definitive linkage. With regard to an association between cadmium and renal or stomach cancer in both cases only a single report exists and in the absence of verification these sites must be considered as less than established. Further epidemiological work is necessary to determine the exact carcinogenic risk of cadmium to humans.

CADMIUM CARCINOGENESIS; ANIMAL STUDIES

Cadmium has been long recognized as a potent carcinogen in rodents. The original efforts in this area started in the early 1960s and the studies of Gunn et al. (1963, 1964) and Roe et al. (1964) are representative of work of this period. Cadmium was first detected as a carcinogen at the site of intramuscular or subcutaneous injection where it forms sarcomas. It was also determined that cadmium was a very effective testicular tumorigen, forming a high incidence of interstitial cell tumors in the rat or mouse. These early efforts also revealed the remarkable ability of zinc to prevent the carcinogenic effects of cadmium in the testes and at the injection site. Zinc treatment was also shown to prevent the chronic degenerative effects of cadmium on the testes.

More recent work has focused on cadmium as a prostatic carcinogen in rats. In this regard two studies (Waalkes et al., 1988, 1989) have clearly shown that systemic cadmium exposure will result in the generation of tumors of the rat prostate in certain specific treatment groups. This includes doses of cadmium that were below the threshold for induction of testicular degeneration and dysfunction (< 5.0 μmol/kg, sc) or when such degeneration is prevented by zinc. Thus it appears that cadmium produces prostatic tumors if testicular support is maintained. In this regard the structural maintenance and function of the prostate is dependent on androgens and will regress upon cadmium induction of testicular degeneration. Likewise most prostatic tumors are androgen dependent. Our recent findings also show that cadmium disposition and retention in the rat prostate is highly dependent on testosterone (Rhodes et al., 1989). The finding of prostatic cancer in cadmium treated rats supports its possible role in human prostatic cancer.

The lung has also recently been established as a target site of cadmium carcinogenesis in the rat (Takenaka et al., 1983), again in support of the role of cadmium in human pulmonary carcinogenesis. Rats exposed to $CdCl_2$ aerosols for 18 months were found to have in excess of 70% lung carcinoma incidence at the highest exposure levels (50 $\mu g/m^3$). Various other forms of cadmium, including the oxide, have now been established as pulmonary carcinogens (Olidiges et al., 1989). Cadmium is effective as a lung carcinogen in both males and females and with continuous (Takenaka et al., 1983) or discontinuous (Olidiges et al., 1989) exposure. Interestingly, as is the case with several other sites of cadmium carcinogenesis, zinc administration will prevent cadmium induced lung tumors (Olidiges et al., 1989).

The testes are a well defined target site of cadmium carcinogenesis in the rat and mouse and a single dose of cadmium (>20 μmol/kg, sc) can result in a high incidence (>80%) of testicular interstitial cell tumors (Gunn et al., 1963, 1964; Roe et al., 1964; Waalkes et al., 1988, 1989). Such tumors are typically benign and it has been thought that they result as a secondary effect of cadmium induction of testicular degeneration and dysfunction, and not due to any specific effects of cadmium. However, recent work indicates that repeated cadmium exposure can result in a relatively very high incidence (>9%) of malignant interstitial cell tumors of the rat testes (Waalkes et al., 1990). Here malignancy is defined by distant metastases to the lung. Considering the rarity of such malignancy in spontaneously occurring interstitial cell tumors (1 in 1500) this indicates that cadmium is in fact having a direct effect on these specific cells. Other rare tumors of the rat testes occur with cadmium exposure including seminomas, a rete testes adenocarcinoma, and a mixed sertoli-Leydig cell tumor (Boorman et al., 1987; Rehm and Waalkes, 1988), although not in sufficiently high numbers to draw a firm conclusion of cause and effect.

Cadmium will induce sarcomas at the site of subcutaneous or intramuscular injection (Gunn et al., 1963; 1964; Poirier et al., 1983; Waalkes et al., 1988, 1989). This is a feature cadmium has in common with many other metals and may be related to the phenomenon of solid state carcinogenesis where tumors form during encapsulation of a chronically irritating implant. Cadmium at the subcutaneous injection site will persist and form a distinct calcified area. Recently it was observed that repeated cadmium exposure caused a marked increase in the rate of metastases of injection site tumors, although not affecting overall incidence (Waalkes et al., 1990). This again would indicate that cadmium is having a direct carcinogenic effect on the cells of this area. It is well know that zinc treatment will prevent the carcinogenic effects of cadmium at the site of injection (Gunn et al., 1963, 1964; Waalkes et al., 1989) even when given by a totally different route. This again points toward direct cellular effects in the area as opposed to non-specific tumor generation.

Cadmium will also affect tumors of the lymph system in rodents. In Swiss mice infected with murine lymphocytic leukemia virus, oral cadmium increased lymphocytic leukemia incidence by 33% (Blakley, 1986). It is thought that cadmium impaired immunosurveillance allowing emergence of the leukemia virus. In rats, oral cadmium causes a dose related increase of up to four fold in leukemia (Waalkes and Rehm, 1990) while dietary zinc deficiency decreased the potency of cadmium in induction of leukemia.

Zinc plays a fundamental role in cadmium carcinogenesis. In lung, testes, and at the injection site, zinc is antagonistic to cadmium carcinogenesis (Gunn et al., 1963, 1964; Waalkes et al., 1989; Olidiges et al., 1989. The prevention by zinc of tumors in so many different target sites of cadmium many point to a basic mechanism of cadmium carcinogenesis. In contrast, zinc treatment facilitates cadmium induced prostatic carcinogenesis (Waalkes et al., 1989). Dietary zinc deficiency also has tissue specific effects on cadmium carcinogenesis. Zinc deficient diets increase the formation of testicular tumors in rats while reducing the incidence of prostatic tumors (Waalkes and Rehm, 1990). Overall the role of zinc in cadmium carcinogenesis appears to be very complex and highly tissue specific.

In summary, cadmium is a potent animal carcinogen and a suspected human carcinogen. Cadmium exposure results in tumors of various tissues in rodents including the lung and prostate, tissues very relevant to human exposure situations. The carcinogenic effects of cadmium are related in a complex fashion to various factors. The ability of zinc to prevent cadmium carcinogenesis in some but not all target tissues may be of great mechanistic importance. Thus the mechanisms of cadmium carcinogenesis are as yet obscure and may well be highly tissue dependent.

REFERENCES

Bako, G., Smith, E.S.O., Hanson, J. & Dewar, R. (1982): The geographical distribution of high cadmium concentrations in the environment and prostate cancer in Alberta. *Can. J. Public Health* 73, 92-94.

Blakley, B.R. (1986): The effect of cadmium and viral-induced tumor production in mice. *J. Appl. Toxicol.* 6, 425-4

Boorman, G., Eustis, S., Waalkes, M.P. & Rehm, S. (1987): Seminomas in rats. In *Pathology of laboratory animals, Vol. 5; genital system*, ed. T.C. Jones, U. Mohr & R.D. Hunt, pp. 192-195. New York: Springer-Verlag.

Friberg, L., Elinder, C.-G., Kjellstrom, T. & Norbert, G.F., eds. (1986): *Cadmium and Health.* Cleveland: CRC Press.

Gunn, S.A., Gould, T.C. & Anderson, W.A.D. (1963): Cadmium-induced interstitial cell tumors in rats and mice and their prevention by zinc. J. Natl. Cancer Inst. 31, 745-759.

Gunn, S.A., Gould, T.C. & Anderson, W.A.D. (1964): Effect of zinc on cancerogenesis by cadmium. *Proc. Soc. Exp. Biol. Med.* 115, 653-657.

IARC (1976). *IARC monographs on the evaluation of carcinogenic risk of chemicals to man; cadmium, nickel, some epoxides, miscellaneous industrial chemicals and general considerations on volatile anesthetics*. Vol. 11, pp. 39-74. Lyon: IARC.

Kazantzis, G., Lam, T.H. & Sullivan, K. (1988): Mortality of cadmium exposed workers. *Scand. J. Work Environ. Health* 14, 220-227.

Lemen, R.A., Lee, J.S., Wagoner, J.K. & Blejer, H.P. (1976): Cancer mortality among cadmium production workers. *Ann. NY Acad. Sci.* 271, 273-279.

Oberdörster, G. (1986): Airborne cadmium and carcinogenesis of the respiratory tract. *Scand. J. Work Environ. Health* 12, 523-529.

Oldiges, H., Hochrainer, D. & Glaser, U. (1989): Preliminary results from a long-term inhalation study with four cadmium compounds. *Toxicol. Environ. Chem.* 23, 35-38.

Poirier, L.A., Kasprzak, K.S., Hoover, K.L. & Wenk, M.L. (1983): Effects of calcium and magnesium acetates on the carcinogenicity of cadmium chloride in Wistar rats. *Cancer Res.* 43, 4575-4581.

Rehm, S. & Waalkes, M.P. (1988): Mixed sertoli-Leydig cell tumor and rete testis adenocarcinoma in rats treated with $CdCl_2$. *Vet. Pathol.* 25, 163-166.

Rhodes, S.W., Wahba, Z.Z., Bare, R.M., Devor, D.E. & Waalkes, M.P. (1989): The effect of testosterone on the distribution of cadmium to the rat prostate. *The Toxicologist* 9, 57.

Roe, F.J.C., Dukes, C.E., Cameron, K.M , Pugh, R.C.D. & Mitchley, B.C.V. (1964): Cadmium neoplasia: Testicular atrophy and Leydig cell hyperplasia and neoplasia in rats and mice following the subcutaneous injection of cadmium salts. *Br. J. Cancer* 18, 674-681.

Takenaka, S., Oldiges, H., König, H., Hochrainer, D. & Oberdörster, G. (1983): Carcinogenicity of cadmium chloride aerosols in Wistar rats. *J. Natl. Cancer Inst.* 70, 367-371.

Waalkes, M.P., Konishi, N., Ward, J.M. & Bare, R.M. (1990): Carcinogenic effects of repeated injections of cadmium in Wistar and Fischer rats. *The Toxicologist*, 10, 22.

Waalkes, M.P. & Oberdörster, G. (1990): Cadmium carcinogenesis. In *Metal Carcinogenesis*, ed. E.C. Foulkes. Boca Raton: CRC Press, in press.

Waalkes, M.P. & Rehm, S. (1990): Effects of dietary zinc deficiency on cadmium carcinogenesis in rats. *Proc. Am. Assoc. Cancer Res.* 31, in press.

Waalkes, M.P., Rehm, S., Riggs, C.W., Bare, R.M., Devor, D.E., Poirier, L.A., Wenk, M.L., Henneman, J.R., & Balaschak, M.S. (1988): Cadmium carcinogenesis in the male Wistar [Crl:(WI)BR] rats: dose-response analysis of tumor induction in the prostate and testes and at the injection site. *Cancer Res. 48, 4656-4663.*

Waalkes, M. P., Rehm, S., Riggs, C. W., Bare, R. M., Devor, D. E., Poirier, L. A., Wenk, M. L. & Henneman, J. R. (1989): Cadmium carcinogenesis in the male Wistar [Crl:(WI)BR] rat: Dose-response analysis of effects of zinc on tumor induction in the prostate and in the testes and at the injection site. *Cancer Res.* 49, 4282-4288.

Webb, M., ed. (1979): In *The chemistry, biochemistry and biology of cadmium*, Elsevier/North-Holland: Amsterdam.

Metal Ions in Biology and Medicine. Eds. Ph. Collery, L.A. Poirier, M. Manfait, J.C. Etienne. John Libbey Eurotext, Paris © 1990, pp. 284-288

Titanium and cancer growth ?

R.H. Barckhaus, P.F. Schmidt, H.J. Höhling

Institute of Medical Physics, University of Münster, Hüfferstr. 68, D-4400 Münster, FRG

Titanium is a widely distributed metal. Clinical studies have shown that titanium in implants and protheses is extremly well tolerated by osseous and soft tissues (Rae 1986b). The results demonstrate an excellent bone adaptation to titanium implants (Steflik et al. 1989). In vitro investigations of titanium and titanium alloys indicate a mild inflammatory potential (Sjöstrand and Rylander 1984, Rae 1986a). The most common findings associated with titanium dioxide exposure are increased pulmonary dust deposition, slight pulmonary fibrosis, and alveolar cell hyperplasia (Trochimowicz et al. 1988). High concentrations of titanium were found in the hilar lymph nodes and in the lung of the human body (Teraoka 1980), in the kidney, spleen, intestine, femur, blood, and subcellularly in the nuclei of cells of rats (Edel et al. 1985). Titanocene dichloride represents an organometallic compound which has proven antitumor activity and shows cytostatic and cytotoxic effects against cancer growth (Köpf-Maier and Martin 1989).

Although the investigations indicate a low toxicity of titanium on both acute and chronic basis there are few reports on the carcinogenicity of titanium and its compounds.

Outstanding results refer to the occurrence of tumors under TiO_2 dust exposure by animals and men, of tumors by titanium compounds injected intramuscularly, and of malignant tumors at the site of pacemaker implantation.

1. Occurrence of tumors under TiO_2 dust exposure (Table 1a)

(a) An experimental study in animals (Trochimowicz et al. 1988)

Rats were exposed to 10, 50, and 250 mg/m^3 of TiO_2 dust for 24 months. A most important finding of the study was except a dose related pigment deposition in the lungs the occurrence of lung tumors in rats exposed 24 months at 250 mg/m^3 of TiO_2. No lung tumors were seen in the 50 or 10 mg/m^3 exposure groups. The incidence of neoplastic lung tumors in rats 24 months is shown in Table 1a.

The squamous cell carcinoma appears to be an unique "experimental" tumor in rats. It appears to have developed in response to an overwhelmed dust-clearance mechanism. The average amount of dust in an overwhelmed lung of a male rat is about 785 mg/lung, that are approximately 31.5% of the weight of the lung. Permissible exposures in the workplace have been maintained typically well below 5 mg/m^3 respirable dust. Lee et al. (1986) have found that TiO_4 exposure led to the development of a few well-differentiated, cystic keratinizing squamous carcinoma.

(b) An epidemiologic study in men (Chen and Fayerweather 1988):

It was studied whether workers who had been exposed to TiO_2 had a significant higher risk of lung cancer than reference groups.

Table 1. List of the occurrence of the different tumors under TiO_2 dust exposure, by titanium compounds injected intramuscularly, and of malignant tumors at the site of pacemaker implantation.

a) Titanium dust exposure	Anaplastic carcinoma	Trochimowicz et al. 1988
	Bronchoalveolar adenoma	Trochimowicz et al. 1988
	Squamous cell carcinoma	Trochimowicz et al. 1988
	Squamous cell carcinoma	Lee et al. 1986
	Lung cancer	Chen and Fayerweather 1988
	Papillary adenocarcinoma	Yamadori et a. 1986
	Osteoblastic osteosarcoma	Barckhaus et al. 1985
b) Titanium injection	Fibrosarcoma	Furst and Haro 1969
	Hepatoma	Furst and Haro 1969
	Malignant lymphoma	Furst and Haro 1969
	Lymphosarcoma	Furst 1971
c) Tumors in pacemaker pockets	Scirrhous carcinoma (2x)	Zafiracopoulos and Rouskas 1974
	Adenocarcinoma	Magilligan and Isshak 1980
	Extramedullarly plasmocytoma	Hamaker et al. 1976
	Malignant fibrous histiocytoma	Fraedrich et al. 1984

A total of 1.576 employees exposed to TiO_2 was observed from 1956 through 1985 for cancer. The observed number of all cancer cases was slightly higher than expected in the TiO_2-exposed cohort (38 observed, 32.6 expected). The expected number based on Du Pont cancer incidence.
The cohort analyses suggest that the risks of developing lung cancer were no higher for TiO_2-exposed employees than for the referent groups.

(c) A case report in man (Yamadori et al. 1986):
An employee was working in the dust of TiO_2 at a factory without a mask. Chest roentgenography revealed bilateral diffuse chadows, diagnosed as having pneumoconiosis. The lymph nodes revealed a metastatic poorly differentiated adenocarcinoma. Depositions of black substances were noted also in the tumor tissue. Histologically, the tumor was identified as a poorly differentiated papillary adenocarcinoma. The black substances were engulfed in the lysosomes of macrophages. X-ray microanalyses showed a high content of titanium . Yamadori et al. (1986) suggested that this case was a rare autopsy case of pneumoconiosis caused by exposure to TiO_2 accompanied by lung cancer. This patient had no exposure to other agents causing pneumoconiosis such as silica. - The lung cancer in this patient was thought to be coincidental.

2. Titanium compounds injected intramuscularly (Table 1b)

The organic compound, titanocene (Furst and Haro 1969) and fine titanium metal powder (Furst 1971) were shown to be carcinogenic when suspended in trioctanoin, a synthetic triglyceride, and administered by intramuscular injection to rats. Fibrosarcomas occurred at the injection sites and the animals developed hepatomas and malignant lymphomas of the spleen.

3. Malignant tumors at the site of pacemaker implantation (Table 1c)

Several cases of malignant tumors related to pacemaker implantation have been described in the literature: scirrhous mammary carcinomas (Zafiracopoulos and Rouskas 1974), adenocarcinoma (Magilligan and Isshak 1980), extramedullarly plasmocytoma (Hamaker et al. 1976), and malignant fibrous histiocytoma (Fraedrich et al. 1984). It is discussed whether the pacemakers had been instrumental in the etiology of the breast cancer by their titanium-covered material or by their elec-

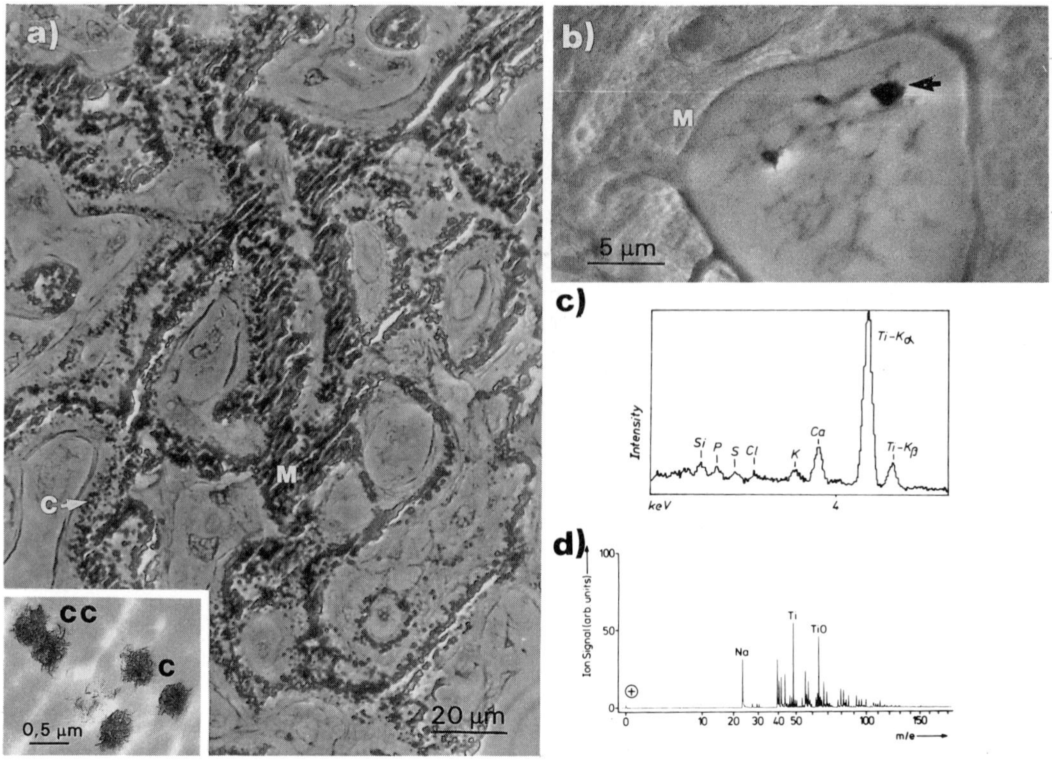

Fig. 1. Fast growing, highly malignant osteoblastic osteosarcoma: a) the tumor cells surrounding by mineral clusters (c) and larger mineralized regions (M); inset: single (c) and coalesced (cc) mineral cluster in the electron microscope; b) osteosarcoma cell with dense microareas(arrow); c) X-ray microprobe analysis shows high titanium contents in the dense regions; d) the mass spectrometric analysis displays both titanium and titanium oxide.

trostimulation. Fraedrich et al. (1984) believe that the malignant fibrous histiocytoma of the lung in a pacemaker pocket surrounding the generator was probably due to oncotaxis and coincidental and not related to material or electrochemical stimulation. The tumor was removed. When the patient died a soft tissue sarcoma was found in the lower lobe of the left lung and an identical tumor mass in the pacemaker pocket.

4. Pigment deposition in the tumor cells (Fig. 1)

TiO_2 dust identified as rutile showed extensive pulmonary deposition (Rode et al. 1981) and the concentrations increased in most tissues especially in spleen, femur, kidney, lymph nodes, and blood, mainly associated with plasma proteins from birth to old age. The long retention of Ti in the body may be due to its ability to form biocomplexes with cellular constituents(Edel et al. 1985). The microscopicallyobserved white pigment seemed black on light and electron microscopy because any material which does not transmit light or electrons will appear black by light and electron microscopy. Electron X-ray microanalysis is an excellent method to investigate the intra- and extracellular distribution of this element. Barckhaus et al. (1985) observed an unusual appearence and characteristic distribution pattern of titanium in the osteoid of a highly malignant typical osteoblastic

osteosarcoma (Fig. 1). The patient had had no titanium exposure beyond the average level. Preoperative chemotherapy with antitumor agent titanocene dichloride had not been administered.
The titanium toxicity is expressed in the formation of stable precipitates or chelates with essential metabolites such as phosphorus (Schroeder 1973). Barckhaus et al. (1985) found titanium in some tumor cells in close proximity of phosphorus. It would be interesting to determine whether this represents a stable titanium-phosphorus chelate complex.

To sum up: Titanium is essentially nontoxic in the amounts and forms which are normally ingested. Toxicity data of titanium are few. But it produces malignant tumors when injected into experimental animals (Furst and Haro 1969; Furst 1971) or in response to an overwhelmed titanium dust clearance mechanism (Trochimowicz et al. 1988). The appearance of a malignant fibrous histiocytoma of the lung in a pacemaker pocket was probably due to oncotaxis or coincidental and not related to titanium covered material or electrochemical stimulation (Fraedrich et al. 1984). Parallel to the experimental tumor increasing depositions of black substances - accumulations of titanium - were noted in the tumor tissues engulfed in the lysosomes of macrophages (Trochimowicz et al. 1988; Yamadori et al. 1986) even without treatment with the titanium containing agent titanocene dichloride (Barckhaus et al. 1985).

REFERENCES

Barckhaus R.H., Schmidt P.F., Roessner A., Timm C., Höhling H.J., Grundmann E. (1985): Electronprobe X-ray microanalysis and laser microprobe mass analysis (LAMMA) of titanium in tumor cells of a typical osteoblastic osteosarcoma. Trace elem. medic. 2, 73-76.

Chen J.L. and Fayerweather W.E. (1988): Epidemiologic study of workers exposed to titanium dioxide. J. Occ. Med. 3o, 937-942.

Edel J., Marafante E., Sabbioni E. (1985): Retention and tissue binding of titanium in the rat. Human Toxicol. 4, 177-185.

Fraedrich G., Kracht J., Scheld H.H., Jundt G., Mulch J. (1984): Sarcoma of the lung in a pacemaker pocket - Simple coincidence or oncotaxis? Thorac. Cardiovasc. Surgeon 32, 67-69.

Furst A. (1971): Trace elements related to specific chronic diseases: Cancer. Geol. Soc. Am. Mem. 123, 1o9-13o.

Furst A. and Haro R.T. (1969): A survey of metal carcinogenesis. Progr. exp. Tumor Res. 12, 1o2-133.

Hamaker W.R., Lindell M.E., Gomez A.C. (1976): Plasmocytoma arising in a pacemaker pocket. Ann. Thorac. Surg. 21, 354-356.

Köpf-Maier P. and Martin R. (1989): Subcellular distribution of titanium in the liver after treatment with the antitumor agent titanocene dichloride. Virchows Archiv B Cell Pathol. 57, 213-222.

Lee K.P., Kelly D.P., Schneider P.W., Trochimowicz H.J. (1986): Inhalation toxicity study on rats exposed to titanium tetrachloride atmospheric hydrolysis products for two years. Toxicol. Appl. Pharmacol. 83, 3o-45.

Magilligan D.J. and Isshak G. (198o): Carcinoma of the breast in a pacemaker pocket - Simple recurrence or oncotaxis? PACE 3, 22o-223.

Rae T. (1986 a): The biological response to titanium and titanium-aluminium-vanadium alloy particles. I. Tissue culture studies. Biomaterials 7, 3o-36.

Rae T. (1986 b): The biological response to titanium and titanium-aluminium-vanadium alloy particles. II. Long-term animal studies. Biomaterials 7, 37-4o.

Rode L.E., Ophus E.M., Gylseth B. (1981): Massive pulmonary deposition of rutile after titanium dioxide exposure . Acta path. microbiol. scand. Sect. A. 89, 455-461.

Schroeder H.A. (1973): Recondite of trace elements . In Essays in toxicology. Vol. 4, ed. W. Hayes Jr., pp 1o7-199. Academic Press New York and London.

Sjöstrand M. and Rylander R. (1984): Enzymes in lung lavage fluid after inhalation exposure to silica dust. Environ. Res. 33, 3o7-311.
Steflik D.E., McKinney R.V. Jr., Koth D.L. (1989): Ultrastructural Comparisons of ceramic and titanium dental implants in vivo: A scanning electron microscopy study. J. Biomed. Mat. Res. 23, 895-9o9.
Teraoka H. (198o): Distribution of 24 elements in the internal organs of normal males and the metallic worker in Japan. Arch. Environ. Health 36, 155-164.
Trochimowicz H.J., Lee K.P., Reinhardt C.F. (1988): Chronic inhalation exposure of rats to titanium dioxide dust. J. Appl. Toxicol. 8, 383-385.
Yamadori I., Ohsumi S., Taguchi K. (1986): Titanium dioxide deposition and adenocarcinoma of the lung. Acta Pathol. Jpn. 36, 783-79o.
Zafiracopoulos P. and Rouskas A. (1974): Breast cancer at site of implantation of pacemaker generator. Lancet 1, 1114.

Metal Ions in Biology and Medicine. Eds. Ph. Collery, L.A. Poirier, M. Manfait, J.C. Etienne. John Libbey Eurotext, Paris © 1990, pp. 289-295

Analysis of the mechanism of methylmercury cytotoxicity

E.J. Massaro[1], R.M. Zucker[2], K.H. Elstein[2]

[1] *Developmental Toxicology Division (MD-67), Health Effects Research Laboratory, U.S. Environmental Protection Agency, Research Triangle Park, NC 27711; and* [2] *NSI Technology Services Corp. – Environmental Services Division, P.O. Box 12313, Research Triangle Park, NC 27709 USA*

It has been reported that methylmercury (MeHg) inhibits mitosis and/or decreases the rate of the cell cycle (Costa et al., 1982; Vogel et al., 1986). Mitotic arrest (the accumulation of cells in the G_2/M phase of the cycle) appears to result from inhibition of microtubule assembly (Miura et al., 1987; Ramel, 1969; Sager & Syversen, 1986; Sager, 1988; Vogel et al., 1985), while decreased cycling rate has been attributed to lengthening of the duration of the G_1 phase as a consequence of inhibition of protein synthesis (Vogel et al., 1986). Whether the duration of other pre-mitotic phases of the cell cycle is altered similarly is not clear. However, Costa et al (1982) have reported that exposure to mercuric chloride induces an S-phase specific block in Chinese hamster ovary cells in vitro.

We have investigated MeHg-induced perturbation of the cell cycle kinetics of the murine erythroleukemic cell (MELC) by flow cytometry (FCM). We observed that, at relatively low levels (2.5 - 7.5 μM), MeHg predominately inhibits progression through the S phase of the cell cycle (in a dose-dependent manner). Accumulation of cells in the G_2/M phase of the cycle also occurs, but to a considerably lesser extent. Light microscopy reveals a dose-dependent increase in incidence of chromosomal aberrations (condensation, pulverization). Higher dose levels (10 - 50 μM) induce chromosomal ring formation and progressive perturbation of the cell membrane/cytoplasm complex. The latter is manifested as increased 90° light scatter [refractive index (Shapiro, 1988)], decreased axial light loss [cell size (Cambier and Monroe, 1983)], simultaneous propidium iodide and carboxyfluorescein fluorescence, and resistance to detergent (NP-40)-mediated cytolysis (Zucker et al., 1989). Our observations indicate that DNA synthesis is the primary target of MeHg cytotoxicity and that apparent targets and degree of cytotoxicity are a complex function of dose.

MATERIALS & METHODS

Cells: Friend murine erythroleukemic cells (MELC: T3CL2: from Dr. Clyde Hutchison, University of North Carolina, Chapel Hill) were grown in suspension culture in RPMI 1640 (GIBCO, Grand Island, NY) supplemented with 10% fetal bovine serum (FBS) and 25 mM HEPES (Sigma #H3375, St. Louis, MO.). Cell density was monitored by Coulter Counter (Model ZBI: Coulter Electronics, Inc., Hialeah, FL) and the cells were passed every two to three days to maintain logarithmic growth.

Viability assay: Viability was determined by the carboxyfluorescein diacetate (CFDA: Molecular Probes, Eugene, OR)/propidium iodide (PI: Sigma #P5264) assay (Shapiro, 1988; Zucker et al., 1989).
Preparation of nuclei for cell cycle analysis: Log phase cells were harvested and washed. Nuclei were isolated by detergent-mediated solubilization of the plasma membrane/cytoplasm complex and stained with fluorescein isothiocyanate (FITC: Sigma #F7250) for protein content and PI for DNA content (Crissman et al., 1986; Zucker et al., 1989).
Flow cytometry: Analyses were accomplished as described previously (Zucker et al., 1989).
MeHg exposure protocol: Methylmercury (II) chloride (Alfa #37123, Danvers, MA), in methanol, was added to logarithmically growing MELC to final concentrations of 0.1, 0.25, 0.5, 1.0, 2.5, 5.0, 7.5, 10, 25, or 50 μM. The methanol concentration of the medium was 0.1% (no effect on viability or growth rate). Duration of exposure was 1, 2, 4 or 6 h. To investigate recoverability from the effects of MeHg exposure, cells were exposed to MeHg for 6 h, washed in prewarmed FBS-supplemented medium and reincubated for 18 h in MeHg-free medium.
Progression assay: The relative rates of movement into/out of the (flow cytometrically defined) compartments of the cell cycle were estimated by a modification of the stathmokinesis assay of Darzynkiewicz et al. (1987), which quantifies the ability of toxicant-exposed cells to accumulate in the G_2/M phase of the cell cycle after treatment with Colcemid. Quantification of the phase distribution of cells was obtained with Multicycle, a cell cycle analysis PC software package (Phoenix Flow Systems, San Diego, CA).
Quantification of the Mitotic Fraction of G_2/M Nuclei: To quantify the percentage of cells in the M phase of the cell cycle, 1×10^6 nuclei/sample were prepared according to the method of Pollack et al. (1984), which allows flow cytometric discrimination of the M subpopulation (Zucker et al., 1988).
Chromosome morphology: Cells (1×10^6/sample) were washed twice in PBS by centrifugation (5 min, 120 x g), resuspended in 10 ml of 75 mM KCl and fixed in methanol-acetic acid (3:1). Chromosome spreads were prepared by centrifugation (~2×10^4 fixed cells, 500 x g/10 min, room temperature) onto glass slides in Leif cytobuckets (Coulter kit #322, Coulter Electronics, Inc., Hialeah, FL), drying at 56°C, and staining with 6% Giemsa (Fisher #SG-28-100). The percentages of normal and abnormal chromosomes were obtained from 200 mitotic figures. The mitotic index was obtained from 500 cells. The data represent the mean ± standard deviation of three experiments.
Data analysis: The data reported are from representative experiments. The experiments were repeated at least 3 times. For each cytometric parameter investigated (PI or FITC fluorescence, 90° light scatter, axial light loss), the distribution or mean of 10^4 events (cells or nuclei) per condition (dose, duration of exposure) or combination of conditions was determined. Data derived from cells exposed to MeHg concentrations below 2.5 μM did not differ from the control condition and are not included.

RESULTS

Exposure of MELC to MeHg ≤ 5μM for ≤ 6 h has no significant effect on viability, 90° light scatter [protein content (Zucker et al., 1988)], or axial light loss; but inhibits rate of growth (Fig. 1) measured as the increase in number of cells/ml following MeHg washout and reincubation for 18 h in fresh MeHg-free medium. Exposure to higher doses results in loss of viability (Table 1).
MeHg alters the percentage distribution of cells across the cell cycle (Figs. 2-6). Following exposure to 5 μM MeHg (no Colcemid), DNA histogram analysis (Fig. 3) reveals depletion of the G_0/G_1 compartment, increase in % cells in S phase (to a relatively steady state), and a slight increase in G_2/M, suggesting some movement of cells out of the S-phase (i.e. a leaky S-phase block). Addition of Colcemid to MeHg-treated cells results in only a moderate increase (compared to

control) in the percentage of G_2/M cells. If the primary effect of MeHg were on microtubule assembly/disassembly (a Colcemid-like effect) and S-phase progression were normal, cells would accumulate in the G_2/M phase of the cycle at the expense of other phases. Apparently, cells treated with MeHg can enter S phase; but the rate at which they traverse this compartment is retarded, resulting in reduction of the rate of influx of cells into G_2. Compared to control cells, exposure to 10 μM MeHg results in an increased percentage of cells in the S phase of the cycle and a slightly decreased percentage in G_0/G_1 and G_2/M. Exposure of such cells to Colcemid has minimal effect on phase distribution, indicating essentially complete cessation of cycling.

Eighteen hours after MeHg washout and reincubation under standard conditions, the DNA histogram of nuclei obtained from MELC exposed to 2.5 μM MeHg appears normal (Fig. 4). The DNA histogram derived from MELC exposed to 5 μM MeHg also reveals considerable recovery (compare Fig. 4 with Fig. 2), although there is still persistent reduction of the rate of S phase traverse as evidenced by the buildup of cells in early and mid S phase and depletion of cells in the late S and G_2/M phases. Colcemid treatment confirms recovery of normal cell cycle kinetics by cells exposed to 2.5 μM MeHg and persistence of S-phase retardation in cells exposed to 5 μM MeHg. Reincubation of MELC exposed to doses $\geq$ 10 μM MeHg reveals greatly perturbed DNA histograms and increasing amounts of debris, indicative of severe, irreversible toxicity.

The time-dependent effects of exposure to 5.0 μM MeHg or 0.2 μg/ml Colcemid on cell cycle progression are compared in Figs. 5 and 6. By inhibiting mitosis, Colcemid exposure (Fig. 6, open symbols) results in a relatively rapid increase in the percentage of cells in the G_2/M phase of the cell cycle at the expense of the G_0/G_1 and S phases. In contrast, the G_2/M compartment of MeHg-exposed cells decreases, then increases only slightly and at a relatively slow rate over the course of the experiment; the S compartment increases to a maximum at 2 hr and remains constant; and the G_0/G_1 compartment decreases to a minimimum at 4 hr. Also, in contrast to Colcemid exposure, the MeHg-exposed cells accumulate maximally in the S phase of the cycle.

To gain insight into the effect of MeHg on the G_2/M phase of the cell cycle, nuclei were prepared from MELC by an isolation procedure that allows flow cytometric discrimination of mitotic nuclei (Pollack et al., 1984; Zucker et al., 1988). On a contour cytogram of 90° scatter versus PI fluorescence (Fig. 7), M-phase nuclei appear as a distinct subpopulation exhibiting decreased 90° scatter and PI fluorescence compared to G_2 nuclei. Following 6 h exposure to concentrations of MeHg $\leq$ 7.5 μM, the percentage of cells in G_2/M remains relatively constant over dose (control - 16.4%; 2.5 μM MeHg - 14.7%; 5 μM MeHg - 18.2%; 7.5 μM MeHg - 19.4%). However, the contribution of the M subcompartment increases considerably. This suggests that, although progression from the S phase into G_2 is retarded, subsequent progression from G_2 into M is not. However, the percentage of cells exhibiting distinct chromosomes does not increase substantially as a function of dose (Table I), suggesting that cells leaving G_2 become arrested in a pre-mitotic phase in which their nuclei exhibit the same biophysical properties as those of M-phase cells; but chromosome condensation is inhibited.

Chromosome analysis reveals a dose-dependent increase in the percentage of condensed and pulverized chromosomes (Table 1) and a modest increase in the mitotic index (considerably less than that caused by Colcemid). At 10 μM MeHg, chromosome spreading was inhibited and the chromosomes of more than half of the mitotic cells appeared in the form of condensed, wreath-like ring structures (Fig. 8). Following exposure to 25 or 50 μM MeHg for as little as 1 h (the shortest time period investigated), all spreads appeared as ring structures.

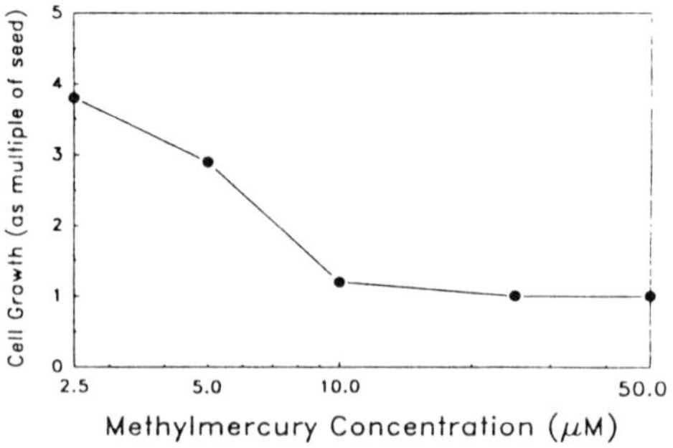

Fig. 1. Rate of MELC growth following MeHg exposure (6h), washout and reincubation (18h) in MeHg-free medium. MELC doubling time after exposure to 2.5 μM MeHg was essentially equal to that of control cells.

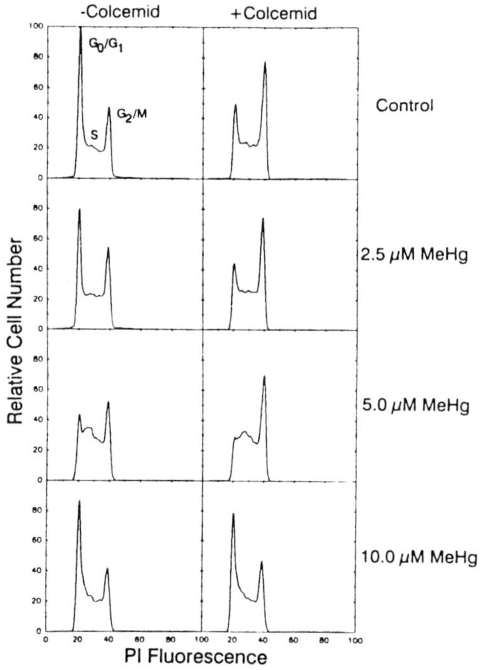

Fig. 2. Representative DNA histograms of nuclei of MELC exposed to MeHg for 6h with or without Colcemid for the last 2h of exposure (progression assay). G_0/G_1 represents the pre-DNA synthetic phase of the cell cycle; S, the phase of DNA synthesis; G_2, the post-synthetic phase preceding mitosis; and M, mitosis. Following exposure to 5.0 μM MeHg, movement of cells through the S phase of the cycle appears to be retarded. At 10 μM MeHg, there is complete cessation of cycling.

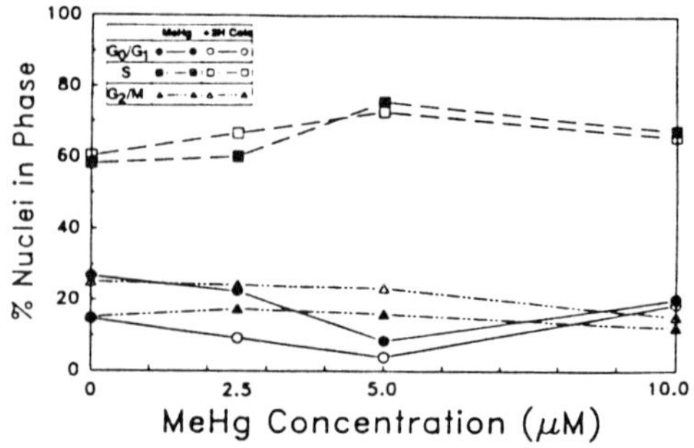

Fig. 3. The percentage of cells in each cell cycle phase was determined by computerized mathematical analysis of the histograms of Fig. 2. Data points at T=0 depict the control condition.

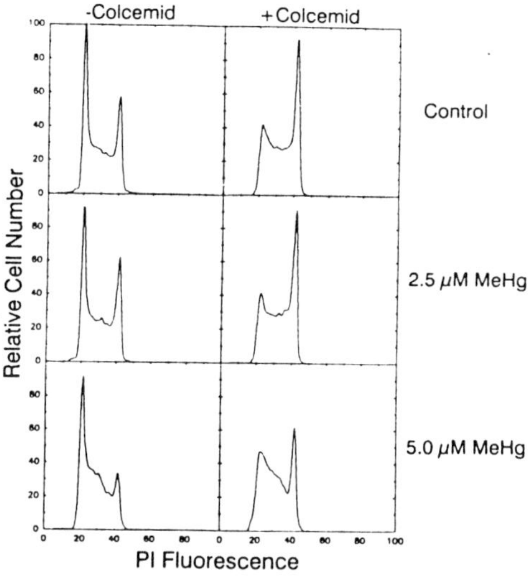

Fig. 4. Representative DNA histograms of nuclei obtained from MeHg-exposed (6h) MELC reincubated in MeHg-free medium for 18 h. Cells recovering from exposure to 5.0 μM MeHg approach normality; but still exhibit retardation of progression into, through and out of the S phase.

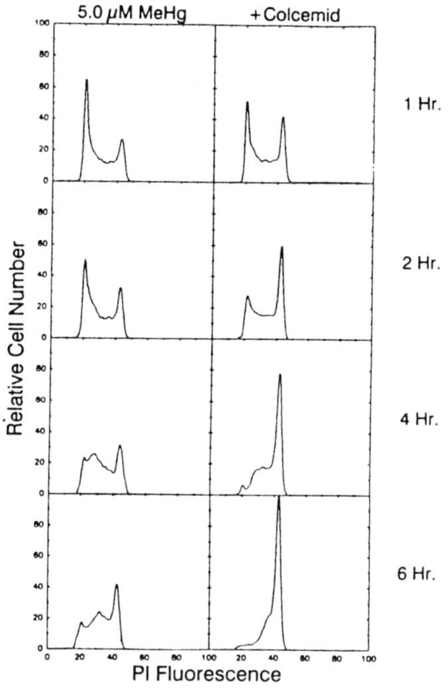

Fig. 5. Exposure to Colcemid (0.2 μg/ml) results in time-dependent accumulation of cells in the G_2/M compartment and a sequential depletion of the G_0/G_1 and S compartments. In contrast, exposure to 5 μM MeHg results in accumulation of cells in S phase and retardation of rate of efflux out of this compartment. As a result, the G_0/G_1 compartment becomes depleted and cells do not accumulate in the G_2/M compartment.

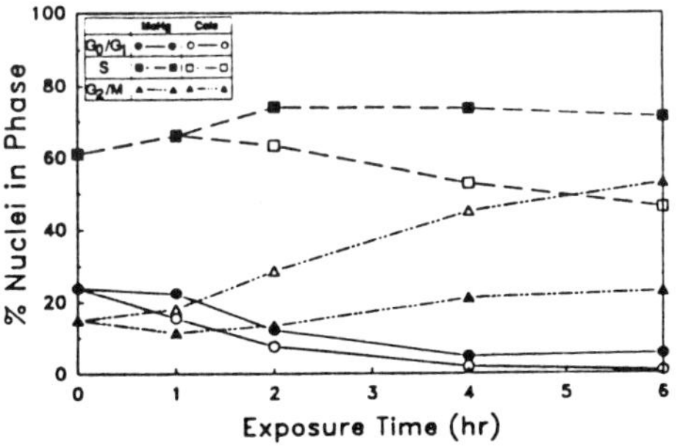

Fig. 6. Computerized mathematical analysis of the histograms of Fig. 5. MELC were exposed to either 5 μM MeHg (filled symbols) or 0.2 μg/ml Colcemid (open sysmbols) as described in the text.

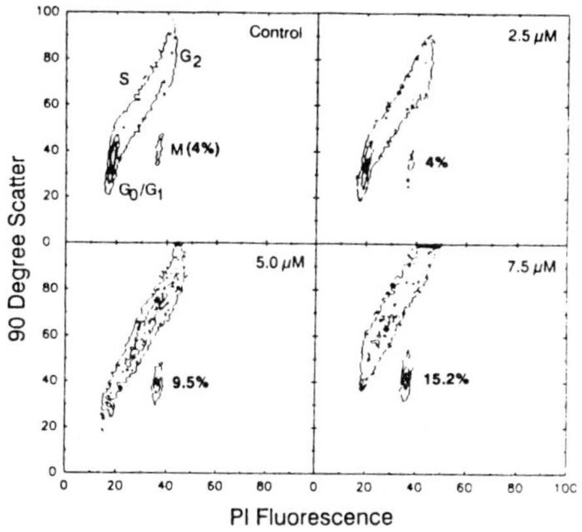

Fig. 7. Contour cytograms of the 90° scatter vs PI fluorescence of nuclei isolated from MeHg-exposed (6h) MELC by treatment with Pollack's buffer (Pollack, et al., 1984). Mitotic nuclei appear as a distinct subpopulation exhibiting decreased 90° light scatter and PI fluorescence. Following MeHg exposure, the relative percentage of this phase increases with dose.

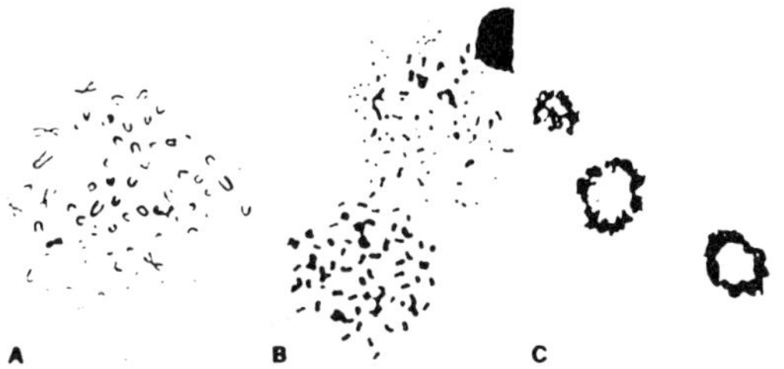

Fig. 8. Photomicrographs (600x) of representative chromosome preparations from cells exposed to MeHg for 6 h. A. Control; B. 10 μM MeHg; C. 25 μM MeHg. The percentage of cells exhibiting chromosomal aberrations increases as a function of dose (see Table I).

Table 1. Cytogenetic Effects of Methylmercury

	%Viability	%Mitotics	Chromosomal Aberrations			
			%Normal	%Condensed	%Pulverized	%Rings
Control	98	3.1 ± 1.0	92 ± 6	5 ± 2	--	3 ± 4
2.5 μM	98	4.2 ± 2.0	84 ± 8	13 ± 8	2 ± 2	2 ± 2
5.0	95	5.0 ± 0.9	74 ± 22	22 ± 15	4 ± 4	--
10.0	14	4.8 ± 3.0	3 ± 4	30 ± 34	12 ± 9	55 ± 41
25.0	3	4.5 ± 0.8	--	--	--	100 ± 0
50.0	5	3.9 ± 1.8	--	--	--	100 ± 0
Colc.	97	41 ± 0	90 ± 4	10 ± 4	--	--

Table I. The mitotic index and percentage of chromosomal aberrations occurring after 6h exposure to 0 - 50 μM MeHg or 0.2 μg/ml Colcemid.

SUMMARY

Flow cytometric analysis of MELC exposed to <7.5 μM MeHg reveals: (1) viability equivalent to that of control cells; (2) a dose-dependent decrease in % G_0/G_1 cells; (3) an increase in % S-phase cells and retardation of S-phase traverse (reduction of rate of DNA synthesis), and (4) a modest increase in % G_2/M cells. If the primary target of MeHg were microtubule assembly/disassembly, increasing dose, below cytotoxic levels, should progressively inhibit mitosis and increase the mitotic index and overall size of the G_2/M compartment. This is not the case. The relative contribution of M-phase cells to the G_2/M compartment increases substantially following exposure to MeHg concentrations > 2.5 μM. However, % M-phase cells exhibiting distinct chromosomes decreases with dose. Apparently, cells leaving G_2 become arrested in a pre-mitotic phase in which their nuclei exhibit biophysical properties similar to those of M-phase cells; but lack condensed chromosomes. This argues against the hypothesis that the mitotic spindle (i.e., microtubule) is the primary target of MeHg, as does our observation of minimal accumulation of G_2/M cells compared with Colcemid treatment. MeHg also induces chromosome aberrations (the incidence and severity of which increase with dose) with exposure at/above 10 μM MeHg resulting in induction of ring structures composed of what appear to be fused chromosomes.

REFERENCES

Cambier, J.C. & Monroe, J.G. (1983): Flow cytometry as an analytical tool·for studies of neuroendocrine function. In Methods in Enzymology, ed. P.M. Conn, Vol. 103, pp. 227.245. New York:Academic Press.

Costa, M., Cantoni, O., Demars, M. & Swartendruber, D.E. (1982): Toxic metals produce an S-phase specific cell cycle block. Res. Comm. Chem. Pathol. Pharmacol. 38, 405-409.

Crissman, H.A., Darzynkiewicz, Z., Tobey, R.A., & Steinkamp, J.A. (1986): Correlated measurements of DNA, RNA and protein in individual cells by flow cytometry. Science 228, 1321-1324.

Darzynkiewicz, Z., Traganos, F., & Kimmel, M. (1987): Assay of cell cycle kinetics by multivariate flow cytometry using the principle of stathmokinesis. In Techniques of Cell Cycle Analysis, eds. J.W. Gray & Z. Darzynkiewicz, pp. 291-336. Clifton, NJ: Humana Press.

Miura, K. & Imura, N. (1987): Mechanism of methylmercury cytotoxicity. CRC Critical Reviews in Toxicology 18, 161-188.

Pollack, A., Moulis, H., Block, N.L., & Irvin III, G.L. (1984): Quantitation of cell kinetic responses using flow cytometric measurements of correlated nuclear DNA and protein. Cytometry 5, 473-481.

Ramel, C. (1969): Methylmercury as a mitosis disturbing agent. J. Japan. Med. Assoc. 61, 1072-1081.

Sager, P.R. & Syversen, T.L.M. (1986): Disruption of microtubules by methylmercury. In: The cytoskeleton: A target for toxic agents, eds. J.W. Clarkson, P.R. Sager, & T.L.M. Syversen, pp. 97-116. 12. New York: Plenum.

Sager, P.R. (1988): Selectivity of methylmercury effects on cytoskeleton and mitotic progression in cultured cells. Toxicol. and App. Pharmacol. 94, 473-486.

Shapiro, H. (1988): Practical Flow Cytometry (Second Edition). New York: Alan R. Liss, Inc.

Vogel, D.G., Margolis, R.L. & Mottet, E.K. (1985): The effects of methylmercury binding to microtubules. Toxicol. App. Pharmacol. 80, 473-486.

Vogel, D.G., Rabinovitch, P.S. & Mottet, N.K. (1986): Methylmercury effects on cell cycle kinetics. Cell Tissue Kinet. 19, 227-242.

Zucker, R.M., Elstein, K.H., Easterling, R.E., & Massaro, E.J. (1988): Flow cytometric discrimination of mitotic nuclei by right-angle light scatter. Cytometry 9, 226-231.

Zucker, R.M., Elstein, K.H., Easterling, R.E., Ting-Beall, H.P., Allis, J.W. & Massaro, E.J. (1989): Effects of tributyltin on biomembranes: Alteration of flow cytometric parameters and inhibition of Na^{+},K^{+}-ATPase two-dimensional crystallization. Toxicol. Appl. Pharmacol. 96, 393-403.

DISCLAIMER

V. Aging

Metal Ions in Biology and Medicine. Eds. Ph. Collery, L.A. Poirier, M. Manfait, J.C. Etienne. John Libbey Eurotext, Paris © 1990, pp. 299-303

Chromium and aging

S. Wallach

Veterans Administration Medical Center Bay Pines, FL 33504 USA; The University of South Florida College of Medicine Tampa, FL 33612 USA

Although chromium (Cr) is classified as a toxic element because of adverse effects from industrial exposure, it is also an essential ultratrace element based on the ability of Cr deficiency (CrD) in experimental animals to impair growth and fertility and cause a diabetic-like state associated with impaired glucose tolerance (IGT), hyperinsulinemia, hypercholesterolemia and enhanced atherogenesis. Human CrD has been indisputably proven only in protein-calorie malnutrition and in patients receiving total parenteral nutrition devoid of Cr supplements, in whom IGT, insulin resistance, peripheral neuropathy and metabolic encephalopathy was reversed by Cr supplementation. CrD also decreases longevity in experimental animals. The hypothesis that CrD can accelerate human aging is supported by data showing a deficit of bioavailable Cr in processed foods and oft-quoted studies claiming a progressive decline in body and organ Cr content from birth onward. The subject of CrD was last reviewed by this author in 1985 (Wallach, 1985). In that review, the frustrations involved in achieving valid Cr analyses, the uncertain estimation of Cr bioavailability, and difficulty in measuring Cr balance were discussed since they impact seriously on the interpretation of the literature relating to Cr nutriture and deficiency.

Considerable progress has now been made in the development of valid Cr measurements and there are three physical methods capable of quantitating the ultratrace amounts of Cr present in biologic fluids and tissues, flameless atomic absorption spectrometry (FAAS), mass spectrometry of stable isotope diluted Cr chelates, and neutron activation analysis. Failure to follow fastidious technical procedures, especially with regard to contamination, can invalidate data obtained with otherwise acceptable methods. The mere use of stainless steel equipment or utensils in sample collection, preparation, or analysis can raise measured Cr levels several-fold. Anderson et al (1983,1985) have indicated the nature of the problem by their summary of plasma and urine Cr levels reported over the past three decades which demonstrate a 1000-fold difference in published values. FAAS based technology is the present "gold standard" for human serum and urine Cr levels. The normal serum Cr level ranges from 0.075 to 0.20 ng/ml and urine Cr excretion does not exceed 300 ng/day. Standardized tissue Cr levels have not as yet been published and it is necessary for now to accept values at the low end of the spectrum of reported values. The issue of bioavailability is equally contentious because of the ability of Cr to exist in foods in three forms, as metallic Cr from contamination, as trivalent Cr^{+3}, and as hexavalent Cr^{+6}. Whether any metallic Cr is converted to bioavailable Cr in the gastrointestinal tract is unknown. Trivalent Cr^{+3} is the biologically active form but is poorly absorbed. Trivalent Cr^{+3} chelated to nicotinic acid and glutathione, so-called glucose tolerance factor (GTF), is better absorbed than Cr^{+3}, and is the form present in Brewer's yeast. Hexavalent Cr^{+6} can find its way into the body via food and inhalation resulting from industrial contamination. There is experimental evidence to suggest some conversion of Cr^{+6} to Cr^{+3} in the acid environment of the stomach and inside cells. Thus, even with accurate and precise measurements of dietary Cr, the degree of bioavailability may be uncertain. Using modern analytic methods, dietary Cr has been stated to range between 25 and 300 ug/day (Gibson, 1985). Elderly patients have been noted to be in Cr balance when ingesting 25 ug/day (Bunker, 1984). Anderson et al (1985a) found the coefficient of Cr absorption to vary inversely with Cr intakes between 10 and 40 ug/day, so that absolute Cr absorption remained approximately the same. The inference to be drawn from such data is that as little as 25 ug/day of Cr ingested as either trivalent Cr^{+3} or GTF may not result in CrD.

Whether CrD occurs commonly in human aging and can only be answered by reliable tissue analyses since it is generally conceded that neither serum nor urine Cr levels are necessarily reflective of body stores (Anderson, 1983). Mostly negative results have been obtained thus far for Cr analyses in relation to age, diabetic status and atheromatous disease. The analytic values are mostly based on obsolete methods which overestimate Cr and the assumption must be made that they are are internally consistent. Body stores of Cr are greatest at birth and decline afterward. There is reasonable evidence for Cr transfer to the placentofetal unit during pregnancy, which may result in maternal Cr depletion in the late parous and post-partum periods (Wallach, 1985). However, after an initial sharp decline of body Cr in the offspring during the first few years of life, there is little further tissue loss of Cr with increasing age in tissues such as hair, lens, aorta, brain, myocardium, liver, pancreas and kidney. However, a significant minority of positive studies also exist. An early study by Schroeder et al (1962) found a progressive decline in liver Cr with increasing age. More recently, Vallerand et al (1984), comparing 7 week and 19 week old rats, noted a trend for Cr loss from myocardium, liver and skeletal muscle and a large 60% decrease in kidney Cr content. Radio-Cr exchange studies in aged rats (Wallach, 1986) are also compatible with, but do not prove, an age-related decline in total body and organ Cr content. The overall data, make it difficult to incriminate Cr deficiency in aging in any direct sense.

Inferential data suggest, however, that CrD may exist in a subset of the adult population and may indirectly influence aging by adverse effects on carbohydrate and lipid metabolism, thereby contributing to degenerative diseases such as diabetes mellitus (DM), coronary artery disease and atherogenesis in general. Supplementation studies in which the impact of Cr addition on various parameters of glucose tolerance (OGTT) and blood lipid levels are summarized in Table 1. Many of these studies indicate improved OGTT and/or insulin action, as well as a decline in total serum cholesterol and/or a rise in HDL levels, in some patients with DM or IGT. However, tissue Cr deficiency has not been uniformly demonstrated in either DM or IGT using newer techniques and recent supplementation studies have had less spectacular results than earlier studies. The role of CrD in atherogenesis is also uncertain. Although there is no unanimity as to serum and tissue Cr levels in atheromatous disease vs controls, Schroeder's (1970) original observation of dramatically decreased aortic Cr content in this condition remains unchallenged. Cr administration to rats and rabbits decreases aortic plaque formation. In Finland, the incidence of coronary artery disease is inversely correlated with the Cr content of the drinking water (Punsar, 1977).

In summary, if Cr deficiency plays an overt role in accelerating human aging, it would appear to do so by enhancing the development and severity of DM and atherogenesis in selected subjects. The extant data, especially those relating to Cr levels in tissues and biologic fluids, are scattered. Until valid Cr data are available for a cross-section of the population, the possibility of subtle adverse effects on aging should not be ignored.

Table 1. Effect of Cr Supplementation on Glucose/Lipid Metabolism

Author, Year	Blood Glucose/OGTT	Plasma Insulin	Total Cholesterol	HDL
Glinnsman, 1966	↓17%*			
Levine, 1968	NC*	NC		
Sherman, 1968	NC			
Liu, 1978	↓7%*	↓28%	↓12%	
Nath, 1979	↓28%*	↓20%	↓12%	
Canfield, 1979	↓9%*	↓20%	↓15%	
Offenbacher, 1980	↓13%	NC	↓12%	
Bialkowska, 1981	↓15%	NC	NC	
Riales, 1981	NC	NC	NC	↑12%
Polansky, 1981	↓15%	NC		
Polansky, 1982	↓15%	NC	NC	
Elwood, 1982			↓8%	↑9%
Grant, 1982	NC(HgbAlc↓17%)		NC	↑36%
Mossop, 1983	↓54%*		NC	↑38%
Uusitupa, 1983	NC	↓14%	NC	NC
Elias, 1984	↓14%	↓39%		
Potter, 1985	NC	NC	NC	NC
Offenbacher, 1985	NC	NC	NC	
Urberg, 1987	NC	NC		

Cr intake 150-1000ug Cr^{+3} or 1.6 - 10g brewer's yeast daily for at least one month. Values are means for the entire group studied. NC: No change; *Contains a subset with improvement noted.

REFERENCES

Anderson, R.A., Polansky, M.M., Bryden, N.A., Patterson, K.Y., Veillon, C., and Glinsmann, W.J. (1983): Effects of chromium supplementation on urinary Cr excretion of human subjects and correlation of Cr excretion with selected clinical parameters. J. Nutr. 113: 276-281.

Anderson, R.A., Bryden, N.A., and Polansky, M.M., (1985): Serum chromium of human subjects: effects of chromium supplementation and glucose[1,2]. Am. J. Clin. Nutr. 41: 571-577.

Anderson, R.A., and Kozlovsky, A.S. (1985a): Chromium intake, absorption and excretion of subjects consuming self-selected diets. Am. J. Clin. Nutr. 41: 1177-1183.

Bialkowska, M., Bruce, A., Magnusson, K., and Lithell, H., (1981): The effect of brewer's yeast on glucose tolerance and serum insulin, triglyceride and cholesterol levels in patients with adult-onset diabetes. Zywienie Czlowieka 8: 129-139.

Bunker, V.W., Lawson, M.S., Delves, H.T., and Clayton, B.E., (1984): The uptake and excretion of chromium by the elderly. Am. J. Clin. Nutr. 39: 797-802.

Canfield, W., (1979): Chromium, glucose tolerance and serum cholesterol in adults. In Cromium in Nutrition and Metabolism, ed. D. Shapcott and J. Hubert, p. 145-161. Amsterdam, Elseview/North Holland.

Elias, A.N., Grossman, M.K., and Valenta, L.J., (1984): Use of the artificial beta cell (ABC) in the assessment of peripheral insulin sensitivity: Effect of chromium supplementation in diabetic patients. Gen. Pharmac. 15: 535-539.

Elwood, J.C., Nash, D.T., and Streeten, D.H.P., (1982): Effect of high-chromium brewer's yeast on human serum lipids. J. Amer. Coll. Nutr. 1: 263-274.

Gibson, R.S., MacDonald, A.C., and Martinez, O.B. (1985): Dietary chromium and manganese intakes of a selected sample of canadian elderly women. Hum. Nutr. Appl. Nutr. 39A: 43-52.

Glinsmann, W.H., and Mertz, W., (1966): Effect of trivalent chromium on glucose tolerance. Metabolism 15: 510-520.

Grant, A.P., and McMullen, J.K., (1982): The effect of brewer's yeast containing glucose tolerance factor on the response to treatment in type 2 diabetes. A short controlled study. Ulster Med. J. 51: 110-114.

Levine, R.A., Streeten, D.H.P., Doisy, R.J., (1968): Effects of oral chromium supplementation on the glucose tolerance of elderly human subjects. Metabolism 17: 114-125.

Liu, V.J.K., and Morris, J.S., (1978): Relative chromium response as an indicator of chromium status. Am. J. Clin. Nutr. 31: 972-976.

Mossop, R.T., (1983): Effects of chromium III on fasting blood glucose, cholesterol and cholesterol HDL levels in diabetes. Centr. Afr. J. Med. 29: 80-82.

Nath, R., Minocha, J., Lyall, V., Sunder, S., Kumar, V., Kapoor, S., and Dhar, K.L., (1979): Assessment of chromium metabolism in maturity onset and juvenile diabetes using chromium-51 and therapeutic response of chromium administration on plasma lipids, glucose tolerance and insulin levels. In Cromium in Nutrition and Metabolism, eds. D. Shapcott and J. Hubert, p. 213. Amsterdam, Elsevier/North Holland.

Offenbacher, E.G., and Pi-Sunyer, F.X., (1980): Beneficial effect of chromium-rich yeast on glucose tolerance and blood

lipids in elderly subjects. Diabetes 29: 919-925.
Offenbacher, E.G., Rinko, C.J., and Pi-Sunyer, F.X., (1985): The effects of inorganic chromium and brewer's yeast on glucose tolerance, plasma lipids, and plasma chromium in elderly subjects. Am. J. Clin. Nutr. 42: 454-461.
Polansky, M.D., Anderson, R.A., Bryden, N.A., Roginski, E.E., Mertz, W., and Glinsmann, W.H., (1981): Chromium supplementation of free-living subjects - Effect on glucose tolerance and insulin. Fed. Proc. 40: 886.
Polansky, M.M., Anderson, R.A., Bryden, N.A., and Glinsmann, W.H., (1982): Chromium (Cr) and brewer's yeast supplementation of human subjects: Effect on glucose tolerance, serum glucose, insulin and lipid parameters. Fed. Proc. 41: 391.
Potter, J.F., Levin, P., Anderson, R.A., Freiberg, J.M., Andres, R., and Elahi, D., (1985): Glucose metabolism in glucose-intolerant older people during chromium supplementation. Metabolism 34: 199-204.
Punsar, S., Wolf, W., Mertz, W., and Karvonen, M.J., (1977): Urinary chromium excretion and atherosclerotic manifestations in two Finnish male populations. Ann. Clin. Res. 9: 79-83.
Riales, R., and Albrink, M.J., (1981): Effect of chromium chloride supplementation on glucose tolerance and serum lipids including high-density lipoprotein of adult men. Am. J. Clin. Nutr. 34: 2670-2678.
Schroeder, H.A., Balassa, J.J., and Tipton, I.H., (1962): Abnormal trace metals in man-chromium. J. Chron. Dis. 23: 941-964.
Schroeder, H.A., Nason, A.P., and Tipton, I.H., (1970): Chromium deficiency as a factor in atherosclerosis. J. Chron. Dis. 23: 123-142.
Sherman, L., Glennon, J.A., Brech, W.J., Klomberg, G.H., and Gordon, E.S., (1968): Failure of trivalent chromium to improve hyperglycemia in diabetes mellitus. Metabolism 17: 439-442.
Urberg, M., and Zemel, M.B., (1987): Evidence for synergism between chromium and nicotinic acid in the control of glucose tolerance in elderly humans. Metabolism 36: 896-899.
Uusitupa, M.I.J., Kumpulainen, J.T., Voutilainen, E., Hersio, K., Sarland, H., Pyorala, K.P., Koivistoinen, P., and Lehto, J.T., (1983): Effect of inorganic chromium supplementation on glucose tolerance, insulin response, and serum lipids in noninsulin-dependent diabetes. Am. J. Clin. Nutr. 38: 404-410.
Vallerand, A.L., Cuerrier, J-P., Shapcott, D., Vallerand, R.J., Gardiner, P.F., (1984): Influence of exercise training on tissue chromium concentrations in the rat. Am. J. Clin. Nutr. 39: 402-409.
Wallach, S., (1985): Clinical and biochemical aspects of chromium deficiency. J. Amer. Coll. Nutr. 4: 107-120.
Wallach, S., and Verch, R.L., (1986): Radiochromium distribution in aged rats. J. Amer. Coll. Nutr. 5: 291-298.

Metal Ions in Biology and Medicine. Eds. Ph. Collery, L.A. Poirier, M. Manfait, J.C. Etienne. John Libbey Eurotext, Paris © 1990, pp. 304-306

Lead poisoning in the aged due to drinking water : clinical and biological characteristics

T. Duriez, P. Kaminsky, M. Duc

Service de Medecine J, C.H.R.U. de Nancy, Hôpitaux de Brabois, 54511 Vandœuvre Cedex, France

Water streaming through standstones and granite rocks takes soft and aggressive characteristics which can induce the corrosion of the canalization. The use of lead piping in water conduits was widespread until 1965, and still persists in old houses. In the Vosges mountains (region of Eastern France), the drinking water pollution by lead induces a chronic intoxication. Elderly people are particularlly exposed and 50 per cent of the patients suffering from plumbism are older than 65 years (Kaminsky et al., 1988).
The aim of this study is to determine the clinical and biological characteristics of chronic lead poisoning in the aged.

1. METHODS

This retrospective study includes all the cases of lead poisoning diagnosed in our Hospital unit between 1976 and 1989. Excessive lead body burden was proven by an increased urinary lead level of over 800ug/day after 1g calcium EDTA provocation test. In accordance with the presence of clinical symptoms and their gravity, three clinical groups were defined: a major clinical form in which the symptoms threatened severely the vital or functional prognosis, a minor and an asymptomatic form. Clinical symptoms were attributed to lead poisoning after their regression following treatment. The hydrous origin was proven after exclusion of accidental or professionnal exposure and by an increased lead water level of greater than 0.05 mg/l (after 10h of water stagnancy).

2. RESULTS

2 a. General considerations

Between 1976 and 1989, 135 cases of lead poisoning due to drinking water were diagnosed. 22 cases presented an intricate pathology, partly explaining the symptoms, and were excluded from the study. Among the remaining 113 patients, 45 were older than 65 years (mean age 72 years, sex ratio=1/1): 17 patients suffered from the major clinical form, 25 from the minor form and 3 were asymptomatic. All the patients came from Vosges mountains. An excessive lead water level was found in 86 per cent (range 100 to 3000 ug/l). The remaining 14 per cent had moved out of the aera of risk or had changed the water canalization

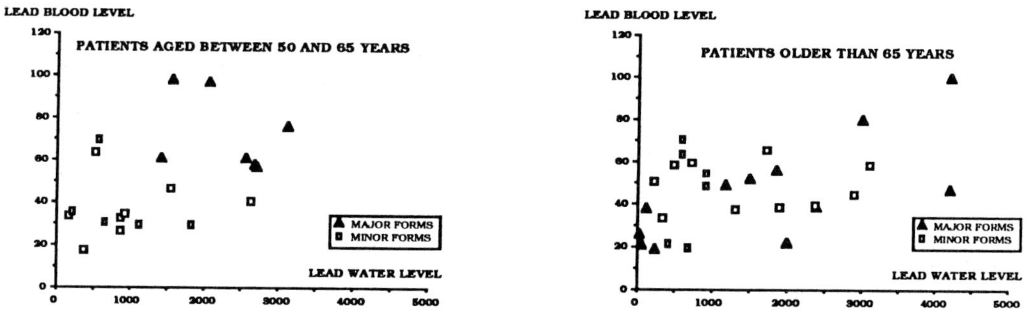

Fig. 1. Relationship between lead blood level and lead water level

during the past five years.

2 b. Clinical findings

Major forms Neurological symptoms are the habitual expression of lead poisoning. They consist of: peripheral neuropathy (n=5), fulgurant pains of lower limbs (n=3), amyotrophic lateral sclerosis-like syndrome (n=2), or troncular palsy. Central manifestations are also possible: dyskinesia or kinetic tremor, cerebellar syndrome or acute encephalopathy. The classical N. Radialis palsy appeared in only one case during the treatment of an encephalopathy. Digestive disturbances are more classical (gastrointestinal colic). A malignant arterial hypertension with renal failure was also found. These severe symptoms are often accompagned by middle clinical signs.

Minor clinical forms Several clinical symptoms can be considerered as minor signs of chronic encephalopathy. Patients with chronic lead poisoning may often complain of asthenia, loss of appetite, headache, irritability, vertigo or rarely static tremor. Some other symptoms probably also related to central manifestations find expression in digestive signs (48%): burning sensation in epigastrium, vague but uncomfortable abdominal feeling, abdominal pain, nausea, diarrhea or constipation. Peripheral sensitive disorders are the most frequent symptoms (60%): aching, burning or shooting pains in the limbs, paresthesia, myalgia or cramps. Weariness and weakness of the limbs are also frequently found. A moderate arterial blood pressure (diastolic pressure range 10-12 and systolic range 16-20 mmHg) is also present in 52 per cent of the patients. To account for the frequency of this symptom in aging, its attribution to plumbism is however not easily verifiable.

Classical symptoms. These symptoms are absent or exceptional in our study: only one case of blue line on gums, one antecedent of gout , one case of sideroblastic anemia and one N.radialis paralysis were found and were never isolated.

2 c. Biology

Toxicological assays

Lead blood levels increase to between 190 and 1000 ug/l. Blood lead is greater than 400 ug/l in only 60 per cent of our patients. No correlation can be found between lead blood level and the clinical severity (Fig 1). Lead level in urine after EDTA provocation test is in the range of 1070-9270 ug/l, without any evident relationship with the clinical form.

Hematological system Discrete anemia is frequently found (about 20%). The protoporphyrin level in erythrocyt is increased in 70 per cent and the activity of the delta aminolevulinic acid (ALA) dehydrase in erythrocyt is decreased in 97 per cent of our cases. ALA and coproporphyrin excretion in urine are however only slightly increased (respectively 14 and 19%) as with ALA blood level (3%). Basophilic stippled red cells are found in only 40 per cent of blood samples. No relationship between all these disorders and the clinical gravity can be shown.

Blood lead level-water pollution relationship In the younger patients, the lead blood level shows a significant correlation with the lead water level ($p<0.001$) and a major clinical form appears only with high blood lead (Fig. 1). The same relationship between hydrous pollution degree and blood lead can be observed in the aged but severe clinical forms are found even with low lead blood level of less than 400 ug/l.

3. DISCUSSION

This study shows the extreme diversity of the clinical symptoms in chronic lead poisoning induced by polluted drinking water. The clinical signs, described in previous (and often old) studies, appear to be very different from the classical pictures observed in acute plumbism (Tsuchiya, 1979; Stainthorpe & Durh, 1974) and we wish to dwell on the atypical appearance of the minor clinical forms. The habitual biological parameters for the screening and follow-up of the professionally exposed patients are often caught out, except protoporphyrin in erythrocyt. The best chronic plumbism screening test is in our opinion the ALA dehydrase activity (Kaminsky et al. 1988). Moreover, this study clearly shows the most important susceptibility of aged patients to lead exposure. Clinical and sometimes severe symptoms can appear even with low lead blood level. In addition, clinical disorders can be found a long time after lead exposure and we have observed two relapses after treatment and verified changing of the lead pipes.

All these statements suggest a diminished detoxication potential in aging and the existence of an endogenous origin of lead. The osseous tissue is known to absorb 90 per cent of the total stock of lead (Tsuchiya, 1979). We believe that the gradual osseous demineralization with ageing induces on the one hand a decreased lead storage capacity and on the other hand, is able to release the toxin from the bone to the soft tissues.

REFERENCES

Kaminsky, P., Leone, J. & Duc, M. (1988): Incidence du saturnisme hydrique dans un service de médecine interne en région de sols acides. *Presse Méd.* 17, 419-422

Stainthorpe,W.W. & Durh, B.S. (1914): Observations on 120 cases of lead absorption from drinking water. *Lancet*, 25, 213-215.

Tsuchiya , K. (1979): Lead. In *Handbook on the Toxicology of metals*, ed. L. Friberg, G.F. Nordberg & V.B. Vouk. pp 451-484. North Holland: Elsevier.

VI. Biological effects

Metal Ions in Biology and Medicine. Eds. Ph. Collery, L.A. Poirier, M. Manfait, J.C. Etienne. John Libbey Eurotext, Paris © 1990, pp. 309-311

Effects of vanadium on the activities of kinases, mutases and phosphatases

G.L. Mendz

School of Biochemistry, The University of New South Wales, P.O. Box 1, Kensington, NSW 2033, Australia

INTRODUCTION

Vanadium elicits a variety of important physiological responses from biological systems; and owing to its high potency as an inhibitor it is a useful probe of biological function. In its higher oxidation states, +5 (vanadate) and +4 (vanadyl), vanadium is known to inhibit or activate a large number of enzymes (Chasteen, 1984). ^{31}P- and ^{51}V-NMR spectroscopy were recently used to study the stimulation of 2,3-bisphosphoglycerate phosphatase by vanadate in intact human erythrocytes (Mendz *et al.*, 1990). In the present work ^{31}P-NMR spectroscopy was employed to investigate *in vitro* and in *haemolysates* the effects of vanadate in the activity of several enzymes that catalyse sugar phosphates and/or high energy phosphates.

MATERIALS AND METHODS

Rabbit muscle phosphoglucose isomerase (EC 5.3.1.9), phosphoglucomutase (EC 5.4.2.2), phosphoglycerate mutase (EC 2.7.5.3), enolase (EC 4.2.1.11) and phosphofructokinase (EC 2.7.1.11) were obtained from Sigma (St. Louis, MO, USA). Enzyme suspensions were prepared in a 0.05 M Tris/0.002 M $MgCl_2$/0.001 M EDTA, pH 7.4 buffer. Ammonium vanadate (NH_4VO_3) (BDH, London, UK) and the metabolites glucose-6-phosphate (Glc6P), glucose-1-phosphate (Glc1P), fructose-6-phospahte (Fru6P), ATP, 3-phosphoglycerate (3PG), glucose-1,6-bisphosphate (Glc1,6P_2), 2,3-bisphosphoglycerate (2,3-DPG) and phosphoenolpyruvate (PEP) (Boehringer-Mannheim, Penzberg, FGR) were prepared in 0.03 M stock solutions and added to the samples in the required amounts.

Fresh human red cells were obtained from the Sydney Blood Bank (NSW, Australia). Erythrocytes were washed five times with isotonic saline (0.154 M NaCl, 277 K, 5 vol). Cell suspensions of low haematocrit were bubbled gently with CO (Matheson, East Rutherford, NJ, USA) for 30 min at 277 K. Lysates were prepared by sonicating cell suspensions kept in ice with two power bursts (40 W, 5 s) of a microtip (Branson, B-15 Sonifier, Danbury, CT, USA). To remove small metabolites from haemolysates, preparations were diluted with 0.14 M KCl (1:1, v/v), and the mixtures were filtered through membranes with a molecular weight cutoff of 2000, until the volume of the retentate was equal to that of the original lysate. The procedure was repeated eight times, the last one employing KCl constituted in ^2H_2O (99.96% ^{2}H, INSE, Lucas Heights, Australia), to provide a deuterium frequency lock for the NMR spectrometer. Retentates were prepared with an equivalent haematocrit of 0.4.

^{31}P NMR chemical shifts are quoted with respect to an internal triethyl phosphate (Et_3P) resonance at 0.44 ppm. Free induction decays were collected using a Varian XL-400 spectrometer operating in the FT mode with quadrature detection. All measurements were carried out at 310 K. The instrumental parameters were: operating frequency 162 MHz, spectral width 5.0 kHz, memory locations 8192, and recycling times between 1 and 5 T_1. Exponential filterings of 2-4 Hz were applied prior to Fourier transformation. Saturation transfer experiments were carried out using a DANTE pulse sequence with a saturating pulse width of 1.3 µs and a delay between pulses of 333 µs. Standard concentration curves were constructed by adding the appropiate metabolite to the samples. The integral of the resonance of the specific metabolite to be measured, relative to the integral of the reference was determined and compared with the appropiate standard curve.

RESULTS

Glucose-6-phosphate isomerase catalyses the exchange reaction between Glc6P and Fru6P. The rate constants (k) characterising the equilibrium reaction are first-order and were determined using saturation transfer techniques. Figure 1 shows the results of experiments carried out in *haemolysates* with 20 mM Glc6P as substrate, and in the presence of different vanadate concentrations. Saturation of the Fru6P resonance (left hand side spectra) results in a decreased intensity in the α- and β-Glc6P peaks. The relationship between k and the reduction in intensity, ΔM, of the peak not undergoing saturation is given by (Brown & Ogawa, 1977):

$$\Delta M = k\ {}^{\mathrm{sat}}T_1\ M_E \qquad (1)$$

where M_E is the equilibrium intensity of the resonance determined from the control experiments (right hand side spectra), and ${}^{\mathrm{sat}}T_1$ is the spin-lattice relaxation time measured during saturation of the exchange partner. The values calculated for the rate constant of the "forward" reaction were 0.124 ± 0.003 and $0.192 \pm 0.004\ s^{-1}$ at 0 and 1.5 mM metavanadate, respectively; and increased linearly with oxyanion concentration. A maximum value of $0.208 \pm 0.005\ s^{-1}$ was measured for the rate constant at 2 mM NH_4VO_3.

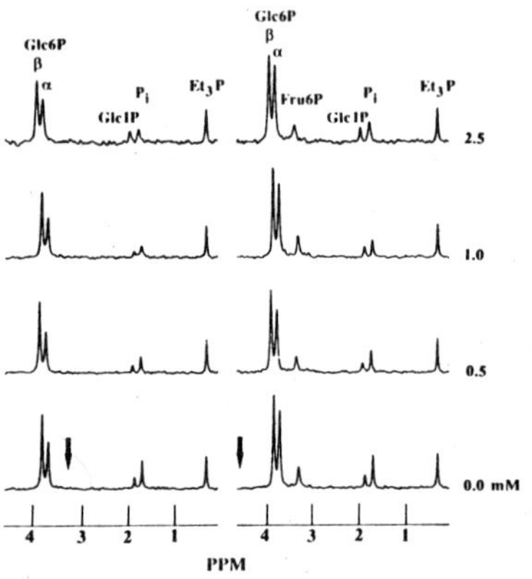

Fig. 1. Saturation transfer of the reaction catalysed by glucose-6-phosphate isomerase in the presence of the concentrations of ammonium vanadate shown on the right of the figure. The positions at which the saturating power was applied are incated by the arrows. Left: saturation at the frequency of the Fru6P resonance. Right: control experiment.

Vanadate was found to stimulate phosphoglucomutase *in vitro*. In enzyme suspensions (0.18 mg/ml) containing Glc1P (25 mM) and Glc1,6P_2 (0.5 mM), the values of the rate constants of the "forward" equilibrium reaction increased linearly from 0.004 ± 0.001 to $0.021 \pm 0.001\ s^{-1}$ between 0 and 2 mM NH_4VO_3.

Magnetisation transfer techniques previously used to measure the rate of exchange of phosphoryl groups between 2PG, 3PG and PEP by the coupled phosphoglycerate mutase-enolase enzyme system *in vitro* (Chapman **et al.**, 1988), were employed to study the effect of metavanadate in the rate constants of the reactions. Substrate was introduced to the reaction by adding 20 mM 3PG to a suspension containing 5 mg of phosphoglycerate mutase and 20 mg of enolase. The exchange rates of the phosphoglycerate mutase reaction increased linearly with oxyanion concentration from $0.043 \pm 0.009\ s^{-1}$ in the absence of vanadate, to $0.124 \pm 0.009\ s^{-1}$ at 3 mM NH_4VO_3. Within experimental error the rate constant for the enolase reaction remained unchanged.

The hydrolysis of 2,3-DPG by phosphoglycerate mutase from an initial concentration of 5 mM was measured *in vitro* in samples containing 0.25 mg of the rabbit muscle enzyme. In the presence of 1 mM 2-phosphoglycolate or metavanadate, the initial rates of decline of 2,3-DPG were 0.032 ± 0.001 and 0.007 ± 0.001 mM/min, respectively.

The effect of metavanadate on the activity of rabbit muscle phosphofructokinase (0.0167 mg/ml) *in vitro* was investigated by adding ATP (7mM) and Fru6P (7mM) to enzyme suspensions, and monitoring the appearance of Fru1,6P_2 in the presence of different concentrations of the oxyanion. Initial rates of formation of the fructose bisphosphate were calculated for the first 18 min of

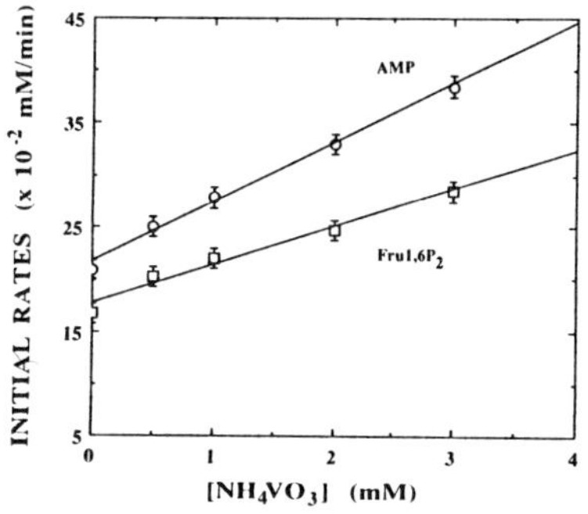

Fig. 2. Dependence of the initial rates of production of β-Fru1,6P$_2$ and AMP on metavanadate concentration in haemolysates.

the time course from good fits of the data to straight lines. The inital rates increased linearly with vanadate concentration up to 0.5 mM. The values calculated for the rates of production of β-Fru1,6P$_2$ (77 per cent of the total) were 0.16 and 0.33 mM/min at 0 and 0.5 mM vanadate, respectively.

Stimulation of phosphofructokinase by vanadate was also observed in *haemolysates* incubated with ATP (10.5 mM) and Fru6P (14 mM). The initial rates of formation of the fructose bisphosphate increased linearly with vanadate concentrations between 0 and 3 mM (Fig. 2). In addition, it was observed a linear increase in the rate of production of AMP with oxyanion concentration (Fig.2). The rates measured for AMP were approximately 20-25 per cent larger than those calculated for β-Fru1,6P$_2$. This difference is in good agreement with the 23 per cent of α-Fru1,6P$_2$ that is also formed. The presence of metavanadate did not affect the rate of production of P$_i$ in haemolysates.

DISCUSSION

Layne & Najjar (1979) reported a 200-fold enhancement of the glucose phosphorylation activity of rabbit muscle phosphoglucomutase *in vitro* in the presence of 1 mM sodium vanadate, and suggested that the oxyanion promotes the transfer reaction probably by assisting in the ionisation of the tyrosine at the active site. This mechanism is consistent with the observation that the rate constants of the phosphoryl transfer between Glc1P and Glc6P increase in the presence of metavanadate. Similar explanations may apply to the other equilibrium reactions whose exchange rates are enhanced by the presence of the oxyanion.

Decavanate is a potent allosteric inhibitor of sheep heart phosphofructokinase *in vitro*, demonstrating that the enzyme is sensitive to inhibition by polyanionic compounds (Choate & Mansour, 1979). The linear dependence of the initial rates of production of Fru1,6P$_2$ and AMP at NH_4VO_3 concentrations between 0 and 3 mM, suggested a direct action of the oxyanion on phosphofructokinase. The stimulation of the enzyme by metavanadate appears to be similar to the action of other known positive effectors such as AMP, cAMP, Fru1,6P$_2$ and Glc1,6P$_2$, which act on allosteric sites different from those for ATP and for other phosphorylated inhibitors.

REFERENCES

Brown, T.R. & Ogawa, S. (1977): ^{31}P-nuclear magnetic resonance kinetic measurements on adenylate kinase. *Proc. Natl. Acad. Sci. USA* 74, 3627-3631.

Chasteen, N.D. (1984): The biochemistry of vanadium. *Structure and Bonding* 53, 105-138.

Chapman, B.E., Stewart, I.M., Bulliman, B.T., Mendz, G.L. & Kuchel, P.W. (1988): ^{31}P Magnetization transfer in the phosphoglycerate-enolase coupled enzyme system. *Biophys. J.* 16, 187-191.

Choate, G. & Mansour, T.E. (1979): Studies on heart phosphofructokinase. *J. Biol. Chem.* 254, 11457-11462.

Layne, P.P. & Najjar, V.A. (1979): Evidence for a tyrosine residue at the active site of phosphoglucomutase and its interaction with vanadate. *Proc. Natl. Acad. Sci. USA* 76, 5010-5013.

Mendz, G.L., Hyslop, S.J. & Kuchel, P.W. (1990): Stimulation of human erythrocyte 2,3-bisphosphoglycerate phosphatase by vanadate. *Arch. Biochem. Biophys.* 276, 160-171.

Metal Ions in Biology and Medicine. Eds. Ph. Collery, L.A. Poirier, M. Manfait, J.C. Etienne. John Libbey Eurotext, Paris © 1990, pp. 312-314

Effects of oral vanadium administration in streptozotocin-diabetic rats

J.L. Domingo, J.M. Llobet, M. Gómez, J. Corbella, C.L. Keen*

*Laboratory of Toxicology and Biochemistry, School of Medicine, University of Barcelona, 43201 Reus, Spain and * Department of Nutrition, University of California, Davis, CA, 95616, USA*

In recent years, various authors have reported that oral administration of sodium orthovanadate (Heyliger et al., 1985; Paulson et al., 1987; Gil et al., 1988), sodium metavanadate (Meyerovitch et al., 1987; Blondel et al., 1989), and vanadyl sulphate (Pederson et al., 1989) in the drinking water of streptozotocin-diabetic rats normalizes the high blood glucose concentration and prevents the decline in cardiac performance due to diabetes. The glucose-lowering effect of vanadium in diabetes is especially relevant since oral administration of insulin in mammals is ineffective. Therefore, the availability of orally administered insulin substitutes would be of great importance in the treatment of diabetics.

On the other hand, the toxic effects of oral vanadium administration are now well-known (Jandhyala & Hom, 1983). In the above reports, vanadium compounds were administered at doses ranging from 0.10 to 1.0 mg/ml. At those, or even lower levels (Parker & Sharma, 1978; Domingo et al., 1985), several signs of toxicity were previously observed in healthy rats. However, in most of the studies on the effects of vanadium in diabetes treatment, the possible toxicity derived from the oral vanadium administration to streptozotocin-diabetic rats was not evaluated (or reported).
In the present investigation, we have compared -under the same experimental conditions- the effectiveness of the administration for four weeks of sodium metavanadate, sodium orthovanadate, and vanadyl sulphate pentahydrate in the drinking water of streptozotocin-diabetic rats to alleviate some symptoms of diabetes.

MATERIALS AND METHODS

Male Sprague-Dawley rats (190-210 g) were obtained from Panlab (Barcelona, Spain). After eight days of adaptation to plastic metabolic cages (day 0) diabetes was induced in 42 animals by two **sc** injections (given on day 0 and 3 at 12.00 hr) of STZ (40 mg/kg) in cold 0.1 M citrate buffer. Rats with glucose levels > 250 mg/dl were considered diabetic. Sodium metavanadate (0.15 mg/ml), sodium orthovanadate (0.23 mg/ml), and vanadyl sulphate pentahydrate (0.31 mg/ml) were dissolved in aqueous solutions of NaCl (80 mM). Vanadium-treated rats were compared to control rats, either diabetic or non-diabetic receiving drinking water containing 80 mM NaCl alone. Body weight, and food and fluid intake were monitored daily, whereas blood glucose

TABLE 1

FLUID AND FOOD INTAKE, VANADIUM INTAKE, WEIGHT GAIN, AND PLASMA GLUCOSE, UREA AND CREATININE CONCENTRATIONS IN VANADIUM TREATED OR UNTREATED DIABETIC AND CONTROL RATS (Mean values ± SEM)

	Fluid intake (ml/kg/day)	Food intake (g/kg/day)	Vanadium intake (mg/kg/day)	Weight gain ($\Delta P/P_o$)	Plasma glucose (mg/dl)	Plasma urea (mg/dl)	Plasma creatinine (mg/dl)
Nondiabetic (untreated control)	144.7±20.7	64.7±2.4	0	0.24±0.023	121.4±3.0	38.2±1.9	0.82±0.12
Diabetic							
Untreated	1106.1±80.2[3]	155.6±6.2[3]	0	0.082±0.016[2]	626.3±30.4[3]	51.0±2.1[3]	0.60±0.07
Vanadyl	363.7±60.9[c,1]	124.4±5.2[3]	22.7±3.8*	-0.032±0.018[a,3]	395.2±74.4[a,2]	80.5±3.6c,3	2.01±0.30b,2
Orthovanadate	244.4±59.7[c]	88.1±4.5[b,3]	15.6±3.7*	-0.048±0.022[b,3]	310.2±67.2[b,1]	66.2±3.5a,3	0.60±0.05
Metavanadate	97.4±45.1c	68.1±1.4[b]	6.1±1.3	-0.072±0.032[b,3]	229.3±35.3[c,1]	71.0±4.2a,3	0.91±0.03a

[1,2,3]Significantly different from nondiabetic (control) rats ($p<0.05$, $p<0.01$, $p<0.001$, respectively).

[a,b,c]Significantly different from untreated diabetic rats ($p<0.05$, $p<0.01$, $p<0.001$, respectively).

*Significantly different with from metavanadate treated rats.

levels were measured weekly. On day 28 of treatment the animals were killed, and vanadium concentrations in liver, kidneys, heart, spleen, pancreas, bone (femur), and muscle were determined by atomic absorption spectrophotometry as described previously (Domingo et al., 1985).

RESULTS AND DISCUSSION

Daily food and fluid intake were significantly reduced in the vanadium-treated animals relative to controls. Also, vanadium treatment reduced the level of hyperglycemia in diabetic rats, with sodium metavanadate being the most effective of the vanadium compounds tested. Paradoxically, however, daily vanadium intake was significantly lower in the animals receiving sodium metavanadate (Table 1). Despite these beneficial effects of oral vanadium administration, several signs of toxicity (some deaths, weight loss, renal impairment) were seen. In addition, vanadium accumulated in all tissues analyzed, which may imply a remarkable risk of toxicity derived from its administration. Bone, kidney and spleen showed the highest vanadium levels. The vanadium accumulation ranked in descending order was $NaVO_3$ > Na_3VO_4 > $VOSO_4$. It must be noted, that for the treatment of diabetes vanadium should be administered repeatedily for an indefinite period.

In view of the concern that the possible use of vanadium compounds as an alternative treatment to insulin in human diabetes, has raised in the last few years, we would like to emphasize that continous oral administration of vanadium compounds at doses which alleviate some symptoms of diabetes may also induce simultaneously various important toxic effects. Therefore, further investigations should clearly evaluate the beneficial and the toxic effects derived from the oral vanadium administration to diabetics.

REFERENCES

Blondel, O., Bailbe, D. & Portha, B. (1989): **In vivo** insulin resistance in streptozotocin-diabetic rats. Evidence for reversal following oral vanadate treatment. Diabetologia **32,** 185-190.

Domingo, J.L., Llobet, J.M., Tomas, J.M. & Corbella, J. (1985): Short-term toxicity studies of vanadium in rats. J. Appl. Toxicol. 5, 418-421.

Gil, J., Miralpeix, M., Carreras, J. & Bartrons, R. (1988): Insulin-like effects of vanadate on glucokinase activity and fructose 2,6-bisphosphatase levels in the liver of diabetic rats. J. Biol. Chem. 263, 1868-1871.

Heyliger, C.E., Tahiliani, A.G. & McNeill, J.H. (1985): Effect of vanadate on elevated blood glucose and depressed cardiac performance of diabetic rats. Science 227, 1474-1477.

Jandhyala, B.S. & Hom, G.J. (1983): Physiological and pharmacological properties of vanadium. Life Sci. 33, 1325-1340.

Meyerovitch, J., Farfel, Z., Sack, J. & Shechter, Y. (1987): Oral administrationof vanadate normalizes blood glucose levels in streptozotocin-treated rats. J. Biol. Chem. 262, 6658-6662.

Parker, R.D.R. & Sharma, R.P. (1978): Accumulation and depletion of vanadium in selected tissues of rats treated with vanadyl sulfate and sodium orthovanadate. J. Environ. Pathol. Toxicol. 2, 235-245.

Paulson, D.J., Kopp, S.J., Tow, J.P. & Peace, D.G. (1987): Effects of vanadate on **in vivo** myocardial reactivity to norepinephrine in diabetic rats. J. Pharmacol. Exp. Ther. 240, 529-534.

Pederson, R.A., Ramanadham, S., Buchan, A.M.J. & McNeill, J.H. (1989): Long-term effects of vanadyl treatment on streptozocin-induced diabetes in rats. Diabetes 38, 1390-1395.

Metal Ions in Biology and Medicine. Eds. Ph. Collery, L.A. Poirier, M. Manfait, J.C. Etienne. John Libbey Eurotext, Paris © 1990, pp. 315-319

Biological effects of occupational and environmental exposure to chromium

S. Anttila

Institute of Occupational Health, Topeliuksenkatu 41 aA, SF-00250 Helsinki, Finland

INTRODUCTION

The adverse effects of chromium (Cr) on human health have been known for a hundred years (Langård, 1990), although Cr also has an essential function in the human organism. Occupational exposure to Cr has been found to cause skin and nasal ulcers, skin allergies, bronchial asthma, and nasal and lung cancers, whereas environmental exposure is usually considered harmless. Several reviews have been published recently concerning epidemiological and experimental evidence of the carcinogenicity of Cr (o.a. Hayes, 1982, 1988; Bianchi & Levis, 1988; Langård, 1990). It is still controversial whether all Cr compounds or only those which are moderately or sparingly soluble and in the hexavalent oxidation state should be regarded as human carcinogens. If all Cr compounds are potentially carcinogenic, also environmental exposure, which applies to a vast number of people, bears a risk. Here environmental exposure to Cr and its possible health effects are discussed more thoroughly than the effects of occupational exposure.

OCCUPATIONAL AND ENVIRONMENTAL EXPOSURE

Occupational exposure to Cr is possible in the metallurgical industry in the production of Cr metal, alloys and ferrochromium steel, in the metal industry for example in stainless steel welding and chrome plating, and in industries producing Cr chemicals and pigments. In addition, Cr compounds are used in various other industries such as in leather tanning and in the production of catalysts, refractory materials, paints, textiles, fungicides, corrosion inhibitors and cement (IARC, 1980).

The amount of Cr occurring naturally in the environment is minimal. Most of the Cr in the environment originates from industrial wastes from the mining of chromite ore, the metallurgical industry, Cr chemical industries, cement producing plants and the burning of coal (IARC, 1980; Cary, 1982). The naturally occurring Cr is probably in the trivalent oxidation state, but it is possible that a part of the Cr added to the atmosphere by man is initially in the hexavalent oxidation state (Cary, 1982). In the waterways and soil Cr is mostly present in the trivalent state, but also hexavalent Cr may exist, depending on pH, organic matter and reducing agents (Cary, 1982).

CHROMIUM AS A HUMAN CARCINOGEN

The first chromium-associated cancer cases which were reported in the literature were a nasal adenocarcinoma of a chrome pigment worker in 1890, and lung cancer in chromate-exposed workers in the production of alizarin dyes in 1911, 1912 and 1935 (Hayes, 1982, 1988; Langård, 1990). Additional cases were found among workers employed in chromate production; also four nasal sinus cancers were described in chromate and chrome pigment workers. Later epidemiological studies have confirmed a two- or three-fold elevated risk of respiratory cancer in Cr chemical production and chromate pigment production. Epidemiological data on the chrome plating industry, ferrochromium production and Cr alloy welding have thus far been inconclusive. Furhermore, exposure to trivalent Cr has not been adequately studied epidemiologically (Hayes, 1982). An excess of cancers of the digestive tract have been reported in Cr chemical and pigment production, in chrome plating and in ferrochromium steel production, but the suggestion of gastrointestinal carcinogenicity has not been consistent in different studies. The evidence collected mostly from case reports suggests an elevated risk of nasal cancer due to Cr chemical exposure. The latency period for Cr-associated lung cancer ranges between 15 and 27 years, and in most studies the increased risk appears within 10-14 years after the initial exposure (Hayes, 1982).

METABOLISM OF CHROMIUM

The main routes of exposure to Cr are the skin, gastrointestinal tract and respiratory tract. The metabolism of Cr is greatly influenced by the route of intake, the compound, the solubility and the oxidation state. Because of the usual route of intake by inhalation and the complex nature of Cr compounds in occupational environments, it is very difficult in animal experiments to simulate the conditions of human exposure. Cr exists mainly in two oxidation states, Cr(III) and Cr(VI). Due to the rapid reduction of hexavalent Cr in the body fluids and tissues, the Cr present in the human body is trivalent. The reduction of Cr(VI) takes place in the secretions of the alimentary tract, in the epithelial-lining fluid and alveolar macrophages in the respiratory tract, in blood circulation mainly in erythrocytes, and also in the parenchymal cells of organs (Petrilli & de Flora, 1988). In the cells, the reduction is possible in the cytosol, endoplasmic reticulum, mitochondria and nuclei. It is partly non-enzymatic, but mainly catalysed by two NADPH-dependent enzymes, DT-diaphorase and cytochrome P450-reductase (Debetto & Luciani, 1988; Petrilli & de Flora, 1988).

In mutagenicity assays, Cr(VI) induces various genotoxic effects in bacteria and mammalian cells, but Cr(III) is inactive. This is due to the rapid diffusion of Cr(VI) through cell membranes, whereas Cr(III) does not permeate membranes in experimental conditions. After the reduction of Cr(VI) inside the cell, the resultant Cr(III) is the ultimate carcinogenic state of Cr, which is able to bind to proteins and DNA. It has recently been found that when suitable ligands make Cr(III) cations diffusible across the plasma membrane, genetic effects are induced by Cr(III) independently of Cr(VI) reduction (Bianchi & Levis, 1988). Inhaled Cr compounds probably occur in particulate form, and it is likely that they are phagosytized by macrophages and epithelial cells in the respiratory tract. Fume particles resulting from stainless steel welding, which produces both Cr(VI) and Cr(III), have been observed by electron microscopy in rat alveolar epithelial cells after inhalation exposure (Anttila, 1986). In their review, Bianchi and Levis (1988) conclude that recent experimental data indicate that chronic low-level exposure to insoluble Cr(III) is not without genotoxic consequences.

CHROMIUM CONTENT OF HUMAN LUNG TISSUE

The respiratory tract is the main target organ for Cr. The determination of pulmonary Cr content may confirm the significant exposure when a Cr-induced respiratory cancer is suspected. Cr cumulates in lung tissue, as the highest Cr concentrations have always been found in lungs in both occupationally exposed and non-exposed persons (Sumino et al., 1975; Hyodo et al., 1980; Kishi et al., 1987). The levels of Cr determined in lung tissue from chromate workers with a probable Cr-associated lung cancer have varied from a few micrograms to one and a half thousand micrograms per gram of dry weight (Tsuneta, 1978; Hyodo et al., 1980; Kishi et al., 1987).

Also the Cr originating from environmental sources accumulates in human lung tissue with age (Schroeder et al. 1962; Kollmeier et al., 1987; Raithel et al., 1988; Pääkkö et al., 1989). Smoking seems to be the most important cause for a high pulmonary Cr content in occupationally unexposed people. Although the traces of Cr in tobacco are very low, the pulmonary Cr concentrations of smokers are up to 2 or 3 times higher than in non-smokers (Raithel et al., 1988; Pääkkö et al., 1989). It is possible that smoking increases the deposition or inhibits the clearance of Cr-containing particles.

The pulmonary Cr content of people living in different geographical areas has been compared, and 3- to 5-fold differences have been found due to local environmental pollution. These differences can not be explained by analytical reasons, because the determinations were done by the same research groups (Kollmeier et al., 1985, 1987; Anttila et al., 1989). Kollmeier et al. (1985, 1987) compared the pulmonary Cr content of people living in the Ruhr District (Bochum), which is defined as a polluted area with high local Cr emissions, with the Cr content of people living in a rural area of West-Germany (Munster). Also Anttila et al. (1989) have compared the pulmonary Cr content of people from the area of Northern Finland (Oulu), where a chromite ore mine and a steel plant produce high Cr emissions, with the Cr of those living in Southern Finland (Helsinki) (see Table 1).

The mean pulmonary Cr content has been higher in lung cancer patients who had not been employed in industries using Cr than in other people with similar smoking habits and living in the same area (Anttila et al., 1989). The exposure of these lung cancer patients may have been partly environmental and partly occupational, caused by unrecognised exposure in construction or industrial work. Also Weber et al. (1983) found higher pulmonary Cr concentrations in lung cancer patients from an area in Belgium, where many foundries and steel plants are located, than in control patients, but occupations and smoking habits were not taken into account in this study (Table 1).

The connection of pulmonary emphysema with high pulmonary Cr may

Table 1. Pulmonary Cr content (mean ± SD ug/g dry weight*) in patients with normal lungs, lung cancer or pulmonary emphysema. The references which report local metallurgical industry or Cr emissions are underlined (n = number of patients).

Reference	Normal lungs	Lung cancer	Emphysema
Weber et al., 1983 Liege, Belgium	2.9 ± 1.8 n = 35	7.4 ± 5.5 n = 21	
Kollmeier et al., 1985 Munster, FRG	0.7 ± 0.5 n = 23		
Kollmeier et al., 1987 Bochum, FRG	3.5 ± 3.7** n = 25		
Raithel et al., 1988 Erlangen, FRG	1.5 ± 2.0 n = 30		
Anttila et al., 1989 Oulu, Finland	1.9 ± 1.3 n = 26	6.4 ± 4.3 n = 53	7.0 ± 4.3 n = 27
Unpublished Helsinki, Finland		2.0 ± 1.9 n = 39	

* The values originally expressed as wet weight multiplied by 5 to attain the value per dry weight.
** Four cases with lung cancer included

be explained by heavy smoking (Table 1) (Anttila et al. 1989). However, as Cr(VI) is a strong oxidant, theoretically it might promote the development of emphysema, because an inbalance between oxidants and antioxidants is considered to contribute to this process of tissue destruction (Cantin & Crystal, 1985). In epidemiological studies, however, no over-expression of chronic obstructive lung disease in Cr-exposed workers has been found.

The different carcinogenic potencies of various Cr compounds make it very difficult to estimate dose-response relations as regards lung cancer. It is not known what amount of Cr in lung tissue should be considered sufficient to reflect a significant exposure for the development of cancer. Tissue analyses may be the most sensitive method to determine the level of environmental exposure and a feasible method to confirm occupational exposure to Cr. The combination of tissue analyses with epidemiological studies on occupationally exposed workers might bring additional information to dose-response estimations.

REFERENCES

Anttila, S. (1986): Dissolution of stainless steel welding fumes in the rat lung: an x ray microanalytical study. Br. J. Ind. Med. 43, 592-596.

Anttila, S., Kokkonen, P., Pääkkö, P., Rainio. P., Kalliomäki, P.L., Pallon, J., Malmqvist, K., Pakarinen, P., Näntö, V. &

Sutinen, S. (1989): High concentrations of chromium in lung tissue from lung cancer patients. Cancer 63, 467-473.
Bianchi, V. & Levis, G. (1988): Review of genetic effects and mechanism of action of chromium compounds. Sci. Total Environ. 71, 351-355.
Cantin, A. & Crystal, R.G. (1985): Oxidants, antioxidants and the pathogenesis of emphysema. Eur. J. Respir. Dis. 66, suppl. 139, 7-17.
Cary, E.E. (1982): Chromium in air, soil and natural waters. In Biological and environmental aspects of chromium, ed. S. Langård, pp. 49-64. Amsterdam: Elsevier Biomedical Press.
Debetto, P. & Luciani, S. (1988): Toxic effect of chromium on cellular metabolism. Sci. Total Environ. 71, 365-377.
Hayes, R.B. (1982): Carcinogenic effects of chromium. In Biological and environmental aspects of chromium, ed. S. Langård, pp. 221-247
Hayes, R.B. (1988): Review of occupational epidemiology of chromium chemicals and respiratory cancer. Sci. Total Environ. 71, 331-339.
Hyodo, K., Suzuki, S., Furuya, N. & Meshizuka, K. (1980) An analysis of chromium, copper, and zinc in organs of a chromate worker. Int. Arch. Occup. Health 46, 141-150.
International Agency for Research on Cancer. (1980): IARC monographs on the evaluation of the carcinogenic risk of chemicals to humans. Some metals and metallic compounds. Lyon: IARC.
Kishi, R., Tarumi, T., Uchino, E. & Miyake, H. (1987): Chromium content of organs of chromate workers with lung cancer. Am. J. Ind. Med. 11, 67-74.
Kollmeier, H., Seemann, J.W., Muller, K.M., Rothe, G., Wittig, P. & Schejbal, V. (1987) Increased chromium and nickel content in lung tissue and bronchial carcinoma. Am. J. Ind. Med. 11, 659-669.
Kollmeier, H., Witting, C., Seemann, J., Wittig, P. & Rothe, R. (1985) Increased chromium and nickel content in lung tissue. J. Cancer Res. Clin. Oncol. 110, 173-176.
Langård, S. (1990) One hundred years of chromium and cancer: A review of epidemiological evidence and selected case reports. Am. J. Ind. Med. 17, 189-215.
Petrilli, F.L. & de Flora, S. (1988): Metabolic reduction of chromium as a treshold mechanism limiting its in vivo activity. Sci. Total Environ. 71, 357-364.
Pääkkö, P., Kokkonen, P., Anttila, S. & Kalliomäki, P.L. (1989): Cadmium and chromium as markers of smoking in human lung tissue. Environ. Res. 49, 197-207.
Raithel, H.J., Schaller, K.H., Reith, A., Svenes, K.B. & Valentin, H. (1988): Investigations on the quantitative determination of nickel and chromium in human lung tissue. Industrial medical, toxicological, and occupational medical expertise aspects. Int. Arch. Occup. Environ. Health 60, 55-66.
Schroeder, H.A., Balassa, J.J. & Tipton, I.H. (1962): Abnormal trace metals in man - chromium. J. Chron. Dis. 15, 941-964.
Sumino, K., Hayakawa, K., Shibata, T. & Kitamura, S. (1975): Heavy metals in normal Japanese tissues. Amounts of 15 heavy metals in 30 subjects. Arch. Environ. Health 30, 487-494.
Tsuneta, H. (1978): Concentration of chromium in the tissues of the respiratory tract among chromate workers with lung cancer. Haigan 18, 341-348.
Weber, G., Robaye, G., Bartsch, P., Collignon, A., Roelandts, I. & Delbrouck-Habaru, J.M. (1983) PIXE using cyclotron and lung cancer pathogenesis. IEEE Transactions on Nuclear Science NS-30, 1313-1315.

Metal Ions in Biology and Medicine. Eds. Ph. Collery, L.A. Poirier, M. Manfait, J.C. Etienne. John Libbey Eurotext, Paris © 1990, pp. 320-322

Clinical aspects of aluminium metabolism and aluminium containing drugs

K. Kisters, C. Spieker, W. Zidek, H.P. Bertram*, F. Wessels**, H. Zumkley†, K.H. Rahn

*Medizinische Poliklinik & * Institut für Pharmakologie der Universität 4400 Münster, Albert-Schweitzer-Str. 33, West Germany; ** San Franziskus Hospital, Essen West Germany*

Summary

Aluminum intoxication is a severe complication in patients suffering from renal failure. The main signs of beginning systemic aluminum intoxication are neurological symptoms accompanied by aggravation of an already existing renal anaemia and/ or osteopathy. Skin lesions and myopathy are also known to occur occasionally. The possible sources for aluminum accumulation in the body stores are aluminum containing drugs. Continuous intake of high doses of aluminum containing drugs are still known to cause elevated aluminum blood and tissue concentrations. In this connection it was interesting that aluminum enters the brain by passing through the blood-brain barrier.
In 10 patients with normal renal function, who got an aluminum containing antacid (2g aluminumhydroxide/d), aluminum brain tissue concentration, determined by atomic absorption spectroscopy, was 1,05±0,31 µg Al/g brain wet weight as compared to 20 contols with 0,58±0,09 µg Al/g brain wet weight significant lower (p<0,01). The results of our study are a warning against an undiscriminating therapy with aluminum containing drugs.

Key words: aluminum, antacid, tissue concentration, renal insufficiency

Introduction

Aluminum concentrations in blood plasma and different tissues are frequently found increased in patients undergoing chronic intermittent hemodialysis treatment (Alfrey et al., 1976; Berlyne, 1979; Kisters et al., 1990; Zumkley et al. 1984). There is evidence that aluminum plays a role in the pathogenesis of some diseases of these patients. Accumulation of aluminum in bone of patients with chronic renal failure was demonstrated by Parsons et al. 1971. Interest in aluminum toxicity in uremic patients was stimulated five years later, when a relation between serum aluminum concentration and dialysis dementia was found (Alfrey et al., 1976).
The exact mechanisms by which aluminum accumulation leads to the symptoms of aluminum intoxication are still unknown.
Continuous intake of high doses of aluminum containing drugs are also known to

cause elevated aluminum blood and tissue concentrations.
Whereas the risks of aluminum accumulation in patients with renal insufficiency is well known, it was interesting whether aluminum is passing through the blood-brain barrier of patients with normal renal function and to which extend aluminum is stored after administration of aluminum containing antacids.

Methods

In 20 patients with normal renal function and without aluminum exposition and in 10 patients with normal renal function, who got an aluminum containing antacid (2g aluminumhydroxide/d) for 14 days, aluminum concentrations were performed in brain tissue. Brain tissue was obtained during brain operation after the patients had given informed consent to the analyzes.
Aluminum concentrations were measured by atomic absorption spectroscopy using the Perkin Elmer apparatus, after tissue specimens were wet ashed (Bertram, 1981).

Results

Figure 1 summarizes the mean values and standard deviations of aluminum concentrations in brain tissue of 20 patients with normal renal function and without aluminum exposition and of 10 patients with normal renal function after an antacid-therapy (2g aluminumhydroxyde/d) for 14 days.

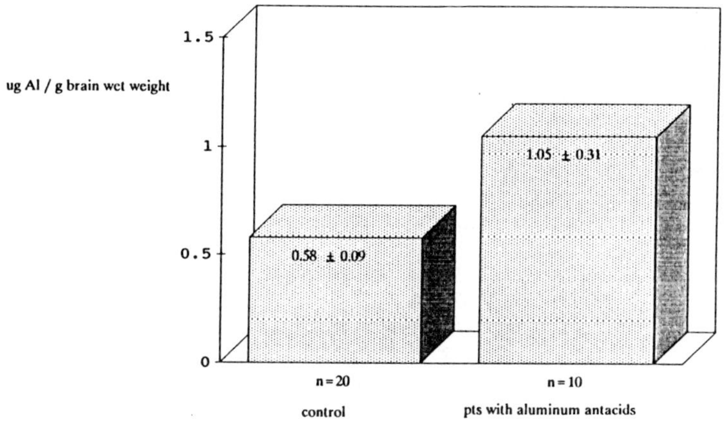

Fig. 1. Mean values and standard deviations of aluminum concentration in brain tissue (in µg Al /g brain wet weight) in 20 controls and in 10 patients with normal renal function after therapy with aluminum containing antacids.

The findings indicate markedly elevated aluminum concentrations in brain tissue ($p<0,01$) of patients with normal renal function after administration of aluminum containing antacids as compared to contols.

Discussion

The results indicate that aluminum concentrations in brain tissue are significantly increased ($p<0,01$) in patients with normal kidney function receiving aluminum containing drugs as compared to controls. In earlier studies it has already been shown, that the continuous intake of high doses of aluminum containing drugs will also lead to elevated aluminum concentrations in the hepatic, rectal, colonic, bone and gastric mucous membrane tissue of patients with normal renal function (Kisters et al., 1990; Zumkley et al., 1988).
How far high aluminum tissue content in brain over prolonged periods of time might lead to local damage has still to be studied in detail.
Whereas the risks of aluminum accumulation in patients with renal insufficiency is well known, our results suggest that also patients with normal renal function might suffer from aluminum accumulation under prolonged therapy with aluminum containing drugs. However, the long-term risks of aluminum load despite normal renal function are not yet assessed in detail. So the results of our study are a warning against an uncritical therapy with aluminum containing drugs.

References

Alfrey, A.C., Le Gendre, G.R., Kaehny, W.D. (1976): Dialysis encephalopathy syndrome. In New Engl. J. Med. 294: 184-188.

Berlyne, G.M. (1979): Aluminum toxicity in man. In Mineral Electrolyte Metab. 2: 71-73.

Bertram, H.P. (1981): Aluminiumbestimmung in Körperflüssigkeiten. In Aluminium in der Nephrologie. Hrsg. : Zumkley, H., Dustri-Verlag, München-Deisenhofen: 1-11.

Kisters, K., Spieker, C., Zidek, W., Fetsch, T., Bertram, H.P., Fromme, H.G., Zumkley, H., Rahn, K.H. (1990): Aluminum concentrations and localisation in gastric mucous membrane before and after therapy with aluminum containing antacids. In Trace elements in medicine (in press).

Parsons, V.R., Davies, C., Googe, C., Ogg, C., Siddiwui, J. (1971): Aluminum in bone from patients with renal failure. In Br. Med. J. 4: 273-275.

Zumkley, H., Spieker, C., Kisters, K., Zidek, W., Fromme, H.G., Losse, H., van Husen, N., Bertram, H.P. (1984): Aluminum concentrations in gastric mucous membrane in renal insufficiency. In Trace Substances in Environmental Health XVIII, ed. Hemphill, D.D., University of Columbia-Missouri: 40-46.

Zumkley, H., Bertram, H.P., Spieker; C. (1988): Klinische Bedeutung des Aluminiumhaushaltes. In Ärztl. Lab. 34: 239-243.

Metal Ions in Biology and Medicine. Eds. Ph. Collery, L.A. Poirier, M. Manfait, J.C. Etienne. John Libbey Eurotext, Paris © 1990, pp. 323-325

Links between aluminium (Al) neurotoxicity and cerebral atrophy (Ca) in regular hemodialysis treatment (RDT) patients : an overview of the pathogenic steps

B. Bocchi, S. Vinci, C. Raimondi, L. Allegri, G.M. Savazzi

Dept. of Internal Medicine and Nephrology, University of Parma 43100, Italy

Recently, brain computerized tomography (CT) has demonstrated CA in a large percentage of RDT subjects (Savazzi et al,1985).Statistical elaboration by discriminant analysis of the various clinical parameters (arterial hypertension, hyperparathyroidism with vascular calcifications, hematocrit, cholesterolemia) all considered possible causes of CA, indicates an etiopathogenic role of the oral elemental Al intake received in the form of $Al(OH)_3$ (Savazzi et al,1985). It is now generally accepted that oral $Al(OH)_3$ intake can be the source of increased brain tissue content of Al in RDT patients (Arieff et al,1979). The role of the metal was also established whatever its source, as an etiological neurotoxic factor in Dialysis Dementia (DD)(Masselot et al,1978). The evidence is consistent with the view that the clinical toxic phenomena associated with long-term oral Al intake in RDT patients includes a non-specific and more subtle form of neurological damage, i.e.CA. Although the mechanism by which Al acts as a neurotoxin has not yet been clearly defined, we can draw a picture of the possible pathogenic links with CA in uremic patients on the basis of experimental and clinical data. Several sites within the CNS could be the potential targets of the Al neurotoxic action. Al is directly implicated in reduced monoaminergic neurotransmitter synthesis by affecting the metabolism of tetrahydrobiopterin, an essential cofactor for the biosynthesis of dopamine, noradrenaline and serotonin.Al inhibition of dihydropteridine reductase, i.e. the regenerating enzyme of the cofactor has been well proved in subjects receiving RDT, and it has been causally related to the neuropsychological impairment observed in such patients (Altmann et al,1987). Al may also restrain cholinergic neurotransmission through the inhibition of choline acetyltransferase (ChAt)activity and synthesis as well as through a depressed choline uptake by nerve terminals (King,1984).Furthermore,Al inhibition of axonal transport could result in a secondary reduction of total ChAt activity, acethylcholinesterase activity and acethylcholine (ACh) levels. Al may affect glutamine decarboxylase activity with reduction in GABA biosynthesis and activity:indeed, a significant reduction in the cortical content of GABA has been observed in DD (Sweeney et al.1985). Al increases the blood-brain-barrier permeability to some peptides and nonpeptide substances including neurotransmitters and uremic toxins; while neurotransmitters (e.g.GABA) could leak out of the brain with a drop in their cerebral concentrations, other substances in the blood could readily enter the brain. In this context, an increased cerebral level of uremic toxins could exert their neurotoxic effect by inhibiting the synthesis of brain substances through interference with enzymatic reactions. A reduction in cerebral glucose utilization with repercussions on vital neuronal functions could be due to the inhibition of cytosolic and mitochondrial hexokinase. In fact, even small amounts of Al effectively compete for Mg binding sites in biological systems; Al inhibites the enzymes for which Mg is a cofactor, for exemple hexokinase and small quantities of Al are sufficient. Mg displacement from ATP, with resulting interference of phosphate transfer and abnormalities of ATP metabolism might explain the increased cerebral ATP and glucose levels as well as the decrease in both the metabolic rate and oxygen consumption observed in the uremic brain (Mahoney et al,1984). Even minimal anomalies in brain utilization of glucose could have weighty repercussions on neurotransmitter metabolism, particularly ACh. Alternative mechanisms might be the Al induced modifications of the steroisomerism of calmodulin with subsequent alteration of calcium-calmodulin interaction and unrestrained calcium influx/release into the neuron (Mahoney et al,1984); because calcium is essential for the function of a large number of intracellular enzyme systems, the disruption of the intraneuronal calcium homeostasis could possibly lead to neuronal damage and death by interfering with any of them. Finally Al has been shown to bind directly to nucleic acids and thus, by interfering with cellular functions such as RNA transcription, may induce hyperproduction of neurofilamentous proteins (NF). Interestingly,segregation and accumulation of NF i.e. Neurofibrillary Degeneration, has been recently found in two patients affected by DD (Scholtz et al,1987). The inappropriate location of NF may have important consequences for cytoskeletal organization and function, including axonal transport: an impairment in cytoplasmic transport of elements essential to neuronal life could conceivably lead to altered trophism with degenerative phenomena and resulting loss of neurons and fibres. Nevertheless, our observations are consistent with the view that the etiopathogenesis of CA in uremia is more likely multifactorial,with few overlapping factors working together in the direction of a progressive parenchimal decay; in this context, in addition to Al intoxication, a paramount etiopathogenic role is probably substained by arterial hypertension.Whatever the case, our present knowledge on the toxicity of Al in the uremic patient clearly warns nephrologists to stop administration of Al phosphate binders in such patients.

REFERENCES

Altmann P., Salihi P.A. Butter K.,Cutler P., Blair J., Leeming R.,Cunnigham J. Marsh F.(1987):Serum aluminium levels and erytrocyte dihydropteridine reductase activity in patients on hemodialysis.N.Engl.J.Med.317: 80-84.

Arieff A.I.,Cooper J.D.,Armstrong D.,Lazarowitz V.C.(1979):Dementia,renal failure and brain aluminium. Ann.Int.Med. 90: 741-747.
King R.G.(1984):Do raised brain aluminium levels in Alzheimer's dementia contribute to cholinergic neuronal deficits? Med.Hypoth. 14: 301-306.
Mahoney C.A.,Sarnacki P.,Arieff A.I.(1984):Uremic encephalopathy:role of brain energy metabolism. Am.J.Physiol. 247: F527-532.
Masselot J.P.,Adhemar J.P.,Jaudon M.C.,Kleinknecht D.,Galli A.(1978): Reversible dialysis encephalopathy:role for aluminium containing gels. Lancet 2:1386-1387.
Savazzi G.M.,Cusmano F.,Degasperi T.(1985):Cerebral atrophy in patients on long-term regular hemodialysis. Clin.Nephrol. 23: 89-95.
Sweeney U.P.,Perry T.L.,Price J.D.E.,Reeve L.E.,Godolphin W.T.,Kish S.T.(1985): Brain γ-aminobutyric acid deficiency in dialysis encephalopathy. Neurology 35: 180-186.
Scholtz C.L.,Swash M.,Bray A.,Kogeorgos T.,Marsh F.(1987):Neurofibrillary neuronal degeneration in dialysis dementia: a feature of Aluminium toxicity. Clin.Neuropathol. 6: 93-97.

Metal Ions in Biology and Medicine. Eds. Ph. Collery, L.A. Poirier, M. Manfait, J.C. Etienne. John Libbey Eurotext, Paris © 1990, pp. 326-328

Effects of aluminium (III) on plasmatic membranes

M. Perazzolo[a], L. Fontana[a], M. Favarato[a], B. Corain[b], G.G. Bombi[b], A. Tapparo[b], M. Nicolini[c], C. Corvaja[d], P. Zatta[e]

University of Padova, [a] Department of Biology, via Trieste 75, I-35121 Padova, Italy; [b] Department of Inorganic, Organometallic and Analytical Chemistry, via Marzolo 1, I-35131 Padova, Italy; [c] Department of Pharmacy, via Marzolo 5, I-35131 Padova, Italy; [d] Department of Physical Chemistry, via Marzolo 1, I-35131 Padova, Italy; [e] CNR, Centro per lo Studio delle Emocianine, via Trieste, 75, I-35121 Padova, Italy

In spite of the large amount of toxicological investigations on aluminum carried out in recent years (Sturman & Wisniewski, 1988), little information is available about the cytotoxicity of the metal (Langui *et al.*, 1988; Shea *et al.*, 1989) and about its effect on the stability of biological membranes (Vierstra & Haug, 1978).

In vivo and *in vitro* experiments with $Al(acac)_3$ (acac = acetylacetonate) carried out on erythrocytes (RBC) in these laboratories (Zatta *et al.*, 1989a; 1989b; Bombi *et al.*, 1990) evidenced an increase of the osmotic fragility and morphological alterations (acanthocytosis, Bessis, 1972).

In the present communication some biochemical and biophysical aspects of aluminum effect on the RBC membrane are discussed. The interaction between membrane constituents and $Al(acac)_3$, $Al(malt)_3$ (malt = maltolate) and $Al(lact)_3$ (lact = lactate) was investigated by SDS-polyacrylamide gel electrophoresis (SDS-PAGE) and by electronic spin resonance (ESR).

MATERIALS AND METHODS. Human, rabbit and rat RBC were obtained by centrifugation followed by washing with 0.9 % NaCl solution; the "ghosts" were obtained as described by Marchesi & Palade (1967).

$Al(lact)_3$ (Fluka) and the spin markers 5-NSA (2-(3-carboxypropyl)-2-tridecyl-4,4-dimethyl-3-oxazolyndinyloxyl) (Sigma) and MAL-6 (2,2,6,6-tetramethylpiperidin-1-oxyl-4-maleimide) (Sigma) were employed as received.

Literature methods were utilized for the preparation and purification of $Al(acac)_3$ (Young & Reynolds, 1946) and of $Al(malt)_3$ (Finnegan *et al.*, 1986), for SDS-PAGE analysis (Montecucco, 1986) and for spin labelling (Markesbery *et al.*, 1980); the aluminum content of ghosts was determined as previously described (Bombi *et al.*, 1990).

Electrophoretic experiments were carried out by a LKB 2001 unit and a Shimadzu CS-930 densitometer; ESR measurements were done on a Bruker ER 200 D spectrometer; micrographs were obtained with a Cambridge Stereoscan 250 SEM.

RESULTS AND DISCUSSION. The treatment of suspended rabbit RBC with $Al(acac)_3$ (concentration range 0.34-8 mM, 2 hours) induced a marked morphological change with production of acanthocytes. This damage was not removed by washing (3 x 5 mL phosphate buffer solution 3.5 mM, pH=7.4) and it was associated to a large increase of Al^{III} concentration in the ghosts (from 0.2 to 42 ppm).

A similar effect was observed on rat RBC, while the effect on human RBC exposed to

8 mM $Al(acac)_3$ was barely detectable. $Al(malt)_3$ and $Al(lact)_3$ do not induce morphological changes on RBC (Zatta *et al.* 1989a; 1989b).

The irreversibility of acanthocyte formation and the high Al^{III} concentration found in ghosts seem to indicate that the effect of $Al(acac)_3$ is due to an interaction with membrane components. This interaction probably does not involves membrane proteins, as no alteration in the SDS-PAGE profile of these proteins was found after $Al(acac)_3$ treatment of RBC.

This hypothesis was confirmed by ESR measurements with MAL-6 (a spin label which binds to proteins through -SH groups): ESR analysis of MAL-6 labelled acanthocyte ghosts from rabbit revealed no alterations of protein configuration (Markesbery et al., 1980). On the contrary, this effect was observed in the membranes of human RBC treated with $Al(acac)_3$, that were not affected from morphologic changes.

The ESR spectra of rabbit acanthocyte ghosts labelled with 5-NSA (a spin label which dissolves in the lipidic bilayer) showed a significant increase of the $2T_{\|}$ value, indicating a decrease of membrane lipid fluidity (Vierstra & Haug, 1978) (Fig. 1).

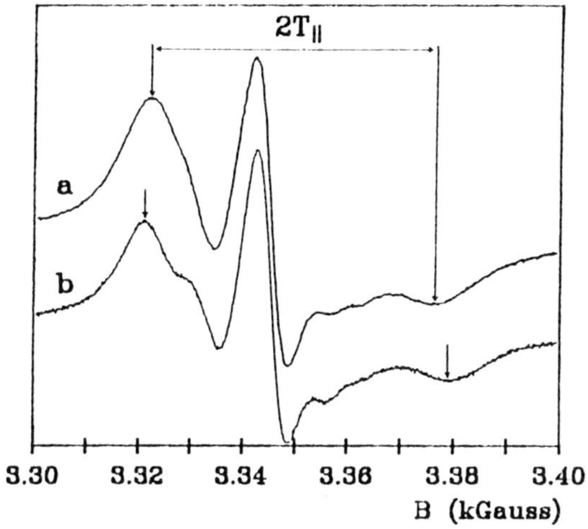

Fig. 1. ESR spectrum of rabbit RBC ghosts labelled with 5-NSA. (a) untreated; (b) treated with 2 mM $Al(acac)_3$.

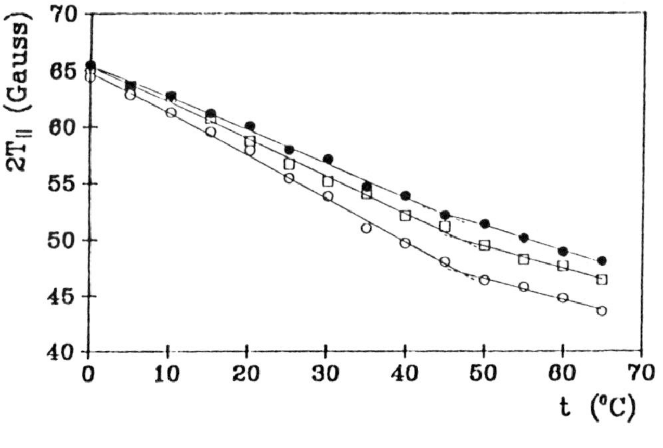

Fig. 2. Temperature effect on $2T_{\|}$ in rabbit RBC ghosts labelled with 5-NSA. ○: untreated; □: $Al(lact)_3$ (4 mM); ●: $Al(acac)_3$ (4 mM). $Al(malt)_3$ (4 mM) behaves as $Al(lact)_3$. Estimated error: ordinate ± 0.2 Gauss; abscissa ± 0.2 °C.

Also experiments in which suspended ghosts from normal rabbit RBC where treated with Al^{III} compounds showed an increase of the rigidity of the lipid matrix. In this case, the effect of $Al(malt)_3$ and $Al(lact)_3$ was similar to that of $Al(acac)_3$, although less marked (Fig. 2).

The membrane rigidity associated with acanthocytosis is probably due to the formation of coordination bonds between Al^{III} and phospholipids. The difference in behaviour between $Al(acac)_3$ and other Al^{III} compounds can be correlated with the high lipophilicity of the former (Tapparo & Perazzolo, 1989), which may promote its penetration into the cell membrane.

The above results represent a model for the deleterious effects that Al^{III} exerts on both the CNS and the peripheral tissues and suggests that its action could be explained, at least in part, as a membrane toxicity.

REFERENCES

Bessis, M. (1972): *Cellules du sang normal et pathologique*, pp. 330-332. Paris: Masson et Cie.

Bombi, G.G., Corain, B., Favarato, M., Giordano, R., Nicolini, M., Perazzolo, M., Tapparo, A. & Zatta, P. (1990): Experimental aluminum dismetabolism in rabbits: effects of hydrophilic and lipophilic complexes. *Environ. Health Persp.* 90: 000.

Finnegan, M.M., Steven, J.R. & Orvig, C. (1986): A neutral water-soluble aluminum complex of neurological interest. *J. Am. Chem. Soc.* 108: 5033-5035.

Langui, D., Anderton, B.H., Brion, J.P. & Ulrich, J. (1988): Effects of aluminium chloride on cultured cells from rat brain hemispheres. *Brain Res.* 438: 67-76.

Marchesi, V.T. & Palade, G.E. (1967): The localisation of Mg-Na-K-activated adenosine triphosphatase on red cell ghost membranes. *J. Cell. Biol.* 35: 385-404.

Markesbery, W.R., Leung, P.K. & Butterfield, A. (1980): Spin label and biochemical studies of erythrocyte membranes. *J. Neurol. Sci.* 45: 323-330.

Montecucco, C. (1986): Hydrophobic photolabelling with ^{125}I-TID of red blood cell membranes. In *Membrane proteins isolation and characterisation*, eds A. Azzi, L. Masotti & A. Vecli, pp. 119-123. Berlin: Springer Verlag.

Sturman, J.A. & Wisniewski, H.M. (1988): Aluminum. In *Metal neurotoxicity*, eds. S.C. Bondy & K.N. Prosad, pp. 61-85. Boca Raton: CRC Press.

Tapparo, A. & Perazzolo, M. (1989): *N*-octanol/water partition coefficients of the acetylacetonate and maltolate complexes of Al(III), Cr(III) and Fe(III) and of aluminum lactate. *Intern. J. Anal. Chem.* 36: 13-16.

Vierstra, R. & Haug, A. (1978): The effects of Al^{3+} on the physical properties of membrane lipids in *Thermoplasma acidophilum*. *Biochem. Biophys. Res. Com.* 84: 138-143.

Young, R.C. & Reynolds, J.P. (1946): Aluminum acetylacetonate. *Inorg. Synth.* 2: 25-26.

Zatta, P., Perazzolo, M. & Corain, B. (1989a). Tris acetylacetonate aluminium(III) induces osmotic fragility and acanthocyte formation in suspended erythrocytes. *Toxicol. Lett.* 45: 15-21.

Zatta, P., Perazzolo, M., Bombi, G.G., Corain, B. & Nicolini, M. (1989b): The role of speciation in the effects of aluminum(III) on the stability of cell membranes and on the activity of selected enzymes. In *Alzheimer's disease and related disorders*, eds. K. Iqbal, H.M. Wisniewski & B. Winblad, pp. 1087-1094. New York: Alan R. Liss, Inc.

Metal Ions in Biology and Medicine. Eds. Ph. Collery, L.A. Poirier, M. Manfait, J.C. Etienne. John Libbey Eurotext, Paris © 1990, pp. 329-331

The role of oral $Al(OH)_3$ ingestion in the development of cerebral atrophy (CA) in patients in regular dialysis treatment (RDT)

G.M. Savazzi, S. Vinci, L. Allegri, B. Bocchi

Department of Internal Medicine and Nephrology, University of Parma, I-43100 Parma, Italy

INTRODUCTION

A sizable number of patients on Regular Dialysis Treatment (RDT) occasionally submitted to Cerebral Computed Tomography (CCT) in our department for diagnostic pourposes had surprisingly shown a sizable incidence of Cerebral Atrophy (CA). Most of the observed CCT abnormalities occured in patients who had been on RDT for long time.The pourpose of this study was to confirm, among the clinical and blood biochemical parameters known to possibly produce brain damage when altered, previous observations indicating an etiopathogenic role of Aluminium (Al) intoxication in the development of CA.

MATERIAL AND METHODS

42 patients in RDT (32 males and 10 females) for not less than 10 years and having less than 50 years of age were investigated for the presence of CA with a CGR-ND-8000 head scanner (All patients received 11 axis cuts of the brain parenchyma, 9 mm thick starting from the skull base; 120 Kv.20 ma; scan time 40 sec.). The results obtained in this series of patients are compared with those of a control group of 55 non-nephropathic patients under age 50, undergoing CCT for diversified reasons and returning compatible with CCT and clinical findings of no evidence of brain pathology. We calculated: a) a mean value of the width in the four larger cortical sulci, constituting an index of cortical atrophy when exceding 3 mm (Bergman et Al.1980, Gildensted 1977): b) the increase of the Evan's ratio indicative of subcortical atrophy when $>$ 0.31 (Brinkmann et Al.1981: Jacob & Levy 1980). The most probable causes of CA were investigated in the last 8 years' anamnestic profile: mean predialysis and postdialysis sistolic and diastolic blood pressure; mean daily $Al(OH)_3$ oral intake; degree of neuropathy as assessed by electromyography - detailed informations about parameters and technics are indicated in earlier published papers (Savazzi et Al.1980)-; blood cholesterol and tryglyceride content; status of arterial vessels as indicated by vascular calcifications. The patients were divided in two groups: positive, showing CA, and negative for CA.Data were statistically considered by the Fisher test (FT) and elaborated

with a CDC 6000 computer for Discriminant Canonic Function Coefficient (DCFC) and Discriminant Analysis.

RESULTS

In the control group of non-nephropatic patients no evidence of CA was found; on the contrary 69% of patients in RDT showed CA (47,6% as cortical atrophy and 21,4 as subcortical damage).The FT and DCFC confirmed a strict correlation of CA with mean blood pressure and mean daily $Al(OH)_3$ intake, while less correlated were mean blood cholesterol and tryglyceride content, vascular calcifications and the degree of neuropathy.The grouping of the patients into CCT positive and CCT negative for CA, related to clinical and biochemical variables processed in Discriminant Analysis, indicated a 92,3% of cases correctly classified, therefore reinforcing the etiopathogenic role of hypertension and $Al(OH)_3$ intake in the production of CA.

DISCUSSION

Statistical evidence of a defined graduation of possible etiologic variables indicates a multifactorial pathogenesis of the CA in RDT patients, in which blood hypertension and Al toxicity are the main factors. It has been clearly shown that Dialysis Dementia - acute Encephalopathy Syndrome - is produced by an accumulation of Al in the brain tissue, Al originating from water rich in $Al(OH)_3$ and not adequately treated by deionization or reverse osmosis for the preparation of dialysis fluids. The cause being identified, those who never met with overt manifestations of dialysis dementia may regard Al toxication as an altogether uncommon phenomenon. But some observations (Kaehny et Al.1962) indicate that the patient on long-term dialysis may develop chronic Al toxication from sources other than untreated water; this resurrects the doubt that CA, as detected by our own and other groups,owes partly to a slow accumulation of Al in the brain tissue, reaching the conclusion that CA, seen in so high a percentage of RDT patients, represents the long-term result of a toxication identical in its etiopathogenesis to that of Dialysis Dementia, only subsymptomatic, due to oral $Al(OH)_3$ assumption and with a chronic course. The existence of Al carriers other that untreated water is indicated by the findings of dialysis centers where the water contained only negligible amount of Al, but there was nevertheless a significant correlation between Al concentration in the gray matter of the patient's brains and the duration of chronic dialysis (Alfrey et Al. 1976): hence of the tendency to hyperphosphatemia and oral dosing with chelating agents. A marked increase of Al concentration in brain tissue was found in hemodialysed patients apparently free of symptoms attributable to organic intellectual deterioration.(McDermott et Al 1977). The neurotoxicity of aluminum,expressed in uremic patients by Dialysis Dementia, is well demonstrated by experimental and clinical observations in situations other than uremia: neurofibrillary degeneration is detected in cerebral areas with high Al content (McDermott et Al.1977) in some but not all animal species: in the rabbit direct exposure of brain tissue to Al salts induces neurofibrillary degeneration, as in the cat, but this does not happen in the rat in spite of the presence of high Al tissue levels.(King et Al.1975). Animal experiments further indicate that Al toxicity may result in irregular distribution of the metal with preference for certain structures of the brain: in the cat, for instance, Al concentrations are higher in frontal and occipital lobes than anywhere else in the gray matter; also, higher concentrations are associated with more extensive neurofibrillary degeneration and a profound disturbance of higher nervous functions (Crapper et Al.1973).

In spite of a variable expression in the different animal species, a significant correlation is found between Al concentration in the brain tissue and morfo-neuronal and functional impairment. This indicates that $Al(OH)_3$ administered orally to RDT patients to control hyperphosphatemia can no longer be tolerated and the high Al concentrations found in the brain tissue of such patients may be regarded as being caused by the medication, possibly with slow but progressive chronic damage,albeit in the absence of clinically apparent intellectual impairment.

REFERENCES

Alfrey A.C.,Le Gendre G.R.,Kaehny W.D.(1976):The dialysis encephalopathy syndrome.N.Engl.J.Med.294; 184-188

Bergman H.,Borg S.,Hindmarsh T.,Idestrom C.M.,Mutzell S.(1980):Computed tomography of the brain and neuropsichological assesment of male alcoholic patients and a random sample. Acta Psych.Scand. 62,Suppl.286, 77-88

Brinkmann S.D.,Sarwar M.,Levin H.S.,Morris H.H.(1981);Quantitative indexes of computed tomography in dementia and normal aging.Neuroradiol.138,89-92

Crapper D.R.,Krishnan S.S.,Dalton A.J.(1973):Brain aluminium distribution in Alzheimer's disease and experimental neurofibrillary degeneration. Science 180, 511-513

Gyldensted C.(1977):Measurements of the normal ventricular system and hemispheric sulci of 100 adults with computed tomography. Neuroradiology 14, 183-193

Jacoby R.J.,Levy R.(1980):Computed tomography in the elderly. 2.Senile dementia:diagnosis and functional impairment.Brit.J.Psych.136, 256-269

Kaehny W.D.,Alfrey A.C.,Holman R.E.,Schorr W.J.(1977):Aluminium transfer during hemodialysis.Kidney Internat. 12, 361-365

King G.A.,De Boni V.,Crapper D.R.(1975):Effect of aluminium upon conditioned voidance response acquisition in the absence of neurofibrillary degeneration Pharmac.Biochem.Behav. 3, 1003-1009

Mc Dermott J.T.,Smith A.I.,Iobal K.,Wisniewski H.M.(1977):Aluminium and Alzheimer's disease.Lancet II,710. 1977

Mc Dermott J.R.,Smith A.I.,Ward M.K.,Parkinson I.S.,Kerr D.N.S.(1978) Brain-aluminium concentration in Dialysis encephalopathy.Lancet i, 901-903

Savazzi G.M.,Cambi V.,Migone L.(1980):The influence of glomerular filtration rate on uremic polyneuropathy. Clin.nephrol. 13, 64-72

Metal Ions in Biology and Medicine. Eds. Ph. Collery, L.A. Poirier, M. Manfait, J.C. Etienne. John Libbey Eurotext, Paris © 1990, pp. 332-335

Bone lead, bone aluminium and renal insufficiency

B. Winterberg[1], R. Fischer[3], R. Korte[3], H. Zumkley[†], H.P. Bertram[2]

[1] Medical Policlinic, University of Münster, F.R.G.; [2] Institute for Pharmacology and Toxicology, University of Münster, F.R.G. and [3] Marienhospital Emsdetten, FRG

Whereas the toxic effects of high doses of lead were already known in antiquity, those of low doses have been recognized only recently (Bellinger et al., 1987; Cullen et al., 1983; Gilfillan, 1965; Neddleman et al., 1984). Lead is a contaminant that is widely distributed in human environment. Individuals with an impaired capacity of eliminating ubiquitous contaminants are at risk of cumulating toxic substances and thus of developing chronic intoxication which must be recognized and treated as early as possible (Alfrey et al., 1976; Winterberg et al., 1986; 1987; Winterberg and Bertram, 1990; Zumkley et al., 1979; Zumkley, 1981).
In this study, the relationship between the degree of elimination impairment (degree of renal insufficiency) and bone lead and bone aluminium was investigated in patients with a normal exposure to lead, out of whom some were continiously exposed to aluminium. None of the patients had gout or lead-induced nephropathy (Batuman et al., 1981; Wedeen and Batuman, 1979).

Patients and Methods

Four different types of patients were included in the study: 8 patients with normal renal function who had to undergo bone biopsy for clarification of other health problems (female n=3, male n=5, mean age 52,4 ± 14,4 years; serum creatinine < 1,3 mg/dl); 8 patients with impaired renal function (female n=4, male n=4; mean age 54,6 ± 15,2 years; mean serum creatinine 5,0 ± 1,62 mg/dl); 14 patients with endstage renal disease (female n=7, male n=7; mean age 52,8 ± 7,6 years; mean time on RDT 4 ± 3 years), and 7 patients with functioning kidney graft (female n=3, male n=4; mean age 38,0 ± 8,7 years; mean serum creatinine 1,9 ± 1,1 mg/dl). Iliac crest biopsy was performed using a Yamshidi needle. In patients with impaired renal function biopsies were carried out as a routine examination to follow up renal osteopathy. Lead and aluminium concentrations were measured by means of electrothermal atomic absorption spectrometry (Perkin Elmer, HGA 500, Perkin Elmer, Norwalk, Connecticut, USA). All materials and containers used were free from lead or aluminium and were pretreated with ultra-clean nitric acid before use (Bertram et al., 1984). All data were expressed in ug/g fresh weight. Statistical data analysis was performed using the nonparametric Kruskal-Wallis test (simple classification).

Results

As is shown in Table 1, patients without renal insufficiency (n=8) and normal exposure to lead had a mean bone lead of 1.63 ± 1.02 ug/g (median 1.15 ug/g). The corresponding values for aluminium were 0.89 ± 0.83 ug/g (median 0.51 ug/g). In patients with impaired renal function (n=8 and a mean serum creatinine of 5.0 ± 1.62 mg/dl, mean bone lead was 2.18 ± 0.98 ug/g (median 1.99 ug/g). Mean bone aluminium was 2.4 ± 1.66 ug/g (median 1.87 ug/g). In the dialysis patients (n=14) mean bone lead and bone aluminium were still more increased (mean bone lead 3.59 ± 2.3 ug/g, median 3.16 ug/g; mean bone aluminium 17.29 ± 11.43 ug/g; median 14.55 ug/g). In the 7 patients with functioning kidney graft and a mean serum creatinine of 1.93 ± 1.1 mg/dl, bone lead and bone aluminium were lower compared to the dialysis patients (bone lead 2.53 ± 1.19 ug/g, median 2.4 ug/g; bone aluminium 10.09 ± 6.61 ug/g, median 6.95 ug/g).
Data analysis by means of the nonparametric Kruskal-Wallis test showed the following results: lead - value of T = 8.64, the corresponding test statistic is chi-squared with f = 3 degrees of freedom (corresponding to p = 0.034). With a significance level of a = 0.05 the null hypothesis is rejected. The corresponding values for aluminium are value of T = 24.8, p = 0.000017, a = 0.001.

Discussion

In this study bone lead and bone aluminium were assessed in 3 patient groups with various degrees of renal insufficiency and in 1 group without renal impairment. All patients were living in a rural area and therefore subjected to low lead exposure. None of the patients had an occupational exposure to lead. The dialysis patients were continuously exposed to aluminium, which was given as a phosphate binder. An indirect measurement of bone lead is done by means of the EDTA-test, i.e. the administration of a chelating agent, which mobilizes lead in the storage tissue and thus causes an increase in blood lead. The chelate-lead complex is eliminated renally. Emmerson (1963) and Behringer et al. (1986) found a positive correlation between serum creatinine and lead excretion in urine after infusion of EDTA in patients with gout and known exposure to lead. A similar correlation was found by Behringer et al. in patients without gout and known exposure to lead. In addition, they found slightly increased blood lead levels in patients with impaired renal function. This was also observed by Szadowski et al. (1969). Van de Vyver et al. (1988) investigated bone lead in dialysis patients and did not find any difference compared to patients without renal insufficiency. None of the groups had an unusual exposure to lead. De Vyver et al. therefore concluded that in people with normal exposure to lead renal insufficiency alone does not cause an increase in bone lead.
On the contrary, our results show that in patients with renal

insufficiency and normal exposure to lead the increase of bone lead depends on the degree of renal impairment. In patients with functioning kidney graft the bone lead depot seems to empty after the re-establishment of renal elimination.
Bone aluminium values were comparable to bone lead values. However, in the dialysis group, bone aluminium was markedly increased compared to the patients without renal insufficiency and those with impaired renal function. This increase is caused by an additional exposure to aluminium due to phosphate binder therapy (Winterberg et al., 1984; Winterberg and Bertram, 1990). Patients with impaired renal function are at particularly high risk of developing chronic intoxication with elements which are preferably excreted in urine. It is therefore important to prevent additional exposure to these elements by prophylactic measures, above all by reducing airborne lead and phosphate binders containing aluminium.

Tab. 1 Bone lead and bone aluminium of all patients' groups in ug/g fresh weight.

Group	N	mean age (years)	Lead ug/g fresh weight Mean	Median	Aluminium ug/g fresh weight Mean	Median
Control (serum creatinine < 1.3 mg/dl)	8	52,4 ± 14,4	1.63 ± 1.02	1.15 (0.72-3.81)	0.89 ± 0.83	0.51 (0.12-2.32)
Patients with impaired renal function (s.creatinine 5.0 ± 1.62 mg/dl)	8	54.6 ± 15.2	2.18 ± 0.98	1.99 (0.72-3.89)	2.4 ± 1.66	1.87 (1.06-6.35)
Dialysis patients	14	52.8 ± 7.6	3.59 ± 2.30	3.16 (1.26-10.3)	17.29 ± 11.43	14.55 (1.77-34.9)
Patients with functioning kidney graft (s. creatinine 1.93 ± 1.11 mg/dl)	7	38.0 ± 8.7	2.53 ± 1.19	2.4 (0.72-4.57)	10.09 ± 6.61	6.95 (3.15-23.0)

References

1. Alfrey A.C., Le Gendre G.R., Kaehny W.D. (1976): The dialysis encephalopathy syndrome. Possible aluminium intoxication. N. Engl. J. Med. 294, 184
2. Batuman V., Maesaka J.K., Haddad B., Tepper E. Landy E. Wedeen R.P. (1981): The role of lead in gout nephropathy. N. Eng. J. Med. 304, 520-523
3. Bellinger D., Levinton A. Waternaux C., Needleman H.L., Rabinowitz M. (1987): Longitudinal analyses of prenatal and postnatal lead exposure and early cognitive development. N. Eng. J. Med. 316, 1037-1043
4. Behringer D., Craswell P., Mohl C., Stoeppler M., Ritz E. (1986): Urinary lead excretion in uremic patients. Nephron 42, 323-329

5. Bertram H.P., Robbers J., Schmidt R. (1984): Multielementanalyse mit ET-ASS im Rahmen der Umweltprobenbank Münster. Fresenius Z. Anal. Chem. 317, 462
6. Cullen M.R., Robins J.M. Eskenazi B. (1983): Acute inorganic lead intoxication. Medicine (Baltimore) 62, 221-247
7. Emmerson B.T. (1963): Chronic lead nephropathy: The diagnostic use of calcium EDTA and the association with gout. Aust. Ann. Med. 12, 310-324
8. Gilfillan S.C. (1965): Lead poisoning and th fall of Rome. J. Occup. Med. 7, 53-59
9. Needleman H.L., Rabinowitz M., Leviton A., Linn S., Schoenbaum S. (1984): The relationship between prenatal exposure to lead and congenital anomalies. JAMA 251, 2956-2959
10. Szadowski D., Schaller K.H., Radunski K. (1969): Das Verhalten des Blutbleispiegels bei einigen internen Krankheiten. Arbeitsmed. Sozialmed. Präventivmed. 4, 54-56
11. Van de Vyver F. L., D'Haese P.C., Visser W.J., Elseviers M.M., Knippenberg L.J., Lamberts L.V., Wedeen R.P., De Broe M.E. (1988): Bone lead in dialysis patients. Kedney Int. 33, 601-607
12. Wedeen P.P., Batuman D.K.V. (1979): Detection and treatment of occupational lead nephropathy. Arch. Intern. Med. 139, 53-57
13. Winterberg B., Lison A.E., Bertram H.P., Spieker C., Kellinghaus H., Zumkley H. (1984): Aluminium intoxication. Treatment with desferrioxamine. Trace Elem. Med. 1, 111-114
14. Winterberg B., Bertram H.P., Lison A.E., Spieker C., Schalthöfer E., Raidt H. Zumkley H. (1986): Desferrioxamine B: aluminium kinetics in patients on regular hemodialysis. Trace Elem. Med. 3, 95-99
15. Winterberg B., Knoll O., Bertram W.P., Zumkley H. (1987): Intoksikatsija aljuminijem - klinitscheskaja problema w. nefrologii. Urologija i Nefrologija 2, 41-44
16. Winterberg B., Bertram H.P. (1990): Aluminium removal in uremic patients after desferrioxamine infusion: is hemofiltration more effective than hemodialysis? Nephron 54, 179 (letter)
17. Zumkley H., Bertram H.P., Lison A.E., Knoll O., Losse H. (1979): Aluminium, zinc and copper concentration in plasma in chronic renal insufficiency. Clin. Nephrol. 12, 18
18. Zumkley H. (1981): Klinische Aspekte des Aluminiumhaushalts. Nieren- und Hochdruckkrankheiten 10, 192

Metal Ions in Biology and Medicine. Eds. Ph. Collery, L.A. Poirier, M. Manfait, J.C. Etienne. John Libbey Eurotext, Paris © 1990, pp. 336-338

Alzheimer disease and dementia syndromes consecutive to imbalanced mineral metabolisms subsequent to blood brain barrier alteration

R. Deloncle[1], O. Guillard[2], P. Turq[3], N. Prulière[3]

[1] *Laboratoire de Chimie Bio-Inorganique, Faculté de Pharmacie, 2 bis Boulevard Tonnellé, 37042 Tours Cedex, France;* [2] *Laboratoire de Biochimie, Institut du médicament, Faculté de Médecine, 34, rue du Jardin des Plantes, 86034 Poitiers Cedex, France;* [3] *Laboratoire d'Electrochimie, Faculté des Sciences Paris VI, 8 rue Cuvier, 75005 Paris, France*

ALUMINIUM, BLOOD BRAIN BARRIER AND ALZHEIMER'S DISEASE

In Alzheimer disease(AD),changes in blood brain barrier(BBB) permeation have been post mortem histologically proved (Alafuzoff et al. 1987) and aluminium's responsibility in this alteration has been suggested (Kim et al.1986,Banks et al.1989). Alteration of BBB does not seem to be the only biological membrane changes in AD (Zubenko et al.1988, Ueda et al.1989). These modifications, possibly genetically explainable, may well be found in biological transfers such as the gastro-intestinal absorption. If a such process is altered, aluminium (Al) transfer to blood can be made easier. This metal will then be able to replace cations such as calcium (Ca) and magnesium (Mg) in vivo by complex competition reactions, and disturb metabolisms in which these alkaline earth cations are involved.

In our laboratory, it has been proven that Al was able as glutamate complex to cross the erythrocyte membrane and the BBB and to be consequently deposited in the brain (Deloncle et al 1990)

CATION METABOLISM DYSFUNCTION BY ALUMINIUM SUBSTITUTICN.

Complex formation can be linked to the ratio of ionic charge to ionic radius; the greater the ratio, the more stable the complex. The previously defined ratios are respectively 2.04 , 2.56 and 5.26 for Ca, Mg and Al. Then, for the same ligand, Ca yields less stable complexes than Mg, themselves less stable than Al relatives . Glutamate complex stability constants are in close agreement with this fact, since log K_1 are respectively 1.43 and 1.9 in KCl 0.1 M for Ca and Mg (Lumb et al. 1953) and log K_1= 15.12 for Al in $NaClO_4$ (Singh et al. 1972).

Alkaline earth cations are involved as labile complexes at various cellular levels; when replaced by aluminium, several disturbances may occur:At bone level, Al substitution for Ca and Mg yields a brittle edifice thas has been confirmed in AD(Buchner & al.1987);at muscle level,such as in myocard, since Ca is involved in muscular contraction, cardiomyopathies described in hemodialyzed patients (Timsit et al. 1986) could well be explained by Al substitution for Ca. The most serious troubles will be noticed at neuronal levels. If Al crosses the BBB as glutamate complex,first target regions in the brain will be essentially glutamatergic ones such as cortical associative area, hippocampus or amygdala(Deutsch et al.1988).Al will be localized there(Santos & al 1987)by binding reactions of glutamate to its receptors and induce cell disturbances by replacing Ca or Mg for example in neuro transmitter release or polymerization-depolymerization process of neurotubules.

If we posit that Al complexes are more stable than Ca or Mg ones, it is conceivable that the resulting structures will be less labile than the Mg-induced one.

Likewise, the stable Al resulting deposits will have analogous antigenic epitopes with the proteins from which they are generated. Similar features are found in AD: paired helical filaments (PHF) and neurofibrillary tangles (NFT) share analogous antigenic structures with normal constitutive proteins of the cytoskeleton (Mori et al. 1987; Mulvihill et al.1989).

Since Al yields stable complexes in biological systems this metal may link with every glutamic or aspartic side carboxylic grouping in protidic sequences. In the genesis of stable hexacoordinated complexes Al may also contract stable links in vivo with strong electron donor groupings such as phosphoric ligands. That would be in close agreement with the brain protein hyperphosphorylation in AD. In this order, the NFT formation may well be explained since the thirteen amino acid phosphorylated repeat motif described by Trojanowski et al.(1989) in neurofilaments, contains two adjacent glutamic groupings. In the same way, the cerebrovascular amyloid protein sequenced by Glenner et Wong (1984) contains a glutamic acid adjacent to an aspartic residue. The presence of two adjacent carboxylic side groupings in protidic sequences allows for stable aluminium complex formation.

ALUMINIUM AND THE GENETIC HYPOTHESIS OF ALZHEIMER BETA AMYLOID DEPOSIT.

Experiments attempting to prove the genetic hypothesis of beta amyloid protein (β AP) deposits have demonstrated that the precursor (β APP) was to be found in membrane-associated proteins in both neural and non neural tissues (Selkoe et al. 1988). Since protidic precursor sequences are rich in aspartic or glutamic residues (Delamarche 1989), we have already explained how self-aggregating amyloid protein could be produced by Al-complex formation. The synthesis of the normal glutamic aspartic rich cell-surface protein, coded on chromosome 21 (St George Hyslop et al. 1987) will then be perturbed. Since the normal protein will not be synthesized, the feedback control would result in such processes as an overproduction of the β APP and a β AP deposit in the brain as has been observed in Alzheimer disease, Down's syndrome and Guam Parkinson Dementia .

PROPOSED MECHANISM FOR ALZHEIMER'S DISEASE

Al accumulation in neurons could well be at the origin of AD, consecutive to biological membrane alteration. In AD, risks factors increase with age ; this feeding is in close agreement with the alteration of biological membranes in the elderly and age-adjusted rates are higher for women, as Ca metabolism is more affected in females than in males, especially following menopausis.

If L-glutamate is present as stable Al chelate inside the neuronal cell, we don't think that this neurotransmitter will be able to fulfill all its biological functions. A very important process in the neuronal cell is ammonia detoxification. It produces L-glutamin, a non toxic metabolite that crosses the BBB on its way out . The consequence in the neuron will be an accumulation of this highly toxic cellular metabolite which will be responsible for cell death. The first neurons destroyed by this process will evidently be glutamatergic ones, but since Al-glutamate crosses the erythrocyte membrane we think that this complex can also transfer through the cell-coat of other neurons. The same destruction process will then affect every neurotransmitter system. Destroyed neurons will be substituted in the brain by a filling tissue such as nevroglie which is in close agreement with the experimentally induced astrocytic proliferation obtained by Brumback et al.(1989) in hepatic encephalopathy.

In conclusion, we think that a mineral imbalance subsequent to BBB alteration is responsible for Alzheimer disease, Down's syndrome and Guam Parkinson Dementia. Ca and Mg are replaced by all metals present in vivo yielding more stable complexes. Blood transported, these metals cross the BBB , such as Al in the form of glutamate complex and generate in vivo stable structures with neuronal and non neuronal proteins . In brain, glutamic acid , in the form of stable metallic complex is then unable to detoxify neuronal ammonia which accumulation is responsible for cell death affecting every neurotransmitter system.

REFERENCES

Alafuzoff I, Adolfsson R, Grundke-Iqbal I, Windblad B.(1987) Blood-brain barrier in Alzheimer's dementia and in non demented elderly. An immunocytochemical study. Acta Neuropathol.; 73: 160-166.

Banks WA, Kastin AJ.(1989) Aluminium induced neurotoxicity. Alteration in membrane function at the blood brain barrier. Neurosci.and Biobehavioral Rev., 13: 47-53.

Brumback RA, Lapham L.(1989) DNA synthesis in Alzheimer type II astrocytosis: The question of astrocytic proliferation and mitosis in experimentally induced hepatic encephalopathy. Arch Neurol. ; 46: 845-848.

Buchner DM, Larson EB.(1987) Falls and fractures in patients with Alzheimer's type dementia. J.A.M.A.; 257(11): 1492-1495.

Delamarche C.(1989)A homologous domain between the amyloid protein of Alzheimer's disease and the neurofilaments subunits. Biochimie 71: 853-856.

Deloncle R, Guillard O, Clanet F, Courtois P, Piriou A.(1990) Aluminium transfer through blood brain barrier as glutamate complex. Possible implication in dialysis encephalopathy. Biol. Trace Elem. Res. in press.

Deutsch SI, Morihisa JM.(1988) Glutamatergic abnormalities in Alzheimer's disease and a rationale for clinical trials with L-glutamate. Clin. Neuropharmacol. 11(1): 18-35.

Glenner GG, Wong CW.(1984) Alzheimer's disease:Initial report of the purification and characterization of a novel cerebrovascular amyloid protein. Biochem. Biophys. Res. Comm. 120(3): 885-890.

Greenamyre JT, Young AB, Penney JB.(1985) A selective loss of glutamate receptors in hippocampus of Alzheimer's disease. Neurology; 35 : S 183.

Kim YS, Lee MH, Winiewski HM.(1986) Aluminium-induced reversible changes in permeability of the BBB to (14 C)-sucrose. Brain Res.; 377:286-291.

Lumb RF, Martell AE.(1953) Metal chelating tendencies of glutamic and aspartic acids. J.Phys.Chem.; 57: 690-693.

Mac Donald TL, Humphreys WG, Martin RB. (1987) Promotion of tubulin assembly by aluminium in vitro. Science; 236: 183-186.

Mori H, Kondo J, Ihara Y.(1987) Ubiquitin is a component of paired helical filaments in Alzheimer's disease. Science; 235: 1641-1644.

Mulvihill P, Perry G.(1989) Immuno affinity demonstration that paired helical filaments of Alzheimer's disease share epitopes with neuro filaments MAP 2 and Tau. Brain Res.; 484: 150-156.

Santos F, Chan JCM, Yang MS, Savory J, Wills MR.(1987) Aluminium deposition, in the central nervous system. Preferential accumulation in the hippocampus in weanling rats. Med. Biol.; 65: 53-55.

Selkoe DJ, Berman-Podlisny M, Joachim CL et al.(1988) B amyloid precursor protein of Alzheimer disease occurs as 110 to 135 kilodaltons membrane-associated proteins in neural and non neural tissues. Proc.Natl.Acad.Sci. USA; 85:7341-734.

Singh MK, Srivastava MN.(1972) Stepwise formation of beryllium and aluminium chelates with aspartic and glutamic acids. J.Inorg Nucl.Chem.; 34: 567-573.

St George-Hyslop PH, Tanzi RE, Polinski RJ et al.(1987)The genetic defect causing familial Alzheimer's disease maps on chromosome 21. Science;235: 885-890.

Timsit F, Galle P, Bourdon R, Merrer J, De Viel E, Hillion D.(1986) Cardiomyopathie aluminique chez un hémodialysé chronique. Nephrol.; 6: 263.

Trojanowski JQ, Schmidt ML, Otvos L et al.(1989) Selective expression of epitopes in multiphophorylation repeats of the high and middle molecular weight neurofilament proteins in Alzheimer neurofibrillary tangles. Ann. Med.; 21: 113-116.

Ueda K, Cole G, Sundsmo M, Katzman R, Saitoh T.(1989) Decreased adhesiveness of Alzheimer's disease fibroblasts . Is amyloid beta protein precursor involved. Annals of Neurol.; 25(3):246-251.

Zubenko GS, Teply I.(1988) Longitudinal study of platelet membrane fluidity in Alzheimer's disease. Biol. Psychiatry; 24: 918-924.

Metal Ions in Biology and Medicine. Eds. Ph. Collery, L.A. Poirier, M. Manfait, J.C. Etienne. John Libbey Eurotext, Paris © 1990, pp. 339-341

Lack of improvement of glucose homeostasis in STZ-diabetic rats after administration by gavage of meta-vanadate

A. Ortega, J.M. Llobet, J.L. Domingo, J. Corbella

Laboratory of Toxicology and Biochemistry, School of Medicine, University of Barcelona, San Lorenzo 21, 43201 Reus, Spain

Vanadium has many insulin-mimetic properties: stimulation of glucose uptake, glycogen synthesis, glycolysis, tyrosine kinase activity, insulin-like growth factor II binding, potassium uptake and initiation of protein degradation (Dubyak & Kleinzeller, 1980). However, the effects of vanadate are different from those of insulin in the stimulatory effects and in the intracellular pathways affected (Mooney *et al.*, 1989). The administration of sodium orthovanadate in the drinking water to streptozotocin(STZ)-diabetic rats reduced the glucose levels and prevented the decline in cardiac performance caused by diabetes (Heyliger *et al.*, 1985). It posed the possibility of dietary vanadate being used as a therapeutic regimen for diabetics. Since then, the effects of metavanadate (Meyerovitch *et al.*, 1987), orthovanadate (Brichard *et al.*, 1988) and vanadyl (Pederson *et al.*, 1989), compounds given in drinking water have been investigated. However, the effects of the discontinuous administration of vanadium (i.e., by gavage) have not yet been evaluated.

On the other hand, the possible toxic effects of oral vanadium administration to STZ-diabetic rats require investigation. Vanadium given to non-diabetic rats at doses lower than those quoted in previous publications has produced renal toxicity and haemopoietic alterations (Domingo *et al.*, 1985; Zaporowska & Wasilewski, 1989). The aim of the present investigation was to evaluate both the therapeutic and toxic effects of the administration of metavanadate ($NaVO_3$) by gavage to STZ-diabetic rats.

MATERIAL AND METHODS

40 male Sprague-Dawley rats, 170-190 g body weight, (Panlab, Barcelona, Spain) were housed in metabolic cages in a light and temperature controlled room, with drinking water and a stock diet *ad lib.* After eight days of acclimatization, diabetes was induced by two subcutaneous injections (given on days 0 and 3, at 12:00 h) of streptozotocin (40 mg/kg in 1 ml) in cold 0.1 M ci- trate buffer (pH 4.5). On day 5, diabetes was confirmed by measuring the plasma glucose concentration. Animals with levels $<$ 8.2 mmol/l were not included in the study.

The 31 diabetic animals were randomly divided into five groups and given by gavage 0, 2.5, 5, 10 or 20 mg/kg/day of $NaVO_3$ disolved in distilled water. The daily dose was divided into two administrations, given at 9:00 and

19:00 hours. Solution concentrations were determined so that a 250-g rat would receive 1 ml. The drinking water of all animals was substituted for solutions of NaCl (80 mM). Both solutions were freshly prepared twice a week. The choice of metavanadate doses was made according to previous results (Domingo *et al.*, 1990), which indicated the efficacy of continuous metavanadate therapy (0.15 ml $NaVO_3$/ml drinking water) in reducing hyperglycemia.

Body weight, food and fluid intake and general clinical changes were monitored daily. Blood glucose levels were measured weekly. Blood was sampled from fed animals at 10:00-11:00 h by retroorbital bleeding. At the end of the ad- ministration period, the plasma levels of glucose, urea, creatinine, gluta- mic-oxaloacetic transaminase (AST), and glutamic-pyruvic transaminase (ALT) were determined. The animals were killed by overexposure to ether and the weight of kidney, liver, heart and spleen measured. Vanadium concentrations were measured in these organs, as well as in pancreas, bone (femur) and muscle (gastrocnemius) by inductively coupled plasma spectrometry, after digestion according to the methods previously described (Ortega *et al.*, 1989).

Table 1. Effects of the administration of $NaVO_3$ by gavage to STZ-diabetic rats during 28 days. Means ± SD.

	Dose (mg/kg/day)				
	0	2.5	5	10	20
% Mortality	0	50	50	43	47
% Wt. change[1]	42 ± 19	13 ± 5	30 ± 7	25 ± 24	-12 ± 9*
Mean water intake	148 ± 65	115 ± 75	132 ± 76	122 ± 79	154 ± 95
Mean food intake	27 ± 9	26 ± 12	28 ± 10	23 ± 8	24 ± 12
Chemistry analyses (day 28)					
Glucose (mmol/l)	**25.0 ± 3.7**	**27.2 ± 6.1**	**23.4 ± 4.7**	**23.9 ± 4.6**	**18.6 ± 6.5**
Urea (mmol/l)	7.92 ± .50	6.90 ± .47*	6.12 ± .10	5.91 ± .68**	6.94 ± 1.5
Creatinine (μmol/l)	66 ± 9	48 ± 18	46 ± 22	54 ± 12	64 ± 4
Relative organ weight[2]					
Kidney	1.00 ± .11	1.24 ± .11*	1.21 ± .05*	1.13 ± .01	1.19 ± .09*
Vanadium concentrations[3]					
Liver	68(1.7)	117(1.1)	419(1.2)**	389(2.5)*	1963(1.8)***
Kidney	94(2.5)	372(1.2)	964(1.3)**	1611(1.4)**	5902(1.6)***
Heart	86(1.7)	80(1.2)	128(1.2)	222(1.1)	635(1.4)***
Spleen	188(1.1)	361(1.1)**	644(1.2)***	1028(1.3)***	2269(1.1)***
Pancreas	265(2.0)	151(1.3)	207(1.4)	255(1.3)	615(1.3)
Bone	115(2.2)	632(2.0)	178(3.3)	1923(1.2)*	4266(1.3)***
Muscle	82(1.5)	55(1.4)	68(1.1)	138(1.2)	281(1.3)**

*P<0.05, **P<0.01, and ***P<0.001, with respect to control. [1]((Weight on d 28 - Weight on d 0) / Weight on d 0)*100. [2]Percent body weight. [3]Geometric mean(geometric standard deviation). The statistical tests were made on the logarithms of the original values.

The data were analyzed using the Bartlett's test for homogeneity of variances, followed either by an ANOVA or a Kruskall-Wallis test. Significant differences between control and treated groups were assessed using the Student's *t* test or the rank sum method (Hayes, 1989).

RESULTS AND DISCUSSION

The glucose levels on day 0 of STZ-diabetic animals were 16.8 ± 4.4 mmols/l, and their weights were 195 ± 26 g. The discontinuous administration of metavanadate to STZ-treated rats did not produce significant reductions of glucose levels in the weekly determinations. Moreover, body weight changes diverge in control rats and in the highest dose group, although water and food intake were not significantly altered. Approximately half of the rats died during the dosage period. In addition, the dose-related decrease in urea as well as the decrease in the relative kidney weights suggest a renal injury, although creatinine levels were not significantly altered. AST, ALT and the relative weights of heart, spleen or liver did not change (data not shown). The accumulation of vanadium in organs and tissues was dose-related and specially prominent in liver, kidney, spleen and bone, as previously described (Domingo *et al.*, 1985).

In conclusion, the discontinuous administration of sodium metavanadate by gavage twice a day to STZ-diabetic rats was not effective in normalising glucose levels, but produced significant toxic effects in the experimental animals, including deaths. This may preclude the use of oral vanadium as a treatment for diabetes.

ACKNOWLEDGEMENTS
We thank Mr. Miguel A. Sánchez-Corral and Mr. Joan Cid for valuable technical assistance.

REFERENCES

Brichard, S.M., Okitolonda, W. & Henquin, J.C. (1988): Long term improvement of glucose homeostasis by vanadate treatment in diabetic rats. *Endocrinology* 123, 2048-2053.
Domingo, J.L., Llobet, J.M., Gómez, M., Corbella, J. & Keen, C.L. (1990): Effects of oral vanadium administration in streptozotocin-diabetic rats. *First International Symposium on Metal Ions in Biology and Medicine* Reims.
Domingo, J.L., Llobet, J.M., Tomás, J.M. & Corbella, J. (1985): Short-term toxicity studies of vanadium in rats. *J. Appl. Toxicol.* 5, 418-421.
Dubyak, G.R. & Kleinzeller, A. (1980): The insulin-mimetic effects of vanadate in isolated rat adipocytes. Dissociation from effects of vanadate as a (Na^+-K^+)ATP-ase inhibitor. *J. Biol. Chem.* 255, 5306-5312.
Gad, S.C. & Weil, C.S. (1989): Statistics for Toxicologists. In *Principles and Methods of Toxicology*, ed. A.H. Hayes, 2nd ed. pp. 435-485. New York, Raven Press.
Heyliger, C.E., Tahiliani, A.G. & McNeill, J.H.(1985): Effect of vanadate on elevated glucose and depressed cardiac performance of diabetic rats. *Science* 227, 1474-1477.
Meyerovitch, J., Farfel, Z., Sack, J. & Shetcher, Y. (1987): Oral administration of vanadate normalizes blood glucose levels in streptozocin-treated rats. *J. Biol. Chem.* 262, 6658-6662.
Mooney, R.A., Bordwell, K.L., Luhowskyj, S. & Casnell, J.E. (1989): The insulin-like effect of vanadate on lipolysis in rat adipocytes is not accompnied by an insulin-like effect on tyrosine phosphorylation. *Endocrinology* 122, 2285-2289.
Ortega, A., Domingo, J.L., Gómez, M. & Corbella, J. (1989): Treatment of experimental acute uranium poisoning by chelating agents. *Pharmacol. Toxicol.* 64, 247-251.
Pederson, R.A., Ramanadham, S., Buchan, A.M.J. & McNeill, H. (1989): Long-term effects of vanadyl treatment on streptozocin-induced diabetis in rats. *Diabetes* 38, 1390-1395.
Zaporowska, H. & Wasilewski, W. (1989): Some selected peripheral blood and haemopoietic system indices in Wistar rats with chronic vanadium intoxication. *Comp. Biochem. Physiol.* 93C, 175-180.

VII. Antidotes

Metal Ions in Biology and Medicine. Eds. Ph. Collery, L.A. Poirier, M. Manfait, J.C. Etienne. John Libbey Eurotext, Paris © 1990, pp. 345-347

Prevention by meso-2,3-dimercaptosuccinic acid (DMSA) of sodium arsenite-induced embryotoxic effects in mice

M.A. Bosque, J.L. Paternain, J.L. Domingo, J.M. Llobet, J. Corbella

Laboratory of Toxicology and Biochemistry, School of Medicine, University of Barcelona, San Lorenzo 21, 43201 Reus, Spain

Arsenic, the twentieth most abundant element of the Earth's crust, is ubiquitous. It is a natural constituent of food, although additional exposure may occur from sources such as herbicides, insecticides, rodenticides, paint pigments, wood preservatives and from the wastes derived from the production of several metals and as a by-product of the uses of fossil fuels (Baxley et al., 1981).

It is known from experimental studies that arsenic in the form of arsenite or arsenate is teratogenic in chicks, golden hamsters, mice and rats (Hood, 1972; Baxley et al., 1981). In previous studies, the effect of chelating agent 2,3-dimercapto-propanol (BAL) on the embryotoxicity and teratogenicity of sodium arsenite was investigated in mice. Treatment with this compound was not effective when exposure to arsenite occurred before BAL was given. Therefore, BAL is unlikely to have a practical beneficial effect on the arsenite exposed conceptus, because it must be administered prior to the teratogen (or perhaps simultaneously with it) to be effective (Hood & Vedel, 1984).

In the present study, meso-2,3-dimercaptosuccinic acid (DMSA) was employed to determine if could prevent arsenite-induced embryotoxic and teratogenic effects. DMSA -a water soluble compound analogous to BAL- has been shown as a potentially useful drug for the treatment of experimental and human poisoning by a number of heavy metals such as lead and mercury (Aposhian, 1983). DMSA has also been effective in the protection of mice against the lethal effects of sodium arsenite (Aposhian et al., 1981), in mobilizing tissue arsenic, and in inducing arsenic excretion (Graziano, 1986; Aposhian & Aposhian, 1989). Moreover, DMSA itself does not possess teratogenic or embriotoxic effects given sc to mice on gestation days 6-15 at the dose levels administered in this study (Domingo et al., 1988).

MATERIALS AND METHODS

On day 10 of gestation, four groups of pregnant Swiss mice received an ip injection of aqueous sodium arsenite ($NaAsO_2$): 12 mg/kg in a volume of 10 ml/kg body weight (Hood, 1972; Hood & Vedel, 1984). Day 10 treatment was expected to result in a maximun level of malformed fetuses, and in a high level of prenatal mortality (Hood, 1972). Subsequently, the animals in the experimental groups received DMSA by sc injection. A series of four injections of DMSA were administered inmediately after $NaAsO_2$ treatment, and at 24, 48 and 72 hr thereafter. DMSA at doses of 0, 80, 160 and 320 mg/kg were dissolved in $NaHCO_3$ solutions and administered at a pH of approximately 7. The animals in the negative control group were pretreated with distilled water, and then posttreated with sodium bicarbonate solutions. Mice were killed on day 18 of pregnancy and examined for anomalies. Live fetuses were observed for gross external defects and weighed. Two-thirds of the available fetuses were cleared and stained with Alizarin red S for their skeletal examinations. Fetuses were not examined for visceral anomalies, since previous results indicated their frequency would be low (Baxley et al., 1981; Hood & Vedel, 1984).

Fetal weight data were analyzed by ANOVA followed by a student-Newman-Keuls multiple range test, and percentage data were compared by the rank sum method of Wilcoxon and Wilcox.

RESULTS AND DISCUSSION

Treatment with sodium arsenite on day 10 of gestation resulted in a high rate of resorptions and dead fetuses (Table 1). Gross and skeletal malformations, predominately exophthalmos, open eyes, anomalous vertebrae, and delayed ossification of parietals and jawbones were also observed following Day 10 treatment with arsenite alone, but there was no significant effect on fetal weight (Table 1). These results agree with those previously reported by Hood (1972) and Hood & Vedel (1984).

When DMSA was given after arsenite administration, it had significant protective effect, allowing enhanced survival of the conceptus. With regard to skeletal malformations, a significant protective effect was also associated with DMSA treatment. No malformed skeletons were observed at 320 mg DMSA/kg. However, although the number of grossly malformed fetuses was significantly decreeased at 160 and 320 mg DMSA/kg, no significant effects were noted at 80 mg/kg.

Relating to the detoxication mechanism of DMSA, it is probable that the drug acts as a chelator of arsenic to obviate its embryotoxic and teratogenic effects. Moreover, DMSA probably enhances the excretion of arsenic into the urine (Graziano, 1986). In contrast to BAL, DMSA offers encouragement with regard to its therapeutic potential for pregnant women exposed to trivalent arsenic.

ACKNOWLEDGMENTS

This work was supported by DGICYT, Spain through Project PB87-0020

REFERENCES

Aposhian, H.V., Tadlock, C.H. & Moon, T.E. (1981): Protection of mice against the lethal effects of sodium arsenite. A quantitative comparison of a number of chelating agents. Toxicol. Appl. Pharmacol. 61: 385-392.

Aposhian, H.V. (1983): DMSA and DMPS-water soluble antidotes for heavy metal poisoning. Annu. Rev. Pharmacol. Toxicol. 23: 193- 215.

Aposhian, H.V. & Apshian, M.M. (1989): Newer developments in arsenic toxicity. J. Amer. Coll. Toxicol. 8: 1297-1306.

Baxley, M.N., Hood, R.D., Vedel, G.C., Harrison, W.P. & Szczech, G.M. (1981): Prenatal toxicity of orally administered sodium arsenite in mice. Bull Environ. Contam. Toxicol. 26: 749-756.

Domingo, J.L., Paternain, J.L., Llobet, J.M. & Corbella, J. (1988): Developmental toxicity of subcutaneously administered meso-2,3-dimercaptosuccinic acid in mice. Fundam. Appl. Toxicol. 11: 715-722.

Graziano, J.H. (1986): Role of 2,3-dimercaptosuccinic acid in the treatment of heavy metal poisoning. Med.Toxicol. 1: 155-162.

Hood, R.D. (1972): Effects of sodium arsenite on fetal development. Bull. Environ. Contam. Toxicol. 7: 216-222.

Hood, R.D. & Vedel G.C. (1984): Evaluation of the effect of BAL (2,3-Dimercapto-propanol) on arsenite-induced teratogenesis in mice. Toxicol. Appl. Pharmacol. 73: 1-7.

Table 1. Effects of DMSA on arsenite-induced embriotoxic and teratogenic effects in mice ($x \pm SD$)

Treatment	0.9% saline + $NaHCO_3$ solution	$NaAsO_2$ 12 mg/kg	$NaAsO_2$ 12 mg/kg + DMSA 80 mg/kg	$NaAsO_2$ 12 mg/kg + DMSA 160 mg/kg	$NaAsO_2$ 12 mg/kg + DMSA 320 mg/kg
Litters (N)	12	12 (5)[c]	12	15	12
Implants per litter	12.7±1.8	12.3±3.2	11.6±1.8	11.5±3.3	12.3±1.5
Dead or resorbed fetuses per litter	0.3±0.2[2]	7.3±5.4[d]	0.8±0.4[2]	0.9±0.6[2]	2.0±0.9[1]
Live fetuses per litter	12.3±1.4[2]	5.0±4.1[d]	10.8±2.1[1]	10.6±2.8[1]	10.3±2.2[1]
Mean fetal weight (g)	1.3±0.1	1.2±0.0	1.3±0.1	1.3±0.1	1.3±0.1
Grossly malformed					
Nº litters affected	1[3]	7[e]	11[e]	13[e]	12[e]
% fetuses affected[a]	0.3[3]	54[e]	45.1[e]	26.8[e,3]	34,1[e,3]
Malformed skeletons					
Nº litters affected	1[3]	5[e]	4[1]	4[1]	0[3]
% fetuses affected[b]	0.4	56.7[e]	10.0[1]	6.1[1]	0.0[3]

[a]Malformed fetuses as a portion of total live fetuses.
[b]Fetuses with malformed skeletons as a proportion of those examined.
[c]Number in parentheses indicates death of dams.
[d,e]Significantly different from untreated controls; $p < 0.05$, $p < 0.01$.
[1,2,3]Significantly different from litters exposed to sodium arsenite controls; $p < 0.05$, $p < 0.01$, $p < 0.001$ respectively.

Metal Ions in Biology and Medicine. Eds. Ph. Collery, L.A. Poirier, M. Manfait, J.C. Etienne. John Libbey Eurotext, Paris © 1990, pp. 348-350

The use of desferrioxamine and other chelating agents in the treatment of aluminium overload in rats

M. Gomez, J.L. Domingo, J.M. Llobet, J. Corbella

Laboratory of Toxicology and Biochemistry, School of Medicine, University of Barcelona, San Lorenzo 21, 43201 Reus, Spain

Aluminium toxicity has been associated with the use of aluminium-contaminated dialysate and aluminium - containing phosphate binders to control hyperphosphatemia in patients receiving chronic haemodialysis (Alfrey et al.,1976). Therefore, aluminium removal is essential in treating patients with aluminium accumulation. Currently, the most effective method has been chelation of aluminium with desferrioxamine (DFOA). The side-effects of DFOA therapy have included hypotension, abdominal pain, gastrointestinal disturbances, auditory neurotoxicity, ocular toxicity, thrombocytopenia, and an increased potential for septicemia (Ackrill & Day, 1985; Olivieri et al., 1986). Due to the complications of DFOA treatment during aluminium chelation, in previous studies we have investigated the efficacy of other therapeutic chelators. Citric, malic and succinic acids were effective in the prevention of acute aluminium intoxication (Llobet et al., 1987; Domingo et al., 1987).

On the other hand, recent studies have demonstrated that ethylenediamine-di-(o-hydroxyphenylacetic acid) (EDDHA) and picolinic acid may also have potential as alternatives to DFOA. The purpose of the present investigation was to compare the effectiveness of repeated i.p. administration of DFOA, EDDHA, citric, malic, succinic and picolinic acids on the distribution and excretion of aluminium after repeated intraperitoneal administration of the metal to rats.

MATERIAL AND METHODS

* Chemicals and animals: Aluminium nitrate nonahydrate and citric acid were obtained from E. Merck (Darmstadt, FRG), desferrioxamine mesylate (DFOA) from Ciba-Geigy (Barcelona, Spain), EDDHA, malic and succinic acids from Sigma Chemical Co. (St Louis, MO, USA). Male Wistar rats (Interfauna Ibérica, Barcelona, Spain) weighing 180-200g were used.

* Distribution and excretion studies: Aluminium as aluminium nitrate nonahydrate was administered i.p. at a daily dose of 51.5 mg/kg (five days/week) for a period of five weeks. The i.p. LD_{50} of aluminium nitrate nonahydrate in rats was 901 mg/kg (Llobet et al.,1987).

One day after the last injection, parenteral therapy with the various chelators or 0.9% saline (control group) was initiated and continued on alternate days (three days/week) for two weeks of treatment. The dose of the chelators was approximately one-fourth of their respective i.p. LD_{50} values. Animals were housed in plastic metabolic cages and urine and faeces collected on days 1, 7 and 14 of treatment.

At the end of the experiment, the rats were sacrificed. The following organs and tissues were removed: liver, spleen, kidneys, brain and bone. The amounts of aluminium collected in the faeces and urine as well as the concentration of aluminium in tissues were determined by atomic absorption spectrophotometry as described previously (Domingo et al., 1988).

RESULTS AND DISCUSSION

The excretion of aluminium into faeces shows that this was the major route of elimination for aluminium in control animals which agrees with our previous studies (Domingo et al., 1988). During the first day of therapy, a significant increase in urinary elimination of aluminium was found in rats treated with DFOA, whereas citric and malic acids significantly increased the faecal elimination of the metal. During successive days (days 7 and 14), DFOA and citric acid sinificantly increased the urinary excretion of aluminium. Only succinic acid significantly increased the amount of aluminium excreted into faeces on days 7 and 14 of treatment (Table 1).

Table 2 summarizes the effect of chelators on the distribution of aluminium in various tissues. The most effective agents for lowering tissue aluminium consistently were DFOA and citric acid, which significantly decreased tissue aluminium concentrations in four of the five tissues studied.

The results of the present investigation show that DFOA and citric acid were the most effective agents of the six chelators tested for decreasing the toxicity of aluminium. DFOA and citric acid significantly increased the urinary aluminium excretion and reduced the concentration of the metal in spleen, kidneys, brain and bone. In the light of these results, citric acid appears to be an alternative to the therapeutic use of DFOA in aluminium toxicity.

REFERENCES

Ackrill, P. & Day, J.P. (1985): Desferrioxamine in the treatment of aluminium overload. Clin. Nephrol. 24: S94-S97.

Alfrey, A.C., LeGendre, G.R. & Kaehny, W.D. (1976): The dialysis encephalopathy syndrome: Possible aluminium intoxication. N. Engl. J. Med. 294: 184-188.

Domingo, J.L., Gómez, M., Llobet, J.M. & Corbella, J. (1988): Comparative effects of several chelating agents on the toxicity, distribution and excretion of aluminium. Human Toxicol. 7: 259-262.

Llobet, J.M., Domingo, J.L. Gómez, M., Tomás, J.M. & Corbella, J. (1987): Acute toxicity studies of aluminium compounds: antidotal efficacy of several chelating agents. Pharmacol. Toxicol. 60: 280-283.

Olivieri, N.F., Buncic, J.R., Chew, E., Gallant, T., Harrison, R.V., Keenam, N. Logan, W., Mitchell, D., Ricci, G., Skarf, B., Taylor, M. & Freedmasn, M.H. (1986): Visual and auditory neurotoxicity in patients receiving subcutaneous deferoxamine infusions. N. Engl J. Med. 314: 869-873.

Table 1. The effects of the chelators on the urinary and faecal excretion of aluminium (µg/kg).

	URINE			FAECES		
	Day 1	Day 7	Day 14	Day 1	Day 7	Day 14
Control	370±23	180±22	180±19	620±20	730±54	430±39
DFOA	1500±200**	700±30***	430±25***	613±23	1200±189	450±51
Citric acid	520±130	550±51**	408±30**	988±40*	1170±93	473±69
Malic acid	400±70	270±32	310±46	973±37*	810±72	392±21
Succinic acid	480±42	363±29	523±69**	809±49	980±102	600±32*
EDDHA	420±68	190±17	283±32	837±63	1150±370	573±43
Picolinic acid	300±23	180±17	176±18	800±42	520±62	290±17

Results are presented as arithmetic means in each group ± S.E.
*p<0.05, **p<0.01, ***p<0.001.

Table 2. The effect of chelators on the concentration of aluminium in various tissues

	Liver	Spleen	Kidneys	Brain	Bone
Control	39.9 3.9	95.5±5.9	14.6±0.8	1.3±0.2	78.8±4.0
DFOA	33.5±4.6	51.5±7.6***	8.5±1.4**	0.5±0.2*	47.6±4.1***
Citric acid	40.9±3.3	64.8±8.4*	9.3±0.9**	0.6±0.2*	58.8±1.5**
Malic acid	44.1±6.0	72.2±11.3	11.2±0.6**	2.3±0.6	81.3±7.6
Succinic acid	31.7±2.9	76.3±3.6*	9.3±1.4**	2.6±1.3	78.1±4.6
EDDHA	49.0±6.8	88.6±6.1	8.0±1.0***	2.3±0.6	82.1±3.7
Picolinic acid	31.3±1.0	77.7±5.4	7.4±0.7***	2.1±1.1	63.0±4.3*

Results are presented as arithmetic means in each group ± S.E.
Concentrations of aluminium are expressed in µg per g of tissue wet weight.
Control rats received 0.9% saline instead of chelators.
*p<0.05, **p<0.01, ***p<0.001.

C

THERAPEUTIC METAL IONS : GALLIUM, GERMANIUM, GOLD, IRIDIUM, LANTHANUM, LITHIUM, MOLYBDENUM, NIOBIUM, PALLADIUM, PLATINUM, RUTHENIUM, SELENIUM, TIN, TITANIUM

I. Platinum

Metal Ions in Biology and Medicine. Eds. Ph. Collery, L.A. Poirier, M. Manfait, J.C. Etienne. John Libbey Eurotext, Paris © 1990, pp. 357-359

Synthesis, molecular structure determination and antitumor activity of Platinum (II) and Palladium (II) complexes with, carrier molecules, derivatives of 3-(2-aminoethylthio)-propionic acide peptides

S. Mylonas[1], S. Caranikas[1], M. Polissiou[2], A. Hatzigiannakou[1], A. Tsiftsoglou[3], I. Christou[3]

[1] *Laboratory of Organic Chemistry, University of Athens, Athens, Greece;* [2] *Laboratory of General Chemistry, Agricultural University of Athens, Athens, Greece;* [3] *Laboratory of Pharmacology, Dpt of Pharmaceutical Sciences, Aristotle University of Thessaloniki, Thessaloniki, Greece*

Despite the widespread use of cis platinum as an anticancer drug there is still scope for improvement (Burchenal et al,1979).In the present work we synthesized a series of cis-platinum analogues, using peptides of 3-(2-aminoethylthio)-propionic acid (AETP) with D- and L-amino esters, as lipophilic carriers. The reason for the use of these peptides as carriers was the strong antitumor activity of the S-2-aminoethyl-L-cystein and S-2-aminoethyl-D, L-penicillamine complexes with Pt(II) and Pd(II) as it's cited in our previous works (Mylonas et al,1982;1983).

We used esters not oly of L-amino acids but also D-amino acids hoping that with increasing of metabolic stability we would have better results. Proteinases generaly distinguish enantiomers: a susceptible peptide bond is often rendered resistant to attack by substitution to one or both amino acid residues, adjoining the susceptible peptide bond by the D-isomer (Witter,1983). For the synthesis of these derivatives of 3-(2-aminoethylthio)-propionic acid (AETP) enough sensitive molecule, as a result of the presence of thio atoms, we chose like protective groups: i) the t-butoxycarbonyl group $(CH_3)_3COCO-$ (BOC) (Moroder et al,1976) and ii) the triphenyl methyl group $(C_6H_5)_3C-$ (Trt) (Zervas and Theothoropoulos, 1956; Barlos et al,1982). The procedure that we have followed for the coupling of the N-BOC-AETP or N-Trt-AETP derivatives with the esters of D- and L- amino acids was mainly the Konig and Geiger's method (1970). The N-BOC group was cleaved with HCl/CH_3OH (Anderson and Mc Gregor,1957) and the N-Trt group with HCl/Et_2O follwed by solvolysis. THe synthesis of complexes compounds was performed by the procedure described previously (Mylonas et al,1982).

The complexes and the free ligands were studied by IR and ^1H NMR spectroscopies and the results in conbination with elemental analysis, have been used to determine the structure of the complexes commpounds. The main point of this analysis was that the ligands behave as bidentates being bound to the metal [Pt (II), Pd (II)] via the sulphur atom and the amino group for each dipeptide. The two chlorine atoms were found at cis-position (Mylonas et al,1982).

The antitumor activity of metallocompounds was determined in cultures K562 (human Leukemia), MEL (mouse Leukemia) and TE-671 (human meduloblastoma). Cis-platin was used as positive control throughout these studies (Table 1).

Table 1. Anti-proliferative effect of amino-acid derivative-complexes with Pt (II), Pd (II) and cis-platin on cultured murine erythroleukemia cells.

DRUG	M = Pt (II)		M = Pd (II)	
	CONC (M)	CELL GROWTH (% OF CONTROL)	CONC (M)	CELL GROWTH (% OF CONTROL)
None		100.00		
cis-Platin	10^{-6}	37.03		
$[(NH_3)_2PtCl_2]$	10^{-5}	6.83		
	10^{-4}	N.D.		
$[AETP\text{-}D\text{-}Val\text{-}OMe]MCl_2$	10^{-6}	90.80	10^{-6}	98.73
	10^{-5}	91.95	10^{-5}	93.67
	10^{-4}	80.45	10^{-4}	56.96
$[AETP\text{-}Gly\text{-}OMe]MCl_2$	10^{-6}	96.77	10^{-6}	98.73
	10^{-5}	92.74	10^{-5}	88.60
	10^{-4}	91.12	10^{-4}	46.83
$[AETP\text{-}L\text{-}Ala\text{-}OMe]MCl_2$	10^{-6}	100.00	10^{-6}	100.00
	10^{-5}	89.65	10^{-5}	93.67
	10^{-4}	67.81	10^{-4}	48.10
$[AETP\text{-}D\text{-}Ala\text{-}OMe]MCl_2$	10^{-6}	94.33	10^{-6}	98.11
	10^{-5}	94.33	10^{-5}	98.11
	10^{-4}	100.00	10^{-4}	94.33
$[AETP\text{-}D\text{-}Leu\text{-}OMe]MCl_2$	10^{-6}	100.00	10^{-6}	100.00
	10^{-5}	87.80	10^{-5}	92.68
	10^{-4}	60.97	10^{-4}	60.97
$[AETP\text{-}L\text{-}Leu\text{-}OMe]MCl_2$	10^{-6}	97.56		
	10^{-5}	80.48		
	10^{-4}	41.46		

The conclusion from these studies are:

1) The complexes of Pd (II) are more strong antiproliferative agents than those of the Pt (II).

2) It isn't clear that the derivatives with D-stereroisomers are more drastic than those of L- in cultures of mouse (MEL).

3) Some complexes gave antineoplastic activity (50% were inhibited the cell growth in relatively high cocentrations (10^{-5} - 10^{-4}). Comparatively the complexes of Pd (II) and Pt (II) are cleary less effective in the face of 100-1000 times from the cis-platin.

ACKNOWLEDGEMENTS

This research was supported by grant from the Ministry of Industry, Energy and Technology, General Secretariat Of Research and Technology.

REFERENCES

Aderson, G. W. and Mc Gregor, A. C. (1957): t-Butyloxycarbonylamino acids and their use in peptide synthesis. J. Amer. Chem. Soc. 79: 6180-6184.

Barlos, K.; Papaioannou, D. and Theothoropoulos, D. (1982): Efficient "on pot" synthesis of N-trityl amino acids. J. Org. Chem. 47: 1324-1326.

Burchenal, J. H.; Kalaher, K.; Dew, K. and Lokys, L. (1979): Rationale for development of platinum analogs. Canc. Treat. Res. 63: 1493-1498.

Konig, W. and Geiger, R. (1970): Eine neue method zur synthese von peptiden: Aktivierung der carboxylgruppe mit dicyclohexylcarbodiimid unter zusatz von 1-hydroxybenzotriazolen. Chem. Ber. 103: 788-798.

Moroder, L.; Hallet, A.; Wunsch, E.; Keller O. and Wersin, G. (1976): Di-tert-butyldicarbonat-ein vorteilhaftes reagenz zur einfuhrung der tert-butyloxycarbonyl - schutzgruppe. Z. Physiol. Chem. 357: 1651-1653.

Mylonas, S.; Valavanidis, A.; Voucouvalidis, V. and Polissiou, M. (1982): Synthesis of S-2-Aminoethyl-L-Cysteine and S-2Aminoethyl-D,L-Penicillamine complexes with Pt (II) and Pd (II). Interpretation of IR and ^{1}H NMR spectra conformational implications. Inorg. Chim. Acta 66: 25-28.

Mylonas, S.; Valavanidis, A.; Voucouvalidis, V. and Polissiou, M. (1983): Antitumor activity of Platinum (II) and Paladium (II) complexes of S-2-Aminoethyl-L-Cysteine and S-2-Aminoethyl-D,L-Penicillamine. Chim.Chron. New Series 12: 165-168.

Witter, A. (1983): Stereoselectivity in peptides. InStereochemistry and Biological Activity of Drugs. Eds E. Ariens, W. Soudijn and P.B.M.W.M Timmermans, pp. 151-159. Oxford: Blackwell Scientific Publications.

Zervas, L. and Theothoropoulos, D. (1956): N-Tritylamino acids and peptides: A new method of peptide synthesis. J. Amer. Chem. Soc. 78: 1359-1363.

Metal Ions in Biology and Medicine. Eds. Ph. Collery, L.A. Poirier, M. Manfait, J.C. Etienne. John Libbey Eurotext, Paris © 1990, pp. 360-362

Cis-Platinum-Guanosine 3',5' cyclic monophosphate complexes. Conformational studies related to cancer

M. Polissiou[1], T. Theophanides[2]

[1] *Agricultural University of Athens, Lab. General Chemistry, Iera Odos 75, 118 55 Athens, Greece;* [2] *National Technical University of Athens, Lab. Physical Chemistry, Group of Radiation Chemistry, 157 73 Zografou Athens, Greece*

Since the work of Derubertis and Zenser (1976) it is well known that guanosine 3',5'-cyclic monophosphate ($cG^{3'5'}p$) stimulates nucleic synthesis in lymphocytes. They have also demostrated that 8-bromo (Br) nucleotides are more potent mitogens than $cG^{3'5'}p$ itself. By contrast, adenosine 3',5'-cyclic monophosphate ($cA^{3'5'}p$) and its 8-bromo derivative are antagonistic in their influence on lymphocyte mitogenesis. On the other hand, increases in $cG^{3'5'}p$ levels have been found after a sigle application of active tumor promoting agent, such as phorbol mirystate acetate (PMA), to mouse skin (Belman et al., 1978). The antagonistic role of $cA^{3'5'}p$ was confirmed by Curtis et al (1974) and Perchollet and Boutwell (1981), who found inhibiton of carcinogenesis by injection of $cA^{3'5'}p$ simultaneously with promoter treatment or topical application.

In order to clarify the differences in the mitogenic effect of $cG^{3'5'}p$ and its derivatives the syn <--> anti conformational changes upon substitution at C_8 by bromine or metalation at N_7 by cis-platinum (cis-DDP) and by hydrated magnesium ion ($Mg(ClO_4)_2).6H_2O$ have been ivestigated by NMR and FT-IR spectroscopies.

Experimental

Guanosine 3',5'-cyclic monophosphate ($cG^{3'5'}p$) sodium salt and 8-bromo guanosine 3',5'-cyclic monophosphate ($Br^8cG^{3'5'}p$) sodium salt were purchased from Sigma Chemical Co. The cis-$Pt(NH_3)_2Cl_2$ was obtained from Engelhard and the $Mg(ClO_4)_2$ from Merk Chemical Co. The method of preparation of the complexes was previously described (Polissiou et al,1985). The NMR spectra were recorded with a Brucker HW-400 spectrometres and the FT-IR spectra on a Digilab FTS-15C/D Fourier Transform infrared interferometer located both at University of Montreal. The chemical shifts are referenced relative to an internal standard sodium 2,2-dimethyl-2-silapentane-5-suffonate (DSS) in D_2O solutions. The samples, in aqueous solution (0.02M), for FT-IR spectra, were freeze-dried before use and KCl pellets were prepared.

Results and Discussion

The interaction of the metal ion with the N_7 site of guanine doesn't seem to affect C_3-endo for $cG^{3'5'}p$ which is charateristic of the cyclic nucleotides (Chwang and Sundaralingam,1974). However, the

Table 1. Proton NMR chemical shifts (δ in ppm) of guanosine 3',5'-cyclic monophosphate sodium salt ($cG^{3'5'}p$) with its derivatives in D_2O solution (0.02M).

compound	H_8	$H_{1'}$	$H_{2'}$	$H_{3'}$	$H_{4'}$	$H_{5'}$	$H_{5''}$
$cG^{3'5'}p$	7.876	5.969	4.760	5.007	4.238	4.485	4.293
Mg^{2+}-$cG^{3'5'}p$	7.883	5.970	4.758	4.996	4.241	4.489	4.295
Δδ*	0.007	0.003	-0.002	-0.011	0.003	0.004	0.002
cis-[Pt^7 $(cG^{3',5'}p)_2]^{2+}$	8.334	5.955	4.662	4.773	4.263	4.508	4.334
Δδ*	0.458	-0.014	0.002	-0.234	0.025	0.023	0.041
$Br^8cG^{3',5'}p$	-	5.993	4.982	5.280	4.188	4.459	4.267
Δδ*		0.024	0.222	0.273	-0.050	-0.026	-0.026

*Δδ is the difference between the sodium salt of $cG^{3',5'}p$ its N_7 metalated or C_8 brominated derivatives.

conformational changes that takes place here is the syn <--> anti equilibrium when the purine base is turned around the glycosyl $C_{1'}$-N_9 bond. Upon substitution of the H_8 proton of guanine by a heavy lagre atom, such as bromine, the equilibrium syn <--> anti is turned into syn almost completely. This is indicated by the lagre change of the chemical shift of the $H_{3'}$ sugar proton together with the change of the $H_{2'}$ sugar proton, while for cis-[platinum^7 $(cG^{3'5'}p)_2]^{2+}$ complexes only the $H_{3'}$ sugar proton shows an upfield shift upon complexation, whereas the $H_{2'}$ sugar proton remains unchanged (Table 1). Furthermore, the FT-IR spectra in the region 800-870 cm^{-1} show characteristic bands for $C_{3'}$-endo, syn (838-869 cm^{-1}) and for $C_{3'}$-endo, anti (807-810 cm^{-1}), (Fig.1), (Polissiou and Theophanides,1987). These bands may be considered as signature bands of the above conformations for guanosine 3',5'-cyclic monophosphate (Nishimura and Tsuboi,1986).

Conclusion

It is clear from this study that syn and anti conformations must be determining factors for the biological activity of $cG^{3',5'}p$ and its derivatives. Thus, the more potent mitogenic effect of 8-bromo derivatives is probably due to its exclusive syn conformation. By contrast, in the case of the $cG^{3',5'}p$ and its complexes with the metals platinum (II) and magnesium (II) ions, well known for their anticarcinogenic effect, the presence of the anti conformation is shown.

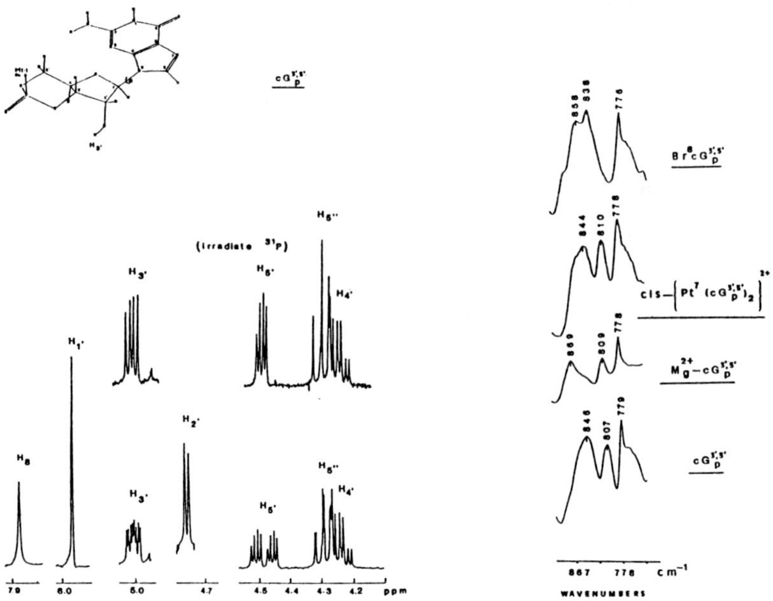

Fig.1. Molecular structure, 400 MHz ^{1}H NMR spectra and part of FT-IR spectra of $cG^{3',5'}p$ with its derivatives.

REFERENCES

Belman, S.; Troll, W.; Garte, S.J. (1978): Effect of phorbol myristate acetate on cyclic nucleotide levels in mouse epidermis. Cancer. Res. 38: 2978-2982.

Chwang, A.K. and Sundaralingam, M. (1974): The crystal and molecular structure of guanosine 3',5'-cyclic monophosphate (cyclic GMP) sodium tetrahydrate. Acta Cryst. B30: 1233-1240.

Curtis, G.L.; Stenbanck, F.; Ryan, L. (1974): Inhibition of skin tumor formation with adenosine 3',5'-cyclic monophosphate in initiation-promotion carcinogenesis. Proc. Am. Assos. Cancer Res. 15: 61.

Derubetis, R.F. and Zenser, T. (1976): Activation of murine lymphocytes by cyclic guanosine 3',5'-monophosphate:specificity and role in mitogen action. Biochim. Biophys. Acta. 428: 91-103.

Nishimura, Y. and Tsuboi, M. (1986): Local conformation and polymorphism of DNA duplexes. In spectroscopy of Biological Systems. Eds R.J.H. Clark and R.E. Ester pp 177-232. John Wiley & Son Ltd.

Perchollet, J.P.; Boutwell R.K. (1981): Effects of 3-isobutyl-1-methylxanthine and cyclic nucleotides on th biochemical processes linked to skin tumor promotion by 12-O-tetradecanoylphorbol-13-acetate. Cancer Res. 41: 3927-3935.

Polissiou, M.; Phan Viet, M.T.; St-Jacques, M. and Theophanides, T. (1985): Binding study of the drug cis-dichlorodiammine platinum (II) to $G^{5'}p$ and $dG^{5'}p$ by high resolution proton and carbon-13 NMR spectroscopy. Inorg. Chim. Acta. 107: 203-210.

Polissiou, M. and Theophanides, T. (1987): NMR and FT-IR conformational studies of 8-substituted guaniné nucleosides and nucleotides and their metal adducts and cancer. Inorg. Chim.Acta. 137: 195-201.

Metal Ions in Biology and Medicine. Eds. Ph. Collery, L.A. Poirier, M. Manfait, J.C. Etienne. John Libbey Eurotext, Paris © 1990, pp. 363-365

Enhancement of cisplatin antitumor activity by ethyl-deshydroxy-sparsomycin

H.P. Hofs[1], D.J.Th. Wagener[1], V. De Valk-Bakker[1], F.C.M. Biermans[2], H.C.J. Ottenheijm[3]

[1] *Division of Medical Oncology, Nijmegen University Hospital, P.O. Box 9101, 6500 HB Nijmegen, The Netherlands;* [2] *Department of Organic Chemistry, University of Nijmegen, Toernooiveld, Nijmegen, The Netherlands;* [3] *Organon International B.V., Oss, The Netherlands*

SUMMARY

Ethyldeshydroxy-sparsomycin (EdSm) was chosen for further preclinical investigation on the synergism between sparsomycins and cisplatin. Combined treatment of EdSm and cisplatin was studied in L1210 leukemia bearing mice. Best results were obtained in the L1210 i.p. tumor model when drug treatment was i.p. on days 1, 5 and 9. Using nontoxic doses of EdSm (5 mg/kg) and cisplatin (3 mg/kg), no differences in antitumor activity were observed after pretreatment, simultaneous or posttreatment of cisplatin with EdSm. Although single drug treatment by EdSm or cisplatin did not generate cures, all schedules of combined treatment resulted in the cure of 4 to 6 mice in each group consisting of six mice.

INTRODUCTION

Combination chemotherapy programs have improved the cure rate of patients with cancer significantly, but in general the success rate is still low. Most drugs currently in use have direct or indirect effects at the DNA level, or the microtubules of the mitotic spindle figure. Bone marrow toxicity is usually dose limiting. Sparsomycins, cytotoxic agents synthesized by our group (Ottenheijm et al, 1981; Broek et al, 1989), act differently by inhibiting protein synthesis. These properties offer innovative possibilities for its use in combination chemotherapy. Because the target of sparsomycins differs from those of other cytostatics, a potentiating effect can be expected by combining them with classical cytotoxic agents.

Studies of sparsomycin (Sm) and cisplatin in L1210 leukemia cells <u>in vitro</u> and <u>in vivo</u> supported this hypothesis (Zylicz et al, 1986; 1987). Treatment of L1210 leukemia cells <u>in vitro</u> with Sm enhanced strongly the cytotoxic effects of cisplatin (Zylicz et al, 1987). <u>In vivo</u> Sm potentiated cisplatin antitumor activity when given 3 to 6 h prior to cisplatin and resulted in prolongation of the median survival time (>60 days) and 66% cures (Zylicz et al, 1986).

Sm-analogs which were <u>in vitro</u> more active than the parent drug were tested <u>in vivo</u> for their antitumor activity in eight murine tumor models (Zylicz et al, 1988). The most active compounds are deshydroxy-Sm (dSm), ethyldeshydroxy-Sm (EdSm) and n-pentyl-Sm (pSm). The most sensitive tumors for sparsomycin analogs appeared to be L1210, P388 and RC carcinoma. Only EdSm showed antitumor activity on s.c. implanted tumors.

RESULTS

Potentiation of cisplatin antitumor activity by Sm and analogs.

In vivo potentiation of cisplatin antitumor activity was studied for Sm and three above mentioned active analogs in s.c. implanted L1210 leukemia (Zylicz et al, 1989). Cisplatin treatment (5 mg/kg) in this tumor model resulted in 156% T/C. Treatment with sparsomycins alone resulted in increased survival time for EdSm only (139%). Although Sm itself was unable to potentiate cisplatin antitumor activity in this model, two analogs, dSm and EdSm, were active. At a dose of 10 mg/kg, EdSm potentiated cisplatin antitumor activity 2.8 times. PSm, the third analog, showed no potentiation of cisplatin antitumor activity in this tumor model. Based on these results EdSm was chosen for further preclinical studies on the synergism between sparsomycins and cisplatin as well as other antitumor agents.

Potentiation of cisplatin antitumor activity by EdSm.

The maximum doses for the combination of EdSm and cisplatin in normal mice, in the absence of toxicity defined as early death, severe loss of body weight induced by cisplatin and severe diarrhoea induced by EdSm, are: EdSm 10 mg/kg with cisplatin 4 mg/kg, and EdSm 5 mg/kg with cisplatin 5 mg/kg.

Table 1: Modulaton of cisplatin (CDDP) antitumor activity by ethyl-deshydroxy-Sm (EdSm).

		[1]MST (DAYS)	[2]LTS
SINGLE DRUG TREATMENT:			
control		10	0
EdSm	5 mg/kg	16	0
cisplatin	3 mg/kg	19	0
COMBINED DRUG TREATMENT:			
EdSm + CDDP:	0 hour	60	4
Pretreatment			
EdSm + CDDP:	-24 hours	60	5
EdSm + CDDP:	- 6 hours	60	5
EdSm + CDDP:	- 3 hours	60	4
Posttreatment			
CDDP + EdSm:	+ 3 hours	60	6
CDDP + EdSm:	+ 6 hours	60	5
CDDP + EdSm:	+24 hours	60	5

[1]MST means median survival time in days after tumor implantation.
[2]LTS means long term survivors (>60 days = Cures).

Combined treatment of EdSm and cisplatin were tested in L1210 leukemia bearing mice. Treatment conditions were investigated with regard to: doses, treatment schedule, time interval between and route of administration of the two drugs, as well as site of tumor inoculation. Best results were obtained in the L1210 i.p. tumor model (table 1). Mice were treated i.p. on days 1, 5 and 9 with cisplatin alone, EdSm alone and a combination of both. No differences in antitumor activity were observed after pretreatment (3, 6 or 24 h), simultaneous treatment or posttreatment (3, 6 or 24 h) of cisplatin with EdSm. Each group consisted of six mice, and in all cases 4 to 6 mice were cured (MST=>60 days). In single drug treatment, EdSm 5 mg/kg (MST=16 days) and cisplatin 3 mg/kg (MST=19 days), no cures were noticed.

The following animal tumor models are under investigation: 1. Murine tumors: B16 melanoma, LL Lewis Lung, RC renal cell carcinoma. 2. Human tumor xenografts: large cell lung cancer, gastric and ovarium cancer.

REFERENCES

Broek, L.A.G.M. van den, Lazaro, E., Zylicz, Z., Fennis, P.J., Missler, F.A.N., Lelieveld, P., Garzotto, M., Wagener, D.J.T., Ballesta, J.P.G., and Ottenheijm, H.C.J. (1989): Lipophilic analogs of sparsomycin as strong inhibitors of protein synthesis and tumor growth: a structure-activity relationship study. J. Med. Chem. 32: 2002-2015.

Ottenheijm, H.C.J., Liskamp, R.M.J., Nispem, S.P. van, Boots, H.A., Tyhuis, M.W. (1981): Total synthesis of the antibiotic sparsomycin, a modified amino-acid monoxodithioacetal. J. Org. Chem. 46: 3273-3283.

Zylicz, Z., Hofs, H.P., and Wagener, D.J.T. (1989): Potentiation of cisplatin antitumor activity on L1210 leukemia s.c. by Sparsomycin and three of its analogs. Cancer Letters 46: 153-157.

Zylicz, Z., Wagener, D.J.T., Rennes, H. van, Kleijn, E. van der, Lelieveld, P., Broek, L.A.G.M. van der, and Ottenheijm, H.C.J. (1988): In vivo antitumor activity of sparsomycin and its analogs in eight murine tumor models. Invest. New drugs 6: 285-292.

Zylicz, Z., Wagener, D.J.T., Rennes, H. van, Wessels, J.M.C., Kleijn, E. van der, Grip, W.J. de, Broek, L.A.G.M. van den, and Ottenheijm, H.C.J. (1987): In vitro modulation of cisplatin cytotoxicity by sparsomycin inhibition of protein synthesis. J.N.C.I. 78: 701-705.

Zylicz, Z., Wagener, D.J.T., Rennes, H. van, Wessels, J.M.C., Kleijn, E. van der, Grip, W.J. de, Ottenheijm, H.C.J., and Broek, L.A.G.M. van den (1986): In vivo potentiation of cis-diamminedichloroplatinum (II) antitumor activity by pretreatment with sparsomycin. Cancer Letters 32: 53-59.

Metal Ions in Biology and Medicine. Eds. Ph. Collery, L.A. Poirier, M. Manfait, J.C. Etienne. John Libbey Eurotext, Paris © 1990, pp. 366-368

In vitro toxicity of cisplatin and carboplatin on tubular cells LLC-PK1 and rat hepatocytes in primary culture

J. Poupon, L. Benel, P. Chappuis, F. Rousselet

Laboratoire de Biochimie Appliquée, Faculté de Pharmacie, 75006 Paris, France

Cisplatin (CDDP) and carboplatin (CBDCA) are two platinum-containing complexes used in chemotherapy in the treatment of human malignancies. Although commonly utilised, the posology of these compounds is limited by their toxicity which is mainly renal fo CDDP and hematologic for CBDCA.

The in vivo studies show a preferential tissular distribution in liver and kidney, but in spite of similar intratissular levels observed in both organs, hepatic toxicity remains low. The in vitro studies on continuous cell lines allowed to clarify the action mechanism of these two products.

Renal toxicity has been studied in vitro using different systems : renal cortical slices (Phelps et al., 1987), isolated tubular cells (Tay et al., 1988), and more recently, $LLC\text{-}PK_1$ cells, a continuous porcine kidney cell line (Montine and Borch, 1988).

However, the exact mechanism of this toxicity remains unclear. Moreover, to the best of our knowledge, no study has been conducted on hepatocytes in culture.

This is the reason why we compared the action of CDDP and CBDCA on two cellular systems : $LLC\text{-}PK_1$ cells and Rat hepatocytes in primary culture in an attempt to :
1) validate a simple cellular system for comparative studies of platinum compounds toxicities, 2) demonstrate clearly a possible difference of action on these two types of cells in order to explain the differences of toxicity observed in vivo.

MATERIALS AND METHODS

The liver was removed from the rats (Sprague-Dawley) and perfused with a collagenase solution. $LLC\text{-}PK_1$ cells were utilised when reaching confluence.

Cells have been incubated 20 hours at 37°C in waymouth medium containing 10 per cent foetal calf serum. Each of the 2 drugs, CDDP and CBDCA, was dissolved in the medium at the following concentrations : CDDP : 0 to 400 µmol/l, CBDCA : 0 to 1600 µmol/l.

After 20 hours, cells viability has been determined by the neutral red uptake method and results expressed as percentages of control cultures (100 per cent). From a graphic representation of viability, we determined a toxic concentration 50 per cent (TC_{50}) defined as the concentration of drug (μmol/l of medium) which produces a 50 per cent decrease of cells viability, and calculated the corresponding cellular platinum level.

Intracellular platinum has been measured by flameless Atomic Absorption Spectrometry and expressed as μmol Pt per gram of cellular Protein.

Synthesis activities of hepatocytes have been inferred from determination of albumin secreted in the medium and cellular glycogen content.

Cellular lysis was estimated by enzymes leakage : AST, ALT and LDH for hepatocytes, ALP, γ-GT and NAG for $LLC\text{-}PK_1$.

Results were expressed as percentages.

RESULTS

Platinum caption by cells increase according to the concentration of the products in the medium. It is similar for both cells systems but six to ten times superior for CDDP than for CBDCA (Fig. 1a and 1b).

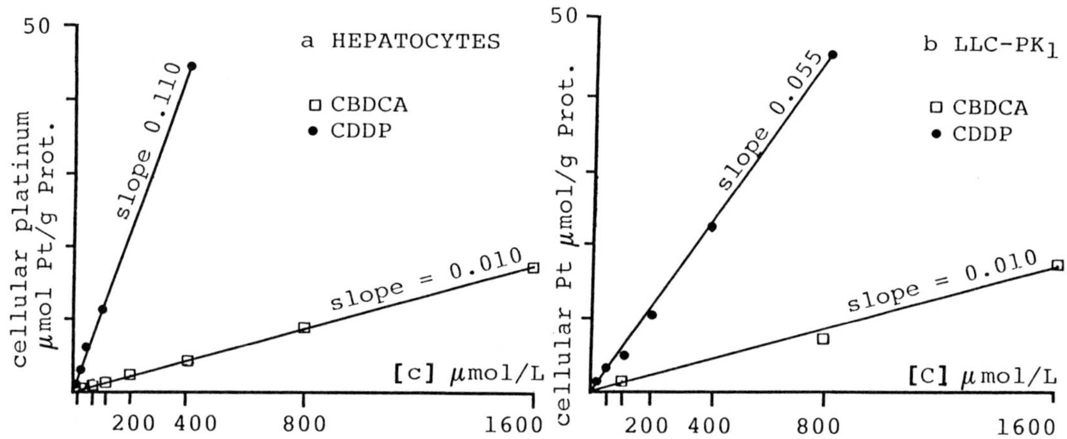

Fig. 1 : Cellular platinum according to the concentration [C] of CDDP and CBDCA on hepatocytes (a) and $LLC\text{-}PK_1$ (b)

Toxic concentrations (Fig. 2) obtained for hepatocytes are close to those obtained for $LLC\text{-}PK_1$, but they are 8 to 9 times superior for CBDCA than for CDDP, as evidenced by their toxic concentrations ratios (CBDCA/CDDP) : 7.7 for hepatocytes and 9.1 for $LLC\text{-}PK_1$. These values are similar to the DL_{50} ratio observed *in vivo* in rats.

It is of importance to notice that these toxic concentrations are achieved at similar cellular platinum levels for both drugs (Table 1).

Table 1 : Toxic concentration (TC_{50}) and cellular platinum on hepatocytes and LLC-PK_1

	HEPATOCYTES		LLC-PK_1	
	TC_{50} µmol/L	cellular Pt µmol Pt/g Prot.	TC_{50} µmol/L	Cellular Pt µmol Pt/g Prot.
CDDP	64	7.0	70	3.9
CBDCA	495	5.2	640	6.5
CBDCA/CDDP	7.7		9.1	

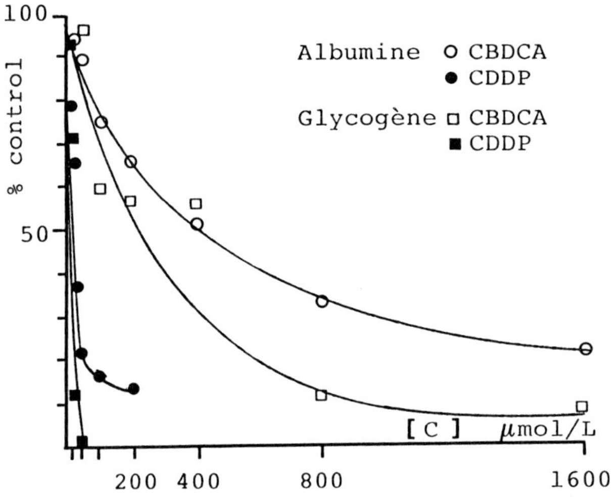

Fig. 2 : Albumin secretion and cellular glycogen of hepatocytes.

The secretion of albumin and the cellular glycogen of hepatocytes proved to be reliable and precocious markers of cellular impairment as shown by Fig. 2.

However, leakage of enzymes show that CDDP and CBDCA have only a minor lytic effect at the concentrations studied (Results not indicated here).

CONCLUSION

Our results do not allow to reproduce the predominance of the renal toxicity of CDDP observed in vivo indicating that kidney toxicity of CDDP might be mostly related to renal physiology. However, they give a fairly good picture of toxicity ratios of cisplatin and carboplatin. For this purpose, Rat hepatocytes in culture prove to be a good and reliable in vitro model.

References

Montine T.J. and Borch R.F., (1988) : Quiescent LLC-PK_1 cells as a model for cis-diaminedichloroplatinum (II) nephrotocixity and modulation by thiol rescue agents. Cancer Res., 48 (21), 6017-6024.

Phelps J.S., Gandolfi A.J., Brendel K. and Dorr R.T., (1987) : Cisplatin nephrotoxicity : in vitro studies with precision-cut Rabbit renal cortical slices. Toxicol. Appl. Pharmacol., 90, 501-512.

Tay L.K., Bregman C.L., Masters B.A. and Williams P.D., (1988) : Effects of cis-diaminedichloroplatinum (II) on Rabbit kidney in vivo and on Rabbit renal proximal tubule cells in cuture. Cancer Res., 48, 2538-2543.

Metal Ions in Biology and Medicine. Eds. Ph. Collery, L.A. Poirier, M. Manfait, J.C. Etienne. John Libbey Eurotext, Paris © 1990, pp. 369-371

Treatment of rat osteosarcoma with phosphonic acid linked platinum complexes *in vivo* and *in vitro*

T. Klenner[1], B.K. Keppler[2], H. Münch[2], P. Valenzuela-Paz[1]

[1] *Institute of Toxicology and Chemotherapy, German Cancer Research Center, INF 280;* [2] *Institute of Inorganic Chemistry, University of Heidelberg, INF 270, 6900 Heidelberg, FRG*

INTRODUCTION:

Osteosarcoma, a malignant disease mostly seen in adolescents, normally has a poor prognosis. The clinical treatment of osteosarcoma patients has made some progress in the past years but remains still unsatisfactory (Winkler et al. 1988). The most important complications are the metastases frequently seen in the lungs (Souhami R.L. 1989). Our model of the intratibially transplanted osteosarcoma of the rat resembles this condition. It is characterized by slow growth rate, dissemination into lungs, kidneys and lymphnodes, bone-matrix formation and sensitivity to different anticancer drugs. The compounds we investigated are platinum complexes linked to phosphonic acids. The phosphonic part of these molecules is thought to act as a carrier to the bone. Experiments were performed in-vivo as well as in-vitro.

MATERIALS AND METHODS:

Our experiments focussed on the evaluation of the following phosphonic-acid linked platinum complexes (Fig. 1):
cis-Diammine-(nitrilotris(methylphosphonato)(2-)-O^1,N^1)platinum II (AMDP),
cis-Cyclohexane-1, 2-diamine- (nitrilotris (methylphosphonato) (2-) -O^1, N^1) platinum II (DADP),
cis-Diammine-((bis(phosphonatomethyl)amino)acetato)(2-)-O^1,N^1)platinum II (DBP),
cis-Cyclohexane-1,2-diamine-((bis(phosphonatomethyl)amino)acetato(2-)-O^1,N^1)platinum II (DDBP),
cis-Diammine-(iminobis(methylphosphonato)(2-)-O^1,N^1)platinum II (IMD),
cis-Cyclohexane-1, 2-diamine-(iminobis (methylphosphonato) (2-)- O^1, N^1) platinum II (DIMD).

The tumor model used was the intraosseously transplantable osteosarcoma of the rat. Transplantation, treatment and evaluation were performed according to previously published methods (Wingen et al. 1986).

The in-vitro cell lines were obtained from a lung metastasis of an osteosarcoma bearing animal using the limited dilution method. The cells were maintained and the experiments evaluated as described earlier (Angres 1989). A quickly growing clone (C25-28h) and a slower growing clone (C36-47h) were compared with respect to their sensitivity to 4 and 24 hours incubation with AMDP, IMD, and DADP.

Fig. 1: Chemical formulae

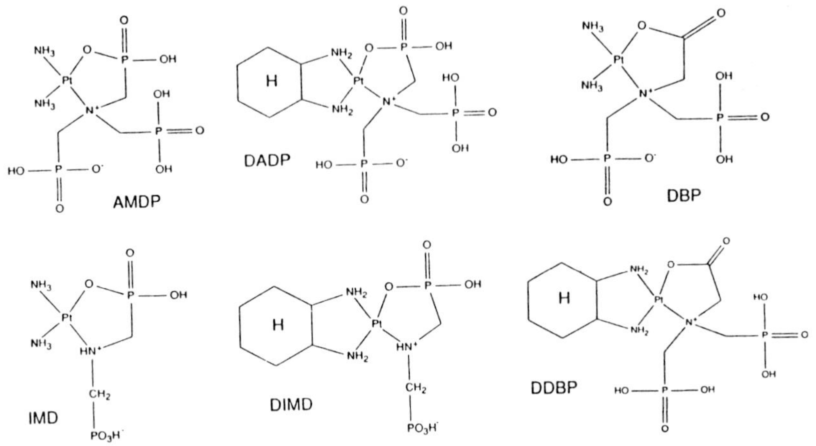

RESULTS AND DISCUSSION:

The results of different in-vivo experiments are compiled in Table 1. They indicate a high anticancer activity as shown in a standstill of tumor growth after three weeks therapy (T/C%) and an increase in life span (ILS). This could be seen with all compounds, except DDBP; the best one, AMDP, with an ILS of more than 130% at 0.396 and 0.6 mmol/kg total dose. The toxicity of the compounds was mainly reversible reduced weight gain (BWD) and mildly reduced kidney function, but no effect on the bone marrow was observed.

Table 1: Results of therapy with cisplatin-linked phosphonates in vivo.

Group	Total dose mg/kg	single dose mmol/kg	T/C% day 42	ILS%	BWD (day 42) male	BWD (day 42) female
AMDP	225.0	0.066	38	+131	- 2	- 29
AMDP	337.5	0.100	29	+145	- 9	- 39
DADP	400.2	0.110	39	+ 89	- 13	- 23
DBP	405.0	0.100	77	+ 98	+ 18	- 9
DBP	607.5	0.150	36	+113	+ 5	- 14
DDBP	300.0	0.088	105	+ 5	+ 3	+ 5
IMD	389.0	0.150	50	+ 65	- 6	- 13
DIMD	464.0*	0.230	56	+ 69	+ 7	+ 3

* therapy lasted 2 weeks

The cisplatin-linked phosphonates are valuable compounds in the treatment of experimental osteosarcoma with respect to tumor inhibition and increase in survival time. Only few side effects were found. So far, the development of cisplatin analogues has focussed on the alteration of the molecule itself, e.g. by introduction of a cyclohexane ring. These compounds were very interesting because they lack the cross resistence to cisplatin (Burchenal et al. 1979). Therefore, the comparison of these new compounds including the cyclohexane derivatives is very important.

The comparison of the sensitivity of a fast growing clone (C25) and a slow growing clone (C36) of a lung metastasis to treatment with AMDP, DADP and IMD in-vitro showed an

important difference. C25 cells were reduced in their growth in a time and concentration dependent manner, but for the lowest concentration and short incubation period the cells recovered and started to grow more rapidly again (Fig. 2). For clone C36 a constantly reduced growth rate was seen after incubation with the same compounds and similar conditions (Fig. 2).
For the future, further experiments on other metal complexes in this model and with a variety of clones are planned.

Fig. 2: Results of incubating clone C25 and C36 with AMDP in-vitro.

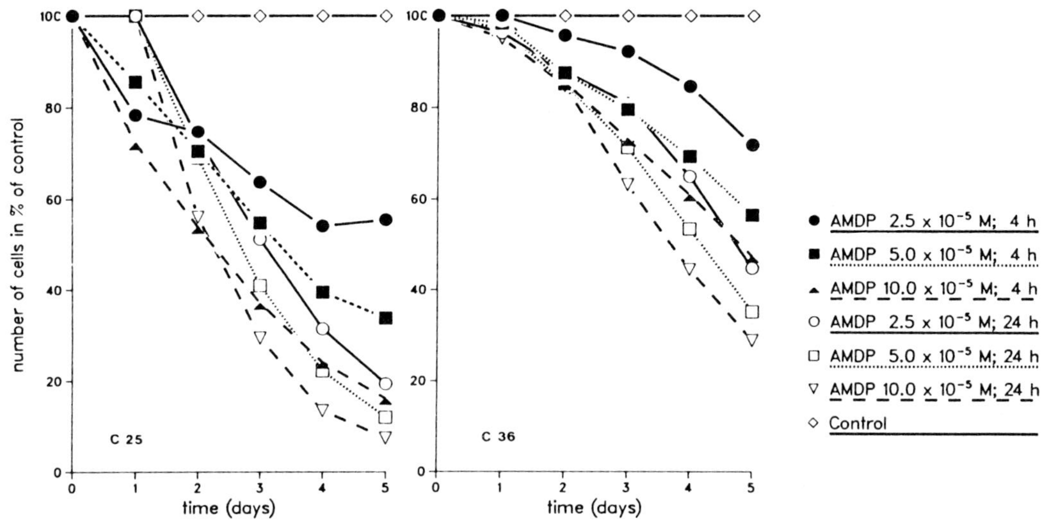

REFERENCES:

Angres G., Scherf H.R. (1989) Different effects of cyclophosphamide in vivo and phosphamide mustard in vitro on two cell clones of chemically induced mammary carcinoma of the rat. J Cancer Res Clin Oncol 115, 203-206

Burchenal H.H., Kalaher K., Dew K., Lokys L. (1979) Rationale for development of platinum analogs. Cancer Treat Rep 63, 1493-1498

Souhami R.L. (1989) Chemotherapy for osteosarcoma. Br J Cancer 59, 147-148

Winkler K., Beron G., Delling G. et al. (1988) Neo-adjuvant chemotherapy of osteosarcoma: results of a randomised cooperative trial (COSS-82) with salvage chemotherapy based on histological tumor response. J Clin Oncol 6, 329

Wingen F., Sterz H., Blum H. et al. (1986) Synthesis, antitumor activity, distribution and toxicity of 4-(4-(Bis(2-chloroethyl)amino)phenyl)-1-hydroxybutane-1,1-bisphosphonic acid (BAD), a new lost derivative with increased accumulation in rat osteosarcoma. J Cancer Res Clin Oncol 111, 209-219

Metal Ions in Biology and Medicine. Eds. Ph. Collery, L.A. Poirier, M. Manfait, J.C. Etienne. John Libbey Eurotext, Paris © 1990, pp. 372-376

A novel series of cis-platinum complexes

T. Theophanides, J. Anastassopoulou, J.Y. Gauthier*, S. Hanessian*

*National Technical University of Athens, Chemical Engineering Department, Laboratory of Radiation Chemistry and Biospectroscopy, Zografou Campus, 157 73 Zografou, Athens, Greece and * Université de Montréal, Department of Chemistry, Montréal, Quebec, Canada H3C 3J7*

Cis-platinum (cis-diamminedichloroplatinum II) is a platinum coordination compound which was shown (Rosenberg et al, 1965) to posses antibiotic and antitumor activity. More studies (Rosenberg et al, 1969) established the antitumor activity of cis-platinum in experimental animal tumors. Clinical trials followed later by several laboratories with cis-platinum and today cis-platinum plays a mojor role in the chemotherapy of several human malignascies (Theophanides, 1981). A large number of platinum complexes have been synthesized and tested for antitumor activity since the initial discovery of platinum ammine compounds. Complexes of the type PtA_2X_2 where X=monodentate or X_2=bidentate anionic ligands and A=monodentate ammines or A_2=bidentate amine ligands have been obtained and structure-activity studies show that these analogs have antitumor activity similar or even superior to cis-platinum (II) (Cleare, 1974). Thus, a more detailed study was undertaken in order to investigate the new analogs of cis-platinum and compare <u>in vitro</u> cytotoxicity to L1210 cells of the new derivatives.

Materials and Methods

The complexes are of the general formula cis-Pt(L)Cl_2, where L= an amine ligand and their preparation was described (Okamoto et al, 1986). The spectroscopic studies concerning some of the complexes have been described briefly (Theophanides & Anastassopoulou, 1989) and will be given fully together with testing and screening data as compared to cis-platinum.

The FT-IR spectra of ligand (L) and the complexes cis-Pt(L)Cl_2 have been recorded with a DIGILAB FTS-15C/D Michelson FT interferometer with resolution 4 cm^{-1}. The spectra were taken in KCl pellets. The interferometer was equipped with a HgCdTe detector (Infrared Associates, New Breenswick, NJ), a KBr beam splitter and a globar source. A published procedure of data collecting and calculating the spectra was used (Tajmir-Riahi & Theophanides, 1983).

The amine ligands that have been used to synthesize the new platinum complexes have been obtained according to Fig.1.

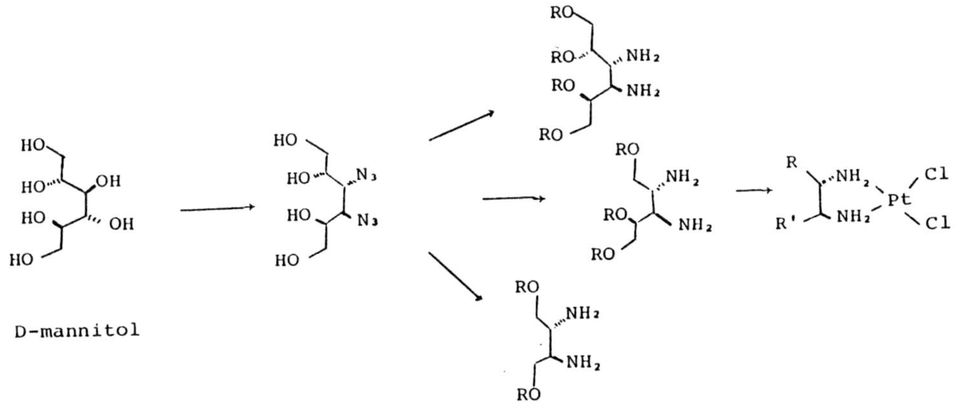

Fig. 1. Here is shown the route of preparation followed to synthesize the complexes. The R and R' groups give various combinations of new compounds.

Several complexes have been prepared with the above method. Some complexes are shown in Fig.2. Both the hydroxy and methoxy series have been synthesized and a great number of these complexes have been tested and screened for antitumor activity. The methoxy series showed lack of cross-resistance with cis-platinum which is a very interesting and novel property of these new cis-platinum analogs.

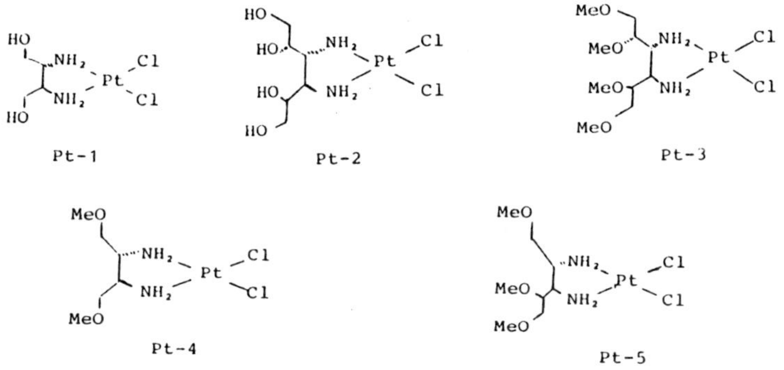

Fig. 2. Some of the new cis-platinum analogs.

Results and Discussion

A representative FT-IR spectrum of the acyclic diol amine ligand is given in Fig.3, where A is the amine ligand and B the complex with the chelating ligand. Upon complexation of the ligand through the NH_2 groups (Theophanides & Anastassopoulou, 1989) the stret-

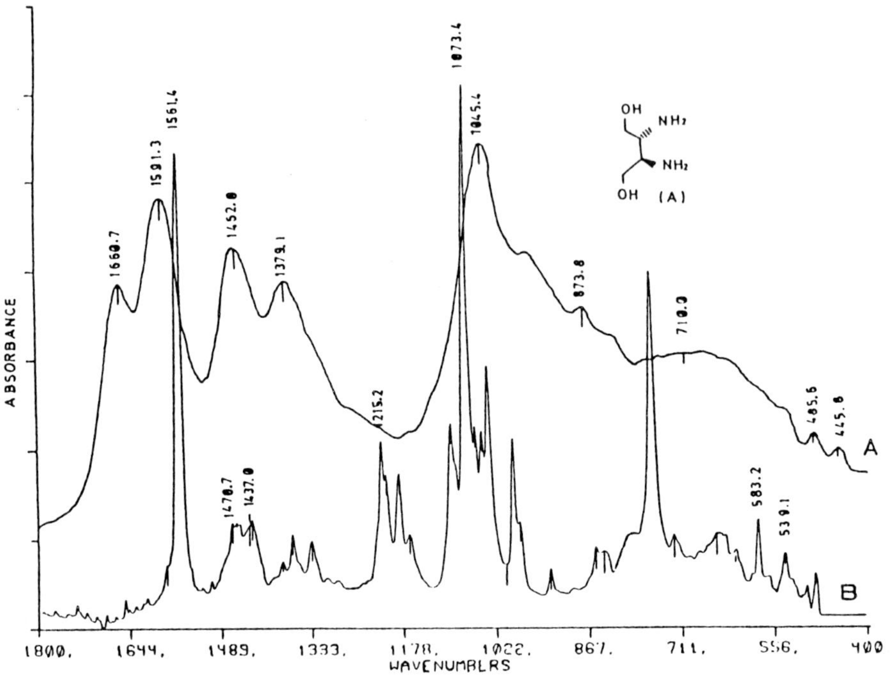

Fig. 3. The FT-IR spectra of the amine ligand 2,3-diamine-2,3-diol (A) and its cis-platinum complex (B).

ching νNH_2 frequencies are shifted to lower frequencies in the region 3500-3200 cm^{-1} by about 200 cm^{-1} and the bending δNH_2 frequency was also shifted to the lower frequencies by about 100 cm^{-1}. Furthermore, the C-O stretching and the C-OH bending frequencies have also been slightly perturbed. These frequencies are normally located in the regions 1100-1000 and 1400-1300 cm^{-1} as strong absorptions (Ballamy, 1975). A small shift of the order of 30 cm^{-1} was observed for the δC-OH bending frequency and an even smaller shift for the νC-O stretching frequency (see Fig.3). The testing and screening data of most of the new complexes were equal or superior to cis-platinum with regard to various data that have been done on the complexes. As an example, the testing status on M5076 sarcoma (IP) solid tumor is given in Table 1. The complete results of testing and the crystallografic data will be published elsewhere (Theopfanides et al, 1990). The *in vitro* cytotoxicity to L1210 cells is given in Table 2. In Table 1, where are given the data on M5076 sarcoma it is shown that the activity ratings of Pt-1, Pt-2, Pt-4 and Pt-5 are very close or similar to those of cis-platinum. The comparison of Table 2 is favorable for the new analogs.

TABLE 1. M5076 SARCOMA (IP)

COMPLEX	DOSE (mg/Kg/inj)	ROUTE	SCHEDULE	TOXIC DEATHS	WEIGHT CHG. (g)	ZT/C (CURES)	ACTIVITY RATING
Pt-1	20.0	IP	D1,5,9	1/10	-1.6	214	+++
Pt-2	15.6	IP	D1,5	0/4	-1.2	207	+++
Pt-3	62.5	IP	D1,5,9	0/4	2.3	180	++
Pt-4	18.6	IP	D1,5,9	0/10	-0.3	253(3/10)	++++
Pt-5	150.0	IP	D1,5,9	0/10	0.2	220(1/10)	++++
CISPLATINUM	8.0	IP	D1,5,9	0/6	-2.0	276(3/6)	++++
	8.0	IP	D1,5,9	1/10	-3.6	301(8/10)	++++

TABLE 2. *In vitro* cytotoxicity to L1210 cells

COMPLEX	IC50 (uM) TEST 1	TEST 2
Pt-1	10.8	7.53
Pt-2	11.9	----
Pt-3	12.9	8.44
Pt-4	10.6	8.95
Pt-5	INACTIVE	INACTIVE
CISPLATINUM	1.45	0.95

* INACTIVE= IC50 greater than 1 ug/ml

Acknowledgements

This research was supported in part by a Grant from the National Institute of Cancer, Canada (TT and SH). We also thank Johnson Matthey Centre for the loan of platinum salts (TT).

REFERENCES

Bellamy, L.J. (1975):The infrared spectra of complex molecules, Vol. 1, 3rd Edition,Chapman and Hall, pp. 122-125, London.

Cleare, M.J. (1974): Coordination compound of the platinum group metals. Their preparative methods and application, Platinum Met. Rev. 18: 122-129

Okamoto, K., Behnam, V., Gauthier, J.-Y., Viet, M.T.P., Polissiou, M., Hanessian, S. and Theophanides, T. (1986):FT-IR and ^{1}H NMR spectroscopic studies of C2'-endo, C3'-endo sugar ring conformations in 5'-GMP and 3'-GMP nucleotides and their platinum complexes. Inorg. Chim. Acta. 123: L3-L5.

Rosenberg, B., Van Camp, L. and Krigas, T. (1965):Inhibition of cell division in E. coli by electrolysis. Products from platinum electrode, Nature, 205: 698-699.

Rosenberg, B., Van Camp, L., Trosko, J. and Mansour, V.H. (1969): Platinum compounds: a new class of potent antitumor agents, Nature, 222: 385-386.

Tajmir-Riahi, H. and Theophanides, T. (1983): Adenosine-5'-monophosphate complexes of Pt(II) and Mg(II) metal ions. Synthesis, FT-IR spectra and structural studies: Inorg. Chim. Acta. 80: 183-189.

Theophanides, T. (1981): Platinum coordination compounds and cancer, Chem. in Canada, 32: 30-32.

Theophanides, T. and Anastassopoulou, J.D. (1989): FT-IR spectra of platinum-sugaramines. A novel series of antitumor platinum drugs. In Spectroscopy of Biological Molecules-State of the Art, eds A. Bertoluzza, C. Fagnano and P. Monti, pp.365-366, Bologna, Societa Editrice, ESCULAPIO.

Theophanides, T. (1990): Synthesis and X-ray analysis of cis-platinum analogs of diaminediols. To be submitted for publication.

Metal Ions in Biology and Medicine. Eds. Ph. Collery, L.A. Poirier, M. Manfait, J.C. Etienne. John Libbey Eurotext, Paris © 1990, pp. 377-379

Clinical pharmacokinetic analys of a five-day continuous infusion of cisplatin in Cancer patients

B. Desoize, F. Marechal, P. Dumont, H. Millart, A. Cattan

Institut J. Godinot, BP 171, 51056 Reims, France

In order to study the pharmacokinetics of Cis-Dichloro-Diammine-Platinum (CDDP), a daily blood sample was taken from a group of cancer patients treated with a 120-hour continuous infusion (CI) of CDDP. The aim of the study was to analyse Total Platinum (TP) and Ultrafilterable Platinum (UP) variations during chemotherapy courses and the search for possible correlations with clinical efficacy and/or toxicity. Correlations between clinical evaluation and pharmacokinetic parameters could lead to adaptation of dosage according to plasma concentration (conc., 1).

PATIENTS AND METHODS

22 kinetics were carried out in 11 patients, 2 of them were previously treated; 7 had non-small cell lung cancers, 2 had carcinomas of unknown origins, 1 had a breast cancer, and 1 had a head & neck cancer. Phase I clinical trials have shown that CI of CDDP can be well tolerated provided that it is accompanied by a 3-litre hyperhydration to prevent renal toxicity (2). Each patient was planned to receive every 28 days (d) a continuous 120-h infusion of 20 mg/m2/d CDDP in 1 l/d saline associated with 50 mg/m2/d Etoposide administered simultaneously as a 120-h CI in 2 l/d saline, using volumetric pumps. The stability of the CDDP etoposide mixture has been established in vitro by Creagan (3). The chronobiology of CDDP (4) led us to withdraw blood samples at a constant hour (8:30 am), in heparinized tubes. The samples were immediately centrifuged and ultrafiltered on Amicon MPS-1 filters to determine TP and UP conc. (5). Platinum assay was performed according to the flameless atomic absorption spectrophotometric technique (6).

Toxicity was evaluated after each course and efficacy after 3 courses according to the WHO criteria (7). The simultaneous infusion of Etoposide precluded to evaluate precisely the proper role of CDDP, nevertheless it can be assumed that most of GI tract and renal toxicities were due to CDDP. The areas under the curve (AUC_{0-120h}) of TP and UP were calculated by the trapezoidal integration method from time 0 to the end of the infusion. The 2-tailed, paired or independent sample, Student t-test was used for the comparison of the means of paired or independent samples.

RESULTS

Correlations between pharmacokinetic parameters : 22 pharmacokinetic studies were available for TP, and 20 for UP. The AUC_{0-120h} of TP was positively correlated with the administered dose of CDDP ($\alpha<0.01$) and with serum creatinine ($\alpha<0.05$). In the same way, the AUC_{0-120h} of UP was correlated with the administered dose of CDDP ($\alpha<0.05$) and with serum creatinine ($\alpha<0.05$). The plasma conc. of TP were correlated ($\alpha<0.05$) with those of UP, and closely so ($\alpha<0.001$) from d3 to d5. The AUC_{0-120h} of TP was closely correlated with the AUC_{0-120h} of UP ($\alpha<0.001$).

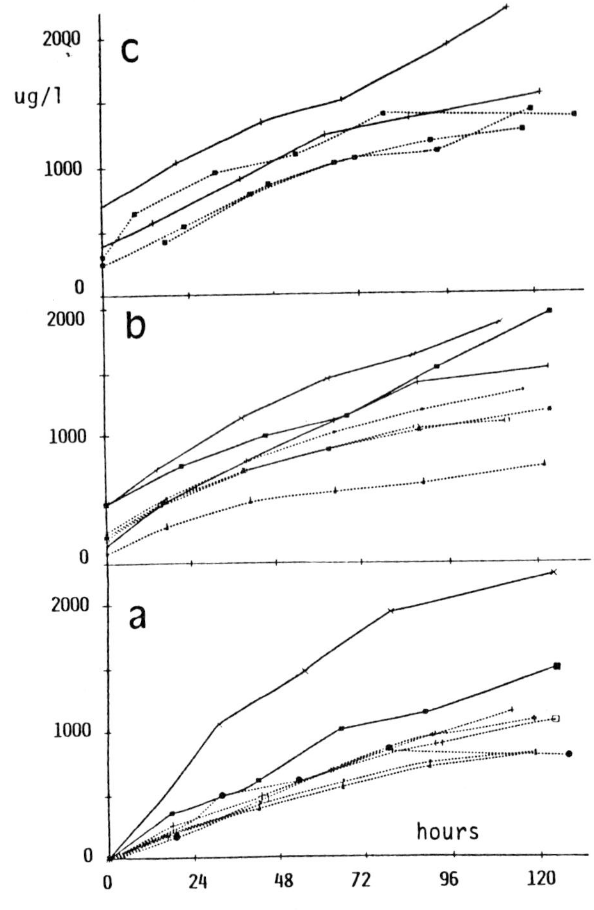

Fig. 1. Plasma conc. of total platinum during the 1st (a), 2nd (b) and 3rd courses (c) in responders (solid lines) and non-responders (dotted lines).

Accumulation of TP and UP : TP conc. increased during the courses. The curves of TP conc. were similar from course to course (fig.1) except that TP conc. did not drop back to 0 after courses. The TP conc. and AUC of the courses n°1 were significantly higher than for the courses n°3 ($\alpha<0.02$). The plasma conc. and AUC of UP did not statistically increase during and between courses

Comparison of pharmacokinetic parameters with clinical parameters : (i) Efficacy : there were two responders and six non-responders, significantly higher levels of TP (AUC_{0-120h} and conc. from d1 to d5) were found amongst responders as compared to non-responders during the 1st course. The mean AUC_{0-120h} of TP were 507 ± 187 $g*l^{-1}*s$ and 251 ± 34 $g*l^{-1}-1*s$, respectively ($\alpha<0.02$), and the mean TP plasma conc. at d5 were 1823 ± 505 $\mu g*l^{-1}$ and 948 ± 167 $\mu g*l^{-1}$, respectively ($\alpha<0.01$). As far as UP was concerned, no statistical difference was found between responders and non-responders.

In the same manner, the pharmacokinetics of the 2nd and 3rd courses were analysed : the AUC_{-120h} and conc. from d2 to d5 of TP were statistically higher ($\alpha<0.02$) amongst the 3 responders of the 2nd course as compared to the 4 non-responders. But no difference was found during the 3rd course. UP conc. was not different between the 2 groups during the 2nd and 3rd courses.

(ii) Toxicity : No renal toxicity was observed, neither the intensity nor the frequency of vomitting increased throughout courses, in contrast with the regular increase in TP conc., and UP to a lesser extent. Nevertheless, the existence of a grade 1-3 digestive intolerance was correlated with a higher AUC_{0-120h} of TP (451 ± 119 vs 324 ± 109 $g*l^{-1}*s$, $\alpha<0.02$) and increased plasma conc. of TP from d1 to d5 ($t<0.05$). Conversely, no statistically significant difference was found in the case of UP.

DISCUSSION

This study was aimed at analysing the first three courses of a 120-hour continuous administration of CDDP. It was not possible to obtain the pharmacokinetics of the 3 courses for all the patients, which hampers the conclusions that can be drawn from our study. TP plasma conc. were correlated with the administered doses of Platinum and with serum creatinine. An accumulation of TP was observed, both within each course, and from one course to the other, as previously reported by Vermorken (8). The accumulation of UP during courses and from one course to the other was less important and not significant, these results were comparable to those of Forastiere (9)

CDDP and Etoposide are amongst the most active drugs in solid tumours, and in particular in non-small cell lung cancer (10). The existence of a synergism between their antineoplastic effects has been shown in vitro (11) and in vivo (12). In this study, the simultaneous infusion of Etoposide precluded a thorough analysis of the clinical efficacy and toxicity. Only digestive and renal toxicities which are linked to CDDP were evaluable. No renal sign of toxicity was observed. GI tract toxicity was correlated with the AUCO-120 of TP and with TP plasma conc. in our study. Maximum plasma levels of TP at the end of the 120-hour CI are reported to be considerably lower than the peak level of a short-time infusion (8), this difference in peak conc. may be responsible for the difference in GI-tract toxicity between bolus and CI, in favour of CI.

TP includes protein-bound Platinum and UP; UP is composed of Platinum bound to low molecular weight molecules (amino acids, glutathione...), free CDDP and metabolite species. UP composition varies during infusion time, even after blood taking, since free CDDP and metabolites still react (8,13,14). Platinum binding is not reversible (14) whereas the chemical nature of free CDDP and its metabolites is, according to Reece et al. (15), "mobile". Therapeutic activity is mainly due to free CDDP and its metabolites (6,13).

In this study, UP was not correlated with clinical data, although it was prepared very carefully. The absence of correlation could be due to the variability of UP composition. Although TP is made of 90 % to 95 % of inactive bound Platinum, TP is correlated to efficacy and toxicity. These positive results have to be taken cautiously owing to our limited set of patients and to simultaneous Etoposide administration. So, our **CONCLUSION** is that no clinical response was observed in the group of patients whose TP conc. and AUC were low. It will be of great interest to confirm these data on a larger series of patients.

REFERENCES

1-FAVRE R, CHARBIT M, RINALDI Y, ILIADIS A, CANO JP, et CARCASSONNE Y (1988) Bull Cancer 75: 541

2-SALEM P, KHALYL M, JABBOURY K and HASHIMI L (1984) Cancer 53: 837

3-CREAGAN ET, RICHARDSON RL and KOVACH JS (1988) J Clin Oncol 6: 1197

4-HECQUET B and SUCCHE M (1986) J Pharmacokinet Biopharm 14: 79

5-Van der VIJGH WJF and KLEIN I (1986). Cancer Chemother Pharmacol 18: 129

6-BANNISTER SJ, CHANG Y, STERNSON LA and REPTA AJ (1978) Clin Chem 24:877

7-MILLER A, HOOGSTRATEN B, STAQUET M and WINKLER A (1981) Cancer 47: 207

8-VERMORKEN JB, Van der VIJGH WJF and KLEIN I (1986) Clin Pharmacol Ther 39: 136

9-FORASTIERE AA, BELLIVEAU JF, GOREN MP, VOGEL CV, POSNER MR and O'LEARY GP (1988) Cancer Res 48: 3869

10-BONI C, COCCONI G, BISAGNI G, CECI G and PERACCHIA G (1989) Cancer 63: 638

11-DURAND RE and GOLDIE JH (1987) Cancer Treat Rep 71: 673

12-SCHABEL FM, TRADER MW, LASTER WR, CORBETT TH and GRISWOLD DP (1979) Cancer Treat Rep 63: 1459

13-DALEY-YATES PT and McBRIEN DCH (1984) Biochem Pharmacol 33: 3063

14-REPTA AJ and LONG DF (1980) In "Cisplatin: current status and new developments" PRESTAYKO et al Ed., Academic Press, New York, p285

15-REECE PA, STAFFORD I, RUSSEL J, KHAN M and GILL PG (1987) Cancer Chemother Pharmacol 20: 26

Metal Ions in Biology and Medicine. Eds. Ph. Collery, L.A. Poirier, M. Manfait, J.C. Etienne. John Libbey Eurotext, Paris © 1990, pp. 380-382

Positron-emission-tomography (PET) for therapy management of patients with advanced cancer of the oro-and Hypopharynx treated with cisplatinum and 5-fluorouracil

U. Haberkorn, L.G. Strauss, M.V Knopp, A. Dimitrakopoulou, A. Schadel, J. Doll, W.J. Lorenz

Institute of Radiology and Pathophysiology, German Cancer Research Center, Im Neuenheimer Feld 280, D-69 Heidelberg, FRG

INTRODUCTION

The systemic chemotherapy of advanced cancer of head and neck is applied either in combination with radiation and/or surgery or as a single measure. Combinations of 5-fluoro-uracil and cisplatinum have proved useful with remission rates between 50 and 80 per cent (Ervin et al., 1985). For the evaluation and individual planing of the chemotherapy it seems useful to gain information about the tumor metabolism during therapy. Ultrasound, computed tomography (CT) and magnetic resonance tomography deliver mainly morphologic data. This is achieved in interpreting parameters like echogenicity or density, proton density or relaxation times. In contrast PET with Fluorine-18-deoxyglucose (FDG) is a specific method, that delivers functional information about glucose metabolism.

PATIENTS AND METHODS

Ten male patients with histological proved tumors of oro-or hypopharynx (9 squamous cell ca., 1 anaplastic ca.) underwent a PET-examination prior and after the first chemotherapeutic cycle with cisplatinum (150 mg/m on day 1) and 5-FU (1000mg/m on day 1-4) Only patients with a tumor or lymph node diameter exceeding 1.5 cm were accepted. Eight milimeter thick continuous sections were acquired in CT (Siemens Somatom DRH). Skin markings were used for correct positioning of PET. Tumor or lymph node volumes were calculated using a region of interest technique. We assumed an exponential function $c = \ln(V0/V1)/t$ with V0 and V1 as the volumina before and after therapy and t as the time interval between V0 and V1 for the estimation of the tumor growth rate. PET-examinations were performed with a two ring detector system (PC2048-7WB Scanditronix). Three PET sections with a thickness of 11 mm were acquired. FDG was used to assess the regional glucose uptake and phosphorylation. While the transport of FDG into the cell is comparable to glucose, FDG is trapped after the phosphorylation. Therefore the distribution of FDG reflects the local glucose accumulation in the tumor. One hour after intravenous administration of 9 to 12 mCi 18FDG, PET images were acquired for 10 minutes. PET images were generated by use of an iterative reconstruction program on a VAX 11/750 (Digital equipment, Maynard, Mass.) computer system. Thereby the image matrix was 128 x 128; for display a interpolation to 256 x 256 was done. The spatial resolution is approximately 5.1 mm. Pixel size in all reconstructed images was 2 mm. Attenuation and scatter correction was done. For quantitative evaluation, regions of interest (ROI) were defined in tumor and soft tissue with a region size exceeding 32 pixels in all cases. The identification of anatomic structures was done by comparing PET-sections with the CT images. FDG uptake was then expressed as the standardized uptake value (SUV): SUV = tissue concentration (nCi/g)/(injected dose [nCi]/body weight [g]).

RESULTS

FDG data were available for 5 tumors and 9 lymph nodes, volumetric data existed for 4 tumors and 7 lymph nodes. The regional FDG metabolism and the changes after therapy are demonstrated in Figure 1. Different lymph nodes in the same patient may show a different FDG uptake pattern. The relation between the change in FDG metabolism and the growth rate is shown in Figure 2. The data demonstrate, that the same change in FDG metabolism results in a higher reduction in tumor volume as compared to the lymph node volume.

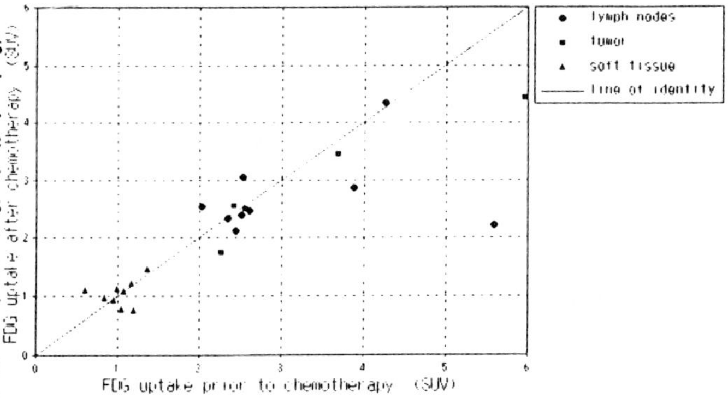

Fig. 1: FDG uptake prior and after chemotherapy.

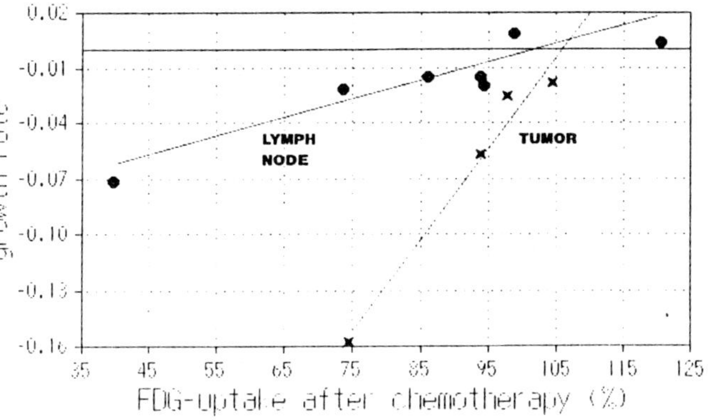

Fig. 2: Changes in FDG uptake and tumor growth rate in response to chemotherapy.

DISCUSSION

The assessment of tumor metabolism in patients undergoing therapy is of importance for the evaluation and individual planning of therapy regimens. This task is a domain of nuclear medicine as compared to morphologic methods provided by other procedures. Gallium scanning has found use in conventional nuclear medicine for treatment control. In a tumor model a dependance of Gallium uptake on the amount of viable cells was observed (Iosilevsky et al., 1985). Bichel (1972) found that a high incorporation of 67Gallium corresponded to a high proliferation rate. However the exact mechanism of Gallium uptake in the cell is not yet known. The incorporation of a transferrin-Gallium-complex is conversly discussed (Higashi et al., 1989). FDG as an analog of glucose is transported like glucose in the cell, then phosphorylated and trapped as FDG-6-phosphate, since there is no significant further metabolism of FDG-6-phosphate. This accumulation in metabolic active cells can be used for the differential diagnosis in patients with probably recurrent colorectal cancer (Strauss et al., 1989). A further application of the PET-FDG method is the evaluation of treatment response measuring glucose metabolism prior and during therapy. Clinical and experimental studies report on a decreasing FDG uptake in tumors treated with radiation therapy (Abe et al., 1986). In our ongoing study, we found, that the FDG uptake prior to chemotherapy was increased in all tumors and lymph nodes. We observed in most cases uniform changes in FDG accumulation during therapy. However a difference in degree was found. Lymph nodes in the same patient can show differences in metabolic activity prior and during therapy. This indicates the heterogeneity of the tumors cell population. There was a linear relation between the changes in metabolism and the growth rate with different regression functions for tumors and lymph nodes. We like to emphasize, that six tumors or lymph nodes had no significant change in tumor metabolism. In this respect it is useful to discuss the results of a flow cytometric study done by Brauneis (1989). They found no significant difference in tumor reaction to therapy in tumors with different proliferation rates. However there was a higher incidence of regional or distant metastases in tumors exhibiting a high proliferation rate. Minn (1988) observed a strong correlation of FDG uptake and the amount of S-phase-cells. This supports the thesis, that FDG indicates the proliferative activity and the incidence of metastatic spread. The FDG uptake has to be seen as an in-vivo-measure for the aggressiveness of a tumor. Using PET it is possible to gain absolute and therefore comparable data about the tumor metabolism prior and after chemotherapy. PET is a useful method for the observation and the improvement of therapeutic measures in patients undergoing systemic chemotherapy.

REFERENCES

Abe, Y., Matsuzawa, T., Fujiwara, T., Fukuda, H., Itoh, M., Yamada, K., Yamaguchi, K., Sato, T. and Ido, T. (1986): Assessment of radiotherapeutic effects on experimental tumors using 18F-2- fluoror-2-deoxy-D-glucose. Eur J Nucl Med 12: 325-328

Bichel, P. and Hansen, H.H. (1972): The incorporation of 67Ga in normal and malignant cells and its dependence on growth rate. Br J Radiol 45: 182-184 Brauneis, J.W., Laskawi, R., Schröder, M. and Göhde, W. (1989): Ergebnisse der Impulscytophotometrie bei malignen Tumoren des Kopf-Hals-Bereiches. HNO 37: 369-372

Ervin, T.J., Clark, J.R. and Weichselbaum, R.R. (1985): Multidisciplinary treatment of advanced squamous carcinoma of the head and neck. Seminars in Oncology 12: 71-78

Higashi, T., Kobayashi, M., Wakao, H. and Jinbu, Y. (1989): The relationship between 67Ga accumulation and ATP metabolism in tumor cells in vitro. Eur J Nucl Med 15: 152-156

Iosilevsky, G., Front, D., Bettman, L., Hardoff, R. and Ben-Arieh, Y. (1985): Uptake of Gallium-67 Citrate and (2- 3H)Deoxyglucose in the tumor model, following chemotherapy and radiotherapy. J Nucl Med 26: 278-282

Minn, H., Joensuu, H., Ahonen, A. and Klemi, P. (1988): Fluorodeoxyglucose imaging: a method to assess the proliferative activity of human cancer in vivo. Comparison with DNA flow cytometry in head and neck tumors. Cancer 61: 1776-1781

Strauss, L.G., Clorius, J.H., Schlag, P., Lehner, B., Kimmig, B., Engenhart, R., Marin-Grez, M., Helus, F., Oberdorfer, F., Schmidlin, P. and van Kaick, G. (1989): Recurrence of colorectal tumors: PET evaluation. Radiology 170: 329-332

II. Gold

Metal Ions in Biology and Medicine. Eds. Ph. Collery, L.A. Poirier, M. Manfait, J.C. Etienne. John Libbey Eurotext, Paris © 1990, pp. 385-389

8-(Thiotheophyllinato) (triphenylphosphine) Gold (I) (tTPau) : a new complex of therapeutic gold ion

M.M. De Pancorbo[1], A. García-Orad[1], M. Paz Arizti[1], J.M. Gutiérrez-Zorrilla[2], E. Colacio[3]

[1] Departamento de Biología Celular; [2] Departamento de Química Inorgánica, Universidad País Vasco, E-48940 Leioa, Vizcaya, Spain; [3] Dpto. Química Inorgánica, Universidad de Granada, Granada, Spain

The biological use of gold compounds as therapeutic agents in various diseases dates as far back as 2500 BC in China. Paracelsus (1493-1541) an outstanding alchemist was convinced that the true of chemistry had to be in the preparation of medicines and not in the making of gold. Half a century on, $AuCl_3$ was recommended as a treatment against leprosy by Bacon (1561-1626). In 1890 Koch observed that the growth of tuberculosis bacilli was hindered by AuCN and it is to this observation that gold pharmacology owes its beginnings. This gave rise, some thirty years later, to potassium dicyanoaurate, $K[Au(CN)_2]$ being used in the treatmente of tuberculosis only to be abandoned later due to its toxicity (Sadler, 1976). When this discovery was used clinically againts skin tuberculosis and syphilis, it met with some success but there were severe toxic side effects. Thiolate ligands (eg. like those found in mercaptopropanol gold) were introduced as a result of the search for lower toxicity compounds between the years 1913-1927. In the period 1925 to 1935 the therapeutic results were uncertain and the problem of toxicity remained. During these years the drugs were readly accepted but later the interest in these drugs diminished and no new gold drugs were discovered. Due to an incorrect assumption that there existed a relationship between chronic polyarthritis and tuberculosis, Landé used gold compounds for the treatment of arthritis in 1927. Chrysotherapy is the term generally used for the application of gold compounds in the treatment of rheumatoid arthritis. The most remarkable drugs accepted in rheumatoid arthritis therapy have been Myocrisin (disodiumthiomalatogold(I)), Solganal (thioglucosegold(I)), SK36914 (chlorotriethylphosphinegold(I)) and Auronafin (2,3,4,6-O-acetyl-1-thio-ß-D-glucopyranosate)(triethylphosphinegold(I) (Ainscough and Brodie, 1985). We owe the discovery of many pharmaceutical properties of metallodrugs to pure chance. Until very recently there were many technical difficulties in inorganic chemistry and this in part has been responsible for the slow progress in the design of these drugs.

The gold atom. The electronic configuration of gold is $[Xe]4f^{14}5d^{10}6s^1$. When gold is in the oxidation state III its electronic configuration is $[Xe]4f^{14}5d^8$. The small covalent radius of Au(III) produces a preference for small non-polarisable ligands such as those of oxygen and nitrogen. Complexes with larger ligand donor atoms are not used as drugs as they tend to be stronger oxidising agents (eg. $H[AuCl_4]$ (a four coordinate square-planar stereochemistry) and $H[AuBr_4]$). Example: when methionine sulphurs ($R-SCH_3$) are found on a protein, they will be oxidised to the sulphoxides ($RS(O)CH_3$) by the $AuCl_4^-$ ion. However if they are not be found, then this ion can react with the NH_2 groups of proteins (Ainscough and Brodie, 1985). Au(III) is isoelectronic with Pt(II), $5d^8$, and also forms strong bonds to nitrogen ligands. Cisplatin, now the most commonly used anticancer agent reacts with DNA via N-7 coordination (Lippard, 1984). It is an interesting speculation as to whether Au(III) could act in a similar way to Cisplatin. But it is highly unlikely that a parallel exists between Au(III) and Cisplatin due to the former being a strong oxidant and highly polarising, although anticancer activity exists in some Au(III) alkyls -eg $[Au(CH_3)_2Cl_2]$ (Sadler et al., 1984). Gold in an oxidising state Au(I) loses the electron $6s^1$ and adquires a configuration of pseudo noble gas. The large Au(I) ion will be stabilised in water by certain large, polarisable ligands and CN^-, thiolates, phosphines and thiosulphates are among these. Ligands not possesing these characteristics would cause a disproportionation reaction to take place very quickly according to the equation:

$$3\ Au(I) \longrightarrow 2\ Au(0) + Au(III)$$

The stereochemistry for Au(I) compounds is linear when it is linked to two ligands, trigonal planar when linked to three and tetrahedral when linked to four (Ainscough and Brodie, 1985)

ANTICANCER GOLD-DRUGS

Characteristics common to gold drugs are their gold content and thiolate or phosphine ligands which inhibit the dispropotionation reaction. Quite a number of gold(I) compounds, with organic groups bonded to the gold by sulfur and/or phosphorus atoms, have been tested *in vitro* on B16 melanoma and P388 leukemia and *in vivo* on P388 leukemia (Mirabelli et al., 1986a; Haiduc and Silvestru, 1989).The compounds with a coordinated phosphine ligand and a thiosugar group connected to gold through sulfur were found to be the most effective. The phosphine groups have proved to be very important because the compounds, lacking these groups, used as antiarthritic drugs do not possess anticancer activity as is the case of Miocrisin and Solganal. Miocrisin, despite having an Au-S bond cannot pass through the cellular barriers, probably due to the presence of carboxilate groups.

Phosphinegold(I) complexes. Gold(I) phosphine complexes are of a lipophilic nature. Auranofin when added to human blood enters the red cells immediately. This is because the lineal sterochemistry S-Au-P inhibits the breaking of its terminal Au-S bonds (Berners-Price and Sadler (1987). This enables Auranofin to produce a membrane blistering on lymphocytes, and in micromolar concentrations is capable of killing tumour cells in vitro, and prolonging the lifespand of mice with P388 leukaemia (Simon et al., 1981; Mirabelli et al., 1985a; Mirabelli et al., 1986a).

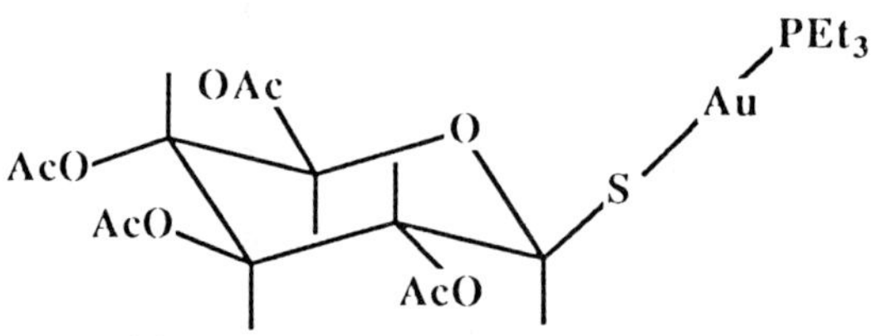

(2,3,4,6,-tetra-*o*-acetyl-1-thio-β-D-glucopyranosato-S)(triethylphosphine)gold(I), (Auranofin)

Triethylphosphine gold (I) complexes, (TEPAu) are highly cytotoxic in vitro and have shown to inhibit DNA, RNA and protein synthesis (Allaudeen et al., 1985; Mirabelli et al., 1985; Mirabelli et al., 1986a,b). But, they result in rapid cell death, thus it seems unlikely that the cytotoxic properties of these compounds can be explained by their effects on macromolecular synthesis. Snyder et al. (1986) have presented data indicating that the first sites likely to be damaged by TEPAu are cellular membranes (Rush et al., 1987a). TEPAu may be acting as a inhibitor, similar to antymicin, in site II of the mitochondrial respiratory chain. The reversal of TEPAu-induced inhibition of mitochondrial respiration and cell lethality by dithiothreitol suggests that mitocondrial thiols may be involved (Rush et al., 1987) The mechanism by which TEPAu induces the collapse of membrane potential may be mediated by a sulfhydril-dependent increase in permeability of the inner mitochondrial membrane to protons (Hoke et al., 1989)

One possibility of achieving broader spectrum of anticancer activity was to reduce the reactivity of Au(I) towards serum components such as albumin (Rosenberg, 1980). Auranofin -as a consecuence of gold binding to the serum proteins shows a cytotoxic potency of only one tenth its normal cytotoxic activity in vitro when foetal calf serum is added to the culture medium (Mirabelli et al. 1985a). Therefore, it would be interesting to design compounds of Au(I) with a low affinity towards the thiols of serum proteins. One way of decreasing the reactivity of Au(I) towards the thiols is by forming bridges of dppe (bis(diphenylfosfinoethane), a ligand which in itself is active against leukemia P388 in mice. One compound of this type is (1)μ-[bis(diphenylphosphine)ethane]-bis[1-thio-ß-Dglucopyranosategold(I)], which contains two 1-thio-ß-D-glucopyranosato groups linked to gold by sulfur (similar to auranofin) and at the same time, bridged by dppe, is more active than auranofin to some tumors (Mirabelli et al., 1986c)

Bis(diphenylphosphine)ethanegold(I) complexes. The diphosphine bridged digold compound, bis(diphenylphosphino)ethane chlorogold(I) [Au(dppe)$_2$Cl] has been found to be eight times more active than the free ligand (dppe) towards B16 melanoma. It has also been found to be active against P388 (ILS = 83%) and L 1210 leukemias (Johnson et al., 1986; Snyder et al., 1986; Berners-Price et al., 1986b). [Au(dppe)$_2$Cl] is often comparable with cisplatin due to its high cytotoxic and because it shows good antitumor activity in a variety of animal tumours. Moreover [Au(dppe)$_2$Cl] and cisplatin administered concurrently act synergistically against moderately advanced P388 leukaemia.

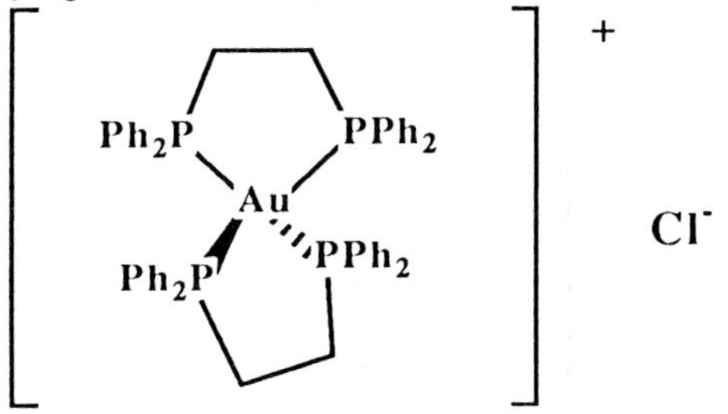

Bis[1,2-bis(diphenylphosphine)ethane]gold(I) chloride, [Au(dppe)$_2$]$^+$

Nevertheless the cytotoxic mechanisms of [Au(dppe)$_2$Cl] and cisplatin are obviusly different because the Au(I) ion is much 'softer' than that of Pt(II) and it is incapable of forming strong bonds to nitrogen ligands and for that reason it is unlikely that the crucial point of attack involves Au binding directly to DNA bases (Berners-Price and Sadler, 1987). X-ray crystallography showed the tetrahedral chelated structure of [Au(dppe)$_2$Cl] to be a ball of phenyl rings enveloping the gold. Furthermore ring closure reactions have been directly observed taking place in human blood plasma forming a chelate (Berners-Price et al., 1986a). When the ligands form strong chelate rings, the cytotoxic activity of gold compounds is at a maximum due to the increased stability introduced by the chelation. For these reasons [Au(dppe)$_2$]$^+$, as was to be expected, was not reactive towards the thiols. This low reactivity towards thiols of [Au(dppe)$_2$Cl] probably makes it possible for the toxic diphosphine to be delivered to targerts within tumor cells by Au(I). Investigations have been carried out on the relation between the structure and the activity of an extensive range of phosphines and their respective metal complexes (Mirabelli et al., 1985b). By means of chelation, metals can protect diphosphines more than monophosphines. Nevertheless, for the phosphine be reactive at the target site, kinetic instability in the metal-phosphine bonds must exist (eg. Au(I), Ag(I) and Cu(I) diphosphine complexes, all of which show good anticancer activity) (Berners-Price et al., 1984). On the other hand [PtCl$_2$(dppe)] is completely inactive (Berners-Price et al., 1985) probably because of the high kinetic stability of its Pt-P bonds. Also active are analogues of Au(I), Ag(I) and Cu(I) complexes with either a three-carbon or a cis-ethene bridge. The replacement of phenyl groups by those of ethlys led to a reduction or total loss of activity. Other analogues such as the diphosphines with alkyl groups are inactive, as are also As or S analogues of dppe.

It is possible that the formation of DNA-protein cross-links and DNA stand breaks may be responsible for the antitumor properties of [Au (dppe)$_2$]$^+$ (Berners-Price et al. 1986b). The existence of DNA single-strand breaks and DNA-protein cross-links in tumor cells probably indicate why nuclear chromatin seems to be a cellular target for [Au (dppe)$_2$]$^+$ (Johnson et al., 1987). It has also been observed that small concentrations of [Au (dppe)$_2$]$^+$ prevent protein synthesis in relation to DNA and RNA synthesis (Hoke et al., 1988). [Au (dppe)$_2$]$^+$ provokes initially a rapid increase in cellular respiration and a decrease in the ATP cellular content. This indicates that mitochondria is a target organ in cytotoxicity. Probably the acceptor properties of O. in the tertiary phosphines are responsible for the above mentioned cytotoxicity. In isolated rat liver mitochondria, state 4 respiration is stimulated by [Au (dppe)$_2$]$^+$ which suggests that it acts as an uncoupler of oxidative phosphorylation (Hoke et al., 1988). Although it was further suggested that during phosphine oxidation, the generation of reactive radical species may play a part in the observed cytotoxic activity of dppe and [Au (dppe)$_2$]$^+$, the consequences of redox reactions in isolated hepatocytes often include GSH oxidation and lipid peroxidation but GSH depletion was not observed before lethal cell injury. Furthemore, no great alterations in the membrane permeability due to phospholipase A2 have been observed. Therefore, these findings would seem to prove that no initial oxidative cell injury is caused by either $[Au\ (dppe)_2]^+$ or the ligand dppe. The mechanism by which $[Au\ (dppe)_2]^+$ causes an alteration of the membrane potential appears to be a selective increased permeability of the inner mitochondrial membrane to cations and protons (Smith et al., 1989)

The triphenylphosphine ligand. Gold(I) complexes of type Ph_3P-AuX, where X = 5-fluorouracil, 6-mercaptopurine, 5-fluorodeoxyuridine or thymidine, also exhibit strong antitumor properties towards several tumor systems (Agrawal et al. 1978). The complex which we are in the process of studing is also of the type Ph_3P-AuX and is the 8-(thioteophyllinatotriphenylphosphine)gold(I) compound (tTPAu) which has an atom of gold(I) coordinated to two ligands. The X-ray crystal analysis has revealed that the gold is bound to the triphenylphosfine ligand through the phosphorous atom and to the 8-tiotheophyllinate ligand through the sulphur atom. The strength of these bonds suggests a low affinity of this compound to the thiol groups giving a lower reactivity with the serum proteins, resulting in a higher probability of arriving at the cells of the affected organs. In adition to that, the tTPAu maximum tolerated dose is 110μM/kg/dayx10, indicating that its organic toxicity is among the lowest when compared to different gold compounds being tested (Mirabelli et al., 1986a)

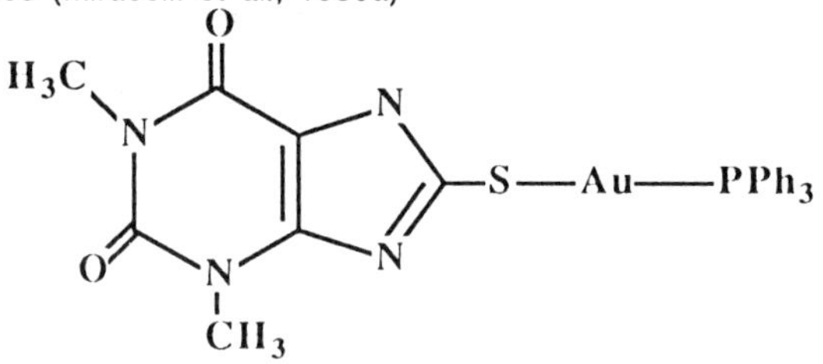

(8-Thiotheophyllinato)(triphenylphosphine)gold(I), (tTPAu)

tTPAu shows a high cytotoxic activity in vitro. Melanome B16F10 cells treated with tTPAu reach very low IC50 (IC50<0.5 μM). This value means a high cytotoxicity for this compound, which allows us to include it among the more active gold compounds. The cytotoxic activity of each ligand is lower than that shown by the complex. This is to be expected since the bond with the atom of gold increases the cytotoxicity of the tertiary phosphines and of the theophylline (Fig. 1).

Table 1. Flux cytometry analysis results (Rh-123 fluorochrome) obtained in L1210 cells treated with tTPAu.

[ ] (μM)	MEAN FALS	*FALS/MEAN FALSc	*IGFL/ IGFLIGFLc	
0	10,5	100,0%	18,3	100,0%
0,6	9,7	92,4%	18,1	98,9%
1,25	10,2	98,4%	17,4	94,8%
5	11,8	113,1%	15,0	82,1%
10	13,1	125,5%	12,6	68,8%

* % relative to control
FALS - Size
IGFL - Green fluorescence

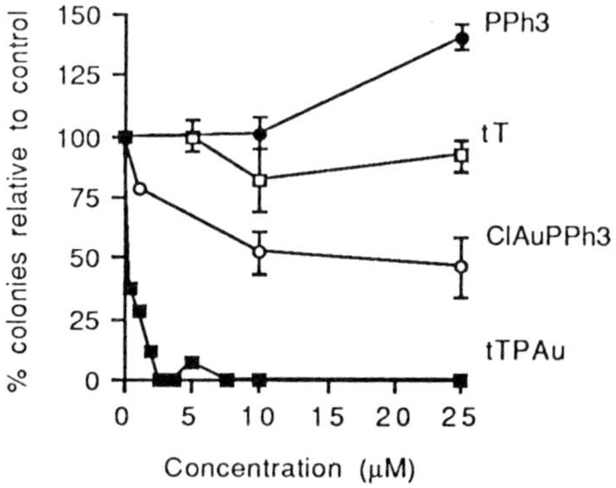

Figure 1. Inhibition of melanome B16F10 colonie formation induced by tTPAu and its free ligands after 2 h of treatment and 72 h of recovery

Melanome B16F10 cell cultures as well as L1210 cell cultures were treated with doses of tTPAu between 0-10 μM. The results using flux cytometry analysis of the cell cycle did not show any differences when compared with the results obtained from the cell culture controls. This indicates that the compound cytotoxic activity is not the result of the interaction between the tTPAu and the chromatine.

It has been reported relative differences in the abilities of mitochondria from normal and cancerous cells to accumulate liphofilic cationic dyes such as rhodamine 123. The demonstration of elevated mitochondrial membrane potential in a variety of cancerous cells, relative to normal cells has been suggested as a rationale for usage of lipophilic cationic compounds as chemotherapeutic agents. Therefore, the selective targeting of mitochondria in cancerous cells might be exploited usefully in an antineoplastic regimen. Gold compounds such as TEPAu y [Au(dppe)$_2$Cl], both having phosphine ligands, affect the mitochondrial activity producing the main cytotoxic action. Therefore, we have studied tTPAu action through mitochondria using flux cytometry analysis. The Rh 123 uptake into the treated cells was significantly lower than the controls. The decrease in the fluorescence was as low as 31.2% at tTPAu 10μM. These results seem to suggest that tTPAu decreases the mitochondrial membrane potential. The mechanism of action proposed in the case of TEPAu and [(dppe)$_2$Au]Cl complexes is by increasing the permeability of the mitochondria membranes which results in an oxidative phosphorylation uncoupling and this appears to be the reason for the observed cytotoxicity. It might be that the tTPAu mechanism is similar to the above, therefore we are programming our work in order to prove this hypothesis.

REFERENCES

Ainscough E.W. and Brodie A.M. (1985) Gold chemistry and its medical applications. Ed. Chem. January: 6-8

Agrawal, K.C., Bears, K.B., Marcus, D. and Jonassen H.B. (1978): Gold triphenylphosphine complexes as a new class of potential antitumor agents. Proc. Am Assoc. Cancer Res. and ASCO: 28

Allaudeen, H.S., Snyder, R.M., Whitman, M.H. and Crooke, S.T. (1985). Effects of coordinated gold compounds on in vitro and in situ DNA replication. Biochem. Pharmacol. 18: 3243-3250

Berners-Price, S.J., Mazid, M.A. and Sadler, P.J. (1984): Stable gold(I) complexes with chelate rings: solution studies of bis(phosphino)ethane complexes and X-ray crystal structure of bis[1,2-bis(diphenylphosphino)ethane]gold(I) hexafluoroantimonate-acetone (1/1). J. Chem. Soc. Dalton Trans., 969-974

Berners-Price, S.J., DiMartino, M.J., Hill, D.T., Kuroda, R., Mazid, M., and Sadler, P.J. (1985): Tertyary phosphine complexes of gold(I) and gold(III) with imido ligands: 1H, 31P and 15NMR spectroscopy, antiinflamatory activity and X-ray crystal structure of (phthalimido(triethylphosphine)gold(I). Inorg. Chem. 24: 3425-3434

Berners-Price, S.J. and Sadler (1986a) Gold(I) complexes with bidentate tertiary phosphine ligands: Formation of annular vs. tetrahedral chelated complexes. Inorg. Chem. 25: 3822-3827

Berners-Price, S.J., Mirabelli, Ch.K., Johnson, R.K., Mattern, M.R., McCabe, F.L., Faucette,. F., Sung, Ch.-M., Mong, S.-M., Sadler, P.J. and Crooke S.T. (1986b): In vivo antitumor activity and in vitro cytotoxic properties of bis[1,2-bis(diphenylphosphino)ethane]gold(I) chloride. Cancer Res. 46: 5486-5493

Berners-Price, S.J. and Sadler P.J. (1987) Phosphines in medicine. Chem. Britain Jun.: 541-544

Haiduc, I. and Silvestru, C. (1989): Rhodium, Iridium, copper and gold antitumor organometallic compounds (Review). In vivo 3: 285-294

Hoke, G.D., Rush, G.G., Bossard, G.E., McArdle, J.M., Jensen, B.D. and Mirabelli, C.K. (1988) Mechanism of alterations in isolated rat liver mitochondrial function induced by gold complexes of bidentate phosphines. J. Biol. Chem. 263: 11203-11210

Hoke G.D., Rush G.F. and Mirabelli C.K. (1989): The mechanism of acute cytotoxicity of triethylphosphine gold(I) complexes. III. Chlorotriethylphosphine gold(I)-induced alterations in isolated rat liver mitochondrial function. Toxicol. App. Pharmacol. 99: 50-60

Jonhson, R.K., Faucette, L.F., McCabe, F.L., Rush, G.F., Goldstein, R.S. and Mirabelli, C.K. (1986): Therapeutic and toxicologic properties of bis[1,2-bis-(diphenylphosphino)ethane gold(I)chloride (SK&F 101722). Proc. of Am Assoc. Cancer Res. 27: 281

Lippard, S.J. (1984) Platinum coordination complexes in cancer chemoteraphy. eds. M.P. Hacker, E.B. Douple and I.H. Krakhoff, pp. 11. Boston: Martinus Nijhoff Publishers.

Mirabelli, C.K., Johnson, R.K., Sung, C.-M., Faucette, L., Muirhead, K. and Crooke, S.T. (1985a) Evaluation of the in vivo antitumor activity and in vitro cytotoxicity properties of Auranofin, a coordinated gold compound, in murine tumor models. Cancer Res. 45: 32-39

Mirabelli, C.K., Johnson, R.K., Bartus, J.O., Sung, C.-M. and Von Hoff, D.D. (1985b) Cytotoxic properties and metal interactions of bis(diphenylphosphine)ethane and related antinecplastic bisphosphines. Proc. Am. Assoc. Cancer Res. 26: 256-265

Mirabelli, C.K., Johnson, R.K., Hill, D.T., Faucette, L.F., Muirhead, K. and Crooke, S.T. (1986a): Correlation of the in vitro and in vivo antitumor activities of gold(I) coordination complexes. J. Med. Chem. 29: 218-233

Mirabelli, Ch.K., Sung, Ch.-M., Zimmerman, J.P., Hill, D.T., Mong, S.-M. and Crooke, S.T. (1986b): Interactions of gold coordination complexes with DNA. Biochem. Pharmacol. 35: 1427-1433

Mirabelli, C.K., Jensen, B.D., Mattern, M.R., Sung, C.-M., Mong, S.-M., Hill, D.T., Dean , S.W., Schein, P.S., Johnson, R.K. and Crooke, S.T. (1986c): Cellular pharmacology of μ-[1,2-bis(diphenylphospino)ethane] bis[1-thio-ß-D-glucopyranosato-S)gold(I)]: a novel antitumor agent. Anti-Canc. Drug Design 1: 223-234

Rosenberg, B. (1980) Cisplatin, current status and new developments. A.W. Prestayko, S.T. Crooke and S.K. Carter eds. New York Academic

Rush, G.F., Smith, P.F., Alberts D.W., Mirabelli C.K., Snyder, R.S., Crooke, S.T., Sowinski, J., Jones, H.B. and Bugelski, P.J. (1987a): The mechanism of acute cytotoxicity of triethylphosphine gold(I) complexes. I. Characterization of triethylphosphine gold chloride-induced biochemical and morphological changes in isolated hepatocytes. Toxicol. App. Pharmacol. 90: 377-390

Rush, G.F., Smith, P.F., Hoke G.D., Alberts D.W., Snyder, R.S. and Mirabelli C.K. (1987b): The mechanism of acute cytotoxicity of triethylphosphine gold(I) complexes. II. Triethylphosphine gold chloride-induced alterations in mitochondrial function. Toxicol. App. Pharmacol. 90: 391-400

Sadler, P.J. (1976): The biological chemistry of gold. Gold Bull. 9: 110-118

Sadler, P.J., Nasr, M. and Narayanan, V.L. (1984): The design of metal complexes as anticancer drugs. In Platinum Coordination Complexes in Cancer Chemotherapy, eds. M.P. Hacker, E.B. Douple and I.H. Krakhoff, pp. 290-304. The Hague: Martinus Nijhoff Publishers.

Simon, T.M., Kunishima, D.H., Vilbert, G.J. and Lorber, A. (1981) Screening trial with the coordinated gold compound Auranofin using mouse lymphocytic leukemia P 388. Cancer Res. 41: 94-97

Smith, P.F., Hoke, G.D., Alberts, D.W., Bugelski, P.J., Lupo, S., Mirabelli, C.K., Rush G.F. (1989) Mechanism of toxicity of an experimental bidentate phosphine gold complexed antineoplastic agent in isolated rat hepatocytes. J. Pharmacol. Exp. Ther. 249: 944-950

Snyder, R.M., Mirabelli, Ch.K., Johnson, R.K., Sung, Ch.-M., Faucette, L.F., McCabe, F.L., Zimmerman, J.P., Whitman, M., Hempel, J.C. and Crooke, S.T. (1986): Modulation of antitumor and biochemical properties of bis(diphenylphosphino) with metals. Cancer Res. 46: 5054-5060

III. Gallium

Metal Ions in Biology and Medicine. Eds. Ph. Collery, L.A. Poirier, M. Manfait, J.C. Etienne. John Libbey Eurotext, Paris © 1990, pp. 393-397

Mechanism of gallium uptake in tumours

R. Sephton, S. De Abrew

Peter Mac Callum Cancer Institute, Little Londsdale St., Melbourne, Australia 3000

The clinical place of gallium, as carrier-free Ga-67, is perhaps best judged by its widespread and continuing use in tumour imaging over two decades (eg Johnson, 1981; Tumeh et al, 1987). Due to a remarkably high preferential uptake, some tumours (principally in the lymphoma categories) are detected with high sensitivity but uptake is lower and variable in other tumour types and their detection is less reliable. Its value in tumour detection is also clouded by its propensity to accumulate in inflammatory lesions and there is a large literature on this clinical application as well, including in AIDS related conditions (eg Hoffer, 1981). Stimulated largely by the observations of preferential tumour uptake, some centres began testing non-radioactive Ga for anti-tumour activity - given as citrated Ga nitrate, in massive (>1g) doses by comparison with the carrier free (<ng) diagnostic dose. Some success was reported in refractory lymphoma and in small cell lung cancer (Foster et al, 1986) but its potential as a single agent appears limited. However Ga can be effective in treatment of cancer related hypercalcaemia and of bone metastases (Warrell & Bockman, 1989).

The considerable previous research on mechanisms of Ga-distribution and pathological uptake was done partly for the purpose of optimizing clinical use, partly perhaps out of some fascination with an unexpected or unknown aspect of tumour biology. A clear unified picture has not yet emerged, certainly in terms of basic questions like - what does high (or low) Ga-67 uptake say about the individual tumour's biology, and is that information clinically exploitable? This paper presents our understanding of distribution and uptake mechanisms, and is concerned largely with the notion of Ga as an iron (Fe) analogue, particularly with the transport protein transferrin (TF) and its interactions with cells. There are alternative hypotheses but none of these has a comparable experimental foundation.

THE ROLE OF TRANSFERRIN IN CELLULAR UPTAKE

The TF mechanism became a major line of research following reports that relatively small amounts of TF added to mouse lymphoid tumour cell cultures caused up to 10-fold increased cellular uptakes of Ga-67 and Fe-59 (Harris & Sephton, 1977). These observations were confirmed and extended by other groups, particularly by Larson et al (1980) who documented what we had only inferred - that Ga and Fe uptakes proceed through interactions of TF with saturable transferrin receptors (TFR) - and they postulated that uptake occurred by receptor-mediated endocytosis of Ga-TF. Coincidently during this period, it was shown that TF was an essential growth factor for all cells (Barnes & Sato, 1980) and that TFR were expressed not only on Hb-synthesizing cells but on all proliferating cells (Trowbridge & Omary, 1981). Several groups published detailed studies which established the uptake sequence for Fe as - binding of Fe-TF to TFR, endocytosis of the entire complex, dissociation of Fe in the acidic (ph 4) endosome, return of the TF-TFR complex to the cell surface where they dissociate leaving the TFR to interact with another Fe-TF and the TF to bind more Fe (eg Klausner et al, 1983).

Thus over the last decade it has become accepted that Fe is essential to cell growth, that expression of TFR is necessary to proliferation but also modulated by availability of Fe (Testa et al, 1984), and that possible anti-tumour strategies might be to interfere with Fe metabolism or to target TFR directly by antibodies or by antibody-toxin conjugates (eg Trowbridge, 1989). Assuming high Ga uptake in tumours depends on abundant, rapid cycling TFR, it may indeed denote rapid proliferation or a large growth function. However uptake and TFR expression also vary in response to Fe availability, decreased in experiments using cultured cells which had been grown up in medium containing hemin (Chitambar & Zivkovic, 1987) and increased in cells which had been grown up in medium containing desferrioxamine (our unpublished data). Of course non-proliferating cells of haemopoietic lineage may also express TFR. Andreesen et al (1987) reported that human monocytes, allowed to mature in vitro, expressed TFR in numbers reaching about 10^6 per cell for the mature macrophage (of the same order as recorded for tumour cell lines). The macrophages also showed high levels of TF-mediated Ga-67 uptake.

Much of the recent work on TF-mediated Ga uptake has in fact concerned its cytostatic activity (eg Chitambar et al, 1988; Hedley et al, 1988). Their view is that Ga inhibits DNA synthesis by blocking cellular incorporation of the Fe required to activate the enzyme ribonucleotide reductase. This puts Ga in close relation to other S-phase inhibitors acting on Fe, like desferrioxamine.

TRANSFERRIN'S ROLE IN DISTRIBUTING GA IN VIVO

The above evidence makes a strong case for TF and TFR as key participants in the cellular uptake mechanism but can we translate this to the whole animal? In normals, at least 50% of available metal binding sites on circulating TF are vacant and following i.v. administration, tracer Ga is circulated largely as the (moderately stable) Ga-TF complex. It is cleared gradually; the Ga dissociating from the TF complex is cleared by extravascular diffusion, renal excretion and by binding to bone mineral. If available TF binding sites are decreased eg by additional Fe (marrow suppression or recent blood transfusion, as clinical examples) or by administration of substantial amounts of any TF-loading metal (as in animal experiments), the rate and magnitude of Ga dissociation and clearance are greatly increased. This includes of course the case of Ga chemotherapy where kinetics and distributions are very different from those seen with tracer amounts.

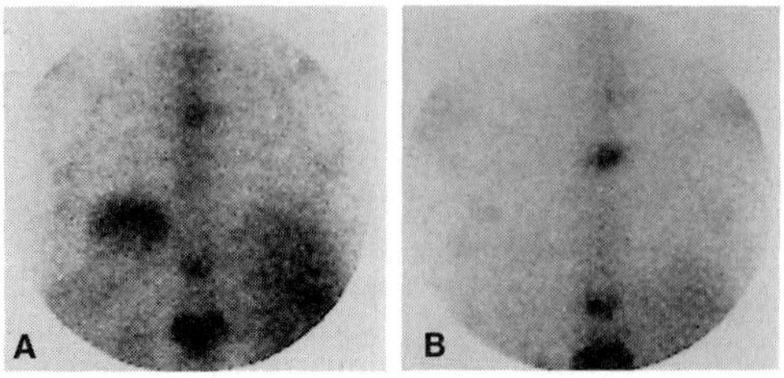

Fig. 1. Posterior images of chest in patient with adenoca L. lung. Tracer Ga-67 shows lung tumour and other soft tissues, plus spine and spinal metastases (A), whereas Ga-67 given with Ga chemo shows only spine and mets (B).

However the notion that TF itself mediates the specific uptake of Ga in high-uptake tumours is controversial for two reasons - 1) other metals including Fe and In show much higher TF-mediated uptakes in cell culture systems, whereas Ga uptake is much higher in tumours in vivo and 2) tumour uptake of Ga remains high in vivo even if the Ga binding capacity of plasma TF is reduced to near-zero, by additional Fe, Sc or other metal. This latter point was the focus of a series of studies by the Oak Ridge group (Hayes, 1983) who injected rodent tumour hosts with saturating doses of Scandium (Sc) and noted that tumour uptake remained high. They proposed that Ga uptake was a TF-independent process involving the diffusion of unbound Ga through hyperpermeable tumour cell membranes. We ourselves had performed detailed studies in mice which showed that hyperferremia accelerated the plasma clearance of Ga and tended to hasten, rather than inhibit, tumour uptake; in fact we showed that tumour visualisation was improved if Fe was given to patients prior to Ga-67 imaging (Sephton & Martin, 1981).

The apparent conflict between the in vivo observations and the TF hypothesis was explored in some detail (Sephton et al, 1982). We suggested that, due to the different stabilities of the metal TF complexes, continuously-dissociating Ga but not Fe or Sc would readily diffuse into tumour interstitial spaces where interstitial TF would then complete the uptake process. This process would be accelerated by loading the plasma with Fe or Sc but in neither case would a moderate excess cause loading of interstitial TF because of the low solubility of hydrolysis products. In support, we showed that TF given as a focal injection into muscle accumulated Ga-67 from the plasma and this accumulation was increased by loading the plasma with Fe or Sc. We also showed that during the first hours post injection, Ga-67 accumulated in tumour could be brought back into the plasma by injecting TF or could be cleared by giving desferrioxamine i.v.; with the passage of time, the accumulated Ga became irreversibly bound.

The foregoing gives plausible but indirect support for the relevance of TF and TFR as key factors in vivo. An interesting recent study which addressed the question more directly showed that Ga-67 in a human melanoma xenograft in athymic mice was greatly reduced by prior injection of an anti-human TFR monoclonal antibody (B3/25) (Ming Chan et al, 1987). The authors were concerned that his particular antibody does not block the binding of TF, but it may interfere with TFR function in vivo.

HOST FACTORS

The possibility of common elements in the uptake mechanisms for tumours and for inflammatory lesions has not to our knowledge been systematically examined. Our approach was to ask whether tumour uptake was somehow dependent on the presence of host inflammatory cells, using host mice which had been pretreated (immunosuppression) such that there would be a reduced traffic of inflammatory cells during the tumour's growth. Details of some of this work were reported elsewhere (Sephton et al, 1990) but Table 1 shows a representative experiment. Tumour uptakes in the RILQ lymphoma were indeed reduced, if the hosts (s.p.f. CBA mice) were given whole body irradiation (WBI) prior to tumour inoculation. Tumour uptakes remained high however if the WBI was countered by giving syngeneic bone marrow cells i.v., to hasten haemopoietic recovery. Similar effects were noted in four of five mouse tumour models tested, the exception being the B16 melanoma in C57/BL. We also noted that prior treatment of hosts with cyclophosphamide produced essentially the same diminution of tumour uptake as did WBI. Figure 2 presents new data showing that cells taken from these lower-uptake RILQ tumours (in the WBI or cyclophosphamide-treated mice) express low numbers of TFR relative to cells from control tumours. The TFR assay was done by cytofluorographic assessment of the cells' capacity to bind fluoresceinated human FeTF. Thus we have an additional confirmation of the role of TFR in Ga uptake, also an important observation that marrow derived host factors seem to influence the expression of TFR within the tumour. It is worth remarking that in these animals pretreated by WBI or cyclophosphamide, the tumours grew normally and appeared on histological sections no different from control tumours (predominantly tumour cells with few macrophages).

Table 1 - Prior WBI in RILQ Hosts

	0		4 Gy -		4 Gy +bm		6 Gy -		6 Gy +bm	
	Ga	Fe	Ga	Fe	Ga	Fe	Ga	Fe	Ga	Fe
Blood	2.0	18	1.9	6.0	2.1	27	0.7	1.9	1.9	23
Liver	3.0 (2.5-3.3)	4.8	3.2 (2.6-3.7)	13.0	3.4 (2.9-4.0)	6.0	2.1 (1.8-2.3)	19	3.5 (3.2-3.7)	8.3
Spleen	2.9 (2.9-3.0)	55	3.0 (2.7-3.5)	27	2.7 (2.3-2.9)	47	2.1 (1.9-2.2)	8.0	4.1 (3.6-4.8)	44
Tumour	11.5 (10.2-12.6)	3.7	5.5 (4.2-6.2)	7.0	10.5 (9.2-11.3)	4.2	2.4 (1.9-3.3)	6.4	7.5 (5.5-8.9)	4.7

Ga-67 and Fe-59 distributions (%A_o per g), at 24h post i.v. injection, 7d post WBI
Mean values for 3 mice per group, and ranges for Ga-67 data.
+bm bone marrow cells injected i.v. at 4h post WBI.

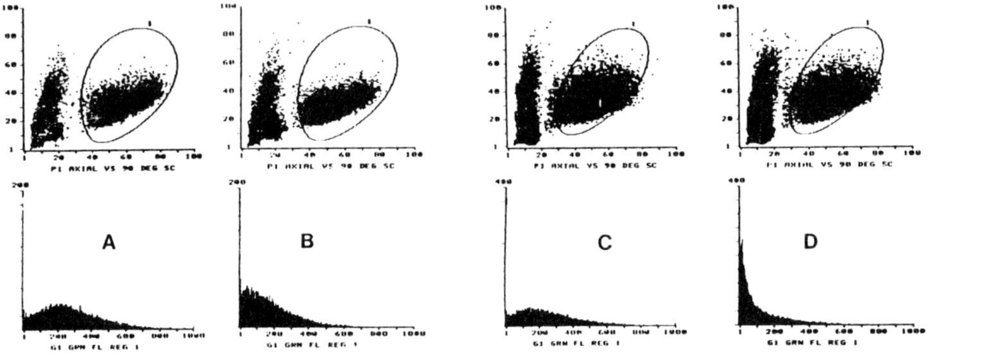

Fig. 2. Binding of TF by cells from RILQ tumours. Lower distributions (green fluorescence) show TF binding for (intact) tumour cells within regions outlined in upper (which indicate all preparations were of comparable quality - regions contain approx. 60% of total counts). Tumours from controls compared against hosts given pre-WBI (A & B), controls compared against hosts given prior cyclophosphamide (C & D).

DISCUSSION

In oncology tumour characterization is complementary to detection, and equally important if it gives good predictive information. A relation between in vivo Ga uptake and TFR has important biological and clinical implications, in terms of current ideas about TFR expression by cycling cells and their Fe needs relative to Fe availability, also in terms of the postulated role for marrow-derived host factors. Recent reports on TF itself as an autocrine growth factor (eg Vostrejs et al, 1988) have other interesting implications. The use of Ga chemotherapy for non-osseous tumours could well be advanced if existing knowledge on two fronts - the essential tumour properties determining uptake and the details of subsequent effects on cell kinetics - were consolidated and applied rationally. As an almost-trivial measure, Ga therapy might be advanced just by imaging beforehand using tracer Ga-67 to predict tumour uptake.

REFERENCES

Andreesen, R., Sephton, R.G., Gadd, S., Atkins, R.C. and De Abrew, S. (1988): Human macrophage maturation in vitro: expression of functional transferrin binding sites of high affinity. Blut 57, 77-83.

Barnes, D. and Sato, G. (1980): Serum free cell culture: A unifying approach. Cell 22, 649-655.

Chitambar, C.R. and Zivkovic, Z. (1987): Uptake of Gallium-67 by human leukemic cells: demonstration of transferrin receptor dependant and independant mechanisms. Cancer Res. 47, 3929-3934.

Chitambar, C.R., Matthaeus, W.G., Antholine, W.E., Graff, K. and O'Brien, W.J. (1988): Inhibition of leukemic HL-60 cell growth by transferrin-gallium: effects on ribonucleotide reductase. Blood 72, 1930-1936.

Foster, B.J., Clagett-Carr, K., Hoth, D., Leyland-Jones, B. (1986): Gallium nitrate: the second metal with clinical activity. Cancer Treat. Rep. 70, 1311-1319.

Harris, A.W. and Sephton, R.G. (1977): Transferrin promotion of Ga-67 and Fe-59 uptakes by cultured mouse myeloma cells. Cancer Res. 37, 3634-3638.

Hayes, R.L. (1983): The interaction of gallium with biological systems. Int. J. Nucl. Med. Biol. 10, 257-261.

Hedley, D.W., Tripp, E.H., Slowiaczek, P. and Mann, G.J. (1988): Effect of gallium on DNA synthesis by human T-cell lymphoblasts. Cancer Res. 48, 3014-3018.

Hoffer, P.B. (1981): Use of gallium-67 for detection of inflammatory diseases. Int. J. Nucl. Med. Biol. 8, 243-247.

Johnson, G.S. (1981): Clinical applications of gallium in oncology. Int. J. Nucl. Med. Biol. 8, 249-255.

Klausner, R.D., Ashwell, G., Van Renswoude, J. et al (1983): Receptor mediated endocytosis of transferrin in K562 cells. J. Biol. Chem. 258, 4715-4724.

Larson, S.M., Rasey, J.S., Allen, D.R., Nelson, N.J., Grunbaum, Z., Harp, G.D., and Williams, D.L. (1980): Common pathway for tumour cell uptake of Ga-67 and Fe-59 via a transferrin receptor. J. Natl. Cancer Inst. 64, 41-53.

Ming Chan, S., Hoffer, P.B., Maric, N. and Duray, P. (1987): Inhibition of Ga-67 uptake in melanoma by an anti human transferrin receptor monoclonal antibody. J. Nucl. Med. 28, 1303-1307.

Sephton, R.G., De Abrew, S. and Hodgson, G.D. (1982): Mechanisms of distribution of gallium in mouse tumour hosts. Brit. J. Radiol. 55, 134-141.

Sephton, R.G. and Martin, J.J. (1981): Ga-67 imaging incorporating administration of iron. Int. J. Nucl. Med. Biol. 8, 341-348.

Sephton, R.G., De Abrew, S. and Hodgson, G.S. (1990): Gallium uptakes in tumours grown in whole body irradiated mice. Proc. 6th Int. Conf. Radiopharmacol. (In press).

Testa, V., Louache, F., Titeux, M., Thomopoulos, P. and Rochant, H. (1985): The iron chelating agent picolinic acid enhances transferrin receptors expression in human erythroleukemic cell lines. Br. J. Haematol. 60, 491-502.

Trowbridge, I.S. and Omary, M.B. (1981): Human cell surface glycoprotein related to cell proliferation is the receptor for transferrin. Proc. Natl. Acad. Sci. 78, 3039-3043.

Trowbridge, I.S. (1988): Transferrin receptor as a possible therapeutic target. In Monoclonal Antibody Therapy: Prog. Allergy 45, ed. H. Waldman, pp. 121-146. Basel: Karger.

Tumeh, S., Rosenthal, D.S., Kaplan, W.D., English, R.J. and Holman, B.L. (1987): Lymphoma: evaluation with Ga-67 SPECT. Radiology 164, 111-114.

Vostrejs, M., Moran, P.L. and Seligman, P.A. (1988): Transferrin synthesis by small cell lung cancer cells acts as an autocrine regulator of cellular proliferation. J. Clin. Invest. 82, 331-339.

Warrell, R.P. Jr and Bockman, R.S. (1989): Gallium in the treatment of hypercalcemia and bone metastasis. In Important Advances in Oncology 1989, eds. De Vita, V.T., Hellman, S. and Rosenberg, S.A., pp.205-220. Lippincott.

Metal Ions in Biology and Medicine. Eds. Ph. Collery, L.A. Poirier, M. Manfait, J.C. Etienne. John Libbey Eurotext, Paris © 1990, pp. 398-402

Role of iron metabolism in carrier-free 67-gallium citrate accumulation by tumor cells

L.J. Anghileri

Biophysics Laboratory (Director : Prof. J. Robert), Faculty of Medicine, University of Nancy, 18, rue Lionnois, 54000 Nancy, France

Nearly all living cells (bacteria, vegetal and animal have an absolute need for iron. A great number of enzymes (electron transport protein, iron flavoproteins, hydroperoxidases, oxygenases, etc) contain iron or their activity depends on the presence of iron. In animal cells the evolution from the G_1 phase into the S phase (phase of DNA synthesis) needs the presence of exogenous or endogenous iron (Baserga, 1981). For this reason iron-binding agents act as inhibitors of DNA synthesis by blocking animal cells in the G_1 phase (Ganeshaguru et al, 1980). The competition for growth essential iron between pathogenic microorganisms and their animal host demonstrates that the host possesses a nutritional immunity based on iron withholding mechanisms, and the microorganism ability to overcome these mechanisms is a very important factor of its virulence (Weinberg, 1978). An important hypoferremia characterizes the mammals with infection, neoplasms or other conditions that stimulate inflammation. This hypoferremia is accompanied by an increased storage of iron in cell ferritin of the reticuloendothelial system (liver and spleen) (Cartwright and Lee, 1971). In patients with malignant tumors the increased iron storage by liver and spleen is concomitant with a rapid clearance of radioiron from plasma, and the rate of disappearance increases with the tumor dissemination (Miller et al, 1956). The shift of iron from plasma to storage seems to be accompanied by de novo synthesis of ferritin (Fairbanks and Klee, 1981), and this ferritin sequesters iron in parenchymal and reticuloendothelial cells to prevent the normal recycling into the plasma. In addition, apo-lactoferrin released by leukocytes can act as iron-sequestering agent, accumulating it at the site of neoplasm or infection (Weinberg 1981). A variety of solid tumors have shown to be surrounded by macrophages containing large amounts of ferritin and iron (Price and Greenfield, 1958; Magnusson et al, 1977). Hyperplastic mammary tissue has shown a greater iron accumulation than the tumor or the normal cells, and that iron content is comparable or greater than that of liver and spleen (Vijayaraghavan and Rivera, 1981). In Hodgkin's disease a large amount of ferritin and iron accumulates surrounding the tumor nodules (Smithyman et al, 1978).

Gallium binds to at least three iron-binding proteins: transferrin (TF), lactoferrin (LF) and ferritin (FT). In vitro studies of ^{67}Ga uptake by Ehrlich ascites tumor cells have demonstrated that iron-residues, 2-4 nitrogen ligands and one bicarbonate ion are involved in the reaction with ferric ion at each binding site. LF seems to have similar binding sites but the iron-binding is more stable to pH variations (Morgan, 1974). At physiological pH the gallium binding to LF is higher than that corresponding to TF (Hoffer et al, 1977). On the other hand, the binding of I-131-LF has been shown to be three times that corresponding to I-131-TF (for DS sarcoma cells) (Anghileri et al, 1989).

Ferric lactate µ mol	µ mol Fe in cells	Percent of ^{67}Ga in cells
0.1	0.026 ± 0.002	12.3 ± 1.26
0.25	0.066 ± 0.005	15.9 ± 1.06
0.5	0.124 ± 0.006	19.1 ± 1.89
1.0	0.296 ± 0.034	21.9 ± 1.54
2.5	1.910 ± 0.241	69.5 ± 5.78
5.0	4.663 ± 0.040	89.1 ± 0.58
0.0	-	4.9 ± 0.15

6 X 10^7 Ehrlich ascites cells in Tyrode medium incubated at 37°C for 30 minutes. Mean value ± SD of five tubes.

Table 1. Ga-67 and Fe-59 taken up by Ehrlich ascites cells

These experimental results appear to support the concept of a double way for 67-Ga uptake by the tumor cell: 1) an increased permeability to 67-Ga citrate, and 2) a binding to the cell through a complex formation with iron-loaded proteins (mainly LF). Four hs after i.v. injection only half of the injected 67-Ga remains in blood. Only 5 to 10 % of the dose is slowly eliminated with a 48 hs half-life (Larson 1978). The fraction rapidly dialyzed by kidneys is mostly gallium citrate (Zivanovic et al, 1978), and the 67-Ga remaining in blood is bound to transferrin and other blood proteins (Hartman and Hayes, 1969). An increased permeability of tumor cell membrane to citrate ions has been demonstrated (Anghileri et al, 1988), and carrier-free 67-Ga seems to follow citrate ions migration. The finding that in presence of gallium-carrier 67-Ga does not penetrate the cell by physical diffusion agrees with the dilution effect of citrate ions. The ultimate destination of 67-Ga citrate on the cell surface and after penetration into the cell seems to be related to Mg2+- and Ca2+-binding sites (Anghileri and Heidbreder, 1977), to hydrolyzed ferric ions, as above reported, and to IBP specially LF because of its highest affinity and its abnormally high concentration in tumors. The 67-Ga-citrate carried over by TF is released at the cell surface and follows the same uptake process as 67-Ga-citrate directly reaching the cell. These two ways of 67-Ga uptake are represented in figure 2.

The results of 67-Ga uptake by tumor cells in the presence of TF and FT disolved in the medium indicate a competition for 67-Ga with the cells, while LF enhances the binding by the cells. Since LF showed a high affinity for tumor cells, and the LF-mediated 67-Ga uptake is not modified by sodium citrate, it can be assumed that the uptake is due to direct binding of LF with the 67-Ga incorporated, pressumably by co-binding with iron.

These experimental results have clearly shown that iron metabolism, directly by hydrolytic ferric ion transformation, or by binding to IBP appears involved in the accumulation of 67-Ga by tumors, and on similar bases in the accumulation of 67-Ga by inflammatory lesions.

saturated lactoferrin (is-LF) in the incubation medium increases by 700 % the uptake, while for iron-low lactoferrin (il-LF), iron-saturated transferrin (is-TF), iron-free transferrin (if-TF) and ferritin (FT) the uptakes are equal or lower than that of control cells (Figure 1). On the other hand, 67-Ga uptake by iron-binding protein in the absence of cells is 7.9 % for (is)LF, 3.0 % for (is)TF, 752 % for (il)LF, 58.5 % for (if)TF and 379 % for FT, respectively of the values corresponding to the tumor cells uptake in the presence of the same amount of iron-binding protein (IBP) and for the same dose of 67-Ga-citrate (Figure 1). Citrate ion is known to inhibit the uptake of 67-Ga-citrate by tumor cells (Anghileri et al, 1988). When the 67-Ga uptake is performed in the presence of 1 mM sodium citrate the control cells show a 60 % decrease, and all the tested IBP but the (is)LF and (il)LF presented a decrease of uptake similar to that of the control cells (Figure 1).

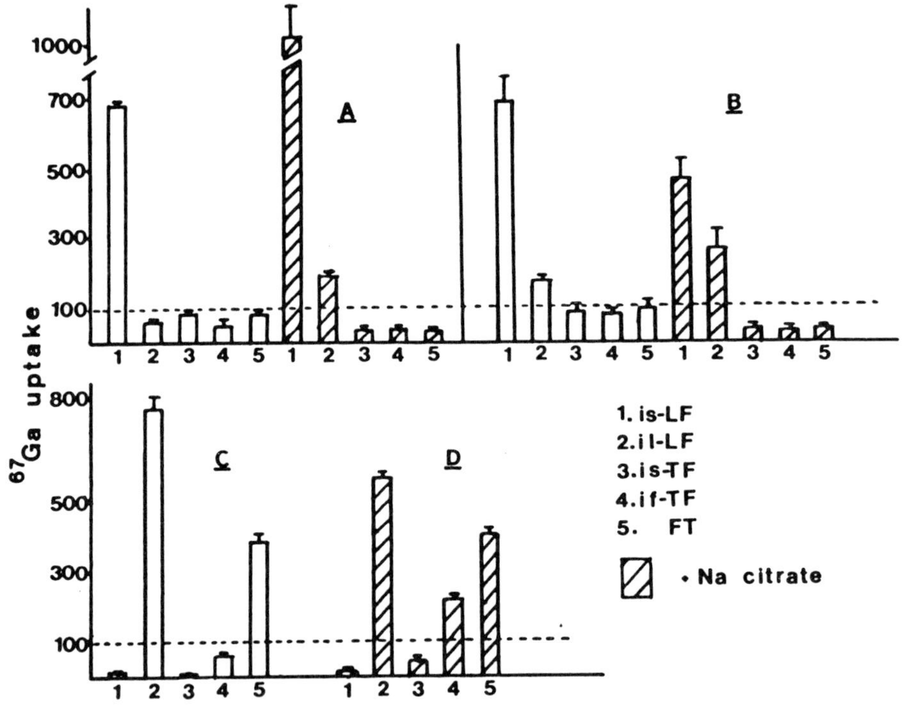

Figure 1. In vitro 67-Ga-citrate uptake.- A: single incubation (as % of cells without IBP); B: preincubation with IBP (as % of cells without IBP); C and D: binding to IBP in absence of cells (as % of 67-Ga taken up by cells plus IBP from the same 67-Ga dose). 1: (is)LF; 2: (il)LF; 3: (is)TF; 4: (if)TF and 5: FT. Shadow bars: in presence of 1 mM sodium citrate. All values are mean ± SD of five tubes.

Hydrolyzed ferric ions are adsorbed or bound to tumor cell surface, and 67-Ga is incorporated to that colloidal iron by the same phenomenon that suggested its use as a scavenging agent to concentrate gallium from very diluted solutions (Wainer, 1934). In the case of table 1 the hydrolyzed iron is pressumably as hydroxi-phosphate, a compound similar to the ferric ions that are present in FT as large numbers of hydrous ferric-oxide-phosphate cores or micelles (Harrison, 1974). The iron-binding is different for TF where three tyrosyl

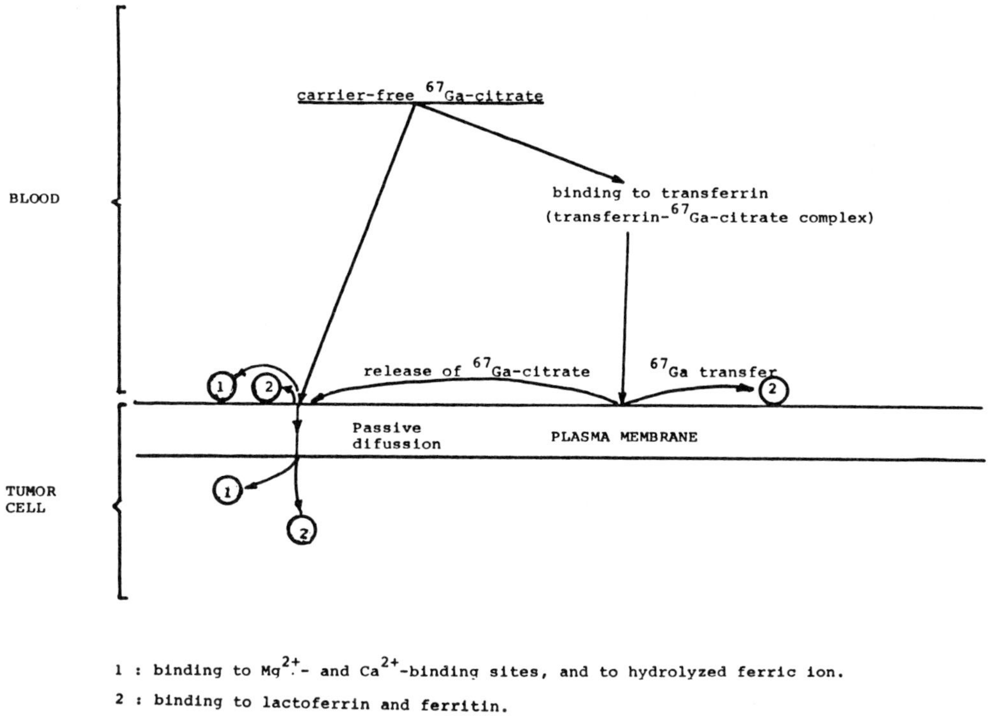

Figure 2. Probable mechanisms of carrier-free 67-Ga-citrate accumulation by tumors.

The ability of tumor cells to produce siderophore-like peptides (Fernandez-Pol, 1978) may be a cellular mechanism to overcome the iron-withholding action of the host. These peptides that solubilize and bind ferric ions may be a way for the tumor to obtain iron from TF that might contain normal or excess amounts of iron. Will be of interest to study the effects of this type of siderophores on 67-Ga-citrate accumulation by tumors.

REFERENCES

Anghileri, L.J., and Heidbreder, M.(1977): On the mechanisms of accumulation of Ga-67 by tumors. Oncology 34: 74-77.

Anghileri, L.J., Chrone-Escanye, M.C., Thouvenot, P., Brunotte, F. and Robert, J.(1988): Mechanisms of gallium-67 accumulation by tumors: Role of cell membrane permeability. J. Nucl. Med. 29: 663-668.

Anghileri, L.J., Thouvenot, P., Crone-Escanye, M.C., Brunotte, F. and Robert, J.(1989): Iron-binding proteins and Ga-67 accumulation by tumor cells. Nucl.-Med. 28: 105-109.

Baserga, R.(1981): The cell cycle. New Engl. J. Med. 304: 453-459.

Cartwright, G.E., and Lee, G.R.(1971): The anaemia of chronic disorders. Br. J. Haematol. 21: 147-152.

Fairbanks, V. F., and Klee, G. G.(1981): Ferritin. Prog. Clin. Pathol. 8: 175-203.

Fernandez-Pol, J. A.(1978): Isolation of a siderophore-like growth

factor from mutants of SV40-transformed cells adapted to Picolinic acid. Cell 14: 489-499.
Ganeshaguru, K., Hoffbrand, A. V., Grady, R.W., and Cerami, A.(1980) Effect of various iron chelating agents on DNA synthesis in human cells. Biochem. Pharmacol. 29: 1275-1279.
Harrison, P. M., Hoare, R. J., Hoy, T. G., and Macara, I. G.(1974): Ferritin and haemosiderin: Structure and function. In Iron in Biochemistry and Medicine, eds A. Jacobs and M. Worwood, p. 80 London, Academic Press.
Hartman, R.E. and Hayes, R. L.(1969): The binding of gallium by blood serum. J. Pharmacol. Exp. Ther. 168: 193-198.
Hoffer, P. B.(1980): Gallium: mechanisms. J. Nucl. Med. 21: 282-285.
Hoffer, P. B., Huberty, J. and Khayam-Bashi, H.(1977): The association of Ga-67 and lactoferrin. J. Nucl. Med. 18: 713-715.
Larson, S. M.(1978): Mechanisms of localization of gallium-67 in tumors. Seminars in Nuclear Medicine 8: 193-203.
Magnusson, G., Flodh, H., and Malmfors, T.(1977): Oncological study in rats of Ferastrol, and iron-poly(sorbitol-gluconic acid) complex, after intramuscular administration. Scand. J. Haematol. (Suppl.)32: 87-98.
Miller, A., Chodos, R. B., Emerson, C. P., and Ross, J. F.(1956): Studies of the anemia and iron metabolism in cancer. J. Clin. Invest. 35: 1248-1262.
Morgan, E.H.(1974): Transferrin and transferrin iron. In Iron in Biochemistry and Medicine, eds. A. Jacobs and M. Worwood, pp. 37-42 London, Academic Press.
Price, V. E. and Greenfield, R. E.(1958): Anemia in cancer. Adv. Cancer Res. 5: 199-284.
Smithyman, A. M., Munn, G., Koziner, B., Tan, C. T. C., and de Sousa M.(1978): Spleen cell populations in Hodgkin's disease. Adv. Exp. Med. Biol. 114: 585-588.
Vijayaraghavan, S., and Rivera, E. M.(1981): Preferential accumulation of iron in hyperplastic tissue of rat mammary gland. Proc. Soc. Exp. Biol. Med. 168: 151-154.
Wainer, E.(1934): The concentration of gallium by means of adsorption on hydrated aluminium and iron oxides. J. Amer. Chem. Soc. 56: 348-350.
Weinberg, E. D.(1978): Iron and infection. Microbiol. Rev. 42: 45-66.
Weinberg, E. D.(1981): Iron and neoplasia. Biol. Trace Elem. Res. 3: 55-80.
Zivanovic, M. J., Taylor, D. M., and McCready, V. R.(1978): The chemical form of gallium-67 in urine. Int. J. Nucl. Med. Biol. 5: 97-98.

Metal Ions in Biology and Medicine. Eds. Ph. Collery, L.A. Poirier, M. Manfait, J.C. Etienne. John Libbey Eurotext, Paris © 1990, pp. 403-405

Protonation of hexamethylenetetramine by $GaCl_3.xH_2O$ and growth inhibition effect on K562 cells

M. Polissiou[1,2], H. Morjani[2], Ph. Collery[3], J.F. Angiboust[2], D. Lamiable[3], M. Manfait[2]

[1] *Laboratory of General Chemistry, Agricultural University of Athens. Iera Odos 75, 118 55 Athens, Greece;* [2] *Laboratoire de Spectroscopie Biomoléculaire, U.F.R. de Pharmacie, 51096 Reims Cedex, France;* [3] *Clinique des Maladies Respiratoires et Laboratoire de Pharmacologie, Hôpital Maison Blanche, 51092 Reims Cedex, France*

Introduction

The antitumor activity of gallium is known from the work of Hart and Adamson (1971). Foster et al. (1986) cited at least five reasons to continue evaluating this antitumor agent : (a) it has been shown to concentrate in human tumors ; (b) it has a broad spectrum of preclinical activity in massive solid tumors ; (c) it does not produce significant myelosuppression in humans making it an attractive candidate for combination chemotherapy ; (d) the risk of renal toxicity can be reduced by 30% according to the mode of drug delivery, without impairing antitumour activity and (e) it has exhibited clinical antitumor activity especially in patients with refractory lymphomas and cancer related hypercalcemia.
Although the exact mechanism of the cytotoxic action of gallium is not known, it is probably related to the metal's ability to interact with plasma proteins (Larson et al., 1980 ; Chitambar and Seligman, 1986), and other molecules of the cell membrane (Anghileri et al., 1987) or to inhibit DNA polymerases (Adamson et al., 1975).
In order to prove the role of the coordinated water molecules of gallium salts, a comparative study in vitro, on K562 cells, was attempted between $GaCl_3.xH_2O$ and the protonated form of hexamethylenetetramine.

Results and Discussion

The protonation of hexamethylenetetramine (HMTA) was performed with $GaCl_3.xH_2O$ (a gift from Laboratoire MERAM, 77020 MELUN cedex) which was added to an absolute ethanol solution of HMTA. This protonated form, as a microcrystalline compound, was isolated, purified and characterized by spectroscopic analysis methods. The absence of gallium ion bound to this organic salt, $C_6H_{12}N_4.HCl$, was checked by 0atomic absorption.
Fourier transform infrared and Raman spectra, in solid state, showed the presence of hydrogen bonds in the range of wavenumbers from 2200 to 2800 cm^{-1} due mainly to the dimerization or polymerization of the protonated form. In parallel, free HMTA showed, in this region, a relatively sharp band centred at 2900 cm^{-1} due only to υ (C-H) vibrations. In the range of wavenumbers from 600 to 1800 cm^{-1}, the infrared spectrum reveals i) a partial removal of degeneracy of the triply degenerate state T_2 of C-N stretches (υ(C-N) = 1000-1030 cm^{-1}, 1220-1270 cm^{-1}) because of the lowering of the symmetry of $C_6H_{12}N_4.HCl$, ii) the appearance of vibrations which, according to the selection rules for free HMTA, were only observable in the Raman spectrum (782, 1300-1450 cm^{-1}) (Bowmaker and Hannan, 1971). Thus, the spectrum obtained in solid state from $C_6H_{12}N_4.HCl$, with a splitting of υ (C-N) bands (811.9, 1006.7, 1238 cm^{-1}) into triplets, suggests that the protonated form of HMTA is polymeric via strong hydrogen bonds (Fig. 1).

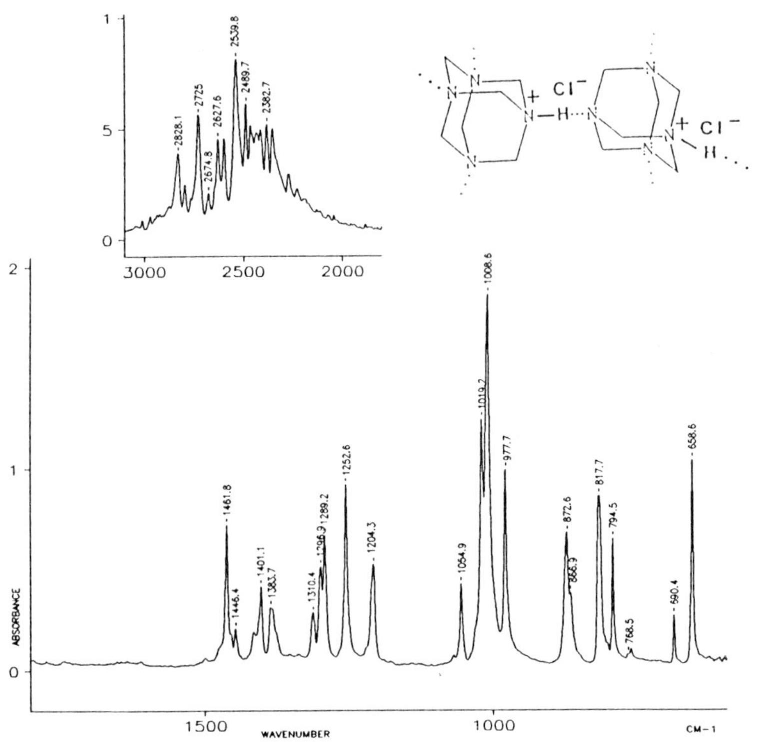

Fig. 1. FT-IR spectrum and structure of protonated HMTA ($C_6H_{12}N_4$.HCl) in KBr disk.

The results are confirmed by ^{1}H-NMR data. In fact a singlet was found for (N^+-CH_2-N) at 4.80 ppm in DMSO, with an additional sharp signal at 8.32 ppm for the (N^+-H) protons. The corresponding chemical shift for (N-CH_2-N) protons of free HMTA was 4.55 ppm (Farminer and Webb, 1976). This protonated form of HMTA was also revealed by electronic impact mass spectrometry.
From the studies of the growth inhibition of K562 cells it was found that HMTA had no effect while protonated HMTA showed an ID 50 of 3.5 X 10^{-5} M. It must be noted that the incubation of K562 cells in the presence of $GaCl_3.xH_2O$ led to a growth inhibition from 4 to 7 X 10^{-5} M according to the hydration degree. Thus it appears, as shown in Fig. 2, that $GaCl_3.xH_2O$ and the $C_6H_{12}N_4$.HCl organic salt show almost the same cytotoxic action (ID 50) on K562 cells.

Conclusion

The similar antitumor activity of these compounds could be related to their ability to interact with other biological molecules of the cell membrane or DNA polymerases, basically by fast proton exchange processes. They might, therefore, be used as agents against some rapidly proliferating tumors in vivo.

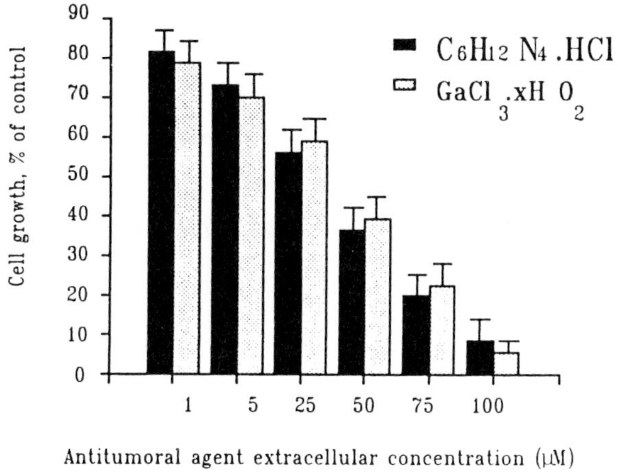

Fig. 2. Growth inhibitory effect of $C_6H_{12}N_4$.HCl and $GaCl_3$.xH_2O on K562 cells. Cells in exponential growth phase at a 50,000/ml density were incubated in the medium containing the appropriate amount of drugs. After 3 days, a triplicate cell counting was performed by phase contrast microscopy. Each point represents the mean and each bar the standard deviation of three experiments.

References

Adamson R.H., Canellos G.P. and Sieber S.M. (1975) : Studies of antitumor activity of gallium nitrate (NSC-15200) and other group IIIa metal salts. Cancer Chemother. Rep. 59, 599-610.

Anghileri L.J., Crone-Escanye M.-C. and Robert J. (1987) : Antitumor activity of Gallium and Lanthanum : Role of cation-cell membrane interaction. Anticancer Res. 7, 1205- 1208.

Bowmaker G.A. and Hannan S.F. (1981) : The vibrational spectra and structure of the bis (hexamethylenetetramine) iodine (I) cation. Aust. J. Chem. 24, 2237-2248.

Chitambar C.R. and Seligman P.A. (1986) : Effect of different Transferrin receptor expression, iron uptake, and cellular proliferation of human leukemic HL60 cells. J. Clin. Invest. 78, 1538-1546.

Farminer A.R. and Webb G.A. (1976) : Structural studies on some 1,2,3,7-tetraaza and 1,3,5-triazaadamantane derivatives. Org. Magn. Resonance. 8, 102-107.

Foster B.J., Clagett-Carr K., Hoth D. and Leyland-Jones B. (1986) : Gallium nitrate : the second metal with clinical activity. Cancer Treat. Rep. 70, 1311-1329.

Hart M.M. and Adamson R.H. (1971) : Antitumor activity and toxicity of salts of inorganic group IIIa metals : aluminum, gallium, indium and thallium. Proc. Natl. Acad.Sci. 68, 1623-1626.

Larson S.M., Rasey J.S., Allen D.R., Nelson N.J., Grunbaum Z., Harp G.D. and Williams D.L. (1980) : Common pathway for tumor cell uptake of gallium-67 and iron-59 via a transferrin receptor. J. natl. Cancer Inst. 64, 41-53.

Metal Ions in Biology and Medicine. Eds. Ph. Collery, L.A. Poirier, M. Manfait, J.C. Etienne. John Libbey Eurotext, Paris © 1990, pp. 406-408

Effect of gallium on the cell cycle of tumor cells in vitro

Y. Carpentier, F. Liautaud-Roger, Ph. Collery*, M. Loirette, B. Desoize, P. Coninx

*Institut Jean-Godinot, BP 171, 51056 Reims Cedex and * Hopital Maison Blanche, 45, rue Cognacq-Jay, 51100 Reims, France*

Gallium, a metal from group III, has been reported to be active against various tumour cells, both in vitro and in vivo (Adamson et al, 1975 ; Foster et al, 1986). Using the murine mammary carcinoma Ca 755, we have shown that $GaCl_3$ is more efficient for tumours in the exponential phase than for those in the plateau phase of growth, i.e. when the growth fraction is low (Carpentier et al, 1987). In vitro, its cytotoxicity has been shown to be time-dependent (Rasey et al, 1982). The study of Ga effect at cellular level and in particular on the cell cycle is therefore of interest. The aim of this study was to determine $GaCl_3$ effects on the growth of L1210 murine leukemia cells in vitro using asynchronous and partially synchronized cells.

MATERIALS AND METHODS

Except where otherwise stated, asynchronous L 1210 cells were cultured in DMEM supplemented with 2.5 % fetal calf serum (FCS) for 24 h-incubations. Cells were numbered using a phase contrast microscope.
In some cases, cells were washed and incubated in culture medium without drug in order to study reversibility, or in semi-solid medium (cloning efficiency, CE) in order to study viability.
Partial synchronization was obtained with a 7.5 h-incubation in 30 nmol/l Colcemid. After washing, these cells were reseeded for various periods of time without drug, with 500 µmol/l $GaCl_3$ (kindly furnished by Meram Lab, France) or 30 nmol/l methotrexate (MTX).
Cell DNA content was assessed by flow cytometry (FCM) using ethidium bromide staining of RNase-treated cells after ethanol fixation. Forward angle scatter (FAS) and DNA histograms of 30000-60000 cells were generated with a Cytofluorograf 50H (Ortho-Instruments) after excitation at 488 nm. Proportions of cells in $G_{0/1}$, S and G_2 + M phases were determined by the Fried method (Fried, 1976)

RESULTS

Cell growth inhibition : Cell growth (CG) was inhibited by 30, 50 and 70 % for 20, 100 and 500 µmol/l $GaCl_3$. A plateau was reached for the highest concentrations.
Cell DNA distribution : An increase in the proportion of cells in the $G_{0/1}$ phase was observed after a 24 h-incubation as seen in Fig. 1A and 1E, while cell size (FAS) was hardly altered (Fig. 1B and 1F). After washing and reincubation for 24 h without drug, cell DNA distribution became similar to that of the controls (Fig. 1C and 1G) although CG inhibition increased from 45 % during the first 24 h-

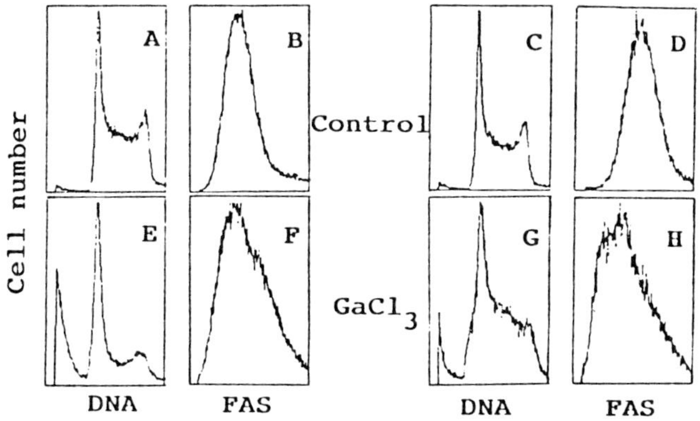

Fig. 1. FCM histograms of Control and 100 µmol/l $GaCl_3$-treated cells. Lefthand part: 24 h-incubation with drug. Righthand part: washing and reincubation for 24 h without drug.

incubation to 72 % during the second one. Moreover, numerous debris appeared as depicted by the FAS histograms (Fig. 1D and 1H).

Concentration related effects : No direct relations were observed when Ga effects on the cell cycle and CE were considered as a function of CG (Fig. 2). A fraction of 25 % $G_{0/1}$ cells was observed from 0 to 20 % CG inhibition and a plateau at 45-50 % $G_{0/1}$ cells appeared beyond 50 % CG inhibition. Similarly, CE was decreased to a plateau of 30 % of controls when CG inhibition was higher than 40 %. For Ga concentrations hardly altering CG, CE was increased by 30 % as compared with the controls (Fig. 2).

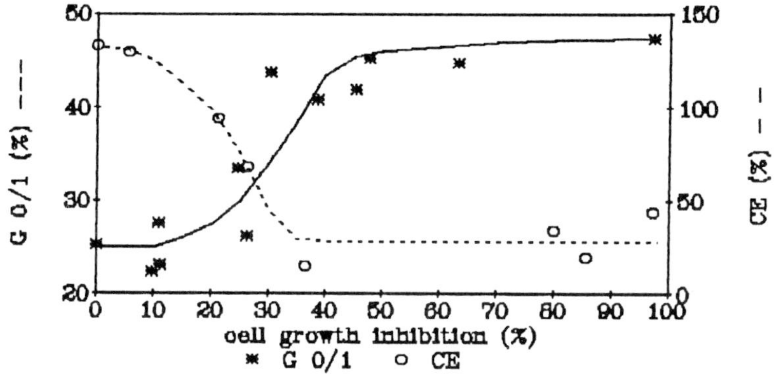

Fig. 2. $G_{0/1}$ and CE versus CG inhibition after 24 h-incubation with different Ga concentrations.

Time effect : No $G_{0/1}$ increase was observed in asynchronous cells up to 10 h-incubations (Fig. 3), i.e. about the length of one doubling time. Using partially synchronized cells, Ga needed the same delay to express cell cycle alterations when the $G_{0/1}$ blockade was quite immediate with MTX (Fig. 3).

DISCUSSION

The antitumour agent Ga was active on L 1210 cell growth for a concentration

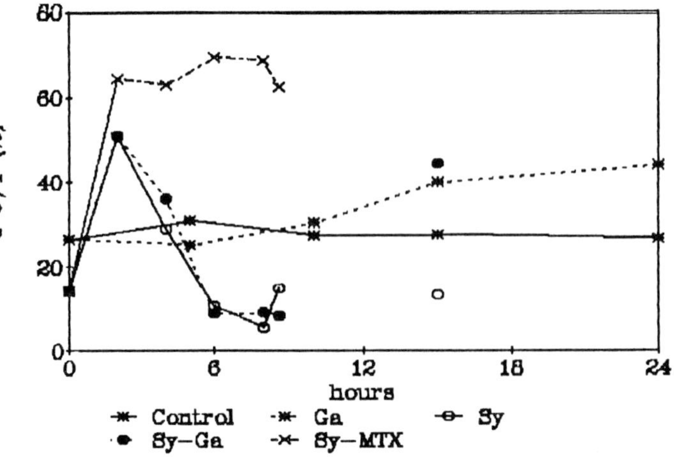

Fig. 3. Percentage of $G_{0/1}$ cells in asynchronous (*) and synchronized (o • x) cultures versus time. $GaCl_3$: 500 µmol/l ; MTX : 30 nmol/l.

similar to that active in a human leukemic cell line (Hedley et al., 1988). On the contrary, cell DNA distribution was differently altered in both cell lines, i.e. an increase of $G_{0/1}$ L1210 cells and an S-phase arrest in human cells.

The Ga dose-effect relations according to different criteria were quite similar in that no effect was observed when CG was decreased by less than 20 % and a plateau was reached for about 45 %, suggesting the presence of both lethally damaged and spared cells. Both cell types were also revealed by the presence of dividing cells 24 h after washing although CG inhibition increased.

$G_{0/1}$ accumulation, rather infrequent when using anticancer agents, was revealed after about one cycle time. This delay was not shortened using synchronized cells whereas cell cycle blockade appears rapidly with usual antitumour drugs such as MTX.

In conclusion, Ga inhibited cell growth by reducing cell viability, induced a delayed accumulation of $G_{0/1}$ cells that was similar with asynchronous or synchronized cells. This indicates that Ga is an unusual antitumour drug.

REFERENCES

Adamson, R., Canellos, G., and Sieber, S. (1975): Studies on the antitumour activity of gallium nitrate (NSC-15200) and other group IIIa metal salts. Cancer Chemother. Rep. 59, 599-610

Carpentier, Y., Liautaud-Roger, F., Labbé, F., Loirette, M., Collery, P., and Coninx, P. (1987): Effect of Gallium at two phases of the Ca 755 tumour growth. Anticancer Res. 7, 745-748.

Foster, B.J., Clagett-Carr, K., Hoth, D., and Leyland-Jones, B.(1986): Gallium nitrate: the second metal with clinical activity. Cancer Treat.Rep. 70,1311-1319.

Fried, J. (1976): Method for the quantitative evaluation of data from flow microfluorometry. Comput. Biomed. Res. 9, 263-276.

Hedley, D.W., Tripp, E.H., Slowiaczek, P., and Mann, G.J. (1988): Effect of gallium on DNA synthesis by human T-cell lymphoblasts. Cancer Res. 48,3014-3018.

Rasey, J., Nelson, N., and Larson, S. (1982): Tumor cell toxicity of stable Gallium nitrate: enhancement by transferrin and protection by iron. Eur. J. Cancer Clin. Oncol. 18, 661-668.

Metal Ions in Biology and Medicine. Eds. Ph. Collery, L.A. Poirier, M. Manfait, J.C. Etienne. John Libbey Eurotext, Paris © 1990, pp. 409-411

Enhancing effects of isoprenoid (L-623) on accumulation of Ga-67 in mice tumor cells

M. Maeda, H. Nihonmatsu, T. Kawagoshi, M. Okamoto, M. Shoji, O. Ogawa*, Y. Furukawa* and T. Honda

*Radioisotope Laboratory, Toyama Medical and Pharmaceutical University, 2630 Sugitani, Toyama 930-01, Japan and * Lederle (Japan), 1-6-34, Kashiwa-cho, Shiki, Saitama 353, Japan*

Ga-67-citrate is clinically widely applied for the imaging of a variety of epithelial and lymphoreticular neoplasms (Edwards et al., 1969). Several local factors influencing Ga-67 uptake have been proposed, including hyperpermeability in tumor bearing areas (Anghileri et al., 1988), exchange of extracellular Ga-67 with the intracellular calcium pool (Anghileri et al., 1986) and mediation by a specific cell surface transferrin receptor (Chan et al., 1987). L-623 is an emulsified preparation which contains one per cent (w/v) of N-solanesyl-N,N' -bis(3,4-dimethoxybenzyl) ethylenediamine malate (N-1379). N-1379 is reported to potentiate the growth inhibition effect of clinically useful antitumor drugs (Ikezaki et al., 1984). The potentiating effect of this drug may be due to an inhibition of the efflux pump for some kinds of antitumor agents (Nakagawa et al., 1986), but other factors are also suggested. This study investigated the effects of L-623 on the accumulation of Ga-67 in tumor cells in vitro and in vivo to improve the clinical usefulness of the radiometal.

Materials and Methods

Reagents Aqueous carrier free Ga-67 solution was used in citrated form at a concentration of 37 MBq/ml. L-623 emulsion was prepared as follows; to the vessel containing 1.0 g of N-1379, 9.2 g of soybean lecithin, 10.0 g of soybean oil and 6.5 g of D-sorbitol was added distilled water to be total volume of 100 ml, and the mixture was homogenized mechanically.

In vitro experimental procedure Three kinds of mouse tumor cell lines, colon 26 (C26), melanoma B16 (B16) and Lewis lung carcinoma (LLC) were used for the experiment. Each tumor cells were suspended in Eagle's MEM containing 10 per cent of FBS, so as to be 10^5-10^6 cells per 0.1 ml of medium. To 16 mm wells of plastic microplates containing 1.0 ml of the medium, added 0.1 ml of cell suspension and L-623 at the final concentration of N-1379 to be 3.82×10^{-5}M. Cells were incubated at 37°C for 60 min in a 5 per cent CO_2 in air. Then 37 kBq of diluted Ga-67 solution (0.03ml) was added in each well, and cells were continued incubation. Cells were harvested with trypsin-EDTA solution at 18 hr, 42 hr and 66 hr after the addition of Ga-67 solution. Collected cells were centrifuged at 2000 rpm for 5 min, supernatant was removed and the residual radioactivity was measured.

In vivo experimental procedure Male 6 wk old CDF1 mice for C26 and BDF1 mice for B16 and LLC were used as the host animals. Mice were administered 10 mg/kg i.v. of N-1379 in the form of L-623 emulsion diluted 10-fold with saline 10 days after the implantation of tumor cells to dorsal subcutaneous region. The tumor was 1.5 or 1.0 g in weight on the day of L-623 administration. Diluted Ga-67 solution was injected i.v. at a dose of 185 kBq per mouse 30 min after the dose of L-623. Blood was collected, and organs and tissues were isolated and weighed periodically, thereafter the radioactivity was measured. Statistical analysis was carried out with Student's t-test.

Results and Discussion

Fig. 1 shows the results of an experiment with cultured mice tumor cells. Ga-67 uptake was increased periodically in both L-623 and control groups of each tumor cell line. The Ga-67 uptake in C26 and B16 cells was 1.5-fold and 1.3-fold higher in L-623 group than in control group after the 66 hr period of exposure, respectively. In LLC cells, Ga-67 accumulated 5.1-fold higher in L-623 group after exposed for 30 hr. Experiments were designed to evaluate the effects of L-623 on the localization of Ga-67 in tumor and other tissues. Results are shown in Table 1. In C26 mice, the accumulation of Ga-67 in tumor tissue was significantly increased in L-623 group at 48 hr ($p<0.05$) and 72 hr ($p<0.01$) after administration. Remaining Ga-67 in the spleen, liver and kidney was also higher in L-623 group of C26 mice at various times examined. Distinct significance between L-623 and control groups was not observed in tumor tissue of B16 and LLC mice (data not shown). Blood levels did not show significant difference between L-623 and control groups of all of the tumor bearing mice.

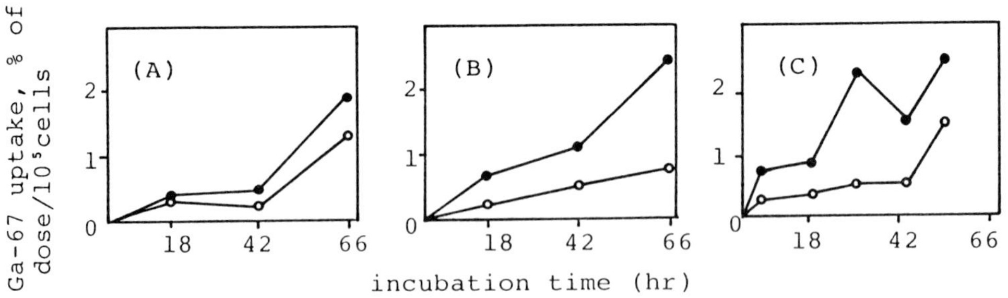

Fig. 1. Effect of L-623 on Ga-67 uptake into cultured tumor cells Tumor cell lines were (A), colon 26; (B), melanoma B16 and (C), Lewis lung carcinoma. —●—, L-623 group; —○—, control group.

Table 1. Effect of L-623 on Ga-67 distribution after administration of Ga-67 to colon 26 mice

	L-623 group			Control group		
	24 hr	48 hr	72 hr	24 hr	48 hr	72 hr
Blood	2.94±0.34	1.10±0.21	0.63±0.01	3.05±0.19	0.89±0.06	0.47±0.05
Tumor	8.53±0.85	7.92±0.49*	6.73±0.25**	7.44±0.16	5.47±1.10	4.30±0.38
Spleen	8.69±1.49*	10.10±1.62**	8.65±0.43**	4.34±0.29	4.46±0.34	3.99±0.19
Liver	7.73±0.41	11.70±1.77*	9.37±0.27*	6.81±0.24	7.35±0.36	7.09±0.70
Kidney	8.52±1.27*	7.78±0.20**	6.57±0.53*	5.43±0.30	4.19±0.45	3.40±0.20
Lung	3.98±0.34*	3.79±0.42	3.44±0.14**	3.24±0.07	4.66±3.50	2.04±0.19
Heart	1.70±0.04	1.66±0.16	1.59±0.12*	1.73±0.26	1.38±0.09	1.16±0.00
Pancreas	2.58±0.23	2.67±0.09	2.76±0.20	2.91±0.20	3.21±0.12	3.13±0.14
Stomach	4.01±0.08	3.63±0.29	3.08±0.38	4.26±0.40	3.59±0.43	3.00±0.24
Muscle	0.73±0.12	0.58±0.17	0.53±0.02	0.85±0.04	0.57±0.07	0.58±0.11
Bone	5.53±0.65	4.63±0.97	8.30±3.59	4.87±0.86	3.23±0.60	11.30±0.90

Each value is expressed as the mean % of dose/g or ml± S.D. for three mice. $*p<0.05$, $**p<0.01$.

The exact mechanism of Ga-67 localization to tumors is still unsettled. It was revealed that the transferrin receptor plays an important role in intracellular incorporation of Ga-67 (Chan et al., 1987). In this study with cultured cells, the uptake of Ga-67 was increased by an addition of L-623 to the medium in every cell line. Furthermore, L-623 potentiated the accumulation of Ga-67 in tumor tissue of C26. Our results may be suggest that L-623 is useful for enhancing the accumulation of Ga-67 in tumor cells. How L-623 affects the Ga-67 localization in tumor cells, promotion of influx or inhibition of efflux, is being studied.

References

Anghileri L.J., Thouvenot P., Brunotte F., et al. (1986): Experimental evidence for the probable involvement of calcium ion transport in 67Ga uptake by tumor cells. Eur. J. Nucl. Med. 12, 179-181.

Anghileri L.J., Thouvenot P., Brunotte F., et al. (1988): Mechanisms of gallium-67 accumulation by tumors: role of cell membrane permeability. J. Nucl. Med. 29, 663-668.

Bell E.G., OMara R.E., Henry C.A., et al. (1971): Non-neoplastic localization of 67Ga-citrate. J. Nucl. Med. 12, 338-339.

Chan S.M., Hoffer P.B., Nada M., et al. (1987): Inhibition of gallium-67 uptake in melanoma by an anti-human transferrin receptor monoclonal antibody. J. Nucl. Med. 28, 1303-1307.

Edwards C. L., Hayes R.L. (1969): Tumor scanning with 67Ga citrate. J. Nucl. Med. 10, 103-105.

Ikezaki K., Yamaguchi T., Miyazaki C., et al. (1984): Potentiation of anticancer agents by new synthetic isoprenoids. I. Inhibition of the growth of cultured mammalian cells. JNCI. 73, 895-901.

Nakagawa M., Akiyama S., Yamaguchi T., et al. (1986): Reversal of multidrug resistance by synthetic isoprenoids in the KB human cancer cell line. Cancer Res. 46, 4453-4457.

Metal Ions in Biology and Medicine. Eds. Ph. Collery, L.A. Poirier, M. Manfait, J.C. Etienne. John Libbey Eurotext, Paris © 1990, pp. 412-414

Effects of $GaCl_3$-CDDP combination on the intratissular concentrations of Ga, Pt, Mg, Fe and Ca in healthy mice

Ph. Collery[1], R. Vistelle[2], F. Arsac[3], F. Habets[3], H. Millart[4], H. Choisy[4]

[1] *Hôpital Maison Blanche, Centre Hospitalier Universitaire Régional, 51092 Reims Cedex, France;* [2] *Laboratoire de Pharmacologie, Faculté de Pharmacie, 51100 Reims, France;* [3] *Laboratoire de Toxicologie, Faculté de Pharmacie, 51100 Reims, France;* [4] *Laboratoire de Pharmacologie, Centre Hospitalier Universitaire Régional, 51092 Reims cedex, France*

INTRODUCTION

Potentiation of a chemotherapy using CDDP (Cis-platinum) by gallium chloride (GaCl3) was demonstrated in a randomized study in cancer patients by Collery et al , 1989. The purpose of the present study was to obtain information about the possible Gallium - Platinum interactions on their intratissular distribution in healthy mice receiving both these drugs, in comparison with mice receiving only GaCl3 or CDDP. The interactions with intratissular concentrations of Mg, Ca and Fe were also evaluated since alterations in these elements have been reported after oral administration of gallium in mice bearing a transplanted C3HBA mammary adenocarcinoma by Collery et al, 1986 and in healthy rats by Vistelle et al,1989.

MATERIAL AND METHODS

45 mice received either 200 mg/24 h GaCl3 (group 2) daily as an intragastric administration for 20 days , or CDDP (3 mg/ 24h) intraperitoneously from days 16 to 20 (group 3), or the combination of GaCl3 for 20 days and CDDP from days 16 to 20 (group 4). An other group received GaCl3 from days 1 to15 and CDDP from days 16 to 20 (group 5). Controls (group 1) were used to determine the influence of Ga and Platinum (Pt) on Mg, Fe and Ca intratissular concentrations . In these 5 groups (n=9), Ga, Pt, Mg, Fe and Ca were assayed by atomic absorption spectrometry in kidneys, spleen, lungs, bone and liver. Results were expressed in µg per gram of wet tissue (ppm). Mann and Whitney test was used for statistical analysis.

RESULTS

Intratissular concentrations of Ga, Pt, Mg, Ca and Fe are shown in table I.

A significant increase in Ga concentrations ($p<0.02$) was observed in the spleen of mice of group 4 (6.45 ± 1.8 µg/g) as compared with group 2 (4.48 ± 1.2 µg/g). This increase was not significant in group 5 (5.76 ± 1.7 µg/g).

A significant increase in Pt concentrations ($p< 0.05$) was noted in the kidney of mice of group 5 (5.90 ± 1.41 µg/g) as compared with those of group 3 (3.85 ± 1.62 µg/g) but not in group 4 (4.42± 1.55 µg/g) as compared with group 3. The difference between groups 4 and 5 was also significant.

In mice receiving Ga without Pt (group 2) the only significant biological modification was a decrease in Mg concentrations in the liver. In contrast, in all groups receiving Pt, significant marked changes were noted such as a decrease in Ca concentrations and an increase of Fe in the liver , an increase in Ca, Mg and Fe concentrations in the spleen and an increase in Fe concentrations in the kidneys .

A significant decrease in the weight of organs / body weight ratio was observed for liver and spleen in all groups receiving CDDP (table II) but not in mice receiving Ga alone. No statistical differences were noted between groups 3, 4 and 5.

Table I : Ga, Pt, Mg, Ca and Fe intratissular concentrations.

Groups : 1 = controls ; 2 = 20 days of GaCl3 oral administration ; 3 = 5 days I.P. injection of CDDP from days 16 to 20 ; 4 = 20 days of GaCl3 oral administration and I.P. injection of CDDP from days 16 to 20 ; 5 = 15 days of GaCl3 oral administration plus I.P. injection of CDDP from days 16 to 20. * p<0.05

	Group 1 m ± SD n = 9	Group 2 m ± SD n = 9	Group 3 m ± SD n = 9	Group 4 m ± SD n = 9	Group 5 m ± SD n = 9
Bone					
Ga (µg/g)		31.2 ± 9.9		34.1 ± 23.8	10.8 ± 3.2
Pt (µg/g)			3.4 ± 1.7	5.5 ± 4.4	2.5 ± 1.3
Mg (mg/g)	2.8 ± 0.4	2.7 ± 0.6	3.0 ± 0.2	2.9 ± 0.5	3.0 ± 0.2
Ca (mg/g)	164 ± 21	153 ± 36	172 ± 12	173 ± 30	176 ± 12
Fe (µg/g)	114 ± 30	124 ± 55	120 ± 20	147 ± 42	129 ± 29
Kidney					
Ga (µg/g)		6.1±3.1		8.4 ± 10	3.5 ± 2.4
Pt (µg/g)			3.9±1.6	4.4 ± 1.6	5.9 ± 1.4
Mg (µg/g)	214 ± 20	289 ± 95	223 ± 37	242 ± 67	252 ± 55
Ca (µg/g)	94 ± 10	96 ± 14	96 ± 12	96 ± 12	116 ± 16*
Fe (µg/g)	126 ± 11	131 ± 20	149 ± 23*	145 ± 32	156 ± 18*
Liver					
Ga (µg/g)		9.2 ± 2.5		10. 5 ± 2.5	8.2 ± 2.7
Pt (µg/g)			5.2 ± 0.8	6.2 ± 1.2	4.9 ± 1.1
Mg (µg/g)	392 ± 166	282 ± 20*	327 ± 82	373 ± 78	328 ± 52
Ca (µg/g)	21 ± 12	18 ± 11	10 ± 3*	11 ± 2*	10 ± 2*
Fe (µg/g)	165 ± 38	158 ± 21	236 ± 20*	241 ± 32*	211 ± 34*
Lung					
Ga (µg/g)		2.4 ± 2.2		2.7 ± 0.9	2.0 ± 1.3
Pt (µg/g)			0.9 ± 0.3	1.0 ± 0.5	0.7 ± 0.4
Mg (µg/g)	257 ± 158	357 ± 121	299 ± 125	254 ± 61	240 ± 55
Ca (µg/g)	103 ± 61	102 ± 63	51 ± 18	52 ± 15	48 ± 14
Fe (µg/g)	168 ± 28	163 ± 29	166 ± 32	176 ± 40	169 ± 27
Spleen					
Ga (µg/g)		4.5 ± 1.2		6.4 ± 1.8	5.8 ± 1.7
Pt (µg/g)			1.9 ± 0.8	1.7 ± 0.6	2.1 ± 0.8
Mg (µg/g)	342 ± 46	345 ± 40	633 ± 357*	603 ± 354	501 ± 227
Ca (µg/g)	27 ± 7	32 ± 8	81 ± 55*	46 ± 17*	46 ± 22*
Fe (µg/g)	639 ± 112	698 ± 101	1192 ± 192*	1237 ± 131*	1270 ± 127*

Table II : weight of organs / body weight (expressed as %) . * p< 0.05

	Group 1 m ± SD n = 9	Group 2 m ± SD n = 9	Group 3 m ± SD n = 9	Group 4 m ± SD n = 9	Group 5 m ± SD n = 9
Kidney	1.33 ± 0.10	1.30 ± 0.08	1.29 ± 0.06	1.33 ± 0.20	1.27 ± 0.06
Liver	5.50 ± 0.49	5.37 ± 0.36	4.77 ± 0.35*	4.52 ± 0.47*	4.62 ± 0.35*
Lung	0.65 ± 0.09	0.62 ± 0.04	0.66 ± 0.09	0.55 ± 0.06	0.69 ± 0.09
Spleen	0.47 ± 0.06	0.48 ± 0.06	0.35 ± 0.06*	0.33 ± 0.05*	0.32 ± 0.04*

DISCUSSION

Noujaim et al. (1981) noted that Pt could interfere with the Ga intracellular uptake . Pt has, in vitro, no effect on Ga binding to transferrin . The decrease of tumor accumulation of Ga observed in a canine tumor model when Ga was administered one week after Pt administration could be related to a decrease in cellular protein synthesis rather than a direct effect on the transport of Ga to the tumor. We also observed an interference between Pt and Ga with an increase in intratissular Ga concentrations in the spleen in healthy mice receiving both Ga and Pt. It is noteworthy that the spleen is the organ in which we observed the most important biological modifications with a marked increase in Mg, Fe and Ca concentrations and a decrease in the ratio of weight of organ/body weight . Mg, Fe and Ca are known to act on the cell cycle , protein and DNA synthesis (Collery et al ,1978 ; Basset et al,1985). As already suggested by Noujaim , the uptake of Ga may be influenced by metabolic conditions.

An increase in Ga tissular uptake has been observed in mice receiving simultaneously Ga and Pt (group 4) in all studied organs. Even though it was only significant in the spleen , the increase also noted in bone, kidney, liver and lung could explain the decrease in serum Ga concentrations observed in cancer patients during the 5 days CDDP infusion by Collery et al,1989 . It was proposed to interrupt Ga treatment during the 5 days of CCDP infusion in order to avoid a renal toxicity due the combination of Ga and Pt . By contrast, an increase in Pt concentrations was noted in the kidney of mice of group 5 and not in those of group 4. To avoid renal toxicity due to the accumulation of Pt it seems better to maintain Ga treatment during CDDP infusion rather than interrupt it. This schedule of treatment permits then an increase in Ga tissular uptake . Our study has to be completed by the comparison of the Ga and Pt concentrations in tumor-bearing mice receiving either Ga or Ga plus Pt .

REFERENCES

Basset , P., Zwiller, J., Revel, M.O., Vincendon, G. (1985) : Growth promotion of transformed cells by iron in serum-free culture. Carcinogenesis 6, 355-359.

Collery, Ph., Coudoux, P., Geoffroy, H. (1978) : Role of magnesium in the development of cancer. Trace Substances in Environmental Health. eds. Hemphill D.D. , Missouri, Columbia, XII, 140-147.

Collery, Ph., Millart, H., Pluot, M., Anghileri, L.J. (1986) : Effects of gallium chloride oral administration on transplanted C3HBA mammary adenocarcinoma : Ga, Mg, Ca and Fe concentrations and anatomopathological characteristics. Anticancer Res.6,1085-1088.

Collery, Ph., Millart, H., Cossart, C., Perdu, D., Vallerand, H., Morel, M., Pechery, C., Etienne, J.C., Choisy, H., Dubois de Montreynaud, J.M. (1989): Potentiation of chemotherapy by gallium . J. Am. Coll. Nutr. 8, 428.

Noujaim, A.A., Terner, U.K., Turner, C.J., Van Nieuwenhuyze, B., Lentle, B. (1981) : Alterations of Gallium-67 uptake in tumors by cis-platinum. Int . J. nucl. Med. Biol. 8, 289-293.

Vistelle, R., Collery, Ph., Millart, H. (1989) : In vivo distribution of gallium in healthy rats after oral administration and its interaction with Fe, Mg and Ca. Trace Elem. Med. 6, 27-32.

Walker, G.M., Duffus, J.H. (1983) : Magnesium as the fundamental regulator of the cell cycle. Magnesium 2, 1-16.

Metal Ions in Biology and Medicine. Eds. Ph. Collery, L.A. Poirier, M. Manfait, J.C. Etienne. John Libbey Eurotext, Paris © 1990, pp. 415-419

A review of the pharmacological and toxicological properties of gallium

J.L. Domingo, J. Corbella

Laboratory of Toxicology and Biochemistry, School of Medicine, University of Barcelona, San Lorenzo 21, 43201 Reus, Spain

Gallium is a metal which belongs to the group IIIa elements of the periodic Table. There are two gallium radioisotopes, ^{67}Ga and ^{68}Ga, possessing nuclear properties that make them attractive for use in nuclear medicine. ^{67}Gallium citrate has long been established as a diagnostic tool for the localization of neoplastic lesions in animals and humans (Edwards & Hayes, 1969; Edwards et al., 1970; Edwards & Hayes, 1970). Nevertheless, the mechanism by which gallium localizes in tumors is largely unknown.

Chemotherapy is a field of cancer treatment which has evolved over the last 40 or 50 years, and its use has escalated enormously in the last 20-25 years (Cleare & Hydes, 1980). Rosenberg et al. (1969, 1970) demonstrated that various salts of platinum possess antitumor activity against leukemia L1210 and sarcoma 180 in mice. Subsequently, the compound cis-diamminedichloroplatinum (II) (DDP) showed therapeutic activity in bladder, head and neck, testicular and other carcinomas (Higby et al., 1975; Rozencweig et al., 1977; Wittes et al., 1977). Nevertheless, nephrotoxicity is one of the primary limitations for the therapeutic administration of cis-platinum in human malignancies (Guarino et al., 1979; Krakoff, 1979; Goldstein et al., 1981). The effectiveness of cis-platinum in neoplastic disease stimulated the investigation of other inorganic compounds. Gallium has been the second metal ion with clinical activity used in cancer treatment (Foster et al., 1986).

Pharmacological and toxicological properties of gallium in cancer treatment

In 1971, the toxicity and antitumor activity of salts of the group IIIa metals aluminium, gallium, indium and thallium were reported for the first time (Hart & Adamson 1971; Hart et al., 1971). All four of the metal salts tested were cytotoxic in tissue culture against Walker 256 carcinosarcoma and leukemia L1210 cells (Hart & Adamson 1971). Significant antitumor activity of gallium nitrate was also reported by Adamson et al. (1973).

Preclinical toxicologic studies in mice, rats, dogs and monkeys indicated that the major toxicity of gallium nitrate was weight loss, pneumonitis, and renal tubular and liver damage. There was some depression of lymphoid cells but no demonstrable bone marrow toxicity (Dudley & Levine 1949; Hart et al.,

1971; Adamson et al., 1975; Newman et al., 1979). During the last two decades, several clinical Phase I studies of gallium nitrate (NSC-15200) have been carried out with different results (Bedikian et al., 1978; Brown et al., 1978; Hall et al., 1979; Krakoff et al., 1979; Kelsen et al., 1980; Samson et al., 1980; Warrell et al., 1983).

Two clinical phase I trials reporting activity in bone and soft tissue sarcomas led to the study of gallium nitrate in advanced soft tissue and bone sarcomas by the Southwest Oncology Group (SWOG) (Brown et al., 1978; Samson et al., 1980). 32 patients with advanced, histologically confirmed soft tissue and bone sarcomas were entered in the study. All patients were hydrated with i.v. fluids followed by the administration of gallium nitrate at a dose of 700 mg/m^2 as a 30-min i.v. infusion. Treatments were administered at 2-week intervals. Gallium nitrate failed in demonstrating activity in these patients (Saiki et al., 1982). Warrell et al. (1983) evaluated the effects of gallium nitrate given as a continous infusion for seven days in patients with advanced malignant lymphoma. Four dose levels which ranged from 200 to 400 mg/m^2/day were studied in 27 patients. Hypocalcemia occurred in two-thirds of patients. Other toxic effects including paresthesiae, diarrhea, and hearing loss were noted in $<5\%$ of patients. Gallium nitrate administered as a continous infusion for seven days at 300 mg/m^2/day was well-tolerated and effective treatment for patients with advanced malignant lymphoma.

In those reported clinical trials, renal toxicity of gallium has hindered the use of efficient doses. Consequently, in recent years, oral administration of gallium has been proposed in order to avoid nephrotoxicity derived from parenteral use of the metal. According to previous experimental results, which showed that gallium administered orally as gallium chloride daily during several months could result in a selective uptake by tumor cells without renal toxicity (Collery et al., 1983; Collery et al., 1984), a clinical trial was proposed with this new form of administration. Thirty cancer patients were treated with gallium chloride (300-800 mg/day) for a median period time of treatment of 4.5 months. No renal toxicity was observed in patients treated for more than 1 year (800 mg/day) (Collery et al., 1985). To determine the influence of the length of treatment on the anatomopathological and biochemical intratumor changes induced by gallium, oral gallium chloride was given to C3HBA mammary adenocarcinoma-bearing mice for a period of either 21 or 42 days. In both cases, 200 mg/kg/24 h was administered. Gallium permits an increase in length of survival with a regression of the tumor growth after continous treatment, but without any diminution of the tumor size (Collery et al., 1986). Due to the lack of renal and hematological toxicity, together with the significant uptake by the tumor, it has been suggested that gallium chloride can be used orally in conjuction with other cytotoxic agents (Collery et al., 1989a; Collery et al., 1989b).

Hypercalcemia is a substantial problem in the clinical management of patients with cancer (Bockman, 1980; Mundy et al., 1984). However, in recent years, it has been demonstrated that gallium nitrate is an effective treatment for cancer-related hypercalcemia (Warrell et al., 1984; Warrell et al., 1986). On the other hand, no significant differences in the magnesium or calcium kidney concentrations were noted when gallium chloride was administered orally to C3HBA mammary adenocarcinoma-bearing mice (Collery et al., 1986).

Experimental toxicity of gallium

Anorexia, decreased weight gain, depressed pupilary reflex, exophthalmos, miosis, and hemorrhages in extremities were the most remarkable physical and clinical signs seen in rats and mice after acute gallium intoxication (Dudley & Levine 1949; Newman et al., 1979; Domingo et al., 1987a). Slight-to-moderate chemical peritonitis, interstitial pneumonitis with slight pleural effusion, and moderate liver and kidney damage were the most conspicuous signs in rats after gallium nitrate administration (Hart et al., 1971).

There is a remarkable lack of information on the reproductive and developmental toxicology of gallium. When gallium sulphate was injected i.v. in increasing amounts into pregnant hamsters on day 8 of gestation, only three embryos from gallium-treated mothers showed mild malformations consisting of one limb bud abnormality, one case of spina bifida, and one mild exencephaly (Ferm & Carpenter, 1970). Subcutaneous gallium sulphate administration (150 mg/kg) in a single dose given during the period of organogenesis showed some teratogenic effects in mice (Caujolle et al., 1972). With regard to the chelating agents to be used as possible antidotes in gallium intoxication, DFOA, citric, succinic, malic, and oxalic acids have been reported to be effective chelators in preventing acute gallium nitrate poisoning (Domingo et al. 1987b).

Gallium in the electronics industry

The current revolution in microelectronics technology has been fueled by the advent of several compound semiconductor materials composed primarily of elements from groups IIIb and Vb of the periodic table (Goering et al., 1988). One of these III-V semiconductor compounds, gallium arsenide (GaAs), is the material of choice in high-frequency microwave and millimeter wave telecommunications and ultrafast supercomputers (Robinson, 1983). With the release of gallium and arsenic from GaAs, one might expect development of their respective signs of systemic toxicity. Notwithstanding, histologic examination of the kidneys of rats at 14 days following intratracheal instillation of 100 mg GaAs/kg did not show any of the signs of renal toxicity reported by Newman et al., (1979) following i.p. administration of gallium nitrate (Webb et al., 1984). Webb et al. (1986) evaluated the relative toxicity of Ga_2O_3, GaAs and As_2O_3 following intratracheal instillations of these particles in rats. They found that the pneumotoxicity ranking in descending order was GaAs> As_2O_3>> Ga_2O_3. The authors suggested that the pathological responses observed in the lung were likely to be primarily due to arsenic.

Inhaled Ga_2O_3 produced cytotoxic, inflammatory, and fibrogenic responses of comparable or greater magnitude than those seen after similar exposures of rats to inhaled quartz particles in other studies (Wolff et al., 1988). In terms of extrapulmonary toxicity, GaAs also produced a marked uroporphyrinuria indicative of arsenic perturbation of the hepatic heme biosynthetic pathway (Woods & Fowler, 1978). Both GaAs constitutive elements, particularly gallium can affect the activity of δ-aminolevulinic acid dehydratase (ALAD) in vitro, suggesting that these moieties may be responsible for the systemic toxicity of GaAs as defined by inhibition of ALAD in blood, kidney, and liver as well as changes in renal ultrastructure (Goering et al., 1988).

REFERENCES

Adamson, R.H., Canellos, G.P. & Sieber, S.M. (1975): Studies on the antitumor activity of gallium nitrate (NSC-15200) and other group IIIa metal salts. Cancer Chemother. Rep. 59: 599-610

Bedikian, A., Valdivieso, M. & Bodey, G. (1978): Phase I clinical studies with gallium nitrate. Cancer Treat. Rep. 62: 1449-1453

Bockman, R.S. (1980): Hypercalcemia in malignancy. Clin. Endocrinol. Metab. 9: 317-333

Brown, J., Santos, E., Rosen, G., Helson, L., Young, C. & Tan, C. (1978): Phase I study of gallium nitrate in patiens with advanced cancer. Proc. Am. Assoc. Cancer. Res. 19: 198

Caujolle, F., Bouissou, H., Gros, S., Maurel, E., Voisin, M.C. & Tollon, Y. (1972): Influence du gallium sur la souris gestante. I. Aspects morphologiques. CR. Soc. Biol. (París) 166: 952-958

Cleare, M.J. & Hydes, P.C. (1980): Antitumor properties of metal complexes. In: Sigel, H. (ed) Metal ions in Biological systems (vol XI). Marcel Dekker, New york pp 1-62

Collery, P.H., Millart, H., Simoneau, J.P., Pluot, M., Halpern, S., Vistelle, R., Lamiable, D., Choisy, H., Coudoux, P. & Etienne, J.C. (1983): Selective uptake of gallium administered orally as chloride, by tumour cells. In: Spitzy, H.K., Karrer, K. (eds) Proceedings of 13th internacional congress of Chemotherapy (part 284), Vienna, pp 35-43

Collery, P.H., Millart,M., Simoneau, J.P., Pluot, M., Halpern, S., Pechery, C., Choisy, H. & Etienne, J.C. (1984): Experimental treatment of mammary carcinomas by gallium chloride after oral administration: intratumor dosages of gallium, anatomapathologic study and intracellular microanalysis. Trace. Elem. Med. 1: 159-161

Collery, P.H., Millart, H., Ferrand, O., Jouet, J.B., Dubois, J.P.,Barthes, G., Pourny, C., Pechery, C., Choisy, H., Cattan, A., Dubois de Montreynaud, J.M. & Etienne, J.C., (1985): Gallium chloride treatment of cancer patients after oral administration: a pilot study. Chemioterapia 2 (suppl): 1165- 1166

Collery, P.H., Millart, H., Pluot, M. & Anghileri, L.J., (1986): Effects of gallium chloride oral administration on transplanted C3HBA mammary adenocarcinoma: Ga, Mg, Ca and Fe concentration and anatomopathological characteristics. Anticancer. Res. 6: 1085-1088

Collery, P.H., Millart, H., Cossart, C., Perdu, D., Vellerand, H., Morel, M., Pechery, C., Etienne, J.C., Choisy, H. & Dubois de Montreynaud, J.M (1989a): Potentiation of chemotherapy by gallium. J. Am. Coll. Nutr. 8: 428

Collery, P.H., Millart, H., Lamiable, D., Vistelle, R., Rinjard, P., Tran, G., Gourdier, B., Cossart, C., Bouana, J.C., Pechery, C., Etienne, J.C., Choisy, H. & Dubois de Montreynaud, J.M. (1989b): Clinical pharmacology of gallium chloride after oral administration in lung cancer patients. Anticancer Res. 9: 353-356

Domingo, J.L., Llobet, J.M. & Corbella, J. (1987a): Acute toxicity of gallium in rats and mice. J. Toxicol. Clin. Exp. 7: 411-418

Domingo, J.L., Llobet, J.M. & Corbella, J. (1987b): Relative efficacy of chelating agents as antidotes for acute gallium nitrate intoxication. Arch. Toxicol. 59: 382-383

Dudley, H.C. & Levine, M.D. (1949): Studies on the toxic action of gallium. J. Pharmacol. Exp. Ther. 95: 487-493

Edwards, C.L. & Hayes, R.L. (1969): Tumor scanning with ^{67}Ga-citrate. J. Nucl. Med. 10: 103-105

Edwards, C.L. & Hayes, R.L. (1970): Scanning malignant neoplasms with gallium^{67}. JAMA 212: 1182-1190

Edwards, C.L., Nelson, B. & Hayes, R.L. (1970): Localization of gallium in human tumors. Clin. Res. 18: 89

Ferm, V.H. & Carpenter, S.J. (1970): Teratogenic and embryopathic effects of indium, gallium and germanium. Toxicol. Appl. Pharmacol. 16: 166-170

Foster, B.J., Clagett-Carr, K., Hoth, D. & Leyland-Jones, B. (1986): gallium nitrate: the second metal with clinical activity. Cancer Treat. Rep. 70: 1311-1319

Goering, P.L., Maropont, R.R. & Fowler, B.A. (1988): Effect of intratracheal gallium arsenide administration on -amnolevulinic acid dehydratase in rats : relationship to urinary excretion of aminolevulinic acid. Toxic. Appl. Pharmacol. 92: 179-193

Goldstein, R.S., Noordeweier, R., Bond, J.T., Hook, J.B. & Mayor G.H. (1981): Cis-dichlorodiammineplatinum nephrotoxicity: time course and dose response of renal function impairment. Toxicol. Appl. Pharmacol. 60: 163-175

Guarino, A.M., Miller, D.S., Arnold, S.T., Pritchard, J.B., Davis, R.D., Urbanek, M.A., Miller,

T.J. & Litterst, C.L. (1979): Platinate toxicity: past, present and prospects. Cancer Treat. Rep. 63: 1475-1483

Hall, S.W. Yeung, K., Benjamin, R.S., Stewart, D., Valdivieso, M., Bedikian, A.Y. & Loo, T.L.(1979): Kinetics of gallium nitrate, a new anticancer agent. Clin. Pharmacol. Ther. 25: 82-87

Hart, M.M. & Adamson, R.H. (1971): Antitumor activity and toxicity of salts of inorganic group IIIa metals: aluminum, gallium, indium and thallium. Proc. Nat. Acad. Sci. USA 68: 1623-1626

Hart, M.M., Smith, C.F., Yancey, S.T. & Adamson, R.H. (1971): toxicity and antitumor activity of gallium nitrate and periodically related metal salts. J. Nat. Cancer Inst. 47: 1121-1127

Higby, D., Wallace, J., Albert, D. & Holland, J. (1975): Diamminodichloroplatinum: a Phase I study showing responses in testicular and other tumors. Cancer 33: 1219-1225

Kelsen, D.P., Alcock, N., Yeh, S., Brown, J. & Young, C. (1980): Pharmacokinetics of gallium nitrate in man. Cancer 46: 2009-2013

Krakoff, I.H. (1979): Nephrotoxicity of cis-dichlorodiammineplatinum (II). Cancer Treat. Rep. 63: 1523-1525

Levenson, S.M., Warren, R.D., Johnston, G.S. & Chabner, R.A. (1976): Abnormal pulmonary gallium accumulation in p.carinii pneumonia. Radiology 19: 395-398

Mundy, G.R., Ibbotson, K.J., D'Souza, S.M., Simpson, E.L., Jacobs, J.W. & Martin T.J. (1984): The hypercalcemia of cancer: clinical implications and pathogenic mechanisms. N. Engl. J.Med. 310: 1718-1727

Newman, R.A., Brody, A.R. & Krakoff, I.H. (1979): Gallium nitrate (NSC-15200) induced toxicity in the rat. A pharmacologic, histopathologic and microanalytical investigation. Cancer 44: 1728-1740

Pearson, R.G. (1963): Hard and soft acids and bases. J. Am. Chem. Soc. 85: 3533-3539

Rasey, J.S., Nelson, N.J. & Larson, S.M. (1982): Tumor cell toxicity of stable gallium nitrate: enhancement by transferrin and protection by iron. Eur. j. cancer Clin. Oncol. 18: 661-668

Robinson, A.L. (1983): GaAs readied for hihg-speed microcircuits. Sciencie 219: 275-277

Rozencweig, M., Von Hoff, D.D., Penta, J.S. & Muggia, F.M. (1977): Clinical status of cis-diamminedichloroplatinum (II). J.Clin. Hemat. Oncol. 7: 672-678

Saiki, J.H., Baker, L.H., Stephens, R.L., Fabian, C.J., Kraut, E.H. & Fletcher, W.S. (1982): Gallium nitrate in advanced soft tissue and bone sarcomas: A southwest oncology group study. Cancer Treat. Rep. 66: 1673-1674

Samson, M.K., Fraile, R.J., Baker, L.H. & O'Bryan, R. (1980): Phase I-II clinical trial of gallium nitrate (NSC-15200). Cancer Clin. Trials. 3: 131-136

Warrell, R.P., Coonley, C.J., Straus, D.P. & Young, C. W. (1983): Treatment of patients with advanced malignant lymphomas using gallium nitrate administration as a seven-day continuos infusion. Cancer 51: 1982-1987

Warrell, R.P., Bockman, R.S., Coonely, C.J., Isaacs, M. & Staszewski, H. (1984): Gallium nitrate inhibits calcium resorptions from bone and is effective treatment for cancer-related hypercalcemia. J. Clin. Invest. 73: 1487-1490

Warrell, R.P., Skelos, A., Alcock, N.W. & Bockman, R.S. (1986): Gallium nitrate for acute treatment of cancer-related hypercalcemia: clinicopharmacological and dose-response analysis. Cancer Res. 48: 4208-4212

Webb, D.R., Wilson, S.E. & Carter, D.E. (1986): Comparative pulmonary toxicity of gallium arsenide, gallium (III) oxide, or arsenic (III) oxide intratracheally instilled into rats. Toxicol. Appl. Pharmacol. 76: 96-104

Wittes, R., Cvitkovic, E., Shah, J., Gerald, F. & Strong, E. (1977): Cis-dichlorodiammineplatinum (II) in the treatment of epidermoid carcinoma of the head and neck. Cancer Treat. Rep. 61: 359-366

Wolff, R.K.,Henderson, R.F., Eidson, A.F., Pickrell, J.A., Rothenberg, S.J. & Hahn, F.F, (1988): Toxicity of gallium oxide particles following a 4-week inhalation exposure. J. Appl. Toxicol. 8: 191-199

Woods, J.S. & Fowler, B.A. (1978): Altered regulation of mammalian hepatic heme biosynthesis and urinary porphyrin excretion during prolonged exposure to sodium arsenate. Toxicol. Appl. Pharmacol. 43: 361-371

Yamauchi, H., Takahaski, K. & Yamamura, Y. (1986): Metabolism and excretion of orally and intraperitoneally administered gallium arsenide in the hamster. Toxicology 40: 237-246

Metal Ions in Biology and Medicine. Eds. Ph. Collery, L.A. Poirier, M. Manfait, J.C. Etienne. John Libbey Eurotext, Paris © 1990, pp. 420-422

Osteosarcoma Diagnosis and therapy respectively with tumour-specific gallium-67 and yttrium-90

S.K. Shukla[a,b], C. Cipriani[a], M. Monteleone[c], U. Tarantino[c], L. Mastidoro[c], S. Hani[d], T. Medici[d], K. Schomäcker[e], R. Münze[e], G. Argiro[a]

[a] *Servizio di Medicina Nucleare, Ospedale S. Eugenio, P. le Umanesimo 10, I-00144 Roma;* [b] *Istituto di Cromatografia, C.N.R., C.P. 10, I-00016 Monterotondo Scalo (Roma); Clinica Ortopedica, 2° Università di Roma, Roma;* [d] *Ambulatorio Veterinario, Via A. Silvani 66, Roma, Italy;* [e] *Zentralinstitut für Kernforschung, Akademie der Wissenschaft der DDR, Rossendorf, Dresden, DDR*

Osteosarcoma is a rare and very variable disease (Taylor et al. 1989), and this seems to be the main cause of poor development of its diagnostic and therapeutic methodologies. Computerized tomography (CT), ultrasound, radionuclidic imaging, laminography, arteriography, and plain radiography are main osteosarcomaimaging techniques, of which CT has become most popular (Levine et al. 1979). Therapeutic procedures include surgery (mainly amputation), chemotherapy and teletherapy but their efficacy is limited and life expectancy of the patient still remains less than 5 years (Bell et al., 1988; MSD Manual, 1988; Taylor et al., 1989). Since osteosarcoma, like other malignant cancers, is a systemic disease producing metastasis in organs far from the site of the primary tumour, only systemic therapy can be effective in its therapy. Radionuclidic diagnosis of osteosarcoma has been carried out by bone imaging with technetium-99m diphosphonates (Rosenberg et al., 1988).with gallium-67 citrate (Bekerman et al., 1985; Kaufman et al., 1977; Lepanto et al., 1976; Levine et al., 1979), and with indium-111-labelled monoclonal antimyosin Fab fragments (Kairemo et al. 1990). Intense concentration of the radionuclide from these radiopharmaceuticals in healthy organs and little uptake in the tumour leads to poor diagnostic results and produces high radiation burden to the patient. We undertook, therefore, a systematic study of the commercially available radiopharmaceuticals for tumour imaging and therapy. In the present articles we report our results with gallium-67 and with yttrium-90, the former used for noninvasive diagnosis and the latter one for noninvasive therapy. Conditions were found under which the radionuclide concentrates only in the tumour and not in healthy orgnas.

Osteosarcoma belongs to the anionic gallium-67 group of our tumour classification (Shukla et al., 1989). Figure 1a shows why gallium-67 scintigraphy with commercial radiopharmaceutical is tumour-nonspecific. In the radiopharmaceutical solution the radionuclide is present in insoluble, easily hydrolysable, and anionic forms, depending on the sodium citrate concentration in the radiopharmaceutical solution (Shukla et al., 1985). When imaging osteosarcoma, only anionic fraction of Ga-67 concentrates in the tumour, other species are taken up by healthy organs. We found conditions under which, the commercial gallium-67 solution can be made chromatographically and electrophoretically pure in anionic form (Fig. 1b).

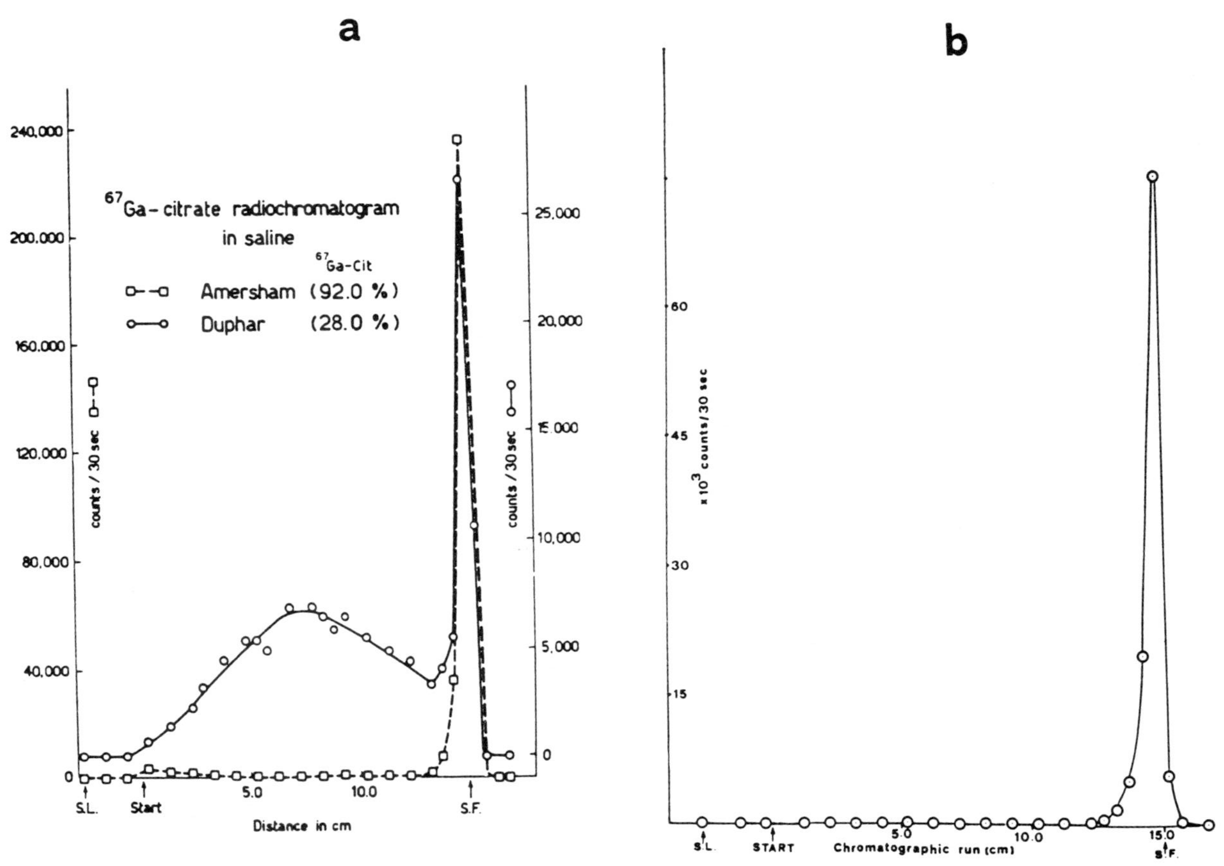

Fig. 1. Radiochromatograms in physiological saline on Whatman 3MM paper strips (3 x 20 cm) of: a) two commercially available gallium-67 citrate solutions; b) osteosarcoma-specific Ga-67 solution prepared by us.

With this osteosarcoma-specific Ga-67, rapid (3 h p.i.) uptake of the radionuclide in the tumour was found. After 24 h the radionuclide disappeared from the blood stream and clean tumour image (Fig. 2a) was obtained. The radionuclide is firmly bound to the tumour which permits follow-up studies for 10 days. The osteosarcoma-specific Ga-67 could image the tumour in all subjects. In present study osteosarcoma was imaged in humans, dogs, and cats.

Since yttrium-90 is trivalent like Ga-67, and emits high energy (2.27 KeV) beta rays, we obtained it also in osteosarcoma-specific form, which was chromatographically and electrophoretically identical to osteosarcoma-specific Ga-67 in solution. The anionic nature of the two radionuclides was obtained in sodium citrate solution, or in presence of other chelating agents;

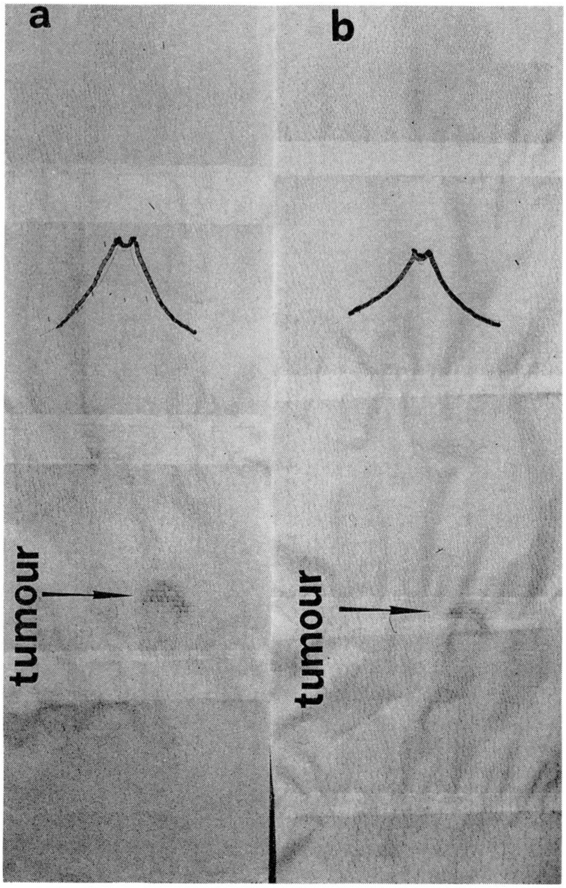

Fig. 2. Total-body scintigram of an osteosarcoma patient injected with osteosarcoma-specific Ga-67 solution: a) image 24 h p.i., b) image 10 days p.i.

OSTEOSARCOMA THERAPY WITH Y-90

Osteosarcoma in cats has been treated with osteosarcoma-specific Y-90. The pain in the tumour was eliminated after the first injection of the radiopharmaceutical and the tumour regressed after 3 administrations.The dose of Y-90 so far given has been 20 MBq/Kg. The studies are being followed.

REFERENCES

Bekerman, C., et al.(1985): The role of Ga-67 in the clinical eveluation of cancer. Semin. Nucl. Med. 15, 72-103.

Bell, R.S., et al. (1988): Timing of Chemotherapy and surgery in a murine osteosarcoma model. Cancer Res. 48, 5533-5338.

Kairemo, K.J.A., et al. (1990): Imaging of soft tissue sarcoma with In-111-labeled monoclonal antimyosin Fab fragments. J. Nucl. Med. 31, 23-31.

Levine, E., et al.(1979): Comparison of computed tomography and other imaging modalities in the evaluation of musculoskeletal tumours. Radiology 131, 431-437.

MSD-Manual der Diagnostik u. Therapie, 4. Aufl.,pp. 132-133. München: Urban & Schwarzenerg.

Taylor, W.F., et al. (1989): Prognostic variables in osteosarcoma: A multi-institute study. J. Natl. Cancer Inst. 81, 21-30.

Metal Ions in Biology and Medicine. Eds. Ph. Collery, L.A. Poirier, M. Manfait, J.C. Etienne. John Libbey Eurotext, Paris © 1990, pp. 423-425

Melanoma early diagnosis and systemic therapy respectively with tumour-specific gallium-67 and with yttrium-90

S.K. Shukla[a,b], C. Cipriani[a], K. Schomäcker[c], L. Taglia[d], S. Marrone[d], A. Muller[e], G. Politano[a]

[a] *Servizio di Medicina Nucleare, Ospedale S. Eugenio, P. le Umanesimo 10, I-00144 Roma;* [b] *Istituto di Cromatografia, C.N.R., C.P. 10, I-00016 Monterotondo Scalo (Roma);* [c] *Zentralinstitut für Kernforschung, Rossendorf, Dresden, D.D.R.;* [d] *Policlinico Veterinario, Via Lucania, 9, Roma;* [e] *Istituto Regina Elena per lo Studio e la Cura dei Tumori, Roma, Italy*

The incidence of malignant melanoma is steadily increasing in every country and more so among fair complexion individuals. For the year 1988, 28,000 new cases were reported in the U.S.A. alone (Mitchell et al., 1988). Melanoma has been found (Carrasquillo et al., 1984) to be a very aggressive disease, once metastasized it responds poorly to conventional treatment. Surgery may be successful in the treatment of primary tumour in its early stage. But, even in those cases we have found many patients to develop metastasis within three to four years. This indicates that melanoma easily disseminates and is resistant to conventional types of therapy, e.g., chemotherapy, teletherapy, or immunotherapy (Carrasquillo et al. 1984; Mitchell et al.,1988). Since melanoma is a widely distributing systemic disease, only systemic diagnostic procedure can give a complete picture of the tumour or can produce its effective therapy. We undertook, therefore, the synthesis and study of the in vivo behaviour of melanoma-specific radiopharmaceuticals and pharmaceuticals. For total-body imaging the trivalent radionuclide gallium-67 (half-life 78 h; gamma emissions: 90, 180, 300 KeV) was selected,which can be detected externally both with a gamma camera or a rectilinear scanner.

MELANOMA IMAGING RADIOPHARMACEUTICALS

1. Gallium-67 citrate (Bekerman et al., 1985)
2. Antimelanoma antibody labelled with I-131 (Larson, 1985), In-111 (Murray et al., 1985), and Tc-99m (Eary et al., 1989).

Total body distribution of the radionuclide after the injection of aboveradiopharmaceuticals showed little affinity of the radionuclide for the tumour and intense uptake in healthy organs, mainly liver, spleen, kidneys.

MELANOMA THERAPY MODES

1. Boron Neutron Capture Therapy (BNCT) (Coderre et al., 1988)
2. Immunotherapy (Mitchell et al. 1988)
3. Chemotherapy (Cummings et al., 1985)
4. Radioimmunotherapy (Carrasquillo et al. 1984; Larson, 1985).

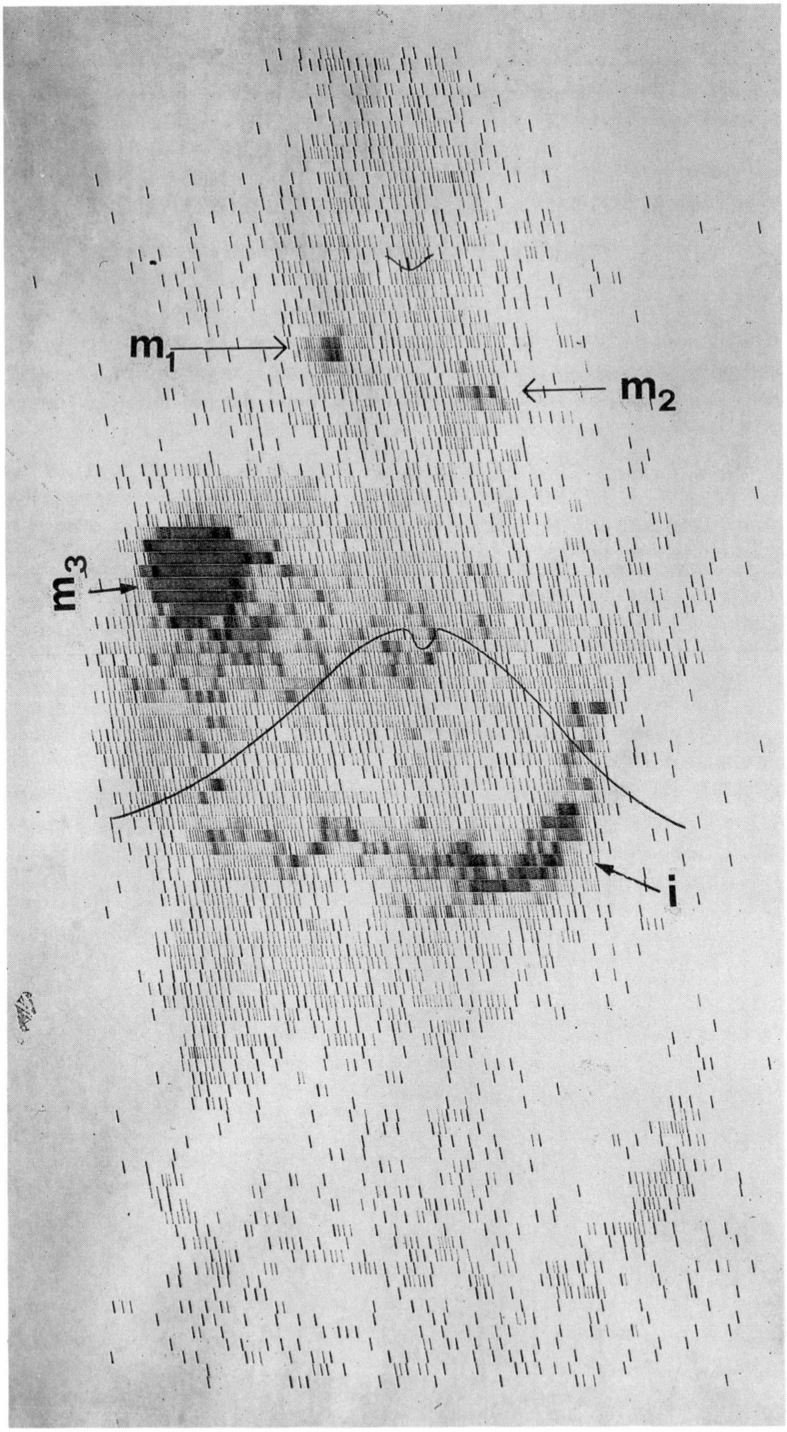

Fig. 1. Total-body Ga-67 distribution in a patient, injected with pure anionic melanoma-specific Ga-67 solution, who had developed metastases in lung (m_1 & m_2) and in liver (m_3) four years after the surgical removal of the primary melanoma at left side of his chest. The activity (i) is due to breakfast, the patient had taken.

Since melanoma belongs to anionic Ga-67 affine group of tumours (Shukla et al., 1989), we injected chromatographically and electrophoretically pure Ga-67 solution for its distribution in the body. So far we have examined about 80 patients in four different clinics in Rome and in Athens and observed tumour-specific uptake of the radionuclide (Fig. 1). The binding of the radionuclide in the primary and secondary tumours is fast, permitting its localization already after 3 h post-injection. After 24 hours and on following days the radionuclidic image of the tumour is clean. The long half-life, 78 h, of Ga-67 makes it possible to follow the course of the tumour for at least 10 days p.i. Due to adsorption of Ga-67 in solid food, its presence in the intestine may be seen in early imaging after the administration of the radiopharmaceutical. The patients are therefore advised not to take solid food at least 12 hours before presenting themselves for diagnosis.

TUMOUR-SPECIFIC MELANOMA THERAPY WITH MELANOMA-SPECIFIC Y-90 AND WITH COLD GALLIUM

Having no antimelanoma antibody so far at our disposal, we have so far examined antitumour activity of melanoma-specific Y-90 and of cold gallium in spontaneous melanoma-bearing cats and dogs. When the tumour was not very big and highly metastasized it disappeared for ever after two to three injections of Y-90 (20 MBq/Kg), or gallium (3 mg/Kg) at intervals of 3 days. Tumour growth was arrested in bigger ones and life span prolonged. No side effects of the therapy was observed.

REFERENCES

Bekerman, C., C., Hoffer, P.B., et al. (1985): The role of Ga-67 in the clinical evaluation of cancer. Semin. Nucl. Med. 15, 72-103.

Carrasquillo, J.A., Krohn, K.A., et al. (1984): Diagnosis of and therapy for solid tumors with radiolabeled antibodies and immune fragments. Cancer Treat. Rep. 68, 317-328.

Coderre, J.A., Kalef-Ezra,J.A., et al. (1988): Boron neutron capture therapy of a murine melanoma. Cancer Res. 48, 6313-6316.

Cummings, F.J., McDonald, C.J., et al. (1985): Melanoma. In: Medical Oncology, eds. P. Calabresi, P. Schein & S.A. Rosenberg, pp. 650-679. New York: MacMillan & Co.

Eary, J.F., Schroff, R.W., et al. (1989): Successful imaging of malignant melanoma with technetium-99m-labeled monoclonal antibodies. J. Nucl. Med. 30, 25-32.

Mitchell, M.S., Kan-Mitchell, J., et al. (1988): Active specific immunotherapy for melanoma: Phase I trial of allogeneic lysates and a novel adjuvant. Cancer Res. 48, 5883-5893.

Murray, J.L., Rosenblum, M.G., et al. (1985): Radioimmunoimaging in malignant melanoma with In-111-labeled monoclonal antibody 96.5. Cancer Res. 45, 2376-2381.

Shukla, S.K., Cipriani, C., et al. (1989): Twenty years of gallium-67 for tumor imaging --- What next ? Preparation of tumour-specific gallium-67 in solution for systemic noninvasive diagnosis. Book of Abstracts of the 20th Annual Meeting of the Australian and New Zealand Society of Nuclear Medicine, p. 27. Christchurch School of Medicine, Christchurch, New Zealand.

Metal Ions in Biology and Medicine. Eds. Ph. Collery, L.A. Poirier, M. Manfait, J.C. Etienne. John Libbey Eurotext, Paris © 1990, pp. 426-431

Gallium a unique anti-resorptive agent in bone : pre-clinical studies on its mechanisms of action

R. Bockman, R. Adelman, R. Donnelly, L. Brody, R. Warrell, K. Jones*

*Department of Medicine, The Hospital for Special Surgery, Cornell University Medical College, 530 East 70th Street, New York, New York and * Brookhaven National Laboratory, Upton, New York, USA*

The discovery of gallium as a new and unique agent for the treatment of metabolic bone disorders was in part fortuitous. Initially studied as a possible antineoplastic agent, attention shifted to its effects on bone and mineral metabolism with the clinical observations that patients receiving pharmacologic levels of gallium nitrate frequently became hypocalcemic. Hypocalciuria had been reported in animals given gallium by bolus injection (Newman et al., 1979), however, the mechanism responsible for this effect was not appreciated. Early studies using explanted fetal rat bones led to the hypothesis that the drug had potent anti-resorptive activity in bone (Warrell et al., 1984). This antiresorptive activity was subsequently confirmed by metabolic balance studies in man (Warrell et al., 1985). Laboratory studies showed gallium would block bone resorption induced by a variety of cytokines and calcitropic hormones (Warrell et al., 1984, Bockman et al., 1987). This anti-resorptive action occurred at dose levels that were without evidence of a cytotoxic effect on bone cells (Cournot-Witmer et al., 1987, Hall & Chambers, 1990). Unique compared to all other clinically useful antiresorptive agents, gallium appears to have favorable affects on new calcium accretion and collagen formation (Bockman et al., 1986, Bockman et al., 1987). This dual action may in part explain gallium's broad efficacy in a variety of metabolic bone disorders compared to other clinically available, purely anti-resorptive agents.

Gallium, listed as a Group III a transitional element, is a near-metal of intermediate mass. In its pure form it has the unique property of being a liquid above 29.8^{0}C. It can have several valance states from +1 to +3, the latter being most common. It forms salts with strong acids and stable coordinated complexes. The arsenide compounds have found wide application in the electronic semi-conductor industry. For pharmaceutical applications, gallium salts are typically suspended in citrate containing solutions to provide a citrated-coordinate complex of sufficient solubility at neutral pH. Other weak chelators such as maltol, provide soluble coordinate complexes. Lecoq de Boisbaudran is credited with discovering this element and naming it for the region of its discovery.

Based on a growing body of laboratory and clinical studies, the physiologic effects of gallium, particularly with regard to bone metabolism, result from a rather complicated series of physical and cellular interactions. When injected parenterally into man or animals, there is rapid clearance of free gallium, ie.

non-complexed gallium through the kidneys. With prolonged administration, a fraction of the injected dose demonstrates a very long plasma half-life. This long plasma half life results for at least two reasons, first there is binding of gallium to large circulating proteins in the blood (mainly transferrin, Clausen et al., 1974). Secondly, there is deposition and slow release of gallium from a large tissue reservoir (probably bone). Bone can accumulate large amount of gallium (Angliheri 1971, Bockman et al., 1986 and Repo et al., 1988). This accumulation in bone is dependent on the dose administered (Warrell et al., 1984 and Bockman et al., 1988). Radiolabeled gallium taken up by the soft tissues can be readily displaced by a "cold" gallium chase. Gallium that has accumulated in bone is not readily displaced by newly administered gallium (Angliheri 1971). In contrast to blood, bone levels of gallium decline very slowly following cessation of therapy, Fig. 1. This localization of gallium in bone has raised many important questions as to the interaction of gallium with bone matrices and its effects on the physical properties of bone as well as on the cellular elements of bone. Several of these issues will now be discussed.

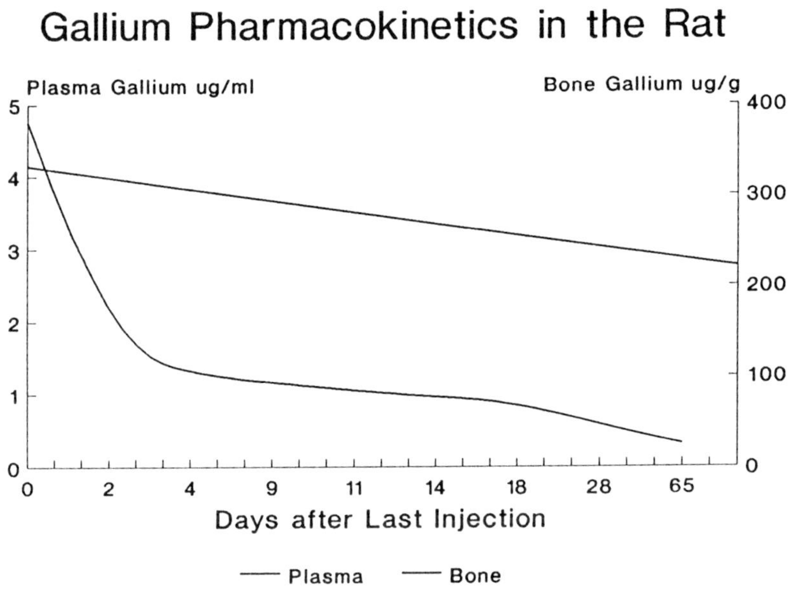

Fig. 1. Fall in Plasma and Bone gallium levels following 7 s.c. injections with gallium nitrate (5 mg/rat) given every other day. Blood and bone samples removed at various times after therapy and analyzed by flameless atomic absorptiometry.

<u>Localization in Bone:</u>

Therapeutic doses of gallium result in trace levels (ppm) accumulating in bone (Bockman et al., 1986, Repo et al., 1988), therefore, the quantification and precise localization of gallium in bone poses significant technical problems. Indeed, it has not been possible to readily detect gallium in bone by the usual ion-probe analysis methods (Cournot-Witmer et al., 1987). To achieve spatial resolution at the microscopic level for selected, naturally occurring trace technique of synchrotron-generated x-ray microscopy. This methodology employs the National Synchrotron Light Source at Brookhaven National Laboratory to provide a dense, collimated beam of x-rays of sufficient energy to permit x-ray fluorescence analyses. The methodology enables one to simultaneously detect zinc, iron, strontium, copper and gallium with high spatial resolution (Bockman et al., 1988). Calcium which is present in relatively abundant amounts, provides an internal standard for the measurements. Comparison of the atomic weight ratios for the naturally occurring trace elements was carried out in the femoral bones of rats who had received 7 mg of elemental gallium over 14 days. Following this treatment the rats showed significant decreases in zinc and iron ratios (27 and 47% respectively) in the metaphyseal region, a site of active bone turnover (Bockman et al., in press). No change in strontium or copper levels were noted. When looking at the microscopic distribution of gallium by XRM, the data show that gallium preferentially accumulates in those regions where active, bone-cell metabolism and thus bone matrix turnover are occurring, ie. the endosteal and periosteal surfaces of bone, Fig. 2. Several cofactor-dependent enzymes known to be critical for normal bone metabolism are concentrated in the regions of gallium accumulation. Gallium-substitution for the naturally occurring trace elements, (iron or zinc) provides a possible mechanism by which to explain many of the biologic effects gallium has in bone. While it is evident that gallium accumulation in bone is associated with displacement of iron and zinc, it has not been demonstrated that gallium actually replaced the naturally occurring trace elements. Newer technological advances in the XRM-methodology have allowed micron level resolution to be achieved; at that level, it may be possible to identify the specific bone cells that are involved with gallium accumulation.

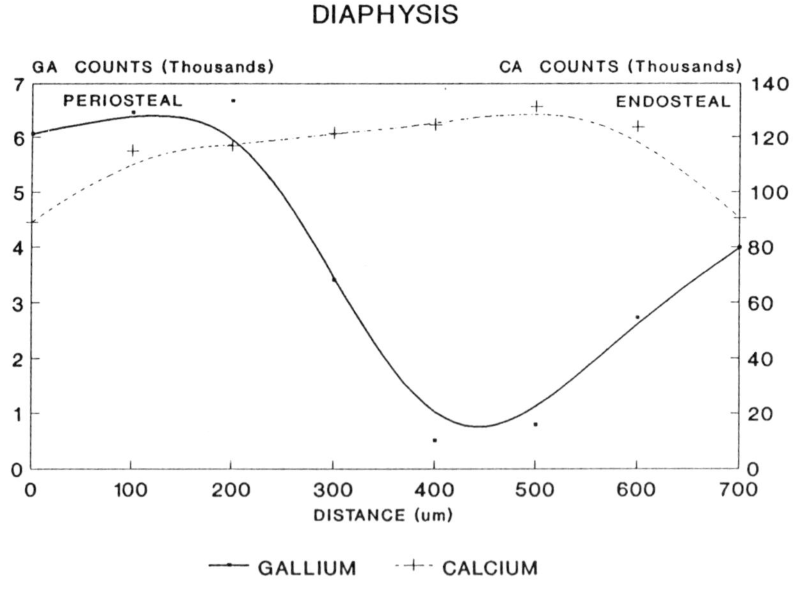

Fig. 2. X-Ray microprobe scan across the diaphyseal cortex from the periosteal to the endosteal surface and analyzed for gallium and calcium fluroescence.

Effects on Bone-Mineral Properties:

The accumulation of trace levels of gallium ion in bone is associated with consistent and reproducible changes in the physical properties of the mineral matrix of bone. X-ray diffraction analysis of bone powder from gallium treated rats shows narrowing of the absorption peak for the long axis of the hydroxyapatite crystals compared to the crystals from untreated, litter-mate controls. Such a change is consistent with an increase in hydroxyapatite crystal perfection or size. Analyses of the gallium-treated bone powder by infrared spectroscopy suggests that bone carbonate content is decreased (Bockman et al., 1989). These changes in physical properties would be anticipated to decrease the solubility of hydroxyapatite. Indeed, when the solubility of hydroxyapatite was tested in mildly acid solutions, the bone powder from gallium treated rats showed significantly slower dissolution rates (Repo et al., 1988).

Other changes in bone mineral properties have been noted. Bone powder particles from the metaphyseal regions of gallium-treated rats showed a shift to higher density, and an increase in calcium and phosphorus content Repo et al., 1988). Injection of radio-labeled calcium into gallium-treated rats, consistently demonstrated more "new" calcium accretion in the treated animals. This finding argues that gallium treatment favors increased mineralization of newly forming bone. This combination of decreased solubility of hydroxyapatite with increased accretion provides a possible explanation for the greater anti-resorptive activity of gallium nitrate that is seen clinically (Warrell et al., 1988 compared to calcitonin and bisphosphonates such as EHDP that have little or no bone-forming activity (Schenk et al 1973).

Cellular Effects

In vitro studies with various explant bone models clearly demonstrate the ability of gallium to prevent osteolysis. Fetal rat long bones well as newborn calvariae prelabeled in utero with ^{45}Ca show dose dependent inhibition of calcium release from tissue explants over a range of 10-200 uM gallium. The effect is independent of the gallium salt used. Gallium is effective in blocking bone resorption induced by PTH-peptides at high (Warrell et al 1984, Hall & Chambers, 1990) or physiologic doses (Stern et al., unpublished), as well as cytokines such as tumor necrosis factor (Bockman et al., 1987). Since these inducers of osteolysis may act via different biochemical pathways (Bockman & Repo 1981), the data suggests gallium acts at a common point (or end-point) to block bone resorption.

In separate, in vivo studies of bone resorption, the effects of gallium were examined in depth. Bone particles from control and gallium treated rats were implanted subcutaneously into rats. Over a three week period, multi-nucleated tartrate-resistant, acid phosphatase staining cells were noted to surround the bone particles. These bone particles were subsequently resorbed. Chronic gallium injections into rats with normal bone powder implants (from non-treated animals), had no effect on the resorption of the implants compared to sham (citrate) injected controls. By contrast, a significant delay in resorption was seen when bone powder taken from gallium treated rats was compared to bone powder from untreated controls. While the gallium-containing bone implants were more resistant to resorption; no differences in the recruitment of the tartrate resistant, acid-phosphatase staining cells with regard to number or appearance was observed. This would suggest that the chemotactic response to the bone particles was not altered by gallium treatment.

Histological sections of bone explants provide an important piece of information. Bone sections from gallium-treated rats show normal osteoclasts apposed to bone with no differences between control and gallium-treated samples. Morphological studies on isolated osteoclasts showed that doses of gallium nitrate as high as it affect their survival on bone slices (Hall & Chambers, 1990). This result is in striking contrast to the devastating cytotoxic effects seen with other therapeutic metals such as platinum and germanium or with mithramycin, (Bockman et al., 1986). Gallium over the therapeutic range (10-100 uM) did not alter DNA or protein synthesis in calvarial explants or in model osteoblast cell lines. In fact, gallium caused a significant enhancement (200-300%) above controls) of the major matrix protein, Type-I collagen (Bockman et al., 1987). This gallium-induced increase in the major protein component of the organic matrix of bone along with the increase in newly formed mineral matrix provides the basis for concluding that gallium treatment actually favors bone formation.

One concern raised by the accumulation of trace amounts of gallium ion and the subtle changes in bone matrix properties is that bone strength may be altered following gallium treatment. A standardized biomechanical test of bone strength (Einhorn et al., 1984) was carried out on rat femurs from rats treated with gallium-nitrate. The rats received 27 and 33 mg of elemental gallium over 9 or 11 weeks before testing was carried out. No changes in bone strength were noted (Adelman et al., 1989, Donnelly et al., unpublished). These data argue that gallium accumulation in bone did not alter bone strength after months of chronic administration.
Reviewing the possible mechanisms of gallium's antiresorptive activity, we propose that:

(1) Gallium accumulates preferentially in the metabolically active regions where bone resorption and formation occur.

(2) Gallium induces changes in the physical properties of newly synthesized bone mineral which renders the mineral less resorbable. (3) Gallium alters the concentration and distribution of naturally occurring trace elements (iron and zinc) in bone. Cofactor changes in key, metal-dependent enzymes responsible for bone turnover could result in altered enzyme function.

(4) Gallium treatment is associated with an increase in the calcium and phosphate content of bone. (5) Cell mediated bone resorption, regardless of the inducing factor can be inhibited by gallium.

(6) Inhibition of bone resorption is not associated with evidence of a cytotoxic effect or a failure to recruit resorptive cells.

(7) Increased collagen production is seen in bone explants treated with gallium.

In summary, gallium is an exciting new therapeutic agent for the treatment of pathologic states characterized by accelerated bone resorption. Compared to other therapeutic metal compounds containing platinum or germanium, gallium affects its antiresorptive action without any evidence of a cytotoxic effect on bone cells. Gallium is unique amongst all therapeutically available antiresorptive agents in that it favors bone formation.

Acknowledgements: This work was supported in part by U.S. Department of Health and Human Service grants CA38645, CA29502; NIH Biotechnology Research Resource Grant P41RR01838 and D.O.E. contract DE-AC02-76CH00016.

References:

Adelman, R.J., Jones, K.W., Donnelly, R., Wright, T.M., Bockman, R.S. (1989): Gallium localization and effect on mechanical bone strength. Trans. Orth. Res. Soc. 15, (2) 415.

Anglihileri, L. (1987): Studies on the accumulation mechanisms of radioisotopes used in tumor diagnostic. Strahlentherapie 142, 456-462.

Bockman, R.S., Bohnsack, R., Warrell, R.P., Jr. (1986): Effect of metal based compounds on bone resorption. Clin. Res. 34, 690A.

Bockman, R.S., Boskey, A.L., Blumenthal, N.C., Alcock, N.W., Warrell, R.P., Jr. (1986): Gallium increases bone calcium and

39, 376-381.
Bockman, R.S., Israel, R., Alcock, N.W., Ferguson, R., Warrell, R.P., Jr. (1987): Gallium nitrate stimulates bone collagen synthesis. Clin. Res. 35, 620.
Bockman, R.S., Repo. M.A. (1981): Lymphokine-mediated bone resorption requires endogenous prostaglandin synthesis. Journal of Experimental Medicine. 154, 529-534.
Bockman, R.S., Repo. M.A., Warrell, R.P., Jr., et al. (1988): X-ray microscopy studies on the pharmaco-dynamics of therapeutic gallium in rat bones. In Sayre D, Howell M, Kirz, J., Rayback, H (eds): X-Ray Microscopy II. Springer Series in Optical. Sciences. New York, Springer-Verlag, vol 56, 391-394.
Bockman, R.S., Repo, M.A., Warrell, R.P., Jr., Israel, R., Gabrilove, J. (1987): Gallium nitrate inhibits bone resorption induced by recombinant human tumor necrosis factor (TNF). Proc. Am. Assoc. Cancer Res. 28, 449.
Clausen, J., Edeling, C.J., Fogh, I. (1974): ^{67}Ga binding to human serum proteins and tumor components. Cancer Res. 34, 1931-1937.
Cournot-Witmer, G., Bourdeau, A., Lieberherr, M., et al. (1987): Bone modeling in gallium-nitrate-treated rats. Calcif. Tissue Int. 40, 270-275.
Einhorn, T.A., Lane, J.M., Burstein, A.H., Kopman, C.R., Vigorita, V.J. (1984): The healing of segmental bone defects induced by demineralized bone matrix. J. Bone & Joint Surgery. 66A, 274-279.
Hall, T.J. and Chambers, T.J. (1990): Gallium inhibits bone resorption by a direct effect on osteoclasts. Bone and Mineral, 8, 211-216.
Newman, R.A., Brody, A.R., Krakoff, I.H. (1979): Gallium nitrate (NSC-15200_ induced toxicity in the rat: A pharmacologic, histopathologic and microanalytical investigations. Cancer 44, 1728-1740.
Repo, M.A., Bockman, R.S., Betts, F., Boskey, A.L., Warrell, R.P., Jr. (1988): Effect of gallium on bone mineral properties. Calcif Tissue Int. 43, 300-306.
Schenk, R., Merz, W.A., Muhlbauer, R., Russell, R.G.E., Fleisch, H. (1973): Effect of EHDP and Cl2 MDP on the calcification and resorption of cartilage and bone in the tibial metaphysis of tats. Calc. Tiss. Res. 11, 196-214.
Warrell, R.P., Jr., Bockman, R.S., Coonley, C.J., Isaacs, M., Staszewski, H. (1984): Gallium nitrate inhibits calcium resorption from bone and is effective treatment for cancer-related hypercalcemia. J. Clin. Invest. 73, 1487-1490.
Warrell, R.P., Jr., Isaacs, M., Coonley, C.J., Alcock, N.W., Bockman, R.S. (1985): Metabolic effects of gallium nitrate administered by prolonged intravenous infusion. Cancer Treat Rep. 69, 653-655.
Warrell, R.P., Jr., Israel, R., Frisone, M., Snyder, T., Gaynor, J.J., Bockman, R.S. (1988): Gallium nitrate for acute treatment of cancer-related hypercalcemia: A randomized double-blind comparison to calcitonin. Ann Intern. Med. 108, 669-674.

Metal Ions in Biology and Medicine. Eds. Ph. Collery, L.A. Poirier, M. Manfait, J.C. Etienne. John Libbey Eurotext, Paris © 1990, pp. 432-436

Gallium for treatment of bone loss in cancer and metabolic bone diseases

R.P. Warrell Jr, R.S. Bockman

Memorial Sloan-Kettering Cancer Center, 1275 York Avenue, New York, NY 10021, USA

The initial observations of hypocalcemia due to pharmacologic uses of gallium nitrate were made in the late 1970's following administration of high-doses as a method of anticancer treatment. An early report indicated that gallium increased urinary excretion of calcium. However, we consistently observed hypocalciuria in our patients who were hypocalcemic (Warrell *et al.*, 1983). This apparent discrepancy initiated a series of studies wherein it was found that: a) administration of gallium nitrate is associated with a profound reduction in urinary calcium excretion (Warrell *et al.*, 1985); and b) the principal mechanism for this effect is a potent reduction in the release of calcium from bone (Warrell *et al.*, 1984). These observations have provided the impetus for a series of clinical studies to evaluate whether gallium-containing compounds could be useful for diseases characterized by loss of bone mineral.

Compelling evidence has been assembled to indicate that gallium is a potent inhibitor of bone resorption and that this agent also acts to maintain or restore calcium content in bone (see Bockman *et al.*, in this volume). These effects suggest that gallium could be useful treatment for a variety of diseases characterized by accelerated bone loss. Such conditions span a spectrum of disorders that range from cancer-related hypercalcemia to senile osteoporosis. In this chapter, we review the clinical evidence which indicates that gallium-containing drugs are useful therapeutic agents for patients with these conditions.

Studies in Hypercalcemia

Gallium nitrate was initially tested in a series of 10 patients with cancer-related hypercalcemia. The drug was administered as a continuous i.v. infusion at a daily dose of 200 mg/m^2 for periods ranging from 5-7 days. All patients responded with a reduction in serum calcium to normal or sub-normal concentrations (Warrell *et al.*, 1984). This study was followed by a dose-ranging study in which 36 infusions were administered to 31 patients. A clear dose-response was observed. Patients treated with the higher dose (200 mg/m^2/day) achieved a superior rate of normocalcemic control compared to the lower dose (100 mg/m^2/day) and this response was sustained for a substantially longer period of time (Warrell *et al.*, 1986).

We then initiated the first randomized double-blind study that compared active methods of treatment in patients with cancer-related hypercalcemia. In that study, hospitalized patients resistant to 2 days of parenteral hydration were randomized to receive either gallium nitrate (200 mg/m^2/day) or high doses of calcitonin (8 IU/kg every 6 hours) for 5 days. Overall, 75% of patients who received gallium nitrate achieved normocalcemia compared to only 27% of patients who received calcitonin ($P < 0.0006$) (Warrell *et al.*, 1988). These response data include only the most stringent criteria (i.e. intent-to-treat, no exclusions, and adjustments of serum calcium for serum albumin). If conventional (but less conservative) criteria are employed as have been used in other studies (i.e. no adjustment of serum calcium and exclusion of patients who died prior to completing treatment), 100% of patients treated with gallium achieved normocalcemia compared to 69% of patients treated with calcitonin ($p < 0.001$).

Duration of normocalcemia is difficult to assess given multiple confounding factors in critically ill patients, especially the use of cytotoxic or other potentially hypocalcemic drugs. If time to recurrence is censored at the time the serum calcium was first above the normal range *or* other treatment was administered (a very conservative method of analysis), the mean duration of normocalcemia was 6 days for patients treated with gallium nitrate and 1 day for patients treated with calcitonin ($p < 0.001$). If the effects of intercurrent treatment are ignored, the duration of normocalcemia was 13+ days for patients treated with gallium nitrate and 2 days for patients treated with calcitonin ($p < 0.01$).

Recent studies have shown that epidermoid (squamous) carcinomas are particularly associated with elevated plasma levels of a protein with biochemical activities identical to parathyroid hormone (PTH) - the so-called "PTH-related protein" (Budayr *et al.*, 1989). In a previous study, gallium nitrate was shown to be highly effective for treatment of hypercalcemia in patients with parathyroid carcinoma, a virulent hypercalcemic syndrome caused by grossly elevated serum levels of PTH (Warrell, Isaacs, *et al.*, 1987). Prior to initiation of the gallium/calcitonin study, patients were stratified by tumor histology since patients with hypercalcemia due to epidermoid carcinomas were believed more likely to have a resistant, "humorally-mediated" hypercalcemia. Patients with epidermoid carcinoma who received calcitonin in that study fared especially poorly, and only 1 of 10 such patients achieved a normal serum calcium. However, the response to gallium nitrate was independent of histology, and equal proportions (75%) of patients with both epidermoid and non-epidermoid tumor types responded. These studies have particular importance since the current generation of bisphosphonates appear to be significantly less effective for hypercalcemic syndromes mediated by the PTH-related protein (Thiebaud *et al.*, 1990).

A recent study has shown exceptional therapeutic superiority of gallium nitrate compared to the bisphosphonate, etidronate (Warrell *et al.*, 1990). A follow-up study that compares the effectiveness of gallium to another bisphosphonate (pamidronate; APD) is underway in France. Additional clinical studies are evaluating dosing regimens that are more convenient than continuous i.v. infusion, as well as low-dose schedules that can be administered chronically to prevent recurrence of hypercalcemia. These clinical studies have confirmed *in vitro* observations that gallium antagonizes bone resorption irrespective of mechanism or cancer type. At present, it appears that gallium nitrate is the *most* effective drug in clinical use for the treatment of cancer-related hypercalcemia.

Studies in Bone Metastases

The possibility of using medical treatment as an adjunct to strengthen bone tissue against erosion from cancerous metastasis has gained increased credence. Ideally, such therapy should not only minimize further bone loss but also restore bone that has been previously eroded. Other drugs, such as bisphosphonates, fluorides and calcitonin, have previously been proposed for this use. The unique dual activity of gallium to both decrease bone resorption and stimulate bone formation has suggested that this agent might be useful for preservation or restoration of bone that has been destroyed due to cancer.

Accelerated bone turnover in patients with bone metastases is frequently associated with increased urinary excretion of calcium and hydroxyproline. In a preliminary study, gallium nitrate was administered by a continuous i.v. infusion daily for 5-7 days to 22 patients with lytic bone metastases. Urinary excretion of calcium and hydroxyproline was determined before and after therapy. Urinary calcium excretion was significantly reduced in 21 of 22 patients. Urinary hydroxyproline excretion was also significantly lowered in patients with high basal levels of excretion (Warrell, Alcock, *et al.*, 1987). Similar findings have been observed by others in patients with prostatic cancer who have dense osteoblastic bone metastases (Scher *et al.*, 1987). These results document that therapy with gallium acutely lowers biochemical parameters associated with accelerated bone turnover. Several key issues are outstanding: first, can these effects be sustained; second, what is the lowest dose at which these effects are observed; third; what treatment schedules are most effective and safe; and fourth, can extended treatment provide an important benefit to patients. Presumably, an effective drug would substantially relieve bone pain, reduce the incidence of pathologic fractures, and prevent hypercalcemia.

Some of the difficulties in performing such studies have been discussed elsewhere (Warrell & Bockman, 1989). In the United States, research has focused on multiple myeloma and breast cancer - the former since it is the prototypic osteolytic disease in cancer, and the latter since it is the most prevalent osteolytic condition. The initial study in myeloma has been completed. In that study, patients who had achieved a "plateau" in their disease with conventional chemotherapy were randomized to receive gallium nitrate or no additional treatment for 6 months along with their chemotherapy. Patients randomized to no gallium were crossed over after 6 months. The principal outcome determinant was the measurement of total-body calcium content using a highly sensitive technique (neutron activation analysis) (Lovett *et al.*, 1989). Patients who received gallium nitrate showed a significant increase in total-body calcium relative to untreated controls. The increase was usually (though not uniformly) observed after 6 months of treatment. Most patients also experienced substantial relief of bone pain while receiving gallium. Clinically apparent pathologic fractures (and a single episode of hypercalcemia) were observed only in patients who were initially randomized not to receive gallium treatment. No important side-effects were observed (Warrell *et al.*, 1989). Some patients with myeloma have remained on gallium nitrate for periods in excess of 3 years. These preliminary results suggest that prolonged treatment with gallium nitrate may be effective in reducing the morbidity associated with bone metastases; however, questions of dose and scheduling are yet to be resolved.

A follow-up study has been initiated in patients with osteolytic metastases from breast cancer whereby patients receiving standard systemic treatment are randomized to 1 of 3 dose schedules of gallium nitrate (administered as a subcutaneous injection) or to no additional therapy. In addition to conventional measures of pain and fracture, quantitative digital computed tomography is being evaluated in that study as a method of quantifying focal changes in lytic bone destruction. Additional long-term multicenter studies are also being conducted in the United States and Europe to broadly test gallium nitrate as an adjuvant to conventional chemotherapy in women with breast cancer.

Studies in Metabolic Bone Diseases

Gallium nitrate has recently been shown to reduce biochemical parameters of disease activity in patients with advanced Paget's disease of bone. Similar to the aforementioned studies in bone metastases, short-term treatment with both moderate-dose i.v. infusions and low-dose subcutaneous injections (0.25-0.5 mg/kg/d [10-20 mg/m^2/d]) were highly effective in reducing urinary excretion of calcium and hydroxyproline, as well as in significantly lowering serum levels of alkaline phosphatase (Warrell, Bockman *et al.*, 1989; Bockman *et al.*, 1989. The findings of activity after high-dose infusions have now been confirmed by other groups (Matcovic *et al.*, 1990). Current studies are investigating remission induction in Paget's disease using low-dose therapy administered for 6 months. Additional studies are underway as a method of treatment for osteoporosis, both for patients who are post-menopausal as well as for patients with corticosteroid-induced disease.

Summary

Gallium nitrate appears to a unique, safe, and highly effective method of reducing accelerated bone loss in patients with cancer and metabolic bone disease.

REFERENCES

Bockman, R. *et al.* (1989): Treatment of Paget's disease of bone with low dose subcutaneous gallium nitrate. J Bone Min Res (Suppl.) 4, S167.

Budayr, A.A. *et al.* (1989): Increased serum levels of a parathyroid hormone-like protein in malignancy-associated hypercalcemia. Ann. Int. Med. 111, 807-812.

Lovett, D. *et al.* (1989): Quantitative assessment of osteolysis in myeloma: a comparative study of technical methods. Blood (Suppl.) 74, 377a.

Matkovic, V. *et al.* (1990): Use of gallium to treat Paget's disease of bone: a pilot study. Lancet 335, 72-75.

Scher H.I. *et al.* (1987): Gallium nitrate in prostatic cancer: evaluation of antitumor activity and effects on bone turnover. Cancer Treat. Rep. 71, 887-893.

Thiebaud, D. *et al.* (1990): Response to retreatment of malignant hypercalcemia with the bisphosphonate AHPrBP (APD): respective role of kidney and bone. J. Bone Mineral Res. 5, 221-226.

Warrell, R.P. *et al.* (1983): Treatment of advanced malignant lymphoma using gallium nitrate administered as a seven day continuous infusion. Cancer (*Phila.*) 51, 1982-1987.

Warrell, R.P. *et al.* (1984): Gallium nitrate inhibits calcium resorption from bone and is effective treatment for cancer-related hypercalcemia. J. Clin. Invest. 73, 1487-1490.

Warrell, R.P. *et al.* (1985): Metabolic effects of gallium nitrate administered by prolonged infusion. Cancer Treat. Rep. 69, 653-655.

Warrell, R.P. *et al.* (1986): Gallium nitrate for acute treatment of cancer-related hypercalcemia: clinicopharmacological and dose-response analysis. Cancer Res. 46, 4208-4212.

Warrell, R.P., Isaacs, M. *et al.* (1987): Gallium nitrate for treatment of refractory hypercalcemia from parathyroid carcinoma. Ann. Int. Med. 107, 683-686.

Warrell, R.P., Alcock, N.W. *et al.* (1987): Gallium nitrate inhibits accelerated bone turnover in patients with bone metastases. J. Clin. Oncol. 5, 292-298.

Warrell, R.P. *et al.* (1988): Gallium nitrate for acute treatment of cancer-related hypercalcemia: a randomized double-blind comparison to calcitonin. Ann. Int. Med. 108, 669-674.

Warrell, R.P. & Bockman, R.S. (1989): Gallium for treatment of hypercalcemia and bone metastasis, in *Important Advances in Oncology, 1989*, eds. V.T. DeVita, Jr., S. Hellman, S.A. Rosenberg. J.B. Lippincott: New York, pp. 205-220.

Warrell, R.P., Lovett, D. *et al.* (1989): Gallium nitrate for prevention of osteolysis in myeloma: a pilot randomized study. Blood (Suppl.) 74, 24a.

Warrell, R.P., Bockman R.S. *et al.* (1989): Biochemical effects of gallium nitrate in Paget's disease of bone. Clin. Res. 37,463A.

Warrell, R.P. *et al.* (1990): Gallium nitrate for treatment of cancer-related hypercalcemia: a randomized double-blind comparison to etidronate. Proc. Am. Soc. Clin. Oncol. (in press).

Metal Ions in Biology and Medicine. Eds. Ph. Collery, L.A. Poirier, M. Manfait, J.C. Etienne. John Libbey Eurotext, Paris © 1990, pp. 437-442

Oral administration of gallium in conjunction with platinum in lung cancer treatment

Ph. Collery, M. Morel, H. Millart, B. Desoize, C. Cossart, D. Perdu, H. Vallerand, J.C. Bouana, C. Pechery, J.C. Etienne, H. Choisy, J.M. Dubois De Montreynaud

Centre Hospitalier Universitaire Régional, 51092 Reims Cedex, France

To avoid the renal toxicity of intravenously administered gallium nitrate, Warrel and al. (7) proposed to give it as a continuous infusion of 300 mg/m2/24 h over 5 or 7 days instead of a bolus infusion of 750 mg/m2 . Daily oral administration of gallium chloride ($GaCl_3$) has also been proposed (1) to prevent this renal toxicity of gallium (Ga).

PHARMACOLOGICAL STUDIES : with oral daily doses of 400 mg of $GaCl_3$, the mean serum Ga concentrations at steady state were of 371 ± 142 µg/l in lung cancer patients (3). In the whole tested population of 45 lung cancer patients with various histological characteristics and at different stages of the disease , serum Ga concentrations increased with the enhancement of doses from 100 to 200 mg/24h and from 200 to 400 mg/24h, but not with higher doses up to 1400 mg/24h. Nevertheless, large individual variations were observed that may be in part due to the staging of cancer : metastatic patients had lower significant Ga serum concentrations than nonmetastatic patients. Besides, tissue Ga concentrations were assayed after death in 2 patients having received prolonged daily oral dosages of $GaCl_3$: Ga concentrations were higher than 10 µg/g in metastases, 3.6 ± 2.9 µg/g in the primary tumor and 2.3 ± 0.9 µg /g in the kidney. In another preliminary pharmacological study (1), tissue Ga concentrations were determined in 2 other patients : the first, who died of a lymphoma of the esophagus after having received $GaCl_3$ for 6 months, had Ga concentrations of 6.7 µg/g in the tumor, 1.95 µg/g in the healthy oesophagus and 2 µg/g in the kidney. In another patient with a non small cell lung carcinoma , the treatment was interrupted 2 months before death. In spite of the absence of Ga treatment during those last 2 months, the Ga concentrations were 7.8 µg/g in the primary tumor, 4.8 µg/g in a cerebral metastasis and 10.1 µg/g in a kidney metastasis, while Ga was undetectable in all healthy tissues. $GaCl_3$, which has neither renal nor hematological toxicity after oral prolonged administration, can be used orally in conjunction with other cytotoxic agents as the significant uptake of Ga by the tumor may potentiate the effect of a conventional chemotherapy. To evaluate the efficacy of oral administration of $GaCl_3$ in potentiating conventional chemotherapy, a randomized study comparing plain chemotherapy with chemotherapy plus 400 mg Ga /24h has been proposed (5).This dosage of 400 mg/24h $GaCl_3$ should not be considered definitive : individual adaptations could be necessary in order to obtain maximum serum Ga concentrations for each patient at steady-state (4) .

A RANDOMIZED CLINICAL TRIAL was then performed to determine whether the addition of gallium chloride (GaCl3) to a CDDP (Cis-platinum)-VP16 (etoposide) combination chemotherapy was superior to this chemotherapy alone. To be eligible, patients had to fulfill the following requirements : they could be treated neither by surgery nor by radiation , they had to have a progressing disease and Karnofsky performance status higher than 40%, they had to have either measurable or evaluable disease and normal renal function (serum creatinine level lower than 150 µmol/l), normal bone marrow reserve (leukocytes more than 4,000/mm3 and platelets more than 100,000/mm3). 18 non small cell lung cancer patients were included in the study.

Patients were randomized into 2 groups

The first one (control group) consisted of 9 patients (2 females, 7 males), treated by CDDP and VP 16, with 3 adenocarcimomas, 4 squamous cell carcinomas and 2 undifferentiated carcinomas, and with a mean age of 61 (range 52-73); 1 patient had a stage 3 and 8 patients had stage 4 cancers

The second one (Ga group) consisted of 9 patients (9 males), treated by CDDP-VP 16 and by GaCl3, with 3 adenocarcinomas and 6 squamous cell carcinomas, and with a mean age of 61 (range 48-73) ; 1 patient had a stage 3 and 8 patients had stage 4 cancers.

GaCl3 was orally administered in the Ga group at the daily dose of 400 mg as soon as their cancers were diagnosed, before starting conventional chemotherapy, and was maintained until progression or toxicity. CDDP was given as a continuous I.V. infusion for 5 days (15 mg/m2/24h) every 28 days with an assay of total and ultrafilterable platinum at 96 and 120 hours of the first and second courses. VP16 was given as a bolus I.V. infusion of 100 mg/m^2 on day 2, 3, 4 of CDDP infusion.

After 3 courses of chemotherapy :

5 partial responses were observed in the 9 patients receiving GaCl3 and none in the other group : the number of objective responses was significantly higher in the Ga group than in the control group according to an exact probability test with $p < 0.05$.

No significant differences were noted for toxicity :

- 2 patients died with a progressing disease before completing 3 courses of chemotherapy in the gallium group, versus 1 in the other
- septicemia was observed in 2 cases (1 in each treated group).
- hematological toxicity was the same in the 2 groups with the following minimum values :
  - control group : red blood cells = 2.7 ± 0.47. 10^9 /l; hemoglobin = 85 ± 13 g/l; leukocytes = 1.9 ± 0.6.10^9 /l and neutrophils = 0.7 ± 0.5.10^9 /l
  - Ga group : red blood cells = 2.8 ± 0.55. 10^9 /l ; hemoglobin = 86 ± 23 g/l ; leukocytes = 2.9 ± 1.3 .10^9 /l and neutrophils = 1.1 ± 1.2.10^9 /l
- a significant and similar decrease in plasma and red blood cell (RBC) magnesium (Mg) was observed in the 2 groups (Wilcoxon test) :
  - control group : plasma Mg significantly decreased ($p<0.02$) from 0.83 ± 0.05 mmol/l (before treatment) to 0.53 ± 0.14 mmol/l, and RBC Mg ($p<0.02$) from 2.56 ± 0.24 mmol/l to 2.01 ± 0.22 mmol/l .
  - Ga group: plasma Mg significantly decreased ($p<0.05$) from 0.80 ± 0.15 mmol/l to 0.58 ± 0.18 mmol/l , and RBC Mg ($p<0.02$, Wilcoxon test) from 2.39 ± 0.29 mmol/l to 1.98 ± 0.40 mmol/l .

A sharp decrease in serum gallium concentrations was observed during the five days of each CDDP-VP16 infusion :

	0 h	96 h	120 h
1 st course:	100 ± 66 µg/l	27 ± 27 µg/l	17 ± 4 µg/l
2 nd course:	133 ± 142µg/l	30 ± 21 µg/l	21 ± 13 µg/l
3 rd course:	188 ± 60 µg/l	41 ± 44 µg/l	15 ± 9 µg/l

It was not possible to give 3 additional courses of chemotherapy in the responders, as a toxicity appeared in 3 patients between the 3rd and the 5th course of chemotherapy with a severe pneumopathy and a cardiac arythmia in 1 patient , 1 septicemia in 1 patient and a renal deficiency in another. The treatment had to be interrupted in these 3 patients so that, if GaCl3 was effective in potentiating the effect of chemotherapy after 3 courses, there was no significant increase in the median survival time (5 months in the control group and 8 months in the Ga group) and no significant difference in the actuarial survival curves . This late toxicity could be due in part to the well-known toxicity of platinum and, indeed, we observed a significant increase in both total and ultrafilterable serum platinum (Pt) concentrations from the 1st to the 2nd course of chemotherapy but without any significant differences between both groups:

- total Pt serum concentrations :
  - first course = 853 ± 247 µg/l
  - second course = 1199 ± 314 µg/l ($p < 0.001$, Mann and Whitney test)
- ultrafilterable Pt plasma concentrations :
  - first course = 49± 29 µg/l
  - second course = 76 ± 43 µg/l ($p < 0.05$, Mann and Whitney test)

To prevent the late toxicity of the CDDP-VP16 combination with $GaCl_3$, an adaptation of CDDP doses has been proposed (6) in order to obtain serum total Pt concentrations yielding a constant area under the curve for each 120 hours infusion.

ADAPTATION OF CDDP DOSES in order to obtain serum total Pt concentrations yielding a constant area under the curve during the 5 days of each CDDP infusion .

A CDDP-VP16 combination chemotherapy was administered as a 5 days continuous infusion every 4 weeks in 9 lung cancer patients with non small cell carcinomas. These patients were also daily treated with oral administration of $GaCl_3$ as soon as their cancers were diagnosed and considered inoperable . In the randomized study, comparing chemotherapy versus chemotherapy plus Ga, a decrease in Ga serum concentrations was noted during the 5 days of CDDP infusion. This drop in Ga concentrations being unexplained, $GaCl_3$ treatment was then interrupted during each course of CDDP-VP16 infusion in half of the patients in order to show a possible difference in toxicity or efficacy between these 2 modes of treatment. The mean age of the patients (all males) was 59 (range: 43 - 70). Anatomopathogical characteristics were : 5 squamous cell carcinomas (1 stage 2, 1 stage 3, and 3 stage 4), 3 adenocarcinomas (1 stage 3 and 2 stage 4) and 1 undifferentiated carcinoma (stage 3) . The Karnofsky performance status was always higher than 50%. During the first course of chemotherapy, CDDP doses were 15mg/m^2/24h and VP 16 doses 25 mg/m^2/24h. Doses were adapted during the following courses of chemotherapy in order to obtain the same area under the curve in total serum Pt concentrations during the 5 days (120 hours) of the infusion . To allow this adaptation, total serum Pt concentrations were determined by AAS at the following hours of each infusion : 0, 16, 24, 40, 64, 72, 88, 96, 112 and 120 . Serum VP 16 concen-

Figure 1 : adaptation of CDDP doses in order to obtain serum total Pt concentrations yielding a constant AUC for each 120 hours infusion :

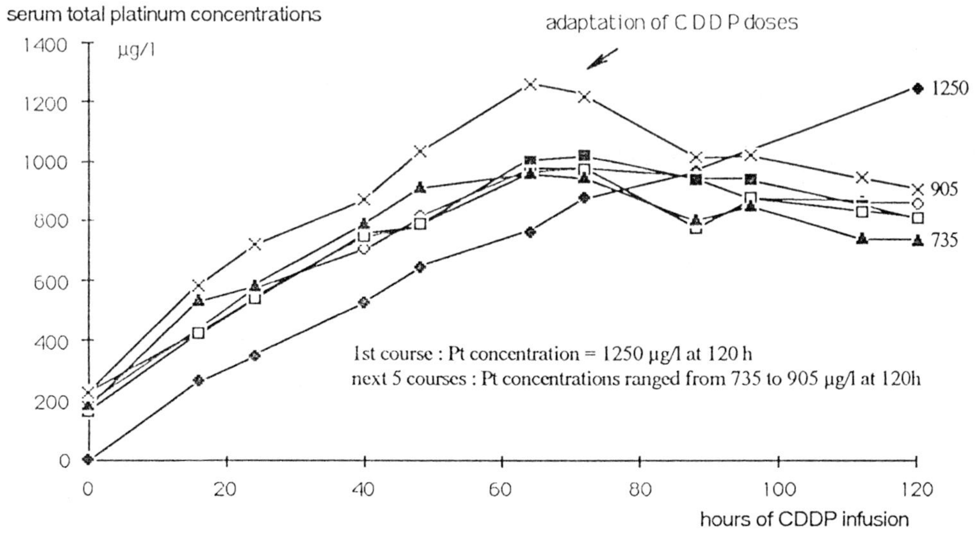

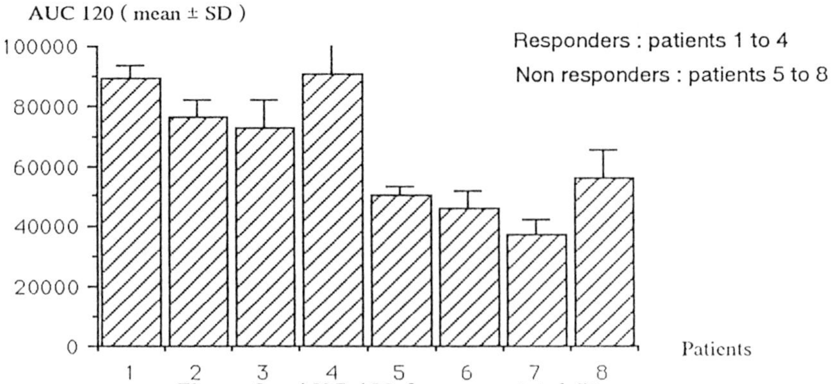

Figure 2 : AUC 120 for serum total Pt

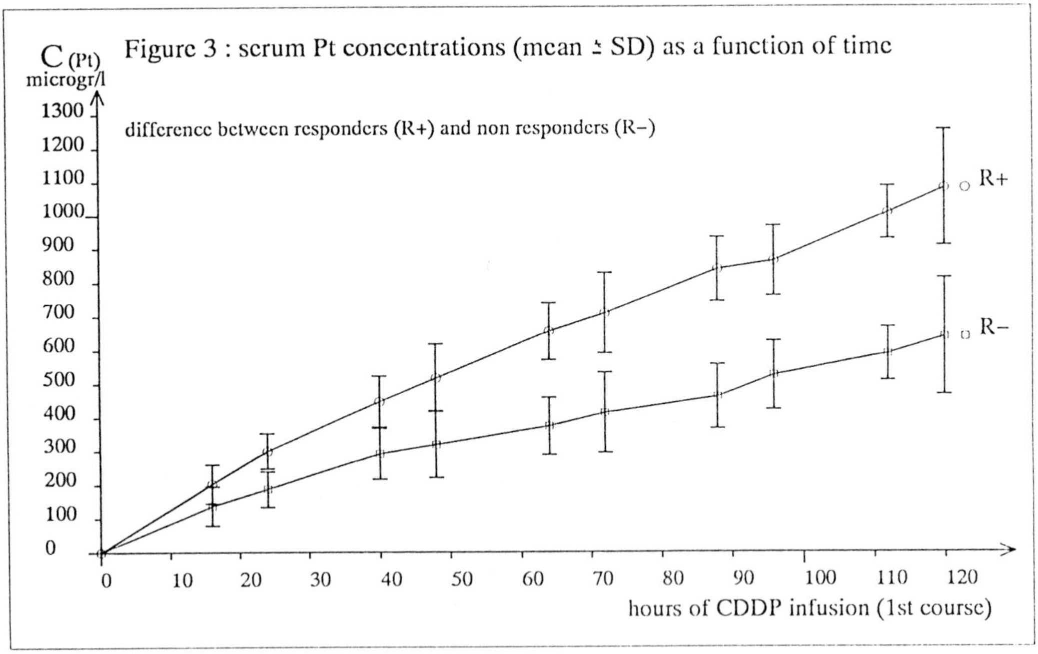

Figure 3 : serum Pt concentrations (mean ± SD) as a function of time trations were determined by HPLC. The area under the curve (AUC) was calculated for time 0 to 64 hours (AUC 64) and time 0 to 120 hours (AUC 120) of each infusion . Results were expressed in $\mu g.l^{-1}.h$ for total serum Pt concentrations and in $mmol.l^{-1}.h$ for VP 16. Doses of CDDP were modified at time 64 hours of the second course as a function of :

- AUC 64 of this second course: $(AUC\ 64)_2$
- AUC 64 of the first course: $(AUC\ 64)_1$
- CDDP dose administered during this second course from 0 to 64 h : $D(0\text{-}64)_2$
- CDDP dose administered during the first course from 0 to 64 h : $D(0\text{-}64)_1$
- CDDP dose administered during the first course from 0 to 120 h : $D(0\text{-}120)_1$

The CDDP dose administered during the second course from 64 to 120 h : $D(64\text{-}120)_2$ was calculated according to the formula : $D(64\text{-}120)_2 = D(0\text{-}64)_2 . \{[D(0\text{-}120)_1 / D(0\text{-}64)_1].[AUC(64)_1 / AUC(64)_2]-1\}$.
Explanations of the formula in (6). CDDP doses were adapted in the following courses according to the same principle.

8 patients were treated by at least 3 courses of chemotherapy (1 patient with many metastases died after the 1st course). Thanks to the adaptation of CDDP doses, the increase in AUC 120 for serum total platinum was only of 20± 10% between the second and the first course. By contrast AUC 64 increased by 59± 24 % between these 2 courses. The increase in AUC 120 was of 5 ± 8 % between the third and the second course and the changes were of -3 ± 9 % for AUC 64. Results for AUC 120 for the 3 first courses (mean ± SD) are summarized in figure 2 :

No toxicity was observed in these 8 evaluable patients after 3 courses of chemotherapy with Ga. 4 objective responses were observed : 3 partial responses and 1 complete response (patients 1 to 4 in the graph) . The mean AUC 120 were 82292 ± 8222 $\mu g.l^{-1}.h$ in responders and 48540 ± 6748 $\mu g.l^{-1}.h$ in non-responders. This difference was significant with $p < 0.05$ (Wilcoxon test).

Individual variations in serum total platinum concentrations could be recorded as from the first course and are shown in figure 3 :

<u>DISCUSSION</u> : According to these individual variations and the significant correlation between the AUC 120 and efficacy, it appears that CDDP doses could not only be given according to body area as usual, but mainly according to pharmacokinetic parameters. On the other hand, the AUC 120 for serum VP16 was 165 ± 38 $\mu g.l^{-1}.h$ in responders versus 108 ± 51 $mmol.l^{-1}.h$ in non-responders. This difference was not significant. There were 3 responders in patients receiving Ga continuously and 1 complete response in 1 of the 4 patients without Ga treatment during CDDP-VP16 infusions. This patient reached serum Ga concentrations as high 4 as 1250 µg/l with the daily dose of 400 mg $GaCl_3$. No obvious interest in interrupting Ga treatment during CDDP infusion has been noted. On the other hand, a correlation between Ga serum concentrations and efficacy having already been demonstrated (2), it could be necessary to optimize the dosage of $GaCl_3$ for each patient in order to obtain the maximum serum Ga concentrations at steady state, these depending on several factors such as the stage of cancer (3,4) and perhaps also on serum transferrin concentrations (4). In the next study, we shall try to obtain an AUC 120 of at least 80000 $\mu g.l^{-1}.h$ for serum total Pt concentrations. The best protocol of treatment for VP16 has not been fully determined but responses have been obtained with a mean AUC of 165 $mmol.l^{-1}.h$ during 120 hours VP16 infusions. The combination of CDDP and VP16 in a 5 days infusion will be associated with $GaCl_3$ continuously administered per os .

<u>CONCLUSION</u> :

Oral daily administration of $GaCl_3$ is able to potentiate the efficacy of a conventional chemotherapy with Cisplatinum and VP16 administered as continuous infusions every 4 weeks. These 2 treatments need to be optimized according to pharmacokinetic parameters in order to obtain the highest serum Ga concentrations at steady state allowed by oral administration, and serum total Pt concentrations yielding an AUC of 82000 $\mu g.l^{-1}.h$ during the120 hours infusion of each course.

REFERENCES:

(1) Collery, Ph., Millart, H., Pourny, C., Choisy, H., Etienne, J.C. (1984) . Pharmacokinetics of gallium after oral administration. Proc. 9th International Congress of Pharmacology (London), abstract 501

(2) Collery Ph., Millart H., Ferrand O., Jouet J.B., Dubois J.P., Barthes G., Pourny C., Pechery C., Choisy H., Cattan A., Dubois de Montreynaud J.M., Etienne J.C. (1985). Gallium chloride treatment of cancer patients after oral administration . A pilot study. Chemiotherapia 4, 1165-1166.

(3) Collery Ph., Millart H., Lamiable D., Vistelle R., Rinjard P., Tran G., Gourdier B., Cossart C., Bouana J.C., Pechery C., Etienne J.C., Choisy H., Dubois de Montreynaud J.M. (1989). Clinical Pharmacology of gallium chloride after oral administration in lung cancer patients . Anticancer Res. 9, 353 - 356.

(4) Collery Ph., Millart H., Lamiable D., Vistelle R., Berthier A., Betbeze P., Cossart C., Mulette T., Masure F., Gourdier B., Pechery C., Choisy H., Etienne J.C., Dubois de Montreynaud J.M. (1989) . Magnesium alterations and pharmacokinetic data in gallium-treated lung cancer patients. Magnesium 8, 56-64.

(5) Collery Ph., Millart H., Cossart C., Perdu D., Vallerand H., Morel M., Pechery C., Etienne J.C., Choisy H., Dubois de Montreynaud J.M.(1989): Potentiation of chemotherapy by gallium . J. Am. Coll. Nutr. 8, 428.

(6) Collery Ph., Millart H., Desoize B., Morel M., Dubois J.P., Choisy H., Etienne J.C., Dubois de Montreynaud J.M. (1989). Optimization of CDDP doses in continuous infusion for 5 days. Proc. third ARTAC Workshop on therapeutic trials in oncology (PARIS). Abstract in Cancer Communication 3, 328-329. Manuscript in press.

(7) Warrell R.P., Coonley C.J., Strauss D.J., Young C.W. (1983) : Treatment of patients with advanced malignant lymphoma using gallium nitrate administered as a seven-day continuous infusion. Cancer 51, 1982 - 1987.

Metal Ions in Biology and Medicine. Eds. Ph. Collery, L.A. Poirier, M. Manfait, J.C. Etienne. John Libbey Eurotext, Paris © 1990, pp. 443-444

Effect of gallium chloride on inflammation and experimental polyarthritis

Ph. Collery[1], P. Rinjard[2], C. Pechery[3]

[1] *Hôpital Maison Blanche, Centre Hospitalier Universitaire Régional, 51092 Reims Cedex, France;* [2] *Centre de Recherche, Coopération Pharmaceutique Française, 77000 Melun, France;* [3] *Laboratoire Meram, 1, avenue de la Libération, 77020 Melun cedex, France*

The antiinflammatory effect of Gallium Chloride (GaCl3) has been tested in rats on Carrageenin Paw Edema , Granuloma Pouch and Adjuvant Polyarthritis .

MATERIAL AND METHODS :

Carrageenin paw edema : 32 Female Rats received 1.10 ml of 1% carrageenin in the right hind leg. A control group of 8 rats was compared with 3 treated groups of 8 rats each according to the dose of GaCl3 : 2,5 mg/kg , 5 mg/kg and 10 mg/kg . GaCl3 was administered intraperitoneously 1 hour before the injection of carrageenin .

Granuloma Pouch : 32 Female Rats received 1 ml of 1% Croton Oil in the dorsal area when anesthezised. A control group of 8 rats was compared with 3 treated groups of 8 rats each. GaCl3 was administered intraperitoneously daily for 7 days . The tested doses were 2,5 mg/kg ; 5 mg/kg and 10 mg/kg .

Adjuvant polyarthritis : Polyarthritis was observed in rats receiving 0.1 ml Mycobacterium Butyricum 0,6% in the plantar region of the right hind- leg . GaCl3 was administered for 21 days either orally or intraperitoneously . Daily tested doses of GaCl3 were 1 mg/kg , 2 mg/kg and 4 mg/kg for IP injections and 50 mg/kg , 100 mg/kg , 200 mg/kg and 400 mg/kg for oral administrations .

Statistical Analysis : treated groups were compared with controls (n = 10 for each group) by Student't test . Results were significant when p was at least <0.05 .

RESULTS

Carrageenin paw edema : the paw swelling was significantly decreased 3 hours after the carrageenin injection in rats receiving 10 mg/kg GaCl3 versus controls. No significant effect was observed in rats treated with lower doses.

Granuloma Pouch : the decrease in the weight of the sac was significant on day 8 for each treated group versus controls . There was no significant decrease in the volume of the fluid within the sac.

Adjuvant polyarthritis : the edema of the left hind- leg was significantly decreased in all treated groups except with 1 mg/kg IP GaCl3. This edema disappeared with IP doses of 4 mg/kg and oral doses of 200mg/kg . The edema of the fore-legs was significantly decreased in all treated groups except with oral doses of 50mg/kg GaCl3 . Lesion nodules at the tails and ears were significantly decreased with IP doses of 4 mg/kg and all oral doses except 50 mg/kg GaCl3.

CONCLUSION :

GaCl3 has an antiinflammatory effect in acute and chronic experimental conditions . Single I.P doses of 10 mg/kg are active against acute inflammation . Daily I.P. injections of 4mg/kg (0.02271 mmol/kg GaCl3) for 21 days or daily oral administration of 100 mg/kg (0.56 mmol/kg) for 21 days are active against the experimental adjuvant polyarthritis . These doses are far from the toxic doses. These results confirm those of Delbarre and Rabaud (1) who observed a significant effect of 0.00266 mmol gallium sulfate , administered subcutaneously for 19 days after the injection of Mycobacterium Tuberculosis, in experimental adjuvant polyarthritis in rats . The efficacy may be in relation with the selective uptake of 67 gallium by rheumatoid joints with clinically active synovitis as observed in patients by I.W. Mc. Call et al . (2).

REFERENCES :

DELBARRE F., RABAUD M. (1976). Prévention chez le rat, de la polyarthrite expérimentale à adjuvant, par un sel de gallium. C.R. Acad. Sci. Paris 283 , 1469 - 1472.

I.W. Mc CALL., SHEPPARD H., HADDAWAY M., PARK W.M., WARD D.J. (1983) . Gallium 67 scanning in rheumatoid arthritis . Br. J. Radiol. 56 , 241 - 243.

IV. New metal complexes

Metal Ions in Biology and Medicine. Eds. Ph. Collery, L.A. Poirier, M. Manfait, J.C. Etienne. John Libbey Eurotext, Paris © 1990, pp. 447-451

Cryptands : coordination chemistry of alkali, alkaline-earth and toxic cations

B. Dietrich

Université Louis Pasteur, Institut Le Bel, 4, rue Blaise Pascal, 67000 Strasbourg, France

The macrobicyclic cryptands **1-11** (Figure) were designed and synthesized by Lehn et al. (Dietrich et al. 1969, 1973 ; Lehn and Montavon, 1972, 1976 ; Vitali, 1980). They display high complexation ability toward many cations and in addition, several cryptands exhibit good complexation selectivity. These aspects have been largely reported and reviewed in several articles (Lehn, 1973, 1978, 1985, 1988 ; Potvin and Lehn, 1987 ; Dietrich, 1984, 1985). The applications of this class of substances are numerous, however we will focuse only on the cation complexation aspects for which stability constant and selectivity are of prime importance (Table).

One of the first applications was the decorporation of radioactive cations using cryptand **3** : $^{85}Sr^{2+}$ (Müller, 1970), $^{224}Ra^{2+}$ (Müller and Müller, 1974 and Müller, 1977), $^{140}Ba^{2+}$ (Müller et al., 1977). Complexation selectivities : Sr^{2+}/Ca^{2+} = 4.10^3, Ra^{2+}/Ca^{2+} = 1.6 10^2, Ba^{2+}/Ca^{2+} = 1.3 10^5 and Sr^{2+}/K^+ = 4.10^2, Ra^{2+}/K^+ = 16, Ba^{2+}/K^+ = 1.3 10^4 are all in favour of the radionuclide (log Ks Ra^{2+} = 6.6). The experiments were performed *in vivo* (on rats) and the excretion of radioactive cations in faeces and urine were greatly enhanced in the presence of the cryptand.

For the use as detoxication agents of heavy metals, several cryptands seem to fulfill the necessary requirements i.e. strong binding of the toxic metals (Cd^{2+}, Hg^{2+}, Pb^{2+}) and weak binding of the biologically essential cations (Na^+, K^+, Mg^{2+}, Ca^{2+}, Zn^{2+}). This is achieved : for Cd^{2+} with cryptands **6-8** (selectivities Cd^{2+}/Ca^{2+} from 10^5 to 10^9, Cd^{2+}/Zn^{2+} from 10^3 to 10^6) ; for Hg^{2+} with cryptands **1-3**, **6-8** (selectivities Hg^{2+}/Ca^{2+} from 10^{13} to 10^{25}, Hg^{2+}/Zn^{2+} from 10^{11} to 10^{19} ; and for Pb^{2+} with cryptands **2**, **3**, **6-10** (selectivities Pb^{2+}/Ca^{2+} from 10^6 to 10^{14}, Pb^{2+}/Zn^{2+} from 10^5 to 10^{10}).

Extraction and transport are also possible with cryptands (Lehn, 1983). These processes have been used with heavy metals and compound **11** (Bacon and Jung, 1985, Bacon and Kirch, 1987) and on Cu^{2+}, Ni^{2+}, Co^{2+}, Hg^{2+}, Pb^{2+} with the photoresponsive cryptand **12** ; the represented trans-form extracts more efficiently than the photoisomerized cis-form (Shinkai, 1982).

It should be noted in the table that the lithium cation is poorly bound (except with **1**). Due to the importance of Li^+ in the treatment of mental illnesses there is a need of receptors or carriers of this cation. It has recently been described (Bencini et al., 1989) that ligand **13** shows high affinity for Li^+, and also seems to show appreciable Li^+/Na^+ selectivity.

Cryptands have also found a large variety of applications in the field of analytical chemistry. For example trace amounts of lead(II) can be analyzed by spectrophotometric (Szczepaniak et Juskowiak, 1982) or fluorometric (Gomis et al., 1985) methods. These determinations require the extraction into an organic solvent of the ion-pair formed between the lead cryptate (with ligand **3**) and a lipophilic anion.

The chromophore containing cryptand **14** (Klink et al., 1983) shows a high complexation selectivity for K^+ and a characteristic color change in its presence ; with the exception of Rb^+ no other alkali or alkaline earth cations give the same coloration upon complexation.The field of chromo- and fluoroionophores has been widely explored (Löhr and Vögtle, 1985).

Cryptands have found further applications in isotopic separation (Heumann, 1985), in rare earth complexation (Bünzli and Wessner, 1984) and in many other fields (Kolthoff, 1979 ; Tagaki and Nakamura, 1986). Cryptands have also been immobilized on polymers leading to resins useful for analytical and preparative separation of cations (Blasius and Janzen, 1981 ; Smid and Sinta, 1984). Finally, it should be mentioned that a recovery process for cryptand **3** from final reaction mixtures has been described (Perdicakis and Bessière, 1986).

Concluding remark : due to space limitation it was not possible to mention the rich field of macrocyclic ligands ; only macrobicyclic cryptands have been considered.

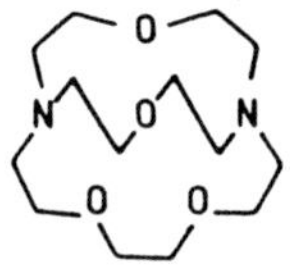

1

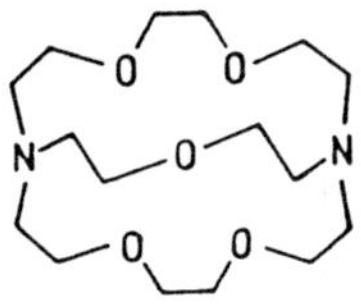

2

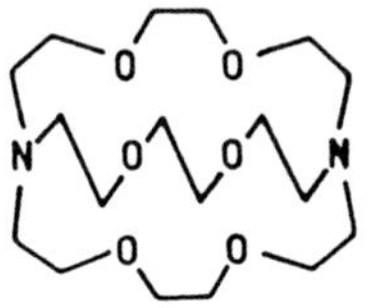

3

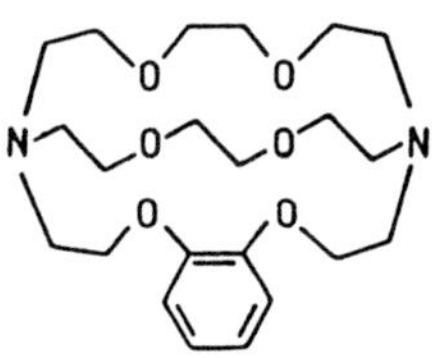

4

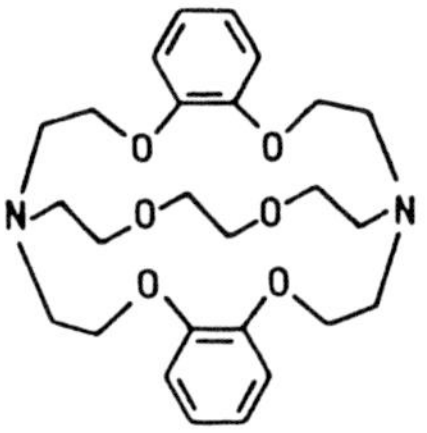

5

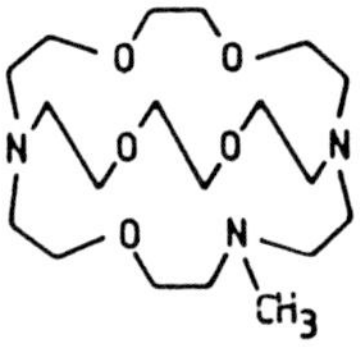

6

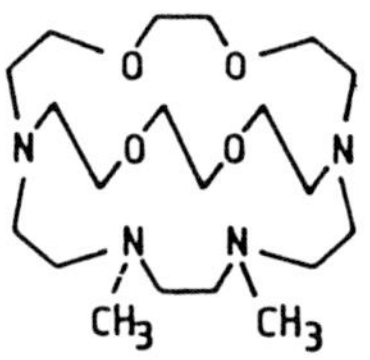

7

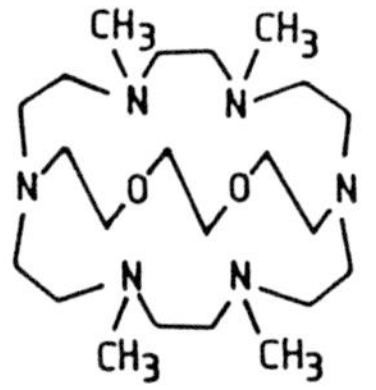

8

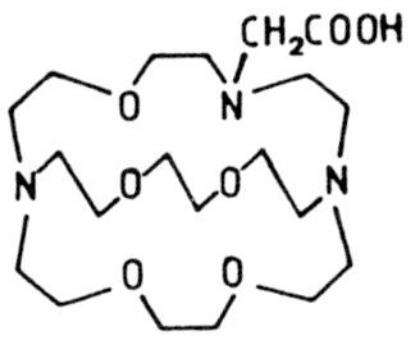

9

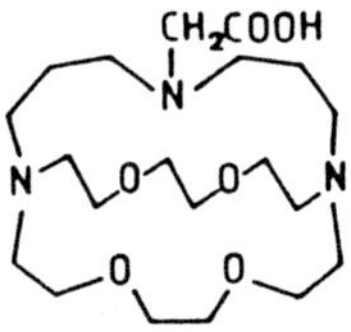

10

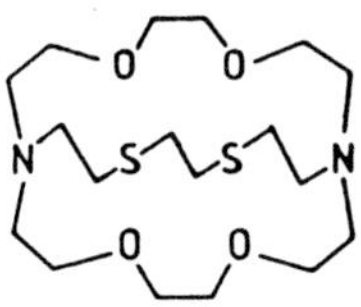

11

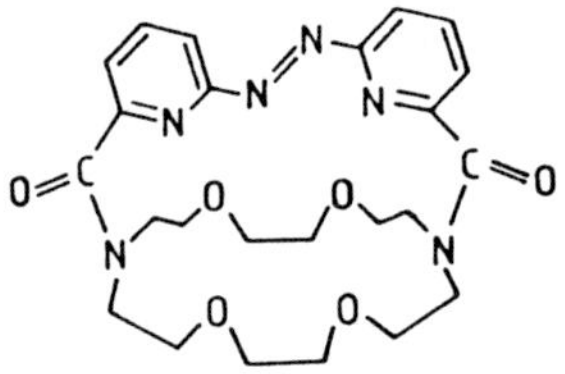

12

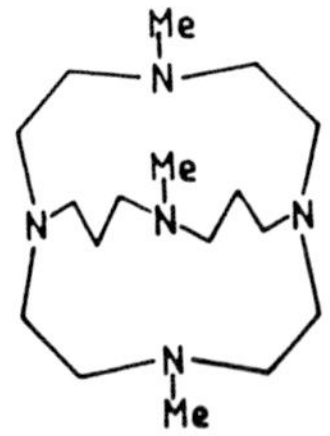

13

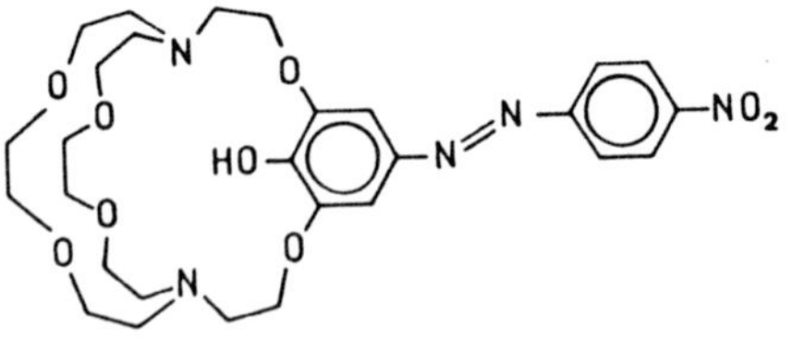

14

FIGURE

TABLE. Stability constants, log K_s, of the cryptate complexes of cryptands **1-10**. Solvent : water or methanol (M).

Ligand / Cations	1	2	3	4^d(M)	5^d(M)	6^e	7^e	8^e	9^f	10^f
Li^+	[a] 5.5	2.5	<2	2.2	2.0	1.5	2.4	-	2.45	1.0
Na^+	3.2	5.4	3.9	7.5	7.6	3.2	2.5	-	3.15	1.4
K^+	<2	3.95	5.4	9.2	8.75	4.2	2.7	1.7	4.0	1.6
Mg^{2+}	2.5	<2	<2	-	-	1.9	2.6	-	2.8	2.3
Ca^{2+}	2.5	6.95	4.4	-	-	4.6	4.3	1.5	6.5	1.3
Sr^{2+}	<2	7.35	8.0	-	-	7.4	6.1	1.5	9.5	3.2
Ba^{2+}	<2	6.3	9.5	-	-	9.0	6.7	3.7	10.55	2.35
Mg^{2+}(M)	[b] 4.75	4.15	4.85	-	-	-	-	-	-	-
Ca^{2+}(M)	5.45	9.3	8.45	7.05	5.95	-	-	-	-	-
Sr^{2+}(M)	4.85	10.6	11.0	10.3	8.85	-	-	-	-	-
Ba^{2+}(M)	5.35	10.05	12.9	11.0	8.85	-	-	-	-	-
Co^{2+}	[c] <4.7	5.4	<2.5	-	-	5.2	4.9	5.2	7.35	5.2
Ni^{2+}	<4.5	4.3	<3.5	-	-	5.0	5.1	5.7	9.20	5.55
Cu^{2+}	7.8	7.55	6.8	-	-	9.7	12.7	12.5	12.75	10.5
Zn^{2+}	<5.3	5.4	<2.5	-	-	6.3	6.0	6.8	8.9	6.1
Cd^{2+}	<5.5	10.05	7.1	-	-	9.7	12.0	10.7	8.6	5.35
Hg^{2+}	16.0	20.0	18.2	-	-	21.7	24.9	26.1	-	-
Pb^{2+}	7.9	13.1	12.7	12.2	10.9	14.1	15.3	15.5	16.65	11.25
La^{3+}	-	6.6	6.45	-	-	-	-	-	-	-
Ag^+	8.5	10.6	9.85	12.0	11.85	10.8	11.5	13.0	-	-
Tl^+	-	-	-	-	-	6.3	5.5	4.1	-	-

Table references : a) Lehn and Sauvage, 1975 ; b) Arnaud-Neu et al., 1986 ; c) Arnaud-Neu et al., 1977 ; for Hg^{2+} Arnaud-Neu et al., 1982 and Anderegg, 1975 ; for La^{3+} Burns and Baes, 1981 ; d) Buschmann, 1985 and 1987 ; e) Lehn and Montavon, 1978 ; f) Vitali, 1980.

REFERENCES

Anderegg, G. (1975) : Thermodynamik der metallkomplexbildung mit polyoxadiazamacrocyclen. *Helv. Chim. Acta* 58, 1218-1225.

Arnaud-Neu, F., Spiess, B. and Schwing-Weill, M.-J. (1977) : Stabilité en solution aqueuse de complexes de métaux lourds avec des ligands diaza-polyoxamacrocycliques. *Helv. Chim. Acta* 60, 2633-2643.

Arnaud-Neu, F., Spiess, B. and Schwing-Weill, M.-J. (1982) : Solvent effects in the complexation of [2] cryptands and related monocycles with transition- and heavy-metal cations. *J. Am. Chem. Soc.* 104, 5641-5645.

Arnaud-Neu, F., Yahya, R. and Schwing-Weill, M.-J. (1986) : pH-metric study of the complexation equilibria of alkaline-earth cations with [1] and [2] cryptands in methanol.*J. Chim. Phys.* 83, 403-408.

Bacon, E. and Jung, L. (1985) : Selective extraction and transport of mercury, through a liquid membrane by macrocyclic ligands. Improvement in the transport efficiency and an approach to physiological systems. *J. Membrane Sci.* 24, 185-199.

Bacon, E. and Kirch, M. (1987) : Competitive transport of the toxic heavy metals, lead, mercury and cadmium by macrocyclic ligands. *J. Membrane Sci.* 32, 159-173.

Bencini, A., Bianchi, A., Ciampolini, M., Garcia-Espana, E., Dapporto, P., Micheloni, M., Paoli, P., Ramirez, J.A. and Valtancoli, B. (1989) : Selective encapsulation of lithium ion by the new azacage 5,12,17, Trimethyl-1,5,9,12,17-penta-azabicyclo [7.5.5] nonadecane (L). Thermodynamic studies and crystal structures of the lithium complex [LiL][BPh_4] and of the monoprotonated salt [HL][Cl].$(H_2O)_3$. *J. Chem. Soc. Chem. Comm.*, 701-703.

Blasius, E. and Janzen K.-P. (1981) : Analytical applications of crown compounds and cryptands. In *Host Guest complex chemistry I*, ed. F. Vögtle. Springer-Verlag Berlin 1981, pp. 163-189.

Bünzli, J.-C. and Wessner, D. (1984) : Rare earth complexes with neutral macrocyclic ligands. *Coord. Chem. Rev.* 60, 191-253.

Burns, J.H. and Baes, C.F. (1981) : Stability quotients of some lanthanide cryptates in aqueous solutions. *Inorg. Chem.* 20, 616-619.

Buschmann, H.-J. (1985) : Komplexierung von Blei(II) durch azakronenether und kryptanden in methanol. *Chem. Ber.* 118, 3408-3412.

Buschmann, H.-J. (1987) : The macrocyclic and cryptate effect. Complex formation of the cryptands (222), (222B), (222BB), and (222CC) with different cations in methanol solutions. *Inorg. Chim. Acta* 134, 225-228.

Dietrich, B., Lehn, J.-M. and Sauvage, J.P. (1969) : Diaza-polyoxa-macrocycles et macrobicycles. *Tetrahedron Lett.* 2885-2888.

Dietrich, B., Lehn, J.-M. and Sauvage, J.P. (1973) : Cryptates : control over bivalent/monovalent cation selectivity. *J. Chem. Soc. Chem. Comm.*, 15-16.

Dietrich, B., Lehn, J.-M., Sauvage, J.P. and Blanzat, J. (1973) : Synthèses et propriétés physiques de systèmes diaza-polyoxa-macrobicycliques. *Tetrahedron* 29, 1629-1645.

Dietrich , B. (1984) : Cryptate complexes. In *Inclusion compounds,* ed. J.L. Atwood, J.E.D. Davies and D.D. MacNicol. Academic Press, London, vol. 2, pp. 337-405.

Dietrich, B. (1985) : Coordination chemistry of alkali and alkaline-earth cations with macrocyclic ligands. *J. Chem. Educ.* 62, 954-964.

Gomis, D.B., Alonso, E.F. and Sanz-Medel A; (1985) : Extractive fluorometric determination of ultratraces of lead with cryptand 2.2.2 and eosin. *Talanta* 32, 915-920.

Heumann, K.G. (1985) : Isotopic separation in systems with crown ethers and cryptands. *Topics in current chemistry* 127, 77-132.

Klink, R., Bodart, D. and Lehn, J.-M. (1983) : Kaliumreagens und Verfahren zur Bestimmung von Kaliumionen. Merck Patent GmbH. Europ. Pat. Apl. 83100281.1.

Kolthoff, I.M. (1979) : Application of macrocyclic compounds in chemical analysis. *Analytical chemistry* 51, 1R-22R.

Lehn, J.-M., Montavon, F. (1972) : Polyaza-macrobicycles et leurs cryptates. *Tetrahedron Lett.*, 4557-4560.

Lehn, J.-M. (1973) : Design of organic complexing agents. Strategies towards properties. *Structure and bonding* 16, 1-69.

Lehn, J.-M. and Sauvage, J.-P. (1975) : [2] -Cryptates : Stability and selectivity of alkali and alkaline-earth macrobicyclic complexes. *J. Am. Chem. Soc.* 97, 6700-6707.

Lehn, J.-M., Montavon, F. (1976) : Ligands polyaza-polyoxa-macrobicycliques : synthèses et complexes metalliques. *Helv. Chim. Acta* 59, 1566-1583.
Lehn, J.-M. et Montavon, F. (1978) : Stability and selectivity of cation inclusion complexes of polyaza-macrobicyclic ligands. Selective complexation of toxic heavy metal cations. *Helv. Chim. Acta* 61, 67-82.
Lehn, J.-M. (1978) : Cryptates : the chemistry of macropolycyclic inclusion complexes. *Acc. Chem. Res.* 11, 49-57.
Lehn, J.-M. (1983) : Chemistry of transport processes. Design of synthetic carrier molecules. *Physical Chemistry of transmembrane ion motions,* ed. Spach, G. Elsevier Science Publishers B.V., Amsterdam, pp. 181-207.
Lehn, J.-M. (1985) : Supramolecular chemistry - receptors, catalysts, and carriers. *Science* 227, 849-856.
Lehn, J.-M. (1988) : Supramolecular chemistry - scope and perspectives, molecules, supermolecules and molecular devices (Nobel lecture). *Angew. Chem. Int. Ed. Engl.* 27, 89-112.
Löhr, H.-G. and Vögtle, F. (1985) : Chromo- and fluoroionophores. A new class of dye reagents. *Acc. Chem. Res.* 18, 65-72.
Müller, W.H. (1970) : Sr-85 decorporation with a cryptating agent. *Naturwissenschaften* 57, 248-249.
Müller, W.H. and Müller, W.A. (1974) : Enhanced $^{224}Ra/^{212}Pb$ excretion provoked by cryptating agents in rats. *Naturwissenschaften* 61, 455.
Müller, W.H. (1977) : Predictibility of the stability constant of a radium-cryptate by means of *in vivo* data from radioactive alkaline earthes. *Strahlentherapie* 153, 570-571.
Müller, W.H., Müller, W.A. and Linzner, U. (1977) : Enhanced $^{140}Ba/^{140}La$ excretion provoked by a cryptating agent in rat. *Naturwissenschaften* 64, 96.
Perdicakis, M. and Bessière, J. (1986) : Recyclage du cryptand (222). *Anal. Letters* 19, 393-401.
Potvin, P.G. and Lehn, J.-M. (1987) : Design of cation and anion receptors, catalysts and carriers. In *Synthesis of macrocycles : the design of selective complexing agents*, ed. R.M. Izatt, J.J. Christensen. John Wiley and Sons, pp. 167-239.
Shinkai, S., Minami, T., Kouno, T., Kusano, Y. and Manabe, O. (1982) : Photoresponsive cryptand with an azapyridine-bridge. *Chemistry Letters,* 499-500.
Smid, J. and Sinta, R. (1984) : Macroheterocyclic ligands on polymers. *Topics in current chemistry* 121, 105-156.
Szczepaniak, W. and Juskowiak, B. (1982) : Spectrophotometric determination of trace amounts of lead(II) by ion-pair extraction with cryptand (2.2.2) and eosin. *Anal. Chim. Acta,* 140, 261-269.
Takagi, M. and Nakamura, H. (1986) : Analytical application of functionalized crown ether-metal complexes. *J. Coord. Chem.* 15, 53-82.
Vitali, P. (1980) : Synthèse et cryptates de complexones macropolycycliques. Thèse de Doctorat d'Etat, Strasbourg.

Metal Ions in Biology and Medicine. Eds. Ph. Collery, L.A. Poirier, M. Manfait, J.C. Etienne. John Libbey Eurotext, Paris © 1990, pp. 452-456

Screening for new metal anti-tumour agents

S.P. Fricker

Johnson Matthey Technology Centre, Blount's Court, Sonning common reading, Berks, RG4 9NH, UK

INTRODUCTION

Strategies for selecting and identifying novel agents for the treatment of neoplastic disease have changed and evolved, often in the light of retrospective analysis of empirical data. Historically these strategies have been predominantly based on in vivo rodent tumour models (Corbett et al, 1987). The earliest rodent models used were the allogeneic transplantable tumours e.g. Yoshida ascites sarcoma, Sarcoma 180 and Erlich ascites carcinoma. These are rapidly growing, undifferentiated cell types often, as in the case of Erlich ascites, with high sensitivity to cytotoxic agents giving a high incidence of false positives. The solid tumours e.g. Sarcoma 180 are susceptible to host weight loss so that apparent drug effects are due to inanition rather than the effect of the drug. These tumour models had limited success and were replaced in the 1950's by syngeneic transplantable murine tumours e.g. L1210, P388, B16. A possible alternative model system is the use of autochthonous tumours. These tumours are induced in the host animal and have the advantage of being histologically representative of the tissue of origin. They have a long latent period and are slow growing and are generally considered to be unsuitable for intensive screening.

Transplantable syngeneic murine tumours have formed the backbone of the majority of screening programmes. These tumours retain their histologic characteristics on transplantation, grow rapidly and repeatably, and give reproducible results with test compounds therefore satisfying many of the requirements of a screening system. The predominant models used have been the P388 and L1210 murine leukaemias. They have been used extensively by many laboratories for the pre-clinical selection of new anti-tumour agents.

One of the most well documented screening programmes is that of the NCI (Goldin et al, 1979, Venditti et al, 1984). In 1966 the L1210 leukaemia was their primary screening model. In 1971 the more sensitive P388 leukaemia replaced L1210 as the primary screen. The screening strategy also incorporated a secondary screen of four murine transplantable tumours and three human xenograft tumours. However the murine leukaemias formed the basis of the screening programme. A retrospective evaluation of total NCI screening between 1955-1982 indicated that of 600,000 compounds screened only

50 cytostatic drugs were in the clinic. An evaluation of the programme between 1976-1982 showed that of 1085 compounds tested 225 showed activity and of these 106 were selected by the L1210 leukaemia. Only 2% of the agents selected by P388 were active against murine solid tumours. The general conclusions that can be drawn from this analysis are **1)** screening stategies based on murine leukaemias select only for anti-leukaemia agents, very few solid tumours show a successful response to chemotheropeutic agents in the clinic. **2)** previous screening strategies only select cytostatic/cytotoxic agents. A novel approach is therefore required for screening for new anti-tumour agents.

ALTERNATIVE SCREENING STRATEGIES

1) Xenografts

The use of xenograftable human tumours into immune deprived athymic mice has enabled the testing of new agents against human solid tumours in an in vivo model system. The disadvantages with this system are that it is technically difficult, slow, expensive and as in all tumour models, the heterogeneity of the tumour cell population is lost. The latter characteristic can be turned to advantage when developing drugs to treat clinically unresponsive tumours and is being exploited in the search for new platinum drugs to treat platinum resistant ovarian tumours (Harrap et al, 1989). Though not practical as a primary screen xenografts are a major advance in secondary screening.

2) Biochemical systems

These have been of limited value being mainly of use for identifying specific classes of agents with known biochemical targets e.g. anti-metabolites. Recent advances in our knowledge of cell signalling systems and related membrane events and the importance of these events in the development of the cancer cell has stimulated the development of novel screening systems in this area.

3) Cell culture systems

Historically cell culture screening systems have been of little value. This has been because the cell lines used have been rapidly proliferating lines with few characteristics of cells of neoplastic disease. Usually the cytotoxicity of test compounds has been evaluated against a single cell line. Such systems bear little relationship to the clinical situation. A re-evaluation of screening models has lead to the development of screening systems utilising a range of cell lines established from human solid tumours with the aim of providing a clinically relevant in vitro screen for new anti-tumour agents (Alley et al 1988, Fricker & Buckley, 1989, Hills et al, 1989).

SCREENING FOR METAL ANTI-TUMOUR AGENTS

The majority of clinically used anti-tumour drugs demonstrate little or no specificity. One striking exception to this is the metal anti-tumour drug cisplatin which has a marked specificity for testicular cancer. The discovery of the platinum anti-tumour drugs has lead to an increasing interest in the development of other metal complexes as anti-tumour drugs. The specificity of cisplatin to testicular cancer also suggests that in principle it may be possible to synthesise other metal complexes that are specific for other tumour types. In order to identify new active drugs a novel approach to screening is required.

To overcome the limitations of conventional in vivo murine tumour screening systems we have established a panel of human tumour derived cell lines as a potential in vitro screen for metal containing anti-tumour drugs. The cell lines chosen (SW403, SW620, SW1116 colon adenocarcinomas; HTI376 bladder carcinoma; HT29/219 rectal carcinoma; ZR-75-1 breast carcinoma; SK-OV-3 ovarian carcinoma) are representative of solid tumours of different tissue origin and clinical responsiveness. The requirements for the screen were **(i)** a simple assay system for cell viability suitable for automation **(ii)** a dosing regimen allowing the development of delayed cytotoxicity and/or recovery from potential lethal damage and **(iii)** a panel of tumour derived cell lines which exhibit a differential response to known anti-tumour agents. The latter is an important concept as different tumours and different cell populations within a tumour respond differently to chemotherapeutic agents. Differential cytotoxicity, as opposed to non-specific cytotoxicity, is an indicator of **(a)** anti-tumour activity and **(b)** selectivity against a tumour cell type.

The dosing regimen adopted is as follows: **(1)** Pre-incubate the cells for 24hr at pre-determined starting concentrations ($5 \times 10^4 - 10^5$ cells/ml) in 96 well micro-titre plates, final volume 200 µl. **(2)** Challenge the cells with the test compound for 2 hr at a concentration range of 0-200 µg/ml. **(3)** Replace the drug with 200µl fresh medium and incubate for a further 72hr. **(4)** Assess cell viability and calculate IC_{50} values (drug concentration giving 50% cell kill) from dose/response curves.

Two assay systems for cell viability have been compared. These were **(1)** MTT assay based on the reduction of the tetrazolium salt 3-(4,5 dimethylthiazol-2-yl)-2,5 tetrazolium bromide by viable cells and **(2)** sulforhodamine B (SRB) protein dye which measures total cell protein. Both assays are colorimetric assays capable of semi-automation using standard microtitre equipment. Both assay systems give similar dose/response curves and IC_{50} values for test compounds. The SRB assay has been adopted for general use as it is a more robust assay with better endpoint stability.

Cisplatin was used to characterise the response of the cell line panel. The seven cell lines exhibit a marked differential response to cisplatin, the three colon lines being particularly insensitive **(Table 1)**. A number of metal complexes reported to have activity in alternative screening models have been tested against the panel **(Table 1)**. An ideal compound for further development would be one that **(a)** exhibits differential cytotoxic response across the cell line panel and **(b)** exhibits a different response to cisplatin. The activity of ruthenium complexes in tumour models has been well documented eg $[RuCl_2((CH_3)_2SO)_4]$ (Giraldi et al 1977). This compound has reported activity against L1210 at high concentrations. It demonstrated marginal activity against most of the cell lines but interestingly exhibited activity against ZR-75-1. Imidazolium complexes of ruthenium have exhibited activity against rodent tumours, including an autochthonous rat colorectal tumour (Keppler 1989). Two of these complexes, when tested against the cell line panel, showed no activity.

Titanocene complexes have exhibited activity in several animal models. Of more interest is their activity against xenografts of human tumours (Kopf-Maier & Kopf 1987) where they have been shown to inhibit tumour growth in a dose-dependent fashion. One representative of this class, $[TiCl_2(C_5H_5)_2]$, exhibited limited activity against the tumour panel with some indication of differential activity. This was not as marked as with cisplatin and the lowest IC_{50} values obtained were against the cisplatin sensitive lines.

One metal complex with good activity was the dirhodium tetraacetate. This was active against all cell lines and demonstrated differential activity with lowest IC_{50} values against the cisplatin sensitive lines. It must be a criterion of any screening system that it can identify structure - activity relationships within a class of compounds. The binuclear rhodium tetracarboxylates have activity in Erlich ascites and L1210 tumours. The results indicate that toxicity and activity increase in the order acetate <propionate < butyrate correlating with increasing hydrophobicity (Howard et al 1977). When tested in the tumour cell panel a marked increase in cytotoxicity was seen on going from acetate to propionate as would be predicted. Both propionate and butyrate were equally toxic over the concentration range tested to the extent that differential activity was lost **(Table 2).** The propionate and butyrate would therefore be regarded as non-specific cytotoxics in this screening system.

CONCLUSIONS

Retrospective analysis of present screening systems indicates that a novel approach to screening is required. The use of a panel of human tumour cell lines is one such approach. The cell lines chosen exhibit differential activity to test compounds and can be used to identify structure - activity relationships within a class of compounds. The use of in vitro human tumour cell lines is a potentially relevant primary screen for novel metal anti-tumour agents. However further investigations, including in vivo secondary screening using xenografts of these cell lines in athymic mice, are required to demonstrate the potential utility of this system.

TABLE 1 IN VITRO CYTOTOXICITY OF PUBLISHED METAL ANTITUMOUR AGENTS

	IC_{50} µg/ml vs CELL LINE						
	SW620 COLON	SW1116 COLON	SW403 COLON	ZR-75-1 BREAST	HT1376 BLADDER	HT29/219 RECTUM	SKOV-3 OVARIAN
$[PtCl_2(NH_3)_2]$	115	140	163	5	9	5	11
$[TiCl_2(Cp)_2]$	130	145	>200	120	130	80	-
$ImH[RuCl_4Im_2]$	INACTIVE	INACTIVE	INACTIVE	INACTIVE	-	INACTIVE	-
$(ImH)_2[RuCl_5Im]$	INACTIVE	INACTIVE	INACTIVE	INACTIVE	INACTIVE	INACTIVE	INACTIVE
$[RuCl_2(dmso)_4]$	120	>200	>200	22	100	-	49
$[Rh_2(OOCCH_3)_4]$	30	18	28	4	10	8	22

Cp = cyclopentadiene
Im = Imidazole
dmso = dimethylsulfoxide

TABLE 2 IN VITRO CYTOTOXCITIY OF A SERIES DIRHODIUM TETRACARBOXYLATES

	IC_{50} µg/ml vs CELL LINE					
	SW620 COLON	SW1116 COLON	SW403 COLON	ZR-75-1 BREAST	HT1376 BLADDER	HT29/219 RECTUM
$[Rh_2(OOCCH_3)_4]$	30	18	28	4	10	8
$[Rh_2(OOCC_2H_5)_4]$	<1	<1	<1	<1	<1	<1
$[Rh_2(OOCC_3H_7)_4]$	<1	<1	<1	<1	<1	<1

REFERENCES

Alley, M.C., Scudiero, D.A., Monks, A., Hursey, M.L., Czerwinski, D.L., Fine. D.L., Abbot, B.J., Mayo, J.G., Shoemaker, R.H., & Boyd, M.R. (1988): Feasibility of drug screening with panels of human tumours cell lines using a microculture tetrazolium assay. Cancer Res. 48, 589-601.

Corbett, T.H., Valeriote, F.A., & Baker, L.H. (1987) Is the P388 murine tumour no longer adequate as a drug discovery model. Invest New Drugs 5, 3-20

Fricker, S.P. & Buckley. R.G. (1989): Characterisation of a human tumour cell line panel for screening of anti-tumour agents in vitro. Biochem. Soc. Trans. 17. 1049-1050

Giraldi, T., Sava, G., Bertoli, G., Mestroni, G. & Zassinovich, G. (1977): Anti-tumour activity of two rhodium and ruthenium complexes in comparison with cis-diamminedichloroplatinum (II). Cancer Res. 37, 2662-2666

Harrap, K.R., Jones, M., Goddard, P.M., Orr, R.M. & Ziddik, Z.H. (1988): Drug resistance as a focus for new drug design. In Mechanisms of drug resistance in neoplastic cells ed.. P.V. Woodley III & K.D. Tew, pp. 307-328. Academic Press Inc.

Hills C.A., Kelland, L.R., Abel, G., Siracky, J., Wibson, A.P. & Harrap, K.R. (1989): Biological properties of ten human ovarian carcinoma lines: calibration in vitro against four platinum complexes. Br. J. Cancer 59, 527-534

Howard, R.A., Sherwood, E., Erck, A., Kimball, A.P. & Bear, J.L. (1977): Hydropholicity of several rhodium(II) carboxylates correlated with their biologic activity. J. Med. Chem. 20, 943-946.

Keppler, B.K., Henn, M., Juhl, U.M., Berger, M.R., Niebel, R.E. & Wagner, F.E. (1989): New ruthenium complexes for the treatment of cancer. Progress in Clinical Biochemistry. 10, 41-70

Kopf-Maier, P. & Kopf, H. (1987): Non-platinum-group metal anti-tumour agents: History, current status and perspectives. Chem. Rev. 87, 1137-1152

Venditti, J.M., Wesley, R.A. & Plowman, J. (1984): Current NCI preclinical anti-tumour screening in vivo: results of tumour panel screening 1976-1982, and future directions. In Advances in Pharmacology and Chemotherapy, Vol 20, ed. S. Garattini, A. Goldin & F. Hawking. pp. 1-20 Orlands F.L.: Academic Press Inc.

Goldin, A., Schepartz, S.A., Venditti, J.M. & Devita, V.T., (1979): Historical development and current strategy of the Natural Cancer Institute drug development programme In Methods in Cancer Research Vol XVI, ed. V.T. Devita Jr. & H. Busch, pp. 165-245 New York: Academic Press Inc.

Metal Ions in Biology and Medicine. Eds. Ph. Collery, L.A. Poirier, M. Manfait, J.C. Etienne. John Libbey Eurotext, Paris © 1990, pp. 457-461

Bis(cyclopentadienyl) complexes of Ti, V, Nb, Mo, Fe, Ge, Sn, as antitumor agents - state of development and perspectives

P. Köpf-Maier

Institut für Anatomie, Freie Universität Berlin, Königin-Luise-Straße 15, D-1000 Berlin 33, West Germany

Investigations of the past twenty years revealed that numerous inorganic and organometallic compounds are characterized by remarkable antitumor activity. Various and structurally different types of cytostatically active compounds were described which either contained transition metals such as platinum, ruthenium, gold, copper, iron, and titanium, or main group elements, e.g., gallium, germanium, and tin. Of these compounds, only the platinum complexes cis-diamminedichloroplatinum ("cisplatin") and diammine(cyclobutane-1,1-dicarboxylato)platinum(II) ("carboplatin") had successfully passed the clinical trials of the phases I to IV during the past years and were introduced into clinic use, both drugs effecting highest therapeutic activity against testicular and ovarian carcinomas and tumors of the head and neck. All other compounds containing transition or main group metals are still in the preclinical stage or just entering early clinical trials.

Regarding the numerous and structurally quite different cytostatic non-platinum-group metal complexes, it becomes obvious that several of them contain aromatic cyclopentadienyl rings as organic ligands. Three different types of organometallic cyclopentadienyl complexes can be distinguished, differing from one another, among other features, by the central metal atom present within the complexes:

i) Neutral and ionic bis(cyclopentadienyl)metal ("metallocene") diacido complexes $(C_5H_5)_2MX_2$, $[(C_5H_5)_2MXL]^+Y^-$ or $[(C_5H_5)_2ML_2]^{2+}(Y^-)_2$ include an early transition metal such as titanium(IV), vanadium(IV), niobium(IV), niobium(V), molydenum(IV), or molybdenum(VI) as central metal atom (Fig. 1).

ii) In ionic metallicenium salts $[(C_5H_5)_2M]^+X^-$, the central metal M is represented by a medium transition metal, e.g., iron(III) (Fig. 1).

iii) Uncharged decasubstituted metallocenes $(C_5R_5)_2M$ ($R = C_6H_5$, $C_6H_5CH_2$) contain a main-group element, e.g., germanium(II) or tin(II), as central atom M (Fig. 1).

Antitumor testing experiments revealed pronounced tumor-inhibiting properties of these compounds against animal ascitic and solid tumors, e.g. Ehrlich ascites tumor, sarcoma 180, B 16 melanoma, colon 38 carcinoma or Lewis lung carcinoma. The more active complexes were then investigated against diverse heterotransplanted human carcinomas. In the course of these studies, best activity was shown for bis(cyclopentadienyl)titanium(IV) ("titanocene") diacido complexes against adenocarcinomas deriving from the stomach, the colon and the lung, (Fig. 2). Most of the investigated twenty gastrointestinal carcinomas were sensitive to the cytostatic action of titanocene compounds and inhibited by 50 to 90 % to 10 to 50 % of the volume of untreated

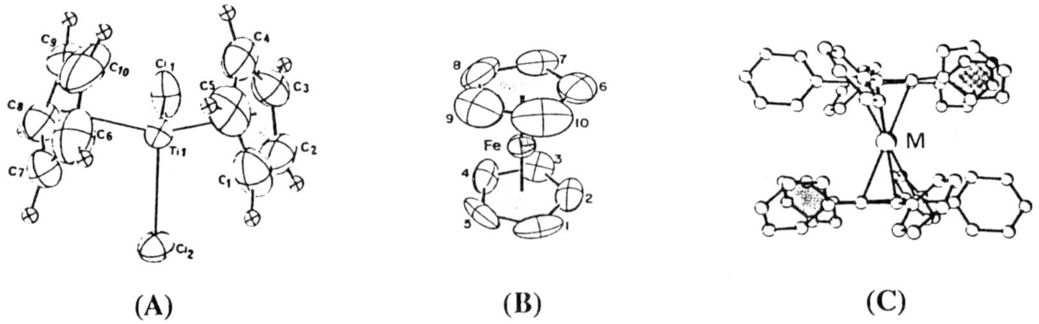

Fig. 1. Molecular structures of typical antitumor metallocene complexes. (**A**) Titanocene dichloride; (**B**) ferricenium cation; (**C**) decaphenylstannocene molecule

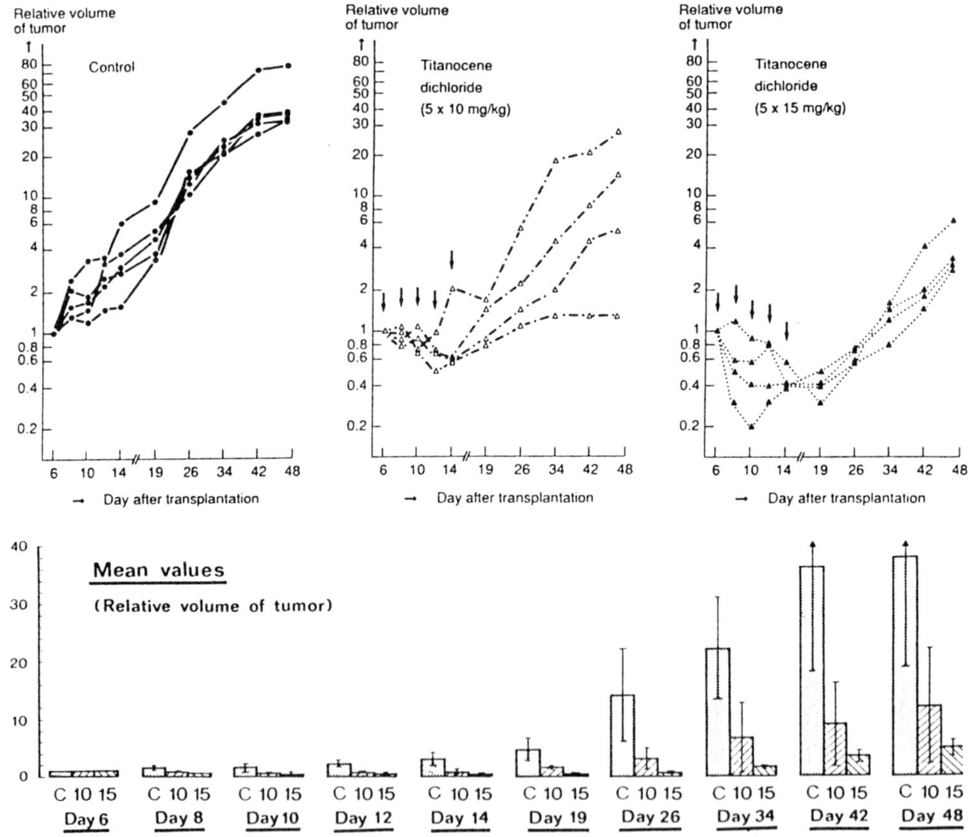

Fig. 2. Growth behavior of the xenografted human colon adenocarcinoma C-Stg 2 under treatment with titanocene dichloride, applied at sublethal doses according to Q2Dx5 (D = 10 or 15 mg/kg) on days 6, 8, 10, 12, 14 after transplantation. Each curve represents the growth curve of an individual tumor.

control tumors. These results are remarkable insofar as it is known from clinical and experimental studies that adenocarcinomas of the colon and the lung do generally not respond to clinical chemotherapy and can be influenced only marginally by established cytostatic drugs including 5-fluorouracil and cisplatin. In this connection it is worth mentioning that another group of inorganic titanium complexes, the neutral bis(benzoylacetonato)titanium(IV) diacido complexes, lacking aromatic cyclopentadienyl ring ligands, but containing titanium as central atom, are also characterized by pronounced antitumor activity against chemically induced autochthonous tumors of the colon, which are highly resistant to the therapeutic action of most cytostatic drugs.

Besides the antitumor spectrum, organometallic bis(cyclopentadienyl)metal ("metallocene") complexes also distinguish from established organic cytostatics and antitumor platinum compounds by a differing pattern of organ toxicity. Following application of therapeutic and toxic doses of metallocene complexes, there was no mentionable depression of bone marrow function, which is the main toxic side effect after treatment with organic cytostatics or carboplatin. The kidneys, which are usually severely damaged following administration of cisplatin, remained unaffected under the influence of metallocene compounds and did not show any functional impairments or structural injuries even after application of toxic (LD_{50}) doses. On the other hand, liver toxicity seemed to be the main and dose-limiting side effect of metallocene compounds. It manifested by elevations of the serum levels of typical enzymes of hepatocytes, such as GLDH, GOT, and GPT, and by the occurrence of single cell necroses within the liver parenchyma. Interestingly, the above mentioned bis(benzoylacetonato)titanium(IV) diacido complexes are characterized by a similar pattern of organ toxicity, which is dominated by dose-limiting hepatotoxicity.

Morphological and cytokinetic studies performed with diverse tumor cell lines in vitro and animal and human tumors in vivo revealed profound cytological and histological changes occurring in sensitive tumors which were treated with titanocene dichloride (Fig. 3). The mitotic activity decreased within 24 h to about 10 to 20 % of control values (100 %), the nuclear chromatin condensed, and the nuclei became bizarrely segmented. In the cytoplasm, lipid droplets and inclusion bodies appeared. These phenomena indicated cytoplasmic degeneration as a consequence of cellular, most probably of nuclear injuries effected by titanocene dichloride. Numerous inflammatory cells invaded the tumor tissue and removed damaged and necrotic tumor cells between days 3 to 6 after substance application. Microanalytical studies performed at the ultrastructural level by use of electron spectroscopic imaging (ESI) illustrated that titanium actually accumulated at first within the nucleus of tumor and liver cells (Fig. 4), both cell types being susceptible to the cytotoxic action of titanocene dichloride. Interestingly, at all sites where Ti was found in traceable amounts within the cells, phosphorus was also enriched in high concentration and similar local distribution as titanium. These results suggest the formation of aggregates between the phosphorus-rich nucleic acids and titanium-containing metabolites following application of titanocene dichloride to mammalian organisms.

In vitro syntheses of model complexes confirmed such an interaction between cyclopentadienly-metal units and nucleic acid components. The results illustrate the possibility of metal (Ti, Mo) nucleobase linkage as well as metal (Mo, V) phosphate interaction. Different types of model compounds were successfully synthetized during the past years:

- A titanocene purinato derivative was isolated where titanium(IV) was bound monofunctionally to the nitrogen-9 atom of the purinato ligand.
- A titanocene theophyllinato derivative was synthetized where the titanium(III) atom was bound via bifunctional chelation to the nitrogen-7 and oxygen-6 atoms of the theophyllinato ligand.
- A trinuclear complex was found with the xanthinato dianion where two bis(cyclopentadienyl)titanium(III) units were bound bifunctionally to both the N-7 and O-6 and the N-1 and O-2 atoms of the xanthinato dianion, the third titanocene center being coordinated by a monodentate dative bond to the N-9 atom.

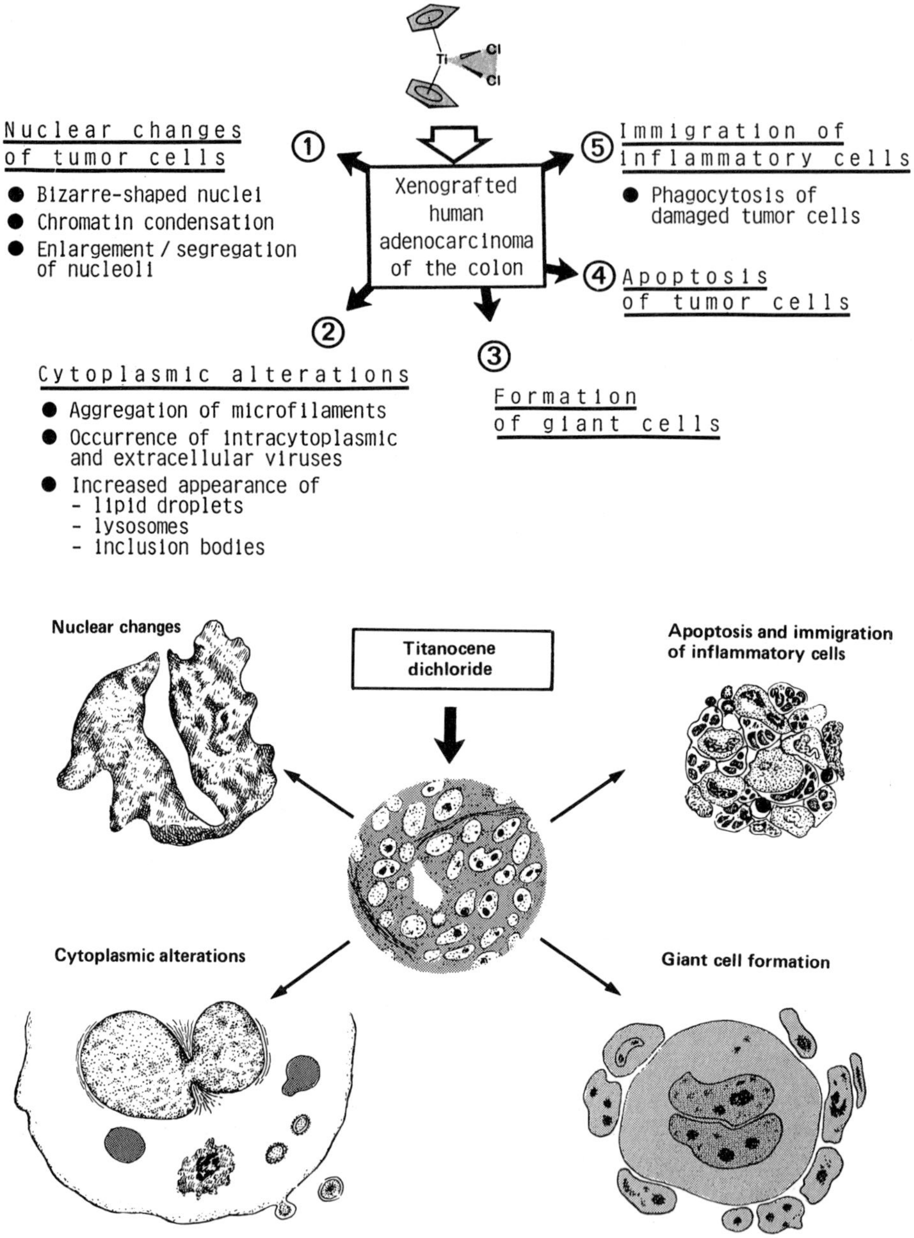

Fig. 3. Cytologic phenomena occurring in the xenografted human colon adenocarcinoma S 90 after treatment with titanocene dichloride in nude mice.

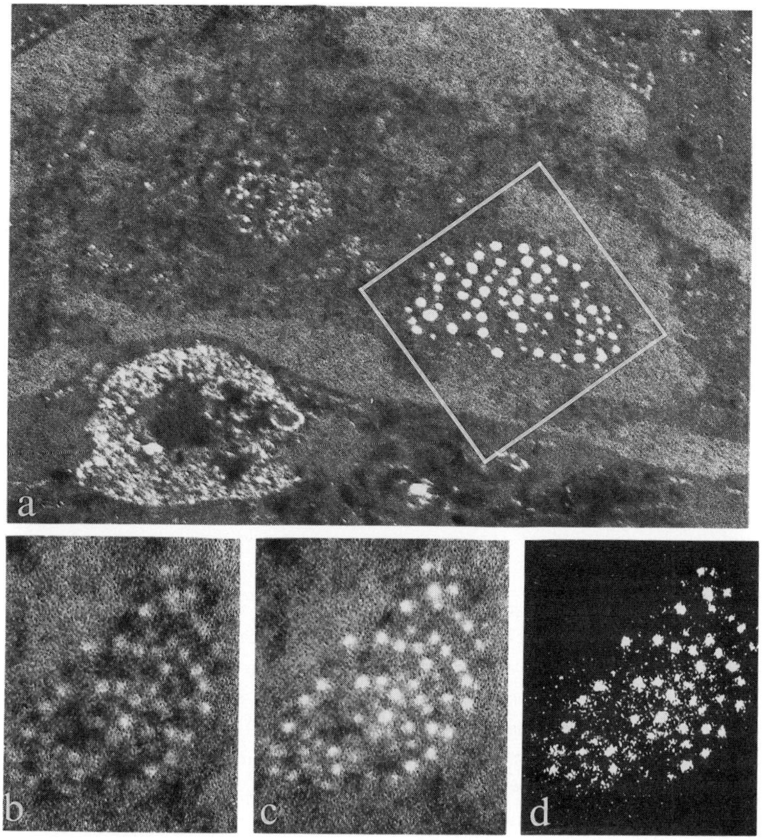

Fig. 4. Phosphorus image of a hepatocyt, taken at 165 eV (**a**), 24 h after application of titanocene dichloride. Note the nuclear granules and the cytoplasmic inclusion body, both containing high phosphorus content. **b-d,** The titanium analysis within the nuclear region marked in **a** (**b**, titanium image at 465 eV; **c**, unspecific underground at 410 eV; **d**, computer-calculated net distribution of Ti in the same area) reveals high titanium concentration in the nuclear granules. x 48,000 (**a**); x 58,000 (**b-d**)

A labile outer-sphere complexation of the bis(cyclopentadienyl)vanadium(IV) moiety to the nucleotide phosphate group of mononucleotides built up in aqueous solution near physiological pH, whereby the nucleotide-nucleotide Watson-Crick base-pairing was obviously not disrupted by the binding of the vanadocene molecule.

- Finally, in the case of bis(cyclopentadienyl)molydenum(IV), a direct O(phosphate)-Mo bond and a N7(adenine)-Mo bond simultaneously occurred within the complex $[(C_5H_5)_2Mo(5'\text{-}dAMP)]$.

All these biological and chemical results qualify organometallic metallocene complexes to enter soon early clinical trials. In preparing this application, a special formulation must be developed during the past months, in order to prevent hydrolysis reactions, to allow a long-term storage of the compounds, and to guarantee a rapid and complete dissolution immediately before application. The results of early clinical studies planned for the next months are awaited to see if they will confirm the toxic features which were found in animals and fundamentally differed from the toxic patterns of organic and inorganic cytostatics and, secondly, if they will actually establish the antitumor activity of metallocene complexes which had shown main therapeutic effect against human lung and gastrointestinal carcinomas xenografted to athymic mice.

Acknowledgement. The author's work on cyclopentadienyl metal complexes was supported by financial grants of the Medac GmbH, Hamburg, the Trude Goerke Heritage Foundation for the benefit of cancer research at the Freie Universität Berlin, and the Deutsche Forschungsgemeinschaft, Bonn.

Metal Ions in Biology and Medicine. Eds. Ph. Collery, L.A. Poirier, M. Manfait, J.C. Etienne. John Libbey Eurotext, Paris © 1990, pp. 462-464

Biological evaluation of new Ir(I) organometallic complexes

D.G. Craciunescu, V. Scarcia*, A. Furlani*, A. Papaioannou*, E. Parrondo-Iglesias

*Department of Inorganic and Analytical Chemistry, Faculty of Pharmacy, University of Madrid, 28040 Madrid 3, Spain and * Institute of Pharmacology and Pharmacognosy, Faculty of Pharmacy, University of Trieste, 34100 Trieste, Italy*

INTRODUCTION

A large series of coordination complexes containing group VIII metals have been tested in these years for their antineoplastic activity in various experimental systems.Nevertheless little researches have been carried out on iridium complexes, even though the limited data available showed that Iridium(I) complexes of mono- and diolefins exhibited interesting antitumour and, in some cases, antimetastatic properties (Sava et al., 1987a; Haiduc & Silvestru, 1989). Such effects did not appear to be related to a strong in vivo toxicity (Sava et al., 1987b).
This finding prompted us to synthesize and test for in vitro cytostatic and in vivo antitumour activity a series of five organometallic Ir(I) complexes of formula $[Ir(CO)_2(L)]^{\circ}$ where L were dithiocarbamate (dtc) or xanthate (xan) derivatives. Moreover, following the discovery of the dual pharmacological properties(antitumour and antitrypanosomic) displayed by a large number of metal complexes (Farrell et al., 1984), we have also assayed these complexes against three trypanosoma strains.

MATERIALS AND METHODS

Synthesis and chemical characterization of complexes

The complexes were obtained by reaction of the dimer $[Ir(CO)_4Cl_2]$ in methanol hot solution with the potassium salts of the respective ligands previously dissolved in hot methanol, in a 1:1 metal:ligand molar ratio. The mixture was stirred vigorously for 3 h at 80-90° C and then kept at room temperature for about 24 h until the complexes precipitated as microcrystalline powders. They were removed by filtration and washed several times with methanol and ether. The yields were approximately 33-35%.
The following ligands were employed: aniline-dtc, N-methylpiperazine-dtc, dicyclohexyl-dtc, buthyl-xan and benzyl-xan. The synthesis of these ligands was previously reported (Craciunescu et al., 1988).The complexes were moderately soluble in dimethylsulfoxide (DMSO), almost insoluble in water, methanol and ethanol.
Elemental analyses (N, O, S, Ir) of the complexes were carried out at Galbraith Laboratories (Knoxville, TN). Infrared spectra for complexes and pure ligands (potassium salts) were also recorded.

In vitro cytostatic activity evaluation

The Ir(I) complexes were assayed in vitro against an established human tumour cell line (KB), following the method previously described (Craciunescu et al., 1987). The compounds were dissolved or suspended, immediately before use, in sterile DMSO. The cytostatic activity was evaluated on the basis of cell growth inhibition in the treated cultures with respect to the controls.

In vivo antitumour activity evaluation

The antitumour in vivo assay was carried out against P388 and L1210 leukaemias and against Ehrlich ascites carcinoma in collaboration with National Cancer Institute (Bethesda, MD), according to methods from literature (Craciunescu et al., 1987; Craciunescu et al., 1985). The compounds were suspended in 5% dextrose in water and administered by a single i.p. injection on day 1 after tumour transplantation (day 0).

Antitrypanosomic activity evaluation

The antitrypanosomic activity was evaluated in collaboration with Institute of Veterinary Medicine (Berlin) against three trypanosoma strains (T.rhodesiense, T.gambiense and T.congolense) in Swiss mice. The compounds, suspended in Tween 80, were administered by a single s.c. injection 60 min after the 10^5 parasite i.p. inoculation. Six animals for each drug dose were used. The doubling of the mean survival time of the treated animals compared to that of the controls and no parasitaemia were the minimum responses for the activity.

RESULTS AND DISCUSSION

Elemental analyses for all complexes gave satisfactory results within $\pm$ 0.4% of the calculated values. The IR spectra of all complexes presented a single band of medium intensity at 985-960 cm^{-1}range, which was assigned to the ν(C$\overset{..}{-}$S) vibrations of the ligands (Forghieri et al., 1983). These values suggested that the dithioligands act as bidentate, since only one band was present, whereas in the case of monodentate coordination a doublet is present (Fabretti et al., 1984). In addiction new bands, absent in the ligand spectra, were observed in the 580-575 cm^{-1}range. They were attributed to the ν (Ir-S) stretching mode. The existence of the CO groups coordinated to Ir(I) appeared to be confirmed by the presence in the IR spectra of complexes of three slowly resolved bands (νC=O) in the range 2100-2040, 2075-2030 and 2015-2000 cm^{-1}respectively.

From results of the in vitro assay, reported in Table 1, it is possible to note that all Ir-complexes displayed very marginal cytostatic effect, the ID_{50}-values (molar concentration of the compound at which the cells show a 50% growth inhibition in relation to the control values) being higher than $5x10^{-5}$ M, upper-limit criterium for significant activity. Table 1 also summarizes the antitumour activity of our compounds against P388 and L1210 leukaemias as well as against Ehrlich ascites carcinoma. Only the $[Ir(CO)_2$(aniline-dtc)] displayed a moderate activity against Ehrlich ascites at 50 mg/Kg dose (1/8 LD_{50}), nevertheless such result was far lower than that of cisplatin that produced a T/C% value of 285 at 6 mg/kg dose (about 1/2 LD_{50}). The other tumour systems appeared no responsive. Consequently, as it has already been pointed out by various authors (Giraldi et al., 1977; Sava et al., 1987a) the activity of the complexes containing metals other than platinum varies depending on the tumour system employed.

The antitrypanosomic activity was found exclusively for $[Ir(CO)_2$(aniline-dtc)], the same complex that appeared effective against Ehrlich ascites, nevertheless no cures (animals surviving for 30 days) were obtained.

Table 1. In vitro cytostatic and in vivo antitumour activity .

Compounds	KB cells (ID_{50})	Dose (mg/Kg)	P388 (T/C%)[a]	L1210 (T/C%)[a]	Ehrlich asc. (T/C%)[a]
$[Ir(CO)_2(aniline\text{-}dtc)]$	$>5x10^{-5}M$	25	-	-	150
		50	110	-	170
		100	130	-	170
		200	140	110	175
$[Ir(CO)_2(N\text{-}methylpiperazine\text{-}dtc)]$	$>5x10^{-5}M$	100	120	-	150
		200	135	119	160
$[Ir(CO)_2(dicyclohexyl\text{-}dtc)]$	$>5x10^{-5}M$	100	117	-	148
		200	125	116	157
$[Ir(CO)_2(buthyl\text{-}xan)]$	$>5x10^{-5}M$	200	120	102	150
$[Ir(CO)_2(benzyl\text{-}xan)]$	$>5x10^{-5}M$	200	120	108	152

a) The antitumour activity was expressed as mean survival time of treated mice x 100/the mean survival time of the controls. T/C% values exceeding 125 for P388 and L1210 and 150 for Ehrlich ascites were taken as indicating effectiveness.

REFERENCES

Craciunescu,D.G., Furlani,A., Scarcia,V., and Doadrio,A. (1985): Synthesis, cytostatic and antitumor properties of new Rh(I) thiazole complexes. Biol. Trace Element Res. 8: 251-261.

Craciunescu,D.G., Scarcia,V., Furlani,A., Doadrio,A., Ghirvu,C., and Ravalico,L. (1987): Cytostatic and antitumour properties for a new series of Pt(II) complexes with cyclopentylamine. In vivo 1: 229-234.

Craciunescu, D.G., Parrondo Iglesias,E., Molina,C., Doadrio,A., gaston de Iriarte, E., and Ghirvu,C. (1988): Organometallic Rh(I) complexes with dual pharmacological effects. An.Real Acad.Farm. 54: 442-466.

Fabretti,A.C., Forghieri,F., Giusti,A., Prèti,C., and Tosi,G. (1984): The syntheses and properties of Cobalt(II), Nickel(II) and Copper(II) complexes with some heterocyclic dithiocarbamates. Inorg.Chim.Acta 86: 127-131.

Farrell,N.P., Williamson,J., and McLaren,J.M. (1984): Trypanocidal and antitumour activity of platinum-metal-drug dual-function complexes. Biochem.Pharmacol. 33: 961-971.

Forghieri,F., Graziosi,G., Preti,C., and Tosi,G. (1983): Cyclic substituted dithio carbamates as ligands. Vanadium(III) and oxovanadium(IV,V) complexes. Transition Met.Chem. 8: 372-376.

Giraldi,T., Sava,G., Bertoli,G., Mestroni,G., and Zassinovich,G. (1977): Antitumour action of two rhodium and ruthenium complexes in comparison with cis-diamminedichloroplatinum(II). Cancer Res. 37: 2662-2666.

Haiduc,I., and Silvestru,C. (1989): Rhodium, iridium, copper and gold antitumor organometallic compounds. In vivo 3: 285-294.

Sava,G., Zorzet,S., Perissin,L., Mestroni,G., Zassinovich,G., and Bontempi,A. (1987a): Coordination metal complexes of Rh(I), Ir(I) and Ru(II):recent advances on antimetastatic activity on solid mouse tumors. Inorg.Chim.Acta 137: 69-71.

Sava,G., Piccini,P., Mestroni,G., Zassinovich,G., and Bontempi,A. (1987b): Inhibitory effects of some organometallic complexes of rhodium(I) and iridium(I) on carrageenin paw edema in rats. In vivo 1: 27-30.

Metal Ions in Biology and Medicine. Eds. Ph. Collery, L.A. Poirier, M. Manfait, J.C. Etienne. John Libbey Eurotext, Paris © 1990, pp. 465-467

New, ionic metallocene complexes of titanium, molybdenum, and niobium as antitumor agents

P. Köpf-Maier, T. Klapötke

Institut für Anatomie, Freie Universität Berlin, Königin-Luise-Straße 15, D-1000 Berlin 33, and Institut für Anorganische und Analytische Chemie, Technische Universität Berlin, Straße des 17. Juni 135, D-1000 Berlin 12, Germany

Cyclopentadienyl metal complexes of the type of $(C_5H_5)_2MX_2$, M representing an early or medium transition metal, e.g. Ti, V, Nb, Mo, are known to exhibit pronounced antitumor activity against experimental and human tumors (Köpf-Maier and Köpf 1988). These compounds contain two acido ligands X in adjacent, "cis-like" position and two aromatic cyclopentadienyl ring ligands in an "open sandwich" geometry. In analogy to antitumor platinum compounds and most other non-platinum-group metal complexes (Rosenberg 1973, Köpf-Maier and Köpf 1988), the mentioned metallocene diacido complexes are uncharged and characterized by *electric neutrality*. As a consequence of this property, metallocene complexes are soluble in pure water or in saline only to a limited extent, on account of which it was necessary to develop a special formulation for the clinical application of metallocene compounds.

In order to avoid these problems deriving from poor water solubility of pharmacologically active compounds, we investigated in the present study the antitumor activity of certain *ionic* metallocene complexes which contain titanium(IV), niobium(V), or molybdenum(VI) as central atom and distinguish from known metallocene compounds by superior water solubility in consequence of their polar, charged and salt-like character. Typical representatives of this new generation of antitumor bis(cyclopentadienyl)metal complexes are the *ionic* titanocene, molybdenocene, and niobocene complexes given in Fig. 1.

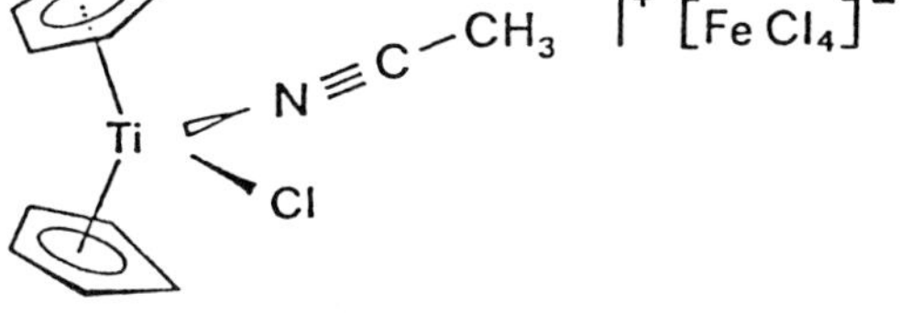

(A)

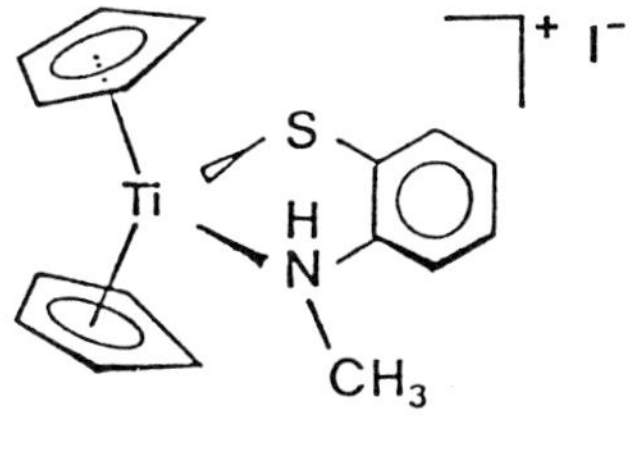

(B)

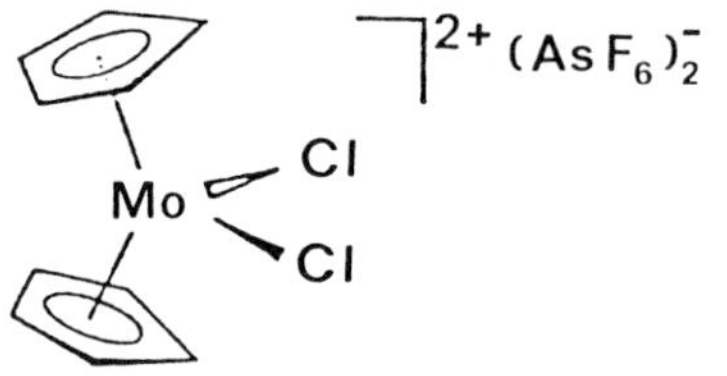

(C)

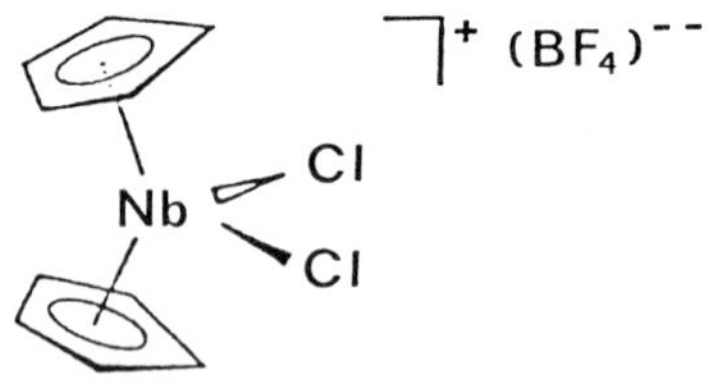

(D)

Fig. 1. Molecular formulae of some ionic metallocene complexes. (**A**) Bis(cyclopentadienyl)acetonitrile(chloro)titanium(IV) tetrachloroferrate(III); (**B**) bis(cyclopentadienyl)-N-methyl-o-aminothiopenolate titanium(IV) iodide; (**C**) bis(cyclopentadienyl)dichloromolybdenum(VI) bis(hexafluoroarsenate(V)); (**D**) bis(cyclopentadienyl)dichloroniobium(V) tetrafluoroborate.

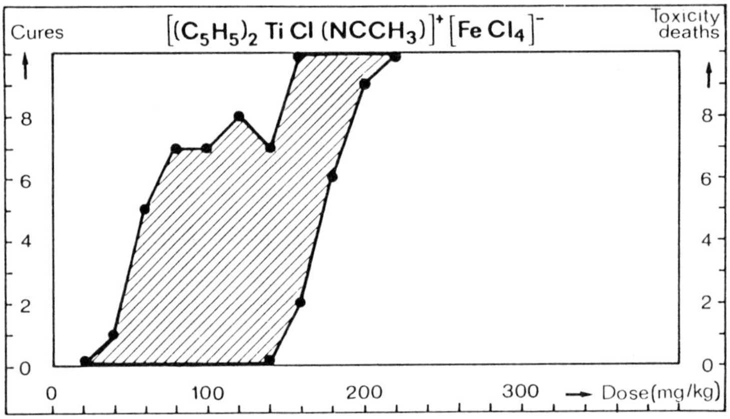

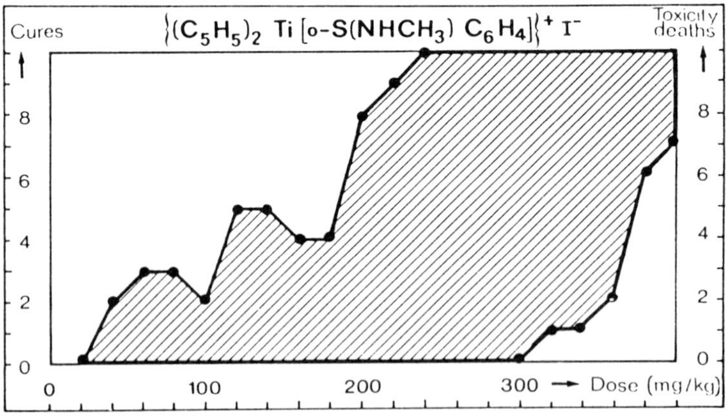

Fig. 2. Dose-activity (left graphs) and dose-lethality (right graphs) relationships of **A** and **B** against fluid Ehrlich ascites tumor. The shaded areas indicate the range of surviving, cured animals.

These compounds were tested against the murine tumor systems Ehrlich ascites tumor, B 16 melanoma, colon 38 carcinoma, and Lewis lung carcinoma. Against these tumors, the compounds exhibited pronounced antitumor activity manifesting in cure rates of 80 to 100 % of animals bearing Ehrlich ascites tumor (Fig. 2) and in growth reductions of animal solid tumors by 50 to 75 % to 25 to 50 % of control tumor size. The compounds **A** and **B** were additionally tested against some xenografted human carcinomas of the colon and the lung. It was shown that both compounds were able to inhibit the growth of the LX lung tumor, one of the standard tumors of the National Cancer Institute, and some colon adenocarcinomas by 50 to 80 % to 20 to 50 % of the size of untreated control tumors (100 %). These growth reductions of colon carcinomas were in a similar range as it was found before following application of the neutral metallocene diacido complex bis(cyclopentadienyl)titanium dichloride, which was shown to inhibit the growth of colon carcinomas in a more pronounced manner than known established cytostatic drugs, especially 5-fluorouracil and cisplatin (Köpf-Maier and Köpf 1988).

These results confirm that *ionic* metallocene complexes are characterized by similarly remarkable antitumor properties against diverse animal and human tumor systems as *neutral* metallocene compounds and, thus, can be considered as a possible "second generation" of antitumor metallocene complexes. Moreover, the findings underline that electric neutrality, which has been postulated to be an essential prerequisite for the achievement of cytostatic activity with antitumor platinum complexes (Cleare et al. 1980), is obviously not imperative in the case of cytostatic metallocene compounds.

References

Cleare, M.J., Hydes, P.C., Hepburn, D.R., and Malerbi, B.W. (1980): Antitumor platinum complexes: structure-activity relationships. In *Cisplatin - current status and new developments*, ed. A.W. Prestayko, S.T. Crooke, and S. K. Carter, pp. 149-170. New York: Academic Press.

Köpf-Maier, P. and Köpf, H. (1988): The antitumor activity of transition and main-group metal cyclopentadienyl complexes. *Struct. Bond.* 70, 103-185.

Rosenberg, B. (1973): Platinum coordination complexes in cancer chemotherapy. *Naturwiss.* 60, 399-406.

V. Ruthenium

Metal Ions in Biology and Medicine. Eds. Ph. Collery, L.A. Poirier, M. Manfait, J.C. Etienne. John Libbey Eurotext, Paris © 1990, pp. 471-475

Metal complexes of ruthenium in cancer chemo -therapy

G. Sava, S. Pacor, F. Bregant, V. Ceschia

Institute of Pharmacology, School of Pharmacy, University of Trieste, via A. Valerio 32, 34127 Trieste, Italy

SUMMARY

Many studies on animal tumors show interesting results for two "families" of ruthenium(III) complexes, namely ammine- and heterocycle- coordinated ruthenium(III) complexes and for few ruthenium(II) compounds with dimethylsulphoxide ligands. Octahedral ruthenium(III) complexes are suggested to be selectively activated to the more reactive ruthenium(II) complexes by the hypoxic environment of solid tumors; the antitumor action of heterocycle-ruthenium(III) complexes can be framed within this hypothesis. A separate mention concerns the antitumor properties of ruthenium(II) complexes with dimethylsulphoxide ligands which inhibit primary tumor growth of solid transplantable tumors but preferentially reduce the formation of their pulmonary metastases. These data indicate ruthenium complexes as a promising tool for the exploration of the potentialities existing within inorganic compounds to add to the panel of anticancer drugs.

INTRODUCTION

Transition metal complexes with metals of group VIII (Platinum group), but also from other groups were synthesized and tested mainly on the neoplastic disease (Cleare & Hydes, 1979; Sava et al., 1987; Williams, 1972). Particular emphasis was given to the examination of the possible antineoplastic properties of ruthenium complexes with a number of biologically active ligands. Chemical interactions and chemical reactivity of ruthenium ions have been studied under different conditions, pointing out their rather high propensity to bind to DNA (Cauci et al., 1987; Clarke, 1979; Clarke, 1983; Clarke et al., 1987).

The interaction of ruthenium ions with DNA is markedly stressed by the studies on the mutagenicity in bacteria (Hansen & Stern, 1984) and by the radiosentizing properties for hypoxic cells (Farrell, 1989). Compounds containing ^{97}Ru or ^{103}Ru radionuclides preferentially distribute in tumor tissue than in normal tissue (for example, the level is 5-fold higher in tumors than in muscle) (Tanabe & Yamamoto, 1975). Tumor disposition is facilitated by transferrin transport (Srivastava et al., 1989). Tumor cells require higher amounts of iron and therefore have a number of transferrin receptors larger than normal tissues. On the basis of the above considerations, the potential usefulness of ruthenium ions in cancer chemotherapy overbearingly emerges.

ANTINEOPLASTIC PROPERTIES

The antineoplastic properties of ruthenium complexes appear to be investigated mainly for two families of ruthenium(III) derivatives, namely the ammino-(tetraammino and pentaammino) ruthenium(III) compounds and the heterocycle-coordinated ruthenium(III) complexes. Indeed, some study was undertaken also with

Table 1. Antineoplastic action of the most representative ruthenium complexes.

COMPOUND	TUMOR INHIBITORY CHARACTERISTICS
$[Ru(III)(NH_3)_5X]^{+2}$	X= Cl, NH_3, purine, pyrimidine, methylxantine: inactive. X= carboxylic acid: moderate to weak antitumor activity in mice bearing P388 lymphocytic leukemia (T/C%= 128-162) (Clarke, 1979). X= ascorbate: active on L1210 lymphoid leukemia (T/C%= 130) and on a melanoma (T/C%= 126) (Clarke, 1988)
$[Ru(III)Im_2Cl_4]^-$ ImH	Active on P388 lymphocytic leukemia (T/C%= 168), B16 melanoma (T/C%= 13-35), S180 sarcoma (T/C%= 45), Walker 256 carcinosarcoma (T/C%= 230) (Keppler et al., 1987) and AMMN induced autochthonous colorectal tumors (T/C%= 18-21) (Garzon et al., 1987). Antitumor activity comparable to that of cisplatin (T/C%= 175 on P388 tumors; T/C%= 54 on sarcoma 180) (Keppler et al., 1987), cyclophosphamide (T/C%= 26-38 on B16 melanoma) (Keppler & Rupp, 1987) and 5-FU (T/C%= 144 on P388 tumors) (Keppler et al., 1987), or much greater than cisplatin (inactive, Keppler et al., 1989) and 5'dFUR (T/C%= 63) (Keppler et al., 1989) on AMMN- induced tumors.
$[Ru(II)(DMSO)_4Cl_2]^{\circ}$	cis-dichloro derivative: significant antitumor activity in mice bearing Lewis lung carcinoma (T/C%= 32), B16 melanoma (T/C%= 37) and MCa mammary carcinoma (T/C%= 55) (Sava et al., 1984). Weak antitumor activity on L1210 lymphoid leukemia (T/C%= 129) and active on Ehrlich ascites carcinoma (T/C%= 13 to cures) (Giraldi et al., 1977). trans-dichloro derivative: active on lung metastases more than on primary tumor growth (T/C%=72 vs 29) and effective in prolonging post-surgical survival time (+50%) in mice bearing Lewis lung carcinoma (Sava et al., 1989). Active on MCa mammary carcinoma after oral administration (T/C%= 53 for the metastatic tumor), on P388 lymphocytic leukemia (T/C%= 131) and inactive on TLX5 lymphoma (Sava et al., 1989a).

For: P388 lymphocytic leukemia, L1210 lymphoid leukemia and Walker 256 carcinosarcoma, T/C% values concern host survival time. For: Lewis lung carcinoma, B16 melanoma, MCa mammary carcinoma, AMMN-induced tumors, S180 sarcoma, Ehrlich ascites carcinoma, T/C% values concern tumor weight.

a dye, ruthenium red, which showed some degree of inhibition of Lewis lung carcinoma (Tsuruo et al., 1972; Anghileri, 1975). Ruthenium(II) complexes with dimethylsulphoxide ligands were also choosen for antitumor activity which resulted particularly interesting on solid tumors than on tumors of lymphoproliferative type (a summary of antitumor properties of the most representative compounds is given in table 1).

Ammine-ruthenium(III) complexes. cis-$Ru(NH_3)_3Cl_3$ can be considered the former amminoruthenium(III) complex studied which resulted active on experimental tumors (T/C%= 189 on P388 lymphocytic leukemia). Similarly to cisplatin, it induces filamentous growth in Escherichia coli (Durig et al., 1976). This effect correlates with the antitumor activity in experimental systems and suggests interactions with mechanisms of cellular replication. A wide number of ammine-ruthenium(III) complexes was synthesized with ligands such as halides, purines and pyrimidines, methylxantines, carboxylic acids and others (Clarke, 1979). The presence of labile ligands does not necessarily bring to antitumor effects. In fact $Ru(III)(NH_3)_5Cl^{++}$, a complex which rapidly replaces the labile Cl ligand for more strongly bonded groups is rather devoid of cytotoxicity and of mutagenicity in vivo and in vitro, althought inhibition of DNA synthesis occurred in vitro on RK cells (Kelman et al., 1977). A discrepancy between antitumor effects in vivo and mutagenicity in vitro is also reported for purine and pyrimidine derivatives which were much more mutagenic than other pentamminoruthenium complexes tested (Clarke, 1979). The only group of these complexes endowed with antitumor activity in the P388 system is that of pentamminoruthenium carboxylates. For these latter complexes: a) the antitumor activity increases with the increasing size of the carboxylic acid; b) the inhibition of DNA synthesis decreases with the increasing size of the carboxylic acid.

Ammine-ruthenium(III) complexes are supposed to be activated to the more toxic ruthenium(II) compounds by the redox potential of the environment. The relevance of the redox potential in activating ruthenium complexes is based upon the observation that: i) tumor tissues have a reducing environment due to the biological characteristics of tumor growth and ii) ruthenium(II), obtained by reduction from ruthenium(III), is several times more reactive than ruthenium(III) itself (Clarke, 1979). Connecting these two events, it can be assumed that ruthenium(III) complexes should permit to be selectively toxic for tumor cells rather than for normal tissues (Clarke, 1979; Clarke, 1989).

Heterocycle-ruthenium(III) complexes. Heterocycle-ruthenium(III) complexes are a series of compounds characterized by: i) the presence of one or two heterocycle ligands such as imidazole, pyrazole and indazole or their methyl substituted derivatives; ii) a generally good hydrosolubility; iii) an ionic structure with one or two negative charges; iiii) a trans- structure when two heterocycles are present (Keppler et al., 1989). On screening tumors, such as P388 lymphocytic leukemia or B16 melanoma, the complex imidazolium bis-(imidazole)tetrachlororuthenate(III) is, at maximum tolerated doses, as active as medium dosages of conventional cytotoxic drugs such as cyclophosphamide, cisplatin and 5-fluorouracil (Keppler & Rupp, 1986; Keppler et al., 1987). Indeed, renal toxicity is rather pronounced as is in cisplatin treated hosts, and is characterized by necrosis, evident also on liver cell parenchyma. Interestingly, the 4-fold increase of the volume of administration significantly reduces the lethality of the given dose (Keppler et al., 1989). The antitumor activity observed on colorectal tumors seems to indicate for these complexes a particular propension to reach the colorectal district (Garzon et al., 1987).

Dimethylsulphoxide-ruthenium(II) complexes. cis- and trans-ruthenium(II)dichlorotetrakisdimethylsulphoxide are the parent compounds of this group which differs from the previous "families" in that they are not influenced by the redox potential of the environment, being rather stable in their ± 2 oxidation state.

According to this consideration, these compounds should exibit a lesser selectivity for tumor tissues. Nevertheless, they exibit different degrees of toxicity on normal tissue, comparable to those shown by heterocycle complexes: marked histopathological alterations at the renal level, a lesser degree of damage on intestinal mucosa and spleen (not resulting from DNA synthesis evaluation) (Giraldi et al., 1977; Sava et al., 1984). cis-$RuCl_2(DMSO)_4$ was choosen because of its similarities with cisplatin. Althought having an octahedral structure, it has: i) two Cl ligands in a planar cis-position, ii) a pronounced affinity for nitrogen-donor ligands, and is: i) neutral, ii) very stable in the +2 oxidation state, iii) rather freely soluble in acqueous solutions. This complex induces lambda profage, produces filamentous growth in Escherichia coli, is selectively toxic for a strain defective in deoxyribonucleic acid repairing system (Monti-Bragadin et al., 1975) and inhibits a number of animal tumors (Giraldi et al., 1977; Sava et al., 1984; Sava et al., 1989b). The antitumor activity is better expressed on a panel of solid tumors suggesting that they are a target better than tumors of lymphoproliferative type. The examination of the differential effects of cis- and trans-$RuCl_2(DMSO)_4$ on primary tumor and on metastasis development has revealed antimetastatic activities superior to the effects on primary tumor growth (Sava et al., 1989b). The comparison between the antitumor effects of cis- and trans-$RuCl_2(DMSO)_4$ indicates the superiority of the latter whose antimetastatic activity is comparable or even greater than that of cisplatin (Sava et al., 1989b). Replacement of Cl with other halides such as Br or I does not increase the overall effect (Sava et al., 1989a).

CONCLUSIONS

Data available indicate the potential usefulness of ruthenium complexes in the field of control of cancer growth. Ruthenium ions can exibit antitumor effects different from those exibited by cisplatin; it appears that colon tumors and lung tumors could be two preferential targets for these drugs. The hypothesis put by Clarke on selective activation of ruthenium(II) toxic complexes from ruthenium(III) inert prodrugs could be of help in preparing new compounds which will be delivered to tumor tissues at the expences of a reduced or even absent host toxicity.

Acknowledgements: This work was supported by a Grant from Italian Ministry of Education (MPI 60%). S. Pacor and V. Ceschia are recipients of Fellowship Grants from Boheringer Mannheim Italia S.p.A. and from Fondazione C. e D. Callerio, respectively.

REFERENCES

Anghileri, L.J. (1975): The in vivo inhibition of tumor growth by ruthenium red: its relationship with the metabolism of calcium in the tumor. Z. Krebsforsch., 83, 213-217.

Cauci, S., Alessio, E., Mestroni, G., & Quadrifoglio, F. (1987): Reaction of cis-$RuII(DMSO)_4Cl_2$ with DNA and with some of its bases in aqueous solution. Inorg Chim Acta, 137, 19-24.

Clarke, M.J. (1979): Oncological implications of the chemistry of ruthenium. In Metal Ions in Biological Systems, ed. M. Dekker, vol. 11, pp. 231-283. New York.

Clarke, M.J. (1983): Ruthenium anticancer agents and relevant reactions of ruthenium-purine complexes. In Platinum, Gold and other chemotherapeutic agents. ACS Symposium series 209, ed. S.J. Lippard, pp. 335-354.

Clarke, M.J., Galang, R.D., Rodriguez, V.M., Kumar, R., Pell, S. & Bryan, D.M.

(1987): Chemical considerations in the design of ruthenium anticancer agents. In Platinum and other metal coordination compounds in cancer chemotherapy, ed. M. Nicolini, pp. 582-600. M. Nijihoff Publishing.
Clarke, M.J. (1989): Ruthenium chemistry pertaining to the design of anticancer agents. In Progress in Clinical Biochemistry and Medicine, vol. 10, pp. 25-39. Springer-Verlag Berlin.
Cleare, M.J. & Hydes, P.C. (1979): Antitumor properties of metal complexes. In Metal Ions in Biological Systems, ed. M. Dekker, vol. 11, pp. 1-62. New York.
Durig, J.R., Danneman, J., Behnke, W.D. & Mercer, E.E. (1976): The induction of filamentous growth in Escherichia coli by ruthenium and palladium complexes. Chem. Biol. Interact. 13, 287-294.
Farrell, N. (1989): Metal complexes as radiosensitizers. In Progress in Clinical Biochemistry and Medicine, vol. 10, pp. 89-109. Springer-Verlag Berlin.
Garzon, F.T., Berger, M.R., Keppler, B.K. & Schmahl, D. (1987): Comparative antitumor activity of ruthenium derivatives with 5'-deoxy-5-fluorouridine in chemically induced colorectal tumors in SD rats. Cancer Chemother. Pharmacol., 19, 347-349.
Giraldi, T., Sava, G., Bertoli, G., Mestroni, G. & Zassinovich, G. (1977): Antitumor action of two rhodium and ruthenium complexes in comparison with cis-diamminedichloroplatinum(II). Cancer Res., 37, 2662-2666.
Hansen, K. & Stern, R.M. (1984): A survey of metal-induced mutagenicity in vitro and in vivo. J. Am College Toxicol., 3, 381-430.
Monti-Bragadin, C., Ramani, L., Samer, L., Mestroni, G. & Zassinovich, G. (1975): Effects of cis-dichlorodiammineplatinum(II) and related transition metal complexes on Escherichia coli. Antimicrob. Agents and Chemother., 7, 825-827.
Kelman, A.D., Clarke, M.J., Edmonds, S.D. & Peresie, H.J. (1977): Biological activity of ruthenium purine complexes. J. Clin. Hematol. Oncol., 7, 274-288.
Keppler, B.K. & Rupp, W. (1986): Antitumor activity of imidazolium-bisimidazole-tetrachlororuthenate(III). J. Cancer Res. Clin. Oncol., 111, 166-168.
Keppler, B.K., Rupp, W., Juhl, U.M., Endres, H., Niebl, R. & Balzer, W. (1987): Synthesis, molecular structure, and tumor-inhibiting properties of imidazolium trans-bis(imidazole)tetrachlororuthenate(III) and its methyl-substituted derivatives. Inorg. Chem., 26, 4366-4370.
Keppler, B.K., Henn, M., Juhl, U.M., Berger, M.R., Niebl, R. & Wagner, F.E. (1989): New ruthenium complexes for the treatment of cancer. In Progress in Clinical Biochemistry and Medicine, vol. 10, pp. 41-69. Springer-Verlag Berlin.
Sava, G., Zorzet, S., Giraldi, T., Mestroni, G. & Zassinovich, G. (1984): Antineoplastic activity and toxicity of an organometallic complex of ruthenium(II) in comparison with cis-PDD in mice bearing solid malignant neoplasms. Eur. J. Cancer Clin. Oncol., 20, 841-847.
Sava, G., Pacor, S., Ceschia, V., Alessio, E. & Mestroni, G. (1989a): trans-Ru(II)dimethylsulphoxides: antineoplastic action on mouse tumours. Pharmacol. Res., 21, 453-454.
Sava, G., Pacor, S., Zorzet, S., Alessio, E. & Mestroni, G. (1989): Antitumor properties of dimethylsulphoxide ruthenium(II) complexes in the Lewis lung carcinoma system. Pharmacol. Res., 21, 617-628.
Sava, G., Zorzet, S., Perissin, L., Mestroni, G., Zassinovich, G. & Bontempi, A. (1987). Coordination metal complexes of Rh(I), Ir(I) and Ru(II): recent advances on antimetastatic activity on solid tumors. Inorg Chim Acta, 137, 69-71.
Srivastava, S.C., Mausner, L.F. & Clarke, M.J. (1989): Radioruthenium-labeled compounds for diagnostic tumor imaging. In Progress in Clinical Biochemistry and Medicine, vol. 10, pp.111-149. Springer-Verlag Berlin.
Tanabe, M. & Yamamoto, G. (1975): Tissue distributions of ^{97}Ru and ^{103}Ru in subcutaneous tumor of rodents. Acta Med. Okayama, 29, 431-436.
Tsuruo, T., Iida, H., Tsukagoshi, S. & Sakurai, Yoshio (1980): Growth inhibition of Lewis lung carcinoma by an inorganic dye, ruthenium red. Gann, 71, 151-154.
Williams, D.R. (1972): Metals, ligands and cancer. Chem. Rev., 72, 203-213.

Metal Ions in Biology and Medicine. Eds. Ph. Collery, L.A. Poirier, M. Manfait, J.C. Etienne. John Libbey Eurotext, Paris © 1990, pp. 476-478

Efficacy of two ruthenium complexes against chemically induced autochtonous colorectal carcinoma in rats

M.H. Seelig[1], M.R. Berger[1], B.K. Keppler[2], D. Schmähl[1]

[1] *Institute of Toxicology and Chemotherapy, German Cancer Research Center Im, Neuenheimer Feld 280, 6900 Heidelberg, F.R.G.;* [2] *Institute of Anorganic Chemistry, University of Heidelberg, Im Neuenheimer Feld 270, 6900 Heidelberg, FRG*

Introduction:

Chemotherapeutic treatment of advanced colorectal cancer is still not very successful, as can be realized by analyzing clinical trials which have been reviewed recently (Arbuck, 1989; Galeano et al., 1990). To find new, more active drugs we tested two new ruthenium–metal–complexes in acetoxymethylmethylnitrosamine (AMMN)–induced colorectal rat carcinoma. This model mimics the human situation with respect to its resistance against many clinically used antineoplastic agents. Previous experiments have shown a high antitumor efficacy of trans–indazolium–bisindazoletetrachlororuthenate (III) [IndH(RuInd$_2$Cl$_4$)] and trans–imidazolium–bisimidazoletetrachlororuthenate (III) [ImH(RuIm$_2$Cl$_4$)] in this model at selected dosages (Berger et al., 1989). This study was undertaken to compare both agents in a dose response study based at equimolar doses to determine the more qualified metal complex in terms of antineoplastic efficacy, toxicity and suitability for administration.

Materials and methods:

130 male Sprague–Dawley rats were induced by intrarectal application of 2 mg/kg AMMN once a week over a period of ten weeks. 15 weeks after beginning of tumor–induction the animals were examined coloscopically and in case of evident tumors distributed to treatment– and control–groups. Treatment started immediately thereafter and was performed by i.v. injection twice a week via the tail vein for ten weeks (see Tab. 1 for experimental details). At the end of treatment the experiment was terminated, the last 20 cm of the gut were dissected, macroscopically visible tumors were measured and their volume was defined by using the formula a × b × c / 2 (for more detailed description see (Berger et al., 1989)). To compare tumor growth inhibition the ratio of mean tumor volumes of treated (T) and control (C) groups was determined according to the formula T/C × 100.

Table 1: Treatment of acetoxymethylmethylnitrosamine – induced colorectal carcinoma in rats with trans – Indazolium – bisindazoletetrachlororuthenate(III) [IndH ($RuInd_2Cl_4$)] and trans – Imidazolium – bisimidazoletetrachlororuthenate(III) [ImH ($RuIm_2Cl_4$)]: experimental design.

Group No.	Animal No.	Treatment	Dosage[a)] (mg/kg)	Dosage[a)] (mMol/kg)	Duration[b)] of therapy (weeks)	Total Dosage (mg/kg)
1	15	Control 1	–	–	0	–
2	20	Control 2	–	–	10	–
3	5	IndH ($RuInd_2Cl_4$)	13	0,021	10	260
4	15	IndH ($RuInd_2Cl_4$)	10	0,016	10	200
5	15	IndH ($RuInd_2Cl_4$)	7,1	0,012	10	142
6	15	IndH ($RuInd_2Cl_4$)	5,1	0,008	10	102
7	15	ImH ($RuIm_2Cl_4$)	7,5	0,016	10	150
8	15	ImH ($RuIm_2Cl_4$)	5,3	0,012	10	106
9	15	ImH ($RuIm_2Cl_4$)	3,8	0,008	10	76

a) Treatment was given twice weekly i.v.

b) Treatment started immediately after endoscopic diagnosis of tumors

Results and discussion:

The mean T/C values of equimolar doses of ImH($RuIm_2Cl_4$) and IndH($RuInd_2Cl_4$) are shown in Fig. 1, A and C. Both compounds caused a maximum tumor growth inhibition of about 75% thus pointing to almost equivalent antineoplastic efficacy. However, a clear dose dependency of T/C values was observed only following ImH($RuIm_2Cl_4$). The non–dose–dependency of tumor growth inhibition of IndH($RuInd_2Cl_4$) probably refers to the fact that the i.v. injection of this less water soluble substance was associated with a high rate of tail vein necroses. Fig. 1, B and D represent the median body weight development during the treatment period indicating the toxic effect of the substances. Here interesting differences were found: Whereas ImH($RuIm_2Cl_4$) caused considerable dose related decreases in body weight (30%, 19% and 9% body weight loss relative to the initial weight, respectively) was administration of IndH($RuInd_2Cl_4$) not related with any gross signs of toxicity (8%, 10% and 2% body weight loss and 1% body weight gain relative to the initial body weight, respectively).

Regarding the much lower toxicity of IndH($RuInd_2Cl_4$) this substance should be preferred for further investigation if a more suited galenic preparation could overcome the observed difficulties in administration.

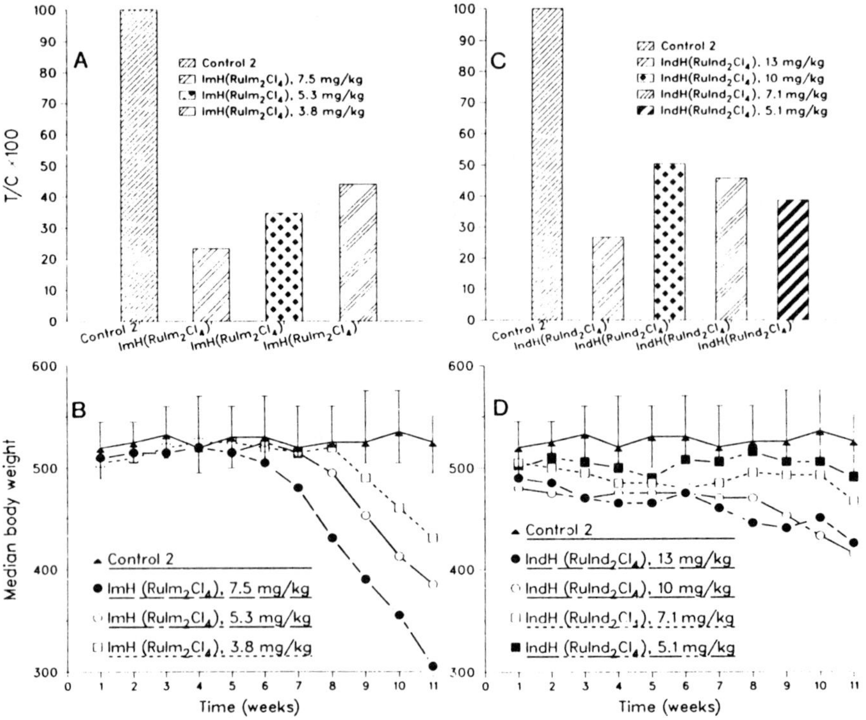

Fig. 1: Mean T/C values (A,C) and body weight development (B,D) of AMMN-induced SD-rats treated with $ImH(RuImH_2Cl_4)$ and $IndH(RuInd_2Cl_4)$.

References:

Arbuck, S.G. (1989): Overview of clinical trials using 5-Fluorouracil and Leucovorin for the treatment of colorectal cancer. Cancer 63, 1036- 1044.

Berger, M.R., Garzon, F.T., Keppler, B.K., Schmähl, D. (1989): Efficacy of new ruthenium complexes against chemically induced autochthonous colorectal carcinoma in rats. Anticancer Res. 9: 761-766.

Galeano, A., Petru, E., Berger, M.R., Schmähl, D. (1990): Use of various cytotoxic regimens in the palliative treatment of colon cancer. TumorDiagnostik & Therapie, in press.

Metal Ions in Biology and Medicine. Eds. Ph. Collery, L.A. Poirier, M. Manfait, J.C. Etienne. John Libbey Eurotext, Paris © 1990, pp. 479-481

Preliminary study of the acute and subacute toxicity of ruthenium-5'-ATP, a possible new antineoplastic agent

J.L. Domingo, A. Ortega, V. Moreno

Laboratory of Toxicology and Biochemistry, School of Medicine, University of Barcelona, San Lorenzo 21, 43201 Reus, Spain

Cis-Pt$(NH_3)_2Cl_2$ (cis-DDP, generic name cisplatin) was the first inorganic compound that demonstrated antitumour activity in a variety of murine tumour models (Rosenberg & Van Camp, 1970). In spite of its beneficial effects, nephrotoxicity is one of the primary limitations for the therapeutic administration of cisplatin (Krakoff, 1979; Goldstein et al., 1981). However, the success of cis-DDP as an antitumour agent has led to a search for other transition metal complexes with similar biological activities, but which are less toxic than cisplatin (Cleare & Hides, 1980; Yasbin et l., 1980). Gallium has been the second metal ion with clinical activity used in cancer treatment (Adamson et al., 1975; Warrell et al., 1983; Foster et al., 1986)

Ruthenium is a platinum group metal which in its lower oxidation states exhibits high affinity for nitrogen ligands. Several investigations suggested that some of its complexes might exhibit antitumour activity (Clarke, 1980). The binding of ammine complexes of both Ru(II) and Ru(III) to nucleotides and nucleic acids is, in many ways, analogous to that of cisplatin (Yasbin et al., 1980; Clarke, 1980; López et al., 1986). Octahedral ruthenium complexes with a number of biological ligands including purine and pyrimidine bases, histidine and cysteine have been synthesized (Clarke, 1980; Yasbin et al., 1980). Although in general these complexes are fairly stable, no really significant antitumour results have been reported yet. Recently, we have synthesized a new water soluble ruthenium compound: ruthenium-5'-ATP (Ru-ATP) (Zato & Moreno, 1990). In the present work, the toxicity of this compound has been investigated in rats previously to test its possible antitumor activity.

MATERIALS AND METHODS

Ru-ATP was synthesized by reaction between methanolic solutions of Na_3ATP and $RuCl_3$. The resulting solution was refluxed with constant stirring for three days. A pale brown soluble water product was isolated, filtered and dried over vacuum (Zato & Moreno, 1990). Ru-ATP was administered to male Sprague-Dawley rats (250-300 g) and male Swiss mice (25-30 g). The animals were supplied by Panlab (Barcelona, Spain) and had free access to a standard pellet diet and tap water. Single doses of Ru-ATP were administered i.m. and i.p. to rats and mice. Solutions were constituted by adding appropiate amounts of Ru-ATP and distilled water, then adding sodium bicarbonate until a pH of 7 was reached.

Solution concentrations were adjusted so that a 300-g rat received 1 ml and a 30-g mouse 0.2 ml. Ten animals per group were used. Survivors were held for a 14-day period of obervation following administration. The LD_{50} values were calculated by the standard method of Litchfield & Wilcoxon.

In a second experiment, four groups of eight rats received eight injections of Ru-ATP alternatively over a 21-day period at doses of 0, 107.5, 215, and 430 mg/kg (approximately 1/20, 1/10 and 1/5 of the i.p. LD_{50} of the Ru-ATP previously determined). Animals were placed in individual metabolic cages. Food and water consumption, as well as body weights, were measured daily. Protein efficiency coefficients were calculated weekly. After 21 days, the animals were exsanguinated from the tail vein and then sacrificed. Haematological examinations included haematocrit, haemoglobin, red and white blood cells, MCV, MCH and MCHC. SGOT, SGPT, total protein, glucose, cholesterol, urea, uric acid, and creatinine were measured according to standard methods. After the sacrifice of the animals, the weight of liver, kidneys, heart, spleen, lungs, brain, testes, suprarenals and thyroid were measured, and macroscopic observations were carried out. The significance of the differences in the results was determined by the Student's t test. A difference was considered to be significant when $P<0.05$.

RESULTS AND DISCUSSION

The i.p. and i.m. LD_{50} values (14 days) for Ru-ATP were respectively: 3060 mg/kg (mice), 2390 mg/kg (rats), and 3180 mg/kg (mice) and 2150 mg/kg (rats). The Ru-ATP acute toxicity was very similar for both ways of administration, being higher for rats than for mice. The majority of deaths occurred during the first three days.

TABLE 1. Haematological and plasma analyses in rats receiving ruthenium-5'-ATP for three weeks. The results are expressed as means ± SE. *P < 0.05, **P < 0.01, ***P < 0.001 by the Student's t test.

	Dose (mg/kg)			
	Control	107.5	215	430
Haematocrit (%)	40.2 ± 1.92	34.4 ± 6.09	32.0 ± 1.26*	34.1 ± 2.96*
Haemoglobin (g/100 ml)	13.9 ± 0.99	12.3 ± 3.04	11.8 ± 0.89	12.2 ± 1.48
Red blood cells (10^6/mm^3)	7.26 ± 0.36	6.21 ± 1.45	5.85 ± 0.41*	5.95 ± 0.44*
White blood cells (10^3/mm^3)	13.6 ± 0.14	7.0 ± 1.35***	9.8 ± 0.25***	10.2 ± 0.40***
SGOT (U/l)	164.1 ± 62.0	155.2 ± 23.9	170.8 ± 21.7	197.6 ± 44.7
SGPT (U/l)	56.3 ± 19.6	48.5 ± 22.5	53.0 ± 24.5	60.3 ± 9.2
Total protein (g/100 ml)	5.5 ± 0.82	5.1 ± 0.28	5.9 ± 0.57	6.2 ± 0.38
Glucose (mg/100 ml)	166.1 ± 18.4	133.9 ± 33.6	155.8 ± 21.1	157.0 ± 49.1
Cholesterol (mg/100 ml)	61.6 ± 9.0	56.6 ± 8.4	73.1 ± 8.3*	78.0 ± 9.6*
Urea (mg/100 ml)	30.7 ± 4.7	27.4 ± 4.2	28.1 ± 6.5	31.0 ± 7.1
Uric acid (mg/100 ml)	3.0 ± 1.2	1.3 ± 0.3*	1.1 ± 0.3*	1.5 ± 0.2*
Creatinine (mg/100 ml)	0.5 ± 0.04	0.5 ± 0.04	0.5 ± 0.06	0.5 ± 0.07

There was no significant effect on the body weight of the rats which received Ru-ATP for three weeks. Moreover, food consumption and drinking water were scarcely affected. However, the accumulated values for the group receiving 430 mg/kg were remarkably lower than the values for the control group. The protein efficiency coefficients were usually lower for the treated animals. No significant or dose-dependent differences between the treated and the control animals were observed in relative organ weights.

Some dose-dependent significant differences in haematocrit, red blood cells and white blood cells can be seen in Table 1. The concentrations of SGOT, SGPT, total protein, glucose, urea and creatinine in plasma were within the normal range for treated and untreated animals. Only cholesterol and uric acid showed slight dose-dependent changes. However, it would seem that Ru-ATP had not any effect on the liver and renal functions (Table 1). Macroscopic observations of the organs examined showed no tissue changes, which can be attributed to Ru-ATP. No specific physical and clinical symptoms or behaviour disturbances could be detected during the period of administration.

Ruthenium compounds have been reported to have a lower toxicity than cis-DDP. This fact would make possible the use of higher concentrations for ruthenium compounds (Yasbin et al., 1980). In the present work, we have administered high doses of Ru-ATP. These doses were generally well tolerated with the possible exception of some hematological parametres. However, due to this low toxicity, we would suggest to undertake systematic studies on the antitumour activity of Ru-ATP.

REFERENCES

Adamson, R.H., Canellos, G.P. & Sieber, S.M. (1975): Studies on the antitumor activity of gallium nitrate (NSC-15200) and other group IIIa metal salts. Cancer Chemother. Rep. 59, 599-610.

Clarke, M.J. (1980). In Metal ions in biological systems, ed. H. Sigel, vol 11 (5) pp. 231-283. New York: Marcel Dekker.

Cleare, M.J. & Hydes, P.C. (1980): Antitumor properties of metal complexes. In Metal ions in biological systems, ed. H. Sigel, vol 11 (1) pp. 1-62. New York: Marcel Dekker.

Foster, B.J., Clagett-Carr, K., Hoth, D. & Leyland-Jones, B. (1986): Gallium nitrate: the second metal with clinical activity. Cancer Treat. Rep. 70, 1311-1319.

Goldstein, R.S., Noordeweier, B., Bond, J.T., Hook, J.B. & Mayor, G.H. (1981): cis-Dichlorodiammineplatinum nephrotoxicity: time course and dose response of renal functional impairment. Toxicol. Appl. Pharmacol. 60, 163-175.

Kelsen, D.P., Alcock, N., Yeh, S., Brown, J. & Young, C. (1980): Pharmacokinetics of gallium nitrate in man. Cancer 46, 2009-2013.

Krakoff, I.H. (1979): Nephrotoxicity of cis-dichlorodiammineplatinum (II). Cancer Treat. Rep. 63, 1523-1525.

Levi, J., Jacobs, C., Kalman, S.M., McTigue, M. & Weiner, M.W. (1980): Mechanism of cis-platinum nephrotoxicity: 1. Effects of sulfhydryl groups in rat kidneys. J. Pharmacol. Exp. Ther. 213, 545-550.

López, R.C., Ruiz, C.M., Craciunescu, D., Doadrio, A., Osuna, A. & Alonso, C. (1986): Studies on the interaction between antitrypanosome cis-DDP analogs and DNA. Chem. Biol. Interact. 59, 99-111.

Rosenberg, B. & Van Camp, L. (1970): The successful regression of large solid sarcoma 180 tumors by platinum compounds. Cancer Res. 30, 1799-1802.

Warrell, R.P. Jr., Coonley, C.J., Straus, D.J. & Young, C.W. (1983): Treatment of patients with advanced malignant lymphoma using gallium nitrate administration as a seven-day continuous infusion. Cancer 51, 1982-1987.

Yasbin, R.E., Matthews, C.R. & Clarke, M.J. (1980): Mutagenic and toxic effects of ruthenium. Chem. Biol. Interact. 31, 355-365.

Zato, E. & Moreno, V. (1990): Complexes of Ru(III) with purine bases. J. Bioinorg. Chem. (in press).

Metal Ions in Biology and Medicine. Eds. Ph. Collery, L.A. Poirier, M. Manfait, J.C. Etienne. John Libbey Eurotext, Paris © 1990, pp. 482-484

Antitumor action of mer-trichlorobis (dimethylsulphoxide) aminoruthenium (III) (BBR2382) in mice bearing lewis lung carcinoma

S. Pacor, G. Sava, F. Bregant, V. Ceschia, E. Alessio*, G. Mestroni*

*Institute of Pharmacology, School of Pharmacy and * Department of Chemical Sciences, University of Trieste, I-34127 Trieste, Italy*

SUMMARY

The differential effects of i.p. treatment of BD2F1 female mice carrying s.c. implants of Lewis lung carcinoma with mer-trichlorobis(dimethylsulphoxide)-aminoruthenium(III), BBR2382, on primary tumor growth and on host survival time, were compared to those of equitoxic doses of cis-dichlorodiammineplatinum (cisplatin) and of imidazoliumbis(imidazole)tetrachlororuthenate ($ImH(RuIm_2Cl_4)$. The results obtained indicate that BBR2382 significantly reduces primary tumor growth by a factor comparable to that of cisplatin but significantly larger than that of $ImH(RuIm_2Cl_4)$. Similar results are obtained in terms of increase of survival time which is prolonged by 33%; this parameter is significantly better for mice treated with BBR2382 than for those treated with cisplatin. These data suggestsah the existence of antimetastatic effects and stress the potential therapeutic usefulness of ruthenium(III)dimethylsulphoxides in cancer treatment.

INTRODUCTION

Ruthenium complexes were shown capable of inhibiting a number of tumors of experimental animals (Clarke, 1989; Garzon et al., 1977; Keppler & Rupp, 1986; Sava et al., 1984). Out of the complexes of Ruthenium(III) studied, interesting results were shown by the use of metal-complexes containing heterocycle ligands such as imidazole, indazole and pyrazole (Keppler et al., 1989). One of them, namely imidazoliumbis(imidazole)tetrachlororuthenate(III), $ImH(RuIm_2Cl_4)$, resulted as active or even more active than cisplatin, cyclophosphamide and 5-fluorouracil against selected mouse tumors (Garzon et al., 1987; Keppler et al., 1987; Keppler & Rupp, 1986).

The aim of the present investigation was grounded upon the evaluation of the antineoplastic activity of a new ruthenium(III) derivative containing dimethylsulphoxide ligands. The compound was prepared taking into account the interesting antitumor properties observed with ruthenium(II) complexes containing dimethylsulphoxides (Sava et al., 1984; Sava et al., 1989) and the hypothesized selective toxicity for tumor tissues attributed to ruthenium complexes at +3 oxidation state (Clarke, 1987). We therefore syntesized and tested mer-trichlorobis(dimethylsulphoxide)amineruthenium(III) (BBR2382) using the Lewis lung carcinoma model.

MATERIALS AND METHODS

Synthesis and animal treatment. The synthesis of BBR2382 was performed as described in the Italian patent 2883M (E. Alessio et al., Complessi di rutenio(III) come agenti antineoplastici, 1989). $ImH(RuIm_2Cl_4)$ was synthesized according to previously reported procedures (Keppler et al., 1987) and cisplatin was obtained from the Drug Synthesis and Chemistry Branch Division of Cancer Treatment NCI, NIH, Bethesda MD, USA. The compounds were dissolved in isotonic saline and administered i.p. in volumes of 0.1ml/10g body weight. The less soluble BBR2382 was sonicated, 10" at maximum power with a MSE sonicator, before use; solutions were kept in cold during sonications. Controls received a similar amount of isotonic saline.

Tumor transplantation and evaluation The Lewis lung carcinoma line used, obtained from the Tumor Repository Bank NCI, NIH, Bethesda MD, USA, is locally maintained according to NCI protocols (Geran et al., 1972). For experimental purposes, the tumor was transplanted into BD2F1 mice purchased from Charles River, Calco (Como), Italy, according to already reported procedures (Sava et al., 1989). Primary tumor growth was evaluated by determining two perpendicular axes (a and b, with a b), by the formula Tumor Volume= $(\pi/6)xa^2xbx$, and was determined on alternate days starting from tumor appareance. In each group the average survival time was determined by recording the life-span of each animal.

RESULTS AND DISCUSSION

The daily dose used for each compound, administered on days 1,5,9,13 following s.c. tumor implantation, is the maximum tolerated dose devoid of lethality; body weight variation, in treated groups vs untreated controls, determined at the end of treatment, is about -2% in all treatment groups.

Table 1. Effects of BBR2382 on the growth of subcutenous tumor: comparison with $ImH(RuIm_2Cl_4)$ and cisplatin .

Compound	Dose mg/kg/day	Primary Tumor Volume (ml) on Day[a] 8	14	21	Survival Time (days)
Controls	-	1,175±195	4,051±478	7,920±785	23.3±2.7
BBR2382	100	407± 60*	1,640±327*	3,518±626*	31.0±3.2°
Cisplatin	4	480±131*	1,036±179*	3,400±579*	25.7±3.0
$ImH(RuIm_2Cl_4)$	40	954±178	3,125±462	6,772±707	23.7±2.1

a: days from s.c. tumor implantation. *: $p < 0.01$ vs controls, Student Newman--Keuls test; °: $p < 0.025$ vs controls, Mann-Whitney U-test.
Each value is the mean±S.E. obtained in groups of 8-10 BD2F1 mice, implanted s.c. with $100mm^3$ of tumor fragments on day 0, which were given i.p. the test compound on days 1,5,9,13.

Data reported in Table 1 indicate that BBR2382 significantly reduces the growth of primary subcutaneous tumor in mice bearing Lewis lung carcinoma by 55-65%, as compared with untreated controls. The reduction is comparable to that obtained with an equitoxic dose of cisplatin (55-75%) and is superior to that caused by $ImH(RuIm_2Cl_4)$ (15-20%), at every day of determination. Correspondingly, the survival time of hosts treated with BBR2382 is significantly prolonged by 33%, compared with untreated controls. This effect is much better than those shown by cisplatin (+10.3%) and by $ImH(RuIm_2Cl_4)$ (+1.3%). BBR2382 also increases the survival time of mice treated after removal of their primary tumor (+32% vs controls). In these conditions, i.p. treatment on days 1,5,9,13 occured after surgical amputation of the left hind leg (referred as day 0), into which 50 mm^3 tumor fragments were inoculated 12 days before (see methods from Sava et al., 1989). This model allows to study the effects of drug treatment on spontaneously formed lung metastases. On this model, similarly to previous findings, the activity of cisplatin (+22%) and that of $ImH(RuIm_2Cl_4)$ (+19%), although better than those reported in Table 1, is less pronounced than that of BBR2382.

These data indicate that BBR2382, a new dimethylsulphoxide ruthenium(III) complex, has a remarkable activity on Lewis lung carcinoma, comparable or even greater than that of cisplatin of a reference ruthenium(III) derivative already shown active in a number of experimental systems (Keppler et al., 1989). Unlike cisplatin and $ImH(RuIm_2Cl_4)$, BBR2382 prolongs the survival time of the treated animals irrespectively of whether primary tumor is left in situ or is surgically removed. This observation, stressing the role of cancer metastasis on host survival, indicates an antimetastatic effect in addition to the effects on primary tumor growth, and this effect seems superior to those of the reference compounds presently employed. Although data from other tumor models are needed, BBR2382 could be identified as the precursor of a series of new ruthenium(III) derivatives to be investigated for their potential usefulness in cancer therapy.

ACKNOWLEDGEMENT: This work was supported by Italian Ministry of Education (MPI 60%) and by Boheringer Mannheim Italia. S. Pacor and V. Ceschia are recipients of fellowship grants from Boheringer Mannheim Italia and from Fondazione C. & D. Callerio, respectively.

REFERENCES

Clarke, M.J. (1989): Ruthenium chemistry pertaining to the design of anticancer agents. In Progress in Clinical Biochemistry and Medicine, vol. 10, pp.25-39. Springer-Verlag Berlin.

Garzon, F.T., Berger, M.R., Keppler, B.K. & Schmahl, D. (1987): Comparative antitumor activity of ruthenium derivatives with 5'-deoxy-5-fluorouridine in chemically induced colorectal tumors in SD rats. Cancer Chemother. Pharmacol., 19, 347-349.

Geran, R.I., Greenberg, N.H., Macdonald, M.M., Schumacher, A.M. & Abbott, B.J. (1972): Protocols for sceening chemical agents and natural products against animal tumors and other biological systems. Cancer Chemother. Rep., pt. 3, 3, 13.

Giraldi, T., Sava, G., Bertoli, G., Mestroni, G. & Zassinovich, G. (1977): Antitumor action of two rhodium and ruthenium complexes in comparison with cis-diamminedichloroplatinum(II). Cancer Res., 32, 2662-2666.

Keppler, B.K. & Rupp, W. (1986): Antitumor activity of imdazolium bis-imidazoletetrachlororuthenate (III). J. Cancer Res. Clin. Oncol., 111, 166-168.

Keppler, B.K., Rupp, W., Juhl, U.M., Endres, H, Niebl, R. & Balzer, W. (1987): Synthesis, molecular structure, and tumor-inhibiting properties of imidazolium trans-bis(imidazole)tetrachlororuthenate(III) andits methyl-substituted derivatives. Inorg. Chem., 26, 4366-4370.

Keppler, B.K., Henn, M., Juhl, U.M., Berger, M.R., Niebl, R. & Wagner, F.E. (1989): New ruthenium complexes for the treatment of cancer. In Progress in Clinical Biochemistry and Medicine, vol. 10, pp.41-69. Springer-Verlag Berlin.

Sava, G., Zorzet, S., Giraldi, T., Mestroni, G. & Zassinovich, G. (1984): Antineoplastic activity and toxicity of an organometallic complex of ruthenium(II) in comparison with cis-PDD in mice bearing solid malignant neoplasms. Eur. J. Cancer Cli. Oncol., 20, 841-847.

Sava, G., Pacor, S., Ceschia, V., Alessio, E. & Mestroni, G. (1989): trans-Ru(II)dimethylsulphoxides: antineoplastic action on mouse tumours. Pharmacol. Res., 21, 453-454.

VI. Selenium

Metal Ions in Biology and Medicine. Eds. Ph. Collery, L.A. Poirier, M. Manfait, J.C. Etienne. John Libbey Eurotext, Paris © 1990, pp. 487-189

The regulatory effect of selenium on the expression of oncogenes associated with proliferation and differentiation on tumor cells

S.Y. Yu, X.P. Lu, S.D. Liao

Cancer Institute, Chinese Academy of Medical Sciences, P.O. Box 2258 Beijing, 100021 P.R. China

The importance of the trace element selenium to human health and its potential as a nutritional anticancer agent has recently been widely recognized. However, relatively little is known about the precise mechanisms involved. In previous report we demonstrated a potential role of selenium to regulate the differentiation and proliferation of tumor cells and induce the cellular and biochemical phenotypes of hepatoma cells toward normal (YU et al, 1988), including decreasing of mitotic index, growth rate, the extent of saturation density, glycolysis, aspartate carbamyl transferase, cGMP, protein kinase A isozyme type I, which are related to cell proliferation, and increasing of adhesiveness of cells, carbamyl phosphate synthetase I, cAMP, protein kinase A isozyme type II, which are associated with cell differentiation. Here we present further evidence for the effect of selenium on the modulation of oncogene expression associated with differentiation and proliferation.

Using dot blot hybridization techniques (White & Bancroft, 1982), we found there was a decrement in the c-myc gene expression and an increase in the c-fos gene expression in *BEL-7402* human hepatoma cells treated with sodium selenite *in vitro* (Fig.1). Densitometric measurement showed that compared to control, treatment of the cells with 1, 2 and 3 parts per million (ppm) sodium selenite for twenty-four hours, the c-myc gene expressions were reduced by 15, 57 and 71 per cent, while the increment of c-fos gene expressions were 1.3, 1.5 and 3 folds, respectively.

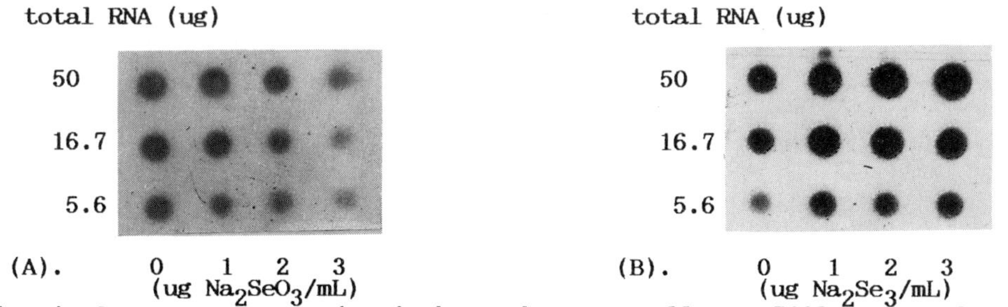

Fig. 1. Oncogene expression in human hepatom cell *BEL-7402* untreated and treated with Na_2SeO_3 for 24 hrs.
(A). c-myc gene expression. (B). c-fos gene expression.

A similar result was obtained *in vivo* experiments. Hepatoma bearing mice were injected intraperitoneally with sodium selenite in dosage of one microgram per gram of body weight daily. Northern blot analysis (Lehrach & Skoultchi, 1984) Showed that the mRNA expression of c-myc oncogene was decreased in hepatoma cells treated with sodium selenite for three days and it was more significantly decreased for four-day treatment. (Fig. 2A)

In hepatoma cells untreated with selenium the c-fos gene expression was undetectable. After selenium treatment for three and four-day. c-fos mRNA was induced significantly and there was a time dependent effect. (Fig. 2B)

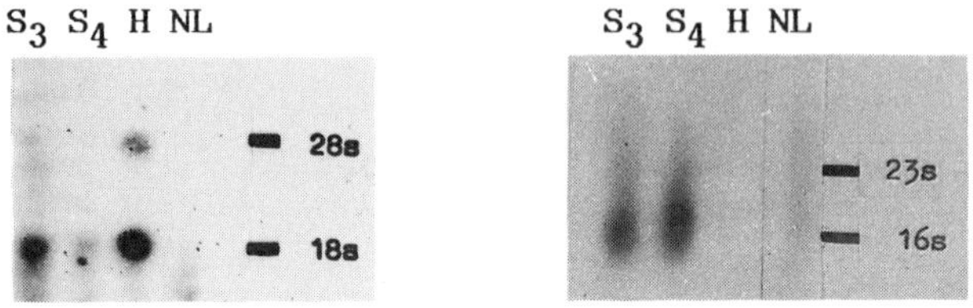

(A). (B).

Fig. 2. Northern blot analysis of oncogene expression with 5ug mRNAs which were prepared from normal liver (NL), hepatoma ascites cells (H), three-day Se-treated hepatoma ascited cells (S3) and four-day Se-treated hepatoma ascites cells (S4) respectively.
(A). c-myc gene expression (B). c-fos gene expression

The myc and fos are nuclear oncogenes. they play important role in regulation of cell division and differentiation. It is most likely that augmented myc gene expression is associated with tumorigenesis and its down-regulation may induce cell to terminal differentiation (Reitsma et al., 1983, Alitalo et al., 1987). Althouth the pattern of fos gene expression is complex, it does appear to show higher expression among some cell types during differdntiation (Sachs 1987, Collins 1987).

The results obtained demonstrate that the c-fos gene expression is stimulated and the expression of c-myc gene is reduced in hepatoma cells treated with selenium. Thus, on molecular level, it is directly attested that the regulatory effects of selenium on the inhibition of proliferation, improvement of differentiation and reduction of malignancy of tumor cells at cellular and biochemical levels are related to the mRNA expression of oncogenes.

REFERENCES

Alitalo, K., Koskinen, P., Makela, T.P., Saksela, K., Sistonen, L. and Wingvist, R. (1987): myc oncogenes: activation and amplification. BBA 907:1-32

Collins, S.J. (1987): The HL-60 promyelocytic leukemia cell line: Proliferation, differentiation, and Cellular Oncogene Expression. Blood 70:1233-1244

Lachman, H.M. and Skoultchi, A.I. (1984): Expression of c-myc changes during differentiation of mouse erythroleukaemia cells. Nature 310:592-594

Reitsma, P.H., Rothberg, P.G., Astrin, S.M., Trial, J., Bar-shavit, Z., Hall, A., Teitebaum, S.L. and Kahn, A.J. (1983): Regulation of myc gene expression in Hl-60 Leukaemia cells by a vitamin D metabolite. Nature 306:492-494
Sachs, L. (1987): Cell Differentiation and Bypassing of Genetic in the Suppression of Malignancy. Cancer Res. 47:1981-1986
White, B.A. and Bancroft, F.C. (1982): Cytoplasmic Dot Hybridization. J. Biol. Chem. 257:8569-8572
Yu, S.Y., Ao, P., Wang, L.M., Huang, S.L., Chen, H.C., Lu, X.P. and Liu, Q.Y. (1988): Biochemical and Cellular Aspects of the Anticancer Activity of Selenium. Biol. Trace Ele. Res. 15:243-255

Metal Ions in Biology and Medicine. Eds. Ph. Collery, L.A. Poirier, M. Manfait, J.C. Etienne. John Libbey Eurotext, Paris © 1990, pp. 490-492

Damage to selenium deficient erythrocyte by oxy-radicals

L.Z. Zhu, M. Gao, J.H. Piao

Institute of Nutrition and Food Hygiene, Chinese Academy of Preventive Medicine, 29 Nanwei Road, Beijing 100050, People's Republic of China

It is well known that selenium is able to protect cells against oxidation via SeGSHPx, thus selenium deficient may be more susceptible to oxy-radicals. So, we have used erythrocytes of selenium deficient rat and human as the model to study oxidation of oxy-radicals and protection of selenium .

For animal experiment, 30 weanling male Wistar rats were divided into two groups of 15 each, one was fed with crops grown in a selenium deficient area in China, another was fed with crops grown in normal region. After three months, blood was sampled by decapitation. Washed erythrocytes were incubated with four oxidants: dihydroxy fumaric acid(DHF) which generates $O_2^{\bar{}}$ automatically, vitamin C, H_2O_2 and t-butylhydroperoxide separately to observe the damage to erythrocyte. Parameters of damage used were changes of haemoglobin and pattern of membrane skeleton protein.

For human experiment, erythrocytes were taken from children aged six to ten living in selenium deficient area and in normal region. Erythrocytes were treated as in amimal experiment.

I. Animal experiment

A. Selenium status and activities of antioxidant enzymes other than GSHPX

Results were shown in Table 1 and Table 2.

Tab. 1 Whole blood selenium level and GSHPx activity in erythrocyte of Se-deficient and Se-adequate rats

Group	Selenium level (ppm)	GSHPx activity (umoles NADPH oxd/min/g Hb)
Se-deficient	0.0131±0.0011[‡] (12)	13.74±0.88 (12)
Se-adequate	0.0746±0.0034 (13)	124.70±6.10 (13)

Mean+SE(n)

Obviously, the difference of selenium status between these two groups was highly significant. Nevertheless no significant difference was found in activities of catalase and superoxide dismutase (Tab. 2, omit).

B. Oxidation of haemoglobin

After erythrocytes were incubated with $O_2^{\bar{\cdot}}$, vitamin C, H_2O_2 or t-BuOOH at 37°C separately, all these four oxidants did not induced haemolysis in our experiment. However, a part of haemoglobin was oxidized in both groups , but the percentages of haemoglobin oxidized by all four oxidants in Se-deficient group were higher than that of se-adequate group (Tab. 3).

Tab. 3 Oxidation of haemoglobin of se-deficient and Se-adequate rats by $O_2^{\bar{\cdot}}$, vit. C, H_2O_2 and t-BuOOH

Group	Percentage of haemoglobin oxidized			
	$O_2^{\bar{\cdot}}$	Vit. C	H_2O_2	t-BuOOH
Se-deficient	9.32±1.35* (12)	11.63±1.04 (8)	22.74±0.67 (12)	8.92±1.59 (12)
Se-adequate	4.20±0.69 (13)	8.77±1.22 (8)	21.10±0.54 (15)	3.67±1.00 (15)
P value	0.01	0.05	0.05	0.01

* Mean+SE(n)

C. Alteration of pattern of erythrocyte membrane protein

Results shown in Figure 1 demonstrated that no significant change in pattern of membrane protein was found, after erythrocytes of Se-deficient and Se-adequate rats were incubated with $O_2^{\bar{\cdot}}$ or vit. C at 37°C. However, when erythrocytes were oxidized with t-BuOOH, percentage of bands 1,2 decreased, and that of band 5 increased in both Se-deficient and Se-adequate groups, but the change of Se-deficient group was more significant than that of Se-adequate group. It means selenium had some protective effect against t-BuOOH oxidation to rat's erythrocyte.

Besides, it is worthy to notice that the consequence of oxidation of these four oxidants to the membrane of erythrocyte was apparently different. The small effect of $O_2^{\bar{\cdot}}$ on the membrane may be due to that it can pass the membrane though band 3 into erythrocyte.

II. Human experiment

A. GSHPx activity in erythrocytes

GSHPx activity in erythrocyte of Se-deficient children (8.61 umoles NADPH oxd/min/g Hb) was significantly lower than that of Se-adequate children (20.22 umoles NADPH oxd/min/g Hb).

B. Oxidation of haemoglobin

Percentages of haemoglobin oxidized by O_2^{-} or vitamin C in Se-deficient children were higher than that of Se-adequate group. But no significant difference between these two groups, when erythrocytes were oxidized with H_2O_2 or t-BuOOH.

C. Alteration of pattern membrane protein of erythrocyte

Results were similar to that of rats.

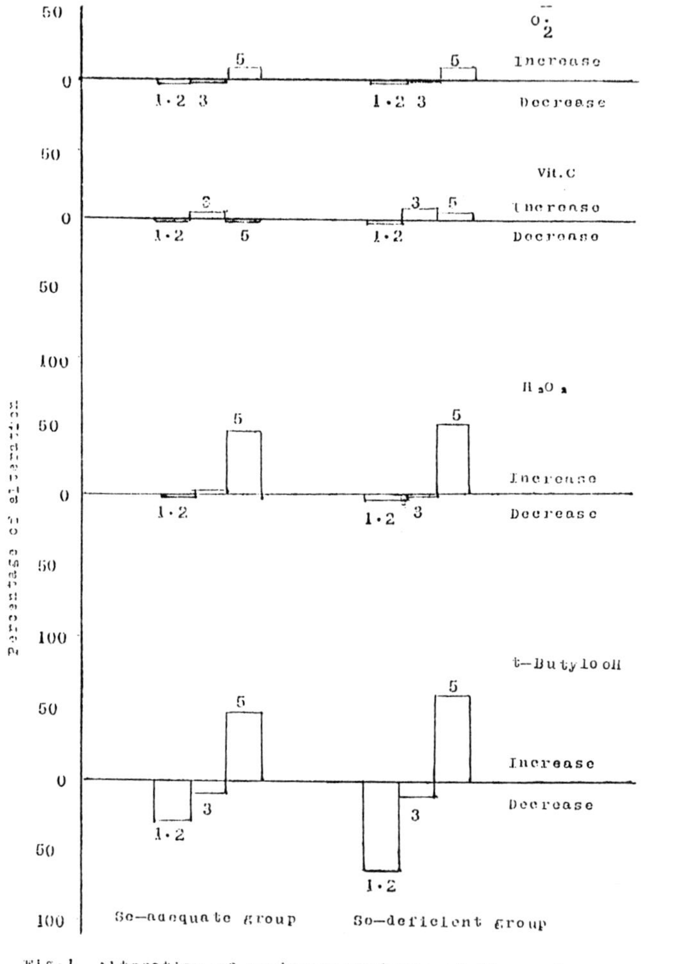

Fig·1 Alteration of membrane proteins of the erythrocytes which were incubated with oxy-radicals at 37℃ in vitro The figures on or under the bars were the numbers of the bands.

Metal Ions in Biology and Medicine. Eds. Ph. Collery, L.A. Poirier, M. Manfait, J.C. Etienne. John Libbey Eurotext, Paris © 1990, pp. 493-496

In vivo and *in vitro* study of the role of selenium in colon carcinogenesis

J.L. Nano, E. Francois, P. Rampal

Laboratoire de Gastroentérologie et de Nutrition, U.E.R. de Médecine, 06034 Nice Cedex, France

Several epidemiological studies have reported an increased incidence of human cancers, and particularly colon cancer, in selenium-deficient regions (Salonen et al., 1984; Scrauzer et al., 1977). Numerous *in vivo* studies have demonstrated significant inhibition of chemical carcinogen-induced tumors in various organs by diet supplementation with sodium selenite (Gregory & Edds,1984; Ip & Daniel,1985; Jacobs 1977; Medina & Shepherd, 1981; Overvad et al., 1985; Thompson & Becci, 1980.)
This study was designed to investigate
1) *In vivo*, prevention of chemically-induced colon cancer in the rat by oral supplementation with Se, and 2) *in vitro*, the effect of the growth of three human colon cancer cell lines.

MATERIALS AND METHODS

1) <u>IN VIVO STUDIES :</u>
Two series of 24 male Sprague Dawley rats seven weeks of age (IFA CREDO, France) were injected weekly with 20 mg DMH (dimethylhidrazine)/Kg body weight for 20 weeks. One group (DMH) was fed with a standard animal food. The drinking water of the other group (DMH+Se) was supplemented with 4ppm Se one week prior to and throughout carcinogen administration. All animals were sacrified by diethyl oxide inhalation the 20th week, and complete autopsies were performed. Tumors were examined after embedding in paraffin and hematoxilin-eosin staining.

2) <u>IN VITRO STUDIES :</u>
a) <u>Cell culture:</u>
HT29 (Fog & Trempe, 1975) and HRT18 (Trompkins et al., 1974) human colon cancer cell lines were kindly provided by J.Marvaldi (ICB, Marseille FRANCE) and Caco2 (Fog et al., 1977) was obtained throught the courtesy of A. Zweibaum (INSERM 178, PARIS, FRANCE). Cells were routinely grown at 37°C in a humidified atmosphere of 5% CO_2 in air in Dulbecco's Modified Eagle Medium (DMEM) (Intermed, FRANCE) supplemented with 10% Fetal Calf Serum (FCS) (Intermed,FRANCE), penicillin (50 U/ml) and streptomycin (50 μg/ml), defined as standard medium. For the experiment, cells were plated at 25 x 10^4 cells/cm2 in 12 well costars plates in standard medium for 20 min with serum-free medium (DMEM/HAM's F12 1/1) supplemented with antibiotics, insulin (5 ug/ml), transferin (5 ug/ml) and EGF (10ng/ml) and incubated in fresh serum-free medium in the absence (control) or presence of 1nM to 1 mM selenium as sodium selenite (Na_2SeO_3) for 72 hours. Cells counts were made using a Coulter Counter after trypsinization.

c) <u>Effect of selenium on the release of ^{51}Cr from preloaded cells :</u>

Growing cells were preincubated for 2 h with serum-free medium containing $Na_2{}^{51}CrO_4$ (40 µCi/ml). The cells were then washed twice with medium and incubated for 72 h with fresh medium in the absence (control) or presence of various selenium concentrations. The medium from each well was then carefully collected, and aliquots were used for determination of the released radioactivity by gamma-emission counting. The cell layer of each well was dissolved in 0.5 ml of O.1 N NaOH, and aliquots of this solution were used for the determination of cellular radioactivity by gamma-emission counting. The fractional release of ^{51}Cr was defined as that portion of total ^{51}Cr appearing in the medium during the second incubation.

RESULTS

1) IN VIVO EXPERIMENTS

a) Weight gain :

The weight gain of rats receiving DMH with and without selenium was 253±23 and 229±19 grams, respectively after 20 weeks (table 1).

b) Tumor incidence and localisation :

Addition of 4 ppm selenium in the drinking water reduced the number of rats developing DMH-induced colon tumors from 22 to 11 in the two groups of 24 animals each. In all, 30 yumors occurred in the DMH-treated animals versus 12 tumors in the DMH + selenium group.. All colon cancers were induced with approximatively the same frequency in the proximal, transverse and distal segments of the colon (table 1)

TABLE 1 : Effects of selenium on body weight, tumor incidence, and location in the presence of DMH.

GROUP	WEIGHT GAIN	COLON TUMORS			TUMOR SITE IN COLON		
	[g]	Total	Average nb/rat	incidence	Proximal	Transverse	Distal
DMH	253±23	30	1.25	22/24	12	8	10
DMH + Se	229±19	12	0.5	11/24	8	0	4

2) IN VITRO EXPERIMENTS :

a) Effect of selenium on cell proliferation :

A selenium concentration close to 10 nM stimulated growth of all cell lines, except TH 29, by 30-60% (Fig 1). At higher levels, selenium inhibited the growth of all the lines studied. DNA synthesis estimated by [3H] thymidine incorporation paralleled the growth curve (results not shown).

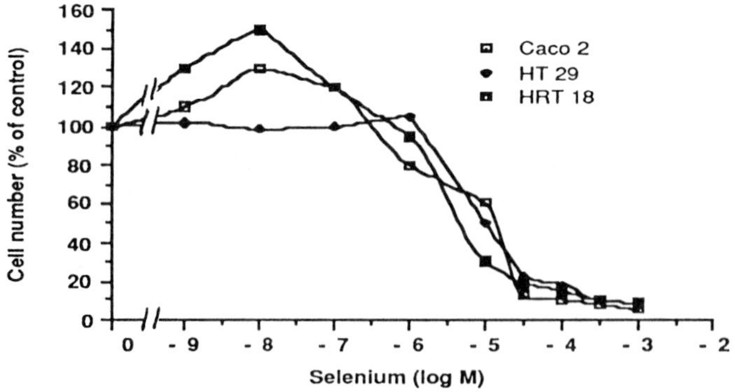

Fig 1 : Effect of selenium on the growth of Caco 2, HT 29 and HRT 18 cell lines. Each value is the mean of triplicate determinations. The SEM was < 10%.

b) Effect of Se on cell viability :

Figure 2 illustrates the viability of Caco 2 cells, measured by the release of radioactive chromium in the presence of increasing concentrations of selenium, and comparison of this parameter with cell proliferation. An approximate shift of a factor of 10 toward higher concentrations occurred between cell viability and growth inhibition. The same situation was true for HT 29 and HRT 18 cells.

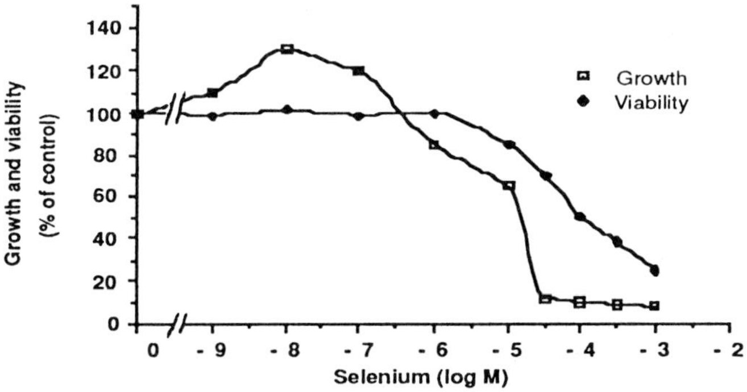

Fig 2 : Effect of selenium on the viability of Caco 2 cell line. Each value is the mean of triplicate determinations. Variations did not exceed 10%.

DISCUSSION :

Following epidemiological reports, numerous experimental studies have suggested that selenium has an inhibitory effect on the proliferation of neoplastic tissues. Most of these studies have been performed in vivo using animals given a carcinogen (Gregory & Edds,1984; Ip & Daniel, 1985; Jacobs, 1977; Medina & Shepherd, 1981; Overvad et al., 1985; Soulier et al., 1981; Thompson & Becci, 1980) and fed a diet with or without selenium supplementation. In our study, we demonstrated that addition of 4 ppm selenium to drinking water reduced the incidence of chemically-induced (DMH) colon cancer in rats by 50%. In order to investigate the mechanism of action of selenium, we evaluated its effects in vitro on human colon cancer cells.

At low concentrations (10 nM), selenium stimulated the proliferation of Caco 2 and HRT 18 cells. Such growth stimulation is in agreement with observations for other cell types (Giasudin & Diplock, 1979 ; McKeehan et al., 1976; Medina & Oborne, 1981; Medina & Oborn, 1984; Webber et al., 1985). This finding is not surprising, because this concentration correponds to that found in fetal calf serum (Price & Gregory, 1982), which is commonly used in cell cultures. Higher selenium concentrations inhibit cell proliferation : the IC_{50} was approximately 10 µM. Human colon cancer cells appear less sensitive to inhibition by selenium than tumoral cells of breast (Medina & Oborn, 1981; Medina & Oborn, 1984), pancreatic (Webber et al.,1985), or cerebral origin (Gruenwedel & Cruikshank, 1979). We also demonstrated that que the rate of ^{75}Se uptake was the same in all of the cell lines studied, although the quantity incorporated differed, and that GSH-Px activity was dependent on the selenium content of the medium (Nano et al. 1989).

A significant shift in cell viability was observed in the presence of selenium, with inhibition of cell proliferation. Indeed, the growth of Caco2 cells incubated in the presence of 10 µM of selenium was inhibited by 50%, whereas 90-100% of the cells were still viable. This observation is similar to that reported by Gruenwedel and Cruikshank (1979), who demonstrated a 50% reduction in DNA, RNA, and protein synthesis in HeLa cells with 10 µM selenium even though their viability was not affected.

In conclusion, the present study shows that selenium prevents chemically-induced colon cancer in vivo and inhibits the growth of human cancer cells in vitro. Inhibition probably occurs at two levels: inhibition of DNA synthesis, which does not affect cell viability, at doses of 0.5-10µM, and cytotoxicity at higher concentrations. These observations merit further exploration concerning the potential usefulness of selenium for colon cancer prevention and treatment.

ACKNOWLEDGMENTS :
The authors are grateful to Nancy Rameau for careful reading of the manuscript. This work was supported by the LABCATAL Laboratories (Montrouge, France).

REFERENCES :

Fogh, J., Fogh, J. M. and Orfeo, T. (1977): One hundred and twenty-seven cultured human tumor cell lines producing tumors in nude mice. J. Natl. Cancer Inst. 59 : 221-226.

Fogh, J. and Trempe, C. (1975): New human tumor cell lines. Plenum Publishing : 115-141.

Giasuddin, A.S.M. and Diplock, A.T. (1979): The influence of vitamin E and selenium on the growth and plasma membrane permeability of mouse fibroblasts in culture. Arch. Biochem. Biophys. 196: 270-280.

Gregory, J. F. III and Edds, G. T. (1984): Effects of dietary selenium on the metabolism of aflatoxin B1 in turkeys. Fd. Chem. Toxic 22/8 : 637-642.

Gruenwedel, D.W. and Cruikshank, M.K. (1979): The influence of sodium selenite on the viability and intracellular synthetic activity (DNA, RNA and protein synthesis) of HeLa S3 cells. Toxicol. Apll. Pharmacol.50: 1-7.

Ip, C. and Daniel, B. F. (1985): Effects of selenium on 7,12 dimethylbenz(a)anthracene-induced mammary carcinogenesis and DNA adduct formation. Cancer Research 45 :61-65.

Jacobs, M. M. (1983): Selenium inhibition of 1,2 Dimethylhydrazine induced colon carcinogenesis Cancer Research 43 : 4458-4465.

McKeehan, W.L., Hamilton, W.G. and Ham, R.G. (1976): Selenium is an essential trace nutrient for growth of WI-38 diploid human fibroblasts. Proc. Natl. Acad. Sci. USA 73: 2023-2027.

Medina, D. and Oborn, C.J. (1981): Differential effects of selenium on the growth of mouse mammary cells in vitro. Cancer Letters 13: 33-344.

Medina, D. and Oborn, C.J. (1984) : Selenium inhibition of DNA synthesis in mouse mammary epithelial cell line YN4. Cancer Letters 44 : 4361-4365.

Medina, D. and Shepherd, F. (1981): Selenium-mediated inhibition of 7,12 dimethylbenz(a)nthracene-induced mouse mammary tumorigenesis. Cancer Letters 2 : 451-455.

Nano, J.L., Czerucka, D., Menguy, F. and Rampal P. (1989): Effect of selenium on the growth of three human colon cancer cell lines. Biol. Trace Element Res. 20: 31-43.

Overlad, K., Thorling, E. B., Bjerring, P. and Ebbesen, P. (1985): Selenium inhibits UV-light-induced skin carcinogenis in hairless mice. Cancer Letters 27 : 163-172.

Price, P.J. and Gregory, E.A. (1982): Relationship between in vitro growth promotion and biophysical and biochemical properties of the serum supplement. In Vitro 18: 576-584.

Salonen, J. T., Alfthan, G., Huttunen, J. K. and Puska, P. (1984): Association between serum selenium and the risk of cancer. Am. J. Epidemiol 120: 342-349.

Schaurzer, G. N., White, D. A., and Schneider, C. J. (1977): Cancer mortality correlation studies-III : statistical associations with dietary selenium intakes. Bioinorg. Chem. 7 : 23-24.

Soullier, B. K., Wilson, P. S. and Nigro, N. D. (1981): Effect of selenium azoxymethane-induced intestinal cancer in rats fed a high fat diet. Cancer Letters 12 : 343-348.

Webber, M.M., Perez-Ripoll, E.A. and James G.T. (1985): Inhibitory effects of selenium on the growth of DU-145 human prostate carcinoma cells in vitro. Biochem. Biophys. Res. Commun. 130 : 603-609.

Metal Ions in Biology and Medicine. Eds. Ph. Collery, L.A. Poirier, M. Manfait, J.C. Etienne. John Libbey Eurotext, Paris © 1990, pp. 497-499

Chemoprevention trial of primary liver cancer with selenium supplementation in Qidong county of China

S.Y. Yu[1], Y.J. Zhu[1], W.G. Li[2], C. Hou[1]

[1] *Cancer Institute, Chinese Academy of Medical Sciences, P.O. Box 2258 Beijing, 100021 P.R. China;*
[2] *Qidong Liver Cancer Institute, Qidong county, Jiangsu province 226200 P.R. China*

Although the potential of Selenium (Se) as a natural anticancer agent is well documented by epidemiologic evidence and laboratory animal model studies, the ultimate measure of the effectiveness of Se in the prevention of cancers will be provided by controlled clinical intervention trials in areas where Se is deficient and there is a prevalence of certain cancers. China offers unique conditions for the studing the cancer-prevention role of selenium. The vast populations of China are generally nonmobile and have their own local dietary pattern. In addition, the characteristic geographical distribution of cancer in China provides an ideal model for elucidating the possible role of Se in human carcinogenesis. Liver cancer is one of the major types of cancer in China. The occurence of liver cancer in China has distinct geographic distribution patterns. For instance, Qidong county, Jiangsu province, has one of highest incidences of this particular cancer. It has a primary liver cancer (PLC) mortality rate of 79.26 for males and 21.57 for females per 100,000 population. It has a population about 1,100,000 living on 1140 square kilometers of land. Even here, the distribution of PLC is uneven throughout the county. Generally, the incidence rates for PLC in the southern part of the county are higher than that in the north. There is a fourfold difference in PLC mortality between the highest and the lowest areas. The characteristic geographical distribution of PLC in Qidong county provides impetus for further investigation of the relationship of Se and liver cancer both in epidemiological studies and in Se-intervention trials using local animals and human population.

GEOGRAPHIC DISTRIBUTION OF PLC AND ITS RELATION TO Se LEVEL IN QIDONG COUNTY.

Epidemiological studies showed an inverse correlation between the age-adjusted PLC incidence by township and the Se content in local grains (rs=-0.63) (Table 1.) or the blood Se levels of residents (rs=-0.72). (Fig.1)

It suggests that low Se status increases risk of PLC and Se supplementation could reduce the PLC incidence.

Table 1. Grain Selenium Content in Regions with Different Age-Adjusted Primary Liver Cancer (PLC) Incidence in Qidong County

PLC incidence per 100,000	No. of townships	Se Content (ppb), M ± S.D.	
		Maize	Barley corn
15-39	8	15.1±4.4 (40)*	42.3±1.9 (40)*
40-49	17	13.5±2.2 (85)	24.7±7.8 (84)
50-59	17	12.7±1.6 (81)	23.5±4.9 (80)
60-	1	9.7±0.083 (5)	18.1±4.54 (5)

* In parenthesis is the number of samples analyzed

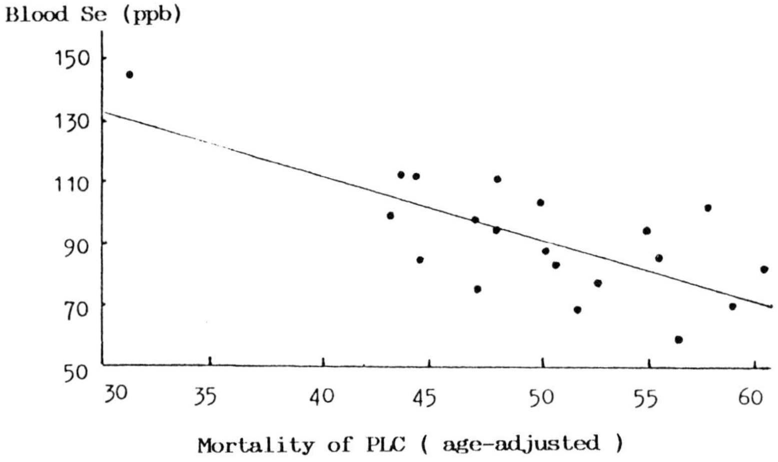

Fig.1. An reverse correlation between the townships' mortality of PLC and blood Se content of its residents.

Se-CHEMOPREVENTION TRIAL IN DUCKS

To test the hypotheses that Se supplementation could counteracts the action of the environmental cancer causing agents or stimuli and reduce the PLC incidence, a four-year field intervention trial on ducks was carried out in an area with low Se level and high PLC incidence both for humans and domestic ducks. Groups of newly-hatched Qidong-ducklings were maintained on local diet for duck-feeding. This type of diet is usually contaminated with aflatoxin (about 15 parts per billion (ppb) on average). Na_2SeO_3 was added to diets at levels of 0, 0.5, 1.0 and 2.0 parts per million (ppm) respectively. Results (table 2.) showed that there is no statistic difference in body weight among the four groups, but the relative liver weight in control group was significantly greater than that of the experimental groups, which was close to the normal(2.0-2.5 per cent). By pathological examination, no tumor was found in any group, but the incidence of precancerous lesions in liver of ducks in control group was much more then that in Se-supplemented groups. It was 25.8 per cent in the control group and 8.3, 5.9, 6.8 per cent in the groups of 0.5, 1.0 and 2.0 ppm Na SeO supplementation, respectively.

Table 2. Preventive Effect of Se Against PLC Incidence of Ducks

No. of ducks	Na_2SeO_3 in diet (ppm)	Blood Se μg/mL	Relative liver weight* %	Preneoplastic incidence** %	Body weight gm
89	0	0.10	3.2	25.8	1200±190
48	0.5	0.18	2.5	8.3	1090±150
51	1.0	0.23	1.9	5.9	1090±180
74	2.0	0.29	2.0	6.8	1030±130

$$^{*}\ \text{Relative liver weight} = \frac{\text{liver weight}}{\text{body weight}} \times 100\%$$

** by histopathological examination

These results greatly enhance the practical significance concerning the reduction of PLC in high risk population by Se supplementation, since the ducks and humans are living in the same environments.

PREVENTION TRIAL OF HUMAN PLC WITH Se SUPPLEMENTATION IN QIDONG COUNTY

An intervention trial of human PLC with supplementation of table salt fortified with 15 ppm Na_2SeO_3 (Se-salt) to the general population of 20847 persons in a township M.Z. at Qidong county has been carring out since 1985. Its neighbourhood townships Y.J. X.Y. J.N. and S.Z. with the similar PLC incidence was served as controls provided with plain table salt. The results are shown in table 3.

Table 3. PLC Incidence of Population with and without Se-supplement

Township	Population (1984)	Age-adjusted incidence of PLC per 100,000					
		Before trial	After trial				
		1972-1984(Avg.)	1985	1986	1987	1988	Avg.
Se-treated							
M.Z.	20847	41.9	38.82	27.99	32.38	27.46	30.41
Non-treated							
Y.J.	28175	41.9	65.04	64.65	54.42	36.51	55.16
X.Y.	30243	43.0	60.23	50.65	48.61	38.75	49.56
J.N.	19186	44.8	58.09	44.19	40.93	52.47	48.92
S.Z.	32020	45.0	51.85	45.99	55.80	53.87	51.88

The average age-adjusted incidence of PLC from 1972 to 1984 was 41.9 per 100,000 for both M.Z. and Y.J. townships. After Se-supplementation in M.Z. the PLC incidence was reduced to 38.8, 28.0, 32.4 and 27.5 per 100,000 for 1985, 1986, 1987 and 1988 respectively. Whereas there was no reduction of PLC incidence in Y.J. as well as in X.Y. J.N. and S.Z. The primary results showed that the measure of Se-salt supplementation is effective, safe and feasible for PLC prevention in large human population.

Metal Ions in Biology and Medicine. Eds. Ph. Collery, L.A. Poirier, M. Manfait, J.C. Etienne. John Libbey Eurotext, Paris © 1990, pp. 500-502

Relationship of serum selenium to tumor activity in acute non lymphocytic leukemia (ANLL) and chronic lymphocytic leukemia (CLL)

Y. Beguin*, V. Bours*, J.M. Delbrouck, G. Robaye, I. Roelandts, G. Fillet*, G. Weber

Departments of Hematology and Experimental Nuclear Physics, University of Liège, IPNE, Sart Tilman B15, 4000 Liège, Belgium*
VB is «Aspirant» and GW «Chercheur Qualifié» of the National Fund for Scientific Research (FNRS), Belgium. Supported in part by grant 4450865 of the Institut Interuniversitaire des Sciences Nucléaires (ISSN), Belgium

Pharmacologic doses of selenium (Se) can reduce tumorigenesis and retard progression of established tumors (Ip, 1986). Several geographic studies have shown an inverse relationship between Se status and cancer mortatility (Clark, 1985) and nested case-control studies have found an inverse relationship between prediagnosis serum Se and the subsequent risk of cancer (Willett et al., 1983). Reduced Se levels have been observed in a variety of cancer patients (Shamberger et al., 1973), but the significance of this finding as a cause or consequence of cancer is poorly understood. We therefore decided to study two groups of leukemic patients to examine the relationships between serum Se, tumor activity, and chemotherapy.

Patients and methods

We studied 70 patients with ANLL who received high-dose induction chemotherapy, consisting of a combination of cytarabine and daunorubicin, mitoxantrone, amsacrine, or etoposide. Complete remission (CR) was defined as a normocellular marrow with < 5% blasts, and a peripheral blood with hemoglobin > 10gr/dl, platelets (PLT) > 100000/µl, and neutrophils > 1500/µl. Partial remission, failure, and death were all classified as failures. Day E refers to the day of evaluation of response to the first course of chemotherapy. Day F refers to the day of final evaluation, i.e. after 2 courses in patients not entering CR after one. Serum samples were obtained before chemotherapy (day 0), and thereafter twice weekly for five weeks. We also studied 47 patients with CLL. Patients were classified into clinical stages 0 through 4 as described (Rai et al., 1975). There were 21 stage 0, 15 stage 1, 12 stage 2, 1 stage 3, and 9 stage 4 patients. For further analysis, stage 3 or 4 patients were assembled in a stages 3-4 group. Eight patients were studied on two occasions, in different stages. Serum selenium was measured by PIXE (Proton-Induced X-Ray Emission) as described previously (Johansson et al., 1970). Normal serum Se levels were determined in 100 healthy subjects. Student's t-test were used to compare two groups, paired Student's t-tests to compare pre- to post-treatment values and ANOVA to compare more than two groups.

Results

Table 1 - Serum Se levels (M±SD) in ANLL patients during induction chemotherapy

Time	Se(μg/ml)		
	All patients (N=70)	CR (N=41)	Failures (N=19)
Pre-treatment	0.082±0.033	0.085±0.032	0.074±0.034
Day 7	0.108±0.037	0.100±0.037	0.131±0.027
Day 21	0.105±0.041	0.105±0.041	0.104±0.041
Day E	0.099±0.032	0.099±0.029	0.100±0.039
Day F	0.100±0.031	0.102±0.029	0.093±0.037
Relapse		0.073±0.027	

ANLL

Pre-treatment serum Se levels were lower in patients than is controls (0.082 ±0.033 μg/ml vs 0.097±0.035 μg/ml, $P<0.01$). Se correlated negatively with the peripheral absolute blast cell count ($R = -0.62$, $P<0.001$) and WBC count ($R = -0.58$, $P<0.001$), bone marrow blast + promyelocyte percentage ($R = -0.41$, $P<0.01$), and serum LDH ($R = -0.53$, $P<0.001$). There was no correlation between serum Se and sex, cytologic sub-typ, infection, age, serum protein, albumin, alkaline phosphatase, fibrinogen, hemoglobin, or platelet count. In multivariate analysis, no other variable added predictive value to the peripheral blast cell count.

On day 7 after chemotherapy (Table 1), Se levels increased significantly ($P<0.01$). The difference between day 7 and day 0 (in μg/ml) correlated with the initial peripheral blast cell count ($R = 0.44$, $P<0.01$). Mean serum Se levels on days 14, 21, 28, E, and F remained comparable to the levels observed in controls. CR was achieved in 49/70 (71%) patients. Failures tended to have higher WBC and blast cell counts and lower pretreatment serum Se levels than CRs (Table 1). However, day 7 Se levels were significantly higher in failures ($P<0.005$). Thereafter Se levels remained stable in CRs while decreasing in failures (Table 1).

CLL

Serum Se levels in CLL patients (0.107±0.034 μg/ml) did not differ from levels in controls (0.097±0.035 μg/ml). Selenium levels were 0.103±0.041 μg/ml, 0.124±0.031μg/ml, 0.105±0.035 μg/ml, and 0.079±0.035 μg/ml, respectively in stages 0, 1, 2, and 3-4 ($P<0.04$). There was a steady decrease in hemoglobin ($P<0.0001$) and PLT count ($P<0.0001$), and an increase in lymphocyte count ($P<0.03$) and LDH ($P<0.04$), from stage 0 to stage 3-4. Serum Se correlated positively with the PLT count ($P<0.01$), and inversely with the lymphocyte count in patients with lymphocytes >20000/mm^3 ($P<0.05$).

Discussion

Reduced Se levels have been reported in patients with solid tumors (Shamberger et al., 1973) but normal levels have been described in a limited number of patients with leukemia (Calautti et al., 1980; McConnell et al., 1975). In patients with ANLL we found decreased Se levels as compared to controls. This could mean that individuals with a low Se status are at increased risk ANLL or that leukemia causes a decline of serum Se. In ANLL patients, nutrition was greatly impaired but Se levels normalized at the time of maximum nutrition impairement.

Several studies reported a trend towards reduced Se levels in more advanced cancer (Dworkin et al., 1988). In ANLL patients, Se levels correlated inversely with measurements of the tumor burden i.e. marrow blast + promyelocyte, LDH, and blood WBC and blast cells. Se increased after chemotherapy, and the increment was proportionnal to the initial burden. Thereafter, Se remained stable in patients who will eventually enter CR. In patients with resistant disease, Se was lower before and increased more after chemotherapy, to fall gradually later on. In CLL patients, serum Se correlated inversely with the lymphocyte count and was significantly reduced in stages 3-4. All these findings are in favor of an inverse relationship between Se levels and disease activity in leukemia. Evidence that some tumors can accumulate Se have been reported (Di Ilio et al., 1987) but this remains to be demonstrated in ANLL and CLL. However, in view of the close relationship between tumor burden and serum Se, as well as the rapid modifications induced by chemotherapy, such a sequestration mechanism appears to be likely. Because of this effect, no conclusion on the relationship between Se status and cancer can be derived from data obtained in patients with active cancer.

References

Calautti, P., Moschini, G., Stievano,B.M., Tomio, L., Calzavara, F. & Perona, G. (1980) : Serum selenium levels in malignant lymphoproliferative diseases. Scand.J.Haematol., 24, 63-66

Clark, L.C. (1985) : The epidemiology of selenium and cancer. Fed.Proc. 44, 2584-2589

Di Ilio C., Del Boccio, G., Casaccia, R., Aceto, A., Di Giacomo, F. & Federici, G. (1987) : Selenium level and glutathione-dependent enzyme activities in normal and neoplastic human lung tissues. Carcinogenesis 8, 281-284

Dworkin, B.M., Rosenthal, W.S., Mittelman, A., Weiss, L., Applebee-Brady, L. & Arlin, Z. (1988) : Selenium status and the polyp-cancer sequence : a colonoscopically controlled study. Am.J.Gastroenterol. 83, 748-751

Ip, C. (1986) : Selenium and experimental cancer. Ann.Clin.Res. 18, 22-29

Johansson, T.B., Akselsson, R., Johansson, S.A.E. (1970) : Elemental trace element analysis at the 10^{-12} level. Nucl.Instrum.Meth. 84, 141-143

McConnell, K.P., Broghamer, W.L., Jr, Blotcky, A.J. & Hurt, O.J. (1975) : Selenium levels in human blood and tissues in health and in disease. J.Nutr. 105, 1026-1031

Shamberger, R.J., Rukovena, E., Longfield, A.K., Tytko, S.A., Deodhar, S. & Willis, C.E. (1973) : Antioxidants and cancer. 1. Selenium in the blood of normals and cancer patients. J.Natl.Cancer Inst. 50, 863-870

Willett, W., Polk, B., Morris, S., Stampfer, M.J., Pressel, S., Rosner, B., Taylor, J.O. Schneider K. & Hames, C.G. (1983) : Prediagnostic serum selenium and risk of cancer. Lancet. ii, 130-134

VII. Titanium, tin, palladium, germanium, lanthanum, Iron

Metal Ions in Biology and Medicine. Eds. Ph. Collery, L.A. Poirier, M. Manfait, J.C. Etienne. John Libbey Eurotext, Paris © 1990, pp. 505-507

Subcellular localization of titanium in the liver and xenografted human tumors after treatment with the antitumor agent titanocene dichloride

P. Köpf-Maier

Institut für Anatomie, Freie Universität Berlin, Königin-Luise-Straße 15, D-1000 Berlin 33, West Germany

Titanocene dichloride, bis(cyclopentadienyl)titanium(IV) dichloride, represents a new-developed organometallic antitumor agent which is cytostatically active against numerous experimental tumors and xenografted human carcinomas deriving from the lung and the gastrointestinal tract (Köpf-Maier, 1989). Investigations using atomic absorption spectroscopy were done to elucidate the organ distribution of titanocene compounds after in vivo application. The experiments revealed that titanium, the central metal atom of titanocene dichloride, or titanium-containing metabolites, respectively, were mainly accumulated within the liver and the intestine of mice, titanium being obviously excreted via the bile and the faeces. Applying this method, titanium-containing species were also detected in transplanted tumors in increasing amounts during several days after substance application (Köpf-Maier et al. 1988).

In the present study, the subcellular localization and distribution of titanium were investigated in the mouse liver and in three human carcinomas xenografted to athymic mice after administration of a single injection of titanocene dichloride in a dose of 80 mg/kg, corresponding to a therapeutic dose of this antitumor agent. The investigations were performed using electron spectroscopic imaging (ESI), a new electronmicroscopical method developed on the basis of electron energy loss spectroscopy (EELS). It allows to analyze and to image the subcellular distribution of light and medium-weighed elements, such as titanium, within ultrathin sections of animal tissues and organs.

In the liver, 24 h after treatment, titanium was mainly accumulated in the cytoplasm of Kupffer and endothelial cells, lining the hepatic sinuses. Additionally, at this time, the element was detectable in smaller amounts in the nucleoli and the euchromatin of hepatocytes, being packaged as granules together with phosphorus and oxygen. One day later, titanium was still present in cytoplasmic inclusions within endothelial and Kupffer cells, whereas in the nucleoli of liver cells only a few deposits of titanium were observable at 48 h. At this time, titanium was mainly accumulated as highly condensed granules in the euchromatin and the perinucleolar heterochromatin of hepatocytes. Moreover, it was incorporated into cytoplasmic lysosomes of liver cells, the so-called peribiliary bodies, which were located near to bile canaliculi and occasionally seen extruding their content into the lumen of bile capillaries.

In the case of the three investigated *human carcinomas* of the lung, the stomach and the colon, all three being sensitive to the therapeutic action of titanocene dichloride, titanium was at first

Fig. 1-3. Sigmoid adenocarcinoma S90 at certain intervals after application of titanocene dichloride (80 mg/kg). The electron spectroscopic images **a** represent survey pictures taken at 150 eV, i.e., just above the phosphorus edge. The images **b-d** document the titanium analyses performed within the marked areas, whereby the pictures **b** and **c** were taken at 465 eV (**b**) and 410 eV (**c**), i.e., just above (**b**) and below (**c**) the titanium absorption edge. The figures **d** illustrate computer-calculated subtractions of the images **b** and **c** and give the net distribution of titanium within the analyzed regions.

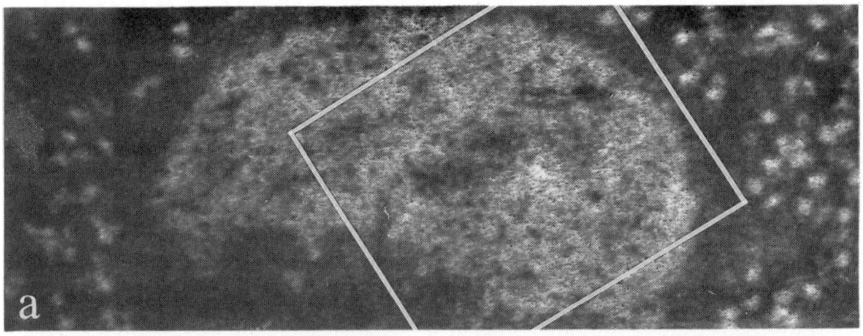

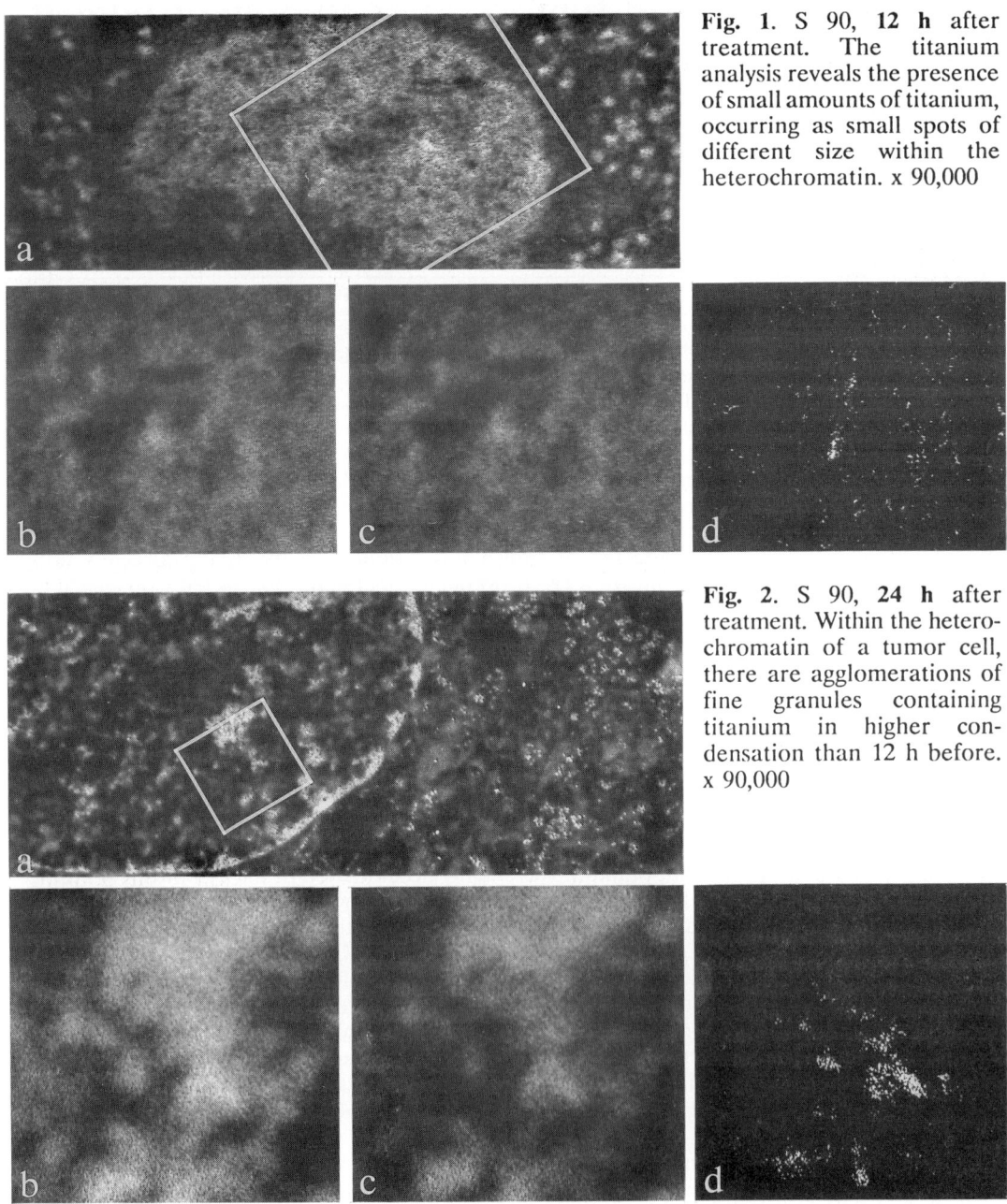

Fig. 1. S 90, **12 h** after treatment. The titanium analysis reveals the presence of small amounts of titanium, occurring as small spots of different size within the heterochromatin. x 90,000

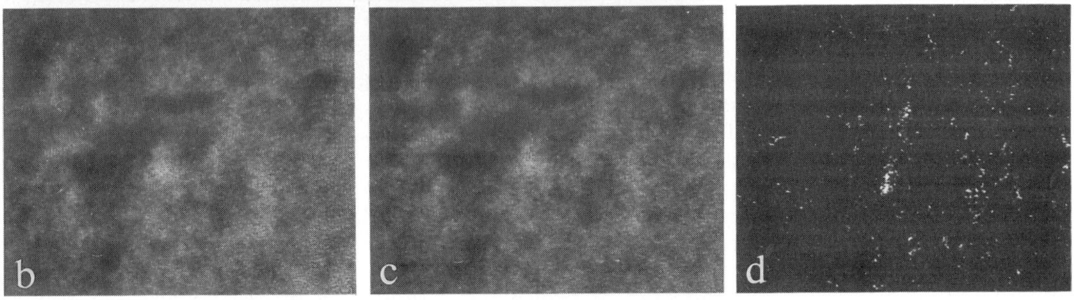

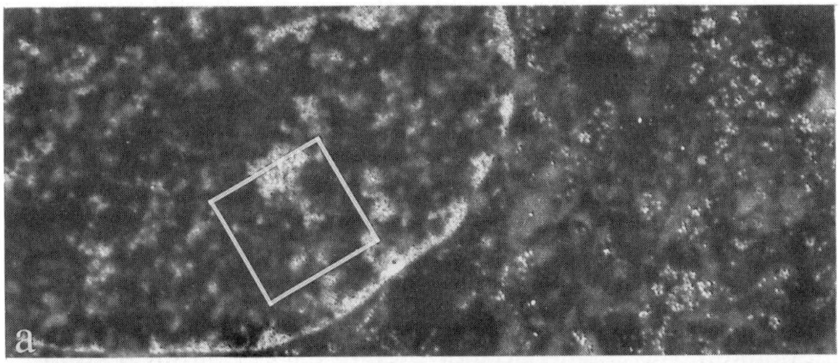

Fig. 2. S 90, **24 h** after treatment. Within the heterochromatin of a tumor cell, there are agglomerations of fine granules containing titanium in higher condensation than 12 h before. x 90,000

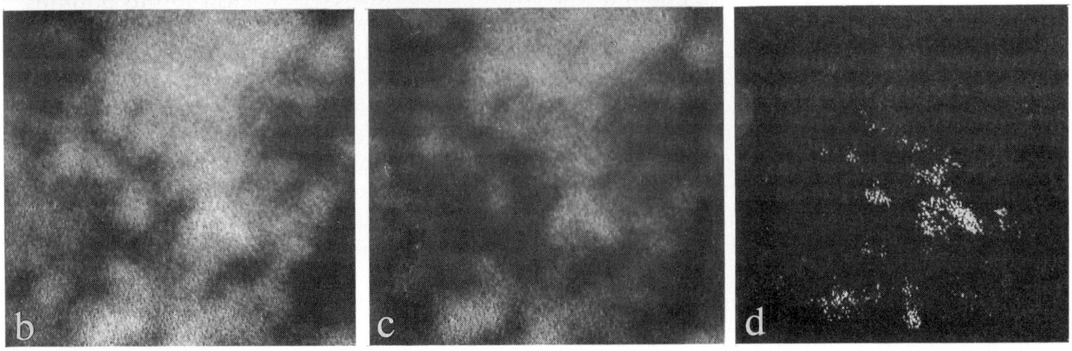

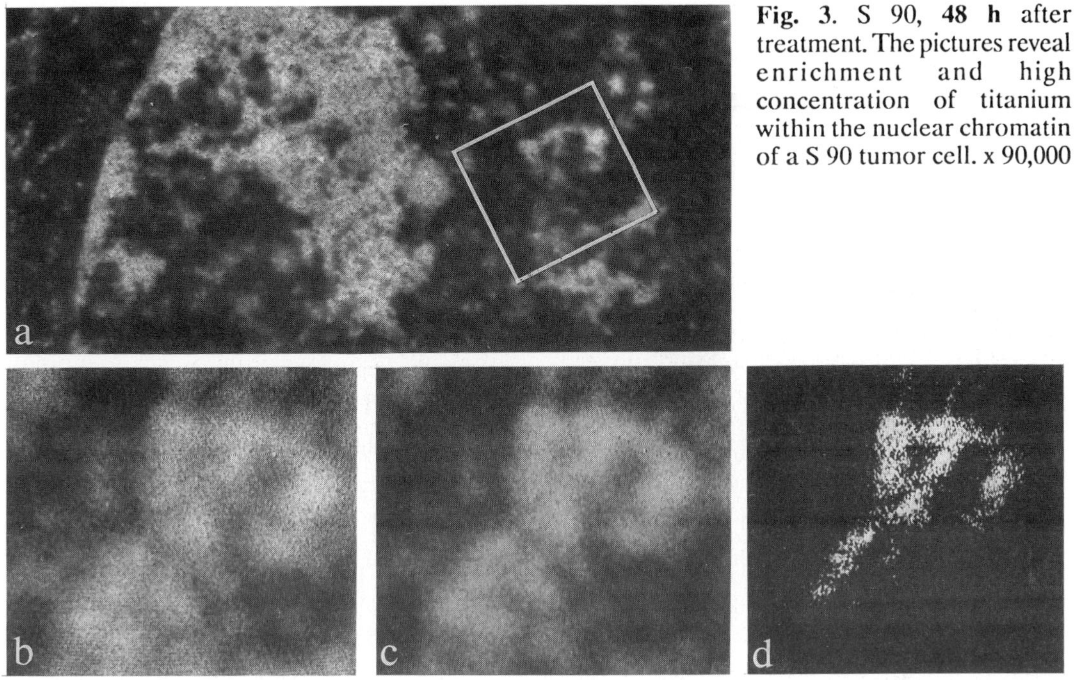

Fig. 3. S 90, **48 h** after treatment. The pictures reveal enrichment and high concentration of titanium within the nuclear chromatin of a S 90 tumor cell. x 90,000

detected within the nucleus and, some hours and days later, it was additionally found incorporated into cytoplasmic lysosomes. As soon as 12 h after application of titanocene dichloride, titanium was traceable in small amounts in the nuclear chromatin of human tumor cells (Fig. 1). During the following intervals, it was then accumulated increasingly in certain areas of the nuclear heterochromatin (Figs. 2, 3), whereby the pattern of enrichment varied from diffuse distribution to spot-like concentration into clear-cut granules. In the case of the L 261 tumor, titanium was additionally detectable in the nucleolus during the first two days after treatment. Generally, maximum concentrations of titanium were attained in the nuclei and nucleoli at 48 h after substance application. Interestingly, phosphorus was always enriched in high concentration in all those regions where titanium was found in traceable amounts. During the following days, titanium was then incorporated in increasing amounts into cytoplasmic lysosomes, which are known to be involved in intracellular digesting processes, titanium being found there in high density at 72 and 96 h. Again, in all lysosomes, where titanium was detected, phosphorus was also enriched in high concentration and similar local distribution as titanium.

These results are in agreement with previous biochemical and cytobiological results pointing to the nucleic acids as probable site of primary intracellular target for titanocene complexes. They confirm the postulated molecular interaction of titanium-containing metabolites deriving from titanocene complexes with nucleic acid molecules, especially with nuclear DNA. Moreover, the findings suggest the formation of aggregates between nucleic acids and titanium-containing metabolites, which are obviously extruded out of the nuclei and incorporated into cytoplasmic lysosomes in order to be degraded and digested there.

References

Köpf-Maier, P. (1989): The antitumor activity of transition and main-group metal cyclopentadienyl complexes. *Progr. Clin. Biochem. Med.* 10. 151-184.

Köpf-Maier, P., Brauchle, U., and Henssler, A. (1988): Organ distribution and pharmacokinetics of titanium after treatment with titanocene dichloride. *Toxicol.* 51, 291-298.

Metal Ions in Biology and Medicine. Eds. Ph. Collery, L.A. Poirier, M. Manfait, J.C. Etienne. John Libbey Eurotext, Paris © 1990, pp. 508-510

Clinical studies with budotitane - A new non-platinum complex for cancer therapy

M.E. Heim[1], H. Bischoff[1], B.K. Keppler[2]

[1] *Onkologisches Zentrum des Klinikums Mannheim, Theodor-Kutzer-Ufer, 6800 Mannheim 1, FRG;*
[2] *Anorganisch-Chemisches Institut der Universität Heidelberg, Im Neuenheimer Feld 270,6900 Heidelberg, FRG*

The inorganic metal complex cis-diamminedichloroplatinum(II), Cisplatin (INN), was introduced into clinical cancer therapy in the late 1970s and, since then, has been of great value in the management of testicular cancer, tumors of the ovaries, cancer of the head and neck, and bladder tumors (Prestayko et al. 1980). Besides cisplatin and its derivatives there are only three metal complexes from the field of inorganic chemistry to have been evaluated in clinical studies for their antitumor activity - the germanium complexes spirogermanium and germanium-132, gallium nitrate, and budotitane (Slavik et al. 1983, Miyao et al. 1980, Hopkins 1981).

Budotitane (INN), Diethoxybis(1-phenylbutane-1,3-dionato)titanium(IV) (Fig. 1), belongs to the class of the bis(β-diketonato) metal complexes. The antitumor activity of this class of complexes was first described in 1982 (Keller et al. 1982). Budotitane has been examined in numerous experimental tumor systems, such as the sarcoma 180 ascitic tumor, the subcutaneously and intramuscularly transplanted sarcoma 180, and the leukemias P 388 and L 1210 (Keppler and Heim 1988, Keppler and Schmähl 1986). Excellent activity was observed in the MAC 15 A colon adenocarcinoma and in the AMMN-induced autochthonous colorectal tumors of the rat. These tumors are of special interest because they resemble the corresponding human tumors both microscopically and macroscopically and thus are highly predictive for the clinical situation (Bischoff et al. 1987). 5-Fluorouracil (5-FU), which is considered the clinical standard therapy in this type of tumors, is at present the only drug to show some activity. Budotitane, however, has been significantly more active in these tumors than 5-FU and cisplatin.

Preclinical toxicity studies with budotitane have shown that the main side-effect is mild hepatotoxicity. Myelosuppression or emesis were not observed (Keppler and Schmähl 1986).

A clinical phase I study with budotitane was started in 1986. The aim of this study was to determine the maximum tolerated dose (MTD), the main toxic side-effects, and the dose-limiting toxicity of this new antineoplastic agent. Patients with histologically confirmed malignant tumors no longer amenable to standard treatments were entered on the study. Other criteria included: age between 18 and 75 years, Karnofsky index of 40 % or better, no major organ dysfunctions, no cytotoxic treatment during the previous 4 weeks, and a life expectancy of at least 3 months.

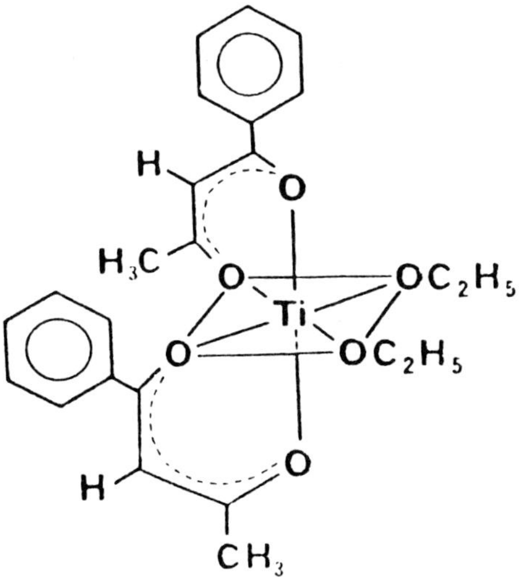

Fig. 1: Budotitane

The first part of the study included the evaluation of budotitane single-dose administration. In the second part, budotitane was given twice a week over four weeks. The MTD was defined as that dose which was not followed by grade 2 toxicity according to WHO criteria or other toxicities of comparable intensity.

Budotitane was applied in the form of a cremophorEL/propylene glycole coprecipitate. The substance was given as a short infusion into a peripheral or a central vein over a period of 30 minutes.

In single-dose administration, budotitane was applied at seven different dose levels: 1, 2, 4, 6, 9, 14, and 21 mg/kg B.W. The first drug-related toxicity was observed at 9 mg/kg B.W., when a patient complained about a reversible impairment of the sense of taste shortly after the infusion. At 14 mg/kg B.W. there was a moderate increase in liver enzymes and LDH. Nephrotoxicity with an increase in urea and creatinine of WHO grade 2 was dose-limiting at the 21 mg/kg B.W. dose level in 2 patients. After some weeks, urea and creatinine values returned to normal.

Nephrotoxicity was accompanied by nausea, weakness, and malaise. Myelosuppression was not observed at any of the tested dose levels.

The MTD of budotitane thus is between 14 and 21 mg/kg B.W. The recommendation for the repeated dose application is 21 mg/kg B.W. (800 mg/m^2 B.S.A.) total dose divided into 8 applications of 100 mg/m^2 B.S.A. twice a week over a 4-week period. So far three patients have been treated at each dose level with 100, 120, and 150 mg/m^2 B.S.A. 8 times in 4 weeks. At these dose levels, there was no hepatotoxicity and no nephrotoxicity, but all patients complained about the described gustatory changes. At a budotitane dose of 14 mg/kg B.W.

the titanium concentration in the serum reached values between 2 and 5 $\mu g/g$, and at a dose of 21 mg/kg B.W. values reached levels between 5 and 12 $\mu g/g$. Titanium could be detected in the serum 7 days and, in one case, even 4 weeks after a single-dose application. In the red blood cells titanium was detected at concentrations of 1 to 2 $\mu g/g$.

The results of the clinical phase I study with budotitane are comparable to what has been found in preclinical experiments. Nephrotoxicity was dose-limiting in the single-dose application, but prophylactic hyperhydration and diuresis may prevent this side-effect. An emetic potential of the drug was not observed except in one patient. In contrast to preclinical toxicity studies, local toxicity could not be observed. The lack of myelosuppression facilitates combination therapy with other anticancer agents. Ageusia could of course not be predicted from animal experiments, and the mechanism of this toxicity is still unclear, but it has turned out to be reversible.

The clinical phase I studies with budotitane have almost been completed by now, and phase II studies are planned. Evidence suggests that the total dose of budotitane can be increased considerably without any severe toxicity. It remains to be seen whether budotitane will develop into what may be an efficient drug against colorectal cancer in humans. The prospects are good.

REFERENCES:

Bischoff, H., Berger, M.R., Keppler, B.K., Schmähl, D. (1987) Efficacy of β-Diketonato Complexes of Titanium, Zirconium, and Hafnium against Autochthonous Colonic Tumors in Rats. J. Cancer Res. Clin. Oncol. 113, 446-450

Hopkins, S.J. (1981) Gallium Nitrate. Drugs of the Future 4, 228-230

Keller, H.J., Keppler, B.K., Schmähl, D. (1982) Antitumor Activity of cis-Dihalogenobis(1-phenyl-1,3-dionato)titanium(IV) against Walker 256 Carcinosarcoma. Arzneim.-Forsch./Drug Res. 32, 806-807

Keppler, B.K., Heim, M.E. (1988) Antitumor-active Bis(β-diketonato) Metal Complexes: Budotitane - A New Anticancer Agent. Drugs of the Future 13, 637-652

Keppler, B.K., Schmähl, D. (1986) Preclinical Evaluation of Dichlorobis(1-phenylbutane-1,3-dionato)titanium(IV) and Budotitane. Arzneim.-Forsch./Drug Res. 36, 1822-1828

Miyao, K., Onishi, T., Asai, K., Tomizawa, S., Suzuki, F. (1980) Toxicology and Phase I Studies on a Novel Organogermanium Compound, Ge-132. In: Nelson, Grassi; Current Chemotherapy and Infectious Disease, 2, 1527-1529

Prestayko, A.W., Crooke, S.T., Carter, S.K. (eds.) (1980) Cisplatin - Current Status and New Developments. Academic Press, New York

Slavik, M., Blanc, O., Davis, J. (1983) Spirogermanium: A New Investigational Drug of Novel Structure and Lack of Bone Marrow Toxicity. Invest. New Drugs 1, 225-234

Metal Ions in Biology and Medicine. Eds. Ph. Collery, L.A. Poirier, M. Manfait, J.C. Etienne. John Libbey Eurotext, Paris © 1990, pp. 511-513

A comparative study between inorganic and organometallic tin dithiocarbamates as cytostatic agents

V. Scarcia, A. Furlani, A. Papaioannou, C. Preti*, V. Cherchi**, G. Fracasso**, D. Marton**, L. Sindellari**

*Institute of Pharmacology and Pharmacognosy, Faculty of Pharmacy, University of Trieste, 34100 Trieste, Italy : * Department of Chemistry, University of Modena, 41100 Modena, Italy and ** Department of Inorganic, Metallorganic and Analytical Chemistry, University of Padova, 35100 Padova, Italy*

INTRODUCTION

The widespread success of platinum compounds in the clinical treatment of certain types of cancer placed the coordination chemistry on the front line in the fight against cancer. Among complexes containing metals other than platinum as antitumour agents, tin-derivatives appeared to play an important role. The interest toward this element was further confirmed by the very high affinity shown by tin for neoplastic tissues and by its relatively low toxicity in mammals (Crowe,1988; Saxena & Huber,1989). Many organotin compounds have been studied for their biological activity, whereas very little is known about the cytostatic effect of tin-inorganic derivatives.
On the basis of these observations, we thought it of interest to extent our research on some inorganic as well organometallic complexes in which tin was present in the two oxidation states (II, IV). The tested compounds were the following:

Y_2SnX_2	Y = DEdtc	X = Cl	$(DEdtc)_2SnCl_2$	DEdtc=diethyldithio-carbamate
	Y = X = DEdtc		$Sn(DEdtc)_4$	
	Y = CH_3	X = DEdtc	$(CH_3)_2Sn(DEdtc)_2$	
	Y = $n\text{-}C_4H_9$	X = DEdtc	$(n\text{-}C_4H_9)_2Sn(DEdtc)_2$	
	Y = DMdtc	X = Cl	$(DMdtc)_2SnCl_2$	DMdtc=dimethyldithio-carbamate
	Y = X = DMdtc		$Sn(DMdtc)_4$	

SnL_2

L =
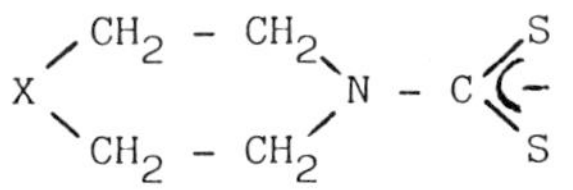

X = CH_2	(Pipdtc)
X = S	(Timdtc)
X = O	(Morphdtc)
X = $H_3C.N$	(CH_3Pzdtc)

MATERIALS AND METHODS

Chemical

Tin(IV) complexes were prepared by stirring either $SnCl_4$ or R_2SnCl_2 (R=Me, n-Bu) with sodium diethyldithiocarbamate, at molar ratio 1:4 or 1:2 in benzene at room temperature, for several hours, with minor medifications of the reported methods (Bonati *et al.*, 1968). The complexes were identified by microanalysis. The i.r.

spectra were recorded in the 4000-400 cm $^{-1}$ range with a Perkin-Elmer 580 spectrophotometer (nujol mull) using KBr discs as well as ^{1}H nmr, ^{13}C nmr and ^{119}Sn nmr spectra were obtained with a Jeol FX 90 Q spectrometer in $CDCl_3$ at room temperature (Sharma et al., 1981).
The synthesis and the i.r. characterization of the complexes of general formula SnL_2 were previously reported (Preti et al., 1980).

In vitro cytostatic activity evaluation
Two established cell lines, an epidermoid human carcinoma (KB) and a murine leukemia (L1210) were used for cytostatic effect evaluation. KB cells were maintained and tested in Eagle's Minimum Essential Medium(MEM) as previously described (Cherchi et al., 1989). The doubling time was ca. 24 h. L1210 cells were maintained as suspension cultures in RPMI 1640 medium supplemented by 10% fetal calf serum. Under these conditions the doubling time was ca. 12 h. Each compound at several concentrations was added to L1210 cells ($2x10^4$ cells/ml) in exponential phase of growth. Cell number was determined at 24 h, 48 h and 72 h after drug administration. In all experiments the compounds were dissolved or suspended, immediately before use,in sterile acetone. The cytostatic activity was evaluated on the basis of cell growth inhibition in the treated cultures with respect to the controls. The ID_{50} values were estimated from dose-response curves and represent the molar drug concentration required to inhibit the replication of neoplastic cells by 50%.

RESULTS AND DISCUSSION

The complexes have been obtained at elevate grade of purity with good yield (about 80%).
^{13}C nmr and ^{119}Sn nmr data were in accordance with ^{1}H nmr (Sharma et al., 1981).
^{13}C nmr spectra of tin(IV) diethyldithiocarbamate derivative showed that N-bonded ethyl groups are equivalent. They showed single signals for all compounds due to chemical equivalence. In the octahedral $Sn(DEdtc)_4$ the ligand exchange between the apical monodentate and the equatorial bidentate dithiocarbamates is too fast, at room temperature, to allow the two types of signals to be detected by nmr.
An only ^{119}Sn resonance sharp signal indicated that just one structure was present in solution, probable hexacoordinated, as in the solid state (Maeda et al., 1966).

The in vitro cytostatic effect of these compounds has been evaluated as a preliminary screening for biological activity (Table 1). Almost all complexes showed significant activity on both cell lines with ID_{50}-values lower than those displayed by cisplatin, determined by us under the same experimental conditions ($0.3x10^{-6}$ M and $0.6x10^{-6}$ M against KB and L1210 cells respectively).
From these results some observations can be inferred. Tin oxidation state appears to influence the cytostatic activity, since tin(IV) complexes showed a more pronounced effect. The organotin(IV) complexes displayed very low ID_{50}-values, that were however similar to those exhibited by the inorganic tin(IV) complexes. This feature was rather unusual since a noteworthy in vitro cytostatic effect for tin inorganic derivatives has been observed rarely (Crowe, 1988). Nevertheless among the inorganic tin(IV) species, it is possible to note a different behaviour, depending on the stoichiometry of the complexes, the 1:4 displaying an activity about three or four times higher than the correspnding 1:2 species.
From these results it seems to arise a possible biological interest toward tin-dithiocarbamates complexes.

Further studies will be carried out for a more complete evaluation of the biological role of these compounds.

Table 1. In vitro cytostatic activity of Sn-complexes.

Compounds	Molar ID_{50} - values	
	KB cells	L1210 cells
$(DEdtc)_2SnCl_2$	0.13×10^{-6}	0.08×10^{-6}
$Sn(DEdtc)_4$	0.05×10^{-6}	0.04×10^{-6}
$(CH_3)_2Sn(DEdtc)_2$	0.07×10^{-6}	0.13×10^{-6}
$(n\text{-}C_4H_9)_2Sn(DEdtc)_2$	0.08×10^{-6}	0.10×10^{-6}
$(DMdtc)_2SnCl_2$	0.14×10^{-6}	0.71×10^{-6}
$Sn(DMdtc)_4$	0.03×10^{-6}	0.11×10^{-6}
$Sn(Pipdtc)_2$	3.34×10^{-6}	1.81×10^{-6}
$Sn(Timdtc)_2$	0.74×10^{-6}	0.73×10^{-6}
$Sn(Morphdtc)_2$	0.43×10^{-6}	0.60×10^{-6}
$Sn(CH_3Pzdtc)_2$	0.13×10^{-6}	0.35×10^{-6}

REFERENCES

Bonati, F., Minghetti, G., and Cenini, S. (1968): Tin(IV) dithiocarbamates and related compounds. Inorg.Chim.Acta 26: 375-378.

Cherchi, V., Faraglia, G., Sindellari, L., Sitran, S., Furlani, A., and Scarcia, V. (1989): Synthesis and in vitro cytostatic activity of platinum(II) nitrate complexes with propan-1-amine. Inorg.Chim.Acta 155: 267-272.

Crowe, A.J. (1988): The antitumour activity of tin compounds. In: Metal-bases anti-tumour drugs, ed. M. Gielen, pp. 103-149. London: Freund Publishing House.

Maeda, Y., Dillard, C.R., and Okawara, R. (1966): Dimethyl diacetate and its configuration. Inorg.Nucl.Chem.Letters 2: 197-201.

Preti, C., Tosi, G., and Zannini, P. (1980): Investigations of Chromium(III), Manganese(III), Tin(II) and Lead(II) dithiocarbamate complexes. J.Mol.Struct. 65: 283-292.

Saxena, A.K. and Huber, F. (1989): Organotin compounds and cancer chemotherapy. Coordination Chem.Rev. 95: 109-123.

Sharma, C.P., Kumar, N., Khandpal, M.C., Chandra, S., and Bhide, V.G. (1981): Studies on the preparation and characterization of bis-dithiocarbamato derivatives of di-n-butyl- and di-n-hexyl Sn(IV). J.Inorg.Nucl.Chem. 43: 923-930.

Metal Ions in Biology and Medicine. Eds. Ph. Collery, L.A. Poirier, M. Manfait, J.C. Etienne. John Libbey Eurotext, Paris © 1990, pp. 514-516

New concepts on the design of palladium (II) complexes as antitumor agents

A. Afcharian[a], J.L. Butour[b], P. Castan[a], S. Wimmer[a]

[a] *Laboratoire de Chimie Inorganique, Université Paul Sabatier, 205 route de Narbonne, Toulouse, France;* [b] *Laboratoire de Pharmacologie et Toxicologie Fondamentales du CNRS, 205 route de Narbonne, Toulouse, France*

The search for antitumor Pt(II) drugs with less side effects and nephrotoxicity compared to cis-Platin has been extensive over the last several years, leading to the design of "second generation" drugs such as "Paraplatin" (Bristol-Myers).

By contrast, the development of palladium anticancer drugs has been disappointing. These compounds have been unsuccessfull probably because their design has been based on the "window of reactivity" usually employed for potential platinum antitumor drugs.

The considerably higher reactivity of palladium complexes compared to the platinum ones implies a suitable choice of the leaving group. If this group is reasonably "non labile", the drug can reach the molecular target (presumably the DNA), if the mechanism of action is the same as for platinum compouds.

We have recently presented (Castan,1990), some results supporting this hypothesis, since we have shown that while Pd(Dach)Cl_2 is inactive, on replacing the chloro groups by a strong chelating ligand, 3-methyl-orotate, we have prepared the complex Pd(Dach)(3-Methyl-Orot). This complex displays a very significant activity. The survival time of mice inoculated with Sarcoma 180 is almost identical, after treatment with Pd(Dach)(3-Methyl-Orot), T/C = 267%, to what it is after treatment with cis-DDP, T/C = 277%. By contrast, the homologous Pt(Dach)(3-Methyl-Orot) is inactive. With platinum as central atom, the two ligands are too strongly bonded to Pt for the complex to be active.

From this preliminary work it appears that:

(i) Obtaining antitumor active palladium complexes is possible.

(ii) Structure-activity studies based on platinum complexes cannot be used as a guide for designing new palladium(II) drugs.

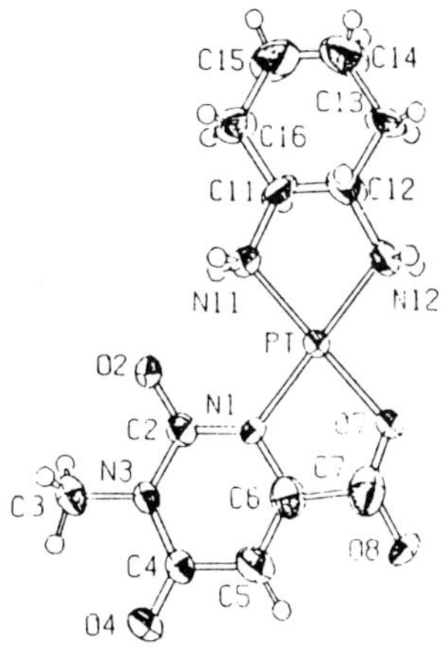

Fig. 1. X-ray structure determination of Pt(Dach)(3-Methyl-Orot)

Beyond platinum and palladium complexes chelating covalently to DNA, aromatic ring stacking plays a major role in the intercalative binding to DNA. Drugs with planar aromatic rings are therefore of prime importance for their antitumor or carcinogenic activities. Platinum(II) complexes containing aromatic ligands coplanar with the metal coordination sphere bind intercalatively to double-stranded DNA, while non coplanar and non aromatic complexes do not.

Some platinum complexes containing heterocyclic ligands such as 2,2'-bipyridine, 1,10-phenanthroline and 2,2',2"-terpyridine have been shown to bind non-covalently to DNA through intercalation by aromatic ring stacking with the base moiety (Howe-Grant,1976).
As part of our study to synthesize palladium complexes with good antitumor activity we have prepared some palladium compounds with the same ligands and compared their ability to intercalate with platinum complexes.

DNA binding constants of the various platinum and palladium complexes were determined by measuring their ability to compete with ethidium bromide (EB) an intercalating agent (Le Pecq, 1967). The experimental conditions have been already reported (Butour, 1985). Scatchard's plots were obtained with Salmon sperm DNA (10 μg/mL) in 0.4 M KNO_3, in the presence of increasing molar Pt or Pd/nucleotide ratios (r_i).
Figure 2 represents typical curves for the competition binding of four different drugs. They allow to determine the apparent binding constants: K' for EB and K_x for Pt or Pd complexes.

The four compounds studied exhibit the same behaviour towards DNA. They intercalate between the base pairs. Nevertheless, the palladium derivatives have a better affinity for the macromolecule than the corresponding platinum derivatives (Table 1). As the ligands bound to the metal are identical, these results suggest that the better affinity of the palladium versus platinum complexes is due, in part, to a different electronic configuration.

	$K_x(M^{-1})$
$(Pt(o\text{-}phen)en)Cl_2$	$0.15x10^{-4}$
$(Pt(4,4'\text{-}DMB)en)Cl_2$	$0.175x10^{-4}$
$(Pd(o\text{-}phen)en)Cl_2$	$0.80x10^{-4}$
$(Pd(4,4'\text{-}DMB)en)Cl_2$	$0.52x10^{-4}$

Table 1

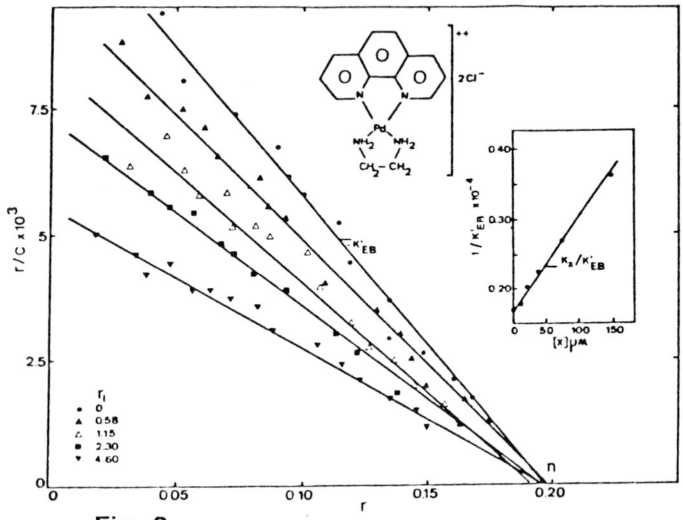

Fig. 2

Abbreviations : en = 1,2-diaminoethane; Dach = 1,2-diaminocyclohexane; 3-Methyl-Orot = 3-methyl-orotate; o-phen = 1,10-phenanthroline; DMB = dimethyl-bipyridine.

References

Butour J.L., Mazard A.M., and Macquet J.P., (1985): Kinetic of the reaction of cis-platinum compounds with DNA in vitro. In Biochem. Biophys. Res. Com. **133**, pp. 347-353.

Castan P., Colacio-Rodriguez E., Beauchamp A.L., Cros S., and Wimmer S., (1990): Platinum and Palladium complexes of 3-methyl orotic acid. A route towards palladium complexes with a good antitumor activity. In Inorg. Biochem. (in the press).

Howe-Grant M., Wu K.C., Bauer W.R., and Lippard S.J., (1976): Binding of platinum and palladium metallointercalation reagents and antitumor drugs to closed and open DNAs. In Biochemistry **15**, pp. 4339-4346.

Le Pecq J.B., and Paoletti C., (1976): A fluorescent complex between ethidium bromide and nucleic acids. In J. Mol. Biol. **27**, pp. 87-106.

Metal Ions in Biology and Medicine. Eds. Ph. Collery, L.A. Poirier, M. Manfait, J.C. Etienne. John Libbey Eurotext, Paris © 1990, pp. 517-519

Inhibitory effect of Germanium (IV) sodium ascorbate on the *in vitro* reverse transcriptase of Human immunodeficiency virus

A.M. Badawi

Department of Application, Petroleum Research Institute, Nasr City, Cairo, Egypt

Many factors involved in host resistance to aquired immune deficiency syndrome (AIDS) might be significantly dependent upon the availability of ascorbate. Ascorbic acid was investigated for its inactivation of bacterial viruses (Murata, 1974; Murata et al., 1975). The mechanism of in vitro viral inactivation is still unclear. It was postulated that free-radical intermediates produced during ascorbate oxidation are the active agents in viral inactivation (Murata, 1975).

Carboxyethyl germanium sesquioxide (Ge-132) was recently disclosed as an oral interferon inducer with antiviral activity (Ishida et al., 1984).

To investigate the possibility that the antiviral activity of ascorbic acid can be potentiated with germanium, it occurred to us to react ascorbic acid with sodium germanate to give the corresponding germanium (IV) sodium ascorbate derivative (GeNaA), $(C_6H_7O_6)_2$ GeO. $2(C_6H_7NaO_6).4\ H_2O$

The resulting compound was submitted to U.S. Army Medical Research Institute of Infectious Diseases, Fort Detrick, Frederick,Maryland,U.S.A., for reverse transcriptase (RT) assay for human immunodeficiency virus (HIV). GeNaA was found to significanlty inhibit in vitro RT suggesting AIDS-retarding activity.

MATERIALS AND METHODS :

Chemicals and radioisotopes. The following compounds were obtained from the indicated sources: leupeptin, Boehringer Mannheim Chemicals; poly (rA).p(dT)$_{12-18}$, 2'-deoxythmidine 5'-triphosphate sodium salt (dTTP), and 2',3'-dideoxythymidine 5'=triphosphate sodium salt (ddTTP), Pharmacia; and {methyl, 1',2'-^{3}H} deoxythymidine 5-triphosphate tetrasodium salt in Tricine buffer (100Ci/mmol), NEN Research Products. Details of synthesising GeNaA will be described elsewhere.

Assay buffer. The assay buffer for Human Immunodeficiency Virus (HIV) RT contians 100 mM Tris-HCl (pH 7.9), 140 mM KCl, 15.5 mM $MgC1_2$, 10 mM β-mercaptoethanol, and 0.625 mM ethyleneglycol-bis (β-aminoethyl ether) N,N,N', N'-tetraacetic acid (EGTA).

Preparation of HIV RT. H9 cells infected with the III-B strain of HIV were grown for at least three days in media. The cells were removed by low speed centrifugation and virus pelleted from the clarified supernatant by centrifugation at 16,000 x g for 2 hours at 4°C. The virus pellet was resuspended in 1/10th of the initial supernatant volume in 10 mM Tris-HCl (pH 7.8), 1 mM ethylenediaminetetraacetic acid (EDTA), 200 mM KC1, 10 mM β-mercaptoethanol, 3 micromolar leupeptin, and 0.5% Triton X-100, dispersed by vortexing, and kept at room temperature for 1 hour to ensure complete disruption of the virus coat. Glycerol is added to the sample to make it 30% in that reagent and the sample stored at -20°C.

Preparation of assay reagents. The template-primer is prepared by dissolving 1 mg of poly $(rA)p(dT)_{12-18}$ in 1 ml of H_2O, heating at 42°C for 5 minutes, and cooling slowly to room temperature. The solution is divided into small aliquots and stored at -20°C.

HIV RT enzyme assays for buffer soluble compound. Ten microliters of a solution containing the test compound 0.5 mg/ml in buffer is added to one well of a 96-well flat bottom microtest plate with lid, followed by 10 microliters of enzyme solution, and finally 30 microliters of a mixture containing the assay buffer, template-primer, and substrate. The final assay for HIV RT contains 59.4 mM Tris-HC1 (pH 7.9), 0.07% Triton X-100, 109 mM KC1, 9.0 mM $MgC1_2$, 7.2 mM β-mercaptoethanol, 0.36 mM EGTA, 0.14 mM EDTA, 0.42 micromolar leupeptin, 6% glycerol, 1 microgram of poly $(rA).p(dT)_{12-18,}$ and 60 micromolar dTTP (4.16 mCi^3H/micromole dTTP). The HIV assay is run for 1 hour at 37°C. It is terminated by the addition of 65 microliters of 0.2 M sodium pyrophosphate. One hundred microliters from each well is spotted on a Whatman GF/A glass microfibre filter and processed. For the HIV assay results are expressed as picocuries of dTTP incorporated per hour. Each sample is assayed in quadruplicate and the average, standard deviation (S.D.), standard error (S.E.), and coefficient of variation (C.V.) are calculated.

HIV RT enzyme assay for DMSO soluble compound. Twenty five microliters of a solution containing the test compound 5.0 mg/ml in DMSO is added to one well of a 96-well flat bottom microtest plate with lid, followed by 10 microliters of enzyme solution, and finally 15 microliters of a mixture containing the assay buffer, template-primer, and substrate. The final assay for HIV RT contains 57.4 mM Tris-HCl (pH 7.9), 0.07% Triton X-100, 2% DMSO, 106.4 mM KC1, 8.68 mM $MgCl_2$, 7.0 mM β-mercaptoethanol, 0.35 mM EGTA, 0.14 mM EDTA, 0.42 micromolar leupeptin, 6% glycerol, 1 microgram of poly $(rA).p(dT)_{12-18,}$ and 60 micromoles dTTP (4.16 mCi^3H/micromole dTTP). The HIV assay is run for 1 hour at 37°C. It is terminated by the addition of 65 microliters of 0.2 M sodium pyrophosphate. One hundred microliters from each well is spotted on a Whatman GF/A glass microfibre filter and processed .

Experimental design. Each screen is set up in the same format. The first four samples are blank filters which have been through the wash procedure. Two sets of controls (OE) which are assays without any enzyme solution are run; one is set at the beginning and one at the end of each screen. Two sets of assays in which no inhibitor was present (OI) are always included. Positive controls with a known RT inhibitor, ddTTP or phosphonoformic acid (PFA), are also included. (If the compound is solubilized in DMSO then the OE, OI, and positive drug controls also contain 2% DMSO). In addition, 5 microliter aliquots of the assay solution are also counted; this could be used to convert the data to picomoles of dTTP incorporated. Screens are done so that the concentration that inhibits the enzyme by 50% (ID_{50}) can be calculated.

The ID_{50} (concentration of GeNaA to which the enzyme is inhibited) was calculated and found as follows:

Exp. No.	Solvent	ID_{50} ug/ml
1	Buffer	99
2	Buffer	50
3	DMSO	20

DISCUSSION

The HIV-RT result summery shows that GeNaA inhibited HIV-RT by 50% at approximately 99 ug/ml, 50 ug/ml and 20 ug/ml during three different experiments. The first ID_{50} value 99 ug/ml appears to be too high and may be dropped, thus the ID_{50} will be average of the other two values: 50 and 20 ug/ml. The investigation presented her demonstrates that GeNaA has a significant inhibitory effect on HIV replication in vitro and might be related to unclear mechanism of action targeted at RT.. Although organic germanium's properties of oxygen modulation are not yet precisely elucidated (Goodman, 1988), the possible oxygenating effect of germanium and(or) dehydroascorbic radical might be responsible for the inhibition of HIV RT enzyme.

ACKNOWLEDGEMENT

The author wish to thank Dr. Michael A. Chirigos, Deputy for Science, U.S. Army Medical Research Institute of Infectious Diseases, for HIV RT assay and for his valuable comment .

REFERENCES

Goodman, S. (1988): Organic germanium oxygen enricher and antioxidant. In Germanium, the Health and Life Enhancer, pp. 32-38. Wellingborough:Thorsons.

Ishida, N., Satoh, H., Suzuki, F. and Miyao, K. (1984): Organo germanium induction of interferon production. United States Patent, No. 4, 473, 581.

Murata, A. (1974): Inactivation of viruses by vitamin C, Mechanism of inactivation. Vitamines 48: 507-511.

Murata, A., Oyadomari, R, Ohashi, T. and Kitagawa, K. (1975): Virus inactivating effect of ascorbic, IV. Mechanism of inactivation of bacteriophage δA containing single-strand DNA by ascorbic acid. J. Nutr. Sci. Vitaminol.21:261-269.

Metal Ions in Biology and Medicine. Eds. Ph. Collery, L.A. Poirier, M. Manfait, J.C. Etienne. John Libbey Eurotext, Paris © 1990, pp. 520-522

Lanthanon Binding to intestinal brush-border membranes

D. Bingham, M. Dobrota

Robens Institute of Health and Safety, University of Surrey, Guildford, Surrey, GU2 5XH, UK

INTRODUCTION

Lanthanide salts are very poorly absorbed from the G.I. tract (Durbin et al, 1955). However, they can seriously affect a number of essential physiological processes most probably because they act as Ca analogues (dos Remedios, 1981; Arvela, 1979). We have examined the binding of lanthanides to cell membranes by employing intestinal brush-border vesicle preparations.

METHODS

Brush-border membrane vesicles were prepared by the method of Simpson and Peters (1984) from the proximal 50% of rabbit intestine. The vesicles were suspended in a buffer containing 0.1M NaCl, 0.1M mannitol, 0.1mM $MgSO_4$ and 20mM HEPES pH 7.4 at a concentration of 2-5mg protein/ml and stored at -70°C. The brush-border vesicle extract was at least 4 fold enriched in the brush-border marker enzyme, maltase, compared to the homogenate.

In vesicle binding experiments, 5µl of vesicles were incubated with 50µl of incubation medium consisting of 0.1M NaCl, 0.1M mannitol and 20mM PIPES pH 6.1. $^{152}EuCl_3$ (Amersham), $EuCl_3$, $Al(NO_3)_3$ and $CaCl_2$ were added to the incubation medium at appropriate concentrations. After incubation at 37°C for 1 minute the vesicles were separated from the medium by filtration through 0.22µM nitrocellulose filters (Millipole GSWP 02500). The filters were immediately washed with 2x5ml ice-cold 0.15M NaCl, 1mM $EuCl_3$, 20mM PIPES pH6.1 (unless otherwise stated).

This procedure was repeated for "blank" samples of 55µl of incubation medium (no vesicles). The blank values of Eu retention on filters were subtracted from results with vesicles.

Samples of vesicles, filtered onto PTFE filters (Millipore, GVWP 02500) were prepared for electron microscopy in order to examine their structure by T.E.M. and also to analyse for elemental composition by EDAX.

RESULTS

TABLE 1. Removal of Eu^{152} from vesicles by different wash solutions

WASH SOLUTION	pH	Eu^{152} BINDING nmol/mg protein	% REMOVAL
1mM $EuCl_3$, 0.15M NaCl	1.8	32.8	64
# 1mM $EuCl_3$, 0.15M NaCl	5.4	70.8	23
0.15M NaCl	5.9	89.2	3
# 0.15M NaCl	6.4	104.7	13 increase
1mM $EuCl_3$, 0.15M NaCl Buffered	6.7	81.1	12
# 1mM $EuCl_3$, 0.15M NaCl Buffered	7.0	91.6	0
1mM DTPA, 0.15M NaCl	6.6	66.2	28
# 1mM DTPA, 0.15M NaCl	6.7	55.2	40

\# Filters presoaked in 1mM $EuCl_3$ prior to filtration of Eu^{152} and vesicles

Vesicles were incubated in 100uM Eu^{152} at pH 7.0 for 1 min.

The effect of various wash solutions on the value obtained for bound Eu^{152} is shown in Table 1. When the wash medium is kept the same but the pH reduced, there is a greater removal of Eu^{152} from the membranes. The presence of the chelating agent DTPA in the wash medium was able to remove bound Eu^{152} without a reduction in pH.

Fig. 1 EUROPIUM-VESICLE AFFINITY

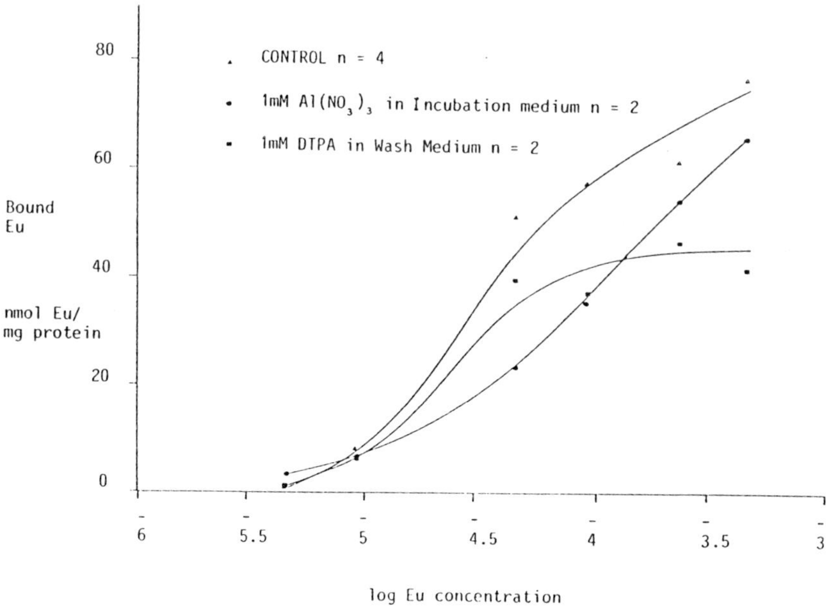

Eu-vesicle affinity curves (Fig. 1) were calculated using the displacement of Eu^{152} from the vesicles by increasing concentrations of $EuCl_3$. Saturation of the vesicles with Eu was not reached at the highest (450μM) concentration of Eu employed. The shape of the saturation curve is only roughly sigmoidal and the complex speciation of europium in solution (eg. carbonates, hydroxides) does not allow Km and maximum binding values to be computed from presently available data. Addition of 1mM Al^{3+} (as $Al(NO_3)_3$) to the incubation medium inhibited some Eu^{3+} binding ($p < 0.05$ at 45μM Eu; Fig. 1). Adding the chelating agent DTPA (1mM) to the wash solution reduced but not significantly bound Eu at higher Eu concentrations.

In another experiment, Ca^{2+} was found not to inhibit Eu^{3+} binding. Vesicles (5μl) were pretreated with incubation medium (40μl) containing 0, 0.1 and 1mM $CaCl_2$ for 1 min. After subsequent addition of Eu^{152} to a final concentration of 500μM, the amount of Eu bound was 156.4 and 156.9 nmol/mg protein for 0.1 and 1mM Ca^{2+} treated vesicles and 147.7 nmol Eu/mg protein in Ca^{2+} free vesicles (means of 2 experiments). Similarly, Eu^{3+} did not inhibit vesicle uptake of ^{45}Ca. The binding of ^{45}Ca to the vesicles was 1.547 nmol/mg prot (at 100μM $CaCl_2$ after 5 mins), 1.050 nmol/mg prot (in the presence of 1mM ATP and 1mM $MgCl_2$) and 1.812 nmol/mg prot (1mM ATP, 1mM $MgCl_2$, 0.1mM $EuCl_3$). In a study of element distribution using EDAX (x-ray) analysis on the electron microscope, Eu was situated on the vesicles though it was not possible to distinguish whether the Eu was membrane bound or intravesicular. Colocalized with the Eu were P, Cl and to some extent Ca.

DISCUSSION

The initial binding of Eu to the vesicles is a very rapid process. The bound Eu is not easily removed by DTPA suggesting that Eu is (a) internalized into the vesicles or (b) very strongly bound to the vesicle exterior. The EDAX analysis found Eu on vesicles associated with P which is most likely inorganic phosphate or phosphate on nucleotides. Alternatively the Eu could be interacting with phosphate groups of the membrane phospholipids (Williams, 1982). Al^{3+} reduces Eu binding by competing for interaction with these groups. In initial in vitro studies, phospholipids and Eu were separated during thin layer chromatography suggesting weak (if any) bonding between them.

The relatively low uptake of Ca^{2+} by vesicles (not involving ATP) was slightly increased by Eu and this may be due to membrane perturbation. Similarly Ca^{2+} had little effect on Eu^{3+} binding under the conditions used here. It may be that lower free concentrations of the ions are required to show competition between them at binding sites. Interestingly, other workers (Stevens and Kneer, 1988) have shown the ability of lanthanides to substitute for Na^{+}, in the cotransport of glucose and amino acids across brush-border membranes suggesting that they may interact with other cations as well as Ca^{2+}. Work with Eu salts is difficult at physiological pH due to their ability to form insoluble species (hydroxides, phosphates). Further work into lanthanide binding and transport across brush-border membranes should be carried out under conditions of carefully controlled and known free ion concentration.

ACKNOWLEDGEMENT

D. Bingham's studentship is supported by Johnson Matthey PLC.

REFERENCES

Arvela, P. (1979). Toxicity of rare-earths. Prog. Pharmacol. 2, 69-114.

Durbin, P.W., Williams, M.H., Gee, M., Newman, R.M. and Hamilton, J.G. (1955). The metabolism of lanthanons in the rat. Proc. Soc. Exp. Biol. Med. 91, 78-85.

dos Remedios, C.G. (1981). Lanthanide ion probes of calcium-binding sites on cellular membranes. Cell calcium 2, 29-51.

Simpson, R.J. and Peters, T.J. (1984). Studies of Fe^{3+} transport across isolated intestinal brush-border membrane of the mouse. Biochimica et Biophysica Acta 772, 220-226.

Stevens, B.R. and Kneer, C. (1988). Lanthanide-stimulated glucose and proline transport across rabbit intestinal brush-border membranes. Biochimica et Biophysica Acta 942, 205-208.

Williams, R.J.P. (1982). The chemistry of lanthanide ions in solution and in biological systems. Structure and Bonding 50, 79-119.

VIII. Radioisotopes, radiosensitizers

Metal Ions in Biology and Medicine. Eds. Ph. Collery, L.A. Poirier, M. Manfait, J.C. Etienne. John Libbey Eurotext, Paris © 1990, pp. 525-529

Metal ions as radiosensitizers

J.D. Anastassopoulou

National Technical University of Athens, Chemical Engineering Department, Laboratory of Radiation Chemistry and Biospectroscopy, Zografou Campus, 157 73 Zografou, Athens, Greece

Living organisms are generally sensitized to the lethal effects of ionizing radiation, if oxygen is present during irradiation (Dowdy et al, 1950). The sensitizing properties of oxygen may be attributed to the ability of this element to react with a radiation-induced free radical, produced in the vital molecule within the living cell, to form a more stable hydroperoxyl radical (Adams, 1970). In the absence of oxygen the free radical initially produced by direct or indirect action of radiation in the vital molecule may undergo back reaction to the original state (Howard-Fladers, 1958). In an attempt to overcome the problem of radioresistance of hypoxic cell a considerable amount of research was focused on organic electron attracting compounds as nitroimidazoles (Brown, 1975). These compounds react by accepting electrons from radical species on DNA causing a fixation of radiation induced damage. Many metal ions and metal complexes show electron affinity properties due to the positive charge of the metal and they become very attractive as possible radiosensitizers. These include complexes of Ag(I), Cu(I), Cu(II), Rh(II), Co(III) and Pt(II). We have investigated Pt(II) because of the antitumor activity of cis-platinum (Theophanides, 1981), which is still used extensively to cure cancers and Mg(II), because it is a non toxic life metal and has been found to be preventive against cancers (Theophanides, 1984, Durlach, 1987)

The cis-Pt(II) complexes show antitumor activity due possibly to their reaction with DNA, at the N(7) position of the guanine residue (Theophanides & Anastassopoulou, 1988 and references therein). For this reason, in order to investigate the role of Pt(II) as radiosensitizer we irradiated aqueous solutions of the complex cis-dichlorodiammineplatinum (cis-platinum) with guanosine-5'-monophosphate disodium salt (5'-GMP=L). As it is shown in Table 1 when deaerated aqueous solutions of the complex, cis-$PtL_2(NH_3)_2$ with Ar, were irradiated 5'-GMP and cis-platinum were released. Platinum is a transition metal quite electrolhyllic and can react with hydrated electrons, produced from radiolysis of water, according to the following reactions (Anastassopoulou & Brekoulakis, 1990):

$$Pt(NH_3)_2L_2 + e^-_{aq} \xrightarrow[+H_2O]{-L} Pt(I)(NH_3)_2L(H_2O) \quad (1)$$

$$Pt(I)(NH_2)_2L(H_2O) \xrightarrow[+H_2O]{-L} Pt(I)(NH_3)_2(H_2O)_2 \quad (2)$$

Pt(I) is not stable and after disproportionation may give Pt(0) and Pt(II) as follows:

$$2Pt(I) \longrightarrow Pt(0) + cis\text{-}Pt(II)(NH_3)_2(H_2O)_2 + 2NH_3 \quad (3)$$

The above results are in agreement with what had been observed by Richmond (1984). He reported that the radiosensitizing activity

Table 1. HPLC data of irradiated cis-$|Pt(NH_3)_2L_2|$ aqueous solutions. Eluent 0.6 M $NH_4H_2PO_4$ (Ph≅6), $|cis\text{-}Pt(NH_3)_2L_2| = 10^{-4}$M.

Time min	Ar, Area Percent 150 Gy	300 Gy	450 Gy	aerated, Area Percent 150 Gy	300 Gy	450 Gy	Product
4.94	9.85	4.40	0.90	----	----	----	cis-Pt
7.27	23.70	40.77	56.27	----	2.54	4.38	5'- GMP
13.56	---	---	---	1.03	1.46	1.99	Gua
14.90	---	---	---	2.85	3.85	3.94	Guo
23.85	39.26	17.01	6.87	69.22	62.85	56.75	complex

of cis-platinum complexes in salmonella typhimurium cells was due to the toxicity of metallic Pt that was released after irradiation in the absence of oxygen.

When aerated aqueous solutions were irradiated, guanine (Gua), guanosine (Guo) and 5'-GMP were released. Under these irradiation conditions (Anastassopoulou, 1990) there were produced O_2^-, HO_2 and OH radicals.

Table 2. HPLC of irradiated $|5'\text{-}GMPNa_2| = 10^{-4}$M aerated aqueous solutions. Eluent 0.6 M $NH_4H_2PO_4$ (Ph≅6).

Time min	Area Percent 150 Gy	300 Gy	450 Gy
7.27	100	76.11	73.54

It is clear that cis-platinum reacted as radiosensitizer in the presence of oxygen as is shown from Tables 1 and 2. The amount of damaged 5'-GMP is greater in the presence of cis-platinum.

Other transition metals, as Co(III) and Fe(III) can also sensitize B. megaterium spores under hypoxic and anoxic conditions. The mechanism of action is not clear and may involve the release of toxic ligands as it was suggested by Teicher et al (1987).

Bhattacharyya et al (1989) have observed that Cu(II) can sensitize in vitro thymine due to the oxidation of the transient hydroxyl adducts of thymine by copper ions and reduction of Cu(II) to Cu(I). Cramp (1967) proposed that the sensitizing activity of Cu(II) might be associated with the formation of Cu(I) which has been shown to be toxic.

In another study of magnesium, a non transition metal, it was observed that Mg(II) ions can sensitize in vitro 5'-GMPNa$_2$ in the absence of oxygen (Anastassopoulou, 1989) and it doubles the radiation effect.

From positive FAB (Fast Atom Bombardment) mass spectrometry it was found that Mg(II) remained fixed on the nucleotide even after irradiation with a dose of 1 kGy, on the guanine residue (Table3).

Table 3. FAB positive mass spectra of $|5'\text{-GMPH}_2\text{-Mg(ClO}_4)_2| = 10^{-4}$ M complexes with possible assignment of fragments. Total dose 1 kGy

Peak Mass	Fragment
152	$[B + H]^+$
228	$[M - B + H]^+$
307	$[M - H_2PO_4 + Mg(OH)]^+$ or $[Guo + Mg(OH) + 2H]^+$
309	$[M - H_2PO_4 + MgOH + 2H]^+$
329	$[2B + Mg + 2H]^+$
364	$[M + H]^+$
369	$[W^* - 2ClO_4 - CH_2H_2PO_4]^+$
399	$[2B + Mg(H_2O)_4]^+$
456	$[M - 2H + Mg(H_2O)_4]^+$
478	$[W - 2HClO_4 + H]^+$
492	$[W - CH_2H_2PO_4 + 4H_2O]^+$
531	$[W - CH_2H_2PO_4 + 2OH]^+$
578	$[W - ClO_4 + 2H]^+$
580	$[W - HClO_4 + 3H]^+$
601	$[W - 4H_2O + 3H]^+$
610	$[W - HClO_4 + 2OH]^+$
713	$[W + 2OH + 3H]^+$
793	$[W + B - 2OH]^+$

* where, M:363, Molecular weight of 5'-GMPH$_2$ and W:676, molecular weight of the comlex, 5'-GMPH$_2$-Mg(H$_2$)$_5$(ClO$_4$)$_2$

This is in accord with the proposal (Theophanides, 1984) that Mg(II) binds to the N(7) position of the 5'-GMP molecule in its reaction with this nucleotide.

Acknowledgments

We wish to thank the Pinawa Research Center for allowing us to do the experiments of HPLC at the invitation of Dr. A. Vikis.

REFERENCES

Adams, G. E. (1970):Molecular mechanisms of cellular radiosensitization. In Radiation Protection and Sensitization, eds H. Moroson and M. L. Quintiliani, pp. 1-14, London, Taylor & Francis.

Anastassopoulou, J. D. (1989): OH radicals as bioactivators. In Spectroscopy of Inorganic Bioactivators. Theory and Applications-Chemistry, Physics, Biology and Medicine, ed T. Theophanides, pp. 273-278, Dordrecht, Kluwer Academic Publishers.

Anastassopoulou, J.D. (1989): Magnesium perchlorate and cis-platinum as radiosensitizers in radiolysis of nucleotides. Magnesium Res.3: 1,(In press).

Anastassopoulou, J.D. and Brekoulakis, J.D. (1990):Radiation chemistry of cis-platinum-nucleotide complexes. To be submitted for publication.

Brown, J.M. (1975): Selective radiosensitization of hypoxic cells of mouse tumors with the nitroimidazoles, metronidazole and Ro-7-0582. Radiat. Res. 64: 633-647.

Bhattacharyya, S.N., Mandal, P.C. and Chakraborty, S. (1989): Copper (II) induced radiosensitization of thymine, Anticancer Res. 9: 1181-1184.

Cramp, W.A. (1967):The toxic action on bacteria of irradiated solutions of copper compounds, Radiat. Res. 30: 221-236.

Dowdy, A.H., Bennett, L.R. and Chastain, S.M. (1950): Protective action of anoxic anoxia against total body roentgen irradiation mammals, Radiology 55: 879-885.

Durlach, J. (1988):Magnesium in clinical practice, John Libbey, London.

Howard-Fladers, P. (1958): Physical and chemical mechanisms in the injurs of cells by ionizing radiations, Advan. Biol. Med. Phys. 6: 552-603.

Richmond, R.C. (1984): Toxic variability and radiation sensitization by dichlorodiammineplatinum(II) complexes in salmonella typhimurium cells, Radiat. Res. 99: 556-603.

Teicher, B.A, Jacobs, J.L., Cathcart, K.N.S., Abrams, M.J., Vollano, J.F. and Picker, D.H. (1987): Radiat. Res. 109: 34-46.

Theophanides, T. (1989): Metal ion-nucleic acid interactions as studied by Fourier Transform Infrared Spectroscopy. In Spectroscopy of Inorganic Bioactivators. Theory and applications- Chemistry, Physics, Biology and Medicine, ed T. Theophanides, pp. 265-272, Dordrecht, Kluwer Academic Publishers.

Theophanides, T. (1984): Metal ions in biological systems. Interl. J. Quantum Chem. XXVI: 933-941.

Theophanides, T, (1981): Platinum coordination chemistry compounds and cancer, Chemistry in Canada 32: 31-32.

Theophanides, T. and Anastassopoulou, J.D. (1988): The reaction of platinum-antitumor compounds with DNA. In Metal-Based Anti-tumor Drugs, ed M.F. Gielen, pp. 151-174, London, Freund Publishing House LTD.

Metal Ions in Biology and Medicine. Eds. Ph. Collery, L.A. Poirier, M. Manfait, J.C. Etienne. John Libbey Eurotext, Paris © 1990, pp. 530-532

Survival of mice bearing a krebs ascitic tumor by means of a protocole based on radioactive copper (^{64} Cu)

S. Apelgot[1], E. Guille[2]

[1] *Institut Curie, Section de Physique et Chimie, 11, rue Pierre et Marie Curie, 75231 Paris Cedex 05;*
[2] *Université Paris Sud, Biologie Moléculaire Végétale (UA 1128) 91405 Orsay, France*

Previous experiments have shown that no cured mice were ever observed when ^{64}Cu (half-life = 12.8 hrs) was injected alone 6 days after the injection in each animal of 5 x 10^5 Krebs ascitic cells and even at day 1 when it is well known that no developing tumor can exist (Fig. 1, curve a). Without treatment, all of these animals died within 12 to 25 days. After the injection of 1 to 6 mCi of ^{64}Cu, only a delay in death was observed, regardless of the ^{64}Cu doses and the injection protocol (Coppey et al., 1982). Moreover, survival seems to depend only on the number of the ascitic cells initially injected (Guillé et al., 1985). These results were puzzling with regard to those of all previous results showing 1) the ^{64}Cu lethal effect for mammalian cells under in vitro growth conditions (Apelgot et al. 1989) and 2) the preferential ^{64}Cu incorporation in the ascitic cells under in vivo conditions (Apelgot and Guillé 1989).

Thus a spectacular difference does exist between the consequences of the same physical event (^{64}Cu transmutation) under in vitro or in vivo conditions: a lethal consequence in the first case but not in the second case even under experimental conditions when no developing tumor can exist. To shed light on these surprising results, we tried to consider, in a new way, the mice injected with the ascitic cells. We took into account the dynamic interactions existing between the injected malignant cells and the cells of the injected mouse, since both kinds of cells were in the same organism (Von Bertalanffy, 1980). Based on these interactions between the two constituent elements of the new system (the mouse and the injected ascitic cells), we hypothesized that if the majority of the injected ascitic cells were eliminated, some of them become slightly modified and consequently accepted, while some of the cells of the injected mouse take on a slightly modified functioning. Along these lines, we called this mouse a "cancer animal" and "cancer-cells" the mouse cells with this modified functioning. These "cancer-cells" are not malignant but have some subtle differences with regard to normal cells (Apelgot and Guillé, 1989).

The difference between the consequences of ^{64}Cu transmutations under in vitro and in vivo conditions can now be understood. ^{64}Cu transmutations induce in the DNA of malignant cells definitive and irreversible modifications, ending in a lethal event under in vitro conditions, but only temporarily injures, under in vivo conditions, because of modifications in the accepted malignant cells. This temporarily injured DNA reorganizes spontaneously towards resistant malignant DNA (Sager et al. 1985; Apelgot and Guillé, 1989). Therefore, an efficient treatment has to act at the same time and in a coordinated manner on the modified malignant cells and on the modified host cells (cancer-cells). An efficient treatment was developed (Table 1) which lasts for 8 days and is based on the use of :

a) $^{64}CuCl_2$ to damage DNA of the modified malignant cells in order to stop temporarily the malignant signals;

b) A radioprotector, glycerol, to decrease the consequences of the irradiation induced in the organism by the particles emitted by ^{64}Cu;

c) The reverse transformation established under in vitro conditions for thioproline (Brugarolas and Gosalvez, 1980) in order to orientate the reorganization of malignant cellular DNA temporarily damaged by ^{64}Cu transmutation towards non-malignant cellular DNA;

d) Metal ions, amino acids, vitamin D_2, thyroxine, dead ^{64}Cu and chelating substances (Table 1) in order to reverse the slightly modified functioning of the host "cancer-cells". These products are able to reinitiate the different living processes in a non-cancerous manner.

Table 1 - The efficient treatment (Apelgot and Guillé, 1989)

Days		Compounds
a	(17 h)	^{64}Cu: 105 µCi + $MgCl_2$: 255 µg + glycerol: 5 µg + aspartic acid: 4.5 µg + arginine - $2H_2O$: 5 µg
a + 1	(17 h)	Dead ^{64}Cu: ~ 2 mCi + $MgCl_2$: 11 µg + $ZnCl_2$: 84 µg
	(8 h)	Hydroquinone - Cu: 15 µg
a + 2	(14-17 h)	Vitamin D_2 (*)
	(19 h)	Thioproline: 0.5 µg
a + 3	(17 h)	Thyroxine: 1.70 µg + $ZnCl_2$: 0.30 µg
a + 4	(17 h)	Fe (SO_4): 38 µg
a + 5	(14-17 h)	Vitamin D_2 (*)
a + 7	(14-17 h)	MoO_3 + $MnCl_2$, $4H_2O$ (*)

"a" corresponds to the day when the treatment began: "a" could be on day 1, 11 or 16 after the injection of the Krebs ascitic cells; the hours between brackets (o'clock time) are only indicative. Compounds marked (*) were given via oral route. For dead ^{64}Cu solutions, see Apelgot and Guillé, 1989.

When this treatment was applied to mice bearing a 1 or 11 day old Krebs ascitic tumor (lag-to exponential tumor growth phase) 90 to 100 % of the treated mice became tumor-free but not when applied to mice bearing a 16 day old tumor (after the exponential tumor growth phase) (Fig. 1).

To understand the difference in the treatment efficiency between 11 and 16 days, we tested the ^{64}Cu incorporation inside the ascitic cells. It was observed that ^{64}Cu incorporation exists on day 11 but no longer on day 16. On day 16, the inefficiency of the treatment must be correlated with the non-incorporation of ^{64}Cu inside the ascitic cells. As the tumor growth is also arrested on day 16, an irreversible stage is reached. A model was developed to explain the results obtained (Apelgot and Guillé, 1989).

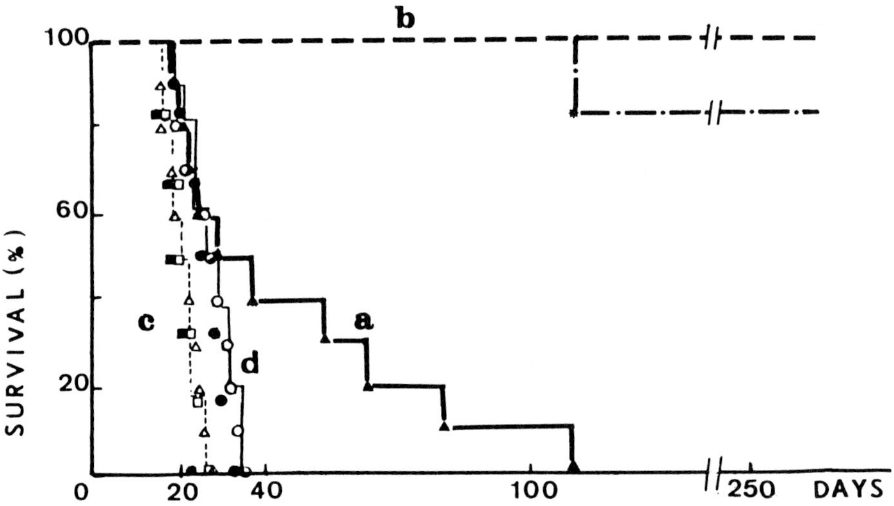

Fig. 1. Survival of mice treated on days 1, 11 or 16.
On day zero, mice were injected (IP) with about 5×10^5 Krebs ascitic cells.
Curve c: Cu or Cu-treatment is applied; ^{64}Cu alone is applied on day 1 (Curve a); the proposed treatment is applied on day 1 or 11 (Curves b) or on day 16 (Curve d).

Finally, to test the consequence of this treatment, some of these tumor-free mice were injected with 5×10^5 ascitic cells. No tumor developed in these tumor-free mice meaning that they also were tumor-resistant. This last result is in agreement with the forecast of the systemic analysis (Von Bartalanffy, 1980) on which the efficient treatment was based.

REFERENCES

Apelgot, S., Coppey, J., Gaudemer, A., Grisvard, J., Guillé, E., Sasaki, I. and Sissoëff, I. (1989): Similar lethal effect in mammalian cells for two radioisotopes of copper with different decay schemes, ^{64}Cu and ^{67}Cu: possible meaning of these new results. Int J. of Radiat. Biol. 55: 365-384.

Apelgot, S. and Guillé, E. (1989): Treatment of mice bearing a Krebs ascitic tumor by means of a protocol based on radioactive copper (^{64}Cu). Anticancer Res.: 9: 941-960.

Brugarolas, A. and Gosalvez, M. (1980): Treatment of cancer by an inducer of reverse transformation. The Lancet I n° 8159: 67-79.

Coppey, J., Grisvard, J., Guillé, E., Sissoëff, I. and Apelgot, S. (1982): Application de l'effet létal de la capture électronique: essai de contrôle par ^{64}Cu de la croissance de la tumeur ascitique de Krebs chez la souris. Bull. Cancer Paris 69: 121-130.

Guillé, E., Sissoëff, I., Apelgot, S., Coppey, J., Grisvard, J. and Fromentin, A. (1985): ^{64}Cu and Krebs ascitic tumor in mice: dose effect relationship. 8th Meeting of the European Association for Cancer Research (Bratislava, Czechoslovakia, 13-15 May 1985).

Sager, R., Gadi, I.K., Stephen, L., and Grabowy, C.T. (1985): Gene amplification: an example of accelerated evolution in tumorigenic cells. Proc. Nat. Acad. Sci. USA 82: 7015-7019.

Von Bertalanffy, L. (1980): Théorie générale des systèmes, Paris, Dunod.

Metal Ions in Biology and Medicine. Eds. Ph. Collery, L.A. Poirier, M. Manfait, J.C. Etienne. John Libbey Eurotext, Paris © 1990, pp. 533-537

IIIA group elements in early diagnosis and follow-up and in effective systemic therapy of cancer : a review of past results and suggestions for future improvements

S.K. Shukla[a,b], C. Cipriani[a]

[a] *Servizio di Medicina Nucleare, Ospedale S. Eugenio, P. le Umanesimo 10, I-00144 Roma;* [b] *Istituto di Cromatografia, C.N.R., C.P. 10, I-00016 Monterotondo Scalo, Roma, Italy*

In the preface to the book "Therapeutic Use of Artificial Radionuclides" P.F. Hahn (1956) wrote: "Ideally, one looks for an agent that will seek out and destroy cancer." The elements,which are capable of producing such agents capable of not only treating cancer but also revealing its presence and growth with time with nuclear medicine procedures, are given in Table 1. Radiopharmaceuticals of gamma emitting radionuclides (of short half-life and gamma energy in the range 130 to 160 KeV) are suitable for tumour localization and its follow up, while those of high energy beta or alpha emitting radionuclides can destroy cancerous tissue noninvasively. For both these clinical applications the elements of the IIIA group have all ideal properties and their clinical use is given in Table 2.

Table 2. IIIA Group elements in tumour diagnosis and therapy

Element	Diagnosis	Therapy
Boron	-	Boron neutron capture therapy (Coderre et al., 1988)
Aluminum	-	Aluminum nitrate (Hart et al., 1971)
Gallium	Ga-67 citrate (Edwards et al., 1969)	Indium nitrate (Hart et al., 1971)
Indium	In-111 chloride Hunter et al., 1969)	Indium nitrate (Hart el al., 1971)
Thallium	Tl-201 chloride (Ando et al., 1987)	Thallic nitrate (Hart et al., 1971)

Gallium, the central element of the group, has chemical properties ideal for its use in the synthesis of radiopharmaceuticals and pharmaceuticals (Shukla et al.,

Table 1. Chemical elements, the compounds of which are being presently used or are under study for systemic diagnosis and therapy of cancer: ☼ diagnostic agent, ✱ therapy agent (radionuclidic or chemotherapeutic).

n=Charge Density

H -0.65																	He 2
Li 1.47	Be 5.71											B✱ 13.04	C -1.54	N 18.75	O -1.51	F✱ -0.75	Ne 0.89
Na 1.03	Mg 3.03											Al✱ 5.88	Si 9.52	P 14.3	S -1.09	Cl✱ -0.55	Ar 0.64
K 0.75	Ca 2.02	Sc 3.70	Ti 5.88	V 8.47	Cr 11.54	Mn 6.06	Fe 4.68	Co 4.76	Ni 2.30	Cu 2.78	Zn 2.70	☼Gd✱ 4.84	Ge 7.54	As✱ 10.9	Se✱ 14.3	Br -0.51	Kr 36
Rb 0.68	Sr 1.79	Y✱ 3.26	Zr 5.06	Nb 7.25	Mo 9.68	☼Tc 5.56	Ru 5.97	Rh 4.41	Pd 6.15	Ag 0.79	Cd 2.06	☼In✱ 3.70	Sn 2.15	Sb 3.95	Te 10.7	☼I✱ -0.45	Xe 54
Cs 0.60	Ba 1.49	La 2.63	Hf 5.13	Ta 7.35	W 9.68	Re 5.56	Os 4.55	Ir 4.88	Pt✱ 2.50	Au 0.73	Hg 1.82	☼Tl✱ 3.16	Pb 1.67	Bi 3.13	Po 8.95	At✱ 11.3	Rn 86
Fr 0.56	Ra 1.40	Ac 2.54	Rf 104	Ha 105	106	107	108	109	110	111	112	113	114	115	116	117	118
119	120	121	154	155	156	157	158	159	160	161	162	163	164	165	166	167	168

Ce 2.80	Pr 2.83	Nd 2.88	Pm 2.83	☼Sm✱ 3.00	Eu 3.06	Gd 4.84	Tb 3.23	Dy 3.26	Ho 3.30	Er 3.37	Tm 3.45	Yb 3.49	Lu 3.53
Th 3.92	Pa 4.08	U 4.12	Np 4.21	Pu 4.30	Am 4.35	Cm 96	Bk 97	Cf 98	Es 99	Fm 100	Md 101	No 102	Lr 103

122	123	124	125	126	…	153

Periodic Table of Elements (☐ so far unknown elements)

1980), and is so far the most widely used, as gallium-67 citrate, in scintigraphic detection of tumours (Shukla et al., 1990), and is also under study as a chemotherapeutic agent (Samson et al., 1980). Our study has therefore been mainly concerned with gallium-67 and cold gallium because the gamma emissions (90, 180, 300 KeV) of Ga-67 permit the radionuclide distribution studies in vivo which helps the control of the tumour-specificity of the synthesized products.

The superiority of radionuclidic therapy over other methodologies was recently discussed by Beierwaltes (1985) who concluded: "Many cancer therapists agree that surgery, teletherapy, and chemotherapy are crude and nowhere near as effective as we all want them to be, and what is really needed is an innovative, effective, and less harmful form of cancer therapy --- radionuclide therapy has begun to demonstrate that potential." Although nothing can be done to make surgery and teletherapy to be less harmful and more effective, chemotherapy, being systemic can be improved by synthesizing tumour-specific pharmaceuticals, and the synthesis of such products is the object of our research.

Recent review on radionuclidic detection and therapy (Jasmin, 1987; Thomlinson, 1987; Mitchell et al., 1988; Ott, 1989) show, however, that the radiopharmaceuticals available so far have little tumour affinity in vivo. Even monoclonal antibodies and their fragments used to target the radionuclide to the tumour fail to do so. We, therefore, investigated first the nature of radionuclide species present in Ga-67 solutions and labelled antibody solutions (Shukla et al., 1980; 1987). We found that commercially available gallium-67 and monoclonal antibodies contain many radionuclide species of cationic, neutral, and even anionic nature. The tumour affinity of radionuclide species was therefore studied first. The tumour-specific species was recognized by total-body distribution studies of the radionuclide species in tumour-bearing subjects. Conditions were found under which the radionuclide can be obtained in chromatographically and electrophoretically pure form in solution.

CLASSIFICATION OF TUMOURS BASED ON THEIR AFFINITY FOR GALLIUM-67 SPECIES IN SOLUTION

Tumour-specific uptake of the radionuclide was observed (Fig. 1) when chromatographically and electrophoretically pure radionuclide in tumour-specific form was administered. Strong binding between the radionuclide and tumour was found permitting the tumour imaging at least for 8 days postinjection.

The affinity of chromatographically and electrophoretically pure Ga-67 species in solution permitted us to classify malignant tumours in following groups (Shukla et al., 1989).

<u>Cationic gallium-67 affine tumours</u>: Hodgkin's disease, lung tumour, neuroblastoma, Morris hepatoma-3924A, ...

<u>Anionic gallium-67 affine tumours</u>: Mammary tumour, melanoma, osteosarcoma, Ewing's sarcoma, fibrosarcoma, ...

Tumour affinity of Ga-67 does not depend on the nature of the ligand. What is important is its ionic nature and purity.

TUMOUR THERAPY WITH TUMOUR-SPECIFIC COLD GALLIUM IN SOLUTION

Cold gallium in tumour-specific form was prepared in sodium citrate solution and

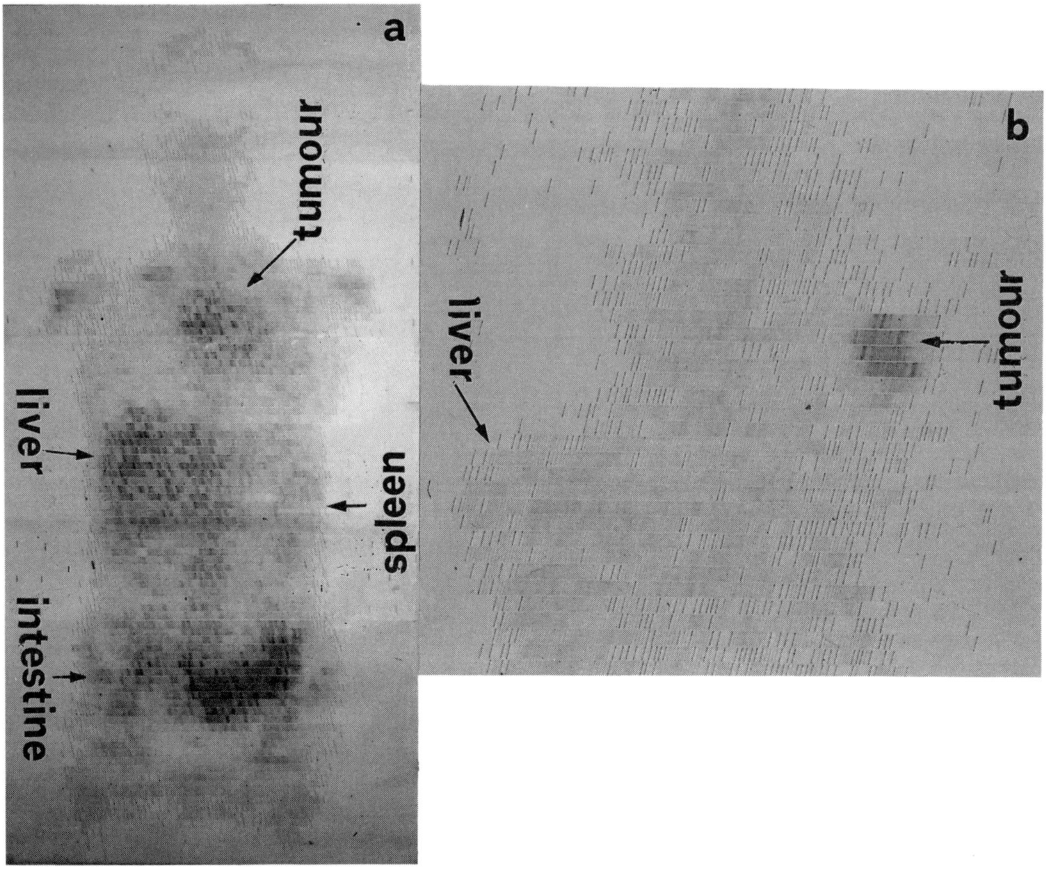

Fig. 1. Total-body distribution of Ga-67 in a lung tumour patient injected with: (a) commercial Ga-67 citrate solution, (b) lung tumour-affine Ga-67 (scintigraphy 24 h p.i. with a rectilinear scanner).

its identity with anionic melanoma-specific Ga-67 was tested chromatographically and electrophoretically.

Antitumour activity of melanoma-specific gallium in solution was studied in melanoma-bearing mice. For both Harding-Passey and B-16 melanoma, melanoma-specific gallium could cure the tumour when its diameter before the therapy was less than 1 cm. Bigger tumours became necrotic and did not further grow with time.When removed surgically after necrosis, the animal lived free from tumour for its normal life.

The antitumour action of our tumour-specific gallium was also seen in a cocker dog. These studies are still being followed.

CONCLUSION

Cancer so far studied in could be classified into: 1. cation affine, 2. anion affine gallium tumours. Tumour specific Ga-67, and cold gallium distribution is obtained with pure tumour-specific gallium species, useful for diagnosis or therapy.

REFERENCES

Ando, A.A., Ando,I., Katayama, M., et al. (1987): Biodistribution of Tl-201 in tumor bearing animals and inflammatory lesion induced animals. Eur. J. Nucl. Med. 12, 567-572.

Coderre, J.A., Kalef-Ezra, J.A., Fairchild, R.G., et al. (1988): Boron neutron capture therapy of a murine melanoma. Cancer Res. 48, 6313-6316.

Edwards, C.L. & Hayes, R.L. (1969): Tumour imaging with Ga-67 citrate. J. Nucl. Med. 10, 103-105.

Hahn, P.F. (1956): Preface. In: "Therapeutic use of artificial radioisotopes", ed. P.F. Hahn, pp. i-iii. New York: John Wiley & Sons, Inc.

Hart, M.M. & Adamson, R.H. (1971): Antitumor activity and toxicity of salts of inorganic Group IIIa metals: aluminum, gallium, indium, and thallium. Proc. Nat. Acad. Sci., USA 68, 1623-1626.

Hunter, W.W. & De Cock, H.W. (1969): Indium-111 for tumour localization. J. Nucl. Med. 10, 343-348.

Jasmin, C. (1987): Le traitement biologique des cancers. La Recherche 18, 754-764.

Mitchell, M.S., Kan-Mitchell, J. Kempf, R.A., et al. (1988): Active specific immunotherapy for melanoma: Phase I trial of allogneic lysates and a novel adjuvant. Cancer Res. 48, 5883-5893.

Ott, R.J.(1989): Nuclear medicine in the 1990s: a quantitative physiological approach. Br. J. Radiol. 62, 421-432.

Samson, M.K., Fraile, R.J., Baker, L.H., et al. (1980): Phase I-II clinical trial of gallium nitrate (NSC-15200). Cancer Clin. Trials 3, 131-136.

Shukla, S.K., Castelli, L., Cipriani, C., et al. (1980): What is wrong in gallium-67 citrate and how to increase its tumour affinity ? Radioakt. Isotope Klin. Forsch. 14, 179- 196.

Shukla, S.K., Cipriani, C., Manni, G.B., et al. (1987): A comparative study of the chemical nature of melanoma imaging radiopharmaceuticals and of the quality of melanoma scans obtained from them. In: "Epidemiology, Prevention, and Diagnosis, eds. K. Lapis & S. Eckhardt, pp. 345-354. Budapest: Akademiai Kiadò.

Shukla, S.K., Xie, H., Hua, R.L., et al. (1990): Hepatis tumour specificity of gallium-67 labelled antihepatoma antibody, hepama-1. Radioakt. Isotope Klin. Forsch. 19, (in press)

Thomlinson, R.H. (1987): Cancer: the failure of treatment. Br. J. Radiol. 60, 735-751.

Metal Ions in Biology and Medicine. Eds. Ph. Collery, L.A. Poirier, M. Manfait, J.C. Etienne. John Libbey Eurotext, Paris © 1990, pp. 538-542

Review of yttrium-90 radiolabeled antibodies

J.E. Crook, Y.C.C. Lee**, L.C. Washburn, T.T.H. Sun, B.L. Byrd, E.C. Holloway, H.S. Julia Ju, and Z. Steplewski*

*The Oak Ridge Associated Universities, Medical Sciences Division, P.O. Box 117, Oak Ridge, Tennessee, U.S.A., 37830 and *The Wistar Institute, Philadelphia, Pennsylvania, U.S.A.; ** Present Address : Bio-decision Clinical Research Institute, 5001 Baum Boulevard, Pittsburgh, Pennsylvania, 15213-1850 USA*

Introduction: The burgeoning development of a large number of monoclonal antibodies occurring in tandem with the synthesis of newer coupling reagents as chelators has heightened interest in a diverse array of radionuclidic metals for potential use in cancer therapy as through radiolabeling of monoclonal antibodies. Yttrium-90 (^{90}Y), a beta emitter, is one such radionuclide that is being increasingly utilized in both the preclinical and clinical areas stimulated perhaps by the discussion of its model characteristics by Wessels and Rogus (1984). Although yttrium-90's potential as an effective radiolabel for many different classes of monoclonal antibodies is great, it possesses important unique features that underscore the necessity of a review of its chemistry, chemical production, physiological distribution, metabolic fate and toxicological effects to ensure its optimum use.

For the past five years, our laboratory has performed extensive chemical and biological investigations focused primarily on ^{90}Y as a radiolabel for the monoclonal antibody, CO17-1A, which has specificity for colorectal and pancreatic carcinomas. In subsequent sections we will review and present our experiences as well as the relevant studies of others.

Chemistry and characteristics: ^{90}Y arises from the decay of its reactor produced parent, strontium-90 (^{90}Sr), in the following manner:

$$^{90}Sr \xrightarrow{\beta^-(0.54\ MeV)} {}^{90}Y \xrightarrow{\beta^-(2.27\ MeV)} \text{Zirconium-90}$$

resulting in the production of a stable end-product. Production of ^{90}Y suitable for clinical radiolabeling studies, i.e. an acceptable level of trace element contamination with material such as Fe^{+3} which competes with ^{90}Y for antibody binding sites and a low ratio of ^{90}Sr to ^{90}Y on the order of μCi to Ci is routinely available (Wike, et al., 1990) using various solvent extractions from a ^{90}Sr cow.

Yttrium-90 possesses a number of features which serve to characterize it as one of the more select radionuclides available for use in radioimmunotherapy. Among its more desirable properties are (1) ready availability from a long-lived parent ($T_{1/2}$ = 28y), (2) decays to a stable daughter, (3) a pure beta emitter, (4) suitable half-life of 64 h and (5) possesses good chelation properties.

Chelating Agents: Since the initial report of Hnatowich, et al., (1983) describing the use of the cyclic anhydride diethylenetriaminepentaacetic acid (DTPA) method to chelate ^{90}Y, progressive and innovative organic synthetic procedures have been developed to accomplish the goal of attaching a chelating agent to the monoclonal antibody which effectively increases the number of carboxylic acid groups available for binding to the radionuclide. As shown in Table 1, the strategy has resulted in increasing the % injected dose/g tissue uptake in the target tissue, i.e., a xenographed tumor in a nude mouse while decreasing the % update by critical organs such as the bone marrow which is known to be radiation dose sensitive.

TABLE 1

COMPARISON OF 48-HOUR TISSUE DISTRIBUTIONS OF ^{90}Y-CO17-1A RADIOLABELED BY FIVE BIFUNCTIONAL CHELATE TECHNIQUES IN FEMALE NUDE MICE BEARING SW 948 COLORECTAL CARCINOMA XENOGRAFTS

Tissue	Cyclic DTPA Anhydride	NCS-Bz-DTPA	Site-Specific NH_2-Bz-DTPA	NCS-Bz-Mx-DTPA	Site-Specific NH_2-Bz-Mx-DTPA
Liver	7.21 ± 0.70	9.13 ± 0.74	9.29 ± 0.51	10.03 ± 0.33	8.27 ± 0.52
Spleen	6.13 ± 1.19	12.03 ± 2.46	12.02 ± 1.61	10.18 ± 1.02	9.45 ± 0.57
Kidney	3.93 ± 0.47	5.49 ± 0.35	4.71 ± 0.20	5.43 ± 0.19	4.18 ± 0.32
Lung	7.96 ± 2.89	5.78 ± 0.43	6.08 ± 1.16	11.13 ± 5.33	4.66 ± 0.58
Muscle	0.78 ± 0.10	1.20 ± 0.08	0.87 ± 0.03	1.11 ± 0.19	0.80 ± 0.06
Heart	2.65 ± 0.18	3.26 ± 0.31	3.47 ± 0.19	3.87 ± 0.10	2.58 ± 0.29
Femur	12.21 ± 1.22	2.67 ± 0.13	2.32 ± 0.13	6.15 ± 0.20	2.47 ± 0.09
Marrow	7.05 ± 1.69	13.74 ± 2.54	10.88 ± 1.28	9.73 ± 1.10	9.16 ± 0.55
Blood	4.73 ± 0.68	8.01 ± 1.15	7.52 ± 0.40	6.39 ± 0.10	6.26 ± 0.93
Tumor	16.63 ± 1.89	24.90 ± 1.35	18.55 ± 1.65	18.18 ± 0.55	18.99 ± 0.96
L. Intestine	1.61 ± 0.18	2.35 ± 0.31	2.70 ± 0.35	2.18 ± 0.09	2.15 ± 0.14
S. Intestine	2.10 ± 0.28	3.50 ± 0.45	2.93 ± 0.22	3.16 ± 0.18	2.93 ± 0.22

As indicated in Table 1, several different approaches were taken to carry out radiolabeling of the monoclonal antibody. For example, the cyclic DTPA anhydride technique brings about formation of an amide bond between a single carboxyl group of the DTPA and an ϵ-lysyl amino group on the monoclonal antibody. Other bifunctional chelate reagents such as the 1-(p-isothiocyanatobenzyl)DTPA (NCS-Bz-DTPA) (Breachbiel, et al., 1986) is conjugated using thiourea links. Potentially, the major drawback to carrying out chelations using this method is that the amide or thiourea linkages may occur on the portion of the antibody instrumental in antigen-antibody binding. Partially to offset this disadvantage, site specific conjugation to the oligosaccharide portion of monoclonal antibodies, located distal to the antigen binding sites, have been reported by Rodwell, et al., (1986).

Physiological Distribution: A variety of organ systems can impact the eventual distribution and metabolic fate of ^{90}Y radiolabeled monoclonal antibodies. Clearly the cardiovascular system plays a major role regardless of whether the route of administration is intravenous or intraperitoneal. The gastrointestinal tract also has a role in the eventual excretion of ^{90}Y labeled monoclonal antibodies since there appears to be a hepatobiliary route of excretion. Interrelated to these two systems is the avidity of ^{90}Y for bone and soft tissue as discussed by Ramsden (1961) since the major route of delivery is via the vascular supply wherein the concentration of the ^{90}Y radiolabeled monoclonal antibody circulating in the blood is limited in one aspect by the rate of excretion. Lee, et al., (1990) has reported on the metabolism and whole-body retention of ^{90}Y-CO17-1A in nude mice. Studies carried out in female nude mice (Fig. 1) clearly demonstrate the operant hepatobiliary circulation wherein 66 per cent of the activity was present in the feces. The per cent whole body retention, measured by counting the bremsstrahlung radiation from ^{90}Y using a geometry independent small-animal whole body counter, decreased by 50 per cent over 18 days indicating that the biological half-life was approximately 21 days for ^{90}Y-CO17-1A labeled by the site-specific p-NH_2-benzyl-Mx-DTPA method.

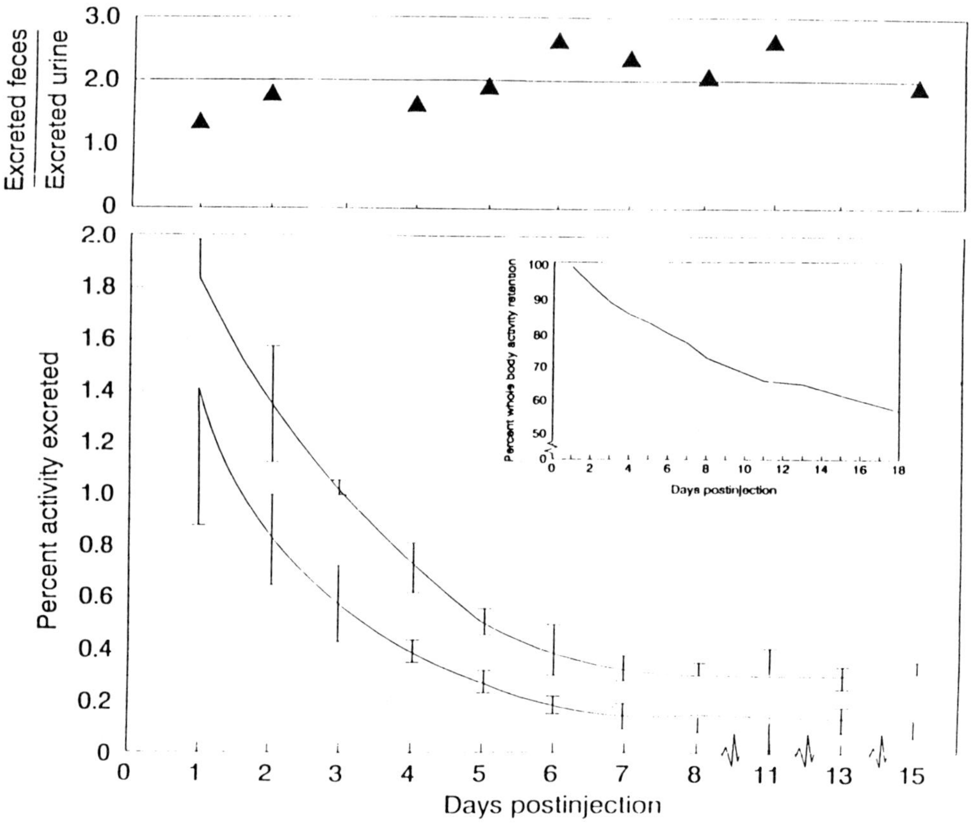

Fig. 1. Excretion of ^{90}Y labeled CO17-1A via feces (open circles) and urine (closed circles). Upper panel demonstrates the constant ratio of feces to urine excretion pattern. Inset demonstrates the decline in whole body activity vs. time.

Other studies by Crook, et al., (1988) have demonstrated the role of age in determining the eventual bone update and consequent radiation dose to the marrow. Measurements of femur uptake of ^{90}Y-CO1701A in groups of nude female mice aged 9-10 weeks and 9.5 months of age revealed a highly statistically significant ($p<0.01$) greater uptake in the younger group. Since uptake was adjusted for body weight and expressed as a per gram value, it appears unlikely to be related solely to bone size. The exact components contributing to the differences have not been fully delineated but it appears that the rapid bone remodeling taking place in the younger group is a significant contributing factor to the noted differences.

Adjuvant treatment: Myelotoxicity induced by the use of monoclonal antibodies radiolabeled with a pure beta emitter such as ^{90}Y is a dose limiting factor for its use in radioimmunotherapy. Washburn, et al., (1990) has proposed that the radiation induced myelosupression could be ameliorated by employing colony stimulating factors (CSF's) such as granulocyte CSF. As demonstrated in Fig. 2, nude mice bearing xenographed human, SW 948, colorectal carcinomas and receiving G-CSF at one of two time intervals clearly demonstrated tumor reduction.

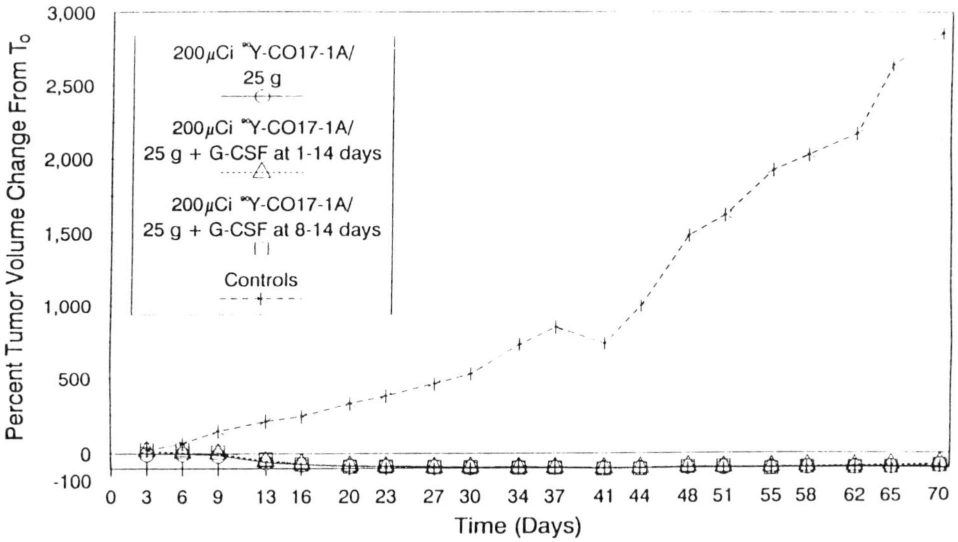

Fig. 2. Effect of adjuvant therapy with G-CSF (3μg/mouse/day) administered at days 1-14 or 8-14 on the growth of SW 948 xenografts in nude mice.

In conclusion, as interest in the general area of radiolabeling of monoclonal antibodies continues to expand, a reasonable forecast is that new, improved and even more specific forms of radioimmunotherapy will be possible.

REFERENCES

Brechbiel, M.W., Gansow, O.A., Atcher, R.W., Schlom, J., Esteban, J., Simpson, D., and Colcher, D. (1987): Synthesis of 1-(p-isothiocyanatobenzyl) derivatives of DTPA and EDTA. Antibody labeling and tumor imaging studies. Inorg. Chem. 25:2772-2781.

Chinol, M. and Hnatowich, D.J. (1987): Generator produced ^{90}Y for radioimmunotherapy. J. Nuc. Med. 28:1465-1469.

Crook, J.C., Washburn, L.C., Lee, Y-C.C., Byrd, B.L., Sun, T.T.H., and Steplewski, Z. (1988): Tissue distribution of ^{90}Y-labeled monoclonal antibody CO17-1A in young and old athymic mice. In Proceedings of the 4th Annual International Conference on Monoclonal Antibody Immunoconjugates for Cancer. eds. R.O. Dillman & I Royston, p.42 (abstract) San Diego.

Hnatowich, D.J., Layne, W.W., Childs, R.L., Lanteigne, D., and Davis, M.A. (1983): Radioactive labeling of antibody: a simple and efficient method. Science 220:613-615.

Lee, Y.C.C., Crook, J.E., Washburn, L.C., Sun, T.T.H., Byrd, B.L., Holloway, E. C., H-S.J., Ju, and Steplewski, Z. (1990): Metabolism and whole body retention of Y-90-CO17-1A in nude mice bearing SW 948 human colorectal carcinoma xenographs. Antibody Immuncon. Radiopharm. 3:51 (abstract).

Ramsden, E. (1961): A review of experimental work on radio-yttrium comprising (1) the tissue distribution, (2) the mechanism of deposition in bone, and (3) the state in blood. Int. J. Rad. Biol. 3:399-410.

Rodwell, J.D., Alverez, V.L., Lee, C., Lopes, A.D., Goers, J.W.F., King, H.D., Powsner, H.J., and McKearn, T.J. (1986): Site-specific covalent modification of monoclonal antibodies: In vitro and in vivo evaluations. Proc. Natl. Acad. Sci. USA 83:2632-2636.

Washburn, L.C., Lee, Y-C.C., Sun, T.T.H., Byrd, B.L., Holloway, E.C., Crook, J.E., Patchen, M. L., MacVittie, T.J., Souza, L.M. and Steplewski, Z. (1990): Adjunctive use of G-CSF with radioimmunotherapy using Y-90-Labeled monoclonal antibodies. J. Nuc. Med.: in press, (abstract).

Wessels, B.W. and Rogus, R.D. (1984): Radionuclide selection and model absorbed dose calculations for radiolabeled tumor associated antibodies. Med. Phys. 11:638-645.

Wike, J.S., Guyer, C.E., Ramey, D.W. and Phillips, B.P. (1990): Chemistry for commercial scale production of yttrium-90 for medical research. Int. J. Appl. Radiation and Isotopes, Part A: in press.

ACKNOWLEDGEMENT

The submitted manuscript has been authored by a contractor of the U.S. Government under contract number DE-AC05-76OR00033. Accordingly, the U.S. Government retains a nonexclusive, royalty-free license to publish or reproduce the published form of this contribution, or allow others to do so, for U.S. Government purposes.

D INTERACTIONS METAL IONS/BIOLOGICAL SUBSTRATES

Metal Ions in Biology and Medicine. Eds. Ph. Collery, L.A. Poirier, M. Manfait, J.C. Etienne. John Libbey Eurotext, Paris © 1990, pp. 545-548

Antagonism or synergy among iron, copper and zinc towards processes involving oxygen radical generation

P. Bienvenu, J.F. Kergonou

Unité de Radiobiochimie, Centre de Recherches du Service de Santé des Armées, 24, avenue des Maquis du Grésivaudan, 38700 La Tronche, France

EARLIER EXPERIMENTS

In an earlier work, we had shown the periodicity of acute intraperitoneal toxicity of metal ions ($LD_{50/30d}$ in Swiss mice), versus atomic number (Bienvenu et al., 1963; Bienvenu, 1965). The respective values for iron, copper and zinc, all expressed as milliatom.g/kg, and ordered following increasing toxicities, were: Fe(II): 0.700±0.080; Fe(III): 0.420±0.010: Zn(II): 0.180±0.010; Cu(II) 0.045±0.003. We have found an inverse relationship between the sum of ionization potentials, (= ΣPi) divided by the oxidation number (v), and log LD_{50} (=log T), for the respective metal ions: ΣPi/v=60.92log T +6.9. A more complex, and not so tight relationship between electronegativity and toxicity was tentatively established. Furthermore, a good correlation was found between relative toxicities and the Irwing-Williams series of stability constants of the divalent metal ions, so that, after Bremner and May (1989) "Zinc(II) can be pushed into metalloproteins by maintaining it at a higher concentration than say copper(II)". Toxicities also paralleled the order of stability constants, as given by BASOLO (cited in: Angelici, 1975). Cu^+ belongs to the soft metal ion class of Pearson, preferentially binding soft ligands, whereas Fe^{3+} is a "hard" ion, reacting with O,N,F ligands, and Fe^{2+}, Cu^{2+} and Zn^{2+} are borderline ions (Angelici, 1975).

GENERAL BEHAVIOUR OF Fe, Cu, Zn, AND INVOLVEMENT IN IRRADIATION

The involvement of copper and iron in active oxygen radicals generation is well documented (Halliwell and Gutteridge,1984),parti--cularly for the "ill-placed" iron (Willson,1984).In this respect, irradiation may trigger an increase in serum iron and copper,whereas zinc remained unaltered,after a small increase (Wood et al.,1970; review:Bienvenu et al.,1990).Chelators,reducing agents,and probably H_2O_2 and $^{\cdot}OH$ may extract iron out of its storage proteins,transfe--rrin and hemosiderin (O'Connell and Peters,1987).Iron overload increased lipid peroxidation,e.g. pentane generation,in irradiated rats (Weiss and Kumar,1988).Copper toxicity might be caused by the ability of some of its complexes to stimulate active oxygen species generation,particularly O_2^- if Cu (I) is involved (Petering and Antho--line,1988).The oxidized forms of iron and copper are the most stable,on a thermodynamic basis,and the oxidation of the former is needed before its incorporation in metalloproteins:copper could be needed for it,and ceruloplasmin,e.g. has ferroxidase properties. Zinc is thought to protect againt free radicals indirectly (Willson, 1989).Zinc antagonizes copper effects (Bremner and May,1989),notably hemolysis and lipid peroxidation (Bettger et al.,1978),and is a therapy in Willson's disease (Cossack,1988).Lipid peroxidation might involve iron redox-cycling,possibly antagonized by zinc Girotti et al.,1985).

RECENT EXPERIMENTS:THERAPY OF RADIATION EFFECTS AND INTERPRETATION

Hereunder,we present the results of tentative therapies,made after irradiation,on OF1 mice weighing circa 26g.

TREATMENT	IRRADIATION	SURVIVAL RATES %	REMARKS
Control	6 Gy	100.0	10 Animals
Control	7 Gy	80.0	"
Control	8 Gy	10.0	"
Control	8 Gy	6.7	15 Animals*
Almitrine	8 Gy	46.7	15 Animals
Zinc chloride	8 Gy	40.0	"
Desferal(R)	8 Gy	20.0	"
Diethyldithio--carbamate	8 Gy	6.7	"

TABLE 1:Survival of mice 10 days after irradiation and treatment; Treatments:1/10 LD_{50} of drug,i.p.route,15min.post-irrad^{n}.;*Sham inj^{n}.

Some of the tested drugs are radioprotective:the dithiocarbamate (Bacq et al.1953)and zinc(Floersheim,1986),but they were administe--red before irradiation,as well as d-penicillamine(Ward et al.1980). Almitrine might be a chelator,and Desferal[(R)] accelerates Fe^{2+} autoxidation(Klebanoff et al.,1989).

REFERENCES

Angelici,R.J. (1975):Stability of coordination compounds,in:Inorganic Biochemistry,ed.Eichhorn,G.L.,pp.63-101.Amsterdam:Elsevier.

Bach,Z.M.et al. (1953):Rayons X et agents de chelation,Bull.Acad.Roy.Med.Belge 18,226.

Bettger,J.W. et al.:Effects of copper and zinc status of rats on erythrocyte sta--bility and superoxide dismutase activity,Proc.Soc.Exp.Biol. 158,279-282.

Bienvenu,P.et al.(1963):Toxicité générale comparée des ions métalliques.Relation avec la classification périodique,C.R.Acad.Sc. 256,1043-1044.

Bienvenu,P. (1965):Toxicité générale aigüe et propriétés physicochimiques des ions métalliques,Pharm.D.Thesis,Lyon.

Bienvenu,P. et al.(1990):Antioxidant effects in radioprotection,in:Antioxidants in therapy and preventive medicine,ed.I.Emerit et al.New York:Plenum;in press.

Bremner,I.& May,P.M.(1989):Systemic interactions of zinc,in:Zinc in Human Biology. ed.C.F.Mills,pp.95-198,Heidelberg:Springer-Verlag.

Cossack,Z.T.(1988):The efficacy of oral zinc therapy as an alternative to penici--llamine for Wilson's disease,N.Engl.J.Med. 318,322-323.

Floersheim,G.L. & P.(1986):Protection against ionising radiation and synergism with thiols by zinc aspartate,Brit.J.Radiol.59:597-602.

Girotti et al.(1985):Inhibitory effect of zinc(II) on free radical lipid peroxi--dation in erythrocyte membranes,J.Free Rad.Biol.Med. 1,395-401.

Halliwell,B.& Gurreridge,J.M.C.(1984):Oxygen toxicity,oxygen radicals,transition metals and disease,Biochem.J.,219,1-14.

Klebanoff,S.J. et al.(1989):Oxygen-based free radical generation by ferrous ions and deferoxamine,J.Biol.Chem.264,19765-19771.

O'Connell,M.J.& Peters,T.J.(1987):Ferritin and hemosiderin in free radical gene--ration,lipid peroxidation and protein damage,Chem.Phys.Lipids,45,241-249.

Petering,D.H.& Antholine,W.E.(1988):Copper toxicity:speciation and reactions of copper in biological systems,Rev.Biochem.Toxicol.9,225-270.

Ward,W.F. et al.(1980):Survival of penicillamine-treated mice following whole-body irradiation,Radiat.Res.,81,131-137.

Weiss,J.F.&Kumar,K.S.(1988):Antioxidant mechanisms in radiation injury and radio--protection,in:Cellular antioxidant defense mechanisms,ed.C.K.Chow,Vol II, pp.163-189.Boca Raton:CRC.

Willson,R.L.(1984):Ill-placed iron,oxygen free radicals and disease:some recent and not so recent radiation studies,in:The Biology and Chemistry of active oxygen,ed.J.V.&W.H.Bannister,pp.238-258.Amsterdam:Elsevier.

Willson,R.L.(1989):Zinc and iron in free radical pathology and cellular control, in:Zinc in Human Biology,ed.C.F.Mills,pp.147-172.Heidelberg:Springer Verlag.

Woods,A.H. et al.(1970):Trace metal behaviour in rabbits after whole-body irradiation,Int.J.Appl.Radiat.Isot.,21,389-394.

Angelici,R.J. (1975):Stability of coordination compounds,in:Inorganic Biochemistry,ed.Eichhorn,G.L.,pp.63-101.Amsterdam:Elsevier.

Bach,Z.M.et al. (1953):Rayons X et agents de chelation,Bull.Acad.Roy.Med.Belge 18,226.

Bettger,J.W. et al.:Effects of copper and zinc status of rats on erythrocyte sta--bility and superoxide dismutase activity,Proc.Soc.Exp.Biol. 158,279-282.

Bienvenu,P.et al.(1963):Toxicité générale comparée des ions métalliques.Relation avec la classification périodique,C.R.Acad.Sc. 256,1043-1044.

Bienvenu,P. (1965):Toxicité générale aigüe et propriétés physicochimiques des ions métalliques,Pharm.D.Thesis,Lyon.

Bienvenu,P. et al.(1990):Antioxidant effects in radioprotection,in:Antioxidants in therapy and preventive medicine,ed.I.Emerit et al.New York:Plenum;in press.

Bremner,I.& May,P.M.(1989):Systemic interactions of zinc,in:Zinc in Human Biology, ed.C.F.Mills,pp.95-198,Heidelberg:Springer-Verlag.

Cossack,Z.T.(1988):The efficacy of oral zinc therapy as an alternative to penici--llamine for Wilson's disease,N.Engl.J.Med. 318,322-323.

Floersheim,G.L. & P.(1986):Protection against ionising radiation and synergism with thiols by zinc aspartate,Brit.J.Radiol.59:597-602.

Girotti et al.(1985):Inhibitory effect of zinc(II) on free radical lipid peroxi--dation in erythrocyte membranes,J.Free Rad.Biol.Med. 1,395-401.

Halliwell,B.& Gurreridge,J.M.C.(1984):Oxygen toxicity,oxygen radicals,transition metals and disease,Biochem.J.,219,1-14.

Klebanoff,S.J. et al.(1989):Oxygen-based free radical generation by ferrous ions and deferoxamine,J.Biol.Chem.264,19765-19771.

O'Connell,M.J.& Peters,T.J.(1987):Ferritin and hemosiderin in free radical gene--ration,lipid peroxidation and protein damage,Chem.Phys.Lipids,45,241-249.

Petering,D.H.& Antholine,W.E.(1988):Copper toxicity:speciation and reactions of copper in biological systems,Rev.Biochem.Toxicol.9,225-270.

Ward,W.F. et al.(1980):Survival of penicillamine-treated mice following whole-body irradiation,Radiat.Res.,81,131-137.

Weiss,J.F.&Kumar,K.S.(1988):Antioxidant mechanisms in radiation injury and radio--protection,in:Cellular antioxidant defense mechanisms,ed.C.K.Chow,Vol II, pp.163-189.Boca Raton:CRC.

Willson,R.L.(1984):Ill-placed iron,oxygen free radicals and disease:some recent and not so recent radiation studies,in:The Biology and Chemistry of active oxygen,ed.J.V.&W.H.Bannister,pp.238-258.Amsterdam:Elsevier.

Willson,R.L.(1989):Zinc and iron in free radical pathology and cellular control, in:Zinc in Human Biology,ed.C.F.Mills,pp.147-172.Heidelberg:Springer Verlag.

Woods,A.H. et al.(1970):Trace metal behaviour in rabbits after whole-body irradiation,Int.J.Appl.Radiat.Isot.,21,389-394.

Metal Ions in Biology and Medicine. Eds. Ph. Collery, L.A. Poirier, M. Manfait, J.C. Etienne. John Libbey Eurotext, Paris © 1990, pp. 549-554

Regulation of gene expression by metals : Zinc finger-loop domains in transcription factors, hormone receptors, and proteins encoded by oncogenes

F.W. Sunderman Jr.

Departments of Laboratory Medicine and Pharmacology, University of Connecticut Medical School, 263 Farmington Avenue, Farmington, Connecticut 06032, USA

SCOPE

This article considers seven principal topics: *(1)* the discovery of 'Zn-fingers' in Transcription Factor IIIA ('TFIIIA') of the African toad, *Xenopus laevis*; *(2)* structure of the finger-loop domains in TFIIIA, a prototype of Zn-finger proteins; *(3)* a classification of Zn-finger proteins, with examples of each class, *(4)* studies of Zn in finger-loop domains and evidence that other metals can sometimes substitute for zinc; *(5)* insights into how Zn fingers of steroid and thyroid receptors specify their respective DNA promotor loci, which are termed 'hormone response elements' *(6)* a résumé of the oncogenes proteins that contain finger-loop domains; and *(7)* the author's proposal that substitutions of foreign metal ions for divalent Zn in finger-loop domains of transcription factors, hormone receptors, and proteins encoded by oncogenes could be a molecular mechanism for genotoxicity, including carcinogenesis and teratogenesis. This article summarizes and brings up to date earlier reviews of this subject (Sunderman and Barber 1988, Sunderman 1990a,b).

TRANSCRIPTION FACTOR IIIA

TFIIIA is a 40 kD Zn-metalloprotein that binds to the internal control region of the gene for 5S RNA in *Xenopus laevis*; TFIIIA also binds to the gene product, 5S rRNA, forming the 7S ribonucleoprotein particles that are abundant in *Xenopus* oocytes (Brown 1984). After the amino acid sequence of TFIIIA was determined by Ginsburg et al (1984), Brown et al (1985) and Miller et al (1985) independently reported that the DNA-binding domain, beginning at residue 15 from the N-terminus, contains tandem repeats of segments of approximately 30 residues. These segments possess pairs of cys and his residues arranged in the following pattern:

1 5	C	S F A D -	C	G A A Y N K N W K L Q A	H	L C K -	H	T G E K - P F P
4 5	C	- K E E G	C	E K G F T S L H H L T R	H	S L - T	H	T G E K - N F T
7 5	C	- D S D G	C	D L R F T T K A N M K K	H	F N R F	H	N I K I C V Y V
1 0 7	C	H F E N -	C	G K A F K K H N Q L K V	H	- Q F S	H	T Q Q L - P Y E
1 3 7	C	P H E - G	C	D K R F S L P S R L K R	H	- E K V	H	A G - - - - Y P
1 6 4	C	K K D D S	C	S F V G K T W T L Y L K	H	V A E C	H	Q D - - - L A V
1 9 5	C	- - - D V	C	N R K F R H K D Y L R D	H	- Q K T	H	E K E R T V Y L
2 2 3	C	P R - D G	C	D R S Y T T A F N L R S	H	I Q S F	H	E E Q R - P F V
2 5 4	C	E H A - G	C	G K C F A M K K S L E R	H	S V - V	H	D P E K R K L -

Figure: Amino acid sequence of TFIIIA of *Xenopus* oocytes, arranged to show the repeated segments.

FINGER-LOOP STRUCTURE

The consensus sequence of the repeated segments of TFIIIA consists of 2 cys residues, separated by 2, 4, or 5 other amino acids, followed by a longer string of 12 amino acids, and then 2 his residues separated by 3 or 4 other amino acids. Each repeat ends in a 'linker string' of 4 to 8 amino acids. Brown et al (1985) and Miller et al (1985) speculated that tetrahedral coordination of a Zn^{2+} ion to the thiol sulfur of the cys residues and the imidazole nitrogen of the his residues might arrange each segment in a finger-like configuration that could bind to DNA in a site-specific fashion. During the past five years, more than one hundred examples of proteins with such Zn-finger domains have been discovered throughout the phyla of living organisms, extending from plants, viruses, bacteria, yeast, insects, and amphibia to mammals, including humans (Sunderman 1990b). NMR studies have elucidated the three-dimensional structure of typical finger-loop domains (Lee et al 1989, Párraga et al 1988). The N-terminal half of each finger-loop consists of two β-sheets linked by a hair-pin turn near the 2 cys residues that are coordinated to the Zn^{2+} ion; the C-terminal half of each finger-loop contains an α-helical segment that includes the 2 his residues that are also coordinated to the Zn^{2+} ion. The Zn^{2+} evidently serves as a strut that maintains the folding of the finger, enabling it to fit into the major groove of double stranded DNA, with the β-sheets facing outside, parallel to the phosphate backbones of the DNA, and the α-helix facing inside, making sequence-specific contacts with the edges of the base-pairs (Lee et al 1989). The linker region between each of the fingers contributes a turn, so that the over-all protein conformation resembles, in certain respects, the helix-turn-helix category of DNA-binding structures.

CLASSIFICATION OF FINGER-LOOP DOMAINS

Three major classes of finger-loop domains have been identified (Sunderman 1990b). Proteins (*e.g.*, TFIIIA) with *Class I* fingers (C-C-H-H), usually have 2 to 5 residues in the knuckles, between the pairs of cys or his residues, and 9 to 20 residues in the central loop. The central loop typically contains conserved hydrophobic residues (phe, tyr, leu, ileu, val) that hypothetically stabilize the structure by forming hydrophobic bonds across the finger, and basic or polar residues (arg, asp, glu, his, thr) that probably interact with DNA. *Class I* proteins, which generally have multiple loops, are exemplified by the proteins encoded by several morphogenic genes of *Drosophila* (*e.g.*, the Krüppel and Hunchback genes, which control initial segmentation of the embryo) (Schuh et al 1986, Tautz et al 1987). Another protein of *Class I* is *E coli* UvrA, the damage-recognition unit of the ABC excision nuclease, which is involved in DNA repair (Navaratnam et al 1989). UvrA has two finger-loops that bind to the damaged DNA that will be excised. The regulatory proteins of mammals in *Class I* include: REX-1, a transcription factor of murine F9 teratocarcinoma cells that is suppressed by retinoic acid (Hosler et al 1989); Egr-1 and EGR-2, mammalian transcription factors that increase following mitogenic stimulation of early growth response genes (Seyfert et al 1989); Sp1, a protein that binds to GC-boxes and activates transcription of many viral and cellular promotors (Kadonaga et al 1988); pGLI, a protein that is expressed in many glioblastomas and childhood sarcomas (Kinzler et al 1988); and pZFY, a Zn-finger protein encoded by the Y chromosome, which may play a role in determining the sex of mammals by regulating testicular development (Mardon and Page 1989).

In *Class II* finger proteins, the Zn-binding site comprises four cys residues. Some of these proteins contain a single finger, as in the transposons and light-responsive chloroplast proteins of various plants, and in numerous regulatory factors of yeast (*e.g.*, GAL4, MAL63) (Vodkin and Vodkin 1989, Johnston 1987, Kim and Michels 1988). Other proteins of this class have two fingers, including GF1 (erythroid growth factor), protein kinase C, and several hormone receptors and related proteins (*e.g.*, the thyroid, glucocorticoid, mineralocorticoid, estrogen, progesterone, androgen, vitamin D, and retinoic acid receptors) (Tsai et al 1989, Ono et al 1989, Berg 1989, Beato 1989, Sluyser 1990)

Class III finger proteins have three cys and one his residues as the Zn^{2+}-binding site, including *Type IIIa* (C-C-C-H), exemplified by coat proteins of the tobacco streak and tobacco rattle viruses (Sehnke et al 1989). *Type IIIb* (C-C-H-C) occurs as one of two finger domains of Rad-18, an error-prone DNA-repair enzyme of yeast, and as both finger domains of human poly(ADP-ribose)-polymerase, which is involved in DNA repair and sister-chromatid exchanges (Chanet et al 1988, Mazen et al 1989). *Type IIIc* (C-H-C-C) occurs, for example, in the second finger-loop of the Rad-18 DNA-repair enzyme, in gene 32 protein of phage T4, and in the human Int-1 oncogene product (Giedroc et al 1986, Rijsewijk et al 1987).

STUDIES OF METALS IN FINGER-LOOP DOMAINS

Several investigators have studied the complexation of metals by TFIIIA isolated from *Xenopus* oocytes. Hanas et al (1983) showed that Zn could be removed from TFIIIA by chelation with EDTA or 1,10-phenanthroline, and that apoTFIIIA did not bind to the 5S RNA gene unless Zn was replaced.

Additions of divalent Co, Ni, Mn or Fe did not restore the binding. Miller et al (1985) demonstrated that TFIIIA contained ~9 atoms of Zn per mole and that chelation of Zn in 7S ribonucleoprotein particles dissociated TFIIIA from 5S RNA, as did exposure of the particles to divalent Cd or Cu, which evidently displaced the native Zn. Diakun et al (1986) showed by EXAFS analysis that Zn in TFIIIA is coordinated to two his and two cys residues, consistent with the predicted finger-loop structure. Studying a synthetic peptide corresponding to the second finger of TFIIIA, Frankel et al (1987) showed that the peptide folded after Zn was added *in vitro* and that it also bound Co, with a characteristic change of absorption spectrum. In a further study, Berg and Merkle (1989) showed that Zn or Mn can displace Co bound to the synthetic peptide.

Studies of viral proteins with finger loop domains show that the apoproteins avidly bind Zn^{2+} and that divalent Cd or Co can offtimes be substituted for Zn (Giedroc et al 1986, Keating et al 1988, Culp et al 1988). South et al (1989) reported NMR analyses of a Cd-substituted nucleic acid binding protein, p7, from HIV-1 virus. The NMR spectra showed that cadmium is coordinated to three cys and one his residue, consistent with the predicted structure. Presence of a finger loop domain in the virus that causes AIDS raises the possibility of chemotherapy by chelation of the intrinsic Zn atom. Various DNA-binding proteins from yeast or *E coli* possess finger-loop domains that coincide with the DNA-binding site (Johnston 1987, Nagai et al 1988, Pan and Coleman 1989, Carey et al 1989). In general, the apoprotein does not bind to DNA; the binding can be restored by addition of Zn^{2+}, and in some cases by the addition of Cd^{2+}, Co^{2+}, or Hg^{2+}. Navaratnam et al (1989) reported that UvrA, the damage recognition subunit of the excision nuclease repair enzyme of *E coli* contains two atoms of Zn/mole, with each Zn atom bound tetrahedrally to four cys residues in a finger-loop domain. Site-directed mutagenesis showed that replacement of one of the cys residues by a his residue partially diminished the complementing activity of UvrA; further diminutions were noted when the cys residue was replaced by serine or alanine residues.

Studies by Kadonaga et al (1987) and Westin and Schaffner (1988) showed that DNA-binding activity of transcription factor Sp1 is prevented by removal of Zn from the finger-loop domains and partially restored by Zn-replacement, but not by additions of divalent Cd, Co, Ni, Cu, or Mn. Freedman et al (1988) showed by EXAFS analysis that the Zn atoms in human glucocorticoid receptor are coordinated to four cys groups, consistent with the predicted structure of the finger-loops. Surprisingly, the affinity of the glucocorticoid receptor for Cd was 200-fold that for Zn, suggesting that Cd might be substituted for Zn the finger-domains of Cd-exposed cells or animals. Mazen et al (1989) studied the two finger loops in poly(ADP-ribose)-polymerase, noting that the native enzyme contains two atoms of Zn per mole, and that the Zn could be exchanged in vitro by divalent Cd or Cu, but not by Fe, Mn, or Ni. Removal of Zn prevented the binding of this DNA-repair enzyme to DNA.

FINGER-LOOP DOMAINS OF HORMONE RECEPTORS AND RELATED PROTEINS

As reviewed by Berg (1989), Beato (1989), and Sluyser (1990), human steroid and thyroid hormone receptors, vitamin D receptor, and retinoic acid receptors have a similar organization. A variable immunogenic region is located at the N-terminus. The DNA-binding 'C' region, which contains two finger-loops, 'CI' and 'CII', is located in the middle of the receptor molecules. The hormone-binding site is located near the COOH-terminus. Binding of the respective hormone to the C-terminal site activates the receptor, possibly by releasing a heat-shock protein, 'HSP-90', that may block nuclear uptake of the receptor. The receptor then binds to a specific palindromic hormone response element (HRE), causing transcriptional activation. Two molecules of the steroid receptor may bind on either side of double-stranded DNA that contains the palindromic HRE, forming a dimer.

Mutagenesis studies by Danielson et al (1989) showed that the sites that confer specificity of binding to the HRE are the two amino acids between the second pair of cys residues in finger-loop 'CI'. There are two families of hormone receptors, the glucocorticoid family, which contains gly and ser residues at these positions, and the estrogen family, which contains glu and gly residues at these positions. Studies by Umesono and Evans (1989) indicate that the five amino acids between the first pair of cys residues in the second finger-loop, 'CII', discriminate between the thyroid and estrogen response elements. Severne et al (1988) showed that the fifth conserved cys residue in the CII finger is nonessential for site specific binking of the glucocorticoid receptor to its response element. Green and Chambon (1989) deduced that the CI finger specifies the hormone response element and that the CII finger stabilizes the protein binding to the DNA, possibly by forming a dimeric complex.

FINGER-LOOP DOMAINS OF ONCOGENE PRODUCTS AND RELATED PROTEINS

Several oncogenes and related proteins have been shown to contain Zn-finger domains. The ErbA oncogene, first found in the avian erythroblastosis virus, encodes a protein with two *Class II* fingers in its

DNA-binding domains (Weinberger et al 1986). ErbA protein, which closely resembles the human thyroid receptor, acts as a constitutive repressor that blocks T3 activation of the hormone response element. Int-1, a mammary tumor associated proto-oncogene, is homologous to the *wingless* gene of *Drosophila* (Rijsewijk et al 1987) and encodes a *Class IIIc* finger protein that is expressed in mid-gestational embryos and adult testis. The Ret oncogene encodes a protein with two *Class I* fingers and one *Class IIIb* finger; this protein, designated rfp, is expressed in a variety of human and rodent cell lines, as well as in mouse embryos and in testis (Takehashi et al 1988). The Evi-1 proto-oncogene is a common site of retroviral insertion in murine myeloid leukemias, especially those that proliferate and differentiate in the presence of interleukin-3. The protein product of Evi-1 contains eight *Class I* fingers and two fingers of *Class IIIb*; these finger domains closely resemble the finger domains of TFIIIA of *Xenopus* (Morishita et al 1988) The human GLI oncogene, which is frequently amplified in glioblastomas and childhood sarcomas, contains five fingers of *Class I*; these finger domains are homologous to the Krüppel gene product of *Drosophila* (Kinzler et al 1988). The NGFI-A gene encodes a protein that is rapidly induced by nerve growth factor in PC12 rat pheochromocytoma cells. This protein has three fingers of *Class I*, which are similar to the fingers of human Sp1 transcription factor (Changelian et al 1989). REX-1 is a gene found in murine F9 teratocarcinoma cells. When the expression of its gene product is suppressed by physiological concentrations of retinoic acid, the teratocarcinoma cells undergo differentiation into a homogeneous population of primitive endoderm cells. The REX-1 protein has three finger-loop domains of *Class I* that resemble the fingers of *Xenopus* TFIIIA (Hosler et al 1989). The 11p13 Wilms' tumor gene, which is homozygously deleted in patients with Wilms' nephroblastoma, contains four finger-loops of Class I (Call et al 1990, Gessler et al 1990). This list of oncogene products and related proteins with Zn fingers supports the premise that finger-loop domains of oncogenes may play an important role in neoplastic transformation.

Zn-FINGERS AS POTENTIAL MOLECULAR TARGETS FOR GENOTOXIC METALS

Substitution of foreign metal ions for Zn in finger-loop domains proteins that regulate gene expression could be a molecular mechanism for metal-induced genotoxicity. The following considerations provide the rationale for this hypothesis: *(1)* Peptide complexes of divalent nickel, cobalt, copper, and cadmium are known to generate oxygen free radicals, such as the hydroxyl radical, in biological systems (Ueda et al 1985, Inoue and Kawanishi 1989). *(2)* Zn^{2+} has an ionic radius of 0.72 Å, which is close to the ionic radii of divalent nickel, cobalt, and copper, suggesting that the latter metals would fit within the Zn binding site. The ionic radii of divalent Cd, Hg, and Pb are larger, but these ions are known to substitute for Zn in proteins and peptides with similar binding sites, such as metallothionein and δ-aminolevulinate dehydratase. *(3)* The foreign metals can mimic zinc by forming tetrahedral coordination complexes with thiol and imidazole groups, so that they might substitute *in vivo* in finger-loop domains, affecting the conformation or stability of the DNA-binding structures. *(4)* By generating oxygen free radicals close to specific gene loci, foreign metal ions in finger loops might cause DNA cleavage or DNA-protein cross-links. In support of this concept, one may note that Cu-complexes are used as DNA-cleavage reagents in hydroxyl radical foot-printing assays. *(5)* By substituting for Zn in finger-loop domains of enzyme complexes that repair DNA damage, foreign metal ions might reduce the fidelity of DNA repair or induce sister chromatid exchanges. *(6)* If the genotoxicity induced by such mechanisms occurs in oncogenes or at other critical regulatory sites, carcinogenesis or teratogenesis could hypothetically result.

REFERENCES

Beato, M. (1989): Gene regulation by steroid hormones. Cell 56: 335-344.

Berg, J.M. (1989): DNA binding specificity of steroid receptors. Cell 57: 1065-1068.

Berg, J.M., Merkle, D.L. (1989): On the metal ion specificity of "zinc finger" proteins. J. Am. Chem. Soc. 111: 3759-3761.

Brown, D.D. (1984): The role of stable complexes that repress and activate eukaryotic genes. Cell 37: 359-365.

Brown, R.S., Sander, C., Argos, P. (1985): The primary structure of transcription factor TFIIIA has 12 consecutive repeats. FEBS Lett. 186: 271-274.

Call, K.M., Glaser, T., Ito, C.Y., Buckler, A.J., Pelletier, J., Haber, D.A., Rose, E.A., Kral, A., Yeger, H., Lewis, W.H., Jones, C., Housman, D.E. (1990): Isolation and characterization of a zinc finger polypeptide gene at the human chromosome 11 Wilms' tumor locus. Cell 60: 509-520.

Carey, M., Kakidani, H., Leatherwood, J., Mostashari, F., Ptashne, M. (1989): An amino-terminal fragment of GAL4 binds DNA as a dimer. J. Mol. Biol. 209: 423-432.

Chanet, R., Magana-Schwencke, N., Fabre, F. (1988): Potential DNA-binding domains in the RAD18 gene product of *Saccharomyces cerevisiae*. Gene 74: 543-547.

Changelian, P.S., Feng, P., King, T.C., Milbrandt, J. (1989): Structure of the NGFI-A gene and detection of upstream sequences responsible for its transcriptional induction by nerve growth factor. Proc. Natl. Acad. Sci. USA 86: 377-381.

Culp, J.S., Webster, L.C., Friedman, D.J., Smith, C.L., Huang, W.J., Wu, F.Y.H., Rosenberg, M., Ricciardi, R.P. (1988): The 289-amino acid E1A protein of adenovirus binds zinc in a region that is important for trans-activation. Proc. Natl. Acad. Sci. USA 85: 6450-6454.

Danielsen, M., Hinck, L., Ringold, G.M. (1989): Two amino acids within the knuckle of the first finger specify DNA response element activation by the glucocorticoid receptor. Cell 57: 1131-1138.

Diakun, G.P., Fairall, L., Klug, A. (1986): EXAFS study of the zinc-binding sites in the protein transcription factor IIIA. Nature 324: 698-699.

Frankel, A.D., Berg, J.M., Pabo, C.O. (1987): Metal-dependent folding of a single zinc finger from transcription factor IIIA. Proc. Natl. Acad. Sci. USA 84: 4841-4845.

Freedman, L.P., Luisi, B.F., Korszun, Z.R., Basavappa, R., Sigler, P.B., Yamamoto, K.R. (1988): The function and structure of the metal coordination sites within the glucocorticoid receptor DNA binding domain. Nature 334: 543-546.

Gessler, M., Poustka, A., Cavenee, W., Neve, R.L., Orkin, S.H., Bruns, G.A.P. (1990) Homozygous deletion in Wilms tumours of a zinc-finger gene identified by chromosome jumping. Nature 343: 774-778.

Giedroc, D.P., Keating, K.M., Williams, K.R., Konigsberg, W.H., Coleman, J.E. (1986): Gene *32* protein, the single-stranded DNA binding protein from bacteriophage T4, is a zinc metalloprotein. Proc. Natl. Acad. Sci. USA 83: 8452-8456.

Ginsberg, A.M., King, B.O., Roeder, R.G. (1984): *Xenopus* 5S gene transcription factor, TFIIIA: characterization of a cDNA clone and measurement of RNA levels throughout development. Cell 39: 479-489.

Green, S., Chambon, P. (1989): Chimeric receptors used to probe the DNA-binding domain of the estrogen and glucocorticoid receptors. Cancer Res. 49S: 2282-2285.

Hanas, J.S., Hazuda, D.J., Bogenhagen, D.F., Wu, F.Y.H., Wu, C.W. (1983): *Xenopus* transcription factor A requires zinc for binding to the 5 S RNA gene. J. Biol. Chem. 258: 14120-14125.

Hosler, B.A., LaRosa, G.L., Grippo, J.F., Gudas, L.J. (1989): Expression of REX-1, a gene containing zinc finger motifs, is rapidly reduced by retinoic acid in F9 teratocarcinoma cells. Mol. Cell. Biol. 9: 5623-5629.

Inoue, S., Kawanishi, S. (1989): ESR evidence for superoxide, hydroxyl radicals and singlet oxygen produced from hydrogen peroxide and nickel(II) complex of glycylglycyl-L-histidine. Biochem. Biophys. Res. Comm. 159: 445-451.

Johnston, M. (1987): Genetic evidence that zinc is an essential co-factor in the DNA binding domain of GAL4 protein. Nature 328: 353-355.

Kadonaga, J.T., Carner, K.R., Masiarz, F.R., Tjian, R. (1987): Isolation of cDNA encoding transcription factor Sp1 and functional analysis of the DNA binding domain. Cell 51: 1079-1090.

Kadonaga, J.T., Courey, A.J., Ladika, J., Tjian, R. (1988): Distinct regions of Sp1 modulate DNA binding and transcriptional activation. Science 242: 1566-1570.

Keating, K.M., Ghosaini, L.R., Giedroc, D.P., Williams, K.R., Coleman, J.C., Sturtevant, J.M. (1988): Thermal denaturation of T4 gene 32 protein: effects of zinc removal and substitution. Biochemistry 27: 5340-5245.

Kim, J., Michels, C.A. (1988): The MAL63 gene of *Saccharomyces* encodes a cysteine-zinc finger protein. Curr. Genet. 14: 319-323.

Kinzler, K.W., Ruppert, J.M., Bigner, S.H., Vogelstein, B. (1988): The *GLI* gene is a member of the *Krüppel* family of zinc finger proteins. Nature 332: 371-374.

Lee, M.S., Gippert , G.P., Soman, K.V., Case, D.A., Wright, P.E. (1989): Three-dimensional solution structure of a single zinc finger DNA-binding domain. Science 245: 635-637.

Mardon, G., Page, D.C. (1989): The sex-determining region of the mouse Y chromosome encodes a protein with a highly acidic domain and 13 zinc fingers. Cell 56: 765-770.

Mazen, A., Menissier-de Murcia, J., Molinete, M., Simonin, F., Gradwohl, G., Poirier, G., de Murcia, G. (1989): Poly(ADP-ribose)polymerase: a novel finger protein. Nucl. Acids Res. 17: 4689-4698.

Miller, J., McLachlan, A.D., Klug, A. (1985): Repetitive zinc-binding domains in the protein transcription factor IIIA from *Xenopus* oocytes. EMBO J. 4: 1609-1614.

Morishita, K., Parker, D.S., Mucenski, M.L., Jenkins, N.A., Copeland, N.G., Ihle, J.N. (1988): Retroviral activation of a novel gene encoding a zinc finger protein in IL-3-dependent myeloid leukemia cell lines. Cell 54: 831-840.

Nagai, K., Nakaseko, Y., Nasmyth, K., Rhodes, D. (1988): Zinc-finger motifs expressed in *E. coli* and folded *in vitro* direct specific binding to DNA. Nature 332: 284-286.

Navaratnam, S., Myles, G.M., Strange, R.W., Sacar, A. (1989): Evidence from extended X-ray absorption fine structure and site-specific mutagenesis for zinc fingers in UvrA protein of Escherichia coli. J. Biol. Chem. 264: 16067-16071.

Ono, Y., Fujii, T., Ogita, K., Kikkawa, U., Igarashi, K., Nishizuka, Y. (1989): Protein kinase C ξ subspecies from rat brain: its structure, expression, and properties. Proc. Natl. Acad. Sci. USA 86: 3099-3103

Pan, T., Coleman, J.E. (1989): Structure and function of the Zn(II) binding site within the DNA-binding domain of the GAL4 transcription factor. Proc. Natl. Acad. Sci. USA 86: 3145-3149.

Párraga, G., Horvath, S.J., Eisen, A., Taylor, W.E., Young, E.T., Klevit, R.E. (1988): Zinc-dependent structure of a single-finger domain of yeast ADR1. Science 241: 1489-1492.

Rijsewijk, F., Schuermann, M., Wagenaar, E., Parren, P., Weigel, D., Nusse, R. (1987): The *Drosophila* homolog of the mouse mammary oncogene int-1 is identical to the segment polarity gene *wingless*. Cell 50: 649-657.

Schuh, R., Aicher, W., Gaul, U., Côté, S., Preiss, A., Maier, D., Seifert, E., Nauber, U., Schröder, C., Kemler, R., Jäckle, H. (1986): A conserved family of nuclear proteins containing structural elements of the finger protein encoded by *Krüppel*, a *Drosophila* segmentation gene. Cell 47: 1025-1032.

Sehnke, P.C., Mason, A.M., Hood, S.J., Lister, R.M., Johnson, J.E. (1989): A "zinc-finger"-type binding domain in tobacco streak virus coat protein. Virology 168: 48-56

Severne, Y., Wieland, S., Schaffner, W., Rusconi, S. (1988): Metal binding 'finger'structures in the glucorticoid receptor defined by site-directed mutagenesis. EMBO J. 7: 2503-2508.

Seyfert, V.L., Sukhatme, V.P., Monroe, J.G. (1989): Differential expression of a zinc finger-encoding gene in response to positive versus negative signaling through receptor immunoglobulin in murine B lymphocytes. Mol. Cell. Biol. 9: 2083-2088.

Sluyser, M. (1990): Steroid/thyroid receptor-like proteins with oncogenic potential: a review. Cancer Res. 50: 451-458.

South, L., Kim, B., Summers, M.F. (1989): ^{113}Cd NMR studies of a 1:1 Cd adduct with and 18-residue finger peptide from HIV-1 nucleic acid binding protein. J. Am. Chem. Soc. 111: 395-396.

Sunderman, F.W., Jr. (1990a): Cadmium substitution for zinc in finger-loop domains of gene-regulating proteins as a possible mechanism for the genotoxicity and carcinogenicity of cadmium compounds. Toxicol. Environ. Chem. 27: 131-141.

Sunderman, F.W., Jr. (1990b): Finger-loop domains and trace metals. In Proceedings of the Second International Conference on Trace Element Research in Humans, ed H. Tomita, Tokyo: Springer-Verlag. (In press).

Sunderman, F.W., Jr., Barber, A.M. (1988): Finger-loops, oncogenes, and metals. Ann. Clin. Lab. Sci. 18: 267-288.

Takahashi, M., Inaguma, Y., Hiai, H., Hirose, F. (1988): Developmentally regulated expression of a human "finger"-containing gene encoded by the 5' half of the r*et* transforming gene. Mol. Cell Biol. 7: 1978-1983.

Tautz, D., Lehmann, R., Schnürch, H., Schuh, R., Seifert, E., Kienlin, A., Jones, K., Jäckle, H. (1987): Finger protein of novel structure encoded by *hunchback*, a second member of the gap class of *Drosophila* segmentation genes. Nature 327: 383-389.

Tsai, S.F., Martin, D.I.K., Zon, L.I., D'Andrea, A.D., Wong, G.G., Orkin, S.H. (1989): Cloning of cDNA for the major DNA-binding protein of the erythroid lineage through expression in mammalian cells. Nature 339: 446-451.

Ueda, K., Kobayashi, S., Morita, J., Komano, T. (1985): Site-specific DNA damage caused by lipid peroxidation products. Biochim. Biophys. Acta 824: 341-348.

Umesono, K., Evans, R.M. (1989): Determinants of target gene specificity for steroid/thyroid hormone receptors. Cell 57: 1139-1146.

Vodkin, M.H., Vodkin, L.O. (1989): A conserved zinc finger domain in higher plants. Plant Mol. Biol. 12: 593-594.

Weinberger, C., Thompson, C.C., Ong, E.S., Lebo, R., Gruol, D.J., Evans, R.M. (1986): The c-*erb*-A gene encodes a thyroid hormone receptor. Nature 324: 641-646.,

Westin, G., Schaffner, W. (1988): Heavy metal ions in transcription factors from HeLa cells: Sp1, but not octamer transcription factor requires zinc for DNA binding and for activation function. Nucl. Acids Res. 16: 5771-5781.

Metal Ions in Biology and Medicine. Eds. Ph. Collery, L.A. Poirier, M. Manfait, J.C. Etienne. John Libbey Eurotext, Paris © 1990, pp. 555-557

How coordination chemistry can help to study the interactions of biological molecules with metal ions

M. Aplincourt, C. Gerard, R. Hugel, J.C. Pierrard, J. Rimbault

Université de Reims Champagne-Ardenne, Faculté des Sciences, Laboratoire de Chimie Minérale, BP 347, 51062 Reims, France

The aim of this short communication is to present ,with some examples, coordination chemistry as a tool for study the interactions in aqueous solution of biological molecules with metal ions.

During the last years, we have investigated the complexation of transition (iron, copper, nickel, cobalt, manganese...) and non-transition metal ions (calcium, magnesium, cadmium, zinc...) with ligands belonging to the following types:

- thiadicarboxylic acids,
- aminoacids derived from cysteine,
- heterocyclics diketones,
- dihydroxybenzoic acids and their derivatives,
- aromatic hydroxamates.

Potentiometry and spectrophotometry are the most convenient and successful methods employed for metal complex equilibrium measurements.

Potentiometric measurements are based on the difference of affinity of the donor ligand atoms for the proton or for the metal cation. The potentiometric studies give a rapid information on the metal-ligand interaction strength and allow to calculate the stability constants of the formation equilibria :

$$ML_{n-1} + L \Leftrightarrow ML_n \quad (K_n = [ML_n] / [ML_{n-1}][L])$$

For example, the reaction of the 2,3 dihydroxybenzoic acid (2,3dhb) with copper(II) and iron(III) is a good illustration of this method (Aplincourt, 1986). The neutralisation curves of 2,3dhb-metal systems (B and C curves) and of the ligand at same concentration (A curve) are reported on figure 1.

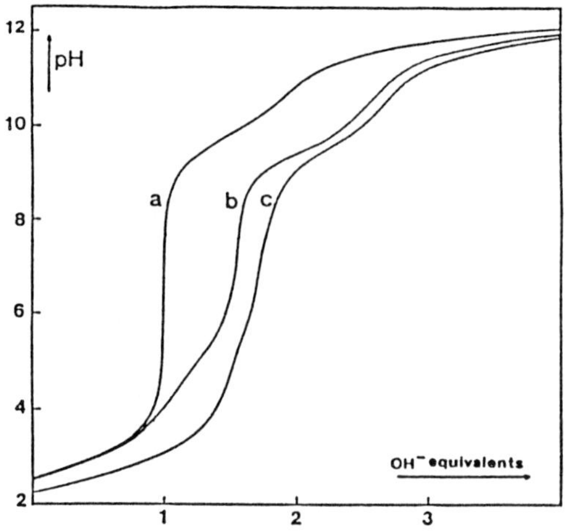

Figure 1: titration curves for 2,3dhb alone (A), 2,3dhb-Cu^{2+} (B), 2,3dhb-Fe^{3+} (C)
Ligand concentration : 10^{-2} $mol.l^{-1}$
Metal concentrations : 2.5 10^{-3} $mol.l^{-1}$

From this results, we observe that :
- in acidic medium (pH<3), the two cations have a very different behaviour. The curve of the 2,3dhb-Fe(III) system is below the one of the ligand alone, indicating an increase of the proton concentration and a complexing reaction.The low pH region of the curve for2,3dhb-Cu(II) system is coincident with the ligand curve and therefore implies no complex formation with the metal ion.
- the interaction ligand-Fe(III) is stronger than the one observed with copper(II). The pH differences between the titration curve of the ligand and the corresponding ones for the 2,3dhb-metal systems increase with the stability of the complexes : Cu < Fe.

The values of the equilibrium constants are calculated from the pH or pM (obtained with selective electrodes) measurements (Martell, 1988).

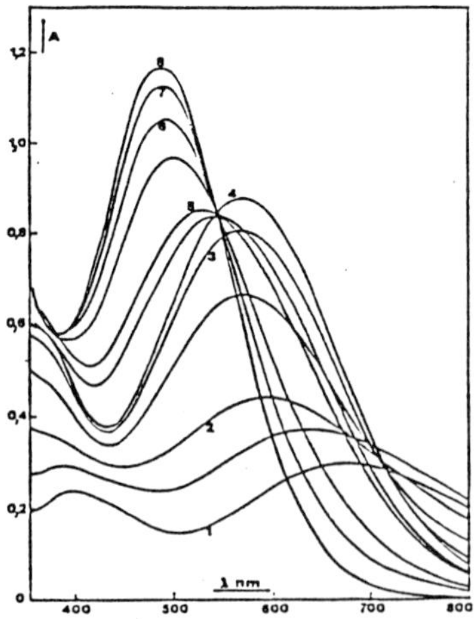

Figure 2: Absorption spectra of the 3,4dhb-Fe(III) system
ligand concentration : 1.94 10^{-3} $mol.l^{-1}$
metal concentration : 2 10^{-4} $mol.l^{-1}$
3 < pH < 9

Spectrophotometric measurements are based on the absorbance differences observed in the visible and ultraviolet region for the free metal ion and the formed complexes. For example, 3,4 dihydroxybenzoic acid react with iron(III), forming strong colored complexes (Gérard, 1987, Aplincourt, 1987). The electronic spectra, corresponding to solutions with constant concentration in metal ion and ligand but with variable pH ($3.3 < pH < 9$), are shown on figure 2. Two isosbestic points are obtained, at 707 and 546 nm, indicating the existence of three absorbing species. The equilibrium domaines being separated, the corresponding constants are calculated from the absorbances.

Potentiometry and spectrophotometry are complementary methods. Their results permit :

- to compare the stability of the complexes formed for several metals with a ligand,
- to study the influence of the susbtituents on the ligand-metal interaction,
- to determine the domaines of existence of the equilibrium species and to calculate their concentration for particular pH values.

REFERENCES

APLINCOURT, M. *et al.*, (1986) : Thermodynamic stability of copper(II) complexes with dihydroxybenzoic acids. Part I. In J. Chem. Research , (S) 134-135, (M) 1253-1273

APLINCOURT, M. *et al.*, (1987) : Thermodynamic stability of copper(II) and iron(III) complexes with dihydroxybenzoic acids. Part III. In J. Chem. Research , (S) 398-399, (M) 3214-3241

GERARD, C. *et al.*, (1987) : Thermodynamic stability of copper(II) and manganese(II)complexes with three dihydroxybenzoic acids. Part II. In J. Chem. Research , (S) 294-295, (M) 2516-2535

MARTELL, A. E. and MOTEKAITIS, R. J., (1988) : The determination and use of stability constants VCH, Verlagsgeselschaft mbH, Weinheim

Metal Ions in Biology and Medicine. Eds. Ph. Collery, L.A. Poirier, M. Manfait, J.C. Etienne. John Libbey Eurotext, Paris © 1990, pp. 558-560

Comparison of the local structures of common (Cu, Fe, Mn, Zn), toxic (Ni, Ag) and therapeutic (Sn) metal complexes formed with carbohydrates and nucleotides

L. Nagy*, T. Yamaguchi, L. Korecz, K. Burger

Department of Inorganic and Analytical Chemistry, A. József University, H-6701 Szeged, P.O. Box 440, Hungary

Introduction

Carbohydrates and related compounds have long been known to form complexes with metal ions. In spite of this, the field of carbohydrate complexes continued to remain largely unexplored. The complexing of metal ions with carbohydrates and their derivatives has received considerable interest only in the past two decades, mainly because of the possible importance of such interactions in a variety of biological processes, e.g. in the binding of metal ions to cell walls, etc. Structural informations - bond distances, bond angles - about transition metal ion - sugar derivative complexes are very difficult to obtaine, because of the lack of suitable single crystals in solid state. In recent years EXAFS (Extended X-ray Absorption Fine Structure) spectroscopy has proved to be succesful in elucidating the short-range order of atoms in solution and in solid state. Our aim is, therefore, to determine the local structure of the sugar complexes of titled metal ions by the EXAFS method. XANES (X-ray Absorption Near Edge Structure), Mössbauer and ESR spectra of the sample were also measured in order to obtain information on the electronic state of the metal ions in the complexes. EXAFS and XANES spectra were measured in transmission mode using BL10B and BL7C stations at the Photon Factory in the Nation Laboratory for High Energy Physics (Japan).

Results and Discussion

Nine nickel(II), five zinc(II), five manganese(II) and three silver(I) complexes of 2-(polyhydroxy-alkil)-thiazolidine-4-carboxylic acid (PHTAc) derivatives have been synthetized in methanolic and alkaline solutions. The EXAFS and XANES spectra supported the distorted octahedral structure of nickel(II) complexes with a mean Ni-O distance of 202 pm irrespective of the conformation of sugar moiety in the molecule. The Ni...C distances in the second and third shells (287 pm and 390 pm) are compatible with values of chelated compounds of nickel(II). The zinc(II) complexes are all four·coordinated with either square planar or tetrahedral structure, except of the L-rhamnose complex which is six coordinated. The mean Zn-O and Zn...C distances in former ones are 203 pm and 290 pm and in latter 210 and

250 pm, respectively (Nagy1 et. al., 1990). All manganese(II)-PHTAc complexes have a slightly distorted octahedral structure with a mean Mn-O distance of 215 pm. The Mn...C distance in the second shell are slightly longer (305 pm) than those of the nickel(II) (280 pm) and zinc(II) (292 pm) complexes. In the silver(I) complexes the metal is coordinated to sulfur and nitrogen of PHTAc. The coordination number is two, while the average Ag-S and Ag-N distance is 216 pm (Nagy2 et al. 1990).
A series of diethyltin(IV) complexes with carbohydrate ligands (aldoses, polyalcohols, sugar acids, sugar amines, and di- and trisaccharides) were prepared. The composition of the complexes were determined by standard analytical methods. The results showed that complexes containing the diethyltin moiety and the carbohydrate ligand in 1:1 ratio are formed. The IR spectra are consistent with the presence of tin-carbohydrate oxygen vibrations. The comparison of experimental quandrupole splitting (PS) with those calculated on the basis of the partial quadrupole splitting (PQS) concept revealed that the compounds are of three types: complexes with central tin(IV) atoms a) in purely trigonal bipyramidal surroundings, b) in purely octahedral surroundings, c) in both octahedral and trigonal bipyramidal arrangements in approximately 1:1 ratio (Nagy3 et al. 1990). The local structures of complexes in solid state have been determined by EXAFS (Nagy4 1990, et al.). Dioxastannolane units are associated into an infinite ribbon polymer, in which the tin is bound by two carbon atoms and by three or four oxygen atoms. Within each unit the average tin-oxygen and tin-carbon bond lengths in the first coordination sphere are 200 pm, while the Sn...C distances in the second shell are 270 pm similar to X-ray diffraction data of analogous complexes.
Three kinds of iron(III) complexes formed with fructose, mannitol and gluconic acid were prepared. The Fe K-edge absorption spectra of the complexes were measured both in aqueous solution and in the solid state to reveal the structure from the analysis of their EXAFS and XANES. It was concluded that the iron(III) sugar complexes have a distorted octahedral structure with a mean Fe-O distance of 195 pm irrespective of the sugar type ligands both in aqueous solution and in the solid state. In the case of iron(III) fructose complex the dimerization of the complex was evidenced and the Fe...Fe distance is 310 pm (Nagy et al. 1989).
Interactions between D-ribose or D-glucosamine and copper(II) ion have been studied by electronic spectra, ESR, and Cu K-XANES and EXAFS measurements. All data have shown that (1:1) copper(II)-sugar complexes are formed in aqueous solution at a high pH. The ESR spectra of the two copper(II) complexes have indicated the presence of a dimeric species via hydroxide oxygen atoms in the solution. It was found that the oxygen coordination geometry (distorted octahedral) around copper(II) ion and the Cu-O bond lengths ($Cu-O_{eq}$ = 191 pm, $Cu-O_{ax}$ = 230 pm) whithin the copper(II)-sugar complexes are not very different from those of the hexaaqua copper(II) ion. EXAFS Fourier transforms have clearly shown a peak ascribed to non-bonded Cu(II)...C interactions, indicating the formation of chelate rings of the sugar ligand around the copper(II) ion. The presence of the second neighbour Cu...Cu interaction was not conclusive from the EXAFS analysis, because the Cu...Cu and Cu...C (271 pm) peaks are probably overlapping in the Fourier transform (Nagy5 et al. 1990).
The Cu-O bond lengths (190 pm) within the square-planar copper(II)-adenosine and uridine complexes in water-DMSO solution at different pHs are not significant different from each others, but shorter than

those within the aqua complex (195 pm). On other hand the Cu-O distance agrees very well with data measured for copper(II)-D-ribose complex in solution, supporting the D-ribose moiety coordination of nucleosides to copper(II) (Yamaguchi et al. 1990).

References

Nagy1, L., Gajda, T., Yamashita, S., Noura, M., Yamaguchi, T. and Wakita, T. (1990): EXAFS and XANES studies of nickel(II) and zinc(II) complexes of some 2-(polyhydroxyalkil)-thiazolidine-4-carboxylic acids. Accepted for publication in Inorg. Chem.

Nagy2, L., Gajda, T., Yamashita, S., Noura, M., Yamaguchi, T. and Wakita, T. (1990) EXAFS and XANES studies of silver(I) and manganese(II) complexes of some 2-(polyhydroxy-alkil)-thiazolidine-4-carboxylic acids. To be published in Inorg. Chem.

Nagy3, L., Korecz, L., Kiricsi, I., Zsikla, L. and Burger, K. (1990): Synthesis, Mössbauer and IR spectroscopic studies and thermal behaviour of diorganotin(IV) complexes with carbohydrate ligands. Struct. Chem. in press

Nagy4, L., Gyurcsik, B., Burger, K., Yamashita, S., Nomura, M., Yamaguchi, T. and Wakita, H. (1990): The local structure of diorganotin(IV) complexes with carbohydrate. To be published in J. Chem. Soc. Dalton Trans.

Nagy, L., Ohtaki, H., Yamaguchi, T. and Nomura, M. (1989): EXAFS study of iron(III) complexes of sugar-type ligands. Inorg. Chim. Acta 159, 201 (1989)

Nagy5, L., Yamaguchi, T., Nomura, M., Páli, T. and Ohtaki, H. (1990): Copper(II) complexes of D-ribose and D-glucosamine. Spectroscopic and EXAFS studies. Accepted for publication in Inorg. Chim. Acta

Yamaguchi, T., Nagy, L., Nomura, M. and Ohtaki, H. (1990): EXAFS and XANES studies of copper(II) complexes formed with adenosine and uridine. Accepted for publication in Bull. Chem. Soc. Jpn.

Metal Ions in Biology and Medicine. Eds. Ph. Collery, L.A. Poirier, M. Manfait, J.C. Etienne. John Libbey Eurotext, Paris © 1990, pp. 561-563

A comparative study of DNA interaction with divalent metal ions. Metal ion binding sites and DNA conformational changes, studied by laser raman spectroscopy

H.A. Tajmir-Riahi*, M. Langlais, R. Savoie

*Département de chimie, Université Laval, Québec (Qué.), Canada G1K 7P4 and * Centre de recherche en biophysique, Université du Québec à Trois-Rivières, Trois-Rivières (Qué.), Canada G9A 5H7*

The stabilizing and destabilizing effects of various metal ions on duplex DNA were quickly recognized in earlier studies, as the melting temperature of DNA, usually determined from the change of optical density at 260 nm in the UV spectrum, was found to depend very strongly on the nature of the metal ion (Eichhorn & Shin, 1968). Divalent metal ions were classified in the decreasing order Mg^{2+}, Co^{2+}, Ni^{2+}, Mn^{2+}, Zn^{2+}, Cd^{2+}, and Cu^{2+} for their relative ability to bind to the phosphate groups rather than to the bases and, consequently, to stabilize the double-helical structure of DNA by neutralizing the negative charges on the polynucleotide backbone. We have recently used Raman spectroscopy to clarify the modes of interaction of Pb^{2+} and Cu^{2+} ions with DNA, the latter having a very high affinity for the bases (Tajmir-Riahi *et al.*, 1988). These results are compared with those obtained with two types of metal ions, Mg^{2+} and Ca^{2+}, which are generally believed to interact exclusively with the phosphate groups of nucleic acids, and with Zn^{2+} and Cd^{2+} ions, which have a higher affinity for the bases and are particularly interesting due to the fact that they can bring about the renaturation of thermally denaturated DNA.

As some of the salts studied in this work, especially at low concentration, did not modify appreciably the Raman spectrum of DNA (4% w/w, 0.1 M DNA(P)), difference spectra obtained by subtracting the spectrum of free DNA from that of the mixtures are presented in order to better show the spectral changes caused by metal-DNA interactions. The subtraction was done using the 786 cm^{-1} band of DNA as an internal standard. Characteristic difference spectra, obtained by subtracting the spectrum of free DNA from those of the polynucleotide in the presence of Mg^{2+}, Ca^{2+}, Zn^{2+}, Pb^{2+}, Cd^{2+}, and Cu^{2+} ions at various metal:DNA(P) molar ratios are reproduced in Fig. 1. The main features in these difference spectra consist of negative or positive peaks, particularly at *ca.* 786, 1093, 1337, 1374, 1488, and 1578 cm^{-1}. The 1093 cm^{-1} band is related to the phosphate groups, whereas the others arise from the nucleic bases.

Metal binding to the PO_2^- groups in DNA is reflected in the Raman spectra by the decrease in intensity of the 1093 cm^{-1} band, which arises from the symmetric PO_2^- stretch. Assuming that the observed intensity decrease of this band is caused by a direct type of interaction (covalent binding) with one particular oxygen atom of the phosphate group, giving a $-P(=O)-O-Metal^+$ type of complex, it is found that the amount of bound metal is comparable for Mg^{2+}, Ca^{2+}, and Cd^{2+} ions, and somewhat higher for the other ions.

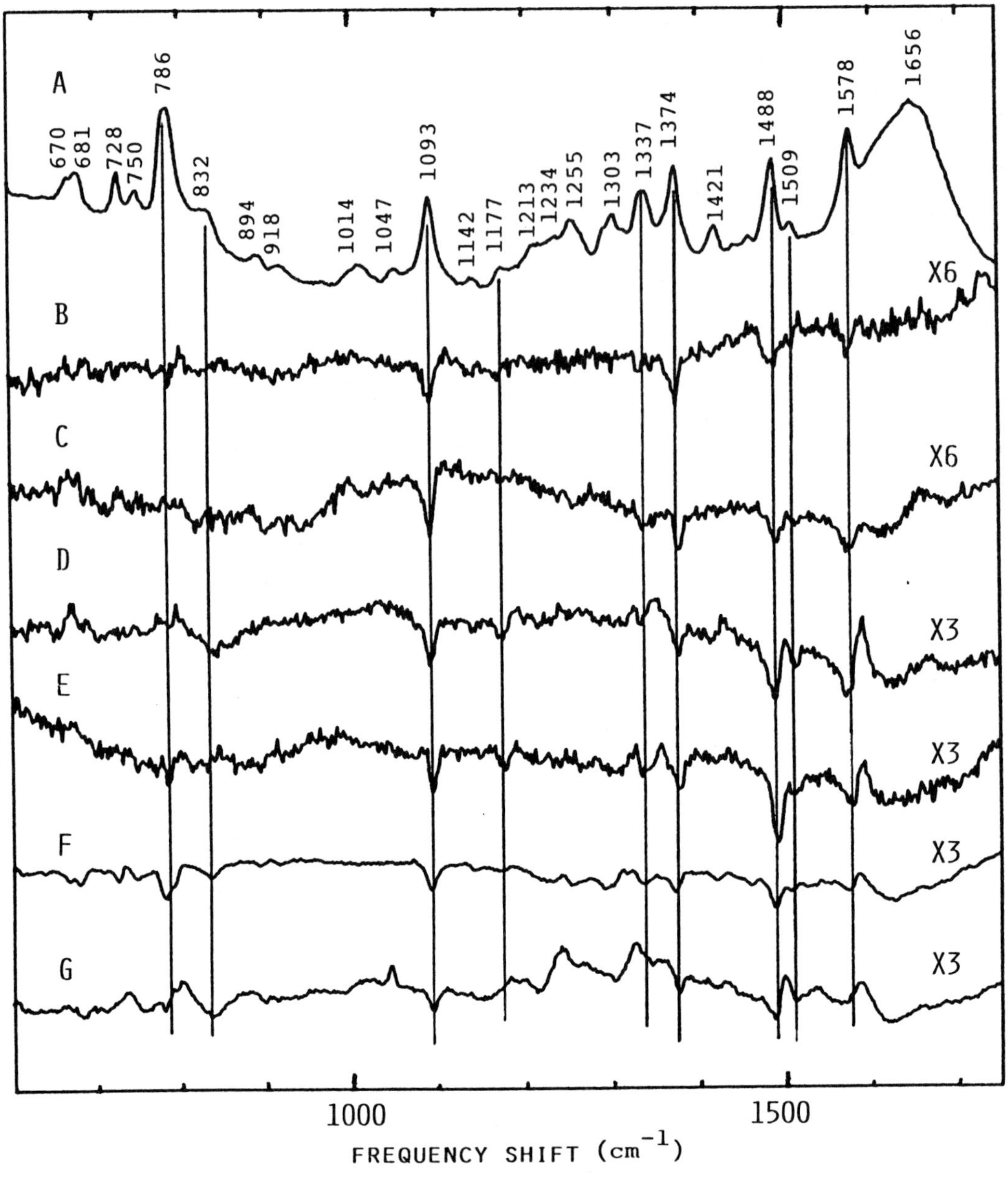

Fig. 1. (A) Raman spectrum of aqueous (4% w/w) calf thymus DNA. Difference spectra obtained from solutions with (B) Mg^{2+} ions at a 1:1 metal:DNA(P) ratio (r = 1), (C) Ca^{2+} (r = 5), (D) Zn^{2+} (r = 4), (E) Cd^{2+} (r = 4), (F) Pb^{2+} (r = 1), and (G) Cu^{2+} (r = 0.5).

The difference Raman spectra of DNA in the presence of very small quantities of divalent ions (metal:DNA ratios ca. 1:10) show negative features in some of the regions where strong DNA bands occur, namely at 1337, 1374, 1488, and 1578 cm^{-1} (Fig. 1). These intensity changes are unexpected for the alkaline-earth ions, as they are not believed to interact with the nucleic bases. Furthermore, they are almost independent of the relative metal concentration. Since the above DNA bands are known to decrease in intensity with increasing base-stacking interactions (Raman hypochromism) (Erfurth & Peticolas, 1975), we believe that the observed spectral modifications are for the most part caused by a structural change in DNA, which occurs at low metal concentration. It is not clear, however, if this change is related to that observed in previous UV and CD spectroscopic studies, which suggested that aqueous DNA undergoes a conformational change from the B form to a "C-like" structure in the presence of divalent metal ions (Zimmer *et al.*, 1974). Our results do not support such a drastic change in structure.

The intensity changes of the bands due to the nucleic bases of DNA in the presence of the transition metal ions are more pronounced than with Mg^{2+} and Ca^{2+} and this is a clear indication of interaction with the bases. The effects on the 1488 and 1578 cm^{-1} bands are a clear indicative of bonding at the N7 (and possibly N1) position of the guanine bases. The involvement of A·T base pairs in the metallation process is also indicated from the changes in the 1509, 1374, and 1177 cm^{-1} bands. The positive peak near 800 cm^{-1} in some of the difference spectra suggests binding at N3 of C, which would be consistent with several models in which interstrand cross-linking has been proposed, such as with the N3 atom of C and N7 of G, to explain the metal-facilitated rewinding of DNA by some metal ions.

The difference spectrum obtained with Cd^{2+} at 5:1 metal:DNA ratio (spectrum not shown) points to partial denaturation of DNA. In particular, a loss in intensity of the 832 cm^{-1} band, characteristic of the B-form conformation, is observed. The situation is somewhat similar to that with Cu^{2+} at a 1:2 metal:DNA(P) ratio, the latter type of ion having a much more pronounced destabilizing effect on double-helical DNA. One major difference, however, is found with Cu^{2+} which, unlike Cd^{2+}, gives a large positive peak near 800 cm^{-1} in its difference spectrum with DNA; this is similar to the Zn^{2+} case, although the effect is more pronounced. This suggests that Cu^{2+} ions, contrary to Cd^{2+}, causes G·C base pairs to pull apart, this being followed by metallation at the N1 position of G and N3 of C. Spectral changes in the 1300-1400 cm^{-1} region, rich in A and T vibrations, and in the 1600-1700 cm^{-1} region (in D_2O solution), where several vibrations of T are active, suggest that base pairs in A·T-rich regions of DNA become separated. Normally protected groups, such as N1 of A, and N3-H and C4=O of T are now available for metal binding and some ions migrate from the phosphate groups to fill these new sites.

REFERENCES

Eichhorn, G.L. and Shin, Y.A. (1968): Interaction of metal ions with polynucleotides and related compounds. XII. The relative effect of various metal ions on DNA helicity. *J. Am. Chem. Soc.* **90**, 7323-7328.

Erfurth, S.C. and Peticolas, W.L. (1975): Melting and premelting phenomenon in DNA by laser Raman scattering. *Biopolymers,* **14**, 247-264.

Tajmir-Riahi, H.A., Langlais, M., and Savoie, R. (1988): A laser Raman spectroscopic study of the interaction of calf-thymus DNA with Cu(II) and Pb(II) ions; metal ion binding and DNA conformational changes. *Nucleic Acids Res.* **16**, 751-762.

Zimmer, Ch., Luck, G., and Triebel, H. (1974): Conformation and reactivity of DNA. IV. Base binding ability of transition metal ions to native DNA and effect on helix conformation with special reference to DNA-Zn(II) complex. *Biopolymers,* **13**, 425-453.

Metal Ions in Biology and Medicine. Eds. Ph. Collery, L.A. Poirier, M. Manfait, J.C. Etienne. John Libbey Eurotext, Paris © 1990, pp. 564-566

Research at the relationship among pH, trace elements (Cd, Zn, Cu) and metallothioneins (MTs)

L.W. Xia, S.X. Liang, C.L. Xiao, X.J. Zhou

Postal Code 410078, Analytical Testing Centre of Hunan Medical University, Changsha, Hunan, P.R. of China

SUMMARY

As is well known, PH has important influence on the physical and chemical properties (such as valence, affinity combination, dissociation, ionigation power and configuration to a great extent) of each element and compound within a system, the possibility, degree, and direction of a chemical reaction (Ling-Wei Xia et., 1988). Therefore, our research is to study the relationship among PH, trace elements and MTs, and go further into the internal relations among PH, trace elements and living phenomena. According to the results of experiments, a new view was put forward that the mechanism of production of MTs in vivo, the amounts of trace elements binding to MTs, and the physiological functions and toxicological efficiency of MTs and trace elements are all in an essential connection with injecting PH. The physiological functions of MT and trace elements in vivo are worthy of further research.

Key words: PH, MT, HPLC, AAS, trace element

Scientists have researched more and more deeply on the biological functions and analytical methods of methallothioneins (MTs) (Chen, R.W. and Ganther, H.E., 1975; Cherian, M.G., 1974; Klauser, S. and Kagi, J.H.R. et al., 1983; Nordberg, G.F. and Nordberg, M. et al., 1972; Piotrowski, J.K. and Bolanoseke, W. et al., 1973; Suzuki, K.T., 1980; Suzuki, K.T. and Sunaga, H. et al., 1983), since Margoshes etc. originally isolated MTs from horse kidney in 1957 (Margoshes, M. and Vallee, B.L., 1957). MTs is an inducible protein of low molecular weight. Some articles reported that MTs in vivo is mainly induced by Cd, Zn, Cu, Fe, etc.. Thereby, the metabolism function and toxicological efficiency of MTs and trace elements (Cd, Zn, etc.) in living body were explained.However, conclusive results are not available (Bremner I. et al., 1975; Chang, C.C. et al., 1980; Foulbes, E.C., 1978; Jones, M.M.et al., 1979; Kiyoshi, Z. et al., 1976; Mcintyre, A.D. et al., 1975; Nordberg, G.F. et al., 1971; Probst, G.S. et al., 1977; Terhaar, C.J. et al., 1965; Webb, M., 1972; 1979; Winge, D.R. et al., 1975).

In order to find out the relations among PH, trace elements and MTs, animal and analytical experiments were carried out in this research. It was shown from the results that both MTs and trace elements (Cd, Zn, etc.) binding to MTs in vivo varied with the injecting PH.

MATERIALS AND METHODS

Healthy Wister rats were elected and randomly divided into 16 groups in experiment. Each group had 15 rats. A animal-model experiment was done as follows: Each rat was injected at abdomen with relevant solution 24h prior to killing. No.1-14 groups were injected with PH 1-14 solution (corresponding to 10^{-1}M, 10^{-2}M, 10^{-3}M, 10^{-4}M, 10^{-5}M, 10^{-6}M, 10^{-7}M hydrochloric acid and 10^{-6}M, 10^{-5}M, 10^{-4}M, 10^{-3} M, 10^{-2} M, 10^{-1} M, 1.0 M sodium hydroxide, respectively) , respectively (0.2ml/100g). The No.15 and No.16 groups were injected with normal saline (0.2ml/100g, PH6.64). Draw blood from aortas, then centrifuged at 4000 revolutions /min. Livers and kidneys were removed and homogenized(B-30 sonifier celldisruptor, Branson Sonic Power Co.). Then, they were ultracentrifuged at 100,000g for 60 min at 4°c (85P-72 automatic preparative ultracentrifuge, Hitachi Co.). MTs in serum and cytosols were determined by HPLC-AAS (Waters 990 HPLC, Waters Co.; Varian 40-P AAS, Varian Co.). Trace elements (Zn, Cd, Cu, Fe) binding to MTs were determined by AAS.

STATISTICAL ANALYSIS

The rank test was used for statistical analysis of all data.

RESULTS AND DISCUSSION

Tab.1, 2, 3 and Fig.1,2 represent the contents of MTs and trace elements at different PH.

Experimental results indicated that injecting only PH solutions could induce the MTs in vivo. The contents of MTs induced varied with the PH. The amounts of MTs induced by PH 1-5 were more larger than those induced by PH 8-14. The content of MTs induced by PH 4-5 was the largest of all. The amount of MTs induced by PH 6-7 was the same as that induced by saline. Especially, the contents of Zn and Cd varied with the contents of MTs induced by different PH. This fact showed that , variation of PH caused the variation of MTs and trace elements (Zn, Cd, etc.) binding to MTs in vivo.

In this paper a new view was put forward that the productive mechanism of MTs in vivo, the amounts of trace elements binding to MTs, the physiological functions and toxicological efficiency of MTs and trace elements are all in an essential connection with injecting PH. PH may be the most important factor of all which influences the health of a living body and has an essential connection with living phenomena. The physiological functions of PH, MTs and trace elements in vivo are worthy of further research.

REFERENCES

Bremner I et al. (1975): Biochem J 149, 733.
Chang, C.C. et al. (1980): Toxicol Appl Pharmacol 55, 94.
Chen, R.W. and Ganther, H.E. (1975): Environ. Physiol. Biochem. 5, 378-388.
Cherian, M.G. (1974): Biochem. Biophys. Res. Commun 1, 920-926.
Foulbes, E.C. (1978): Toxicol Appl Pharmacol 45(2), 505.
Jones, M.M. et al. (1979): Toxicol Appl Pharmacol. 49,41.
Kiyoshi Z et al. (1976): Biol Abstr 62(12):69874.
Klauser, S. and Kagi, J.H.R. et al. (1983): Biochem. J. 209,71-80.
Ling-Wei Xia et al. (1988): Cancer Letters, 41, pp91-97.
Ling-Wei Xia et al. (1988): Trace Element in Medicine, Vol 5, No.3 pp93-96.
Margoshes , M. and Vallee, B.L. (1957): J. Amer. Chem. Soc. 79, 4813-4814.
Mcintyre, A.D. et al. (1975): Ecolog Toxicol Res 177.
Nordberg, G.F. et al (1971): Acta Pharmacol Toxicol 30: 289.
Nordberg, G.F. and Nordberg, M. et al. (1972): Biochem. J. 126, 491-498.
Piotrowski, J.K. and Bolanoseke, W. et al. (1973): Acta Biochim, Pol, 20, 207.
Probst, G.S. et al. (1977): Toxicol Appl Pharmacol 39, 51.
Suzuki, K.T. (1980): Anal Biochem 102, 31-34.
Suzuki, K.T. and Sunaga, H. et al (1983): J. Chromatogr., 281, 159-166.
Terhaar, C.J. et al. (1965): Toxicol Appl Pharmacol 7: 500.
Webb, M. (1972): Biochem Pharmacol 21 (20): 2767.
Webb, M. (1979): in the chemistry, biochemistry and biology of cadmium (Webb, M. ed.) pp 195-226, E sevier-Northholl and Amsterdam.
Winge, D.R. et al. (1975): Arch Biochem Biophys 170:242.

*This study was a part of the research on the connections of the effects of PH and trace element and their mechanisms on living body. Because of the limited space, tables, figures and detail discussion are all omitted.

Metal Ions in Biology and Medicine. Eds. Ph. Collery, L.A. Poirier, M. Manfait, J.C. Etienne. John Libbey Eurotext, Paris © 1990, pp. 567-569

Interactions of 2-Hydroxypyridine N-Oxyde with biological cations (Ca^{2+}, Mg^{2+}, Zn^{2+}, Mn^{2+}...)

M.F. Deida, J.C. Pierrard, J. Rimbault

Université de Reims Champagne-Ardenne, Faculté des Sciences, Laboratoire de Chimie Minérale, BP 347, 51062 Reims, France

Ligands used in treatment of iron overload are of three kinds :hydroxamates, catecholates and hydroxypyridones. This third type of iron chelators are aromatic hydroxamates.

2-hydroxypyridine N-oxide (hpno) has been shown previously to have interesting antimicrobial activity (Albert, 1956). Kontoghiorghes (1986, 1987) has studied the ability of hpno derivatives to mobilise iron from transferrin and ferritin.

We report here a potentiometric study in aqueous solution of the complexing properties of hpno with four metal cations of biological interest (Ca, Mg, Mn, Zn) and with Ni, known for its toxicologic properties.

Protometric studies (glass electrode) were performed at 25°C in $NaClO_4$ solution of 1 $mol.l^{-1}$ ionic strength.

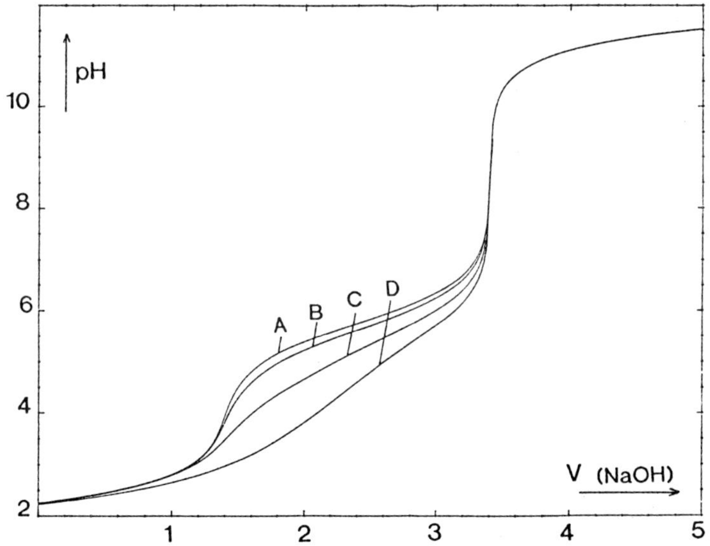

Figure 1: Titration curves for hpno (A), hpno-Ca^{2+}(B), hpno-Mn^{2+}(C), hpno-Ni^{2+}(D), C_L= 8.10^{-3} $mol.l^{-1}$, R = 4

The acidity constant of hpno has been determined from titration of ligand solutions, the concentrations variying from 1.91 10^{-3} to 7.54 10^{-3} mol.l^{-1}. The mean value of pK_a is 5.75 ± 0.01 (95 % confidence).

The hpno-metal(II) systems were studied by varying the ratio $R = C_L / C_M$ (total ligand : total metal concentrations).

The neutralisation curves of hpno-metal(II) systems (B, C and D curves) with R = 4 and of ligand at same concentration (A curve) are reported on figure 1. The difference of pH between the titration curve of the ligand and these obtained in presence of metal increase with the stability of the formed complexes : Ca < Mn < Ni.

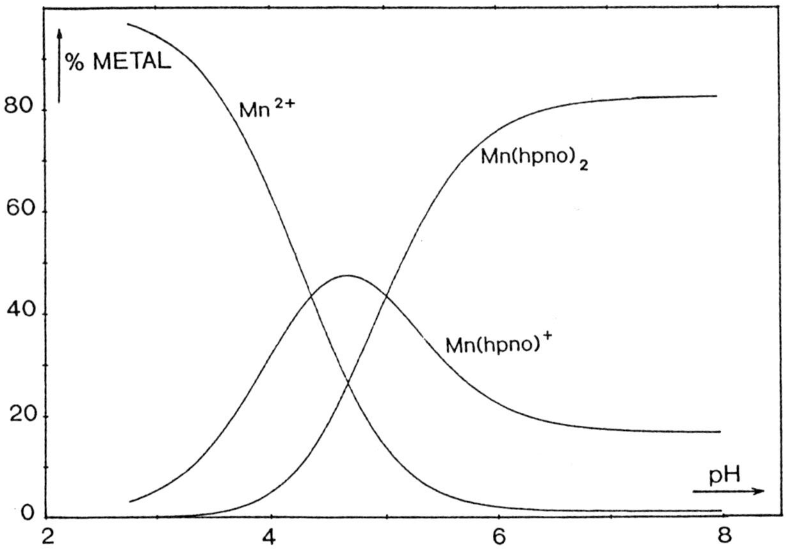

Figure 2: Distribution curves for hpno-Mn^{2+} system
C_L = 8.08 10^{-3} mol.l^{-1}, C_M = 2.01 10^{-3} mol.l^{-1}

The distribution curves corresponding to the hpno-Mn(II) system (Figure 2) show the domaines of existence of three species : Mn^{2+}, $Mn(hpno)^+$ and $Mn(hpno)_2$. This last complex is predominant at biological pH.

The values of the constants corresponding to the equilibria equation are listed for the five studied metals in the table 1 :

$$ML_{n-1}^{(3-n)} + HL \Leftrightarrow ML_n^{(2-n)} + H^+ \qquad K_n$$

metal	$-\log K_1$	$-\log K_2$	$-\log K_3$
Ca^{2+}	3.14 (±0.05)		
Mg^{2+}	2.78 (±0.04)	3.24 (±0.09)	
Mn^{2+}	2.14 (±0.03)	2.64 (±0.06)	
Zn^{2+}	0.84 (±0.06)	1.68 (±0.03)	2.96 (±0.05)
Ni^{2+}	0.82 (±0.03)	1.73 (±0.02)	2.85 (±0.04)

Table 1 : Equilibrium constants of metal complexes with 2 hydroxypyridine N-oxide

The stability of the complexes varies with the nature of the metal ion in the following order :

$$Ca^{2+} < Mg^{2+} < Mn^{2+} < Zn^{2+} \simeq Ni^{2+}$$

REFERENCES

Albert, A. *et al.*, (1956): The influence of chemical constitution on bacterial activity. Part VIII.
In Brit. J. exp. Path., 37, 500-511

Kontoghiorghes, G.J.(1986): Iron mobilisation from lactoferrin by chelators at physiological pH.
In Biochim. Biophys. Acta, 882, 267-270

Kontoghiorghes, G.J.(1987):2 hydroxypyridine-N-oxides: effective new chelators in iron mobilisation.
In Biochim. Biophys. Acta, 924, 13-18

Metal Ions in Biology and Medicine. Eds. Ph. Collery, L.A. Poirier, M. Manfait, J.C. Etienne. John Libbey Eurotext, Paris © 1990, pp. 570-572

Comparative effects of a carcinogenic (As) and an anticancer (Ga) metals on the transfer through the human amnion. Relationship with Mg. Ultrastructural and electrophysiological studies

M. Bara, A. Guiet-Bara, J. Durlach*, Ph. Collery**

*Laboratory of Biology of Reproduction, University P.M. Curie, 7 Quai Saint-Bernard, 75252, Paris Cedex 05, * CHU Cochin Port-Royal Paris, ** CHU Reims, France*

Among the targets of carcinogenic and anticancer metals, the membrane represents a major site of action. During the pregnancy, the major part of the transfer of nutriments between the mother and the fetus, is assumed by the placenta and the fetal membranes which may be a target for cancerous substances and for anticancer agents. In the human amniotic membrane, the transfer of monovalent cations is assumed by two pathways (Guiet-Bara and Bara, 1984) paracellular and cellular structurally located in intercellular spaces, microvilli and podocytes. In this ultrastructural and electrophysiological study, the effects of a carcinogenic (Arsenic, As) and an anticancer (Gallium, Ga) metals, on the amniotic transfer, are compared. The relationship with Mg ions which may produce an anticancer role or a carcinogenic effect is observed.

MATERIAL AND METHODS

- Strips of human amnion, isolated from the placental zone of the amniotic sac, were obtained after eight normal deliveries at term and immediately transferred in Hanks' solution at 37±1°C and pH 7.4.
- Ultrastructural study: The ultrastructure of the human amniotic epithelial cells was studied by the transmission electron microscopy (TEM). After rinsing in Hanks' solution, the amnion was cut into squares approximately 1-2mm wide and immersed in 2 per cent glutaraldehyde in 0.1 M cacodylate buffer for 60 minutes, rinsed in the same buffer at 0.2 M and post-fixed in an osmium (1per cent)- cacodylate (0.2M) mixture during 1 hour at 4°C. Then, the prelevements were dehydrated in graded ethanol and embedded in epon. All sections were stained with both uranyl acetate and lead citrate before examination with a 201 Philips electron microscope.
The results of TEM were analysed by a stereological method using the point counting method (Gundersen and Osterby, 1981) which indicates the ratio between the volume (or area) of intercellular space (R1), microvilli (R2), podocytes (R3), nucleus (R4) and the volume (or area) of the cell.
- Electrophysiological study: The transamniotic conductance Gt was measured by observing the transamniotic potential difference when a direct current was passed across the whole tissue. The ionic fluxes across the amnion (F1: mother to fetus flux, F2: fetus to mother flux) were measured according to the method of Bara et al. (1985a).
- Statistics: The results were expressed as mean ± 1SD. Statistical comparisons were done by conventional paired data analysis.

RESULTS

1) Ultrastructural study

As and Ga, added at 1 mM/1, in the bathing medium decrease R1, increase R2 aad have no effect on R3. As and Ga have a different effect only on R4: As increases the nucleus volume and Ga has no effect. The addition of Mg (1mM) induces a reduction of R4 increased by As and an increase of R4 no changed by Ga.

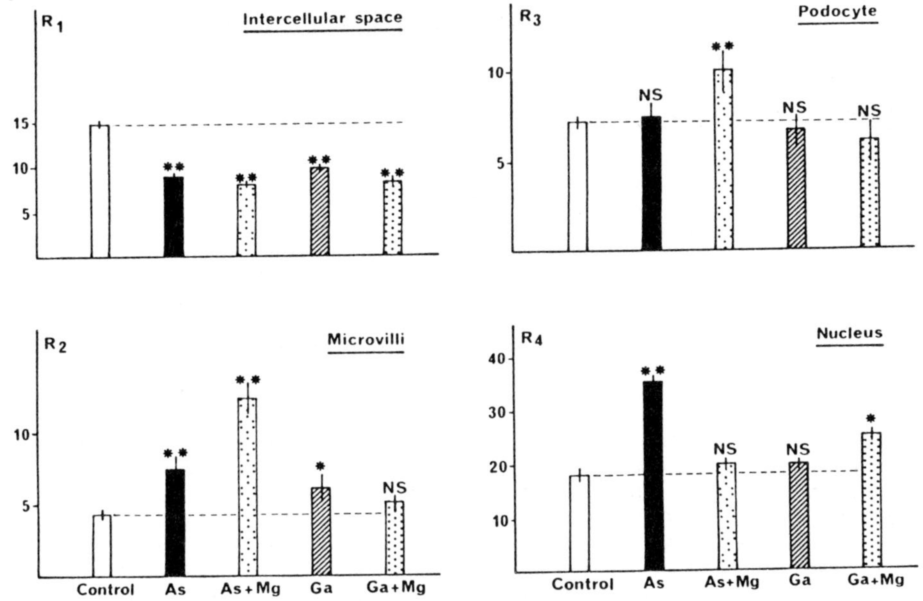

Fig.1. Histogram of the values of the ratio between the volume of intercellular space (R1), microvilli (R2), podocytes (R3), nucleus (R4) and the cell volume reference (*$p<0.05$, **$p<0.01$, NS: no significant)

2) Electrophysiological study

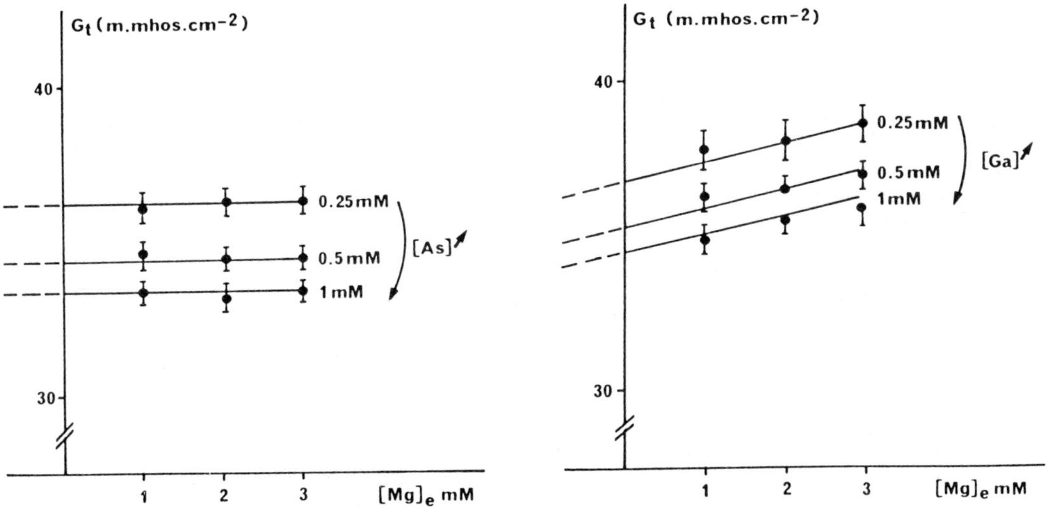

Fig.2. Dixon's curves of the effects of Mg after addition of As and Ga, on the maternal side, on the transamniotic conductance Gt

As and Ga, added on the maternal or the fetal sides have the same effects: reduction of the transamniotic conductance in the mother to fetus and in the fetus to mother ways, reduction of F1 and F2, but the ratio F1/F2 becomes constant (Bara et al., 1985b; 1988). The representation of the Dixon's curves shows no antagonism between Mg and As and a non-competitive inhibition between Mg and Ga (Fig.2).

DISCUSSION AND CONCLUSION

The ultrastructural study indicates that As and Ga have an identical effect on the characteristics of transfer epithelia: intercellular spaces, microvilli and podocytes. At 1 mM/l, As and Ga decrease the volume of intercellular spaces with regard to the cell volume: there is a reduction of the paracellular route of transfer. On the other hand, the microvilli volume is increased and the podocyte volume unchanged: As and Ga increase the cellular route of transfer in the fetus to mother way, without affecting the cellular route in the mother to fetus way. The decrease of the paracellular route is also observed with electrophysiological study because Gt, F1 and F2 are reduced after addition of As and Ga. Moreover, the comparison indicates that a carcinogenic (As) and an anticancer (Ga) metals have the same noxious effect on the ionic transfer. The membrane external sites do not seem to be the target of their cancerous and anticancerous action. The stereological method shows that As and Ga have a different action on the nucleus volume which is increased by As and unchanged by Ga. An antagonistic action of Mg towards As is observed only on the nucleus volume which found again its value in control medium. There is any antagonism between Mg and Ga.
The analyze of the Dixon's curves shows no antagonism between Mg and As (Durlach et al., 1986),i.e., Mg ions are bound to the same sites as As ions but are unable to remove these ions from their external sites. Moreover, there is non-competitive inhibition between Mg and Ga,i.e., Mg and Ga are fixed on different sites of transfer on the membrane.
This study has shown, on the human amniotic membrane, that a carcinogenic (As) and an anticancer (Ga) metals have identical noxious effect on the characteristics of ionic transfer. On the other hand, As increases the nucleus volume unchanged by Ga. Mg ions have an antagonistic effect towards As only on the nucleus volume.

REFERENCES

Bara, M., Guiet-Bara, A., and Durlach, J. (1985a): Monovalent cations transfer through isolated human amnion: a new pharmacological model. Meth. Find. Exptl. Clin. Pharmacol. 7, 209-216.

Bara, M., Guiet-Bara, A., Collery, P., and Durlach, J. (1985b): Gallium action on the ionic transfer through the isolated human amnion.I. Effect on the amnion as a whole and interaction between gallium and magnesium. Trace Elem. Med. 2, 99-103.

Bara, M., Guiet-Bara, A., and Durlach, J. (1988): Relationship between Mg, carcinogenic metals and ionic permeability of the human amnion. Mag. Res. 1, 231-237.

Durlach, J., Bara, M., Guiet-Bara, A., and Collery, P. (1986): Relationship between magnesium, cancer and carcinogenic or anticancer metals. Anticancer Res. 6, 1353-1362.

Guiet-Bara, A., and Bara, M. (1984): Cellular and shunt conductance of human isolated amnion.I. Effect of ouabain, DNP, amiloride and fructose. Bioelectr. Bioenerg. 13, 39-47.

Gundersen, H.J.G., and Osterby, R. (1981): Optimizing sampling efficiency of stereological studies in biology: or "Do more as well!". J. Microsc. 121, 65-73.

Metal Ions in Biology and Medicine. Eds. Ph. Collery, L.A. Poirier, M. Manfait, J.C. Etienne. John Libbey Eurotext, Paris © 1990, pp. 573-575

Adsorption of Mg^{2+}, Ca^{2+} and Mn^{2+} bilayer vesicles constituted of phosphatidic acid, phosphatidylglycerol and phosphatidylserine

M. Fragata, F. Bellemare

Centre de recherche en photobiophysique, Université du Québec à Trois-Rivières, 3351, boul. des Forges, C.P. 500, Trois-Rivières (Québec), G9A 5H7, Canada

Adhesion of biological membranes has long been known to be of general occurrence in a wide variety of cell structures such as, for example, the stacked arrangement of the membranes of visual cells, the Golgi apparatus membranes, or the chloroplast thylakoids. Adhesion is also important in induction phenomena brought about by transient merging of membranes during embryo development. It is clear, in addition, that a defective function of the mechanisms underlying adhesion may lead to pathological states. Divalent cations are among the chemical species which may control membrane adhesion. We undertook a study of this question. Here, we report the adsorption characteristics of Mg^{2+}, Ca^{2+} and Mn^{2+} to bilayer vesicles constituted of anionic lipids, i.e., phosphatidic acid (PA), phosphatidylglycerol (PG) and phosphatidylserine (PS).

METHODS

Experimental Procedures. The phospholipid vesicles were prepared according to methods described previously (see, e.g., Fragata *et al.*, 1985). The binding of the cations to the phospholipid bilayers was achieved by incubating aliquots of the cations solutions with a known volume of vesicles suspended in 0.1 M NaCl, 0.01 Tris-HCl buffer (pH 8.0) to give the final desired concentration of cation. The vesicles solutions were then dialyzed against several changes of 0.01 M Tris-HCl buffer (pH 8.0) which did not contain NaCl. The dialysis time was evaluated as the time required to totally deplete of salts a dialysis bag containing divalent cations at concentrations identical to those present in the lipid-containing bags. At the end of the dialysis period the cation and phosphate contents of the lipid membranes were determined. The cation content was determined by atomic absorption spectroscopy with a Perkin-Elmer, model AA-1278 apparatus. The lipid content was evaluated by phosphate analysis using a modified procedure of Bartlett (see Fragata *et al.*, 1985).

Treatment of Data. Owing to the rather low rate of divalent cation diffusion through the phospholipid bilayers in the time scale of our experiments (see discussions in Papahadjopoulos *et al.*, 1977) the binding data obtained at cation concentrations below the coagulation threshold (CT) (0.1, 0.3 and > 7 mM in PA, PS and PG vesicles, respectively) are corrected for the effective surface of the lipid vesicles, that is to say the surface which is accessible to lipid-cation interactions. The correction factor is about 1.5, i.e., the average outside/inside surface ratio of small, unilamellar vesicles (Huang, 1969). Above CT the correction is not necessary since the aggregation-fusion process involves lysis followed by re-organization of the bilayers. This makes most, if not all the phospholipids available for binding with the divalent cations.

A second correction concerns the ionic strenght of the vesicles solutions. In fact, only 35-45% of the initial concentration of divalent cations is dissociated (see Skoog & West, 1976) and therefore capable of complex formation with the polar head groups of the phospholipids. The activity coefficient (γ) is obtained from the Debye-Hückel expression (cf. Skoog & West, 1976)

$$\log \gamma = (-\,0.5085\, Z^2 \mu^{1/2})/(1 + 0.3281 a \mu^{1/2}) \qquad (1)$$

where Z is the charge of the ion, μ the ionic strength and a the effective diameter of the hydrated ion in Å units. The ionic activity $A = \gamma C$, where C is the molar concentration of free cation, is then used to calculate the association constant (K) of the divalent cations and the number of binding sites per phospholipid molecule (n). To this end, we applied the inverse law of mass action

$$1/\bar{v} = 1/n + 1/nKA \qquad (2)$$

where $\bar{v}$ is the number of cations bound per lipid molecule.

RESULTS AND DISCUSSION

We found that in PA vesicles the number of divalent cations bound per phospholipid decreases in the sequence Ca^{2+}, Mn^{2+}, Mg^{2+}. In PS and PG vesicles the decrease is in the sequence Mn^{2+}, Ca^{2+}, Mg^{2+}. We also studied the variation of n and K in PA, PS and PG bilayers (see Table 1). Table 1 shows that the number of binding sites per phospholipid is in geneal higher in PA than in PS or PG vesicles. It is seen also that the affinity (K) of the divalent cations for the phospholipids is much higher in PA. Of special interest is the fact that n is smaller in PG than in PS bilayers, whereas under the same conditions K is higher in PG.

To explain the differences in the affinities of the divalent cations bound to PA, PG and PS (Table 1) we consider two models, i.e., the inner-sphere complex formation (see, e.g., Lau *et al.*, 1981) and the "cage-like effect" model. The inner-sphere complex model predicts the insertion of the ligand into the first coordination sphere of the divalent cation thereby increasing chemical stability. Given the chemical and geometrical characteristics of the polar head group of PA as well as the observed high values of K, we interpret the binding of Mg^{2+}, Ca^{2+} and Mn^{2+} to the phospholipid as the formation of a complex of the inner-sphere type. On the other hand, the adsorption of the divalent cations to PG and PS bilayers can be best explained by a "cage-like effect" capable of inducing the cations to be entrapped in a region of the lipid vesicles (coordination pocket, see Holwerda *et al.*, 1981) limited between the

PO- region and the layer constituted of glyceryl or serine residues of PG and PS, respectively. Obviously, this would impede the cations of being dialyzed during the desalting procedures described above. In this framework the differences between PG (K ~ 3 000-8 000 M^{-1}) and PS vesicles (K ~ 1 700-3 100 M^{-1}) can be explained either as a loss of bond strength between the divalent cations and the PO- group of PS brought about by the proximity of electronic charges in the carboxyl and amine groups, or as interactions between the cations and the glyceryl OH-groups of PG which do not take place in PS. The latter hypothesis is supported by recent FT-IR work on the binding of Mg(II) and Ca(II) to the O-atoms of D-glucose (Tajmir-Riahi, 1988).

Table 1. K and n of divalent cations adsorbed to phospholipid bilayers.

Cation	PA		PG		PS	
	K(M^{-1})	n	K(M^{-1})	n	K(M^{-1})	n
Mg^{2+}	12 800	0.75	3 000	0.12	1 700	0.25
Ca^{2+}	28 100	0.93	8 000	0.13	2 000	0.57
Mn^{2+}	50 200	0.65	3 900	0.23	3 100	1.15

Acknowledgement. Supported by N.S.E.R.C. Canada and F.C.A.R. Québec.

REFERENCES

Fragata, M., El-Kindi, M., and Bellemare, F. (1985): Mixing of single-chain amphiphiles in two-chain lipid bilayers. 2. Characteristics of chlorophyll *a* and α-tocopherol incorporation in unilamellar phosphatidylcholine vesicles. *Chem. Phys. Lipids* 37, 117-125.

Holwerda, D.L., Ellis, P.D., and Wuthier, R.E. (1981): Carbon-13 and phosphorus-31 nuclear magnetic resonance studies on interaction of calcium with phosphatidylserine. *Biochemistry* 20, 418-428.

Huang, C.-H. (1969): Phosphatidylcholine vesicles. Formation and physical characteristics. *Biochemistry* 8, 344-352.

Lau, A., McLaughlin, A., and McLaughlin, S. (1981): The adsorption of divalent cations to phosphatidylglycerol bilayer membranes. *Biochim. Biophys. Acta* 645, 279-292.

Papahadjopoulos, D., Vail, W.J., Newton, C., Nir, S., Jacobson, K., Poste, G., and Lazo, R. (1977): Studies on membrane fusion. III. The role of calcium-induced phase changes. *Biochim. Biophys. Acta* 465, 579-598.

Skoog, D.A., and West, D.M. (1976): Fundamentals of Analytical Chemistry, 3rd ed., Holt, Rinehardt & Winston, New York.

Tajmir-Riahi, H.-A. (1988): Interaction of D-glucose with alkaline-earth metal ions. Synthesis, spectroscopic, and structural characterization of Mg(II)- and Ca(II)-D-glucose adducts and the effect of metal-ion binding on anomeric configuration of the sugar. *Carbohydrate Res.* 183, 35-46.

Author Index

Reproduction photomécanique
IMPRIMERIE LOUIS-JEAN
BP 87 — 05003 GAP Cedex
Tél. : 92.51.35.23
Dépôt légal : 315 — Mai 1990
Imprimé en France